T0314173

CLASSIFICATION ANALYSIS OF DNA MICROARRAYS

Wiley Series in

Bioinformatics: Computational Techniques and Engineering

Bioinformatics and computational biology involve the comprehensive application of mathematics, statistics, science, and computer science to the understanding of living systems. Research and development in these areas require cooperation among specialists from the fields of biology, computer science, mathematics, statistics, physics, and related sciences. The objective of this book series is to provide timely treatments of the different aspects of bioinformatics spanning theory, new and established techniques, technologies and tools, and application domains. This series emphasizes algorithmic, mathematical, statistical, and computational methods that are central in bioinformatics and computational biology.

Series Editors: **Professor Yi Pan** pan@cs.gsu.edu **Professor Albert Y. Zomaya** zomaya@it.usyd.edu.au

---

Knowledge Discovery in Bioinformatics: Techniques, Methods and Applications / Xiaohua Hu & Yi Pan

Grid Computing for Bioinformatics and Computational Biology / Albert Zomaya & El-Ghazali Talbi

Analysis of Biological Networks / Björn H. Junker & Falk Schreiber

Bioinformatics Algorithms: Techniques and Applications / Ion Mandoiu & Alexander Zelikovsky

Machine Learning in Bioinformatics / Yanqing Zhang & Jagath C. Rajapakse

Biomolecular Networks / Luonan Chen, Rui-Sheng Wang, & Xiang-Sun Zhang

Computational Systems Biology / Huma Lodhi

Computational Intelligence and Pattern Analysis in Biology Informatics / Ujjwal Maulik, Sanghamitra Bandyopadhyay, & Jason T. Wang

Mathematics of Bioinformatics: Theory, Practice, and Applications / Matthew He & Sergey Petoukhov

Introduction to Protein Structure Prediction: Methods and Algorithms / Huzefa Rangwala & George Karypis

Mathematical and Computational Methods in Biomechanics of Human Skeletal Systems: An Introduction Jiri Nedoma & Jiri Stehlik

Rough-Fuzzy Pattern Recognition: Applications in Bioinformatics and Medical Imaging / Pradipta Maji & Sankar K. Pal

Data Management of Protein Interaction Networks / Mario Cannataro & Pietro Hiram Guzzi

Classification Analysis of DNA Microarrays / Leif E. Peterson

CLASSIFICATION ANALYSIS OF DNA MICROARRAYS

LEIF E. PETERSON
Director, Center for Biostatistics, The Methodist Hospital Research Institute, Houston, Texas
Associate Professor of Public Health, Weill Cornell Medical College, Cornell University, New York

Cover design: John Wiley & Sons, Inc.
Cover illustration: © Carlos Olivares/iStockphoto

Published by John Wiley & Sons, Inc., Hoboken, New Jersey
Published simultaneously in Canada

Library of Congress Cataloging-in-Publication Data:

Peterson, Leif E.
Classification analysis of DNA microarrays / Leif E. Peterson
p. cm.
Includes bibliographical references and index.
ISBN 978-0-470-17081-6 (cloth)

Printed in the United States of America

10 9 8 7 6 5 4 3 2 1

To Susan and Evan

CONTENTS

Preface xix

Abbreviations xxiii

1 Introduction 1

1.1 Class Discovery 2
1.2 Dimensional Reduction 4
1.3 Class Prediction 4
1.4 Classification Rules of Thumb 5
1.5 DNA Microarray Datasets Used 9
References 11

PART I CLASS DISCOVERY 13

2 Crisp *K*-Means Cluster Analysis 15

2.1 Introduction 15
2.2 Algorithm 16
2.3 Implementation 18
2.4 Distance Metrics 20
2.5 Cluster Validity 24
2.5.1 Davies–Bouldin Index 25
2.5.2 Dunn's Index 25
2.5.3 Intracluster Distance 26
2.5.4 Intercluster Distance 27
2.5.5 Silhouette Index 30
2.5.6 Hubert's Γ Statistic 31
2.5.7 Randomization Tests for Optimal Value of K 31
2.6 V-Fold Cross-Validation 35
2.7 Cluster Initialization 37

2.7.1 *K* Randomly Selected Microarrays 37
2.7.2 *K* Random Partitions 40
2.7.3 Prototype Splitting 41
2.8 Cluster Outliers 44
2.9 Summary 44
References 45

3 Fuzzy *K*-Means Cluster Analysis 47

3.1 Introduction 47
3.2 Fuzzy *K*-Means Algorithm 47
3.3 Implementation 49
3.4 Summary 54
References 54

4 Self-Organizing Maps 57

4.1 Introduction 57
4.2 Algorithm 57
4.2.1 Feature Transformation and Reference Vector Initialization 59
4.2.2 Learning 60
4.2.3 Conscience 61
4.3 Implementation 63
4.3.1 Feature Transformation and Reference Vector Initialization 63
4.3.2 Reference Vector Weight Learning 66
4.4 Cluster Visualization 67
4.4.1 Crisp *K*-Means Cluster Analysis 67
4.4.2 Adjacency Matrix Method 68
4.4.3 Cluster Connectivity Method 69
4.4.4 Hue–Saturation–Value (HSV) Color Normalization 69
4.5 Unified Distance Matrix (*U* Matrix) 71
4.6 Component Map 71
4.7 Map Quality 73
4.8 Nonlinear Dimension Reduction 75
References 79

5 Unsupervised Neural Gas 81

5.1 Introduction 81
5.2 Algorithm 82
5.3 Implementation 82

5.3.1 Feature Transformation and Prototype Initialization 82
5.3.2 Prototype Learning 83
5.4 Nonlinear Dimension Reduction 85
5.5 Summary 87
References 88

6 **Hierarchical Cluster Analysis** **91**

6.1 Introduction 91
6.2 Methods 91
6.2.1 General Programming Methods 91
6.2.2 Step 1: Cluster-Analyzing Arrays as Objects with Genes as Attributes 92
6.2.3 Step 2: Cluster-Analyzing Genes as Objects with Arrays as Attributes 94
6.3 Algorithm 96
6.4 Implementation 96
6.4.1 Heatmap Color Control 96
6.4.2 User Choices for Clustering Arrays and Genes 97
6.4.3 Distance Matrices and Agglomeration Sequences 98
6.4.4 Drawing Dendograms and Heatmaps 104
References 105

7 **Model-Based Clustering** **107**

7.1 Introduction 107
7.2 Algorithm 110
7.3 Implementation 111
7.4 Summary 116
References 117

8 **Text Mining: Document Clustering** **119**

8.1 Introduction 119
8.2 Duo-Mining 119
8.3 Streams and Documents 120
8.4 Lexical Analysis 120
8.4.1 Automatic Indexing 120
8.4.2 Removing Stopwords 121
8.5 Stemming 121
8.6 Term Weighting 121
8.7 Concept Vectors 124

8.8 Main Terms Representing Concept Vectors 124
8.9 Algorithm 125
8.10 Preprocessing 127
8.11 Summary 137
References 137

9 **Text Mining: *N*-Gram Analysis** **139**

9.1 Introduction 139
9.2 Algorithm 140
9.3 Implementation 141
9.4 Summary 154
References 156

PART II **DIMENSION REDUCTION** **159**

10 **Principal Components Analysis** **161**

10.1 Introduction 161
10.2 Multivariate Statistical Theory 161
10.2.1 Matrix Definitions 162
10.2.2 Principal Component Solution of **R** 163
10.2.3 Extraction of Principal Components 164
10.2.4 Varimax Orthogonal Rotation of Components 166
10.2.5 Principal Component Score Coefficients 168
10.2.6 Principal Component Scores 169
10.3 Algorithm 170
10.4 When to Use Loadings and PC Scores 170
10.5 Implementation 171
10.5.1 Correlation Matrix **R** 171
10.5.2 Eigenanalysis of Correlation Matrix **R** 172
10.5.3 Determination of Loadings and Varimax Rotation 174
10.5.4 Calculating Principal Component (PC) Scores 176
10.6 Rules of Thumb For PCA 182
10.7 Summary 186
References 187

11 **Nonlinear Manifold Learning** **189**

11.1 Introduction 189
11.2 Correlation-Based PCA 190
11.3 Kernel PCA 191
11.4 Diffusion Maps 192

11.5 Laplacian Eigenmaps 192
11.6 Local Linear Embedding 193
11.7 Locality Preserving Projections 194
11.8 Sammon Mapping 195
11.9 NLML Prior to Classification Analysis 195
11.10 Classification Results 197
11.11 Summary 200
References 203

PART III CLASS PREDICTION 205

12 Feature Selection 207

12.1 Introduction 207
12.2 Filtering versus Wrapping 208
12.3 Data 209
12.3.1 Numbers 209
12.3.2 Responses 209
12.3.3 Measurement Scales 210
12.3.4 Variables 211
12.4 Data Arrangement 211
12.5 Filtering 213
12.5.1 Continuous Features 213
12.5.2 Best Rank Filters 219
12.5.3 Randomization Tests 236
12.5.4 Multitesting Problem 237
12.5.5 Filtering Qualitative Features 242
12.5.6 Multiclass Gini Diversity Index 246
12.5.7 Class Comparison Techniques 247
12.5.8 Generation of Nonredundant Gene List 250
12.6 Selection Methods 254
12.6.1 Greedy Plus Takeaway (Greedy PTA) 254
12.6.2 Best Ranked Genes 258
12.7 Multicollinearity 259
12.8 Summary 270
References 270

13 Classifier Performance 273

13.1 Introduction 273
13.2 Input–Output, Speed, and Efficiency 273
13.3 Training, Testing, and Validation 277

13.4 Ensemble Classifier Fusion 280
13.5 Sensitivity and Specificity 283
13.6 Bias 284
13.7 Variance 285
13.8 Receiver–Operator Characteristic (ROC) Curves 286
References 295

14 Linear Regression 297

14.1 Introduction 297
14.2 Algorithm 299
14.3 Implementation 299
14.4 Cross-Validation Results 300
14.5 Bootstrap Bias 303
14.6 Multiclass ROC Curves 306
14.7 Decision Boundaries 308
14.8 Summary 310
References 310

15 Decision Tree Classification 311

15.1 Introduction 311
15.2 Features Used 314
15.3 Terminal Nodes and Stopping Criteria 315
15.4 Algorithm 315
15.5 Implementation 315
15.6 Cross-Validation Results 318
15.7 Decision Boundaries 326
15.8 Summary 327
References 329

16 Random Forests 331

16.1 Introduction 331
16.2 Algorithm 333
16.3 Importance Scores 334
16.4 Strength and Correlation 338
16.5 Proximity and Supervised Clustering 342
16.6 Unsupervised Clustering 345
16.7 Class Outlier Detection 348
16.8 Implementation 350
16.9 Parameter Effects 350
16.10 Summary 357
References 358

17 *K* **Nearest Neighbor** **361**

17.1 Introduction 361
17.2 Algorithm 362
17.3 Implementation 363
17.4 Cross-Validation Results 364
17.5 Bootstrap Bias 369
17.6 Multiclass ROC Curves 373
17.7 Decision Boundaries 374
17.8 Summary 377
References 378

18 **Naïve Bayes Classifier** **379**

18.1 Introduction 379
18.2 Algorithm 380
18.3 Cross-Validation Results 380
18.4 Bootstrap Bias 384
18.5 Multiclass ROC Curves 386
18.6 Decision Boundaries 386
18.7 Summary 389
References 391

19 **Linear Discriminant Analysis** **393**

19.1 Introduction 393
19.2 Multivariate Matrix Definitions 394
19.3 Linear Discriminant Analysis 396
19.3.1 Algorithm 397
19.3.2 Cross-Validation Results 397
19.3.3 Bootstrap Bias 401
19.3.4 Multiclass ROC Curves 402
19.3.5 Decision Boundaries 403
19.4 Quadratic Discriminant Analysis 403
19.5 Fisher's Discriminant Analysis 406
19.6 Summary 411
References 412

20 **Learning Vector Quantization** **415**

20.1 Introduction 415
20.2 Cross-Validation Results 417
20.3 Bootstrap Bias 417
20.4 Multiclass ROC Curves 426

20.5 Decision Boundaries 428
20.6 Summary 428
References 430

21 **Logistic Regression** **433**

21.1 Introduction 433
21.2 Binary Logistic Regression 434
21.3 Polytomous Logistic Regression 439
21.4 Cross-Validation Results 443
21.5 Decision Boundaries 444
21.6 Summary 444
References 447

22 **Support Vector Machines** **449**

22.1 Introduction 449
22.2 Hard-Margin SVM for Linearly Separable Classes 449
22.3 Kernel Mapping into Nonlinear Feature Space 452
22.4 Soft-Margin SVM for Nonlinearly Separable Classes 452
22.5 Gradient Ascent Soft-Margin SVM 454
22.5.1 Cross-Validation Results 455
22.5.2 Bootstrap Bias 457
22.5.3 Multiclass ROC Curves 465
22.5.4 Decision Boundaries 465
22.6 Least-Squares Soft-Margin SVM 465
22.6.1 Cross-Validation Results 470
22.6.2 Bootstrap Bias 477
22.6.3 Multiclass ROC Curves 477
22.6.4 Decision Boundaries 477
22.7 Summary 481
References 483

23 **Artificial Neural Networks** **487**

23.1 Introduction 487
23.2 ANN Architecture 488
23.3 Basics of ANN Training 488
23.3.1 Backpropagation Learning 493
23.3.2 Resilient Backpropagation (RPROP) Learning 496
23.3.3 Cycles and Epochs 496
23.4 ANN Training Methods 497
23.4.1 Method 1: Gene Dimensional Reduction and Recursive Feature Elimination for Large Gene Lists 497
23.4.2 Method 2: Gene Filtering and Selection 502

23.5 Algorithm 502
23.6 Batch versus Online Training 504
23.7 ANN Testing 504
23.8 Cross-Validation Results 504
23.9 Bootstrap Bias 506
23.10 Multiclass ROC Curves 506
23.11 Decision Boundaries 513
23.12 RPROP versus Backpropagation 513
23.13 Summary 522
References 522

24 Kernel Regression 525

24.1 Introduction 525
24.2 Algorithm 527
24.3 Cross-Validation Results 527
24.4 Bootstrap Bias 528
24.5 Multiclass ROC Curves 536
24.6 Decision Boundaries 537
24.7 Summary 540
References 542

25 Neural Adaptive Learning with Metaheuristics 543

25.1 Multilayer Perceptrons 544
25.2 Genetic Algorithms 544
25.3 Covariance Matrix Self-Adaptation–Evolution Strategies 549
25.4 Particle Swarm Optimization 556
25.5 ANT Colony Optimization 560
25.5.1 Classification 560
25.5.2 Continuous-Function Approximation 562
25.6 Summary 567
References 567

26 Supervised Neural Gas 573

26.1 Introduction 573
26.2 Algorithm 574
26.3 Cross-Validation Results 574
26.4 Bootstrap Bias 582
26.5 Multiclass ROC Curves 582
26.6 Class Decision Boundaries 584
26.7 Summary 586
References 588

27 **Mixture of Experts** 591

27.1 Introduction 591
27.2 Algorithm 595
27.3 Cross-Validation Results 596
27.4 Decision Boundaries 597
27.5 Summary 597
References 599

28 **Covariance Matrix Filtering** 601

28.1 Introduction 601
28.2 Covariance and Correlation Matrices 601
28.3 Random Matrices 602
28.4 Component Subtraction 608
28.5 Covariance Matrix Shrinkage 610
28.6 Covariance Matrix Filtering 613
28.7 Summary 621
References 622

APPENDIXES 625

A **Probability Primer** 627

A.1 Choices 627
A.2 Permutations 628
A.3 Combinations 630
A.4 Probability 632
A.4.1 Addition Rule 633
A.4.2 Multiplication Rule and Conditional Probabilities 634
A.4.3 Multiplication Rule for Independent Events 635
A.4.4 Elimination Rule (Disease Prevalence) 636
A.4.5 Bayes' Rule (Pathway Probabilities) 637

B **Matrix Algebra** 639

B.1 Vectors 639
B.2 Matrices 642
B.3 Sample Mean, Covariance, and Correlation 647
B.4 Diagonal Matrices 648
B.5 Identity Matrices 649
B.6 Trace of a Matrix 650

B.7 Eigenanalysis 650
B.8 Symmetric Eigenvalue Problem 650
B.9 Generalized Eigenvalue Problem 651
B.10 Matrix Properties 652

C Mathematical Functions **655**

C.1 Inequalities 655
C.2 Laws of Exponents 655
C.3 Laws of Radicals 656
C.4 Absolute Value 656
C.5 Logarithms 656
C.6 Product and Summation Operators 657
C.7 Partial Derivatives 657
C.8 Likelihood Functions 658

D Statistical Primitives **665**

D.1 Rules of Thumb 665
D.2 Primitives 668
References 678

E Probability Distributions **679**

E.1 Basics of Hypothesis Testing 679
E.2 Probability Functions: Source of *p* Values 682
E.3 Normal Distribution 682
E.4 Gamma Function 686
E.5 Beta Function 689
E.6 Pseudo-Random-Number Generation 692
E.6.1 Standard Uniform Distribution 692
E.6.2 Normal Distribution 693
E.6.3 Lognormal Distribution 694
E.6.4 Binomial Distribution 695
E.6.5 Poisson Distribution 696
E.6.6 Triangle Distribution 697
E.6.7 Log-Triangle Distribution 698
References 698

F Symbols And Notation **699**

Index **703**

PREFACE

Classification analysis is a long-established technique that originated from the areas of statistics and pattern recognition, and is a broad rubric that subsumes *class discovery* and *class prediction*. Earlier forms of class discovery methods of statistical origin include natural grouping techniques such as hierarchical cluster analysis, K-means cluster analysis, and Gaussian mixture models employing the expectation–maximization algorithm. Conversely, pattern recognition-based class discovery techniques include self-organizing maps (Kohonen networks), neural gas, and fuzzy K-means cluster analysis. Class prediction methods that originated in statistics include linear discriminant analysis and logistic regression, while prototype learning, artificial neural networks, and swarm intelligence are more popular in pattern recognition. The main difference between the statistical and pattern recognition approaches is clear; the statistical methods tend to depend more on large sample Gaussian inference, while pattern recognition approaches tend to depend more on distribution-free heuristics employed in machine learning, evolutionary algorithms, or computational intelligence methods.

This book introduces the reader to a variety of statistical, machine learning, and computational intelligence classification algorithms that have been in use for several decades, as well as more recently developed algorithms based on fuzzy methods (soft computing), evolutionary algorithms, and swarm intelligence. Dimensional reduction and text mining for concept and document clustering are also covered to introduce the reader to information retrieval.

The layout of the book is divided into three parts. Part I, on class discovery, includes Chapters 1–9, which address various techniques for identifying the cluster structure of a dataset. Part II, on dimensional reduction, includes Chapters 10 and 11, which focus on linear and nonlinear approaches for reducing the dimensions of a dataset. Part III, on class prediction, covers Chapters 12–28, which present numerous approaches for predicting class membership of test objects after algorithm training is performed. There are also five appendixes, which cover probability, matrix algebra, mathematical functions, statistical primitives, and probability distributions. These are followed by a glossary of the symbols and notation presented throughout the book.

Chapter 1 summarizes class discovery, novel diagnostic classes, comorbidity and disease overlap, outliers and heterogeneity, class prediction, rules of thumb, and descriptions of the nine microarray datasets used throughout this book. Chapter 2 introduces a crisp K-means cluster analysis algorithm, distance metrics, cluster validity to determine the optimal number of clusters, and cluster initialization. Chapter 3 describes use of fuzzification to develop membership functions, which enable the presentation of cluster weights for each microarray. Chapter 4 covers unsupervised cluster analysis using Kohonen networks [self-organizing maps (SOMs)] and the numerous uses of SOM for understanding cluster structure. The fundamental components of self-organizing maps such as neighborhood functions, best-matching units, component maps, and U matrices are also discussed. Chapter 5 introduces prototype learning and the exploration of the cluster structure of data through neural adaptive learning with prototypes. Chapter 6 covers agglomerative clustering, correlation and distance-based agglomeration, dendograms, and heatmaps. Chapter 7 addresses Gaussian mixture models and the expectation–maximization (EM) algorithm for clustering. Chapter 8 develops document and concept clustering via text mining. Methods discussed include stopping, stemming, hash tables, inverse document frequency, and concept vectors. Chapter 9 extends text mining with the use of N grams.

Chapter 10, which discusses linear dimensional reduction by eigenanalysis of the gene-by-gene or array-by-array correlation matrix, develops the concepts of principal component score coefficients, loadings, and principal component scores. Chapter 11 is presented as a means of distance metric learning and dimensional reduction from a nonlinear standpoint, and includes kernel principal component analysis (PCA), diffusion maps, Laplacian eigenmaps, local linear embedding, locality preserving projections, and Sammon mapping.

Chapter 12 presents various methods used for filtering genes in order to develop "optimal" gene lists. A review of variable types (continuous, nominal, ordinal, etc.) is provided, as well as several 2- and k-sample parametric and nonparametric statistical tests for identifying best-ranked genes. A sequential forward–reverse sequential floating method known as "greedy plus takeaway" (greedy PTA) is also introduced, which forms the basis for optimal gene sets used throughout the book. Chapter 13 reviews computational efficiency, confusion matrices and accuracy calculations, cross-validation, bootstrapping, ensemble classifier fusion, random oracles, sensitivity and specificity, receiver–operator characteristic (ROC) curves, and area under the curve (AUC). Chapter 14 presents a matrix algebra approach to multivariate regression in which dependent variables for each class are binarized in the form $y = \pm 1$ and regressed on feature values in order to determine classification decision rules. Chapter 15 discusses decision tree classification, along with parent and daughter nodes, node splitting, terminal nodes, and stopping criteria. Chapter 16 discusses random forests as an extension of decision trees and also discusses feature importance scores, strength and correlation, proximity, unsupervised clustering, and outlier detection. Chapter 17 explores one of the most basic supervised classification methods, in which decision rules are determined from nearest neighbors and applied to test objects to predict class membership. Chapter 18 presents a probabilistic approach that uses Bayes' theorem to formulate class membership predictions. Chapter 19 discusses multivariate definitions, covariance matrices, quadratic discriminants, and Fisher's discriminant function. Chapter 20 presents supervised prototype learning to assign class labels to test objects based on the class of the nearest prototype. Chapter 21 describes binary and polytomous logistic regression. Maximum likelihood theory for the logistic regression model is first developed and then hinged to decision rules for test object class membership prediction. Chapter 22 focuses on linear separability, kernel mapping, hard and soft margins, gradient ascent, and least-squares support vector machines. Chapter 23 covers backpropagation learning, RPROP learning, cycles and epochs, training methods, training with K-means cluster analysis and principal components analysis, recursive feature elimination, gene list generation, and decision rules for class membership assignment. Chapter 24 addresses the use of kernels in linear regression for decision rule generation. Chapter 25 shows how to use metaheuristics such as genetic algorithms, covariance matrix self-adaptation, particle swarm optimization, and ant colony optimization for training a multilayer perceptron (neural network) for class prediction. Chapter 26 discusses the supervised form of neural gas, and describes a distance ranking approach for the neighborhood function. Chapter 27 covers gating probabilities, expert networks, and the EM algorithm for minimizing class prediction error. Chapter 28 covers random matrix theory and methods for filtering covariance (correlation) matrices.

The majority of chapters on class prediction in Part III present results of cross-validation based on fold values of 2, 5, and 10, and bootstrap bias, in which accuracy is shown as a function of increasing bootstrap sample size, multiclass receiver operator characteristic curves and area under the curve as a function of sample size, and class decision boundaries based on posttraining class prediction of arrays and 90,000 reference vectors of a trained 300×300 self-organizing map.

LEIF E. PETERSON

ABBREVIATIONS

AAO	all at once
ACO	ant colony optimization
AID	automatic interaction detection
ANN	artificial neural network(s)
APP	all possible pairs
AQE	average quantization error
ARD	automatic relevance detection
AUC	area under (the) curve
BI	business intelligence
BLOG	binary logistic regression
BMU	best matching unit
CART	classification and regression tress
CDF/PDF	cumulative/probability distribution (density) function
CCAAT	cytidine–cytidine–adenosine–adenosine–thymidine
CKM	crisp *K* means
CMF	covariance matrix filtering
CMSA	covariance matrix self-adaptation (CMSA-ES = CMSA evolution strategies)
CV	cross-validation (CV − 1 = leave-one-out cross-validation)
DM	diffusion map
DTC	decision tree classification
EM	expectation–maximization
EMV	ensemble majority voting (EMWV = ensemble weighted majority voting)
FDA	Fisher's discriminant analysis
FDR	false discovery rate
FKM	fuzzy *K* means
FP/FN	false positive/negative (FPR/FNR = false positive/negative rate; TP/TN = true positive/negative, TPR/TNR = true positive/negative rate)
FWER	familywise error rate
GA	genetic algorithm

GMM	Gaussian mixture model
GOE	Gaussian orthogonal ensemble
GOF	goodness of fit
HCA	hierarchical cluster analysis
HDF	Hierarchical data format
HSV	hue–saturation–value
ICA	independent component analysis
IRLS	iteratively reweighted least squares
KDE	kernel density estimation (KDPCA = kernel density PCA)
KNN	*K* nearest neighbor
KREG	kernel regression
LDA	linear discriminant analysis
LEM	Laplacian eigenmap(s)
LL	loglikelihood
LLE	local linear embedding
LOG	logistic regression
LOOCV	leave-one-out cross-validation
LPP	locality preserving projection(s)
LREG	linear regression
LVQ	learning vector quantization
MCMC	Markov chain Monte Carlo
MLP	multilayer perceptron
MOE	mixture of experts
MSE/MST	mean-square error/total
MSPC	mutative strategy parameter control
NBC	naïve Bayes classification
NG	neural gas (SNG/UNG = supervised/unsupervised NG)
NLML	nonlinear manifold learning
OOA	one against all
OOB	out of (the) bag
ORC	outlier removal clustering
PC	principal component [PCA = principal component(s) analysis]
PDLO	principal direction linear oracle
PSO	particle swarm optimization
PTA	(greedy) plus takeaway
PV	predictive value
QDA	quadratic discriminant analysis
RBF	radial basis function
RF	random forest(s)
RFE	recursive feature elimination
RGB	red-green-blue
RMT	random matrix theory
ROC	receiver–operator characteristic
ROI	return on investment

RPROP	resilient (back)propagation
SAM	statistical analysis of microarrays
s.d.	standard deviation
SMILES	simplified molecular input line entry system
SOM	self-organizing map
SRBCT	small round blue-cell tumor
SSE/SST	sum-of-squares error/total
SVMGA	gradient ascent support vector machine
SVMLS	least-squares support vector machine
UNG	unsupervised neural gas

CHAPTER 1

INTRODUCTION

Classification analysis of DNA microarrays has grown rapidly since the introduction of microarrays in 1995 [1]. Original methodological research covered cluster analysis, which spanned into principal and independent component analysis, and more recently, support vector machines. Current approaches include deterministic annealing, particle swarm optimization, ant colony optimization, and random linear oracles. Future growth areas in classification will likely address parallelization of multiple random linear oracles for ensemble classifier fusion, as a way to overcome diversity problems with today's ensemble methods.

This book discusses classification analysis of DNA microarrays. Topics are introduced in an order that increases with the complexity of the method covered. This chapter introduces the reader to the terminology and rules of thumb used in classification analysis in general, and for DNA microarrays specifically. As the ideas are presented, linkage of topics with the specific information presented in later chapters will become apparent. This chapter is divided into three main parts: class discovery, dimensional reduction, and class prediction. Class discovery (see Section 1.1) is commonly viewed as a hypothesis-generating step that forms an essential part of *knowledge discovery*. Here, the basic assumption is that there is little information about the cluster structure of the data. Analogously, when new molecular methods are used, the goal is often to look for new patterns among the objects (e.g., microarrays) if the research project represents a novel application of laboratory methods for which there is no precedent. The part on dimensional reduction (Section 1.2) offers an approach for collapsing dimensions into a smaller set of features that describe a majority of variation and informativeness of the original larger set. Later, class prediction (Section 1.3) is introduced as a method for training and testing with a learned procedure in order to predict class labels of new unknown objects. Obviously, the learning parameters and predictive

Classification Analysis of DNA Microarrays, First Edition. Leif E. Peterson.

accuracy change as new unknowns are added to the system, and this *concept drift* problem is a greater concern when operationalizing a method in the real world. At this point, we would expect a classification system to perform quite well during external validity tests to classify objects obtained after training and testing, or objects omitted from training that are part of the original training set.

1.1 CLASS DISCOVERY

In today's molecular and genomic world, the first step in analyzing a new dataset should be to perform class discovery from a knowledge discovery approach. You can get more "bang for the buck" or a greater *return on investment* if you exploit newly generated data from the laboratory by trying to determine whether there are new diagnostic classes present, or patients (animals) that don't fit in with the assumed cluster structure. This problem can be partitioned into three categories.

Novel Diagnostic Classes. Never before has there been a better chance to identify new categories of a disease that were previously unknown or unobserved. By augmenting clinical data with DNA microarray data and employing knowledge discovery methods, you may be able to discern new patterns in the data that have never been seen before. Some of the new patterns of objects (patients, animals) may reveal clusters that suggest a new diagnostic class, while others may reflect objects that don't fit in with the assumed structure. Far too often investigators with little experience in linear and nonlinear class discovery methods are unaware of the rich repertoire of methods available for identifying new patterns in a set of data. Other times, there is such a strong focus on hypothesis-driven methods that there is little time available to perform knowledge discovery.

Comorbidity and Overlap. Comorbidity relates to phenotype and represents overlap of disease or diagnostic classes within a disease. Elderly patients presenting with pneumonia may have cardiovascular disease and electrolyte imbalance. In behavioral genetics, one of the best known examples of comorbidity from a biomolecular perspective occurs with alcohol and nicotine dependence [2]. Additionally, it is also well recognized that bipolar disorder, alcoholism, and stress reactivity are comorbid [3]. Depression and comorbidity with epilepsy is another area where DNA microarrays have been employed [4]. Use of class discovery methods for identifying the presence of comorbidity when using DNA microarrays in animal studies or in clinical research may identify new regions of phenotypic overlap.

Outliers and Heterogeneity. Occasionally there are objects that do not cluster with the majority of data and either form their own small groups or act singly as *outliers* due to their unique feature characteristics. Such objects seldom fit in with other objects in the major clusters or major diagnostic categories and are misclassified in class prediction models. Misclassification is common when using genomic data to cluster objects into the diagnostic categories from which objects were originally drawn. Misclassification can be caused by systematic error in sample collection, use of nonstandard buffers and laboratory methods, lack of equipment calibration, intrinsic errors in instrumentation ($1/f$ pink noise, electronic fluctuations, etc.), or genetic and environmental determinants of disease. There is no a guarantee that all objects will fall into discrete known categories, and this should never be assumed.

Table 1.1 lists several class discovery methods described in this book. Crisp and fuzzy K-means cluster analyses (CKM, FKM) are partitional clustering methods, which group objects together in K clusters. The optimal number of clusters is determined by using *cluster validity*. Self-organizing maps (SOMs) provide an unsupervised method that incorporates a neighborhood function to reward learning. Unsupervised neural gas (UNG) uses prototype learning and a punishment-reward learning method for deriving cluster structure. Hierarchical cluster analysis (HCA) is an *agglomerative* cluster method that starts with individual objects, and adds together objects having similar profiles. Divisive cluster analysis works the other way, starting with the entire set of objects as one cluster, and ending up with smaller clusters at the end. The Gaussian mixture model (GMM) method uses the

TABLE 1.1 Unsupervised Class Prediction Methods

Method	Remarks
Crisp K-means cluster analysis (CKM)	Iterative reduction of object-cluster distance
Fuzzy K-means cluster analysis (FKM)	Object-specific membership function for each class
Self-organizing maps (SOM)	Partitions objects using neighborhood functions
Unsupervised neural gas (UNG)	Partitions objects with rank methods
Hierarchical cluster analysis (HCA)	Natural grouping of objects
Gaussian mixture models (GMM)	Exploits the EM algorithm for training

expectation-maximization (EM) algorithm to determine the cluster structure of microarrays.

1.2 DIMENSIONAL REDUCTION

Gene expression datasets are notorious for having a very large number of features (genes). In most cases, not all the features are needed for class discovery and class prediction. Instead, it is possible to reduce the number of features to a lower number of dimensions that can retain the informativeness about the cluster structure of microarrays while explaining a majority of variation in the original dataset. Linear dimensional reduction in the form of principal components analysis (PCA) is first covered in Part II. This is followed by a chapter on nonlinear manifold learning (NLML) for embedding a lower-dimensional map into the original higher-dimensional sample space.

1.3 CLASS PREDICTION

Class prediction methods address the ability that a classifier can learn information from the features of objects, and then make an accurate prediction to assign objects to their true class. This requires knowledge of the true class of each object, and tabulation of this knowledge is commonly referred to as a *truth table*. During *unsupervised* class prediction, the class prediction method does not consider the misclassification error during the learning stage. In *supervised* class prediction, misclassification error is monitored during the learning process, and parameters are updated to achieve better prediction accuracy.

Supervised Methods. The following supervised classification methods (classifiers) are discussed in this book: linear regression (LREG), decision tree classification (DTC), random forests (RF), *K*-nearest neighbor (KNN), naïve Bayes classifier (NBC), linear discriminant analysis (LDA), quadratic discriminant analysis (QDA), Fisher's discriminant analysis (FDA), learning vector quantization (LVQ1), logistic regression (LOG), polytomous logistic regression (PLOG), gradient ascent support vector machines (SVMGA), least-squares support vector machines (SVMLS), artificial neural networks (ANN), kernel regression (KREG), genetic algorithms (GA), covariance matrix self-adaptation (CMSA), particle swarm optimization (PSO), ant colony optimization (ACO), supervised neural gas (SNG), and mixture of experts (MOE). Table 1.2 lists the classifiers used and some remarks regarding their applications.

TABLE 1.2 Supervised Class Prediction Methods

Classifier	Remarks
Linear regression (LREG)	For linearly separable classes
Decision tree classification (DTC)	Classification component of CART[a]
Random forests (RF)	Possibly the least generalization error
K-nearest neighbor (KNN)	Instanced-based learning, "lazy learner"
Naïve Bayes classifier (NBC)	Bayesian classifier, assumes feature independence
Linear discriminant analysis (LDA)	Assumes equal covariance matrices
Quadratic discriminant analysis (QDA)	Assumes unequal covariance matrices
Fisher's discriminant analysis (FDA)	Reduced rank discriminants
Learning vector quantization (LVQ1)	Hebbian nearest prototype learning
Logistic regression (LOG)	2-class maximum likelihood
Polytomous logistic regression (PLOG)	K-class maximum likelihood
Gradient ascent support vector machines (SVMGA)	L_1 soft norm, convex
Least-squares support vector machines (SVMLS)	L_2 soft norm, strictly convex (unique solution)
Artificial neural networks (ANN)	Massively parallel connectionist model
Kernel regression (KREG)	Distance weighted regression
Genetic algorithms (GA)	Genetic selection, crossover, mutation
Covariance matrix self-adaptation (CMSA)	Evolutionary strategy
Particle swarm optimization (PSO)	Swarm intelligence
Ant colony optimization (ACO)	Swarm intelligence with pheromone
Supervised neural gas (SNG)	Prototype learning with error minimization
Mixture of experts (MOE)	Gated mixtures of experts and EM algorithm

[a]Classification and regression trees (algorithm).

1.4 CLASSIFICATION RULES OF THUMB

There are many rules of thumb in machine learning, and classification analysis has its own special rules that have taken form over the last several decades. These are enumerated below, and should be followed to the extent possible because they form a foundation for this book.

1. **Microarrays are objects; genes are features or attributes.** The data used in classification analysis commonly fall into three categories: features, objects, and class labels. *Features* are, by definition, the attributes or characteristics of each object, usually obtained by some sort of measurement or assessment. *Microarrays* are defined as the objects or instances, each of which have features. Thus, an *object* usually consists of a vector of feature values. Objects can also have a *class label*, which for all objects makes up what is called the truth table.
2. **Class discovery does not use class labels.** Class discovery attempts to arrange objects so that a cluster structure is discernible and can provide new insight into the distribution of data in the *sample space*. The fundamental reason for using class discovery is to employ a class discovery technique to partition or arrange objects without considering their true class labels.
3. **Class discovery is commonly independent of feature selection.** The first step in any classification analysis is to determine what features should be used. Class discovery may or may not use all of the available features. However, it is more common to use all of the available features for class discovery, since feature selection is usually done to optimize performance (accuracy) of a class prediction analysis. Expert judgement for selecting "relevant" features is probably the most reasonable reason for preselecting features prior to class discovery analysis. With DNA microarrays it is very typical to cluster the objects using all genes in order to reveal patterns in heterogeneity or new clusters from heatmaps. Certainly, if feature selection is performed for class prediction, then the true classes of objects should be observable in heatmaps generated from the selected features. Thus, the only rationale for selecting features for class discovery is to ensure that only features that are relevant to the data and research questions being addressed are used. For this reason, class discovery is commonly performed independently from feature selection.
4. **Class prediction requires known class labels.** Knowing the truth table of class labels for objects is required for class prediction. Machine learning via training and testing of objects is guided from the truth table of known true objects' classes, and prediction accuracy is hinged to the truth table.
5. **Class prediction commonly employs feature selection.** Feature selection is performed in class prediction in order to increase computational efficiency and optimize class prediction accuracy. Computational efficiency is increased when there are fewer features used, and classification accuracy improves when an optimal set of features is selected. This is a basic tenet of statistical modeling, where the best variables are selected when performing function approximation or

developing a predictive model. The most common methods of feature selection are filtering and wrapping. Filtering selects features using scores independently of the classification procedure, while wrapping incorporates the feature selection process directly into the classification analysis. Use of parametric and nonparametric statistical tests to identify features that are significantly different across class labels independently and prior to statistical modeling would be a form of filtering. Stepwise regression, however, involving feature selection during the optimization process is a form of wrapping. It warrants noting that wrapping, or the selection of features during the classification process, can result in biased predictions since the features selected are specific to the classifier used. Occasionally all the available features are used because either they are relevant to the research question or the investigators chose not to perform feature selection.

6. **Dimensional reduction is typically performed on features.** Dimensional reduction methods are typically used for reducing the number of features involved in a "high-dimensional" classification analysis. DNA microarrays often result in the *small-sample problem,* where the number of features greatly outweighs the number of objects (i.e., $p \gg n$). Principal component analysis (PCA), K-means cluster analysis, and nonlinear manifold learning can be used to reduce the number of features down to a manageable number. An important distinction between dimensional reduction and feature selection is that the former will result in a reduced set of dimensions that are not the same as any of the original features. Any attempt to make inferences on the original features will require a mapping transformation back to the original feature space. Nevertheless, there is great utility in the reduced feature dimensions, since they represent the major source of variation in the *feature space.* Objects can also be used for dimensional reduction. K-means cluster analysis is performed when generating *prototypes* for learning vector quantization and *centers* for radial basis function networks and kernel regression. The number of nodes used in self-organizing maps can also be determined by running K-means cluster analysis on objects, and then setting the nodes equal to the number of centers generated. Any of the dimensional reduction methods described in this book can be used on a larger set of features to obtain reduced features that are then input into class discovery or class prediction algorithms. Class prediction results after using SOM, UNG, and NLML are provided in the respective chapters for these reduction techniques.

7. **Dimensional reduction on objects is common in cluster analysis.** K-means cluster analysis, neural gas, learning vector quantization (LVQ), and radial basis function networks all use methods that

develop *centers* representing multiple groups of objects. In *K*-means cluster analysis the centers have the same dimensions as the number of features and are determined with the average feature values of objects assigned to a given cluster. Neural gas and LVQ use Hebbian learning to derive cluster centers, and in radial basis function networks, the network input node values are based on the distance between each object and every center. What is important is that the user knows that dimensional reduction is being carried out on either the object space or the feature space.

8. **Classification is performed on objects.** It is more common to perform class discovery than class prediction on features, because the use of class labels for features is not nearly as popular as using class labels for objects. For DNA microarrays, class discovery is performed on genes to determine clusters of coregulation or like families of proteins; however, it is less popular to have a research goal to "better predict" gene class membership on the basis of the some truth table.
9. **No single feature set is the best: Ugly Duckling Theorem.** The Ugly Duckling Theorem states that classification always assumes inherent bias, and learning is therefore impossible without bias [5]. If *S* represents a swan and *D* represents a duckling, there is no difference between the ugly duckling and two swans when lined up spatially as *SSD, SDS, DSS*. Therefore, for a specific classifier, there is no reason why a particular set of features should be favored over another.
10. **No single classifier is the best: No-Free-Lunch Theorem.** The No-Free-Lunch Theorem states that "For any two learning algorithms, independent of class priors and the number of objects: There is no difference in expected error when averaged over all target functions, or averaged over all priors" [6]. Hence, if there are *X* reasons why the first learning algorithm outperforms the second, then there are *X* different reasons why the second outperforms the first. Learning is impossible without assumptions, so any observed superiority of one classifier over another is due to the nature of the problem. Overall, the No-Free-Lunch Theorem provides justification that no one classifier is better than another.
11. **Simple is better: Occam's Razor.** The Occam's Razor Theorem states that simpler is better; thus, do not use classifiers that are overly complex. However, a dilemma arises because the No-Free-Lunch Theorem states that simple algorithms should not be favored over complex ones. The Occam's Razor viewpoint about the assumption that accuracy always increases with a higher number of objects is likely not true.

12. **Machine learning is not statistical analysis.** Statistical analysis is based on human interaction, and is usually geared toward inferential hypothesis testing; Type I and Type II errors; sample size and statistical power; normality and distributional assumptions; data range, scale, and transformation; and modeling to fit data. On the other hand, most machine learning methods for classification focus more on establishing consistent objectivity over thousands (millions) of repeated procedures, conducting large-scale repetitive analyses that are humanly impossible in the context of performing them manually, and obtaining high performance and reproducibility consistently in regions of complex decision boundaries where humans break down. In addition, more recent research into machine learning and classification has focused on classifier fusion, diversity, robotics, and automation via expert systems.
13. **Machine learning is not computational intelligence.** Classification with machine learning techniques encapsulates classical statistical methods such as logistic regression and discriminant analysis, Hebbian learning in learning vector quantization, winner-take-all and punishment-reward approaches used in self-organizing maps, as well as kernel methods used in radial basis function networks and support vector machines. Computational intelligence involves three areas, namely, artificial neural networks, soft (fuzzy) computing, and evolutionary algorithms. Artificial neural networks are massively parallel connectionist machines whose constituent perceptrons mimic the neuron in the brain. Soft computing or fuzzy methods exploit uncertainty to make sense of ambiguous data. We have shown on more than one occasion that feature fuzzification can improve classification performance. Evolutionary algorithms such as genetic algorithms mutate and exchange chromosomal regions of object data to arrive at an optimal classification configuration.

1.5 DNA MICROARRAY DATASETS USED

Data used for classification analysis in this book were originally available in C4.5 format from the Kent Ridge Biomedical Data Set Repository (`http://sdmc.i2r.a-star.edu.sg/rp`). At present, many of the datasets used are available at the BRB-ArrayTools Data Archive for Human Cancer Gene Expression [7]. The two-class pediatric brain cancer data consisted of 60 arrays (21 failures, 39 survivors) with expression for 7129 genes [8]. The two-class adult prostate cancer dataset consisted of 102 training arrays (52 tumor, and 50 normal) with 12,600 features. The original report for the prostate data supplement was published by Singh et al. [9]. Two breast cancer datasets were used. The first had two classes and consisted of 15 arrays for eight BRCA1-positive women and seven BRCA2-positive women with expression profiles of 3170 genes [10], and the second was also a two-class set including 78 patient arrays and 24,481 features (genes) consisting of 34 cases with distant metastases who relapsed ("relapse") within 5 years after initial diagnosis and 44 disease-free ("nonrelapse") cases for more than 5 years after diagnosis [11]. Two-class expression data for adult colon cancer were based on the paper published by Alon et al. [12]. The dataset contains 62 arrays based on expression of 2000 genes in 40 tumor biopsies ("negative") and 22 normal ("positive") biopsies from nondiseased colon biopsies from the same patients. An adult two-class lung cancer set including 32 arrays [16 malignant pleural mesothelioma (MPM) and 16 adenocarcinoma (ADCA)] of the lung with expression values for 12,533 genes [13] was also considered. Two leukemia datasets were evaluated; one was a two-class dataset with 38 arrays (27 ALL, 11 AML) containing expression for 7129 genes [14], and the other consisted of three classes for 57 pediatric arrays for lymphoblastic and myelogenous leukemia (20 ALL, 17 MLL, and 20 AML) with expression values for 12,582 genes [15]. The Khan et al. [16] dataset on pediatric small round blue cell tumors (SRBCTs) had expression profiles for 2308 genes and 63 arrays constituting four classes [23 arrays for Ewing sarcoma (EWS), 8 arrays for Burkitt's lymphoma (BL), 12 arrays for NB-neuroblastoma (NB), and 20 arrays for rhabdomyosarcoma) (RMS)].

The entire gene sets listed Table 1.3 were rarely used for unsupervised or supervised classification runs in this book. Instead, gene filtering was

TABLE 1.3 Datasets Used for Classification Analysis

Cancer site	Class	Arrays	Features	Reference
Brain	2	60 (21 failures, 39 survivors)	7,129	8
Prostate	2	102 (52 tumor, 50 normal)	12,600	9
Breast	2	15 (8 BRCA1, 7 BRCA2)	3,170	10
Breast	2	78 (34 relapse, 44 nonrelapse)	24,481	11
Colon	2	62 (40 negative, 22 positive)	2,000	12
Lung	2	32 (16 MPM, 16 ADCA)	12,533	13
Leukemia	2	38 (27 ALL, 11 AML)	7,129	14
Leukemia	3	57 (20 ALL, 17 MLL, 20 AML)	12,582	15
SRBCT	4	63 (23 EWS, 8 BL, 12 NB, 20 RMS)	2,308	16

applied to reduce the number of genes down to a workable level for which there was reduced redundancy and correlation, less noise, and more parsimony. Example 6 (in Chapter 12) describes the main method used for gene filtering.

We did not simulate data for analysis in this book, mainly because the goal was to investigate the characteristics of various classifiers and influence of sample size, statistical significance of features selected, standardization, and fuzzification of features on performance for empirical data. By limiting the coverage to only empirical data, we ensured that the results presented are generalizable to the data considered.

REFERENCES

[1] M. Schena, D. Shalon, R.W. Davis, P.O. Brown. Quantitative monitoring of gene expression patterns with a complementary DNA microarray. *Science* **270**:467–470, 1995.

[2] N. Ait-Daoud, G.A. Wiesbeck, P. Bienkowski, M.D. Li, R.H. Pfutzer, M.V. Singer, O.M. Lesch, B.A. Johnson. Comorbid alcohol and nicotine dependence: From the biomolecular basis to clinical consequences. *Alcohol Clin. Exp. Res.* **29**(8):1541–1549, 2005.

[3] H. Le-Niculescu, M.J. McFarland, C.A. Ogden, Y. Balaraman, S. Patel, J. Tan, Z.A. Rodd, M. Paulus, M.A. Geyer, H.J. Edenberg, S.J. Glatt, S.V. Faraone, J.I. Nurnberger, R. Kuczenski, M.T. Tsuang, A.B. Niculescu. Phenomic, convergent functional genomic, and biomarker studies in a stress-reactive genetic animal model of bipolar disorder and co-morbid alcoholism. *Am. J. Med. Genet. B. Neuropsychiatr. Genet.* **147**(2):134–166, 2008.

[4] S. Koh, R. Magid, H. Chung, C.D. Stine, D.N. Wilson. Depressive behavior and selective down-regulation of serotonin receptor expression after early-life seizures: Reversal by environmental enrichment. *Epilepsy Behav.* 10(1):26–31, 2007.

[5] S. Watanabe. *Pattern Recognition: Human and Mechanical.* Wiley, New York, 1985.

[6] D.H. Wolpert, W.G. Macready. No free lunch theorems for optimization. *IEEE Trans. Evolut. Comput.* **1**:67, 1997.

[7] Y. Zhao, R. Simon. BRB ArrayTools data archive for human cancer gene expression: A unique and efficient data sharing resource. *Cancer Inform.* **6**:9–15, 2008.

[8] S.L. Pomeroy, P. Tamayo, M. Gaasenbeek, L.M. Sturla, M. Angelo, M.E. McLaughlin, J.-Y.H. Kim, L.C. Goumnerovak, P.M. Blackk, C. Lau, J.C. Allen, D. ZagzagI, J.M. Olson, T. Curran, C. Wetmore, J.A. Biegel, T. Poggio, S. Mukherjee, R. Rifkin, A. Califanokk, G. Stolovitzkykk, D.N. Louis, J.P. Mesirov, E.S. Lander, T.R. Golub. Prediction of central nervous system embryonal tumour outcome based on gene expression. *Nature* **415**(6870):436–442, 2002.

[9] D. Singh, P.G. Febbo, K. Ross, D.G. Jackson, J. Manola, C. Ladd, P. Tamayo, A.A. Renshaw, A.V. D'Amico, J.P. Richie, E.S. Lander, M. Loda, P.W. Kantoff, T.R. Golub, W.R. Sellers. Gene expression correlates of clinical prostate cancer behavior. *Cancer Cell* **1**(2):203-209, 2002.

[10] I. Hedenfalk, D. Duggan, Y. Chen, M. Radmacher, M. Bittner, R. Simon, P. Meltzer, B. Gusterson, M. Esteller, M. Raffeld, Z. Yakhini, A. Ben-Dor, E. Dougherty, J. Kononen, L. Bubendorf, W. Fehrle, S. Pittaluga, S. Gruvberger, N. Loman, O. Johannsson, H. Olsson, B. Wilfond, G. Sauter, O-P. Kallioniemi, A.A. Borg, J. Trent. Gene-expression profiles in hereditary breast cancer. *N. Engl. J. Med.* **344**:539-548, 2001.

[11] L.J. van't Veer, H. Dai, M.J. van de Vijver, Y.D. He, A.A. Hart, M. Mao, H.L. Peterse, K. van der Kooy, M.J. Marton, A.T. Witteveen, G.J. Schreiber, R.M. Kerkhoven, C. Roberts, P.S. Linsley, R. Bernards, S.H. Friend. Gene expression profiling predicts clinical outcome of breast cancer. *Nature* **415**:530-536, 2002.

[12] U. Alon, N. Barkai, D.A. Notterman, K. Gish, S. Ybarra, D. Mack, A.J. Levine. Broad patterns of gene expression revealed by clustering of tumor and normal colon tissues probed by oligonucleotide arrays. *Proc. Natl. Acad. Sci. USA* **96**(12):6745-6750, 1999.

[13] G.J. Gordon, R.V. Jensen, L.L. Hsiao, S.R. Gullans, J.E. Blumenstock, S. Ramaswamy, W.G. Richards, D.J. Sugarbaker, R. Bueno. Translation of microarray data into clinically relevant cancer diagnostic tests using gene expression ratios in lung cancer and mesothelioma. *Cancer Res.* **62**(17):4963-5967, 2002.

[14] T.R. Golub, D.K. Slonim, P. Tamayo, C. Huard, M. Gaasenbeek, J.P. Mesirov, H. Coller, M. Loh, J.R. Downing, M.A. Caligiuri, C.D. Bloomfield, E.S. Lander. Molecular classification of cancer: Class discovery and class prediction by gene expression. *Science* **286**:531-537, 1999.

[15] S.A. Armstrong, J.E. Staunton, L.B. Silverman, R. Pieters, M.L. den Boer, M.D. Minden, S.E. Sallan, E.S. Lander, T.R. Golub, S.J. Korsmeyer. MLL translocations specify a distinct gene expression profile that distinguishes a unique leukemia. *Nature Genet.* **30**(1):41-47, 2001.

[16] J. Khan, J.S. Wei, M. Ringner, L.H. Saal, M. Ladanyi, F. Westermann, F. Berthold, M. Schwab, C.R. Antonescu, C. Peterson, R.S. Meltzer. Classification and diagnostic prediction of cancers using gene expression profiling and artificial neural networks. *Nature Med.* **7**:673-679, 2001.

PART I

CLASS DISCOVERY

CHAPTER 2

CRISP K-MEANS CLUSTER ANALYSIS

2.1 INTRODUCTION

Classification analysis often requires objects to be partitioned into groups of objects with like similarity. Unsupervised *class discovery* is one particular example of this, where the natural grouping of objects based on characteristics or features of the objects is the focus. Here, the goal is to learn or discover new classes of objects that heretofore are unknown for various reasons. There are two types of unsupervised cluster algorithms: partitioning and hierarchical. Nonhierarchical partitioning algorithms partition the objects (features) into a finite set of disjoint subsets, each of which has a *center* based on the average feature values within the cluster. Hierarchical cluster algorithms order the objects (features) into a hierarchically nested sequence, or tree structure, for which the result is a *dendogram* that contains a root, branches, and leaves.

Crisp K-means (CKM) cluster analysis is a *partitional* clustering method that is commonly used for class discovery [1]. One of the best ways to learn about the clustering of objects into K groups is to apply CKM cluster analysis to a set of objects, and then determine

1. The size of (i.e., number of objects in) each cluster
2. The mean feature values of objects in each cluster
3. Whether there are clusters containing one object (outliers)
4. The amount of space between the cluster mean vectors (centers) (intercluster distance)
5. The degree of heterogeneity among feature values within clusters (intracluster distance)

Classification Analysis of DNA Microarrays, First Edition. Leif E. Peterson.

Crisp K-means cluster analysis is used not only as a means to an end for learning about object characteristics in class discovery but also as a basis for many *class prediction* methods. In such cases, CKM cluster analysis is typically used to start an analysis for which sample space partitioning is required prior to learning or training. For example, in radial basis function networks (RBFs), the analysis starts by grouping all the objects into K groups, each of which has a mean vector or *center*. Centers are also sometimes called *prototypes*. The Euclidean distance between each input object and each center is then determined, and the farther each object is from a center, the less weight given to the object during training. Likewise, K-means centers can be used to train an artificial neural network as an alternative to performing principal components analysis for feature reduction prior to network training. Instead of using the entire $n \times p$ matrix for network training, K centers are first extracted from the input data to obtain a much smaller $K \times p$ matrix, which is then used for training.

The examples presented above describe how CKM cluster analysis can be used to partition the object space in RBF networks, and for feature reduction prior to artificial neural network training. A more comprehensive list reveals that in class prediction, CKM cluster analysis is used to obtain centers (mean vectors) for RBF networks, reduce input training data used in ANNs, obtain centers for LVQ, or obtain centers for kernel regression.

Contemporary research involving CKM for DNA microarrays includes classification [2], grid methodology for transcription factor binding sites [3], gene annotation and modular gene networks [4], difference-based clustering of short time-course data [5], gene neighborhoods [6], visualization of data under metabolic context [7], and recursive cluster elimination [8]. More recent research includes studies on transcription factor families [9], general applications of CKM [10], text mining biomedical literature for gene-gene relationships [11], and spot image quality with microarray [12].

In the following sections, we introduce the reader to the CKM algorithm, cluster validity, and randomization tests. Several cluster initialization methods are also introduced, followed by the source code. The clustering method known as *K-medoids*, which forces data objects to serve as centers [13], is not addressed in this chapter.

2.2 ALGORITHM

Let $\mathbf{x}_i = (x_{i1}, x_{i2}, \ldots, x_{ip})$ be the set of feature values for object $\mathbf{x}_i$, and $\mathbf{v}_k = (v_{k1}, v_{k2}, \ldots, v_{kp})$ be the set of average feature values for the center vector $\mathbf{v}_k$ $(k = 1, 2, \ldots, K)$. The center vector $\mathbf{v}_k$ is a p-vector representing the average feature values for objects in cluster k. CKM begins by first specifying the total number of center vectors $\mathbf{v}_k$ to be used. Then *cluster initialization* is done to generate the initial set of cluster centers, which are needed during iterative training. This chapter discusses three methods for cluster initialization: (1) K random samples, (2) K random partitions, and (3) prototype splitting. In method 1, K input samples are randomly selected to directly serve as the cluster centers $\mathbf{v}_k$. Method 2 consists in randomly assigning objects to K clusters and determining the cluster centers $\mathbf{v}_k$ by averaging feature values of objects that were randomly assigned to each cluster. Method 3 consists assignment of all objects to a single cluster ($K = 1$), calculation of the cluster center $\mathbf{v}_1$, and then identification of the object with the greatest distance from $\mathbf{v}_1$. This object then becomes the center $\mathbf{v}_2$. This is continued until $k = K$ clusters are formed.

We will focus on K random partitions cluster initialization (Method 2) for introducing CKM. After objects are randomly assigned to K clusters, individual elements of the center vector $\mathbf{v}_k$ for cluster k are found with the formalism

$$v_{kj} = \frac{\sum_{i=1}^{n_k} x_{ij}}{n_k}, \tag{2.1}$$

where n_k is the number of arrays (objects) in cluster k. In the machine learning literature, Equation (2.1) can be specified numerous in ways

$$\begin{aligned} v_{kj} &= \frac{\sum_{i=1}^{n_k} x_{ij}}{|k|} \\ &= \frac{\sum_{\mathbf{x}_i \in k} x_{ij}}{|k|} \\ &= \frac{\sum_{i=1}^{n} x_{ij} I(\mathbf{x}_i \in k)}{|k|} \\ &= \frac{\sum_{\mathbf{x}_i \in k} x_{ij}}{n_k}, \end{aligned} \tag{2.2}$$

where $I(\mathbf{x}_i \in k) = 1$ if object $\mathbf{x}_i$ is in cluster k and 0 otherwise, and $|k|$ is the *cardinality* or number of arrays in cluster k. Note that $I(\mathbf{x}_i \in k)$ must be used when the upper bound of summation is the total number of input arrays n rather than n_k, the number of arrays in cluster k. Another notation commonly used for the centers is

$$\mathbf{v}_k = \frac{\sum_{\mathbf{x}_i \in k} \mathbf{x}_i}{n_k}, \tag{2.3}$$

and this formalism is used extensively in the development of fuzzy K-means cluster analysis.

Once the initial cluster centers are specified, a process of iterative learning is implemented in which the distance between each array and each center is

Algorithm 2.1: Crisp K-Means Cluster Analysis

```
Data: Input arrays x_i (i = 1, 2, ..., n)
Result: K clusters of arrays
Initialize cluster centers v_k with, e.g., K random partitions (Algorithm 2.4)
for t ← 1 to #Iteration do
    numchanged ← 0
    for i ← 1 to n do
        for k ← 1 to K do
            d(x_i, v_k^(t)) = ||x_i − v_k^(t)||
        endfor
        k = arg min_l {d(x_i, v_l^(t))}
        Assign memb(i) ⇝ k
        if membold(i) ≠ memb(i) then
            numchanged+ = 1
        endif
        Assign membold(i) ⇝ k
    endfor
    if numchanged = 0 then
        exit for
    endif
    for k ← 1 to K do
        v_k^(t+1) = (Σ_{x_i∈k} x_i)/n_k
    endfor
endfor
```

determined. Here, we use the Euclidean distance between array $\mathbf{x}_i$ and center $\mathbf{v}_k$, defined as

$$d(\mathbf{x}_i, \mathbf{v}_k) = \|\mathbf{x}_i - \mathbf{v}_k\| = \sqrt{(x_{i1} - v_{k1})^2 + (x_{i2} - v_{k2})^2 + \cdots + (x_{ip} - v_{kp})^2}. \quad (2.4)$$

Arrays are assigned to the cluster having the closest center, and center means are then recalculated using the newly assigned arrays. This iterative process is repeated until there are no arrays remaining for which another cluster center is closer. This can also be accomplished until the center updates are below some threshold level of error ϵ. Algorithm 2.1 lists the computational steps for CKM cluster analysis.

2.3 IMPLEMENTATION

Implementing CKM requires specification of the arrays, genes (features), and number of clusters K. In addition, the user must specify which cluster initialization to use to generate the initial cluster centers. A computational shortcut that is taken to obviate recalculation of the average feature values over all arrays in a cluster is to keep a sum vector to which feature values of appended arrays are added, and from which feature values of removed arrays are subtracted. We also keep track of the number of arrays in each cluster after arrays are appended and removed during each iteration. The last step at the end of each iteration is to calculate the feature-specific means within each cluster using the sum vector and sample size for each cluster. The following source code shows how CKM works, and assumes that cluster initialization has already been performed:

```
For iter = 1 To iterations
  numchanged = 0
  For i = 1 To M
    MinDis = 1.0E+30
    MaxDis = -1.0E+30
      For k = 1 To numclusters
        sum = 0
        For j = 1 To N
          diff = xmatrix(i, j) - v(k, j)
          sum += diff * diff
        Next j
        dist = Math.Sqrt(sum)
        If dist < mindis Then
          mindis = dist
          leastc = k
        End If
      Next k
      clustmemb(i) = leastc
      If clustmemb(i)  <> oldclustmemb(i) Then numchanged += 1
        clustsize(leastc) += 1
        clustsize(oldclustmemb(i)) -= 1
        For j = 1 To N
          vsum(leastc, j) += xmatrix(i, j)
          If clustsize(leastc) > 0 Then v(leastc, j) = _
          vsum(leastc, j) /  clustsize(leastc)
          vsum(oldclustmemb(i),j) -= xmatrix(i, j)
          If clustsize(oldclustmemb(i)) > 0 Then _
          v(oldclustmemb(i), j) = _
          vsum(oldclustmemb(i), j) / clustsize(oldclustmemb(i))
        Next j
      End If
  Next i
  If numchanged = 0 Then Exit For
  For i = 1 To M
    oldclustmemb(i) = clustmemb(i)
  Next i
Next iter
```

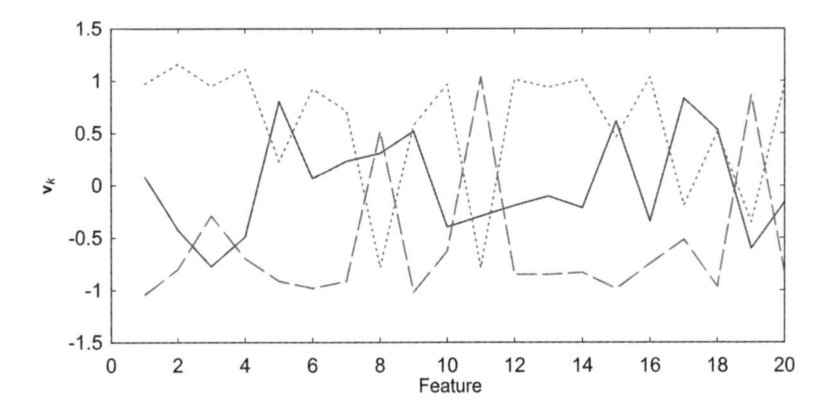

FIGURE 2.1 Cluster centers when $K = 3$ for the three-class MLL_Leukemia data with $n = 57$ arrays and $p = 20$ filtered features. Feature values mean-zero standardized.

Example 2.1 The three-class MLL-Leukemia data were used for developing $K = 3$ clusters. Figure 2.1 shows the values of the cluster centers. Good separation between the cluster centers is observed mostly because of the preselection of $p = 20$ features that strongly discriminate the three diagnostic classes.

2.4 DISTANCE METRICS

A fundamental calculation for classification is based on *resemblance coefficients* or distance metrics, which form a matrix of all pairwise comparisons of *similarity* or *dissimilarity* between arrays (objects). There is greater similarity between the objects with larger similarity coefficients or smaller dissimilarity coefficients. In CKM, the resemblance coefficients computed always compare similarity or dissimilarity between each array and a cluster center, or mean vector of feature values for arrays assigned to a cluster. The notation for the objects is to use $\mathbf{x}_i$ for the ith array and $\mathbf{v}_k$ for the kth cluster center. During iterations, arrays are assigned to the cluster (center) that is the closest, and then the old and new cluster centers are recalculated because the centers for a cluster losing an array and cluster gaining an array will change. In the remainder of this section, we introduce some of the most commonly used similarity and dissimilarity coefficients.

Pitman Correlation. Correlation is a similarity coefficient. A computationally efficient version of the correlation coefficient is Pitman correlation [14], which is a sum product of elements in the two arrays being compared, given in the form

$$d(\mathbf{x}_i, \mathbf{v}_k) = \sum_{j=1}^{p} x_{ij} v_{kj}. \tag{2.5}$$

For the following distance calculations, let `xmatrix(i,j)` be the jth element of array $\mathbf{x}_i$ and `v(k,j)` be the jth element of cluster center $\mathbf{v}_k$. The Pitman correlation between array $\mathbf{x}_i$ and cluster center $\mathbf{v}_k$ is

```
sum = 0
For j = 1 To p
  sum += xmatrix(i, j) * v(k, j)
Next j
dist = sum
```

Pitman correlation is one of the fastest calculations for correlation, and should be implemented when a fast calculation is needed.

Pearson Product Moment Correlation. The sample correlation between arrays $\mathbf{x}_i$ and $\mathbf{v}_k$ is the ratio of their covariance to the product of their standard deviations, given as

$$r = \frac{s_{\mathbf{x}_i \mathbf{v}_k}}{s_{\mathbf{x}_i} s_{\mathbf{v}_k}} = \frac{\sum_{j=1}^{p} (x_{ij} - \bar{x}_i)(v_{kj} - \bar{v}_k)}{\sqrt{\sum_{j=1}^{p} (x_{ij} - \bar{x}_i)^2 \sum_{j=1}^{p} (v_{kj} - \bar{v}_k)^2}}. \tag{2.6}$$

The correlation coefficient is scale-independent and dimensionless, with range $-1 \leq r \leq 1$. When $r \to 1$, there is a tight fit between $\mathbf{x}_i$ and $\mathbf{v}_k$ irrespective of the slope. However, as $r \to 0$, the tightness between $\mathbf{x}_i$ and $\mathbf{v}_k$ weakens. As $r \to -1$, there is a negative correlation, implying that as one variable goes up, the other one goes down. A much simpler formula for correlation involves the crossproduct between the Z scores for $\mathbf{x}_i$ and $\mathbf{v}_k$

$$z_{ij} = \frac{x_{ij} - \bar{x}_i}{s_{x_i}}, \tag{2.7}$$

$$z_{kj} = \frac{v_{kj} - \bar{v}_k}{s_{v_k}}, \tag{2.8}$$

in the form

$$r = \frac{\sum_{j=1}^{p} z_{ij} z_{kj}}{p - 1}. \tag{2.9}$$

In terms of source code, if we assume that `z(i, j)` is the jth element of vector $\mathbf{z}_i$ and `z(k, j)` is the jth element of vector $\mathbf{z}_k$, then the correlation between $\mathbf{z}_i$ and $\mathbf{z}_k$ would be

```
sum = 0
For j = 1 To p
  sum += z(i, j) * z(k, j)
Next j
dist = sum / (p - 1)
```

Euclidean Distance. The Euclidean distance is one of the most popular dissimilarity coefficients and is usually the default choice of many analysts:

$$d(\mathbf{x}_i, \mathbf{v}_k) = \sqrt{\sum_{j=1}^{p} (x_{ij} - v_{kj})^2}. \tag{2.10}$$

With the use of $\mathbf{x}_i$ and cluster center $\mathbf{v}_k$, the Euclidean distance is

```
sum = 0
For j = 1 To p
  diff = (xmatrix(i, j) - v(k, j))
  sum += diff * diff
Next j
dist = Math.Sqrt(sum)
```

City Block (Manhattan) Distance. The city block (Manhattan) distance is another dissimilarity coefficient closely related to the Euclidean distance, since both the Euclidean and Manhattan are Minkowskian r distances, with $r = 2$ and $r = 1$, respectively. The city block distance is defined as

$$d(\mathbf{x}_i, \mathbf{v}_k) = \sum_{j=1}^{p} |x_{ij} - v_{kj}|. \tag{2.11}$$

The city block distance between array $\mathbf{x}_i$ and cluster center $\mathbf{v}_k$ is shown below:

```
sum = 0
For j = 1 To p
  sum += Math.Abs(xmatrix(i, j) - v(k, j))
Next j
dist = sum
```

Other names for Manhattan and Euclidean distance are the L^1 and L^2 distance, respectively, defined with the general equation for L^m, constructed with

$$d(\mathbf{x}_i, \mathbf{v}_k) = \left(\sum_{j=1}^{p} (x_{ij} - v_{kj})^m \right)^{1/m}. \tag{2.12}$$

Chebyshev Distance. Another choice of dissimilarity coefficients is the Chebyshev distance, functionally expressed as

$$d(\mathbf{x}_i, \mathbf{v}_k) = \max_j \{|x_{ij} - v_{kj}|\}. \tag{2.13}$$

The Chebyshev distance between array $\mathbf{x}_i$ and cluster center $\mathbf{v}_k$ is

```
maxd = -1.0E+30
For j = 1 To p
  If Math.Abs(xmatrix(i, j) - v(k, j)) > maxd Then
     maxd = Math.Abs(xmatrix(i, j) - v(k, j))
  End If
Next j
dist = maxd
```

Canberra Distance. Finally, the Canberra distance is another dissimilarity coefficient, given as

$$d(\mathbf{x}_i, \mathbf{v}_k) = \sum_{j=1}^{p} \frac{|x_{ij} - v_{jk}|}{|x_{ij} + v_{kj}|}. \tag{2.14}$$

Computationally, the Canberra distance between array $\mathbf{x}_i$ and cluster center $\mathbf{v}_k$ is

```
sum = 0
For j = 1 To p
  sum += Math.Abs((xmatrix(i, j) - v(k, j)) / _
     (xmatrix(i, j) + v(k, j)))
Next j
dist = sum
```

Tanimoto Distance. Tanimoto distance is a similarity metric often used in information retrieval and is applied to discrete binary vectors consisting of only zeros and ones. Let vector $\mathbf{x}_1 = (1,1,1,0,1,0,1)$ and $\mathbf{x}_2 = (0,0,1,0,0,1,1)$. Then the number of ones in $\mathbf{x}_1$ is $n(\mathbf{x}_1) = 5$, while the number of ones in $\mathbf{x}_2$ is $n(\mathbf{x}_2) = 3$. The number of ones shared by both $\mathbf{x}_1$ and $\mathbf{x}_2$ is $n(\mathbf{x}_1 \cap \mathbf{x}_2) = 2$. The Tanimoti distance is

$$d(\mathbf{x}_1, \mathbf{x}_2) = \frac{n(\mathbf{x}_1 \cap \mathbf{x}_2)}{n(\mathbf{x}_1) + n(\mathbf{x}_2) - n(\mathbf{x}_1 \cap \mathbf{x}_2)}. \tag{2.15}$$

For our example, $d(\mathbf{x}_1, \mathbf{x}_2) = 2/(5+3-2) = \frac{1}{3}$. A continuous version of Tanimoto distance is given in the functional form

$$d(\mathbf{x}_1, \mathbf{x}_2) = \frac{\mathbf{x}_1'\mathbf{x}_2}{\|\mathbf{x}_1\| + \|\mathbf{x}_2\| - \mathbf{x}_1'\mathbf{x}_2}, \tag{2.16}$$

where $\mathbf{x}_1'\mathbf{x}_2$ is the dot product of $\mathbf{x}_1$ and $\mathbf{x}_2$, and $\|\cdot\|$ is the Euclidean norm of a vector defined in (B.62).

The distance metrics described above all vary in their numerical efficiency and cluster results. The Pitman correlation is by far the most efficient, while the Chebyshev and Manhattan distances follow. There are no rules of thumb for employing one over the other, and the choice for a metric is based on

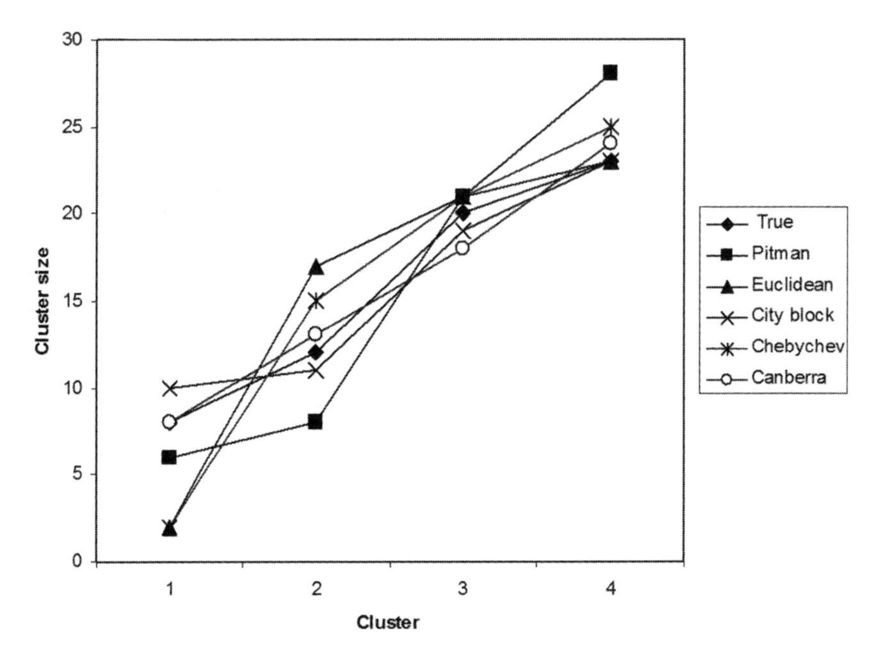

FIGURE 2.2 Plot of cluster size for various distance metrics when $K = 4$ is specified using four-class SRBCT data. Four random partitions are used for cluster initialization.

the particular application. Certainly, a very fast screener for correlation-type measures is approximated by the Pitman correlation, while the remaining are typical dissimilarity metrics. If the goal is to cluster arrays whose profiles have similar additive and multiplicative translations, then Pitman (or correlation) should be the choice. However, if the intent is to cluster arrays whose profiles have similar magnitude from the origin, then the remaining metrics (Euclidean, Manhattan, Chebyshev, and Canberra) would be of interest.

Example 2.2 We used the four-class SRBCT data consisting of $n = 63$ arrays and $p = 23$ features selected with a greedy plus-takeaway 1 method (see Table 12.9) for CKM with $K = 4$. Figure 2.2 shows cluster sizes for various distance metrics when K random partitions was used for cluster initialization. Overall, the Canberra distance resulted in cluster sizes that are closest to cluster sizes based on the true class labels of the 63 arrays. This was followed by the city block distance metric.

2.5 CLUSTER VALIDITY

Partitional cluster analysis assumes that the number of disjoint subsets in the underlying array (feature) space is finite. The clusters resulting from an analysis, however, are not always reliable, so we must evaluate quantitatively whether the imposed structure meets the assumptions for the specified number of K clusters. Usually, we need to ask the following three questions regarding the partitioning:

1. Is there a natural grouping of arrays (features) that represents an underlying cluster structure?
2. What is the appropriate number of clusters in the structure?
3. For a given number of clusters, is the structure partitioning statistically significant?

To answer these questions, we must first assume that there are at least two clusters in the structure. The entire dataset is seldom viewed as being a single cluster since it is being partitioned into smaller groups. An ideal cluster structure will reveal clusters whose centers are far apart and whose assigned arrays are all close in proximity. The more disjoint the clusters are and the less overlap, the greater the chance that the specified number of clusters is the optimal choice [15]. We discuss four cluster validity indices: the Davies-Bouldin index [16], Dunn's index [17], the silhouette index [18], and Hubert's Γ statistic [19].

2.5.1 Davies-Bouldin Index

Let $\mathbf{x}_i$ be an array and K be the number of clusters $\omega_1, \omega_2, \ldots, \omega_K$. After a CKM run with K clusters, the Davies-Bouldin (DB) index for K clusters is given as

$$\mathrm{DB}(K) = \frac{1}{K} \sum_{\ell=1}^{K} \max_{\ell \neq m} \left\{ \frac{d(\omega_\ell) + d(\omega_m)}{d(\omega_\ell, \omega_m)} \right\}, \tag{2.17}$$

where $d(\omega_\ell)$ and $d(\omega_\ell)$ are *intracluster* distances and $d(\omega_\ell, \omega_m)$ is the *intercluster* distance. Intracluster distance $d(\omega)$ and intercluster distance $d(\omega_\ell, \omega_m)$ are described in the following sections. If all the clusters are well separated and compact, then the maximum ratio with small intracluster distances $[d(\omega_\ell) + d(\omega_m)]$ in the numerator and large intercluster distances $d(\omega_\ell, \omega_m)$ in the denominator will result in small values of DB(K). Therefore, the smallest value of DB(K) as a function of K will reflect the optimal number of clusters. Note that DB(K) is undefined for $K = 1$.

2.5.2 Dunn's Index

Dunn's index is similar to the Davies-Bouldin index. Following a CKM run with K clusters, Dunn's index for K clusters is functionally expressed as

$$D(K) = \min_{k} \left\{ \min_{\ell \neq k} \left\{ \frac{d(\omega_k, \omega_\ell)}{\max_{m} \{ d(\omega_m) \}} \right\} \right\}. \tag{2.18}$$

If clusters are well separated and compact, the resulting intercluster distance $d(\omega_k, \omega_\ell)$ in the numerator will be large and intercluster distance $d(\omega_m)$ in the denominator small, causing $D(K)$ to exhibit large positive values at the optimum value of K. Like the Davies-Bouldin index, Dunn's index is also undefined for $K = 1$.

2.5.3 Intracluster Distance

There are various of methods for calculating intracluster and intercluster distances for both the Davies-Bouldin and Dunn indices. We employ some indices from the Bolshakova-Azuaje family of cluster quality indices [20]. For intracluster distance, there are three choices. First is the *complete diameter*, given as

$$d_1(\omega_\ell) = \max_{j>i} \{d(\mathbf{x}_i, \mathbf{x}_j)\} \qquad \mathbf{x}_i, \mathbf{x}_j \in \omega_\ell. \tag{2.19}$$

The actual distance between two arrays is calculated with the Euclidean distance defined in (2.4). The *average diameter* is expressed as

$$d_2(\omega_\ell) = \frac{1}{|\omega_\ell|(|\omega_\ell| - 1)/2} \sum_{j>i} d(\mathbf{x}_i, \mathbf{x}_j) \qquad \mathbf{x}_i, \mathbf{x}_j \in \omega_\ell, \tag{2.20}$$

where $|\omega_\ell|$ is the *cardinality* or number of arrays in cluster ω_ℓ. The *centroid diameter* is calculated using the relationship

$$d_3(\omega_\ell) = \frac{2}{|\omega_\ell|} \sum_{\mathbf{x}_i \in \omega_\ell} d(\mathbf{x}_i, \mathbf{v}_\ell), \tag{2.21}$$

where $\mathbf{v}_\ell$ is the cluster center consisting of mean feature values of arrays in cluster ω_ℓ.

The source code for calculating intracluster distances is shown below. The array `wcdist(i,k)` is used for storage of the distances, where i is the type of specific intracluster distance and k is the cluster. The array elements and the intra-cluster distances are

`wcdist(1,k)` $= d_1(\omega_\ell)$
`wcdist(2,k)` $= d_2(\omega_\ell)$
`wcdist(3,k)` $= d_3(\omega_\ell)$:

```
'within-class distances
For k = 1 To numclusters
    sumclus = 0
    For i = 1 To M
        If clustmemb(i) = k Then sumclus += 1
    Next i
    clussampsize(k) = sumclus
```

```
    maxdist = -1.0E+30
    sumdist = 0
    For i = 1 To M - 1
      For j = i + 1 To M
        If clustmemb(i) = k AndAlso clustmemb(j) = k Then
          sum = 0
          For feature = 1 To N
            diff = xmatrix(i, feature) - xmatrix(j, feature)
            sum += diff * diff
          Next feature
          If sum > maxdist Then maxdist = sum
          sumdist += Math.Sqrt(sum)
        End If
      Next j
    Next i
    wcdist(1, k) = maxdist
    wcdist(2, k) = sumdist / (clussampsize(k) *
                   (clussampsize(k) - 1) / 2)
    sumdist = 0
    For i = 1 To M
      If clustmemb(i) = k Then
        sum = 0
        For feature = 1 To N
          diff = xmatrix(i, feature) - v(k, feature)
          sum += diff * diff
        Next feature
        sumdist += Math.Sqrt(sum)
      End If
    Next i
    wcdist(3, k) = 2 / clussampsize(k) * sumdist
    sumsamp2centerdist(k) = sumdist
Next k
```

2.5.4 Intercluster Distance

For intercluster distance, there are six indices, the first three of which are similar to those used in hierarchical cluster analysis. The *single linkage* distance represents the shortest distance between any of the arrays in clusters ω_ℓ and ω_m:

$$d_1(\omega_\ell, \omega_m) = \min_{i,j} \left\{ d(\mathbf{x}_i, \mathbf{x}_j) \right\} \qquad \mathbf{x}_i \in \omega_\ell, \mathbf{x}_j \in \omega_m. \tag{2.22}$$

Next, the *complete linkage* distance is the maximum distance between arrays in clusters ω_ℓ and ω_m

$$d_2(\omega_\ell, \omega_m) = \max_{i,j} \left\{ d(\mathbf{x}_i, \mathbf{x}_j) \right\} \qquad \mathbf{x}_i \in \omega_\ell, \mathbf{x}_j \in \omega_\ell, \tag{2.23}$$

while the *average linkage* distance is

$$d_3(\omega_\ell, \omega_m) = \frac{1}{|\omega_\ell||\omega_m|} \sum_{i,j} d(\mathbf{x}_i, \mathbf{x}_j) \qquad \mathbf{x}_i \in \omega_\ell, \mathbf{x}_j \in \omega_m. \tag{2.24}$$

One of the most commonly used intercluster distances is the *centroid* linkage distance, expressed as follows:

$$d_4(\omega_\ell, \omega_m) = d(\mathbf{v}_\ell, \mathbf{v}_m). \tag{2.25}$$

The *average to centroids* linkage distance is given as

$$d_5(\omega_\ell, \omega_m) = \frac{1}{|\omega_\ell| + |\omega_m|} \left(\sum_i d(\mathbf{x}_i, \mathbf{v}_\ell) + \sum_j d(\mathbf{x}_j, \mathbf{v}_m) \right) \qquad \mathbf{x}_i \in \omega_\ell, \mathbf{x}_j \in \omega_m, \tag{2.26}$$

and the *Hausdorff distance* is expressed as

$$d_6(\omega_\ell, \omega_m) = \max \left\{ d(\omega_\ell, \omega_m), d(\omega_m, \omega_\ell) \right\}, \tag{2.27}$$

where

$$d(\omega_\ell, \omega_m) = \max_{\mathbf{x}_i \in \omega_\ell} \left\{ \min_{\mathbf{x}_j \in \omega_m} \left\{ d(\mathbf{x}_i, \mathbf{x}_j) \right\} \right\}, \tag{2.28}$$

and

$$d(\omega_m, \omega_\ell) = \max_{\mathbf{x}_i \in \omega_m} \left\{ \min_{\mathbf{x}_j \in \omega_\ell} \left\{ d(\mathbf{x}_i, \mathbf{x}_j) \right\} \right\}. \tag{2.29}$$

In the source code below, the array `bcdist(,,)` stores the intercluster distances after calculations. In tabular form, the array elements and distances are stored as follows:

`bcdist(1,k,l)` $= d_1(\omega_k, \omega_\ell)$
`bcdist(2,k,l)` $= d_2(\omega_k, \omega_\ell)$
`bcdist(3,k,l)` $= d_3(\omega_k, \omega_\ell)$
`bcdist(4,k,l)` $= d_4(\omega_k, \omega_\ell)$
`bcdist(5,k,l)` $= d_5(\omega_k, \omega_\ell)$
`bcdist(6,k,l)` $= d_6(\omega_k, \omega_\ell)$:

```
'Between cluster distances
For k = 1 To numclusters - 1
  For l = k + 1 To numclusters
    mindist = 1.0E+30
    For i = 1 To M
      For j = 1 To M
        If clustmemb(i) = k AndAlso clustmemb(j) = l Then
          sum = 0
          For feature = 1 To N
            diff = xmatrix(i, feature) - xmatrix(j, feature)
```

```
          sum += diff * diff
        Next feature
        If sum < mindist Then mindist = sum
      End If
    Next j
  Next i
bcdist(1, k, l) = mindist
Next l
For l = k + 1 To numclusters
  maxdist = -1.0E+30
  For i = 1 To M
    For j = 1 To M
      If clustmemb(i) = k AndAlso clustmemb(j) = l Then
        sum = 0
        For feature = 1 To N
          diff = xmatrix(i, feature) - xmatrix(j, feature)
          sum += diff * diff
        Next feature
        If sum > maxdist Then maxdist = sum
      End If
    Next j
  Next i
  bcdist(2, k, l) = maxdist
Next l
For l = k + 1 To numclusters
  sumdist = 0
  For i = 1 To M
    For j = 1 To M
      If clustmemb(i) = k AndAlso clustmemb(j) = l Then
        sum = 0
        For feature = 1 To N
          diff = xmatrix(i, feature) - xmatrix(j, feature)
          sum += diff * diff
        Next feature
        sumdist += Math.Sqrt(sum)
      End If
    Next j
  Next i
  bcdist(3, k, l) = sumdist / (clussampsize(k) * clussampsize(l))
Next l
For l = k + 1 To numclusters
  sumdist = 0
  sum = 0
  For feature = 1 To N
    diff = c(feature, k) - c(feature, l)
    sum += diff * diff
  Next feature
  sumdist = Math.Sqrt(sum)
  bcdist(4, k, l) = sumdist
Next l
For l = k + 1 To numclusters
  bcdist(5, k, l) = (sumsamp2centerdist(k) + sumsamp2centerdist(l))_
```

```
          / (clussampsize(k) + clussampsize(l))
  Next l
Next k

Hausdorff distance
   For k = 1 To numclusters
     For l = 1 To numclusters
         maxdist = -1.0E+30
         For i = 1 To M
            If clustmemb(i) = k Then
                 mindist = 1.0E+30
                 For j = 1 To M
                     If clustmemb(j) = l AndAlso outlier(j) = 0 Then
                       sum = 0
                       For feature = 1 To N
                           diff = xmatrix(i, feature)
                              - xmatrix(j, feature)
                           sum += diff * diff
                       Next feature
                       If sum < mindist Then mindist = sum
                 End If
               Next j
               If mindist > maxdist Then maxdist = mindist
             End If
         Next i
         bcdist(6, k, l) = maxdist
       Next l
     Next k
     For k = 1 To numclusters
         For l = 1 To numclusters
             If k <> l Then
                bcdist(6, k, l) = Math.Max(bcdist(6, k, l),
                   bcdist(6, l, k))
              End If
         Next
       Next
```

2.5.5 Silhouette Index

The silhouette index [18] measures the degree of membership of arrays within their assigned clusters. After a CKM run with K clusters, the silhouette index for K clusters is defined as

$$s(i) = \frac{b(i) - a(i)}{\max\{a(i), b(i)\}}. \tag{2.30}$$

The parameter $a(i)$ reflects the average distance between array $\mathbf{x}_i$ and other arrays in its assigned cluster ω_ℓ, and is defined as

$$a(i) = \frac{1}{|\omega_\ell|} \sum_{j \neq i} d(\mathbf{x}_i, \mathbf{x}_j) \qquad \mathbf{x}_i \in \omega_\ell, \mathbf{x}_j \in \omega_\ell. \tag{2.31}$$

The average distance between array $\mathbf{x}_i$ in cluster ω_ℓ and arrays in the closest neighboring cluster ω_m is determined by the functional form

$$b(i) = \min_m \left\{ \frac{1}{|\omega_m|} \sum_j d(\mathbf{x}_i, \mathbf{x}_j) \right\} \qquad \mathbf{x}_i \in \omega_\ell, \mathbf{x}_j \in \omega_m. \tag{2.32}$$

The ratio $s(i)$ ranges from -1 to 1 and requires at least two clusters in a structure. In theory, if all of the arrays in a cluster have the same feature values, then the average intracluster distance $a(i) = 0$, resulting in a numerator of $b(i) - 0 = b(i)$, denominator of $\max\{0, b(i)\} = b(i)$, and ratio $s(i)$ of unity. However, if $a(i) > b(i)$, the numerator $b(i) - a(i)$ and $s(i)$ become negative, indicating that $\mathbf{x}_i$ is misclassified. In the event that there is no real cluster structure present, both $a(i)$ and $b(i)$ will be similar causing $s(i)$ to approach zero. Taken singly, $s(i)$ by itself only reflects the cluster support for $\mathbf{x}_i$, so the average silhouette index is determined as follows:

$$\bar{s} = \frac{1}{n} \sum_{i=1}^{n} s(i). \tag{2.33}$$

Rousseuw has suggested that strong evidence of cluster structure occurs if $0.7 < \bar{s} \leq 1$, reasonable evidence when $0.5 < \bar{s} \leq 0.7$, weak evidence for $0.25 < \bar{s} \leq 0.5$, and no support for structure if $\bar{s} \leq 0.25$ [18].

2.5.6 Hubert's Γ Statistic

Hubert's Γ statistic is another common cluster validity index [21]. Let ω_ℓ be the class of array $\mathbf{x}_i$ (i.e., $\mathbf{x}_i \in \omega_\ell$) and ω_m be the class of array $\mathbf{x}_j$. The normalized $\hat{\Gamma}$ is expressed as

$$\hat{\Gamma} = \frac{1}{n(n-1)/2} \sum_{i=1}^{n-1} \sum_{j=i+1}^{n} \frac{(d(\mathbf{x}_i, \mathbf{x}_j) - \overline{d_{\mathbf{x}_i\mathbf{x}_j}})(d(\omega_\ell, \omega_m) - \overline{d_{\omega_\ell\omega_m}})}{s_{\mathbf{x}_i\mathbf{x}_j} s_{\omega_\ell\omega_m}}, \tag{2.34}$$

where $d(\mathbf{x}_i, \mathbf{x}_j)$ is the interarray distance, $\overline{d_{\mathbf{x}_i\mathbf{x}_j}}$ is the average interarray distance, $d(\omega_\ell, \omega_m)$ is the intercluster distance, $\overline{d_{\omega_\ell\omega_m}}$ is the average intercluster distance, $s_{\mathbf{x}_i\mathbf{x}_j}$ and $s_{\omega_\ell\omega_m}$ are the standard deviation of the between array and intercluster distances, respectively [22].

2.5.7 Randomization Tests for Optimal Value of *K*

The optimal value of K for a CKM cluster run K_{opt} can be determined using a statistical randomization test, otherwise known as a *permutation test* [23]. When $K = K_{\text{opt}}$, the assignment of arrays to the various clusters will yield the

Algorithm 2.2: Cluster Validity Permutation Test

Data: Input arrays $\mathbf{x}_i (i = 1, 2, \ldots, n)$
Result: K_{opt}, optimal number of clusters
Set $K_{\max} = \sqrt{n}$
for $K \leftarrow 1$ **to** $K_{\max}$ **do**
  Initialize cluster centers $\mathbf{v}_k$ using K random partitions (Algorithm 2.4)
  Run CKM (Algorithm 2.1)
  Calculate cluster validity score V_K
  for $b \leftarrow 1$ **to** 50 **do**
    Permute array values within each feature
    Determine $V_K^{(b)}$
  endfor
  Determine the average μ for all values of $V_K^{(b)}$
  Determine the standard deviation σ for all values of $V_K^{(b)}$
  Determine $z_K = (V_K - \mu)/\sigma$
endfor
 $K^* = \arg\max_K \{|z_K|\}$
$K_{\text{opt}} = 1/(\#\text{ of } V_K) \sum_{l=1} K_l^*$

most statistically significant cluster structure. To begin, let V_K be the *observed* value of a validity score [e.g., $\bar{s}$, DB(K), $D(K)$, or NMH(K)] for a specific value of K, and let $V_K^{(b)}$ be the *permutation* value of the score at the bth permutation when array values are permuted (randomly shuffled) within each feature. Permutations are performed repeatedly B times ($b = 1, 2, \ldots, B$), resulting in B values of $V_K^{(b)}$. A standard normal Z score can be defined for the test statistic in the form

$$z_K = \frac{V_K - \mu}{\sigma}, \tag{2.35}$$

where V_K is the observed validity score determined for K clusters, μ is the average of all B values of $V_K^{(b)}$, and σ is the standard deviation of $V_K^{(b)}$. For a given cluster validity score, the greatest absolute value $|z_K|$ over values of K is termed K^*, which reflects a cluster structure with the least intracluster distance and greatest intercluster distance. When V_K is determined with varying combinations of inter- and intracluster distances, we calculate K_{opt} as the average value of K^*. Algorithm 2.2 lists the computational steps.

Example 2.3 We used the four-class SRBCT data consisting of $n = 63$ arrays and $p = 23$ features selected with a greedy plus takeaway-1 method (see Table 12.9) for $K = 2, 3, \ldots, 8$. Table 2.1 lists the average silhouette index $\bar{s}(K)$ and Davies-Bouldin index DB(K) for various combinations of

TABLE 2.1 $|z_K|$ Scores for the Silhouette Index $\bar{s}$ and Davies-Bouldin Score for Various Intra/Intercluster Distances[a]

	K						
Index[a]	2	3	4	5	6	7	8
$\bar{s}$	4.552	2.074	5.895	5.301	6.287	**7.399**[b]	4.971
DB(1, 1)	**7.413**	0.425	0.836	0.975	0.827	1.932	0.956
DB(1, 2)	**6.013**	2.574	1.466	1.574	1.006	0.46	0.341
DB(1, 3)	**6.142**	2.587	1.447	1.638	0.981	0.805	0.418
DB(2, 1)	0.685	0.909	1.701	**2.105**	1.003	3.065	3.366
DB(2, 2)	**4.986**	1.313	2.033	2.874	1.667	0.878	2.123
DB(2, 3)	**4.376**	1.261	1.831	2.773	1.523	1.833	2.431
DB(3, 1)	1.225	0.853	0.716	0.736	0.579	**3.688**	1.653
DB(3, 2)	**9.101**	1.473	3.303	4.476	3.909	2.565	4.62
DB(3, 3)	**6.941**	1.366	2.683	4.296	3.744	4.21	5.318
DB(4, 1)	0.414	0.958	1.213	1.288	1.512	**3.78**	2.679
DB(4, 2)	**6.149**	1.365	3.597	4.559	3.496	3.349	4.452
DB(4, 3)	**5.278**	1.316	3.251	4.494	3.477	3.832	4.664
DB(5, 1)	**2.624**	0.849	0.256	0.519	0.039	0.653	1.724
DB(5, 2)	**3.008**	0.095	1.706	0.261	0.09	1.035	1.034
DB(5, 3)	2.065	0.185	**2.14**	0.468	0.333	0.993	0.74
DB(6, 1)	1.583	0.823	1.901	2.704	1.823	2.96	**3.168**
DB(6, 2)	2.056	1.16	2.188	3.434	2.394	0.826	**3.497**
DB(6, 3)	1.717	1.152	2.083	3.437	2.245	1.294	**3.592**

[a]DB(1, 1) denotes the Davies-Bouldin score for intercluster distance $d_1(\omega_\ell, \omega_m)$ and intracluster distance $d_1(\omega_\ell)$.
[b] Bold entries in each row represent K^*, the value of K for which $|z_K|$ is the greatest.
Notation: In each row (in this and subsequent tables), K^* is denoted in bold typeface. K_{opt} (not listed) is the average of K^* over all rows.

inter- and intracluster distances. The boldfaced values in each row of the table represent K^*, the greatest values of $|z_K|$ for the given validity score. Table 2.2 lists Dunn's index $D(K)$ for the same inter- and intracluster distances used for results in Table 2.1, and Table 2.3 provides results for the normalized Hubert's $\hat{\Gamma}$, NMH(K). Figure 2.3 shows z_K scores for permutation test results using the four-class SRBCT data (see Table 12.9). The average of K^* for all cluster validity scores was 4, suggesting the most appropriate cluster structure is based on $K_{\text{opt}} = 4$ (see Algorithm 2.2). A word of caution is in order when partitioning arrays into the original clusters prior to estimation of V_K and $V_K^{(b)}$ during permutations:

1. The cluster validity scores V_K require at least two arrays in each cluster. If there is only one array in a cluster, there will never be an occasion when $i \neq j$ for intraclass distance calculations.

2. When permuting array values within each feature during permutations, there is an additional risk of failure to assign any arrays to a cluster. Permuting array values within features can result in fewer clusters than K, since by default clustering is not stratified to ensure a minimum of two arrays per cluster. If $K = 8$, the CKM procedure will initially assign arrays into eight clusters, but after permutation the array space may warrant only seven clusters. Having arrays assigned to fewer clusters than K will wreak havoc on the intra- and intercluster

TABLE 2.2 $|z_K|$ Values for Dunn's Index for Various Intra/Intercluster Distances

	K						
Index	2	3	4	5	6	7	8
D(1, 1)	**2.942**	0.692	2.144	0.141	0.102	0.016	0.404
D(1, 2)	**2.757**	2.915	1.778	0.641	0.705	0.226	0.932
D(1, 3)	2.754	**2.935**	2.278	0.524	0.646	0.226	0.918
D(2, 1)	1.184	**1.431**	0.75	1.215	0.247	0.726	0.221
D(2, 2)	**3.893**	1.602	0.405	0.45	1.102	1.205	0.333
D(2, 3)	**3.753**	1.537	0.01	0.931	1.103	1.205	0.309
D(3, 1)	**1.56**	1.492	1.223	0.774	0.671	1.001	0.011
D(3, 2)	**4.014**	0.164	0.024	0.227	0.142	0.551	1.054
D(3, 3)	**3.744**	2.546	0.976	0.209	0.143	0.551	1.113
D(4, 1)	0.816	1.754	**2.462**	2.338	1.836	2.547	0.656
D(4, 2)	**4.05**	3.648	1.812	1.91	1.823	2.425	0.008
D(4, 3)	**3.844**	2.333	3.451	2.889	2.166	2.425	0.051
D(5, 1)	**2.695**	1.226	0.304	0.076	0.204	0.325	0.708
D(5, 2)	1.665	0.756	1.359	1.919	1.661	1.249	**2.166**
D(5, 3)	1.56	0.269	2.059	2.264	1.927	1.249	**2.396**
D(6, 1)	**2.176**	1.425	1.678	0.597	0.89	0.613	0.196
D(6, 2)	**2.094**	1.177	0.864	0.354	0.305	1.074	0.394
D(6, 3)	**2.026**	1.368	1.529	0.075	0.527	1.074	0.369

TABLE 2.3 $|z_K|$ Scores for Normalized Hubert's Index for Various Intercluster Distances

	K						
Index	2	3	4	5	6	7	8
NMH(1)	2.05	2.008	1.772	0.578	0.706	0.832	**3.524**
NMH(2)	2.05	2.406	2.338	0.734	0.652	0.8	**4.027**
NMH(3)	2.05	2.08	2.686	0.422	0.952	0.7	**3.477**
NMH(4)	2.05	2.233	2.38	0.691	0.653	0.959	**3.624**
NMH(5)	2.05	2.258	3.285	0.074	1.158	0.463	**3.71**
NMH(6)	0.654	0.856	0.62	0.505	0.418	0.269	**0.345**

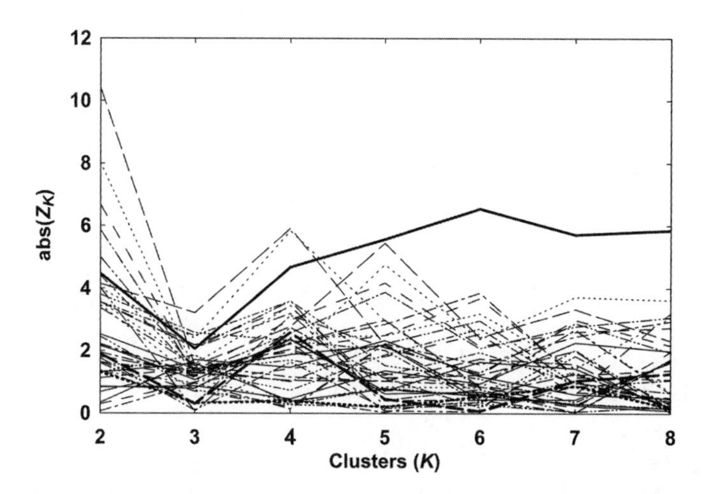

FIGURE 2.3 Z scores for permutation tests of cluster validity scores of the four-class SRBCT gene expression data. Each line represents $|z_K|$ for a particular cluster validity test as a function of cluster size K. The average value of K_{opt} for the cluster validity scores was 4.

distance calculations. Finally, as K increases, there is a greater risk of assignment of fewer arrays into the clusters. The severity of this problem will increase during permutations. As long as the number of arrays assigned to each cluster is greater than or equal to 2, there will be no problems during calculation of V_K or $V_K^{(b)}$.

The difficulties mentioned above can be prevented by requiring the assignment (sp) of at least two arrays to every cluster. If fewer than two arrays are assigned to any cluster, then repeat the cluster partitioning again or the permutation until this criterion (sp) is met. We additionally assume (sp) that a reasonable approximation to $K_{\max} = \sqrt{n}$, where n is the number of arrays being partitioned. There is no problem using $K_{\max} = n/3$ or $K_{\max} = n/2$. The only difficulty with the latter is that if, for example, $n = 1000$, then $K_{\max} = \frac{1000}{2} = 500$. The cost of CKM alone (without permutation testing) for $K_{\max} = 500$ would be very expensive computationally. The additional costs for permutation testing of $n = 1000$ and, for example, $B = 200$ iterations would severely increase the run-time. There may nevertheless be valid reasons for using values of $K_{\max}$ greater than $\sqrt{n}$, so it becomes an issue of choice and computational speed.

2.6 *V*-FOLD CROSS-VALIDATION

One of the most accurate techniques for identifying the number of clusters in a dataset is V-fold cross-validation (VFCV) [24,25]. VFCV starts by partitioning the dataset $\mathcal{D}$ into V mutually exclusive (disjoint) partitions $\mathcal{D}_1, \mathcal{D}_2, \ldots, \mathcal{D}_V$. For a fixed value of K (i.e., number of clusters), the procedure is performed V times, each time using the training set $\mathcal{D} - \mathcal{D}_v$ to generate K cluster centers. The array-to-center distance is then determined for arrays omitted from training in test set $\mathcal{D}_v$ using the closest center, and this is repeated for each test set. An estimate of the average array-to-center distance for the full dataset $\mathcal{D}$ is

$$\widehat{d(\mathbf{x}_i, \mathbf{v}_k)} = \frac{1}{n_v} \sum_{v}^{V} \sum_{\mathbf{x}_i \in \mathcal{D}_v} d(\mathbf{x}_i, \mathbf{v}_k) \qquad k = \underset{m, \mathbf{x}_i \in \mathcal{D}_v}{\arg\min} \{d(\mathbf{x}_i, \mathbf{v}_m)\}, \tag{2.36}$$

where k is the center closest to array $\mathbf{x}_i$ in test set $\mathcal{D}_v$.

Example 2.4 The three-class MLL_Leukemia and four-class SRBCT data with a reduced number of features were employed in VFCV. Each dataset was partitioned into $V = 10$ folds, and the number of clusters K varied from two to eight. Figure 2.4 shows the good agreement between the location of the knee in the curves of the estimated mean array-to-center distance $\widehat{d(\mathbf{x}_i, \mathbf{v}_k)}$ and number of clusters, which happens to be in agreement with the true number of class labels for these datasets. Overall, VFCV is one of the more reliable cluster validity methods and is our most preferred method. An alternative to VFCV is the gap statistic method [26,27], which compares observed cluster quality measures with results from permuted data.

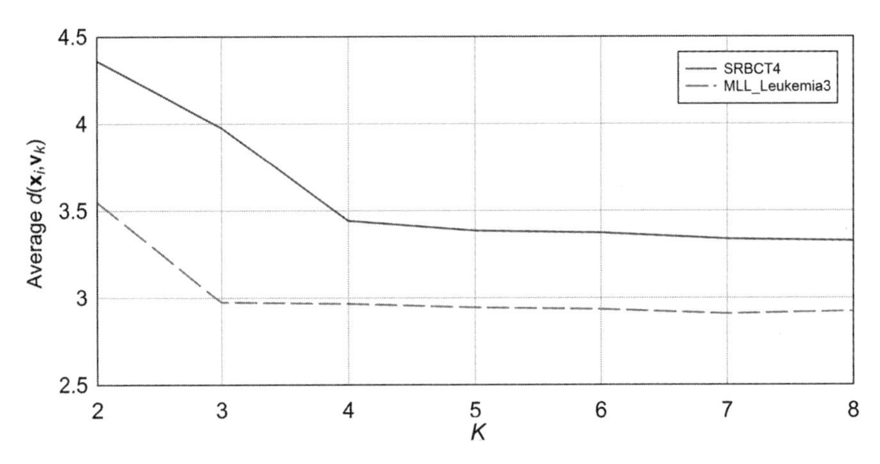

FIGURE 2.4 *V*-fold cross-validation cluster validity run ($V = 10$) results for the four-class SRBCT and three-class MLL_Leukemia data. Average array-to-cluster center distance as a function of number of *K* clusters reflects good agreement with true number of array classes. Features selected using greedy PTA.

2.7 CLUSTER INITIALIZATION

There are many ways of arranging and selecting arrays prior to initiation of CKM. The initialization method used can also have a profound effect on the outcome and performance of CKM. We begin this section by introducing three commonly used cluster initialization methods:

1. Randomly selecting K arrays as the initial cluster centers
2. Randomly assigning all arrays into K partitions
3. Splitting off arrays with the greatest error to create a new cluster center.

2.7.1 *K* Randomly Selected Microarrays

This choice of cluster initialization, which is a divisive form of clustering, first.... assigns all arrays to a single cluster $k = 1$. Next, K arrays are chosen randomly from the arrays in cluster 1, which become the new cluster centers. At this point $k = K$, and the normal iterations are carried out until no arrays migrate into other clusters. The rationale for assigning all arrays into a single cluster at the outset is to ensure that they have a starting cluster from which to migrate during training iterations. Algorithm 2.3 lists the computational steps for randomly selecting K arrays.

There really are no advantages in using this method, since the assignment of initial cluster centers from randomly chosen arrays can result in sparse clusters that, initially, are less populated than others. Whenever a cluster initialization forces small clusters to rely on in-migration of arrays from other larger clusters, there is a potential for reduced *exchangeability*. As such, arrays migrate less and are constrained to exposure to other arrays and their centers. In light of these circumstances, this cluster initialization method would not be our default first choice. Below is an example source code listing for selecting randomly K arrays to serve as cluster centers:

```
ReDim v(numclusters, N)
'Fill vector ranvec() with array id numbers
For i = 1 To M
    ranvec(i) = i
Next i
'Permute ranvec
Nperm = M
Do While Nperm >= 1
    Mperm = Math.Round((Nperm - 1) * Rnd()) + 1
    buffer = ranvec(Nperm)
    ranvec(Nperm) = ranvec(Mperm)
    ranvec(Mperm) = buffer
    Nperm = Nperm - 1
Loop
'Assign all arrays to cluster 1
```

```
For i = 1 To M
    clustmemb(i) = 1
    oldclustmemb(i) = 1
Next i
clustsize(numclusters) = M
'Set prototypes v equal to the first k random arrays
For k = 1 To numclusters
    For j = 1 To N
        v(k, j) = xmatrix(ranvec(k), j)
    Next j
    If k > 1 Then
        clustmemb(ranvec(k)) = k
        oldclustmemb(ranvec(k)) = k
    End If
Next k
'Prototypes v (centers) now ready for iterations
```

Algorithm 2.3: ***K*** **Randomly Selected Arrays**

Data: Input arrays $\mathbf{x}_i$ $(i = 1, 2, \ldots, n)$
Result: K cluster centers $\mathbf{v}_k$
Initialize *ranvec*(n) 'array ID vector
for $i \leftarrow 1$ **to** n **do**
| $ranvec(i) = i$
endfor
N = n
while $N \geq 1$ **do**
| $M = (N - 1)U(0, 1) + 1$
| $buffer = ranvec(N)$
| $ranvec(N) = ranvec(M)$
| $ranvec(N) = buffer$
| $N = N - 1$
endw
for $i \leftarrow 1$ **to** n **do**
| $memb(i) = 1$
| $membold(i) = 1$
endfor
for $k \leftarrow 1$ **to** K **do**
| **for** $j \leftarrow 1$ **to** p **do**
| | $\mathbf{v}(k, j) = x(ranvec(k), j)$
| **endfor**
| **if** $k > 1$ **then**
| | $memb(ranvec(k)) = k$
| | $memold(ranvec(k)) = k$
| **endif**
endfor

Algorithm 2.4: ***K*** **Random Partitions**

Data: Input arrays $\mathbf{x}_i$ ($i = 1, 2, \ldots, n$)
Result: K cluster centers $\mathbf{v}_k$
Initialize *ranvec*(n) 'array ID vector
$k = 0$
'Fill *ranvec*() with consecutive class numbers

```
for i ← 1 to n do
    k = 0
    for j ← 1 to c do
        k+ = 1
        if k > n then
            exit for
        endif
        ranvec(k) = j
    endfor
endfor
N = n
while N > 1 do
M = (N − 1)U(0, 1) + 1
buffer = ranvec(N)
ranvec(N) = ranvec(M)
ranvec(N) = buffer
N = N − 1 do
endw
for i ← 1 to n do
    memb(i) = ranvec(i)
    membold(i) = ranvec(i)
    clussize(ranvec(i))+ = 1
endfor
'Calculate prototype averages for randomly assigned class
for j ← 1 to p do
    for k ← 1 to K do
        for i ← 1 to n do
            if memb(i) = j then
                vsum(k, j)+ = xmatrix(i, j)
            endif
        endfor
        if clussize(k) > 0 then
            v(k, j) = vsum(k, j)/clustsize(k)
        endif
    endfor
endfor
```

2.7.2 *K* Random Partitions

This method is one of the most common choices for CKM. During initialization arrays are uniformly randomly assigned to K clusters. When iterations start, arrays freely migrate back and forth into other clusters for which the array-to-center distance is least. At the end of training, there are K clusters, with the appropriate arrays having the least distance to the centers. Algorithm 2.4 lists the computational steps for K random partitions.

It warrants noting that, in order to achieve uniform cluster assignments, we initially assign cluster numbers based on repeated increments from $1 - K$, and then permute these over the arrays to achieve randomness. We have found this initialization approach to be much more reliable than randomly selecting cluster assignments from the range $1 - K$. Unlike the other two cluster initialization methods, this method is by far the best choice for cluster validity score computation, since there are usually many well-populated clusters at the end of training. Having well-populated clusters is very appropriate for cluster validity score calculation. There is also more exchangeability of arrays, where there is a greater chance that each array will be evaluated for swapping with another cluster. The source code for K random partitions is listed below:

```
ReDim v(numclusters, N)
ReDim vsum(numclusters, N)
ReDim clustsize(numclusters)
'Fill ranvec() with consecutive values of clusters (i.e., if K=3,
'then 1,2,3,1,2,3,... until filled
j = 0
For i = 1 To M
  For k = 1 To numclusters
      j += 1
      If j > M Then GoTo continue2
      ranvec(j) = k
  Next k
Next i
continue2:
'Permute ranvec with class numbers
Nperm = M
Do While Nperm >= 1
  Mperm = Math.Round((Nperm - 1) * Rnd()) + 1
  buffer = ranvec(Nperm)
  ranvec(Nperm) = ranvec(Mperm)
  ranvec(Mperm) = buffer
  Nperm = Nperm - 1
Loop
'Fill memberships with randomly assigned class numbers
For i = 1 To M
  clustmemb(i) = ranvec(i)
  oldclustmemb(i) = ranvec(i)
```

```
  clustsize(ranvec(i)) += 1
Next
'Calculate prototype averages for randomly assigned class
For j = 1 To N
  For k = 1 To numclusters
      For i = 1 To M
        If clustmemb(i) = k Then vsum(k, j) += xmatrix(i, j)
      Next i
      If clustsize(k) > 0 Then v(k, j) = vsum(k, j) / clustsize(k)
  Next k
Next j
'Prototypes v (centers) now ready for iterations
```

2.7.3 Prototype Splitting

This choice of CKM initialization, which is a divisive form of clustering, first assigns all arrays to cluster $k = 1$, determines the center of this cluster, and then finds the single array having the greatest distance (error) from the cluster center. At each iteration, the cluster with the greatest intracluster distance is identified. Within this cluster, the array with the greatest distance to the cluster center is identified. Feature values for this array are then used as the center for new $(k+1)$th cluster. Iterations are performed until $k = K$. Algorithm 2.5 lists the prototype splitting algorithm.

A major disadvantage of prototype splitting is that the array identified for creating the new cluster is an often an outlier. After setting the center values for the new cluster (i.e., the feature values of the outlier array), very few arrays may migrate to the newly formed cluster because it is too far away. Having a single array in a cluster is not amenable to cluster

Algorithm 2.5: Prototype Splitting

Data: Input arrays $\mathbf{x}_i$ $(i = 1, 2, \ldots, n)$
Result: K cluster centers $\mathbf{v}_k$
for $k \leftarrow 1$ **to** $K - 1$ **do**
  if $k = 1$ **then**
    Assign all arrays to first cluster
    Calculate first prototype $\mathbf{v}_1$ for all arrays in cluster $k = 1$
  endif
  Find cluster with maximum error, $l = \arg\max_{\ell}\{\sum_i d(\mathbf{x}_i, \mathbf{v}_\ell)\}$
  Identify array having greatest $j = \arg\max_{i}\{d(\mathbf{x}_i, \mathbf{v}_l)\}$
  Set prototype $\mathbf{v}_{k+1}$ equal to array $\mathbf{x}_j$
  Set membership of array $\mathbf{x}_j$ to cluster $k + 1$
endfor

validity determination, since two or more arrays are needed for intracluster calculations. Analogously, there is less exchangeability of arrays during training iterations, since the new cluster centers are commonly extreme. However, the advantage of prototype splitting is that outlier arrays can be detected. The listing below provides source code for the prototype splitting method. The entire section of code resides in a do loop including the learning iterations, and the number of clusters `numclusters` ramps up from 1 to K for each run of the do loop:

```
If numclusters = 1 Then
    ReDim v(numclusters, N)
    ReDim vsum(numclusters, N)
    ReDim clustsize(numclusters)
    'Randomize first (single) prototype v
    For i = 1 To M
        clustmemb(i) = 1
        oldclustmemb(i) = 1
    Next i
    clustsize(numclusters) = M
    'Calculate first (single) prototype as
    'centroid for all objects, where c=1
    For j = 1 To N
        For k = 1 To numclusters
            For i = 1 To M
              If clustmemb(i) = k Then vsum(k, j) += _
                xmatrix(i, j)
            Next i
            v(k, j) = vsum(k, j) / clustsize(k)
        Next k
    Next j
End If
'find cluster with maximum error (maxerrorc)
ReDim cluserror(numclusters)
MinClusError = 1.0E+30
MaxClusError = -1.0E+30
For k = 1 To numclusters
    cluserror(k) = 0
    For i = 1 To M
       If clustmemb(i) = k Then
          sum = 0
          If DistanceFunc = 1 Then 'Pitman
            For j = 1 To N
              sum += xmatrix(i, j) * v(k, j)
            Next j
            cluserror(k) += sum
          End If
       End If
    Next i
    If DistanceFunc = 1 Then 'Pitman
        If cluserror(k) < mincluserror Then
            mincluserror = cluserror(k)
```

```
            maxerrorc = k
        End If
    End If
    If DistanceFunc > 1 Then
        If cluserror(k) > maxcluserror Then
            maxcluserror = cluserror(k)
            maxerrorc = k
        End If
    End If
Next k
'find feature(array) vector with greatest distance
'within cluster having maximum error
MinDis = 1.0E+30
MaxDis = -1.0E+30
For i = 1 To M
   If clustmemb(i) = maxerrorc Then
      sum = 0
      If DistanceFunc = 1 Then 'Pitman
         For j = 1 To N
           sum += xmatrix(i, j) * v(maxerrorc, j)
         Next j
         dist = sum
      End If
      If DistanceFunc = 1 Then
         If dist < mindis Then
           mindis = dist
           maxi = i
         End If
      End If
      If DistanceFunc > 1 Then
        If dist > maxdis Then
          maxdis = dist
          maxi = i
        End If
      End If
   End If
Next i
numclusters = numclusters + 1
If numclusters > maxc Then
  numclusters -= 1
  Exit Do
End If
ReDim Preserve v(numclusters, N)
ReDim Preserve vsum(numclusters, N)
ReDim Preserve clustsize(numclusters)
For j = 1 To N 'sets new prototype equal to record with greatest error
  v(numclusters, j) = xmatrix(maxi, j)
  vsum(numclusters, j) = xmatrix(maxi, j)
Next j
clustmemb(maxi) = numclusters 'set memberhsip in new prototype
clustsize(numclusters) = 1
clustsize(numclusters - 1) -= 1
```

Algorithm 2.6: Outlier Removal

```
for i ← 1 to n do
    k = min_m{d(x_i, v_m)}
    d_i = d(x_i, v_k)
endfor
d_max = max_i{d_i}
for i ← 1 to n do
    o_i = d(x_i, v_k)/d_max
    if o_i > 0.95 then
        x_i ⇝ Outlier
    endif
endfor
```

2.8 CLUSTER OUTLIERS

Outliers are defined as arrays having extreme distances to their assigned (closest) centers. The operational idea of extreme can be regarded, for example, as arrays whose distances exceed the 95th percentile of distance values for all arrays. Outlier removal clustering (ORC) is a useful method for outlier removal, which improves CKM [28]. ORC is applied during each learning iteration and any time an outlier is detected, it is removed from training during all subsequent iterations. Algorithm 2.6 lists the stepwise procedures used in ORC.

2.9 SUMMARY

This chapter introduces the reader to crisp K-means (CKM) cluster analysis, which is by far one of the most commonly used methods in machine learning. CKM is not a means to an end, that is, to create clusters of arrays for the sake of clustering arrays. Rather, CKM is a useful tool for generating prototypes for arrays in a known true class prior to kernel regression, learning vector quantization (LVQ), and artificial neural network (ANN) classification analysis. CKM can be viewed as a method for grouping arrays with similar characteristics together in clusters, or as a dimensional reduction technique and can be employed on features as well. However, this would most likely be done to reduce a large number of genes down to a manageable number of centers, say, $K = \sqrt{p}$, in order to minimize bias due to the curse of dimensionality (i.e., small sample problem). Knowledge of how to implement CKM based on methods described in this chapter should provide the reader with a strong skill set for addressing today's challenges in unsupervised cluster analysis.

REFERENCES

[1] S. Lloyd. Least squares quantization in PCM. *IEEE Trans. Inform. Theory* **28**:128–137, 1957.

[2] Y.Y. Leung, C.Q. Chang, Y.S. Hung, P.C. Fung. Gene selection for brain cancer classification. *Proc. IEEE Eng. Med. Biol. Soc.* **1**:5846–5849, 2006.

[3] E. van der Wath, L. Moutsianas, R. van der Wath, A. Visagie, L. Milanesi, P. Lio. Grid methodology for identifying co-regulated genes and transcription factor binding sites. *IEEE Trans. Nanobiosci.* **6**(2):162–167, 2007.

[4] M. Shiga, I. Takigawa, H. Mamitsuka. Annotating gene function by combining expression data with a modular gene network. *Bioinformatics.* **23**(13):i468–i478, 2007.

[5] J. Kim, J.H. Kim. Difference-based clustering of short time-course microarray data with replicates. *BMC Bioinform.* **8**:253, 2007.

[6] C. Huttenhower, A.I. Flamholz, J.N. Landis, S. Sahi, C.L. Myers, K.L. Olszewski, M.A. Hibbs, N.O. Siemers, O.G. Troyanskaya, H.A. Coller. Nearest neighbor networks: Clustering expression data based on gene neighborhoods. *BMC Bioinform.* **8**:250, 2007.

[7] M. Weniger, J.C. Engelmann, J. Schultz. Genome expression pathway analysis tool-analysis and visualization of microarray gene expression data under genomic, proteomic and metabolic context. *BMC Bioinform.* **8**:179, 2007.

[8] M. Yousef, S. Jung, L.C. Showe, M.K. Showe. Recursive cluster elimination (RCE) for classification and feature selection from gene expression data. *BMC Bioinform.* **8**:144, 2007.

[9] K. Ushizawa, T. Takahashi, M. Hosoe, H. Ishiwata, K. Kaneyama, K. Kizaki, K. Hashizume. Global gene expression analysis and regulation of the principal genes expressed in bovine placenta in relation to the transcription factor AP-2 family. *Reprod. Biol. Endocrinol.* **5**:17, 2007.

[10] L. Fu, E. Medico. FLAME, a novel fuzzy clustering method for the analysis of DNA microarray data. *BMC Bioinform.* **8**:3, 2007.

[11] Y. Liu, S.B. Navathe, J. Civera, V. Dasigi, A. Ram, B.J. Ciliax, R. Dingledine. Text mining biomedical literature for discovering gene-to-gene relationships: A comparative study of algorithms. *IEEE/ACM Trans. Comput. Biol. Bioinform.* **2**(1):62–76, 2005.

[12] R. Nagarajan, C.A. Peterson. Identifying spots in microarray images. *IEEE Trans. Nanobiosci.* **1**(2):78–84, 2002.

[13] L. Kaufman, P. Rousseeuw. *Finding Groups in Data: An Introduction to Cluster Analysis*. Wiley, New York, 1990.

[14] E.J.G. Pitman. Significance tests which may be applied to samples from any population. *J. Roy. Stat. Soc. Suppl.* **4**:119–130, 1937.

[15] N. Speer, C. Spieth, A. Zell. Biological cluster validity indices based on the gene ontology. *Lecture Notes Comput. Sci.* **3646**:429–439, 2005.

[16] D. Davies, D. Bouldin. A cluster separation measure. *IEEE Trans. Pattern Analy. Machine Intell.* **1**(2):224-227, 1979.

[17] J. Dunn. Well separated clusters and optimal fuzzy partitions. *J. Cybernet.* **4**:95-104, 1974.

[18] P. Rousseuw. Silhouettes: A graphical aid to the interpretation and validation of cluster analysis. *Comput. Appl. Math.* **20**:53-65, 1987.

[19] M. Gonzalez Toledo. *A Comparison in Cluster Validation Techniques*. Master's Thesis, Univ. Puerto Rico-Mayaguez campus, 2005.

[20] N. Bolshakova, F. Azuaje. Cluster validation techniques for genome expression data. *Signal Process.* **83**(4):825-833, 2003.

[21] L. Hubert, P. Arabie. Comparing partitions. *J. Classif.* **2**:193-218, 1985.

[22] M. Halkidi, Y. Batistakis, M. Vazirgiannis. Cluster validity methods: Part i. *SIGMOD Record* **31**(12):40-45, 2002.

[23] R. Pearson, T. Zylkin, J. Schwaber, G. Gonye. Quantitative evaluation of clustering results using computational negative controls. In M. Berry, ed., *Proc. 4th SIAM Int. Conf. Data Mining*, Lake Buena Vista, Florida, 2004, pp.188-199.

[24] T. Hill, P. Lewicki. *Statistics: Methods and Applications*. Statsoft Inc., Tulsa, OK, 2006.

[25] Q. Tan, Z. Zhuang, P. Mitra, C. Lee Giles. A clustering-based sampling approach for refreshing search engine's database. *Proc. 10th Int. Workshop on the Web and Databases (WebDB 2007)*, Beijing, 2007.

[26] T. Hastie, R. Tibshirani, J. Friedman. *The Elements of Statistical Learning: Data Mining, Inference and Prediction*. Springer-Verlag, New York, 2001.

[27] S. Ma, J. Huang. Clustering threshold gradient descent regularization: With applications to microarray studies. *Bioinformatics* **23**(4):466-472, 2007.

[28] V. Hautamäki, S. Cherednichenko, I. Kärkkäinen, T. Kinnunen, P. Fränti. Improving K-means by outlier removal. *Lecture Notes Comput. Sci.* **3540**:978-987, 2005.

CHAPTER 3

FUZZY K-MEANS CLUSTER ANALYSIS

3.1 INTRODUCTION

Fuzzy logic provides a mixture of methods for flexible information processing of ambiguous data [1-3]. In this chapter, fuzzy K-means cluster (FKM) analysis is introduced. FKM was introduced by Dunn [4] and developed by Bezdek [5,6]. FKM provides a capability for generating microarray-specific *membership* probabilities in the interval [0,1]. Microarrays are allowed to have membership in many clusters, with varying degrees of membership [7]. The closer the cluster membership function is to 1, the stronger the degree of membership in a particular cluster. There have been many fundamental developments in fuzzy cluster analysis, including fuzzy learning vector quantization [8], fuzzy decision trees [9], data mining applications [10], rough-fuzzy artificial neural networks [11], evaluation of influential data [12], fuzzy relational classification [13], and generalized weighted conditional fuzzy clustering [14]. Research employing FKM with DNA microarrays to date has addressed spot-based algorithm [15], prognostic factors in cancer [16], data normalization [17], timecourse data [18], cluster reproducibility [19], kernel principal components [20], tumor classification [21], and straightforward application of FKM [22].

3.2 FUZZY K-MEANS ALGORITHM

The equations used for defining CKM in the previous chapter made no mention of the underlying minimization problem inherent in CKM's development. Looking at it from another perspective, the goal of CKM is to find a

Classification Analysis of DNA Microarrays, First Edition. Leif E. Peterson.

microarray partitioning scheme, which minimizes the objective function

$$\zeta(\mathbf{U}) = \sum_{i=1}^{n} \sum_{k=1}^{K} (u_{ik}) \|\mathbf{x}_i - \mathbf{v}_k\|^2, \tag{3.1}$$

where $\mathbf{x}_i$ is the ith microarray, $\mathbf{v}_k$ is the center for cluster k, and u_{ik} is the *membership function* subject to the constraint $\sum_{k=1}^{K} u_{ik} = 1$. The center for cluster k is computed as

$$\mathbf{v}_k = \frac{\sum_{i=1}^{n} (u_{ik}) \mathbf{x}_i}{\sum_{i=1}^{n} (u_{ik})}, \tag{3.2}$$

and the cluster membership function, as

$$u_{ik} = \begin{cases} 1 & \text{if } \|\mathbf{x}_i - \mathbf{v}_k\|^2 < \|\mathbf{x}_i - \mathbf{v}_j\|^2 \quad \forall j \neq k. \\ 0 & \text{otherwise} \end{cases} \tag{3.3}$$

Again, let $\mathbf{X} = \{\mathbf{x}_1, \mathbf{x}_2, \ldots, \mathbf{x}_n\}$ be a set of microarrays $(i = 1, 2, \ldots, n)$, $\mathbf{V} = \{\mathbf{v}_1, \mathbf{v}_2, \ldots, \mathbf{v}_k\}$ be a set of prototypes or "centers," and $\mathbf{U}$ be an $n \times K$ matrix. The objective function that is minimized in FKM clustering is

$$\zeta_m(\mathbf{U}) = \sum_{i=1}^{n} \sum_{k=1}^{K} (u_{ik})^m \|\mathbf{x}_i - \mathbf{v}_k\|, \quad m > 1 \tag{3.4}$$

where $\mathbf{x}_i$ is the ith microarray, $\mathbf{v}_k$ is the center for cluster k, u_{ik} is the cluster membership function subject to the constraint $\sum_{k=1}^{K} u_{ik} = 1$, and $m > 1$ is the fuzzifier. As $m \to 1$, the classifier becomes a crisp K-means classifier (CKM), and therefore, the usual range for FKM cluster analysis is $1 < m \leq 5$. By default, we start with a value of $m = 1.2$. The center for cluster k is computed as

$$\mathbf{v}_k = \frac{\sum_{i=1}^{n} (u_{ik})^m \mathbf{x}_i}{\sum_{i=1}^{n} (u_{ik})^m}, \tag{3.5}$$

where the cluster membership function is expressed as

$$u_{ik} = \left[\sum_{j=1}^{K} \left(\frac{\|\mathbf{x}_i - \mathbf{v}_k\|}{\|\mathbf{x}_i - \mathbf{v}_j\|} \right)^{1/(m-1)} \right]^{-1}. \tag{3.6}$$

Algorithm 3.1 lists the computational steps for FKM cluster analysis.

Example 3.1 Figure 3.1 illustrates the membership functions for FKM cluster analysis of the SRBCT data ($n = 63$ arrays, $p = 23$ filtered genes) using $K = 4$ and $m = 1.2$. As one can observe, microarrays were assigned to clusters for which the membership function value was the greatest among the $K = 4$ membership values.

Algorithm 3.1: Fuzzy K-Means Cluster Analysis

Data: Set of n input microarray $\mathbf{x}_i$ $(i = 1, 2, \ldots, n)$
Result: K clusters of microarrays
Select K and ϵ
Initialize the set of K prototypes $\mathbf{V} = \{\mathbf{v}_1, \mathbf{v}_2, \ldots, \mathbf{v}_K\}$
Specify the fuzziness parameter m in the interval $[2, 5]$
Specify ϵ
for $t \leftarrow 1$ ***to*** *#Iteration* **do**
  for $k \leftarrow 1$ ***to*** K **do**
    for $i \leftarrow 1$ ***to*** n **do**
      $u_{ik}^{(t)} = \left[\sum_{j=1}^{K} \left(\|\mathbf{x}_i - \mathbf{v}_k\| / \|\mathbf{x}_i - \mathbf{v}_j\|\right)^{1/(m-1)}\right]^{-1}$
    endfor
    $\mathbf{v}_k^{(t)} = \sum_{i=1}^{n} (u_{ik}^{(t)})^m \mathbf{x}_i / \sum_{i=1}^{n} (u_{ik}^{(t)})^m$
  endfor
  $E^{(t)} = \sum_{k=1}^{K} \|\mathbf{v}_k^{(t)} - \mathbf{v}_k^{(t-1)}\|^2$
  If $t > N$ or $E^{(t)} \leq \epsilon$ then **exit for**
endfor

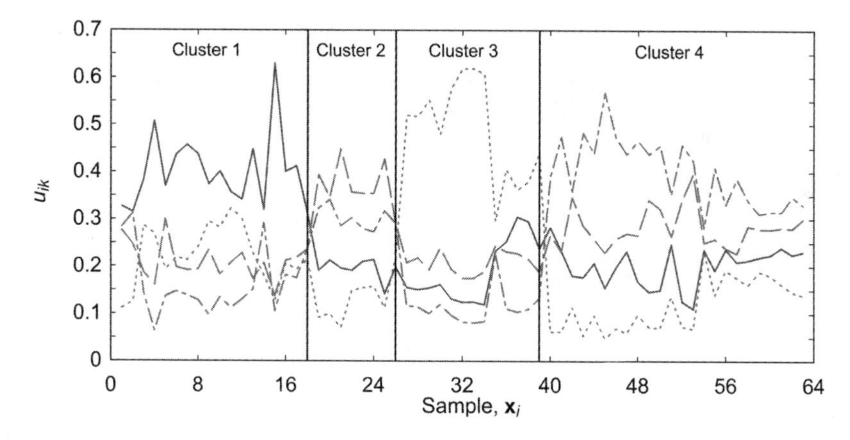

FIGURE 3.1 Membership function values u_{ik} for FKM when $K = 4$ and $m = 1.2$ for $n = 63$ microarrays in the SRBCT dataset. Feature values mean-zero standardized.

3.3 IMPLEMENTATION

A fundamental requirement of FKM cluster analysis is that the K cluster centers $\mathbf{v}_k$ be generated via cluster initialization prior to application of FKM. Once generated, cluster centers and microarrays are fed to the FKM routine for iterative learning. The source code listing below outlines the core methods used for FKM cluster analysis:

```
For iter = 1 To iterations
  ReDim vsum(numclusters, N)'reset (clear) array
  For j = 1 To N
    For k = 1 To numclusters
      denom = 0
      For i = 1 To M
        vsum(k, j) += (u(i, k) ^ power) * xmatrix(i, j)
        denom += u(i, k) ^ power
      Next i
      v(k, j) = vsum(k, j) / denom
    Next k
  Next j
  E = 0
  For i = 1 To M
    For k = 1 To numclusters
      u(i, k) = 0
    Next k
  Next i
  For i = 1 To M
    For k = 1 To numclusters
      sum = 0
      If DistanceFunc = 2 Then 'Euclidean
        For j = 1 To N
          diff = (xmatrix(i, j) - v(k, j))
          sum += diff * diff
        Next j
        d_ik = Math.Sqrt(sum)
      End If
      For l = 1 To numclusters
        sum = 0
         If DistanceFunc = 2 Then 'Euclidean
          For j = 1 To N
            diff = (xmatrix(i, j) - v(l, j))
            sum += diff * diff
          Next j
          d_il = Math.Sqrt(sum)
        End If
        u(i, k) += (d_ik / d_il) ^ (1 / (power - 1))
      Next l
      u(i, k) = 1 / u(i, k)
    Next k
  Next i

  For k = 1 To numclusters
    sum = 0
```

```
    For j = 1 To N
      If v(k, j) <> 0 AndAlso vold(k, j) <> 0 Then
        diff = (v(k, j) - vold(k, j))
        sum += diff * diff
      End If
    Next j
    eucdis = Math.Sqrt(sum)
    E += eucdis
  Next k
  If E < 0.05 Then Exit For
  For k = 1 To numclusters
    For j = 1 To N
      vold(k, j) = v(k, j)
    Next j
  Next k
  For i = 1 To M
    For k = 1 To numclusters
      uold(i, k) = u(i, k)
    Next k
  Next i
Next iter
```

Example 3.2 FKM was implemented using the four-class SRBCT data consisting of $n = 63$ microarrays and $p = 23$ features selected with a greedy plus-take-away (PTA)-1 method (see Table 12.9). The display of cluster membership values was achieved by first performing a descending sort of the K membership values within each microarray, and then plotting a boxplot with the K groups of cluster membership values disregarding which of the K memberships the microarrays were assigned to. For example, if there are $K = 4$ clusters, there will be four cluster membership functions for each microarray and each membership value will represent a specific cluster. Consider microarray $\mathbf{x}_1$ with cluster membership values $\mathbf{u}_1 = \{2.6, 1.3, 6.1, 3.4\}$ for clusters $\boldsymbol{\omega} = \{1, 2, 3, 4\}$, and microarray $\mathbf{x}_2$ with cluster membership values $\mathbf{u}_2 = \{7.2, 3.5, 0.8, 2.2\}$ for clusters $\boldsymbol{\omega} = \{1, 2, 3, 4\}$. After sorting cluster membership values within the microarrays, we get $\mathbf{u}_1 = \{6.1, 3.4, 2.6, 1.3\}$ without regard to cluster assignments for microarray $\mathbf{x}_1$ and $\mathbf{u}_2 = \{7.2, 3.5, 2.2, 0.8\}$, disregarding cluster assignments for microarray $\mathbf{x}_2$. Construction of the boxplot would involve using values 6.1 and 7.2 in the first category (variable), 3.4 and 3.5 in the second boxplot category, and so on.

Figure 3.2 shows the sorted cluster membership values for the four-class SRBCT data using $K = 10$ and $m = 1.2$ [panel (a)] and $m = 2$ [panel (b)]. For all of the 63 microarrays, clearly for $m = 1.2$ stronger cluster membership is preferred for a unique cluster. However, when $m = 2$, there are not nearly as many preferential assignments of microarrays to a unique cluster. A similar trend is shown in Figure 3.3 for FKM ($K = 10$) using the three-class leukemia

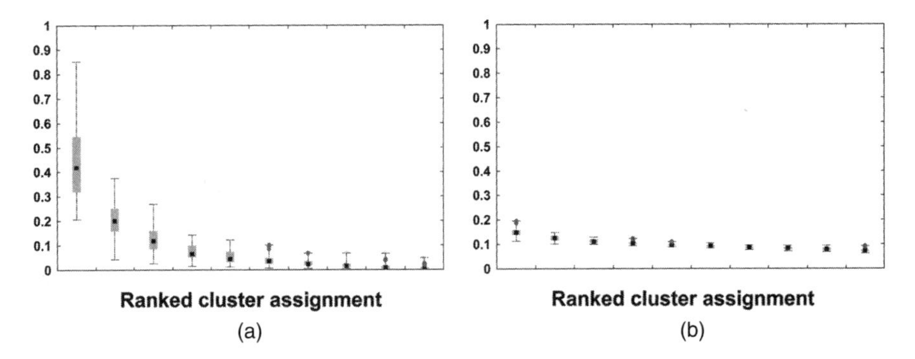

FIGURE 3.2 Sorted microarray-specific cluster membership values u_{ik} for the four-class SRBCT data using ($K = 10$) clusters and $m = 1.2$ (a) and $m = 2$ (b). Cluster membership values for the $K = 10$ clusters were sorted in descending order within each microarray, and are shown in the plots in descending order from left to right.

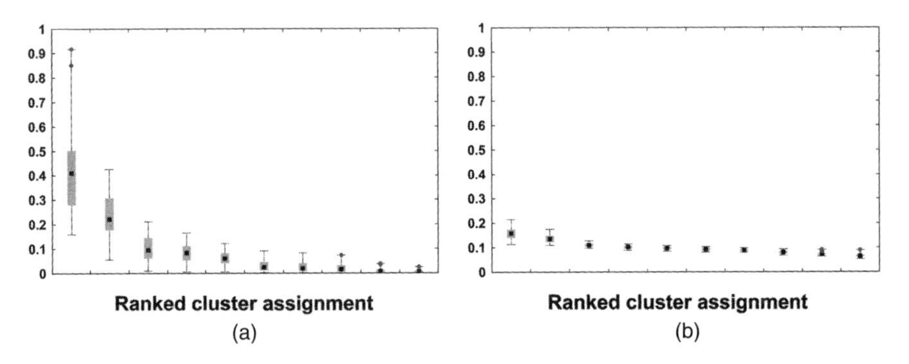

FIGURE 3.3 Sorted microarray-specific cluster membership values u_{ik} for the three-class leukemia data using ($K = 10$) clusters and $m = 1.2$ (a) and $m = 2$ (b). Cluster membership values for the $K = 10$ clusters were sorted in descending order within each microarray, and are shown in the plots in descending order from left to right.

data consisting of $n = 57$ microarrays and $p = 13$ features (Table 12.9). These results suggest that, for these data, as $m \to 5$ there is little chance for cluster membership values to uniquely load on one particular cluster for larger values of m in the range $1 < m \leq 5$. The advantages of FKM are immediately apparent in Figures 3.2 and 3.3, since microarrays are assigned primarily to a single cluster having the greatest membership value, and then to remaining clusters based on monotonically decreasing membership values. In short, every microarray is assigned to a cluster, and there is little confusion regarding the degree of relatedness with a certain cluster center.

This obviously does not rule out the chance for assignment of an outlier microarray to the "wrong" cluster. Cluster validity scores (see previous chapter) can be determined once microarrays are assigned to clusters with the greatest membership value.

An advantage of FKM over CKM is that it assigns each microarray to a cluster by use of K membership values per microarray. Recall that CKM assigns each microarray to only one cluster. The choice of the fuzziness parameter m also depends on the dataset and prespecified value of K, although this is not always straightforward. Dembele and Kastner [22] reported results for applying FKM on genes rather than microarrays, and suggested that m varies with each dataset and that the commonly used value $m = 2$ is not always the best choice. Our findings that $m = 1.2$ resulted in greater first and second membership values when compared with $m = 2$ is in agreement with results of Kim et al. [17], who showed similar results for DNA microarrays. Another advantage of FKM is realized when clusters overlap, and there is a requirement to assign microarrays to more than one cluster. Missing data have also been exploited using FKM [23].

The effect of distance metric on FKM is also of interest, so we ran FKM on the four-class SRBCT data, but instead of using Euclidean distance, we

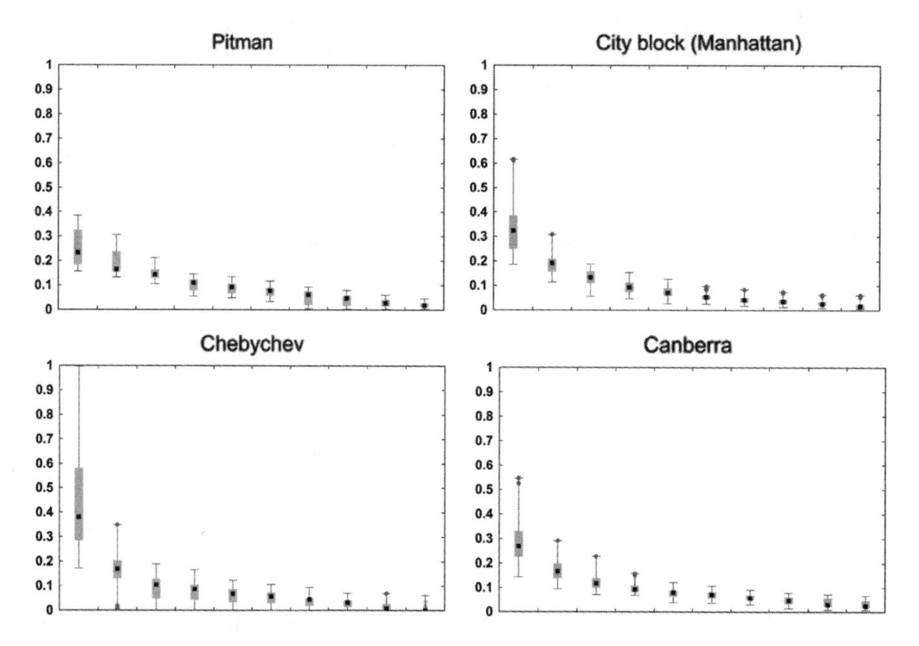

FIGURE 3.4 Cluster membership values u_{ik} comparing FKM results using $K = 10$ and $m = 1.2$ for the SRBCT data with Pitman correlation, city block (Manhattan) distance, Chebyshev distance, and Canberra distance.

used Pitman correlation, city block (Manhattan), Chebyshev, and Canberra. Results are shown in Figure 3.4 and suggest that following Euclidean distance (Figure 3.2), the greatest cluster membership values were observed for Chebyshev distance, followed by city block, Canberra, and Pitman correlation in decreasing order. In conclusion, FKM can increase the clinical significance of microarrays in clusters consisting of microarrays, and biological significance of genes in clusters of genes, although we did not evaluate the latter premise.

3.4 SUMMARY

We introduced the reader to FKM cluster analysis in this chapter and reported on the effects of the fuzzification parameter m on the resulting cluster membership values. The optimal values of m depend on characteristics of the sample and feature space, and in general lower values of $m < 1.5$ tend to result in greater membership in the first and second membership values. We also showed that for the same fuzziness parameter value, specifically, $m = 1.2$, we observed no significant differences for the resulting cluster membership values.

Fuzzy K-means cluster analysis has been successfully applied to a wide range of applications and is now considered to be an established clustering methodology. Development and refinement of FKM cluster methods continue to grow as the utility of soft computing with fuzzy methods increases.

REFERENCES

[1] D. Dubois, H. Prade. *Fundamentals of Fuzzy Sets*. Kluwer, Boston, 2000.

[2] G. Klir, B. Yuan. *Fuzzy Sets and Fuzzy Logic*. Prentice-Hall, Upper Saddle River, NJ, 1995.

[3] S. Pal, P. Mitra. *Pattern Recognition Algorithms for Data Mining: Scalability, Knowledge Discovery and Soft Granular Computing*. Chapman & Hall, Boca Raton, FL, 2004.

[4] J. Dunn. A fuzzy relative of the isodata process and its use in detecting compact well separated clusters. *J. Cybernet.* **3**(3):32–57, 1974.

[5] J. Bezdek, C. Coray, R. Gunderson, J. Watson. Detection and characterization of cluster substructure 1—linear structure: Fuzzy c-lines. *SIAM J. Appl. Math.* **40**(2):339–357, 1981.

[6] J. Bezdek. *Pattern Recognition with Fuzzy Objective Function Algorithm*. Plenum, New York, 1981.

[7] V. Delport, D. Liesch. Fuzzy c-means algorithm for codebook design in vector quantization. *Electron. Lett.* **30**(13):1025–1026, 1994.

[8] N. Karayiannis, J. Bezdek. An integrated approach to fuzzy learning vector quantization and fuzzy c-means clustering. *IEEE Trans. Fuzzy Syst.* **5**(4):622–628, 1997.

[9] S. Mitra, K. Konwar, S. Pal. Fuzzy decision tree, linguistic rules and fuzzy knowledge-based network: Generation and evaluation. *IEEE Trans. Syst. Man, Cybernet.* **32**(4):328–339, 2002.

[10] S. Mitra, S. Pal, P. Mitra. Data mining in soft computing framework: A survey. *IEEE Trans. Neural Networks* **13**(1):3–14, 2002.

[11] S. Pal, S. Mitra, P. Mitra. Rough-fuzzy MLP: Modular evolution, rule genration, and evaluation. *IEEE Trans. Knowl. Data Eng.* **15**(1):14–25, 2003.

[12] H. Imai, A. Tanaka, Miyakoshi. A method of identifying influential data in fuzzy clustering. *IEEE Trans. Fuzzy Syst.* **6**(1):90–101, 1998.

[13] M. Setnes, R. Babushka. Fuzzy relational classifier trained by fuzzy clustering. *IEEE Trans. Syst. Man, Cybern.* **29**(5):619–625, 1999.

[14] J. Leski. Generalized weighted conditional fuzzy clustering. *IEEE Trans. Fuzzy Syst.* **11**(6):709–715, 2003.

[15] A. Daskalakis, D. Cavouras, P. Bougioukos, S. Kostopoulos, P. Georgiadis, I. Kalatzis, G. Kagadis, G. Nikiforidis. Genes expression level quantification using a spot-based algorithmic pipeline. *Proc. IEEE Eng. Med. Biol. Soc.* **2007**:1148–1151, 2007.

[16] T. Czernicki, J. Zegarska, L. Paczek, B. Cukrowska, W. Grajkowska, A. Zajaczkowska, K. Brudzewski, J. Ulaczyk, A. Marchel. Gene expression profile as a prognostic factor in high-grade gliomas. *Int. J. Oncol.* **30**(1):55–64, 2007.

[17] S.Y. Kim, J.W. Lee, J.S. Bae. Effect of data normalization on fuzzy clustering of DNA microarray data. *BMC Bioinform.* **7**: 134, 2006.

[18] M.E. Futschik, B. Carlisle. Noise-robust soft clustering of gene expression time-course data. *J. Bioinform. Comput. Biol.* **3**(4):965–988, 2005.

[19] N.R. Garge, G.P. Page, A.P. Sprague, B.S. Gorman, D.B. Allison. Reproducible clusters from microarray research: Whither? *BMC Bioinform.* **6**(Suppl 2): S10, 2005.

[20] Z. Liu, D. Chen, H. Bensmail, Y. Xu. Clustering gene expression data with kernel principal components. *J. Bioinform. Comput. Biol.* **3**(2):303–316, 2005.

[21] J. Wang, T.H. Bo, I. Jonassen, O. Myklebost, E. Hovig. Tumor classification and marker gene prediction by feature selection and fuzzy c-means clustering using microarray data. *BMC Bioinform.* **4**: 60, 2003.

[22] D. Dembele, P. Kastner. Fuzzy C-means method for clustering microarray data. *Bioinformatics* **19**(8):973–980, 2003.

[23] M. Sarkar, T.Y. Leong. Fuzzy K-means clustering with missing values. *Proc. AMIA* (American Medical Informatics Association) *Symp.* Washington, DC, 2001, pp. 588–592.

CHAPTER 4

SELF-ORGANIZING MAPS

4.1 INTRODUCTION

The self-organizing map (SOM) was developed by Kohonen for mapping a high-dimensional feature space into a low-dimensional space. The SOM is one of the most popular neural network models used in exploratory data analysis for class discovery [1]. SOMs are usually two-dimensional grids consisting of a set of neurons or *nodes* arranged in rows and columns [2]. Underlying each node is a *reference vector* or *prototype* having the same number of features as input objects. Figure 4.1 shows a 2D SOM with 25 nodes and the associated p-dimensional reference vectors. In the schematic, microarray $\mathbf{x}_i$ is closest to the reference vector in row 4 and column 4, which happens to be node $m = 19 = 5(4-1)+4$, based on the relationship $m = \#\text{cols} * (\text{row} - 1) + \text{col}$. The reference vector that a microarray is closest to is called the *best matching unit*.

The SOM is an *associative learning* algorithm that accomplishes its learning in a fashion similar to that of artificial neural networks. For each input microarray, a nearest prototype is identified. Weights of the winning node (reference vector) are updated through a gradient descent approach and weight updates are applied to a lesser degree to nearby nodes that reside in the same *neighborhood*. Two-dimensional (2D) SOMs have been employed in various DNA microarray studies [3-5]. Applications include methodologic developments with independent component analysis [6], infectomics [7,8], molecular classification [9], component plane presentation of data [10], cellular differentiation [11], metastasis [12], and time-course of inflammation [13].

4.2 ALGORITHM

Let $\mathbf{x}_i \in \mathbb{R}^p$ be the ith microarray $(i = 1, 2, \ldots, n)$; $\mathbf{v}_m \in \mathbb{R}^p$ $(m = 1, 2, \ldots, M)$, a p-dimensional $(j = 1, 2, \ldots, p)$ reference vector; and $\mathbf{r}_m \in \mathbb{R}^2$, a 2D fixed *map*

Classification Analysis of DNA Microarrays, First Edition. Leif E. Peterson.

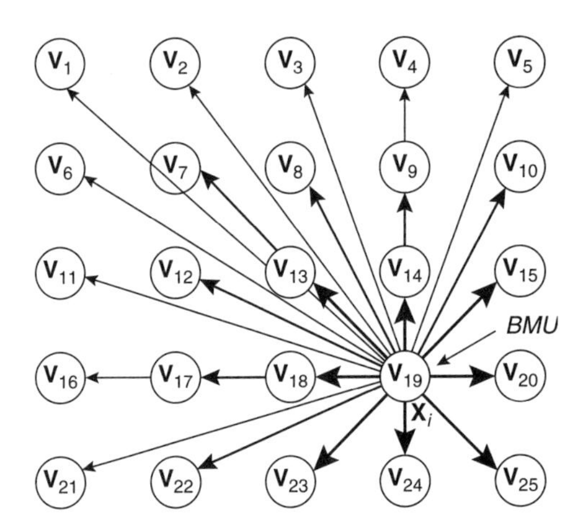

FIGURE 4.1 Structure of a 2D self-organizing map. The SOM shown has $M = 25$ nodes, each of which is associated with a p-dimensional reference vector $\mathbf{v}_m$, whose length is the same size as the p-dimensional input microarrays $\mathbf{x}_i$. Microarray $\mathbf{x}_i$ shown in the map is closest to node $m = 19$, which is the best matching unit (BMU). Line thickness reflects the strength of update according to neighborhood function (4.7) using distance from BMU.

location vector associated with $\mathbf{v}_m$. The map location vector $\mathbf{r}_m$ has coordinates k and l representing the row and column of the two-dimensional (2D) map. The simplest way to design a symmetric 2D SOM of M nodes is to set the number of rows and columns equal to $\sqrt{M}$. If $\sqrt{M}$ does not result in an integer and has a decimal fraction, then round either down or up using the floor-ceiling function, and then square the result. For example, with 500 nodes ($M = 500$), $\sqrt{(500)} = 22.36$. If rounding down, $[\text{Floor}(\sqrt{500})]^2 = 22^2 = 484$. On the other hand, $[\text{Ceiling}(\sqrt{500})]^2 = 23^2 = 529$. Thus, you would use $M = 484$ nodes arranged into 22 rows and 22 columns in the map, or $M = 529$ nodes arranged into 23 rows and 23 columns. Obviously, this approach is intended for designing symmetric maps for which the number of rows equals the number of columns. This may seem overly simplistic; however, keeping the SOM symmetric has its advantages.

Reference Vectors. Each reference vector $\mathbf{v}_m$ in a SOM consists of two subvectors: (1) a p-dimensional vector of learning weights that is of the same length as the number of input features and (2) a fixed 2D vector $\mathbf{r}_m$ of map locations, where r_{1m} is equal to the row and r_{2m} is equal to the column of the node on the 2D map grid. Assume a $M = 100$ node map arranged as a 2D 10×10 grid. The values of map locations in row 1 are $\mathbf{r}_1 = (1,1)$, $\mathbf{r}_2 = (1,2)$, ..., $\mathbf{r}_{10} = (1,10)$; in row 2, $\mathbf{r}_{11} = (2,1)$, $\mathbf{r}_{12} = (2,2)$, ..., $\mathbf{r}_{20} = (2,10)$, and in the last (10th) row are $\mathbf{r}_{91} = (10,1)$, $\mathbf{r}_{92} = (10,2)$, ...,

$\mathbf{r}_{100} = (10,10)$. The structure of reference vectors for a typical 2D SOM are as follows:

$\mathbf{v}_1 =$	v_{11}	v_{21}	v_{31}	$\cdots$	v_{j1}	$\cdots$	v_{p1}	r_{11}	r_{21}
$\mathbf{v}_2 =$	v_{12}	v_{22}	v_{32}	$\cdots$	v_{j2}	$\cdots$	v_{p2}	r_{12}	r_{22}
$\mathbf{v}_m =$	v_{1m}	v_{2m}	v_{3m}	$\cdots$	v_{jm}	$\cdots$	v_{pm}	r_{1m}	r_{2m}
$\mathbf{v}_M =$	v_{1M}	v_{2M}	v_{3M}	$\cdots$	v_{jM}	$\cdots$	v_{pM}	r_{1M}	r_{2M}

As we shall see later in this chapter, the weight values v_{jm} are updated during the learning process, whereas the map location vector $\mathbf{r}_m$ remains fixed and never changes.

4.2.1 Feature Transformation and Reference Vector Initialization

Raw feature values, when input directly into an SOM for training, are usually inappropriate for neural learning. Most neural adaptive learning systems require a narrow range of input values such as $[-1,1]$, [0,1], or standard normal variates distributed as $\mathcal{N}(0,1)$. The appropriate transformation of feature values prior to SOM training will ensure an optimal solution for understanding the structure of the data.

In addition to the appropriate input feature values, the reference vectors employed for neural adaptive learning must be of the same range and scale as the input features in order to optimize learning. There are several techniques for transforming input features and initializing reference vectors.

Mean-zero standardization is one of the most popular methods. First, the input feature values are standardized in the form

$$x_{ij} = \frac{(x_{ij} - \mu_j)}{\sigma_j}, \tag{4.1}$$

where μ_j is the mean value of feature j over all microarrays and σ_j is the standard deviation of feature j based on all microarrays. With mean-zero standardization of input microarrays, the reference vectors are also initialized using randomly generated standard normal variates that are distributed $\mathcal{N}(0,1)$. This ensures that the range and scale of the input features and reference vectors are the same at the beginning of training.

Another type of initialization normalizes feature values of reference vectors in the range [0,1]. For normalization, feature values of the input values need to be normalized first. These are given in the form

$$x_{ij} = \frac{x_{ij} - \min\{x_{ij}\}}{\Delta x_j}, \tag{4.2}$$

where $\min\{x_{ij}\}$ is the minimum value of feature j over all microarrays, and

$$\Delta x_j = \max\{x_{ij}\} - \min\{x_{ij}\} \tag{4.3}$$

is the range of feature j over all microarrays. Once the input features of $\mathbf{x}_i$ have been normalized, feature values of reference vectors can be set in the range [0,1] using uniform random quantiles, $U(0,1)$.

A third technique is to use the raw feature input values, and randomly select values within the range of each feature for initializing reference vectors. For example, the jth feature value for a reference vector can be set to

$$v_{jm} = \min\{x_{ij}\} + U(0,1)\Delta x_j. \tag{4.4}$$

When using this approach, you must ensure that the range and scale of all the transformed feature values are similar. Otherwise, gradient descent learning can fail.

4.2.2 Learning

The SOM learning process involves an iterative process during which the reference vectors learn the patterns among microarrays within a neighborhood. The neighborhood size decreases monotonically with the number of training iterations. The degree of learning decreases monotonically via controlled reductions in a learning rate, up to a point where the change among learned patterns in the reference vectors becomes negligible. Initially, each microarray is selected and the closest reference vector $\mathbf{v}_c$ is determined, using the relationship

$$c = \arg\min_m(\|\mathbf{v}_m - \mathbf{x}_i\|). \tag{4.5}$$

Using the closest reference vector, the neighboring reference vectors are updated according to the relationship

$$\mathbf{v}_m = \mathbf{v}_m + \alpha(t)h_{cm}[\mathbf{x}_i - \mathbf{v}_m], \tag{4.6}$$

where $\alpha(t)$ is the learning rate at the tth iteration ($t = 1, 2, \ldots, T$), the *neighborhood function* is

$$h_{cm} = \exp\left(-\frac{\|\mathbf{r}_c - \mathbf{r}_m\|^2}{2\beta(t)^2}\right), \tag{4.7}$$

$\|\mathbf{r}_c - \mathbf{r}_m\|$ is the Euclidean distance between the 2D map locations of reference vectors $\mathbf{v}_c$ and $\mathbf{v}_m$ and $\beta(t)$ is the width of the neighborhood size at iteration t [14]. Note that $\|\mathbf{r}_c - \mathbf{r}_m\|$ is a straightforward calculation of distance using the row and column numbers of the two nodes. For example, in Figure 4.1, the best matching unit (BMU) is node $c = 19$ whose map location is $\mathbf{r}_{19} = (4, 4)$. The map location for node $m = 6$ is given as $\mathbf{r}_6 = (2, 1)$ and is farther from the BMU than node $m = 18$ for which $\mathbf{r}_{18} = (4, 3)$ since $\|\mathbf{r}_{19} - \mathbf{r}_6\| = \sqrt{(4-2)^2 + (4-1)^2} = 3.6$ and $\|\mathbf{r}_{19} - \mathbf{r}_{18}\| = \sqrt{(4-4)^2 + (4-3)^2} = 1$. The thickness of lines with arrows in Figure 4.1 reflects the general idea of the strength of update by the neighborhood function h_{cm}.

Suitable choices for the learning rate and width are

$$\alpha(t) = \alpha_0 \left(1 - \frac{t-1}{T}\right), \tag{4.8}$$

and

$$\beta(t) = \beta_0 \left(1 - \frac{t-1}{T}\right). \tag{4.9}$$

Reliable results for SOM are commonly obtained using an initial value of $\alpha_0 = 0.05$. Kohonen recommended using an initial value of β_0 equal to at least the column (row) width of the entire SOM grid. We have found that for square grids, the use of $\beta_0 = \sqrt{M}$ works well. During training, Villmann et al. have recommended keeping $\beta(t) > 0.35$ to prevent overfitting [15]. Finally, $T = 100,000$ training iterations should be used.

The preceding calculations at iteration t are repeated for the remaining microarrays, each time looping over all M reference vectors to find the closest reference vector. This is then carried out over T total iterations. Algorithm 4.1 lists the pseudocode for the SOM algorithm.

Figure 4.2 shows the weight values of $M = 100$ 2D reference vectors during 50 learning iterations. We used 100 2D input microarrays $\mathbf{x}_i$ with fixed feature values of (1,1), (1,2), ..., (10,10) for training. Random initialization of reference vector values $\mathbf{v}_m$ was in the range [0,1]. As the competitive learning process progressed, reference vector weights converged to the values of the input microarrays.

4.2.3 Conscience

There is a possibility that several reference vectors will overrepresent the input data because of the initial weight values. This can be prevented by use of *conscience*, which increases the opportunity for reference vectors to represent equal input information [16]. This is accomplished via a bias term B_m and

Algorithm 4.1: Self-Organizing Map

Data: p-dimensional input microarrays $\mathbf{x}_i$ ($i = 1, 2, \ldots, n$), p-dimensional reference vectors $\mathbf{v}_m$ ($m = 1, 2, \ldots, M$), 2D nodes $\mathbf{r}_m$
Result: Organized 2D map
$\alpha_0 = 0.05$
$\beta_0 = \sqrt{M}$ for a square map
Initialize M reference vectors $\mathbf{V} = \{\mathbf{v}_1^{(0)}, \mathbf{v}_2^{(0)}, \ldots, \mathbf{v}_M^{(0)}\}$
$T = 100,000$
for $t \leftarrow 1$ **to** T **do**
  $\alpha(t) = \alpha_0\{1 - [(t-1)/T]\}$
  $\beta(t) = \beta_0\{1 - [(t-1)/T]\}$
  for $i \leftarrow 1$ **to** n **do**
    Determine $c = \arg\min_m \|\mathbf{x}_i - \mathbf{v}_m^{(t-1)}\|$
    for $m \leftarrow 1$ **to** M **do**
      $h_{cm} = \exp\left[\left(\|\mathbf{r}_c - \mathbf{r}_m\|^2\right)/2\beta(t)^2\right]$
      $\mathbf{v}_m^{(t)} = \mathbf{v}_m^{(t-1)} + \alpha(t)h_{cm}[\mathbf{x}_i - \mathbf{v}_m^{(t-1)}]$
    endfor
  endfor
endfor

winning frequency F_m for each reference vector. Determination of the closest reference vector near each microarray is adjusted using the relationship

$$c = \arg\min_m\{\|\mathbf{x}_i - \mathbf{v}_m^{(t-1)}\| - B_m\}. \tag{4.10}$$

During initialization, $B_m = 0$ and $F_m = 1/M$. During training iterations, the bias and frequency of winning are updated as

$$B_m = \gamma^{(t)}\left(\frac{1}{M} - F_m\right), \tag{4.11}$$

where $\gamma^{(t)}$ is a coefficient updated at each iteration, which is set to, for example, 4 initially. Updates during training are as follows

$$\begin{aligned} F_c^{(t)} &= F_c^{(t-1)} + \beta(1 - F_c^{(t-1)}) \\ F_m^{(t)} &= F_m^{(t-1)} + \beta(0 - F_m^{(t-1)}), \qquad \forall m \neq c \end{aligned} \tag{4.12}$$

where β is a small positive fraction such as 0.1. The SOM algorithm with conscience updates is listed in Algorithm 4.2.

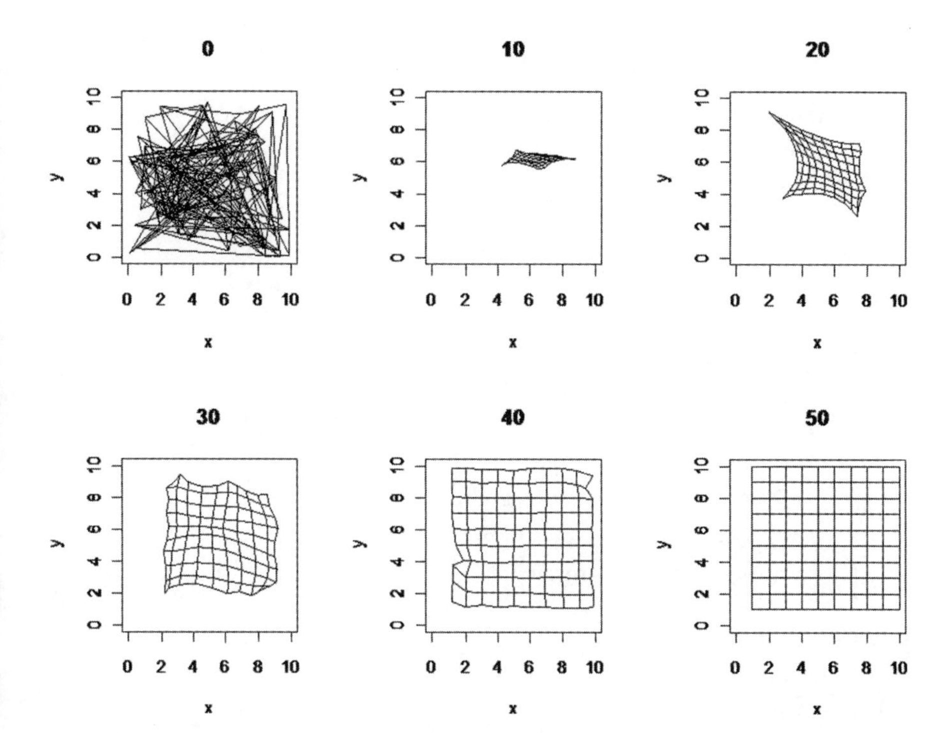

FIGURE 4.2 Weight values of 100 reference vectors vectors $\mathbf{v}_m$ based on 2D weights (x, y) updated during competitive learning iterations of a 10×10 map. One hundred 2D input microarrays $\mathbf{x}_i$ with fixed values of (1,1), (1,2), ..., (10,10) were used for training. Bold values shown above panels are the number of iterations.

4.3 IMPLEMENTATION

4.3.1 Feature Transformation and Reference Vector Initialization

Mean-Zero Standardization. The most common feature transformation for SOM is mean-zero standardization. First, calculate the mean and standard deviation of each feature. Next, standardize each microarray's feature value by subtracting the respective feature mean and divide the feature standard deviation, using the following code:

```
For j = 1 To NumFeatures
  Sum = 0
  For i = 1 To NumObjects
    Sum += x(j, i)
  Next i
  avg = Sum / NumObjects
```

Algorithm 4.2: Self-Organizing Map with Conscience

Data: p-dimensional input microarrays $\mathbf{x}_i$ ($i = 1, 2, \ldots, n$), p-dimensional reference vectors $\mathbf{v}_m$ ($m = 1, 2, \ldots, M$), 2D nodes $\mathbf{r}_m$
Result: Organized 2D map
$\alpha_0 = 1$, $\gamma_0 = 4$, $\beta = 0.1$
Initialize bias $B_m = 0$ ($m = 1, 2, \ldots, M$), $F_m = 1/M$
for $t \leftarrow 1$ **to** T **do**
  $\alpha(t) = \alpha_0\{1 - [(t-1)/T]\}$
  $\gamma(t) = \gamma_0\{1 - [(t-1)/T]\}$
  for $m \leftarrow 1$ **to** M **do**
    $B_m = \gamma^{(t)}[(1/M) - F_m]$
  endfor
  for $i \leftarrow 1$ **to** n **do**
    Determine $c = \arg\min_m\{\|\mathbf{x}_i - \mathbf{v}_m^{(t-1)}\| - B_m\}$
    Set $F_c^{(t)} = F_c^{(t-1)} + \beta(1 - F_c^{(t-1)})$
    Set $F_m^{(t)} = F_m^{(t-1)} + \beta(0 - F_m^{(t-1)}), \forall m \neq c$
    for $m \leftarrow 1$ **to** M **do**
      $\mathbf{v}_m^{(t)} = \mathbf{v}_m^{(t-1)} + \alpha(t)\exp\{-[\|\mathbf{r}_c - \mathbf{r}_m\|/\beta(t)^2]\}[\mathbf{x}_i - \mathbf{v}_m^{(t-1)}]$
    endfor
  endfor
endfor

```
  Sum = 0
  For i = 1 To NumObjects
    Sum += (x(j, i) - avg) * (x(j, i) - avg)
  Next i
  sigma = Math.Sqrt(Sum / (NumObjects - 1))
  For i = 1 To NumObjects
    x(j, i) = (x(j, i) - avg) / sigma
  Next i
Next j
```

Next, the array of reference vectors is dimensioned according to the number of features and number of nodes: `refvec(NumFeatures,Nodes)`. Now that the input feature values are standardized, the reference vector values are initialized by use of standard normal variates, using the following code:

```
For i = 1 To nodes
  For j = 1 To NumFeatures
    refvec(j, i) = StdNorm()
  Next j
Next i
```

Normalization. The following code shows how to normalize the feature values of the input data to the range [0,1]. The vectors `FeatureMin`, `Featuremax`, and `FeatureDelta` contain the minimum (maximum) and range of each feature value among all microarrays used in training. The vectors are zeroed out and then set equal to their respective values:

```
For j = 1 To NumFeatures
    FeatureMin(j) = 1.0E+30
    FeatureMax(j) = -1.0E+30
Next j
For j = 1 To NumFeatures
    For i = 1 To NumObjects
        If x(j, i) < FeatureMin(j) Then FeatureMin(j) = x(j, i)
        If x(j, i) > FeatureMax(j) Then FeatureMax(j) = x(j, i)
    Next i
    FeatureDelta(j) = FeatureMax(j) - FeatureMin(j)
   Next j
```

The feature values for each microarray are then normalized to the range [0,1] using the minimum value and range for each feature:

```
ReDim refvec(NumFeatures, nodes)
For i = 1 To NumObjects
    For j = 1 To NumFeatures
        x(j, i) = (x(j, i) - FeatureMin(j)) / FeatureDelta(j)
    Next
    If FeatureDelta(j) = 0 Then x(j, i) = Rnd()
Next i
```

The reference vector array is then initialized by setting the reference vector values equal to random uniform variates, as follows:

```
For k = 1 To nodes
    For j = 1 To NumFeatures
        refvec(j, k) = Rnd()
    Next j
Next k
```

Random Feature Values. If the raw input feature values are left intact and used during training, the reference vector value initialization should then be based on random feature values that are within the range of the observed values. Unless the input feature value range and scale are uniform across all features, this technique is not recommended for use. Nevertheless, the code below shows how reference vector values can be randomly chosen in the range of feature values, assuming that the input feature values are left intact and are not changed:

```
For k = 1 To nodes
  For j = 1 To NumFeatures
```

```
    refvec(j, k) = FeatureMin(j) + Rnd() * FeatureDelta(j)
  Next j
Next k
```

By initializing reference vector values with the same range and scale as the input feature values, we can prevent overpopulating the SOM with reference vectors whose starting values are irrelevant to the sample space and can increase the chances of identifying clusters that represent the microarrays. If this approach is not followed, there is an increased chance that longer runs with a greater number of iterations will be needed to adequately train the SOM.

4.3.2 Reference Vector Weight Learning

Neural adaptive learning for updating reference vector elements is quite straightforward. The outer loop is based on the iterations, with the next loop for each microarray. For each microarray, the BMU is identified, and the next loop updates each node using the BMU to determine the distance between the BMU and each node $\|\mathbf{r}_c - \mathbf{r}_m\|$, which is squared and rendered negative, divided by $2\beta(t)^2$, and then exponentiated to obtain the neighborhood function in (4.7). The following code provides the necessary calculations for updating reference vectors:

```
alpha = 0.05
beta = NumMapRows
For iter = 1 To somiterations
  AQE = 0
  betat = beta * (1 - (iter - 1) / somiterations)
  alphat = alpha * (1 - (iter - 1) / somiterations)
  For k = 1 To NumObjects
    lowestdist = 1.0E+30
    ransamp = (NumObjects - 1) * Rnd() + 1
    'ransamp = k ' use when NumObjects is small
    For i = 1 To nodes
      Sum = 0
      For j = 1 To NumFeatures
        diff = x(j, ransamp) - refvec(j, i) 'ransamp was k
        Sum += diff * diff
      Next j
      eucdist = Math.Sqrt(Sum)
      If eucdist < lowestdist Then
        lowestdist = eucdist
        lowestnode = i
      End If
    Next i
    AQE += lowestdist
    For i = 1 To nodes

     Sum = 0
      For j = 1 To 2
```

```
        diff = node(j, i) - node(j, lowestnode)
        Sum += diff * diff
      Next j
      eucdist = Math.Sqrt(Sum)
      For j = 1 To NumFeatures
        refvec(j, i) += alphat * Math.Exp(-eucdist * eucdist / _
                        (2 * betat * betat)) * _
                        (x(j, ransamp) - refvec(j, i))
      Next j
    Next i
  Next k
Next iter
```

Note that in the above code, arrays were randomly drawn for training, which is our preferred approach for large-sample problems. This also requires reducing the number of objects sampled. In the event that all arrays are to be used during training, then simply set the number of objects equal to the total number of arrays and use the line `ransamp=k`. Once adaptive learning is complete and the reference vectors are updated, the distance between nodes can be used for component maps, U-matrix calculations, and visualization. Dimensional reduction of an input feature set can be utilized with a trained SOM by using the row and column number of the BMU for each microarray, in order to acquire new x and y values for use in future classification analysis.

4.4 CLUSTER VISUALIZATION

After SOM training, each node in the 2D map will be represented by a p-feature reference vector whose weights were iteratively trained with associative learning. A variety of cluster boundary detection methods have been proposed for SOM. In the following sections, we will discuss some of the more common varieties.

4.4.1 Crisp *K*-Means Cluster Analysis

It may be preferable to assign each reference vector to a cluster and show these clusters in the 2D map by assigning various cluster-specific colors to the reference vectors. The location of microarrays on the 2D map can then be represented by tagging each microarray with its class label, and drawing the microarray labels at the location of the BMUs. Naturally, the number of clusters used for the reference vector space (and sample space) depends on the a priori number of clusters chosen or the optimal numbers of clusters based on cluster validity measures (see Section 2.5). Figure 4.3 shows an example cluster image for a 2D SOM showing class labels and locations of microarrays for the four-class SRBCT data. CKM cluster analysis was used with $K = 4$.

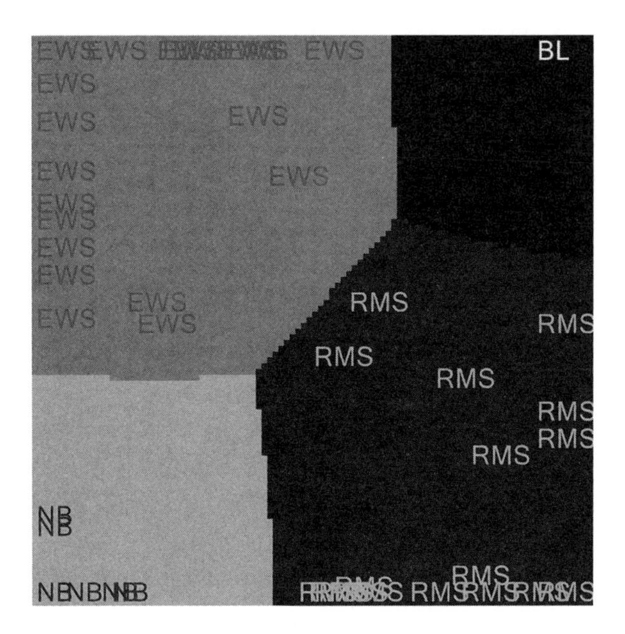

FIGURE 4.3 Cluster image of reference vectors based on CKM cluster analysis with $K = 4$. Input microarray location and class labels are listed at the location of the closest reference vector for each microarray.

4.4.2 Adjacency Matrix Method

When the number of microarrays greatly outweighs the number of nodes in an SOM (i.e., $n >> M$), the adjacency matrix method can be used [17]. Let $\mathbf{A}$ be an $M \times M$ adjacency matrix, where M is the total number of nodes (reference vectors) in an SOM. Each element of $\mathbf{A}$ is equal to

$$a_{jk} = n_j + n_k, \tag{4.13}$$

where n_j is the number of microarrays for which reference vector $\mathbf{v}_j$ is the best matching unit (BMU), and n_k is the number of microarrays for which $\mathbf{v}_k$ is the second BMU. The strength of connection between reference vectors is represented by the symmetric connection strength matrix $\mathbf{C}$, with elements for the connection strength between $\mathbf{v}_j$ and $\mathbf{v}_k$ determined as

$$c_{jk} = c_{kj} = a_{jk} + a_{kj}. \tag{4.14}$$

This approach is appropriate when $n >> M$ and therefore would be much less informative for DNA microarray data, which is commonly plagued by the small-sample-size problem, $n << p$. Nevertheless, it is worth mentioning in this chapter as an alternative to the other cluster approaches, especially for readers interested in image segmentation.

4.4.3 Cluster Connectivity Method

The cluster connectivity method was proposed by Merkel and Rauber [18]. This method determines every pairwise distance d_{jk} between neighboring reference vectors j and k, and assigns colors to lines connecting all possible neighboring nodes based on either mean distance ($\mu \pm \sigma$) or thresholds (e.g., t_1, t_2, t_3). Nodes with similar distances in the hierarchy of distances (range) should belong to the same cluster, whereas nodes linked by heterogeneous distance values are within or are proximal to cluster boundaries. Colors can be assigned by using either thresholds or cutpoints defined on average distance μ and standard deviation, σ. Smaller between-node distances reflect greater similarity, and therefore the following color schemes can be implemented. When thresholds based on quartiles are used, the connection colors for the line joining nodes j and k, that is, line $\overline{JK}$, are

$$\text{Color}(\overline{JK}) = \begin{cases} \text{red} & d_{jk} < t_1 \\ \text{green} & t_1 \leq d_{jk} < t_2 \\ \text{blue} & t_2 \leq d_{jk} < t_3 \\ \text{white} & d_{jk} \geq t_3 \end{cases} \tag{4.15}$$

or when average [standard deviation (s.d.)] are used

$$\text{Color}(\overline{JL}) = \begin{cases} \text{red} & d_{jk} < \mu - \sigma \\ \text{green} & \mu - \sigma \leq d_{jk} < \mu \\ \text{blue} & \mu \leq d_{jk} < \mu + \sigma \\ \text{white} & d_{jk} \geq \mu + \sigma. \end{cases} \tag{4.16}$$

Figure 4.4 shows an example cluster visualization using the connection method with t_1, t_2, and t_3 based on the quartiles of all internode distances for an SOM run for which there are seven known classes. The clusters are represented by regions where the combination of vertical, horizontal, and diagonal internode connection lines is the most frequent.

4.4.4 Hue-Saturation-Value (HSV) Color Normalization

Another approach to cluster identification via color usage is to create three p-dimension vectors $\mathbf{c}_r$, $\mathbf{c}_g$, and $\mathbf{c}_b$ for the p features in an SOM analysis [19].

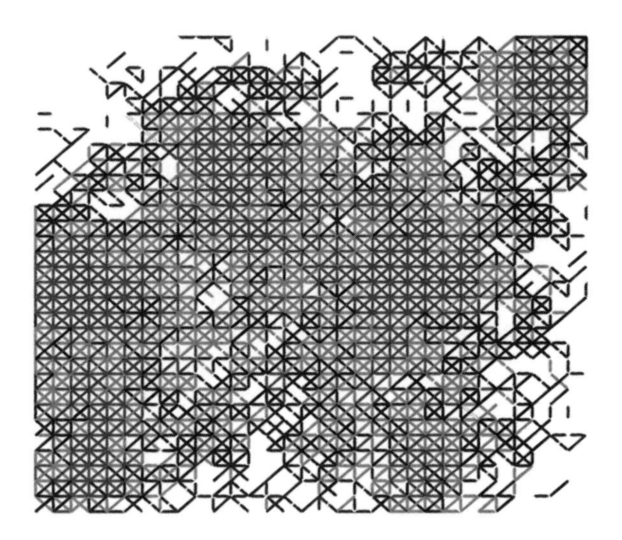

FIGURE 4.4 SOM cluster visualization of a seven-class dataset using connection method.

The elements in these vectors are denoted $c_{j,r}$, $c_{j,g}$, $c_{j,b}$. The color assigned to vector elements is equal to the HSV color value at $(j/p)360^\circ$ transformed into RGB color. For example, if there are $p = 6$ features in an analysis, then $c_{2,r} = 0$, $c_{2,g} = 255$, and $c_{2,b} = 0$ since the HSV color at $120 = \frac{2}{6}360^\circ$ is pure green. The fourth element in the color vectors would be set to $c_{4,r} = 0, c_{4,g} = 0$, and $c_{4,b} = 255$ because at $240 = \frac{4}{6}360^\circ$ the HSV color is pure blue. After SOM training, the RGB colors assigned to reference vector (node) m are

$$\begin{aligned}\text{Red} &= \frac{\sum_{j=1}^{p} c_{j,r} v_{jm}}{\sum_{j=1}^{p} c_{j,r}}\\ \text{Green} &= \frac{\sum_{j=1}^{p} c_{j,g} v_{jm}}{\sum_{j=1}^{p} c_{j,g}}\\ \text{Blue} &= \frac{\sum_{j=1}^{p} c_{j,b} v_{jm}}{\sum_{j=1}^{p} c_{j,b}}.\end{aligned} \tag{4.17}$$

Figure 4.5 illustrates the colors assigned to reference vectors in a 1000-node SOM for which there are seven known classes in the data. One of the drawbacks of this method is that if there are more than, say, ~5 features and there is overlap in the feature-specific values of trained reference vectors, there is a tendency for the image to appear "muddy." A muddy image is the result of homogeneity among the RGB values assigned to a node as in

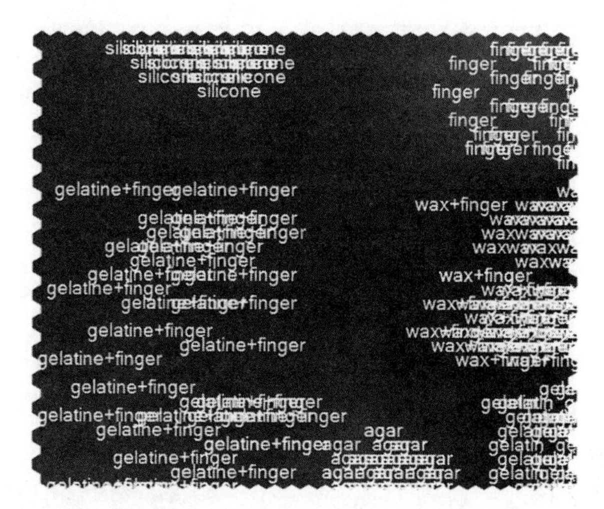

FIGURE 4.5 SOM cluster visualization of a seven-class dataset using the HSV color normalization method.

(4.17), such as the color RGB = (140,132,151) instead of heterogeneity of a particular primary red color, for example, RGB=(170,20,17). Our preliminary work using the HSV normalization method for cluster visualization is quite promising, and we believe that there is great potential for this method to emerge as one of the better methods.

4.5 UNIFIED DISTANCE MATRIX (*U* MATRIX)

The results of an SOM run also commonly provide a unified distance matrix, or *U matrix* to reveal the distance between reference vectors (nodes). Figure 4.6 shows the *U* matrix after an SOM run of the four-class SRBCT dataset. Each node in the 2D grid is assigned a color that represents the distance between its neighboring nodes. Large distances between reference vectors are represented by red and short distances, by blue. Class labels for the $n = 63$ microarrays in the SRBCT dataset are shown in regions of blue, where the distance between reference vectors is lowest after training.

4.6 COMPONENT MAP

The values of each feature within various clusters are shown using a component map. Component maps reflect the variation of feature values for the reference vectors after learning. Again, microarray class labels are drawn

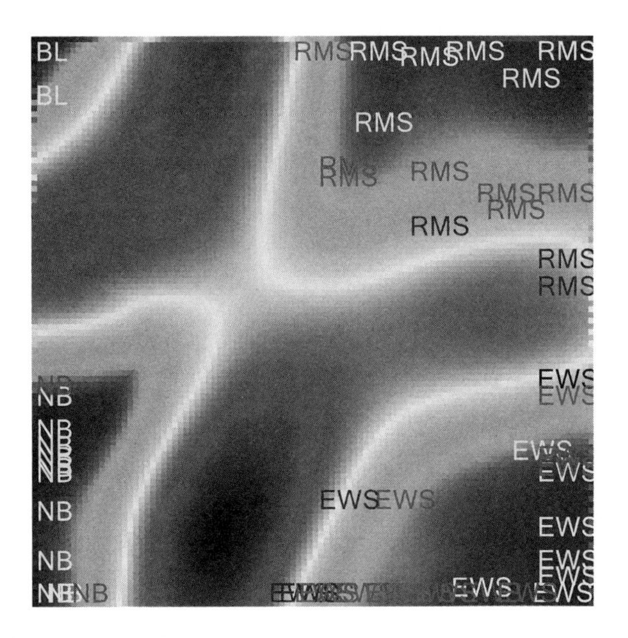

FIGURE 4.6 *U* matrix for four-class SRBCT data. The tubular structure (in red) resembling a deep valley is based on greater distances between reference vectors. Regions with similar shorter distances between reference vectors are shaded in blue. (*See insert for color representation of the figure.*)

near each microarray's BMU (closest reference vector), and the color of the node reveals the value of the particular feature shown in the map. For example, Figure 4.7 shows three component maps for the plasminogen activator gene, CCAAT displacement protein, and translocation protein 1, which are significant class predictors of SRBCT. One can note that the expression of plasminogen activator [panel (a)] is enhanced in rhabdomyosarcoma (RMS) cases that cluster in the upper-right corner of the map. In panel (b), the component map of the CCAAT displacement protein is displayed, whose expression value is enhanced in Burkitt's lymphoma (BL) cases in the upper left of the map and RMS cases in the upper right corner. Finally, the component map for translocation protein 1 is illustrated in panel (c), and its expression is notably enhanced on the left side of the map, where BL and neuroblastoma (NB) cases cluster. By viewing several component maps together, we can identify novel patterns in the feature space that otherwise might be overlooked when viewing only summary statistics, cluster heatmaps, or line plots. A considerable degree of information can be gleaned from component maps, since the associations between features can provide new hypotheses for gaps in knowledge, and establish new leads for future experiments.

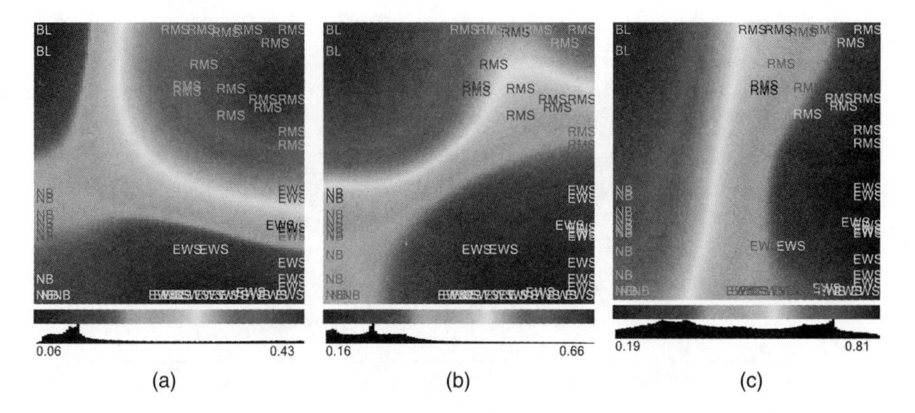

FIGURE 4.7 SOM component maps for three genes in the four-class SRBCT dataset. Expression of plasminogen activator (a) is enhanced in only rhabdomyosarcoma (RMS) cases, CCAAT gene (b) is enhanced in both Burkitt's lymphoma (BL) and RMS, and translocation protein 1 (c) is enhanced in both BL and neuroblastoma (NB). Red signifies enhanced expression, while blue signifies suppressed expression. (*See insert for color representation of the figure.*)

The presentations in Figure 4.7 merely show the feature-specific values of all reference vectors once training is completed. Another approach is to show the contribution of each reference vector feature value as a proportion of the local distance with the relationship

$$f_j = \frac{|v_{jl} - v_{jm}|}{\|\mathbf{v}_l - \mathbf{v}_m\|}. \tag{4.18}$$

Plots of f_j can be constructed for each component (feature) using the feature-specific values of reference vector features.

4.7 MAP QUALITY

The average quantization error (AQE) was introduced [19] for assessing the quality of a map using the distance between all of the training microarrays and their BMUs. Functionally, the AQE is composed as

$$\text{AQE} = \frac{1}{n}\sum_{i=1}^{n} \|\mathbf{x}_i - \mathbf{v}_c\|, \tag{4.19}$$

which is used to determine map *goodness* [20], in the form

$$g(\text{AQE}) = \frac{1}{1 + \text{AQE}^2}, \tag{4.20}$$

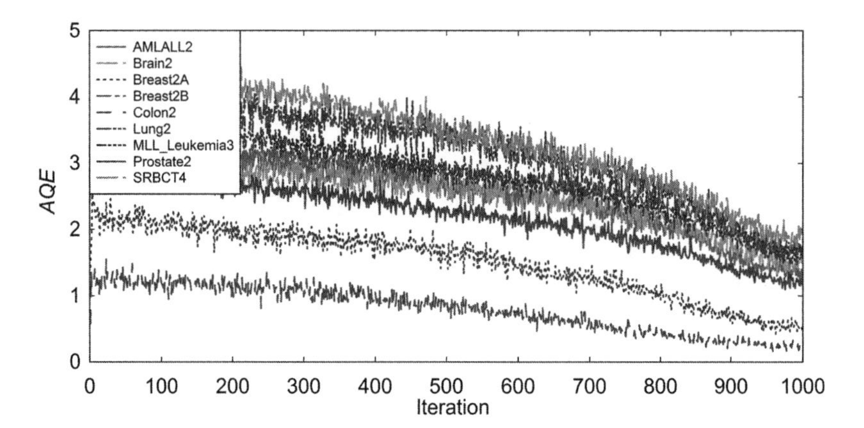

FIGURE 4.8 AQE for all nine datasets based on mean-zero standardized input features originally selected with greedy PTA and reference vectors initialized using standard normal variates, $\mathcal{N}(0, 1)$.

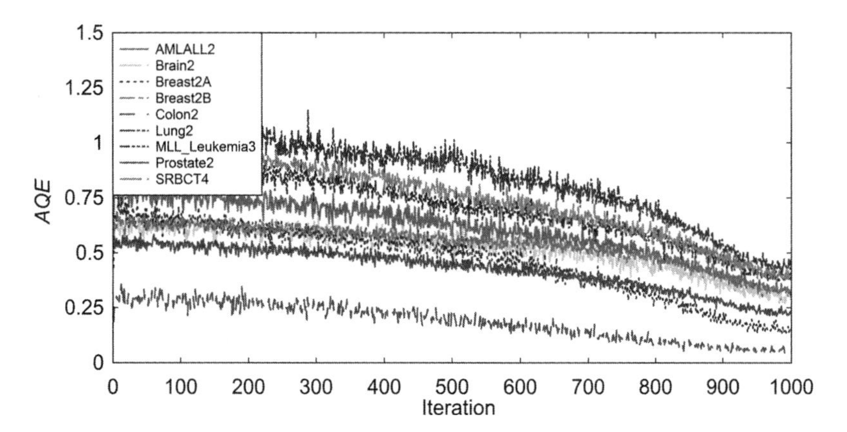

FIGURE 4.9 AQE for all nine datasets based on normalized input features originally selected with greedy PTA and reference vectors initialized using random uniform variates $U(0, 1)$.

for which the beneficial properties are $g(\text{AQE} \rightarrow \infty) = 0$ to $g(0) = 1$. Figure 4.8 illustrates the AQE as a function of SOM training iteration for all nine datasets whose input feature were based on greedy PTA. Input feature values were mean-zero standardized, and reference vector values were initialized using standard normal variates $\mathcal{N}(0,1)$. Figure 4.9 illustrates results for the same data but with normalization of input feature values and reference vector initialization using random uniform variates $U(0,1)$. In both plots, the AQE is observed to stabilize prior to 1000 training iterations.

4.8 NONLINEAR DIMENSION REDUCTION

In addition to being a prototype-based vector quantization method, SOM is a nonlinear mapping algorithm. Because of the 2D mapping characteristics of SOMs, it is possible to gain insight into properties of the multidimensional feature space. Nonlinearity is a property of the data and the classifier applied to a given dataset. Stated formally, *nonlinearity* is defined as the probability that a microarray between two arbitrary microarrays of similar classes is assigned a different class label when uniformly and linearly interpolated using the same classifier. Nonlinearity is independent of the true class labels, and independent of the performance of the classifier used. However, it does depend on the training data used.

The associative neural learning performed by SOMs can preserve nonlinearities in the p-dimensional feature space when mapped into a 2D feature space. Recall that there is nothing linear about an SOM; gradient descent is used with a connectionist model to apply weight updates to a set of reference vectors, in a neighborhood centered around input microarrays, until small weight changes are observed. SOMs can be used for dimension reduction in the same way that principal component analysis (PCA) and crisp K-means (CKM) cluster analysis are used. For most biomedical datasets, PCA can often describe a majority of variance using two principal components (PCs). The disadvantage with PCA, however, is that the PCs themselves are formed as *linear combinations* of the eigenvectors extracted from the scale-invariant $p \times p$ correlation matrix $\mathbf{R}$. SOM, on the other hand, can reliably map nonlinearities in the p dimensions onto the 2D grid of nodes. Once an SOM is trained, the p-dimensional feature vector for each microarray is replaced with the row and column number of the microarray's BMU. As an example, assume that a nine-feature microarray vector is considered. After training, it is discovered that the row-column coordinate of a microarray's BMU is (3,1). The microarray's nine-dimensional feature values are replaced with the row and column numbers of the BMU.

Example 4.1 SOMs can be used to reduce multiple dimensions of microarrays down to 2D prior to classification analysis. Table 4.1 lists classification accuracy for eight classifiers using datasets listed in Table 1.3 whose features were generated with greedy PTA (see Table 12.9). A 32×32 2D SOM ($M = 1024$ nodes) was employed for dimensional reduction of the features identified with greedy PTA for the nine datasets. Initialization of the SOM involved mean-zero standardization of input features or normalization in the range $[0, 1]$, and starting values $\beta_0 = 32 = \sqrt{1024}$, $\alpha_0 = 0.05$, and 50 iterations. The SOM output included a 2D map location vector $\mathbf{r}_m$ for each input microarray containing the integer values k and l which represent the row and column of the reference vector $\mathbf{v}_m$ closest to microarray $\mathbf{x}_i$. The two features from SOM analysis were used with the same classifiers. Regarding the SRBCT dataset, Figure 4.10 shows the 2D feature values obtained by using mean-zero standardization of input feature values and standard

TABLE 4.1 Classification Accuracy for Eight Classifiers Applied to Nine DNA Microarray Datasets[a]

Dataset	5NN	NBC	LDA	SVMGA	SVMLS	KREG	ANN	PLOG
AMLALL2	1.00	0.95	1.00	1.00	1.00	0.89	1.00	1.00
Brain2	0.73	0.83	0.94	0.46	0.67	0.78	0.94	0.90
Breast2	0.93	1.00	0.97	1.00	1.00	0.90	1.00	1.00
Breast2B	0.81	0.83	0.95	0.90	0.68	0.81	0.97	0.97
Colon2	0.80	0.85	0.89	0.72	0.73	0.86	0.94	0.92
Lung2	1.00	1.00	0.88	0.94	1.00	1.00	1.00	0.97
MLL_Leukemia3	0.96	1.00	1.00	0.87	0.87	0.89	1.00	0.96
Prostate2	0.89	0.93	0.98	0.89	0.93	0.96	0.98	0.98
SRBCT4	0.99	0.98	1.00	0.97	0.99	0.99	1.00	1.00

[a] Without use of SOM prior to classification. Ten 10-fold cross-validation used for determining performance.

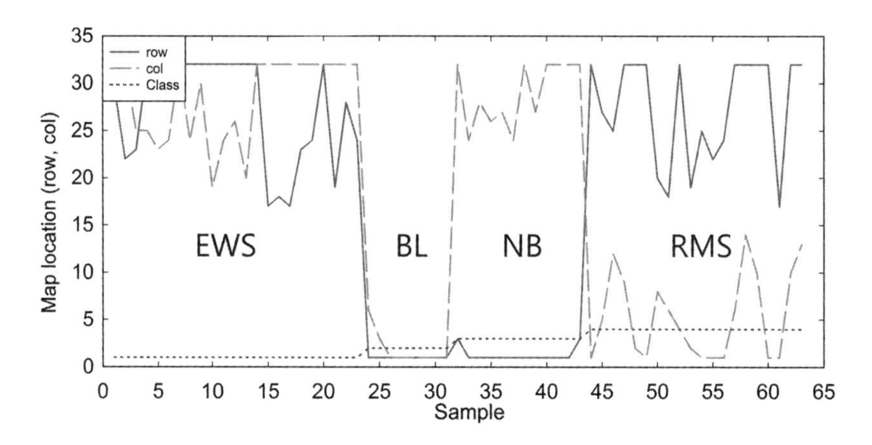

FIGURE 4.10 2D feature values obtained from 32×32 SOM ($M = 1024$) trained using mean-zero standardized input feature values and reference vectors initialized with standard normal variates distributed $\mathcal{N}(0, 1)$. Training dataset based on SRBCT data with $n = 63$ microarrays and $p = 23$ features obtained with greedy PTA. There are four classes of arrays: 23 arrays for Ewing sarcoma (EWS), 8 arrays for Burkitt's lymphoma (BL), 12 arrays for neuroblastoma (NB), and 20 arrays for rhabdomyosarcoma (RMS).

normal variates distributed $\mathcal{N}(0,1)$ for reference vector initialization, and Figure 4.11 shows the 2D feature values obtained by using normalization of input feature values and uniform random variates for reference vector initialization. Table 4.2 lists classification results using the two features identified with SOM and mean-zero standardization of SOM input features.

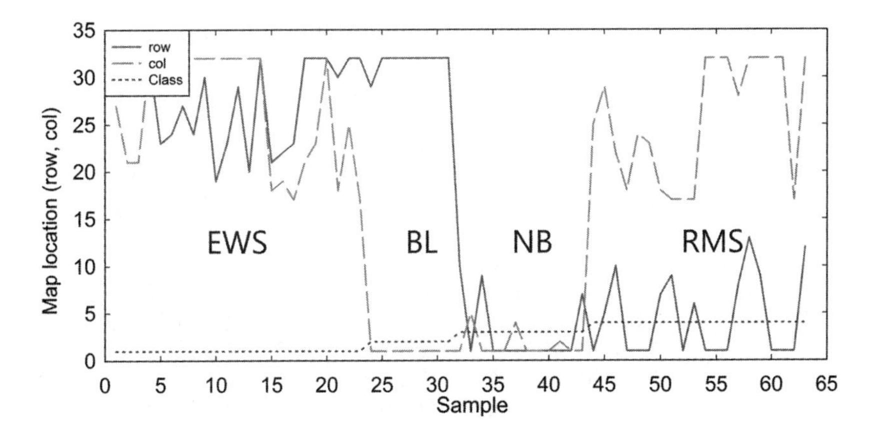

FIGURE 4.11 2D feature values obtained from 32 × 32 SOM ($M = 1024$) trained using normalized input feature (range [0,1]) values and reference vectors initialized with uniform variates distributed $U(0, 1)$. Training dataset based on SRBCT data with $n = 63$ microarrays and $p = 23$ features obtained with greedy PTA. Four classes of arrays: There are 23 arrays for Ewing sarcoma (EWS), 8 arrays for Burkitt's lymphoma (BL), 12 arrays for neuroblastoma (NB), and 20 arrays for rhabdomyosarcoma (RMS).

TABLE 4.2 Classification after Dimensional Reduction Using a 32 x 32 Node SOM[a]

Dataset	5NN	NBC	LDA	SVMGA	SVMLS	KREG	ANN	PLOG
AMLALL2	1.00	0.88	0.94	1.00	0.97	0.81	1.00	1.00
Brain2	0.87	0.84	0.90	0.71	0.81	0.82	1.00	0.87
Breast2	1.00	0.97	0.99	1.00	1.00	1.00	1.00	1.00
Breast2B	0.84	0.85	0.86	0.81	0.75	0.85	0.75	0.85
Colon2	0.89	0.85	0.87	0.78	0.74	0.88	1.00	0.87
Lung2	1.00	1.00	1.00	0.98	1.00	1.00	1.00	1.00
MLL_Leukemia3	0.98	0.88	1.00	0.96	0.96	1.00	0.99	0.95
Prostate2	0.94	0.95	0.94	0.83	0.84	0.94	1.00	0.94
SRBCT4	1.00	0.90	1.00	1.00	1.00	1.00	1.00	1.00

[a]With mean-zero standardization of SOM input features. Ten 10-fold cross-validation used for determining performance.

Table 4.3 gives classification results using the two features identified with normalization of SOM input features. For the nine datasets, Table 4.4 lists the mean classification accuracy for various classifiers using reduced feature dimensions via SOM. On average, SOM with mean-zero standardization

TABLE 4.3 Classification after Dimensional Reduction Using a 32 × 32 node SOM[a]

Dataset	5NN	NBC	LDA	SVMGA	SVMLS	KREG	ANN	PLOG
AMLALL2	1.00	0.95	0.95	1.00	0.98	0.83	1.00	0.99
Brain2	0.80	0.80	0.82	0.74	0.68	0.75	1.00	0.82
Breast2	0.99	0.91	1.00	1.00	1.00	1.00	1.00	1.00
Breast2B	0.81	0.82	0.83	0.77	0.67	0.83	0.76	0.82
Colon2	0.87	0.80	0.86	0.70	0.80	0.85	0.96	0.87
Lung2	1.00	0.98	1.00	0.98	1.00	1.00	1.00	1.00
MLL_Leukemia3	0.99	0.91	0.99	0.99	0.97	0.98	0.99	0.99
Prostate2	0.94	0.94	0.94	0.76	0.88	0.94	1.00	0.92
SRBCT4	1.00	0.90	1.00	1.00	1.00	1.00	1.00	1.00

[a]With normalization of SOM input features. Ten 10-fold cross-validation used for determining performance.

TABLE 4.4 Average Classifier Performance for 9 Datasets with and without (wo) Dimensional Reduction of Input Features to 2D with SOM

Method	5NN	NBC	LDA	SVMGA	SVMLS	KREG	ANN	PLOG
wo/SOM	0.901	**0.930**	**0.957**	0.861	0.874	0.898	**0.981**	**0.967**
SOM(std)	**0.947**	0.902	0.944	**0.897**	**0.897**	**0.922**	0.971	0.942
SOM(norm)	0.933	0.890	0.932	0.882	0.887	0.909	0.968	0.934

yielded greater performance when compared with SOM based on normalization. Specifically, SOM based on mean-zero standardization improved results for K-nearest neighbor ($K = 5$), gradient descent, and least-squares support vector machines (SVMGA, SVMLS), and kernel regression (KREG).

Investigation of the effects of SOM preprocessing entails intense focus on the field of classification. Results shown for this example are encouraging, and would suggest that SOM preprocessing can be considered as an adjunct to other unsupervised methods of dimensional reduction such as PCA and K means. Results from other groups are noteworthy. A hybrid SOM-SVM approach resulted in improved zebrafish gene expression results [21]. Markey et al. [22] also achieved improved classification results using a SOM-ANN approach.

The main attraction for using SOMs is that they are a powerful approach for visualizing high-dimensional data, and employ adaptive learning methods suitable for environments where microarrays are not linearly separable [23]. There are no rules of thumb for the expected performance improvement when preprocessing with an SOM, mostly because of the variation in effects of classifier parameter values, and variation in decision boundaries that ultimately depend on the number and properties of the features.

REFERENCES

[1] T. Kohonen. *Self-Organizing Maps*. Springer, Berlin, 2001.

[2] P. Demartines, J. Herault. Vector quantization and projection neural network. *Lecture Notes Comput. Sci.* **686**:328–333, 1993.

[3] P. Tomayo, D. Slonim, J. Mesirov, Q. Zhu, S. Kitareewan, E. Dmitrovsky, E.S. Lander, T.R. Golub. Interpreting patterns of gene expression with self-organizing maps: Methods and applications to hematopoietic differentiation. *Proc. Natl. Acad. Sci. USA* **96**:2907–2912, 1999.

[4] V. De Bruyne, F. Al-Mulla, B. Pot. Methods for microarray data analysis. *Methods Mol. Biol.* **382**:373–391, 2007.

[5] P. Toronen, M. Kolehmainen, G. Wong, E. Castren. Analysis of gene expression data using self-organizing maps. *FEBS Lett.* **451**(2):142–146, 1999.

[6] A. Dragomir, S. Mavroudi, A. Bezerianos. Som-based class discovery exploring the ICA-reduced features of microarray expression profiles. *Compar. Funct. Genomics* **5**(8):596–616, 2004.

[7] A. Jong, C.H. Wu, W. Zhou, H.M. Chen, S.H. Huang. Infectomic analysis of gene expression profiles of human brain microvascular endothelial cells infected with Cryptococcus neoformans. *J. Biomed. Biotechnol.* **2008**: 375620, 2008.

[8] X. Huang, L.D. Hazlett. Analysis of Pseudomonas aeruginosa corneal infection using an oligonucleotide microarray. *Invest. Ophthalmol. Vis. Sci.* **44**(8):3409–3416, 2003.

[9] D.G. Covell, A. Wallqvist, A.A. Rabow, N. Thanki. Molecular classification of cancer: unsupervised self-organizing map analysis of gene expression microarray data. *Mol. Cancer Ther.* **2**(3):317–332, 2003.

[10] L. Xiao, K. Wang, Y. Teng, J. Zhang. Component plane presentation integrated self-organizing map for microarray data analysis. *FEBS Lett.* **538**(1–3):1171–1124, 2003.

[11] G.R. Burton, Y. Guan, R. Nagarajan, R.E. McGehee Jr. Microarray analysis of gene expression during early adipocyte differentiation. *Gene* **293**(1–2):21–31, 2002.

[12] J.J. Chen, K. Peck, T.M. Hong, S.C. Yang, Y.P. Sher, J.Y. Shih, R. Wu, J.L. Cheng, S.R. Roffler, C.W. Wu, P.C. Yang. Global analysis of gene expression in invasion by a lung cancer model. *Cancer Res.* **61**(13):5223–5230, 2001.

[13] M.R. Saban, H. Hellmich, N.B. Nguyen, J. Winston, T.G. Hammond, R. Saban. Time course of LPS-induced gene expression in a mouse model of genitourinary inflammation. *Physiol. Genomics* **5**(3):147–160, 2001.

[14] E. Berglund, J. Sitte. The parameterless self-organizing map algorithm. *IEEE Trans. Neural Networks* **17**(2):305–316, 2006.

[15] T. Villmann, F. Schleif, M. Kostrzewa, A. Walch, B. Hammer. Classification of mass-spectrometric data in clinical proteomics using learning vector quantization methods. *Brief. Bioinform.* **9**(2):129–143, 2008.

[16] D. DiSieno. Adding conscience to competitive learning. *Proc. Int. Conf. Neural Networks*, IEEE Press, Piscataway, NJ, 1988, vol. 1, pp. 117–124.

[17] K. Taşdemir, E. Merényi, Data topology visualization for the self-organizing map. *European Symposium on Artificial Neural Networks (ESANN'06)*, Bruges, Belgium. 2006

[18] D. Merkl, A. Rauber. Alternative ways for cluster visualization in self-organizing maps. *Proc. of the Workshop on Self-Organizing Maps (WSOM'97)*, pp. 106–111. 1997.

[19] J. Schatzmann. *Using Self-Organizing Maps to Visualize Clusters and Trends in Multidimensional Datasets*. Dept. Computing, Data Mining Group, final year individual project report, Imperial College of London, 2003.

[20] J. Vesanto. SOM-based data visualization methods. *Intell. Data Anal.* **3**(2)111–126, 1999.

[21] W. Wu, X. Liu, M. Xu, J.R. Peng, R. Setiono. A hybrid SOM-SVM approach for the zebrafish gene expression analysis. *Genomics Proteomics Bioinform.* **3**(2):84–93, 2005.

[22] M.K. Markey, J.Y. LO, G.D. Tourassi, C.E. Floyd. Self-organizing map for cluster analysis of a breast cancer database. *Artif. Intell. Med.* **27**(2):113–127, 2003.

[23] J. Mao, A.K. Jain. Artificial neural networks for feature extraction and multivariate data projection. *IEEE Trans. Neural Networks* **6**(2):296–317, 1995.

CHAPTER 5

UNSUPERVISED NEURAL GAS

5.1 INTRODUCTION

Unsupervised neural gas (UNG) is a Hebbian learning approach to class discovery in the field of exploratory data analysis. During early development of UNG, applications included time series prediction [1], handwritten signature verification [2], and learning NG networks with time-constant parameters [3].

Several applications have employed UNG with microarray data. Chelloug et al. clustered microarray data using UNG within an amorphous computing paradigm [4]. Other studies [5,6] investigated growing cell structure and growing NG methods, and observed that UNG runtimes are shorter than those obtained by other unsupervised methods and result in better cluster structures when impurities are present.

Comparison studies have been carried out for several unsupervised approaches. A study based on functional MRI [7] using unsupervised clustering and independent component analysis (ICA) revealed that ICA extracted features rather well for a small number of components but was limited by the linear mixture assumption. On the other hand, unsupervised cluster analysis, including UNG, outperformed ICA with regard to classification results but required longer processing times. In another study on unsupervised class discovery [8], the efficiency and power of several cluster analysis techniques were investigated for several functional MRI datasets. Cluster methods investigated included hierarchical, neural gas, self-organizing maps, hard and fuzzy competitive learning, *K*-means cluster maximin analysis, FCM and FKM, and competitive learning. Results indicated that UNG and *K*-means cluster analysis performed significantly better than all the other methods. The authors concluded that for fMRI, UNG seemed to be the best choice given its stability of results achieved, correct classification of activated pixels [true positives (TPs)], and minimization of misclassification of inactivated pixels [false positives (FPs)]. Ogura et al. [9] developed a topology representing

Classification Analysis of DNA Microarrays, First Edition. Leif E. Peterson.

a network based on UNG for classification of 3D protein images taken by cryoelectronmicroscopy. Their results were superior to those obtained from multivariate statistical analysis and self-organizing maps (SOM), especially for very noisy 3D protein images.

5.2 ALGORITHM

Through a series of iterations, the prototypes learn the patterns among microarrays within a neighborhood and migrate toward naturally occurring clusters of input microarrays. Let $\mathbf{x}_i$ be the ith microarray $(i = 1, 2, \ldots, n)$ and $\mathbf{v}_m$ $(m = 1, 2, \ldots, M)$ be a prototype with the same number of attributes as input microarrays. During learning, each microarray is selected and the prototypes $\mathbf{v}$ are rank-ordered according to their distance from the selected microarray. The locations of all prototypes $\mathbf{v}_m$ are updated according to the relationship

$$\mathbf{v}_m = \mathbf{v}_m + \alpha(t) h_\lambda(\mathbf{v}_m, \mathbf{x}_i)[\mathbf{x}_i - \mathbf{v}_m], \tag{5.1}$$

where the neural gas *neighborhood function* is functionally expressed as

$$h_\lambda(\mathbf{v}_m, \mathbf{x}_i) = \exp\left(\frac{-R_m}{\lambda(t)}\right), \tag{5.2}$$

where $\alpha(t)$ is the learning rate at the tth iteration $(t = 1, 2, \ldots, T)$, $[\mathbf{v}_m - \mathbf{x}_i]$ is the vector difference between prototype $\mathbf{v}_m$ and microarray $\mathbf{x}_i$, and $R_m(m = 1, 2, \ldots, M)$ is the rank index of prototype $\mathbf{v}_m$ based on distance from microarray $\mathbf{x}_i$, where $R_1 = 0$ is the rank closest prototype to $\mathbf{x}_i$, $R_2 = 1$ is the rank of the second closest prototype to $\mathbf{x}_i$, and so on. Suitable choices for the learning rate and width are

$$\alpha(t) = \alpha_i \left(\frac{\alpha_f}{\alpha_i}\right)^{t/T}, \tag{5.3}$$

and

$$\lambda(t) = \lambda_i \left(\frac{\lambda_f}{\lambda_i}\right)^{t/T}. \tag{5.4}$$

Martinetz et al. [1] suggested using values of $\alpha_i = 0.5$, $\alpha_f = 0.005$, $\lambda_i = 10$, $\lambda_f = 0.01$, $T = 100{,}000$.

5.3 IMPLEMENTATION

5.3.1 Feature Transformation and Prototype Initialization

Feature values input into an UNG algorithm need to be transformed to the range and scale that will be used during training. This leads to using either

Algorithm 5.1: Unsupervised Neural Gas

Data: p-dimensional input microarrays $\mathbf{x}_i$ ($i = 1, 2, \ldots, n$), M prototypes $\mathbf{V} = \{\mathbf{v}_1, \mathbf{v}_2, \ldots, \mathbf{v}_M\}$
Result: p-dimensional prototypes
Specify the number of prototypes M
Initialize $\alpha_i = 0.5$, $\alpha_f = 0.005$, $\lambda_i = 10$, $\lambda_f = 0.01$, $T = 100,000$
Initialize M prototypes $\mathbf{V} = \{\mathbf{v}_1^{(0)}, \mathbf{v}_2^{(0)}, \ldots, \mathbf{v}_M^{(0)}\}$
for $t \leftarrow 1$ **to** T **do**
  $\alpha(t) = \alpha_i(\alpha_f/\alpha_i)^{t/T}$
  $\lambda(t) = \lambda_i(\lambda_f/\lambda_i)^{t/T}$
  for $i \leftarrow 1$ **to** n **do**
    Determine prototype ranks $R_1 < R_2 < \cdots < R_M$ based on ascending $d(\mathbf{v}_m, \mathbf{x}_i)$ (*Note*: $R_1 = 0, R_2 = 1, \ldots, R_M = M - 1$)
    for $m \leftarrow 1$ **to** M **do**
      $h_\lambda(\mathbf{v}_m, \mathbf{x}_i) = \exp(-R_m/\lambda(t))$
      $\mathbf{v}_m = \mathbf{v}_m + \alpha(t)h_\lambda(\mathbf{v}_m, \mathbf{x}_i)[\mathbf{x}_i - \mathbf{v}_m]$
    endfor
  endfor
endfor

mean-zero standardization or normalization into the range [0,1]. There really are no general rules of thumbs on which method is best for feature transformation, since results will depend on the data used. The prototypes also need to be initialized via mean-zero standardization or normalization, to ensure similarity of the range and scale to those used for feature value transformation.

5.3.2 Prototype Learning

Algorithm 5.1 lists the pseudocode for the UNG algorithm. Unlike the SOM algorithm, when processing each microarray during training, UNG first sorts in ascending order the node-microarray distance, and then calculates the neighborhood function using the ranked nodes. The implementation of UNG is given in the following source code:

```
For iter = 1 To ngiterations
  For k = 1 To NumObjects
    lowestdist = 1.0E+30
    icol = 0
    For j = 1 To nodes
      icol = icol + 1
      Sum = 0
      For i = 1 To NumFeatures
        Sum += (x(i, k) - refvec(i, j)) ^ 2
```

```
        Next i
        eucdist = Math.Sqrt(Sum)
        arr(icol) = eucdist
        ndx(icol) = icol
        If eucdist < lowestdist Then
          lowestdist = eucdist
          lowestnode = j
        End If
      Next j
      descending = False
      Call sort(arr, ndx, icol, descending)
      alphat = 0.5 * (0.005 / 0.5) ^ (iter / ngiterations)
      lambda = 10 * (0.01 / 10) ^ (iter / ngiterations)
      For j = 1 To nodes
        For i = 1 To NumFeatures
          refvec(i, ndx(j)) = refvec(i, ndx(j)) + alphat * _
                              Math.Exp(-(j - 1) / Lambda) * _
                              (x(i, ranpoint) - refvec(i, ndx(j)))
        Next i
      Next j
    Next k
Next iter
```

Figure 5.1 shows an example plot of UNG learning, where five randomly positioned prototypes move toward the centers of five clusters. Only 15 iterations were used for this five-cluster training example.

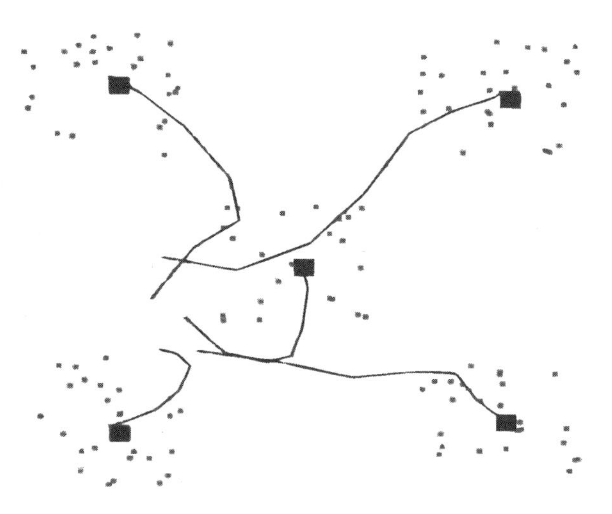

FIGURE 5.1 Neural gas learning for a dataset with five clusters using five prototypes. After 15 iterations the prototypes become located in the middle of the clusters.

5.4 NONLINEAR DIMENSION REDUCTION

Unsupervised neural gas can be used to preprocess feature values of microarrays before they are employed in routine classification. Results of feature dimensional reduction when using UNG is in the form of Euclidean distance $d(\mathbf{x}_i, \mathbf{v}_m)$ between each microarray and prototype used.

Example 5.1 Neural gas was employed for dimensional reduction of the $n = 63$ microarray SRBCT dataset for which $p = 23$ features were originally identified using greedy PTA. A total of $M = 4$ prototypes were used since the SRBCT data has four known class labels. Initialization of UNG involved mean-zero standardization of input features or normalization in the range $[0, 1]$. Regarding the SRBCT dataset, Figure 5.2 shows the 4D feature values obtained by using mean-zero standardization of input feature values and standard normal variates distributed $\mathcal{N}(0, 1)$ for prototype initialization, and Figure 5.3 shows the 4D feature values obtained by using normalization of input feature values and uniform random variates for prototype initialization.

Example 5.2 In this example, we employed UNG prior to classification in order to reduce the dimensions of each dataset, where the number of prototypes (dimensions) used was set equal to the true number of classes (Ω)

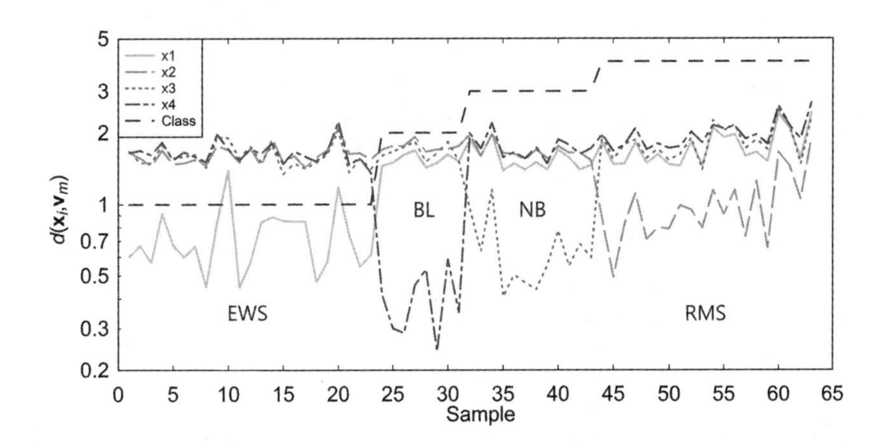

FIGURE 5.2 4D feature values obtained from four-prototype UNG using mean-zero standardized input feature values and prototypes initialized with standard normal variates distributed $\mathcal{N}(0, 1)$. Training dataset based on SRBCT data with $n = 63$ microarrays and $p = 23$ features obtained with greedy PTA. There are four array classes: 23 arrays for Ewing sarcoma (EWS), 8 arrays for Burkitt's lymphoma (BL), 12 arrays for neuroblastoma (NB), and 20 arrays for rhabdomyosarcoma (RMS).

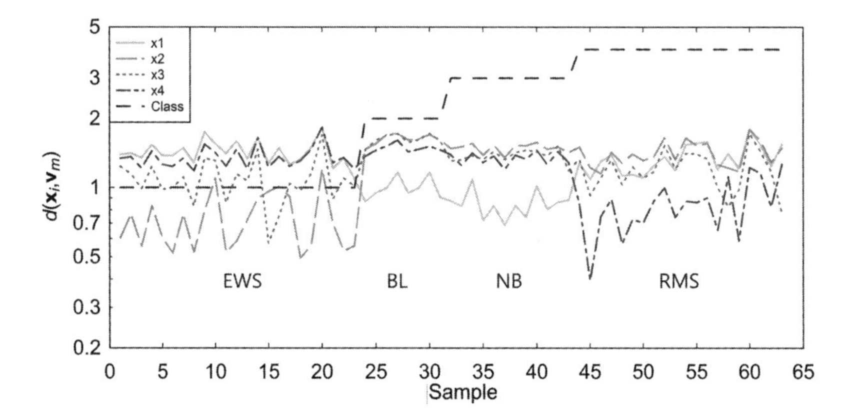

FIGURE 5.3 4D feature values obtained from four-prototype UNG using normalized input feature values (range [0, 1]) and prototypes initialized with uniform variates $U(0, 1)$. Training dataset based on SRBCT data with $n = 63$ microarrays and $p = 23$ features obtained with greedy PTA. There are four array classes: 23 arrays for Ewing sarcoma (EWS), 8 arrays for Burkitt's lymphoma (BL), 12 arrays for neuroblastoma (NB), and 20 arrays for rhabdomyosarcoma (RMS).

in each dataset. Table 5.1 lists classification accuracy for the eight classifiers without dimensional reduction by UNG, using datasets listed in Table 1.3 whose features were generated with greedy PTA (see Table 12.9). Table 5.2 lists classification results after dimensional reduction with UNG and mean-zero standardization of input features. Table 5.3 gives classification results after dimensional reduction with UNG and normalization of input features.

TABLE 5.1 Classification Accuracy for Nine Classifiers Applied to Nine DNA Microarray Datasets[a]

Dataset	5NN	NBC	LDA	SVMGA	SVMLS	KREG	ANN	PLOG
AMLALL2	1.00	0.95	1.00	1.00	1.00	0.89	1.00	1.00
Brain2	0.73	0.83	0.94	0.46	0.67	0.78	0.94	0.90
Breast2	0.93	1.00	0.97	1.00	1.00	0.90	1.00	1.00
Breast2B	0.81	0.83	0.95	0.90	0.68	0.81	0.97	0.97
Colon2	0.80	0.85	0.89	0.72	0.73	0.86	0.94	0.92
Lung2	1.00	1.00	0.88	0.94	1.00	1.00	1.00	0.97
MLL_Leukemia3	0.96	1.00	1.00	0.87	0.87	0.89	1.00	0.96
Prostate2	0.89	0.93	0.98	0.89	0.93	0.96	0.98	0.98
SRBCT4	0.99	0.98	1.00	0.97	0.99	0.99	1.00	1.00

[a]Without use of UNG dimension reduction prior to classification analysis. Ten 10-fold cross-validation used for determining performance.

TABLE 5.2 Classification after Dimensional Reduction of Greedy PTA Features to Ω Dimensions for Each Dataset Using UNG with Input Feature Mean-Zero Standardization[a]

Dataset	5NN	NBC	LDA	SVMGA	SVMLS	KREG	ANN	PLOG
AMLALL2	1.00	0.99	1.00	1.00	0.97	1.00	1.00	1.00
Brain2	0.72	0.63	0.68	0.50	0.69	0.65	0.55	0.69
Breast2A	1.00	0.94	1.00	1.00	1.00	1.00	1.00	1.00
Breast2B	0.75	0.74	0.80	0.66	0.59	0.81	0.84	0.79
Colon2	0.85	0.81	0.85	0.75	0.72	0.85	0.86	0.85
Lung2	1.00	1.00	1.00	1.00	1.00	1.00	1.00	1.00
MLL_Leukemia3	1.00	0.95	0.98	0.98	0.98	0.98	1.00	0.97
Prostate2	0.62	0.66	0.69	0.65	0.55	0.69	0.90	0.69
SRBCT4	1.00	0.88	1.00	1.00	1.00	1.00	1.00	1.00

[a]Ten 10-fold cross-validation used for determining performance.

TABLE 5.3 Classification after Dimensional Reduction of Greedy PTA Features to Ω Dimensions Using UNG with Input Feature Normalization[a]

Dataset	5N	NBC	LDA	SVMGA	SVMLS	KREG	ANN	PLOG
AMLALL2	1.00	0.99	1.00	1.00	0.96	1.00	1.00	1.00
Brain2	0.82	0.68	0.74	0.71	0.58	0.79	0.70	0.81
Breast2A	1.00	0.91	1.00	1.00	1.00	1.00	0.84	1.00
Breast2B	0.81	0.78	0.85	0.79	0.70	0.85	0.85	0.83
Colon2	0.84	0.80	0.87	0.79	0.76	0.87	0.87	0.86
Lung2	1.00	0.99	1.00	0.98	0.95	1.00	1.00	1.00
MLL_Leukemia3	0.98	0.95	0.98	0.97	0.97	0.96	1.00	1.00
Prostate2	0.92	0.89	0.93	0.85	0.71	0.93	0.91	0.92
SRBCT4	0.93	0.82	0.94	0.91	0.84	0.84	0.89	0.92

[a]Ten 10-fold cross-validation used for determining performance.

Average classifier accuracy listed in Table 5.4 shows that feature reduction with SOM and mean-zero standardization resulted in greater performance improvement when compared with classifier performance after UNG feature dimensional reduction.

5.5 SUMMARY

The UNG method has proven useful as an unsupervised cluster analysis technique that generates prototype vectors whose values are similar to *K*-means

TABLE 5.4 Average Classifier Performance with and without Feature Reduction with SOM and UNG Prior to Classification

Method	5NN	NBC	LDA	SVMGA	SVMLS	KREG	ANN	PLOG
wo/SOM or UNG	0.901	**0.930**	**0.957**	0.861	0.874	0.898	**0.981**	**0.967**
SOM(std)	**0.947**	0.902	0.944	**0.897**	**0.897**	**0.922**	0.971	0.942
SOM(norm)	0.933	0.890	0.932	0.882	0.887	0.909	0.968	0.934
UNG(std)	0.882	0.844	0.889	0.838	0.833	0.887	0.906	0.888
UNG(norm)	0.922	0.868	0.923	0.889	0.830	0.916	0.896	0.927

cluster centers. Results of dimensional reduction and follow-on analysis using reduced feature sets vary according to the data being used. UNG is a type of neural adaptive learning method that can exploit nonlinearities in the cluster structure of data. Therefore, when UNG is used for unsupervised cluster analysis in the presence of non-linearity in the data, results are likely to be different from K-means cluster analysis, which tends to be linear. UNG results are also likely to be different from PCA, which is based on linear combinations of standardized feature values (or eigenvector values) and transformed eigenvalues.

REFERENCES

[1] T. Martinetz, S. Berkovich, K. Schulten. Neural gas network for vector quantization and its application to time-series prediction. *IEEE Trans. Neural Networks* **4**(4):558–568, 1993.

[2] B. L. Zhang, M.Y. Fu, H. Yan. Handwritten signature verification based on neural gas based vector quantization. *Proc. 14th Int. Conf. Pattern Recognition*, IEEE Press, Piscataway NJ, 1998, vol. 2, pp. 1862–1864.

[3] B. Fritzke. *A Growing Neural Gas Network Learns Topologies*. Advances in Neural Information Processing Systems series, vol. 7, MIT Press, Cambridge, MA, 1995.

[4] S. Chelloug, S. Meshoul, M. Batouche. Clustering microarray data within amorphous computing paradigm and growing neural gas algorithm. *Lecture Notes Comput. Sci.* **4031**:809–818, 2006.

[5] K. Jackson, I. Koprinska. DNA microarray data clustering using growing self organizing networks. *Proc. 9th Int. Conf. Neural Information Processing (ICONIP'02)*, IEEE Press, Piscataway, NJ, 2002, vol. 2, pp. 805–808.

[6] D. Tasoulis, V. Plagianakos, M. Vrahatis. Unsupervised clustering in mRNA expression profiles. *Comput. Biol. Med.*, **36**(10):1126–1142, 2005.

[7] A. Meyer-Baese, A. Wismueller, O. Lange. Comparison of two exploratory data analysis methods for fMRI: Unsupervised clustering versus independent component analysis. *IEEE Trans. Inform. Technol. Biomed.* **8**(3):387–398, 2004.

[8] E. Dimitriadou, M. Barth, C. Windischberger, K. Hornik, E. Moser. A quantitative comparison of functional MRI cluster analysis. *Artif. Intell. Med.* **31**(1):57–71, 2004.

[9] T. Ogura, K. Iwasaki, C. Sato. Topology representing network enables highly accurate classification of protein images taken by cryo electron-microscope without masking. *J. Struct. Biol.* **143**(3):185–200, 2003.

CHAPTER 6

HIERARCHICAL CLUSTER ANALYSIS

6.1 INTRODUCTION

Hierarchical cluster analysis (HCA) is by far one of the most popular unsupervised class discovery methods for DNA microarray data [1-7]. Demonstrated HCA applications include genomewide characterization of mRNA transcripts during the cell cycle of the budding yeast *Saccharomyces cerevisiae* [8], investigations on multiple sclerosis [9], prostate cancer [10], breast cancer [11], renal cell carcinoma [12], and response to ionizing radiation [13]. In short, microarrays have the distinct advantage over manual blotting methods of providing large quantities of information on transcripts not previously associated with drugs [14], stressors [15], or disease.

6.2 METHODS

6.2.1 General Programming Methods

Cluster analytic methods start with an $n \times n$ array-by-array symmetric Euclidean distance matrix **D** or correlation matrix **R** matrix as the proximity matrix. Let us assume that **R** is chosen as the proximity matrix. To save memory, the lower triangular of **R** is copied to into a $(n+1)((n+1)-1)/2 + (n+1)$ vector in the following manner:

$$\underset{n\times n}{\mathbf{R}} = \begin{pmatrix} r_{21} & & \\ \vdots & \vdots & \\ r_{n1} & r_{n2} & \cdots \end{pmatrix} \tag{6.1}$$

Classification Analysis of DNA Microarrays, First Edition. Leif E. Peterson.

$$\bar{\mathbf{R}} = \begin{pmatrix} r_{21} \\ r_{31} \\ r_{32} \\ r_{41} \\ r_{42} \\ r_{43} \\ \vdots \\ r_{n1} \\ r_{n2} \\ r_{n(n-1)} \end{pmatrix}. \tag{6.2}$$

Storage of only the lower triangular of **R** (not including the ones) reduces the number of stored elements by $(n^2 - n)/2$. The amount of memory saved by this method is listed for several n-symmetric matrices in the following table:

n	Elements eliminated	MBytes memory saved[a]
100	4,950	0.019
1,000	499,500	1.998
10,000	49,995,000	199.98
100,000	4,999,950,000	19,999.8

[a] Assumes 4-byte single-precision floating point format.

6.2.2 Step 1: Cluster-Analyzing Arrays as Objects with Genes as Attributes

***Standardizing Genes (Attributes) in Rows of* Y.** First, the data matrix **Y** are read from disk. With arrays as the objects to cluster, standardization is performed using the gene mean vector $\bar{\mathbf{y}}$, consisting of gene-specific means (i.e., $\bar{y}_1, \bar{y}_2, \ldots, \bar{y}_p$) and standard deviations (of rows), defined as

$$s_j = \left(\frac{\sum_{i=1}^{n} (y_{ji} - \bar{y}_j)^2}{n-1} \right)^{1/2}. \tag{6.3}$$

The resulting data matrix and mean and standard deviation vectors are arranged as follows:

	Array					
Genes	1	2	$\cdots$	n	$\bar{\mathbf{y}}$	$\mathbf{s}$
1	y_{11}	y_{12}	$\cdots$	y_{1n}	$\bar{y}_1$	s_1
2	y_{21}	y_{22}	$\cdots$	y_{2n}	$\bar{y}_2$	s_2
$\vdots$	$\vdots$	$\vdots$	$\ddots$	$\vdots$	$\vdots$	$\vdots$
p	y_{p1}	y_{p2}	$\cdots$	y_{pn}	$\bar{y}_p$	s_p

Next, elements of the standardized data matrix **Z** are determined using the row (gene)-specific means and standard deviations in the form

$$z_{ji} = \frac{y_{ji} - \bar{y}_j}{s_j}. \tag{6.4}$$

After rowwise standardization of attributes (genes), the array mean vector $\bar{\mathbf{z}}$, consisting of column (array)-specific means (i.e., $\bar{z}_1, \bar{z}_2, \ldots, \bar{z}_n$) and standard deviations (of columns) are calculated using the equation

$$s_i = \left(\frac{\sum_{j=1}^{p} (z_{ji} - \bar{z}_i)^2}{p - 1} \right)^{1/2}. \tag{6.5}$$

The column (array)-specific means and standard deviations of **Z** are shown in the following form:

	Array			
Genes	1	2	$\cdots$	n
1	z_{11}	z_{12}	$\cdots$	z_{1n}
2	z_{21}	z_{22}	$\cdots$	z_{2n}
$\vdots$	$\vdots$	$\vdots$	$\ddots$	$\vdots$
p	z_{p1}	z_{p2}	$\cdots$	z_{pn}
$\bar{\mathbf{z}}$	$\bar{z}_1$	$\bar{z}_2$	$\cdots$	$\bar{z}_n$
$\mathbf{s}$	s_1	s_2	$\cdots$	s_n

Correlating Arrays (Objects) in Columns of Z. Column (array)-specific means and standard deviations are then combined to estimate the correlation between each column (array) k and l with the formalism

$$r_{kl} = \frac{\sum_{j=1}^{p} \left[\left(z_{jk} - \bar{z}_k \right) / s_k \right] \left[\left(z_{jl} - \bar{z}_l \right) / s_l \right]}{p - 1}. \tag{6.6}$$

The lower triangular of the $n \times n$ (array-by-array) matrix **R** is used and is copied to a vector.

At the first step in the clustering algorithm, the lowest (most negative) correlation coefficient r_{rs} between arrays R and S is found, and then R and S are "fused" to form a new cluster U. In the remaining steps, the distance between U and all other clusters V is determined in the functional form

$$r_{uv} = \frac{|r|}{|u|} r_{rv} + \frac{|s|}{|u|} r_{sv}, \tag{6.7}$$

where $|r|$, $|s|$, and $|u|$ are the number of objects (i.e., cardinality), in the clusters and r_{rv} and r_{sv} are the joining distances determined from the previous step. This procedure is performed recursively until all clusters have joined and there is one remaining value of r_{rs}.

Once clustering of the n arrays is complete, the columns representing arrays in the original **Y** matrix are reordered according to the agglomeration of arrays during clustering—which results in a new column-reordered matrix termed **Y***. The new **Y*** matrix is then used for gene clustering.

6.2.3 Step 2: Cluster-Analyzing Genes as Objects with Arrays as Attributes

The reordered data matrix **Y*** now appears as follows:

	Array			
Genes	1	2	$\cdots$	n
1	$y_{11}*$	$y_{12}*$	$\cdots$	$y_{1n}*$
2	$y_{21}*$	$y_{22}*$	$\cdots$	$y_{2n}*$
$\vdots$	$\vdots$	$\vdots$	$\ddots$	$\vdots$
p	$y_{p1}*$	$y_{p2}*$	$\cdots$	$y_{pn}*$

Standardizing Array (Attributes) in Columns of* Y*.** Standardization of **Y is performed using the array mean vector $\bar{\mathbf{y}}^*$, consisting of array-specific means and standard deviations (of columns) based on

$$s_i^* = \left(\frac{\sum_{j=1}^{p} (y_{ji}^* - \bar{y}_i^*)^2}{p-1} \right)^{1/2}. \tag{6.8}$$

The column (array)-specific means and standard deviations of **Y*** are shown in the following form:

	Array			
Genes	1	2	$\cdots$	n
1	y_{11}^*	y_{12}^*	$\cdots$	y_{1n}^*
2	y_{21}^*	y_{22}^*	$\cdots$	y_{2n}^*
$\vdots$	$\vdots$	$\vdots$	$\ddots$	$\vdots$
p	y_{p1}^*	y_{p2}^*	$\cdots$	y_{pn}^*
$\bar{\mathbf{y}}^*$	$\bar{y}_1^*$	$\bar{y}_2^*$	$\cdots$	$\bar{y}_n^*$
$\mathbf{s}^*$	s_1^*	s_2^*	$\cdots$	s_n^*

Elements of the standardized data matrix **Z** are

$$z_{ji} = \frac{y_{ji}^* - \bar{y}_i^*}{s_i^*}, \tag{6.9}$$

and the means and standard deviations of rows of **Z** are determined and given in the following arrangement:

	Array					
Genes	1	2	$\cdots$	n	$\bar{\mathbf{z}}$	$\mathbf{s}$
1	z_{11}	z_{12}	$\cdots$	z_{1n}	$\bar{z}_1$	s_1
2	z_{21}	z_{22}	$\cdots$	z_{2n}	$\bar{z}_2$	s_2
$\vdots$	$\vdots$	$\vdots$	$\ddots$	$\vdots$	$\vdots$	$\vdots$
p	z_{p1}	z_{p2}	$\cdots$	z_{pn}	$\bar{z}_p$	s_p

Correlating Genes (Objects) in Rows of Z. Row(gene)-specific means and standard deviations of **Z** are then combined to estimate the correlation between each row(gene) k and l as

$$r_{kl} = \frac{\sum_{i=1}^{n} [(z_{ki} - \bar{z}_k)/s_k]\,[(z_{li} - \bar{z}_l)/s_l]}{n-1} \tag{6.10}$$

Only the lower triangular of the $p \times p$ (gene-by-gene) matrix **R** is used, which is copied to a vector. Clustering of the p genes then begins. After clustering is complete, there will be c major clusters that were formed from clusters of size greater than 1.

Algorithm 6.1: Hierarchical Cluster Analysis (HCA)

Result: Dendograms for arrays and genes
Identify quantiles of expression values for heatmap colors
if *Cluster arrays* **then**
  Determine between-array distance matrix **D**
  Determine array agglomeration sequence
endif
if *Cluster genes* **then**
  Determine between-gene distance matrix **D**
  Determine gene agglomeration sequence
endif
Draw dendogram from agglomeration sequence
Draw heat map

6.3 ALGORITHM

The stepwise procedures used for HCA are listed in Algorithm 6.1.

6.4 IMPLEMENTATION

6.4.1 Heatmap Color Control

The first task in HCA is to sort in ascending order the expression values for all genes from all arrays, and then identify expression values at either percentiles (`colorscalemethod = 1`) or uniformly spaced intervals (`colorscalemethod = 0`) over the range of expression. These values are then saved in the vector called `colorscaledecile()` and are used for assigning colors to cells in the heatmap. We have learned that 22 uniformly spaced cutpoints in the range of expression works well for a variety of color options:

```
Dim arr() As Double
Dim ndx() As Integer
Dim delta As Double
maxx = -1.0E+30
minx = 1.0E+30
ReDim arr(numGenes * NumArrays)
ReDim ndx(numGenes * NumArrays)
icol = 0
For j = 1 To numGenes
  For k = 1 To NumArrays
    icol += 1
    arr(icol) = zmatrix(k, j)
```

```
    ndx(icol) = icol
    If zmatrix(k, j) > maxx Then maxx = zmatrix(k, j)
    If zmatrix(k, j) < minx Then minx = zmatrix(k, j)
  Next k
Next j
descending = False
Call sort(arr, ndx, icol, descending)
globalmedian = arr(icol / 2)
If ClusDataStd = 0 Then
  globalmin = minx
  globalmax = maxx
End If
colorscaledecile(1) = arr(1)
If colorscalemethod = 1 Then
  For j = 2 To 22
    i = (j * icol / 22)
    colorscaledecile(j) = arr(i)
  Next j
End If
If colorscalemethod = 0 Then
  delta = (maxx - minx) / 21
  For j = 1 To 22
    colorscaledecile(j) = minx + (j - 1) * delta
  Next j
End If
```

6.4.2 User Choices for Clustering Arrays and Genes

Dendograms can be generated for arrays and/or genes. Distances are calculated first for arrays, and then genes. The subroutine `clusdist` is called, and when `clusterchoice = 1`, array distances are used to determining the agglomeration sequence reflecting how arrays are joined together. When `clusterchoice = 2` and `clusdist` is called, the between-gene distances are calculated and used for generating the agglomeration sequence of genes; the most common approach is to obtain dendograms for both the arrays and genes. The code below lists how the subroutine `clusdist` is called for clustering arrays and clustering genes. The output of the `clustdist` is the agglomeration sequence containing the order in which arrays and genes are joined, which is needed for generating dendograms:

```
If ClusterArrays = True Then
  ClusterArrays = 1
  If ClusterGenes = False Then
    ClustersBoth = 0
    ClusterGenes = 0
  End If
```

```
  If ClusterGenes = True Then
    ClustersBoth = 1
    ClusterGenes = 1
  End If
  ClusterChoice = 1
  ncol = NumArrays
  nrow = NumGenes
  If ClusDataStd = 1 Then Call ClusStandardize(zmatrix,
   NumArrays, NumGenes)
  Call clusdist(zmatrix, ClusterChoice)
End If
If ClusterGenes = True Then
  ClusterGenes = 1
  If ClusterArrays = False Then
    ClustersBoth = 0
    ClusterArrays = 0
  End If
  If ClusterArrays = True Then
    ClustersBoth = 1
    ClusterArrays = 1
  End If
  ClusterChoice = 2
  nrow = NumArrays
  ncol = NumGenes
  If ClusDataStd = 1 Then Call ClusStandardize(zmatrix,
   NumArrays, NumGenes)
  Call clusdist(zmatrix, ClusterChoice)
End If
```

6.4.3 Distance Matrices and Agglomeration Sequences

The heart of the HCA algorithm is the process by which the arrays and genes are agglomerated based on distance calculations. The `clusdist` subroutine listed below performs the distance calculations using (6.7). Note that the type of distance can be specified, where `clusdistfunc = 1` for Euclidean distance and `clusdistfunc = 2` for one minus correlation, $1 - r$. In addition, note that the square root of the intermediate distance variable `dis` is omitted for Euclidean distance calculations, since this improves performance. The jagged array `dist()()` is used to store the distance results, based on work done by use of the variable `dis`:

```
Sub clusdist(zmatrix, clusterchoice)

'Calculate distance matrix
upperarrays = nrow
For i = 2 To ncol
```

```
  For j = 1 To i - 1
    ReDim Preserve dist(i)(j)
    dis = 0
    For m = 1 To upperarrays
      array = m
      If clusterchoice = 1 Then
        row1 = i
        row2 = j
        col1 = array
        col2 = array
      End If
      If clusterchoice = 2 Then
        row1 = array
        row2 = array
        col1 = i
        col2 = j
      End If
      If clusdistfunc = 1 Then dis += (zmatrix(row1, col1) - _
        zmatrix(row2, col2)) * (zmatrix(row1, col1) - _
        zmatrix(row2, col2))
      If clusdistfunc = 2 Then
        If clusterchoice = 1 Then dis += (zmatrix(row1,
           col1) - _
          arraymean(row1)) * (zmatrix(row2, col2) - arraymean
            (row2))
        If clusterchoice = 2 Then dis += (zmatrix(row1,
           col1) - _
          genemean(col1)) * (zmatrix(row2, col2) - genemean
            (col2))
      End If
    Next m
    If clusdistfunc = 1 Then dist(i)(j) = dis
    If clusdistfunc = 2 Then
      If clusterchoice = 1 Then dis /= ((upperarrays - 1) * _
        arraysd(i) * arraysd(j))
      If clusterchoice = 2 Then dis /= ((upperarrays - 1) * _
        genesd(i) * genesd(j))
    End If
  If clusdistfunc = 2 Then dist(i)(j) = 1 - dis
  Next j
Next i

'Perform agglomeration by filling arraydendogram(,) and
'genedendogram(,) arrays
For m = 1 To ncol - 1
  lowest = 8.0E+28
  For i = 2 To ncol
    For j = 1 To i - 1
```

```
      If colskip(j) = 0 Then
        If dist(i)(j) < lowest Then
          lowest = dist(i)(j)
          lastj = j
          lasti = i
        End If
      End If
    Next j
  Next i
  rowid = lasti
  colid = lastj
  dist(lasti)(lastj) = 9.0E+29
  colskip(colid) = lastj
  colString = clustord(colid)
  rowString = clustord(rowid)
  If lowest = 9.0E+29 Then Exit For
    If clusterchoice = 1 Then
      arraydendogram(m, 1) = clusterid(rowid)
      arraydendogram(m, 2) = clusterid(colid)
      arraydistance(m) = lowest
    End If
    If clusterchoice = 2 Then
      genedendogram(m, 1) = clusterid(rowid)
      genedendogram(m, 2) = clusterid(colid)
      genedistance(m) = lowest
    End If
    joinstring = rowString & "," & colString
    rowsize = clustsize(rowid)
    colsize = clustsize(colid)
    clustord(rowid) = joinstring
    clustsize(rowid) += clustsize(colid)
    'Estimate new distances.  One of them is the lowest
    'distance between all nodes (selected above) and the other
    'is from the remaining distances that need updating and
    'are looped through.
        fac3 = 1 / (rowsize + colsize)
    'As an example, consider the lowest distance to occur
    'between nodes r and s (arrays r and s) consisting of n_r
    'and n_s elements, respectively.   Nodes r and s are then
    '''fused'' to form a new node u with n_u= n_r + n_s elements.
    'Next, distances between the newly formed node u
    '(comprised of nodes r and s) and all other nodes v are '
    'calculated as
    'ClusJoinFunction = 1 (single linkage):
    'D(u,v) = min[D(r,v),D(s,v)]
    'ClusJoinFunction = 2 (UPGMA):
    'D(u,v) = [(D(r,v)*n_r + D(s,v)*n_s)]/n_u  (UPGMA)
    'ClusJoinFunction = 3 (complete linkage):
```

```
    'D(u,v) = max[D(r,v),D(s,v)] (complete linkage)
    r = rowid
    c = colid
    For j = 1 To ncol - 1
      If j = r OrElse j = c Then GoTo nextrec
        If colskip(j) = 0 Then
          If j < r Then
            rowsel1 = r
            colsel1 = j
            rowsel3 = r
            colsel3 = j
          End If
          If j > r Then
            rowsel1 = j
            colsel1 = r
            rowsel3 = j
            colsel3 = r
          End If
          If j < c Then
            rowsel2 = c
            colsel2 = j
          End If
          If j > c Then
            rowsel2 = j
            colsel2 = c
          End If
          If ClusJoinFunction = 1 Then dist(rowsel3)(colsel3)
             = _ minimum(dist(rowsel1)(colsel1), dist(rowsel2)
             (colsel2))
          If ClusJoinFunction = 2 Then dist(rowsel3)(colsel3)
             = _ fac3 * ((dist(rowsel1)(colsel1) * rowsize + _
            dist(rowsel2)(colsel2) * colsize))
          If ClusJoinFunction = 3 Then dist(rowsel3)(colsel3)
             = _ maximum(dist(rowsel1)(colsel1), dist(rowsel2)
             (colsel2))
          dist(rowsel2)(colsel2) = 9.0E+29
      End If
    nextrec:
    Next j
    clusterdis(colid) = lowest
    clusterdis(rowid) = lowest
Next m
joinstring = joinstring.Trim
Dim buffer() As String
buffer = joinstring.Split(",")
For j = 1 To UBound(buffer) + 1
  If j < ncol Then cluster(j) = CInt(buffer(j - 1))
  If j = ncol AndAlso InStr(buffer(j - 1), ",") = 0 Then _
```

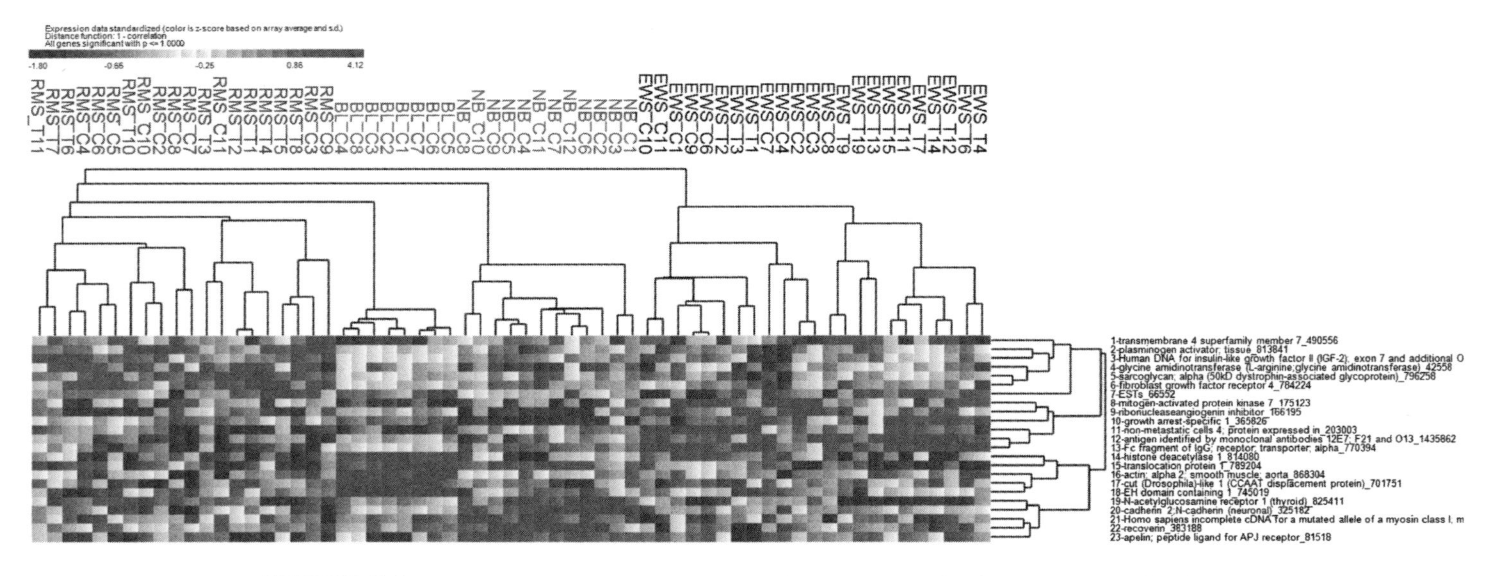

FIGURE 6.1 Hierarchical cluster analysis (HCA) heatmap for four-class SRBCT data showing dendograms for arrays and genes. Data standardized over genes, and one minus correlation ($1 - r$) used for agglomeration. (*See insert for color representation of the figure.*)

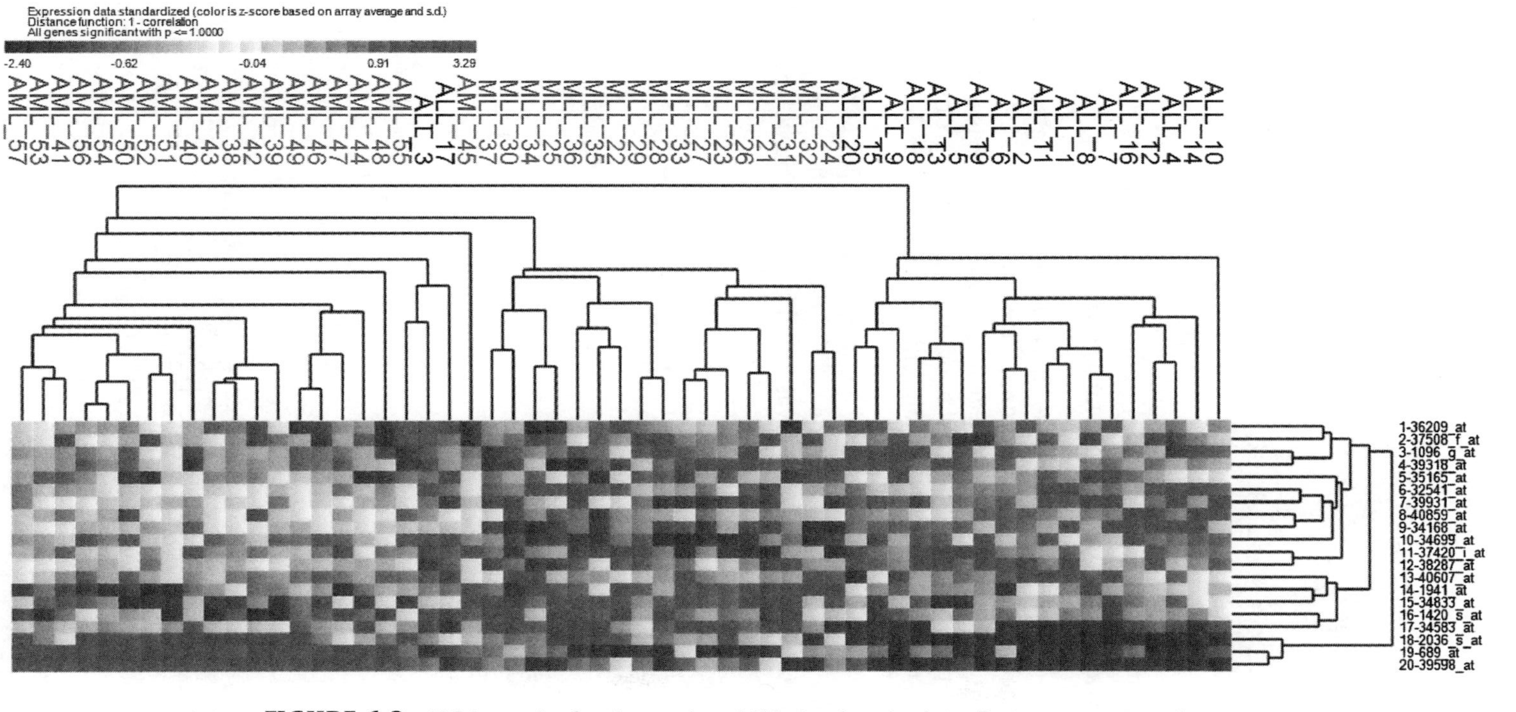

FIGURE 6.2 HCA results for three-class MLL_Leukemia data. Data mean-zero standardized over arrays and genes and $1 - r$ used for agglomeration. (*See insert for color representation of the figure.*)

```
    cluster(j) = CInt(buffer(j - 1))
  If j = ncol AndAlso InStr(buffer(j - 1), ",") > 0 Then _
    cluster(j) = CInt(VB.Left(buffer(j - 1), _
    InStr(buffer(j - 1), ",")))
Next j

Call drawclusterpic(zmatrix, ipca, endgene, startgene)

End Sub
```

6.4.4 Drawing Dendograms and Heatmaps

Drawing dendograms is one of the most challenging programming feats, requiring painstaking, time-consuming efforts to obtain code that performs satisfactorily. Programming the heat map components is less demanding in terms of programming effort. Dendogram and heatmap drawing is performed by the `drawclusterpic` algorithm, which is too lengthy to list here.

Example 6.1 HCA was run on the SRBCT dataset with $n = 63$ arrays and $p = 23$ genes. The agglomeration was based on 1-correlation, and both the

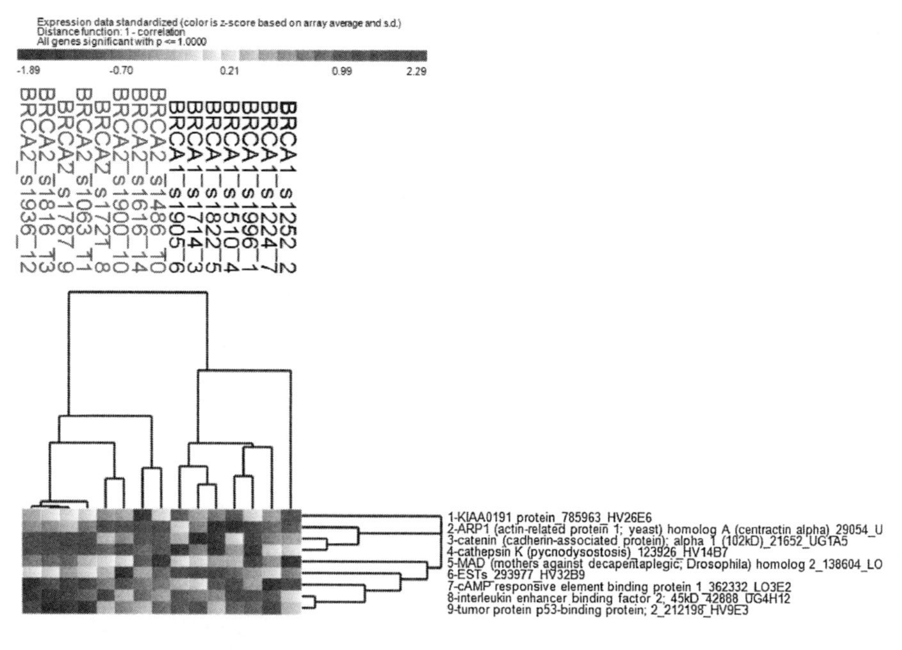

FIGURE 6.3 HCA results for the 2-class breast cancer dataset. Data mean-zero standardized and $1 - r$ for agglomeration.

arrays and genes were clustered to generate dendograms for each. Figure 6.1 shows that HCA results based on the $p = 23$ genes selected using greedy PTA results in four clusters, which represent the true microarray classes (EWS, BL, NB, and RMS), which were expected. The heatmap shows the variation in expression among the $p = 23$ genes over the microarrays, which again reflects the discriminatory value of the genes selected with greedy PTA. Figures 6.2 and 6.3 show HCA results for the three-class MLL_Leukemia and two-class breast cancer dataset for BRCA1- and BRCA2-positive cases. On both these figures, clear separation between tumor diagnostic class is observed, which, again, is expected because the genes selected with greedy PTA were strongly discriminating for class separation.

REFERENCES

[1] M. Schena, D. Shalon, R.W. Davis, P.O. Brown. Quantitative monitoring of gene expression patterns with a complimentary DNA microarray. *Science* **270**: 467–470, 1995.

[2] J.L. DeRisi, V.R. Iyer, P.O. Brown. Exploring the metabolic and genetic control of gene expression on a genomic scale. *Science* **278**:680–686, 1997.

[3] J.J. Chen, R. Wu, P.C. Yang, J.Y. Huang, Y.P. Sher, M.H. Han, W.C. Kao, P.J. Lee, T.F. Chiu, F. Chang, Y.W. Chu, C.W. Wu, K. Peck. Profiling expression patterns and isolating differentially expressed genes by cDNA microarray system with colorimetry detection. *Genomics* **51**:313–324, 1998.

[4] E. Southern, K. Mir, M. Shchepinov. Molecular interactions on microarrays. *Nat. Genet.* **21**(Suppl):5–9, 1999.

[5] D.D.L. Botwell. Options available-from start to finish-for obtaining expression data by microarray. *Nat. Genet.* **21**(Suppl):25–32, 1999.

[6] P.O. Brown, D. Botstein. Exploring the new world of the genome with DNA microarrays. *Nat. Genet.* **21**(Suppl):33–37, 1999.

[7] M.B. Eisen, P.T. Spellman, P.O. Brown, D. Botstein. Cluster analysis and display of genome-wide expression patterns. *Proc. Natl. Acad. Sci.* **95**:14863–14868, 1998.

[8] R. Cho, M.J. Campbell, E.A. Winzeler, L. Steinmetz, A. Conway, L. Wodicka, T.G. Wolfsberg, A.E. Gabrielian, D. Landsman, D.J. Lockhart, R.W. Davis. A genome-wide transcriptional analysis of the mitotic cell cycle. *Mol. Cell.* **2**:65–73, 1998.

[9] L.W. Whitney, K.G. Becker, N.J. Tresser, C.I. Caballero-Ramos, P.J. Munson, V.V. Prabhu, J.M. Trent, H.F. McFarland, W.E. Biddison. Analysis of gene expression in multiple sclerosis lesions using cDNA microarrays. *Ann. Neurol.* **46**:425–428, 1999.

[10] L. Bubendorf, M. Kolmer, J. Kononen, P. Koivisto, S. Mousses, Y. Chen, E. Mahlamki, P. Schraml, H. Moch, N. Willi, A.G. Elkahloun, T.G. Pretlow,

T.C. Gasser, M.J. Mihatsch, G. Sauter, O.P. Kallioniemi. Hormone therapy failure in human prostate cancer: Analysis by complimentary DNA and tissue microarrays. *J. Natl. Cancer Inst.* **91**:1758–1764, 1999.

[11] T.A. Lehman, B.G. Haffty, C.J. Carbone, L.R. Bishop, A.A. Gumbs, S. Krishnan, P.G. Shields, R. Modali, B.C. Turner. Elevated frequency and functional activity of a specific germ-line p53 intron mutation in familial breast cancer. *Cancer Res.* **60**(4):1062–1069, 2000.

[12] H. Moch, P. Schraml, L. Bubendorf, M. Mirlacher, J. Kononen, T. Gasser, M.J. Mihatsch, O.P. Kallioniemi, G. Sauter. High-throughput tissue microarray analysis to evaluate genes uncovered by cDNA microarray screening in renal cell carcinoma. *Am. J. Pathol.* **154**(4):981–986, 1999.

[13] S.A. Amundson, M. Bittner, Y. Chen, J. Trent, P. Meltzer, A.J. Fornace. Fluorescent cDNA microarray hybridization revels complexity and heterogeneity of cellular genotoxic stress responses. *Oncogene* **18**:3666–3672, 1999.

[14] C. Debouck, P.N. Goodfellow. DNA microarrays in drug discovery and development. *Nat. Genet.* **21**(Suppl):48–50, 1999.

[15] R.J. Collier, C.M. Stiening, B.C. Pollard, M.J. VanBaale, L.H. Baumgard, P.C. Gentry, P.M. Coussens. Use of gene expression microarrays for evaluating environmental stress tolerance at the cellular level in cattle. *J. Anim. Sci.* **84**:E1–E13, 2006.

CHAPTER 7

MODEL-BASED CLUSTERING

7.1 INTRODUCTION

Model-based clustering makes the assumption that a finite mixture of probability distributions is responsible for generating the data under consideration. Model-based cluster analysis is an extension of K-means cluster analysis, in which maximum likelihood estimation is used. Determination of the optimal cluster structure, that is, the number of clusters, for a dataset using model-based cluster analysis reduces to the evaluation and comparison of how a selected mixture of probability density functions can best describe the cluster structure.

This chapter addresses the finite *Gaussian mixture model* (GMM), which consists of several unknown clusters and their density components. To date, GMMs have advanced microarray research considerably. Rasmussen explored modeling and visualizing uncertainty in gene expression clusters using Dirichlet mixture models [1]. Microarray image analysis has focused on straightforward GMM analysis of DNA microarray images [2] and fuzzy Gaussian mixture models [3]. Other approaches have addressed data quality [4], inference of gene pathways using mixture Bayesian networks [5], probe-level uncertainty [6], and clustering genes based on quantitative phenotype [7].

Assume that a set of training data consists of the vector pairs $(\mathbf{x}_1, \mathbf{y}_1)$, $(\mathbf{x}_2, \mathbf{y}_2)$, ..., $(\mathbf{x}_n, \mathbf{y}_n)$ $(i = 1, 2, \ldots, n)$, where $\mathbf{x}_i$ represents a p-dimensional input vector and $\mathbf{y}_i$ represents a p-dimensional unobservable output vector. In addition, assume that the underlying data are generated from Ω $(\omega = 1, 2, \ldots, \Omega)$ unknown clusters. For each input vector $\mathbf{x}_i$, we randomly choose a mean vector $\boldsymbol{\mu}_\omega$ and covariance matrix $\boldsymbol{\Sigma}_\omega$ for one of the clusters with probability π_ω $(\sum_\omega^\Omega \pi_\omega = 1)$. Given these definitions, the outcome vector $\mathbf{y}_i$ represents a random variable with multivariate normal density

$$f(\mathbf{y}_i; \boldsymbol{\mu}_\omega; \boldsymbol{\Sigma}_\omega) = \frac{1}{(2\pi)^{p/2}\sqrt{|\boldsymbol{\Sigma}_\omega|}} e^{-1/2}(\mathbf{x}_i - \boldsymbol{\mu}_\omega)\Sigma_\omega^{-1}(\mathbf{x}_i - \boldsymbol{\mu}_\omega). \qquad (7.1)$$

Classification Analysis of DNA Microarrays, First Edition. Leif E. Peterson.

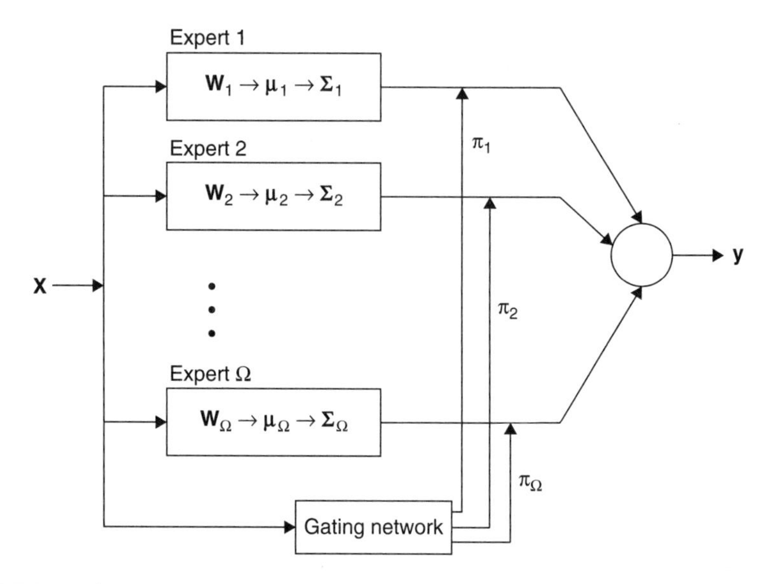

FIGURE 7.1 Structure of Gaussian mixture model (GMM) with gating network.

Figure 7.1 illustrates the GMM learning network with an expert and gating probability $\boldsymbol{\pi}_\omega$ for each cluster. The unknown output vectors $\mathbf{y}_i$ originate from a mixture of densities formed by the sum product of probabilities from experts times the probabilities of the experts, given as

$$p(\mathbf{y}_i; \boldsymbol{\mu}_\omega; \boldsymbol{\Sigma}_\omega) = \sum_{\omega}^{\Omega} \pi_\omega f(\mathbf{y}_i; \boldsymbol{\mu}_\omega; \boldsymbol{\Sigma}_\omega). \tag{7.2}$$

Figure 7.2 shows the expert network responsible for obtaining the mean vector elements for cluster ω. Within each expert, the cluster-specific mean vector and covariance matrix are calculated. For each feature, the cluster-specific mean vector elements are

$$\mu_{j\omega} = \sum_{i=1}^{n} w_{i\omega} x_{ij}, \tag{7.3}$$

where the weights, or *responsibilities,* are determined using an iterative procedure described later.

Now let us consider augmenting the vector $\mathbf{x}_i$ with a class (cluster) label z_ω so that the unobservable outcome vector becomes $\mathbf{y}_i = (\mathbf{x}_i, z_{i\omega})$. By definition, the class labels are missing, and $\mathbf{z}_{i\omega} = 1$ if microarray $\mathbf{x}_i$ is in class

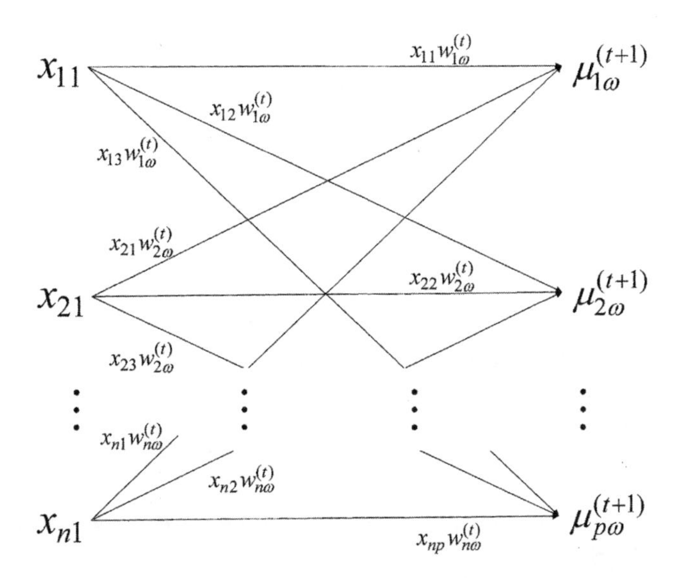

FIGURE 7.2 Expert weight learning network for determining the mean vector $\boldsymbol{\mu}_\omega$ for cluster ω at the $(t+1)$th iteration. Vector elements of $\boldsymbol{\mu}_\omega$ are $(\mu_{1\omega}, \mu_{2\omega}, \ldots, \mu_{p\omega})$ representing the $j = 1, 2, \ldots, p$ features.

ω and $\mathbf{z}_{i\omega} = 0$ otherwise. The incomplete likelihood function is

$$L(\mathbf{y}_i|\boldsymbol{\mu}_\omega; \boldsymbol{\Sigma}_\omega) = \prod_{i=1}^{n}\prod_{\omega=1}^{\Omega}[\pi_\omega f(\mathbf{y}_i; \boldsymbol{\mu}_\omega; \boldsymbol{\Sigma}_\omega)]^{z_{i\omega}}, \tag{7.4}$$

and the corresponding incomplete loglikelihood is given as

$$\log(L(\mathbf{y}_i|\boldsymbol{\mu}_\omega; \boldsymbol{\Sigma}_\omega)) = \sum_{i=1}^{n}\sum_{\omega=1}^{\Omega} z_{i\omega}[\log \pi_\omega + \log f(\mathbf{y}_i; \boldsymbol{\mu}_\omega; \boldsymbol{\Sigma}_\omega)]. \tag{7.5}$$

The mixture likelihood in (7.4) is based on incomplete data because the class labels $\mathbf{z}_i$ are missing values. The expectation-maximization (EM) algorithm [8] can be used to maximize the incomplete loglikelihood in (7.5). The only way to apply the EM algorithm is to evaluate $\mathrm{E}(\mathbf{z}_i|\mathbf{x}_i)$, since $\mathbf{y}_i$ is fixed through the parameters $\boldsymbol{\mu}_\omega$ and $\boldsymbol{\Sigma}_\omega$. Knowing that $\mathbf{z}_i$ is dependent on $\mathbf{x}_i$ through $\mathbf{y}_i$, we can use Bayes' theorem to solve for the probability that $\mathbf{x}_i$ belongs to cluster ω with the functional form

$$\begin{aligned}\mathrm{E}(z_{i\omega}|\mathbf{y}_i) &= P(z_{i\omega} = 1|\mathbf{y}_i)\\ &= \frac{P(z_{i\omega} = 1)P(\mathbf{y}_i|z_{i\omega} = 1|)}{\sum_{\omega}^{\Omega} P(z_{i\omega} = 1)P(\mathbf{y}_i|z_{i\omega} = 1|)}\end{aligned}$$

$$= \frac{\pi_\omega f(\mathbf{y}_i; \boldsymbol{\mu}_\omega; \boldsymbol{\Sigma}_\omega)}{\sum_\omega^\Omega \pi_\omega f(\mathbf{y}_i; \boldsymbol{\mu}_\omega; \boldsymbol{\Sigma}_\omega)}$$

$$= w_{i\omega}. \tag{7.6}$$

7.2 ALGORITHM

The iterative training steps necessary for the EM algorithm are as follows:

1. *Initialize.* First, obtain the starting values for $\boldsymbol{\mu}_\omega$ by using K-means cluster analysis with Ω clusters. Determine the cluster prior probabilities as $\pi_\omega = n_\omega/n$, where n_ω is the number of microarrays assigned to cluster ω during K-means cluster analysis. The initial cluster-specific covariance matrices can be determined by using microarrays assigned to each cluster and the mean vectors obtained from K-means cluster analysis.
2. *E Step.* Calculate the cluster-specific multivariate normal density for each microarray using $\mathbf{x}_i$, the current mean centers, covariance matrices, and determinant of each covariance matrix with the functional form

$$f(\mathbf{y}_i; \boldsymbol{\mu}_\omega; \boldsymbol{\Sigma}_\omega) = \frac{1}{(2\pi)^{p/2}\sqrt{|\boldsymbol{\Sigma}_\omega|}} e^{-1/2}(\mathbf{x}_i - \boldsymbol{\mu}_\omega)\boldsymbol{\Sigma}_\omega^{-1}(\mathbf{x}_i - \boldsymbol{\mu}_\omega). \tag{7.7}$$

Determine the microarray-specific expert network weights used for obtaining the mean vectors and covariances, using the form

$$w_{i\omega} = \frac{\pi_\omega f(\mathbf{y}_i; \boldsymbol{\mu}_\omega; \boldsymbol{\Sigma}_\omega)}{\sum_\omega^\Omega \pi_\omega f(\mathbf{y}_i; \boldsymbol{\mu}_\omega; \boldsymbol{\Sigma}_\omega)}. \tag{7.8}$$

Calculate the loglikelihood $\log(L(\mathbf{y}_i|\boldsymbol{\mu}_\omega; \boldsymbol{\Sigma}_\omega))$ in (7.5), which should increase after each tth iteration, that is

$$\log(L(\mathbf{y}_i|\boldsymbol{\mu}_\omega; \boldsymbol{\Sigma}_\omega))^{t+1} > \log(L(\mathbf{y}_i|\boldsymbol{\mu}_\omega; \boldsymbol{\Sigma}_\omega))^{t}, \tag{7.9}$$

and determine the difference $\delta = \log(L(\mathbf{y}_i|\boldsymbol{\mu}_\omega; \boldsymbol{\Sigma}_\omega))^{t+1} - \log(L(\mathbf{y}_i|\boldsymbol{\mu}_\omega; \boldsymbol{\Sigma}_\omega))^{t}$. When $\delta \leq 10^{-6}$, exit the algorithm.
3. *M Step.* Update the cluster prior probabilities in the form

$$\pi_\omega = \frac{1}{n}\sum_{i=1}^{n} w_{i\omega}, \tag{7.10}$$

and each cluster-specific mean vector using

$$\boldsymbol{\mu}_{\omega} = \frac{\sum_{i=1}^{n} w_{i\omega}\mathbf{x}_i}{\sum_{i=1}^{n} w_{i\omega}}. \quad (7.11)$$

Array elements of the cluster-specific $p \times p$ covariance matrix $\boldsymbol{\Sigma}_{\omega}$ are determined using the functional form

$$s_{jk\omega} = \frac{\sum_{i=1}^{n} w_{i\omega}(x_{ij} - \mu_{j\omega})(x_{ik} - \mu_{k\omega})}{\sum_{i=1}^{n} w_{i\omega}}. \quad (7.12)$$

The EM algorithm repeats steps 2 and 3 above (E step, M step, E step, M step, etc.) until the change in the loglikelihood is negligible. It is important to track the value of the loglikelihood during iterations to ensure that it increases at each step. Using the algorithm outlined above, the first time the loglikelihood is calculated it will be equal to zero, and negative thereafter. Loglikelihood values should always be negative real numbers (they are logarithms of probabilities) since a perfect fit of data with a likelihood of one is rarely observed. Increasing values of the log-likelihood will always be more positive, and appear, for example, as −210.8, −190.2, −185.3, −182.1, −180.7, −180.5, and so on, as the algorithm goes through iterations.

7.3 IMPLEMENTATION

K-means cluster analysis is initially run to obtain the starting cluster centers. Training microarrays are then assigned to the closest center, and the covariance matrix for each cluster is determined using microarrays assigned to the particular clusters. The value of K is set to the desired number of clusters, which is stored in the `NumGMMClusters` variable. The first call is to the `kmeanscluster()` subroutine, which determines the centers for the K clusters. Next, the `means(,)` array is filled with cluster center values, and then used to obtain the covariance matrix `Sc(,,)` for microarrays in each cluster. The source code for these steps is as follows:

```
Call kmeanscluster(NumGMMClusters)

For c = 1 To NumGMMClusters
  For j = 1 To NumFeatures
    means(j, c) = Centers(j, c)'from K-means
  Next j
Next c
Dim sampsizeperclass(NumGMMClusters) As Integer
For i = 1 To NumObjects
  sampsizeperclass(clustassigned(i)) += 1
  'Note: the clustassigned() arrays is determined
```

```
  'in the kmeanscluster subroutine, and contains
  'the cluster id each microarray is closest to
Next
For c = 1 To NumGMMClusters
  pi(c) = 1 / NumGMMClusters
Next
'Covariance matrix for each cluster
For c = 1 To NumGMMClusters
  div1 = sampsizeperclass(c) - 1
  ReDim ycent(sampsizeperclass(c), NumFeatures)
  For j = 1 To NumFeatures
    For k = 1 To NumFeatures
      sum = 0
      For i = 1 To NumObjects
        If clustassigned(i) = c Then
          sum += (xmatrix(i, j) - means(j, c)) * (xmatrix(i, k) - _
            means(k, c))
        End If
      Next i
      sum /= div1
      Sc(c, j, k) = sum
    Next k
  Next j
Next c

For iter = 1 To numiterations
'E-step
  For c = 1 To NumGMMClusters
    totclustdist(c) = 0
  Next c
  For c = 1 To NumGMMClusters
    ReDim S(NumFeatures, NumFeatures)
    For i = 1 To NumFeatures
      For j = 1 To NumFeatures
        S(i, j) = Sc(c, i, j)
      Next j
    Next i
    Call eigen1(NumFeatures, S, 0)
    For j = 1 To NumFeatures
      CovEigVal(j) = EigVal(j, j)
    Next j
    mineigval = 1.0E+30
    maxeigval = -1.0E+30
    For j = 1 To NumFeatures
      If CovEigVal(j) < mineigval Then mineigval = CovEigVal(j)
      If CovEigVal(j) > maxeigval Then maxeigval = CovEigVal(j)
    Next
    condnumb = maxeigval / mineigval
    condnumvec(iter, c) = condnumb
    determ = 1
    For j = 1 To NumFeatures
      If CovEigVal(j) > 0 Then determ *= CovEigVal(j)
```

```
    Next j
    determvec(iter, c) = determ
    lndeterm(c) = Math.Log(determ)
    For i = 1 To NumObjects
      ReDim ycent(NumFeatures, 1)
      ReDim ycenttran(1, NumFeatures)
      ReDim ycenttransinv(1, NumFeatures)
      If iter = 0 Then
        For j = 1 To NumFeatures
          ycent(j, 1) = ybar(j, c) - xmatrix(i, j)
          ycenttran(1, j) = ycent(j, 1)
        Next j
      End If
      If iter >= 1 Then
        For j = 1 To NumFeatures
          ycent(j, 1) = xmatrix(i, j) - means(j, c)
          ycenttran(1, j) = ycent(j, 1)
        Next j
      End If
      Call MATMUL(ycenttran, UTU, ycenttransinv, 1, NumFeatures,_
         NumFeatures)
      Call MATMUL(ycenttransinv, ycent, chisq, 1, NumFeatures, 1)
      distance = Convert.ToDouble(chisq(1, 1))
      p(i, c) = -Math.Log(6.283185307 ^ (NumFeatures / 2)) - _
         Math.Log(Math.Sqrt(determ)) - 0.5 * distance
      totclustdist(c) += distance
    Next i
  Next c
  loglik = 0
  For i = 1 To NumObjects
    For c = 1 To NumGMMClusters
      loglik += w(i, c) * (p(i, c) + Math.Log(pi(c)))
    Next c
  Next i
  loglikvec(iter) = loglik
  For i = 1 To NumObjects
    For c = 1 To NumGMMClusters
      p(i, c) = Math.Exp(p(i, c))
    Next c
    sum = 0
    For c = 1 To NumGMMClusters
      sum += p(i, c)
    Next c
    For c = 1 To NumGMMClusters
      p(i, c) /= sum
    Next c
  Next i
  For i = 1 To NumObjects
    denom = 0
    For c = 1 To NumGMMClusters
      denom += pi(c) * p(i, c)
    Next c
```

```
    For c = 1 To NumGMMClusters
      w(i, c) = (pi(c) * p(i, c)) / denom
    Next c
  Next i

'M-step

  For c = 1 To NumGMMClusters
    sum = 0
    For i = 1 To NumObjects
      sum += w(i, c)
    Next i
    pi(c) = sum / NumObjects
  Next c
  sum = 0
  For c = 1 To NumGMMClusters
    sum += pi(c)
  Next c
  For c = 1 To NumGMMClusters
    pi(c) /= sum
  Next c
  For c = 1 To NumGMMClusters
    denom = 0
    For i = 1 To NumObjects
      denom += w(i, c)
    Next i
    For j = 1 To NumFeatures
      numer = 0
      For i = 1 To NumObjects
        numer += w(i, c) * xmatrix(i, j)
      Next i
      means(j, c) = numer / denom
    Next j
    For j = 1 To NumFeatures
      For k = 1 To NumFeatures
        sum = 0
        For i = 1 To NumObjects
          sum += w(i, c) * (xmatrix(i, j) - means(j, c)) * _
            (xmatrix(i, k) - means(k, c))
        Next i
        Sc(c, j, k) = sum / denom
      Next k
    Next j
  Next c
Next iter
```

Example 7.1 A dataset with three Gaussian clusters was simulated with mean values of $\boldsymbol{\mu} = (0, 0)$, $\boldsymbol{\mu} = (0, 1)$, and $\boldsymbol{\mu} = (0, -1)$ for the x and y values, standard deviation of unity, and correlation of 0.9 or -0.9 between x and y. A total of 1500 microarrays was generated (500 each cluster). Figure 7.3 illustrates the loglikelihood value as a function of training iterations for the GMM algorithm, which reached maximum stable value of −3186.28 within 30 iterations. Figure 7.4 shows the cluster-specific covariance matrix determinants and conditions numbers. Results indicate that a slight increase in collinearity occurred as the cluster-specific determinants reached values of 0.026, 0.027, and 0.027 by 30 iterations. The condition number of a matrix is the ratio of the largest to smallest eigenvalue, and therefore will increase with the multicollinearity or correlation among features. The condition numbers of the cluster-specific covariance matrices decreased with increasing multicollinearity among features, and for this run stabilized at values of 46.16, 47.25, and 48.25 for the three covariance matrices.

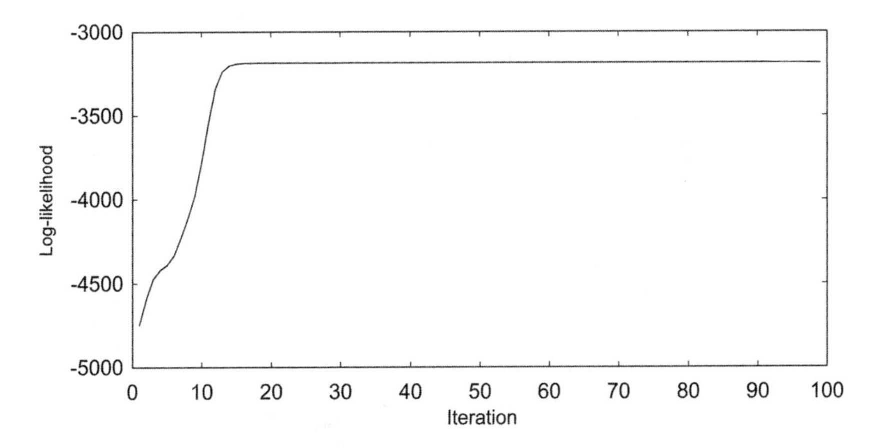

FIGURE 7.3 GMM loglikelihood value as a function of training itertation.

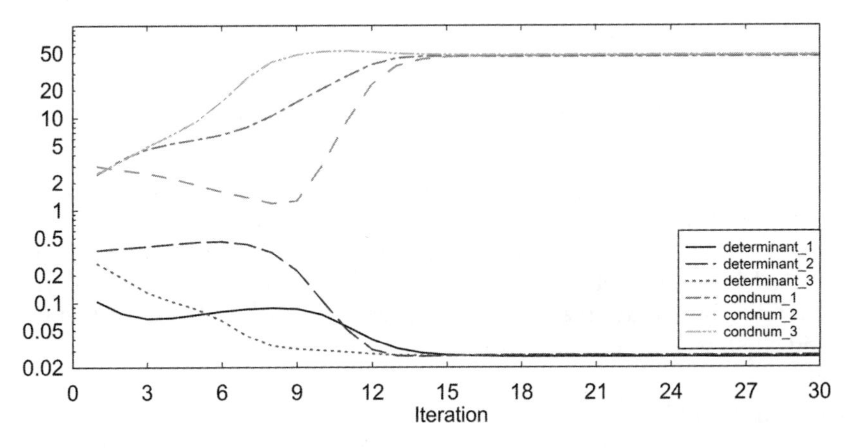

FIGURE 7.4 Cluster-specific covariance matrix determinants and condition numbers.

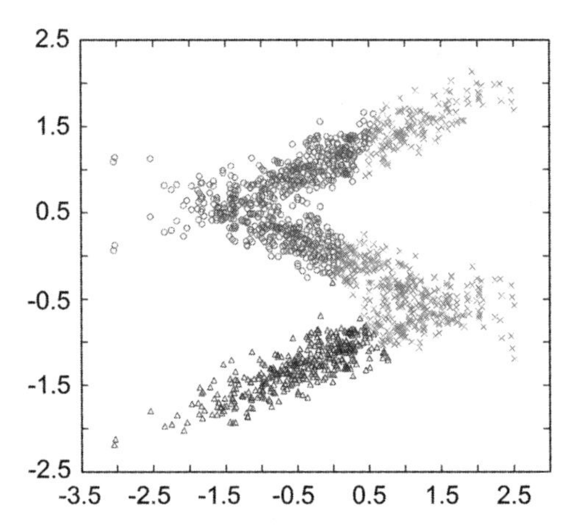

FIGURE 7.5 K-means cluster analysis ($K = 3$) results for 3 Gaussian clusters. (*See insert for color representation of the figure.*)

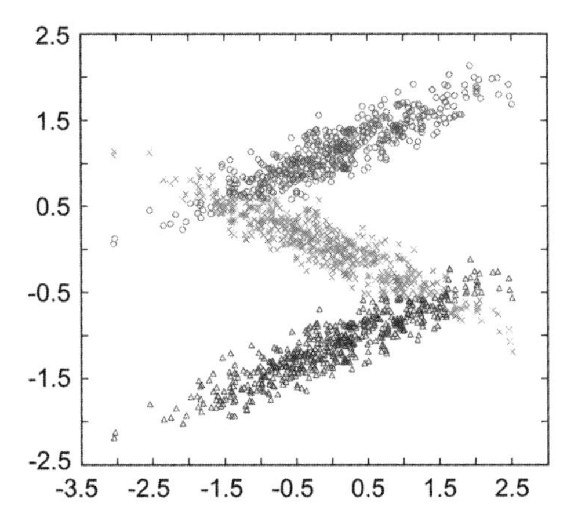

FIGURE 7.6 GMM cluster analysis results for three Gaussian clusters. (*See insert for color representation of the figure.*)

Cluster analysis results based on GMM were quite promising. Figure 7.5 shows that K-means cluster analysis using $K = 3$ clusters failed to correctly assign group labels for microarrays in the three Gaussian clusters. However, the GMM results portrayed in Figure 7.6 indicate very clear separation between cluster assignments and microarrays in each of the three Gaussian-based clusters.

7.4 SUMMARY

Gaussian mixture modeling is one of the most robust parametric unsupervised methods for cluster analysis. As shown in the example, it is clear that

GMM properly identifies microarrays in Gaussian clusters. A disadvantage of GMM, however, is that it is a parametric method that assumes a Gaussian distribution in the feature space. Not all data follow a distribution that can be modeled appropriately with GMM, and like all parametric methods, GMM has its limitations. Mixture-model-based cluster analysis has branched off into other areas of development, including nonparametric Bayesian model analysis based on Dirichlet mixture models [1], fuzzy Gaussian mixture models [3], and genetic algorithm (GA)-based EM approaches to GMM [9]. Outlier detection [11] and GMM's started as early as 1980 [10], and various methods have been introduced since then [12,13].

REFERENCES

[1] C.E. Rasmussen, B.J. de la Cruz, Z. Ghahramani, D.L. Wild. Modeling and visualizing uncertainty in gene expression clusters using Dirichlet process mixtures. *IEEE/ACM Trans. Comput. Biol. Bioinform.* **6**(4):615–628, 2009.

[2] K. Blekas, N.P. Galatsanos, A. Likas, I.E. Lagaris. Mixture model analysis of DNA microarray images. *IEEE Trans. Med. Imaging* **24**(7):901–909, 2005.

[3] E.I. Athanasiadis, D.A. Cavouras, P.P. Spyridonos, D.T. Glotsos, I.K. Kalatzis, G.C. Nikiforidis. Complementary DNA microarray image processing based on the fuzzy Gaussian mixture model. *IEEE Trans. Inform. Technol. Biomed.* **13**(4):419–425, 2009.

[4] B.E. Howard, B. Sick, S. Heber. Unsupervised assessment of microarray data quality using a Gaussian mixture model. *BMC Bioinform.* **10**:191, 2009.

[5] Y. Ko, C. Zhai, S. Rodriguez-Zas. Inference of gene pathways using mixture Bayesian networks. *BMC Syst. Biol.* **3**:54, 2009.

[6] X. Liu, K.K. Lin, B. Andersen, M. Rattray. Including probe-level uncertainty in model-based gene expression clustering. *BMC Bioinform.* **8**:98, 2007.

[7] Z. Jia, S. Xu. Clustering expressed genes on the basis of their association with a quantitative phenotype. *Genet. Res.* **86**(3):193–207, 2005.

[8] A.P. Dempster, N.M. Laird, D.B. Rubin. Maximum likelihood from incomplete data via the EM algorithm. *J. Roy. Stat. Soc.* **39**:1–38, 1977.

[9] F. Pernkopf, D. Bouchaffra. Genetic-based EM algorithm for learning Gaussian mixture models. *IEEE Trans. Pattern Anal. Machine Intell.* **27**(8):1344–1348, 2005.

[10] M. Aitkin, G.T. Wilson. Mixture models, outliers, and the EM algorithm. *Technometrics* 22:325–331, 1980.

[11] V. Barnett, T. Lewis. *Outliers in Statistical Data.* Wiley, New York, 1994.

[12] J.D. Banfield, A.E. Raftery. Model-based Gaussian and non-Gaussian clustering. *Biometrics* **49**:803–821, 1993.

[13] D.W. Scott. Outlier detection and clustering by partial mixture modeling. *Proc. 2004 COMPSTAT Symp.*, Physica-Verlag, Berlin, 1980, pp. 453–465.

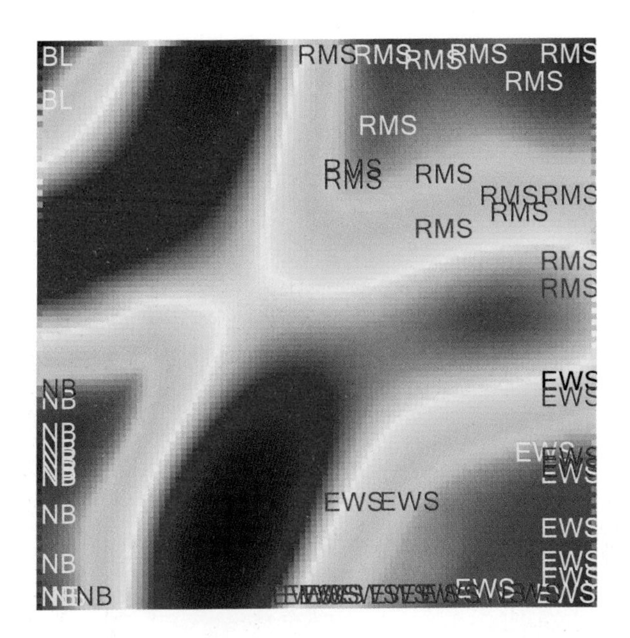

FIGURE 4.6 *U* matrix for four-class SRBCT data. The tubular structure (in red) resembling a deep valley is based on greater distances between reference vectors. Regions with similar shorter distances between reference vectors are shaded in blue.

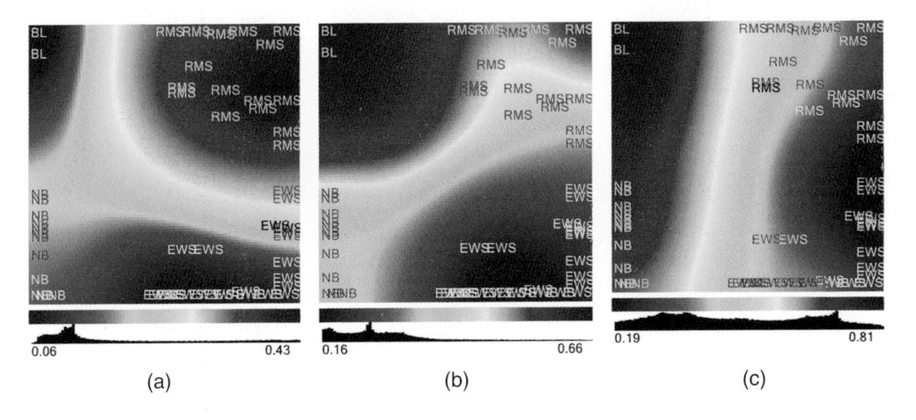

FIGURE 4.7 SOM component maps for three genes in the four-class SRBCT dataset. Expression of plasminogen activator (a) is enhanced in only rhabdomyosarcoma (RMS) cases, CCAAT gene (b) is enhanced in both Burkitt's lymphoma (BL) and RMS, and translocation protein 1 (c) is enhanced in both BL and neuroblastoma (NB). Red signifies enhanced expression, while blue signifies suppressed expression.

Classification Analysis of DNA Microarrays, First Edition. Leif E. Peterson.

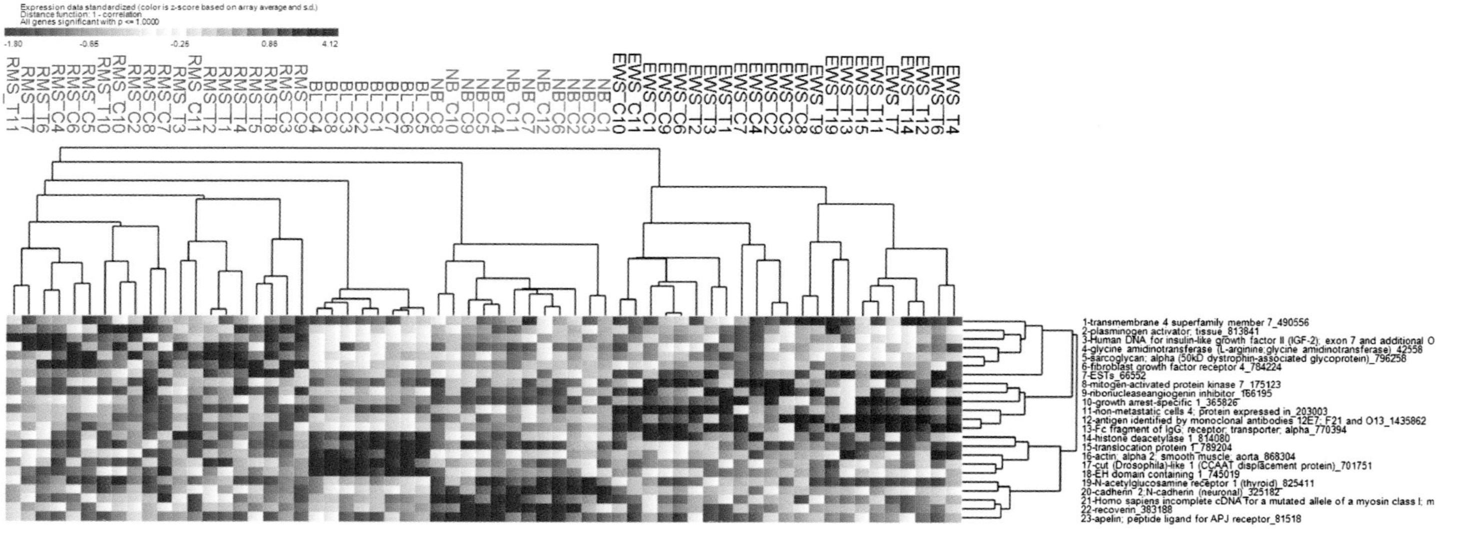

FIGURE 6.1 Hierarchical cluster analysis (HCA) heatmap for four-class SRBCT data showing dendograms for arrays and genes. Data standardized over genes, and one minus correlation $(1 - r)$ used for agglomeration.

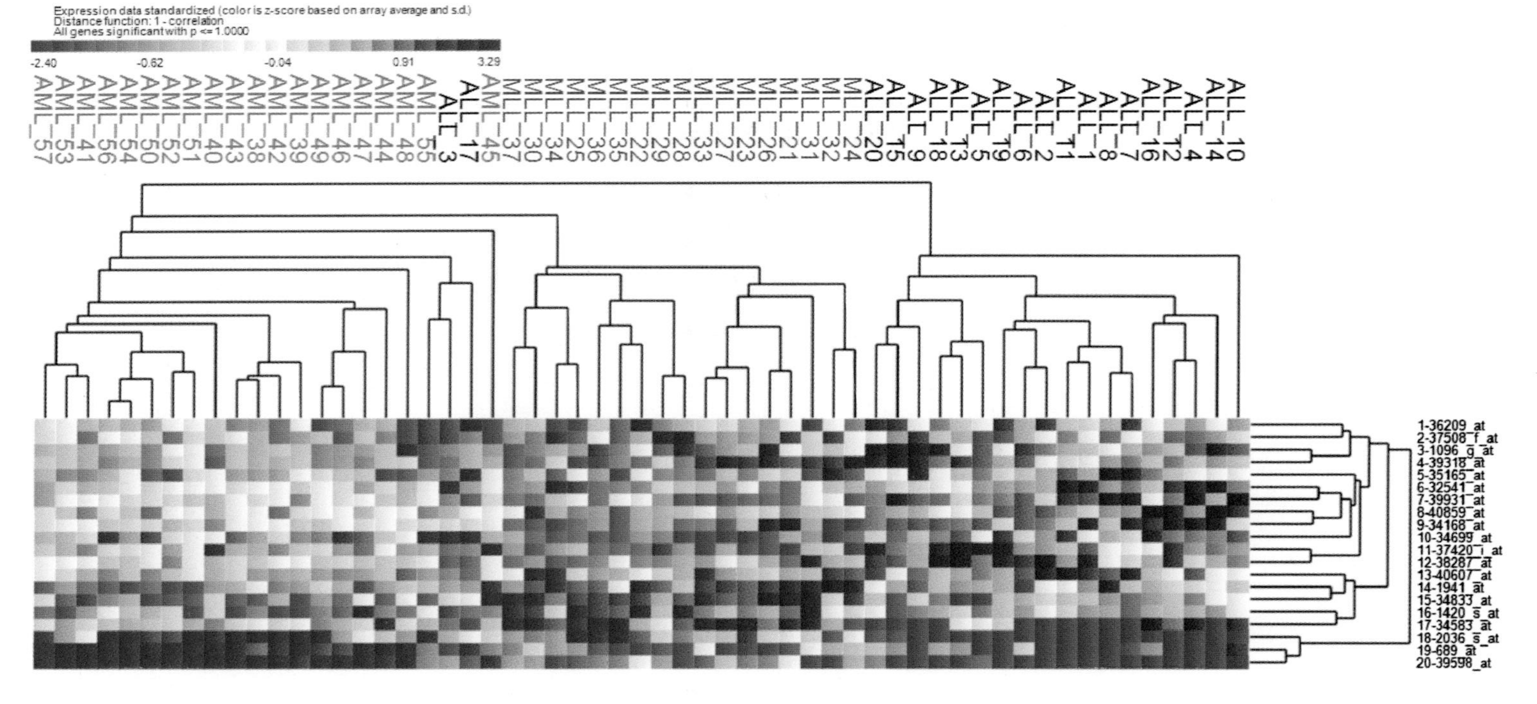

FIGURE 6.2 HCA results for three-class MLL_Leukemia data. Data mean-zero standardized over arrays and genes and $1 - r$ used for agglomeration.

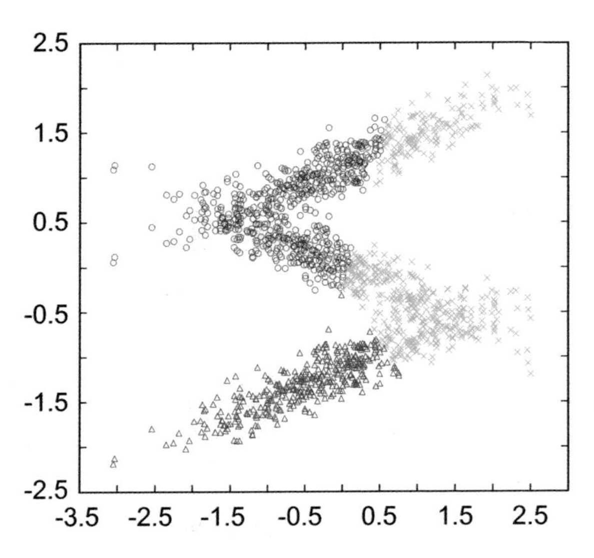

FIGURE 7.5 K-means cluster analysis ($K = 3$) results for 3 Gaussian clusters.

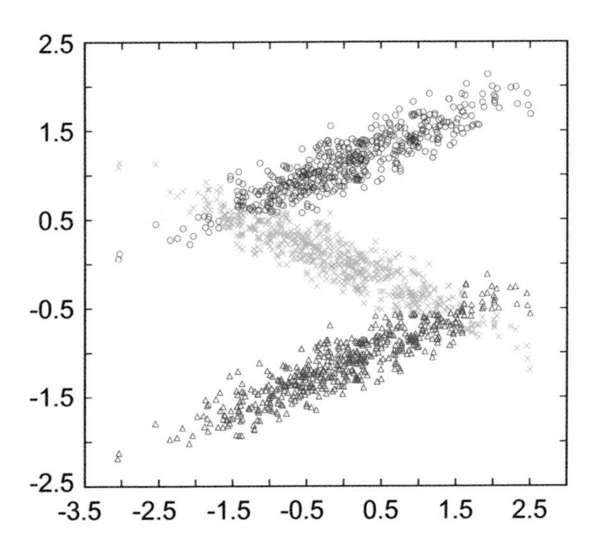

FIGURE 7.6 GMM cluster analysis results for three Gaussian clusters.

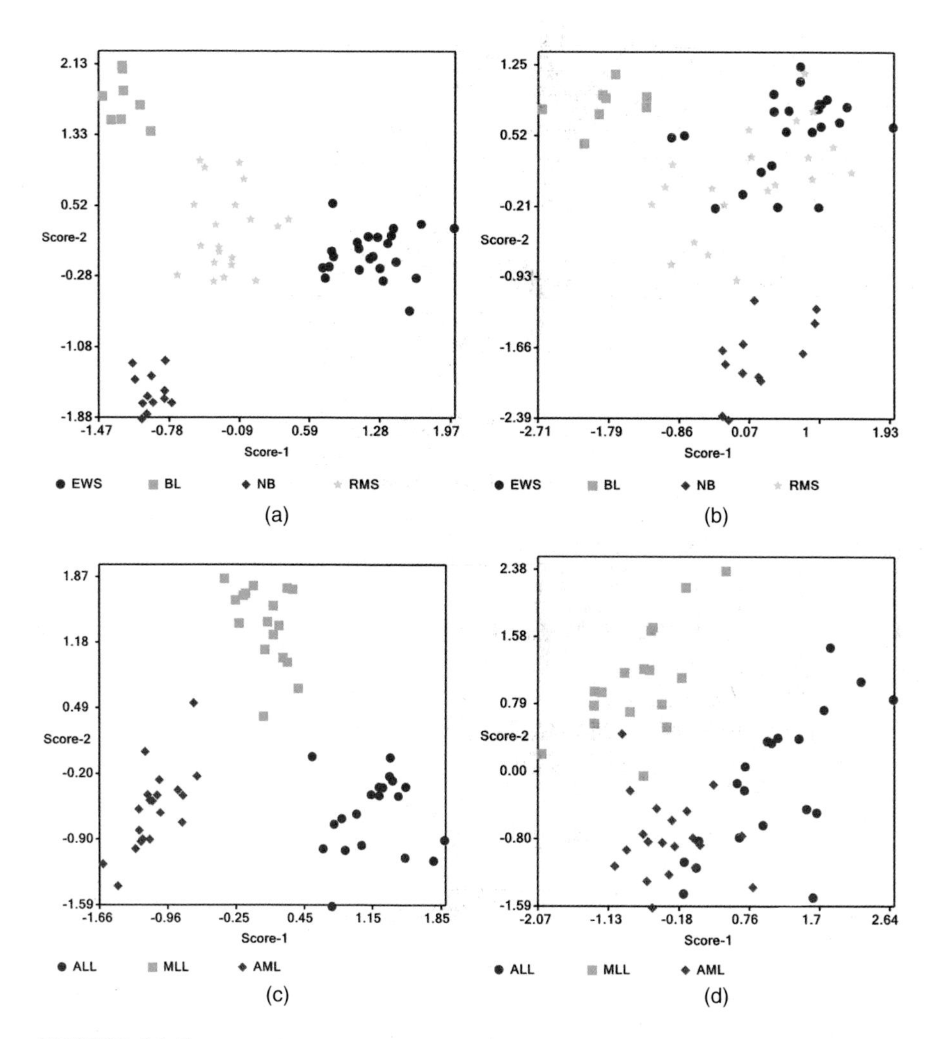

FIGURE 10.5 2D PC, score plots from PCA of each dataset's gene-by-gene correlation matrix. Upper panels contain PC scores for the four-class SRBCT arrays ($n = 63$) determined without (a) and with (b) use of Varimax rotation. Lower panels show PC scores for the three-class MLL_Leukemia arrays ($n = 57$) derived without (c) and with (d) Varimax rotation.

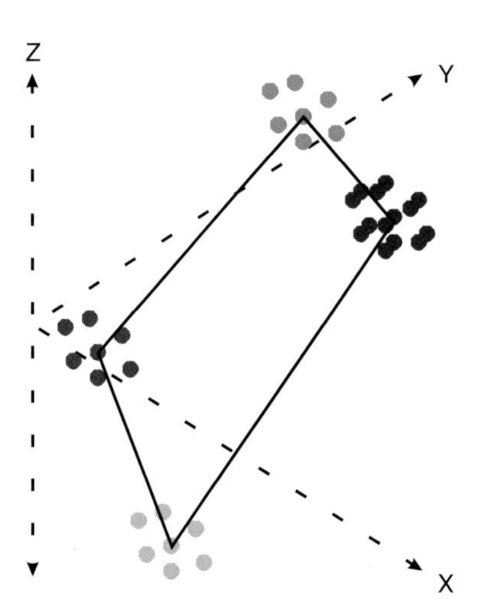

FIGURE 11.1 2D manifold (plane) embedded in a 3D space.

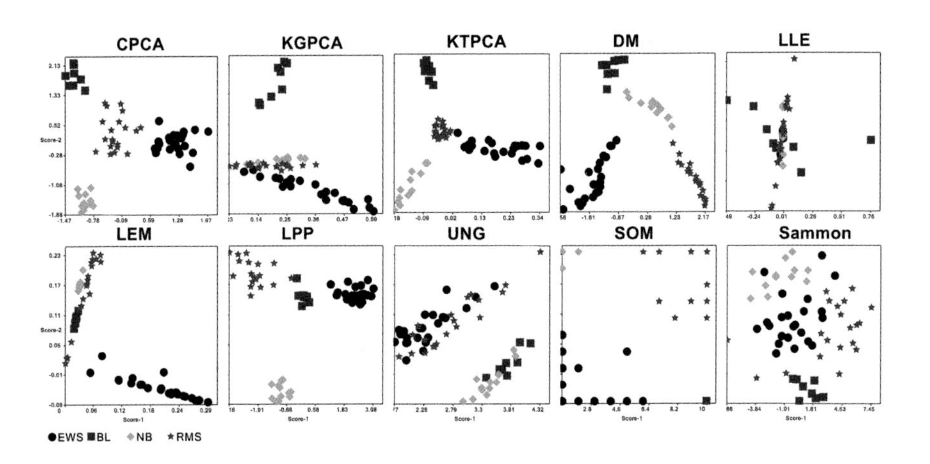

FIGURE 11.3 Cluster structure of four-class SRBCT dataset based on the 2D feature set from NLML.

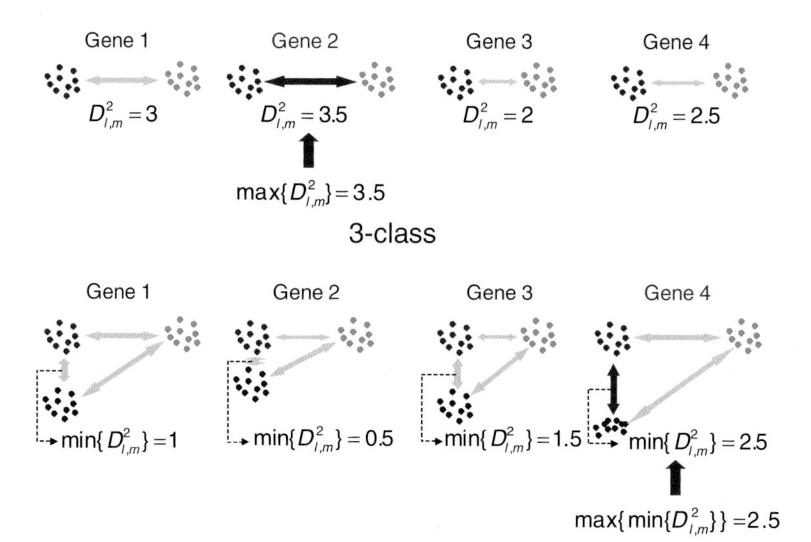

FIGURE 12.10 Schematic showing the Mahalanobis distance D criterion for greedy PTA gene selection. Dots represent arrays, and the distance between groups of arrays represents the Mahalanobis distance. During each forward step in the model for two-class problems the gene resulting in the greatest between-class Mahalanobis distance is selected, whereas in three-class problems the gene resulting in the greatest minimum between-class Mahalanobis distance is added to the model. The process of selecting only genes that progressively widen the between-cluster distance is called "greedy" hill climbing.

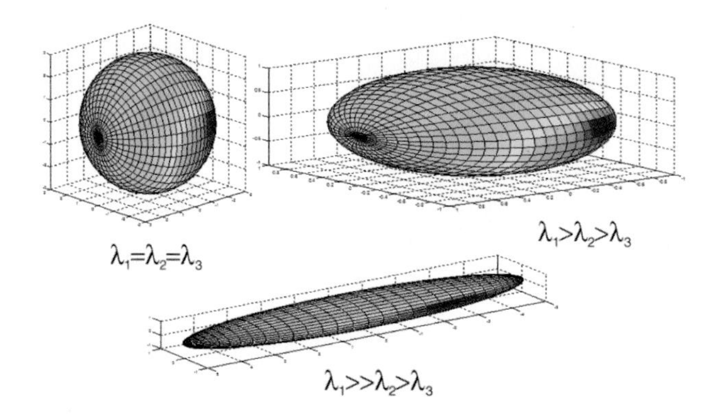

FIGURE 12.14 Contours of equal density for a 3D vector showing the 3-ellipse in 3-space.

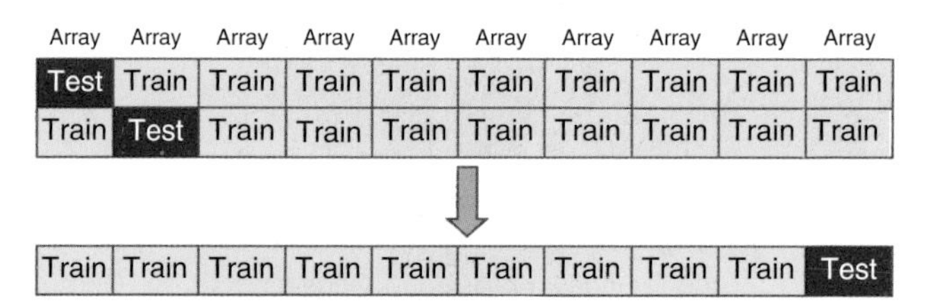

FIGURE 13.2 Workflow for LOOCV.

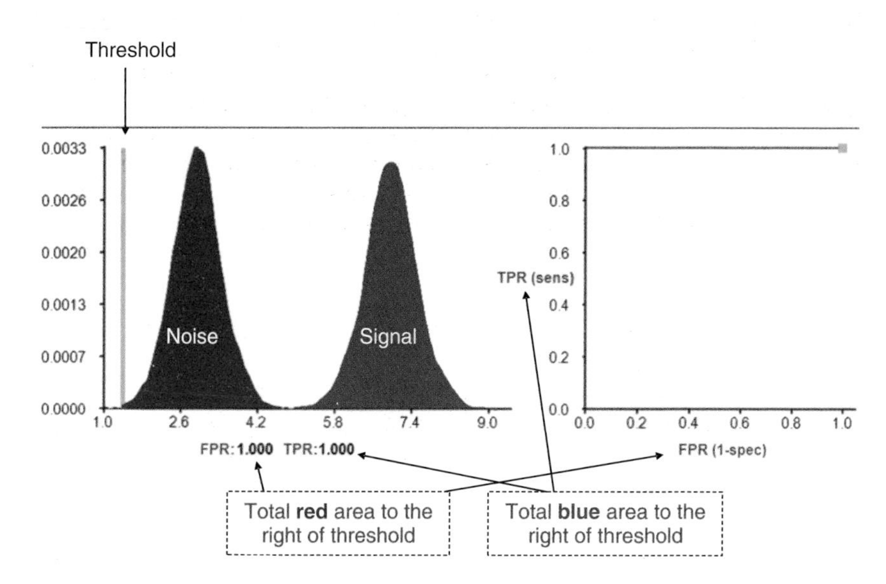

FIGURE 13.7 FPR and TPR of ROC curve when threshold value of x_m^* is small. The noise (null) distribution of classifier accuracy (a) is based on training and testing with permuted class labels, while the signal (alternative) distribution of accuracy (b) is based on training and testing with the true class labels.

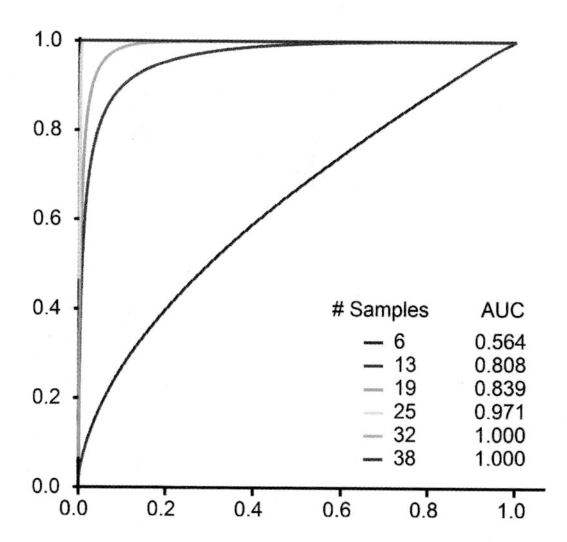

FIGURE 13.14 ROC curves as a function of bootstrapped sample sizes for 5NN classifier run on the four-class SRBCT data with 23 features selected using greedy PTA. The legend key "# Samples" represents the number of arrays randomly selected.

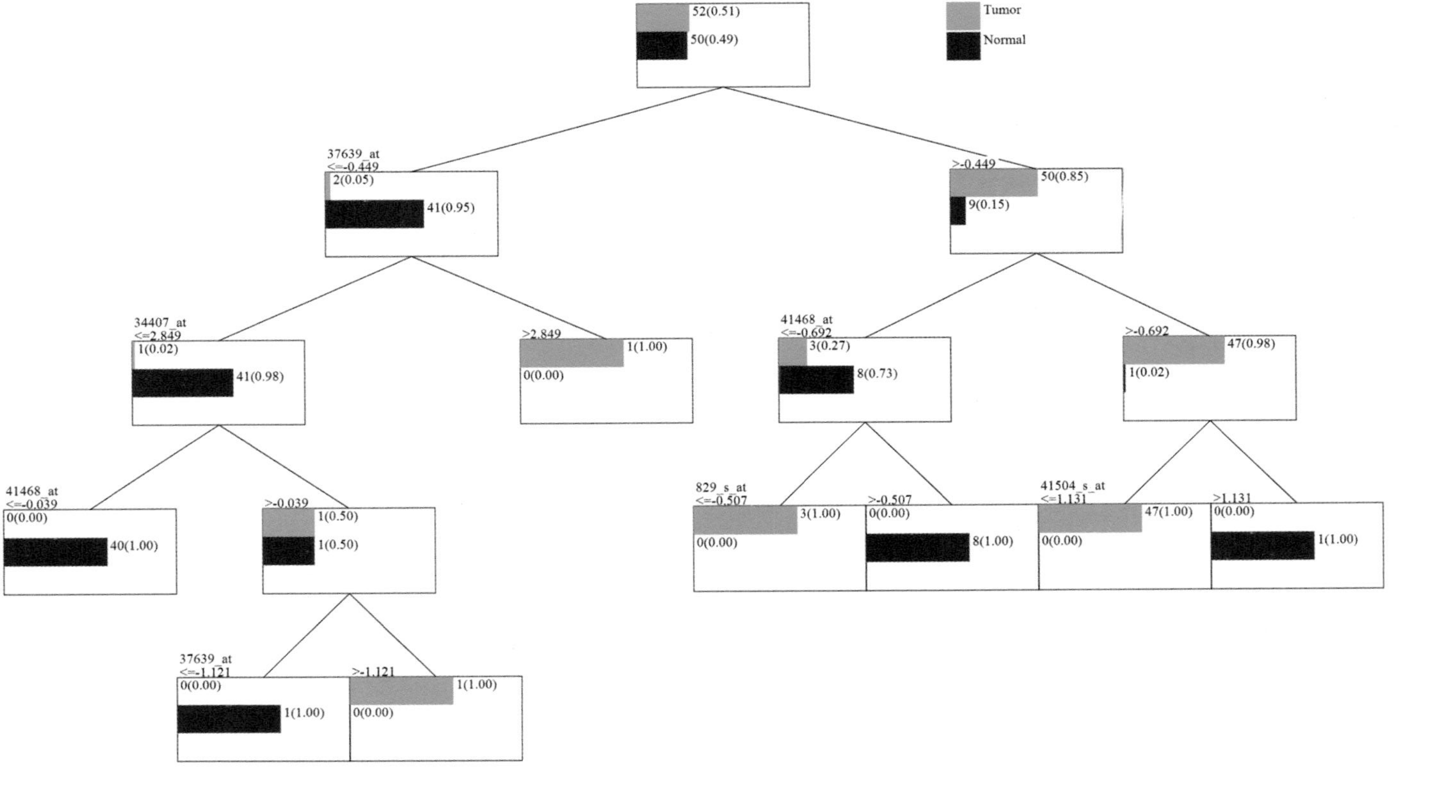

FIGURE 15.3 Decision tree classification (DCT) analysis of two-class prostate cancer dataset with 102 total arrays, partitioned into 52 (51%) tumor arrays and 50 (49%) normal arrays. Results reveal a five-level tree having eight terminal nodes with 100% class purity.

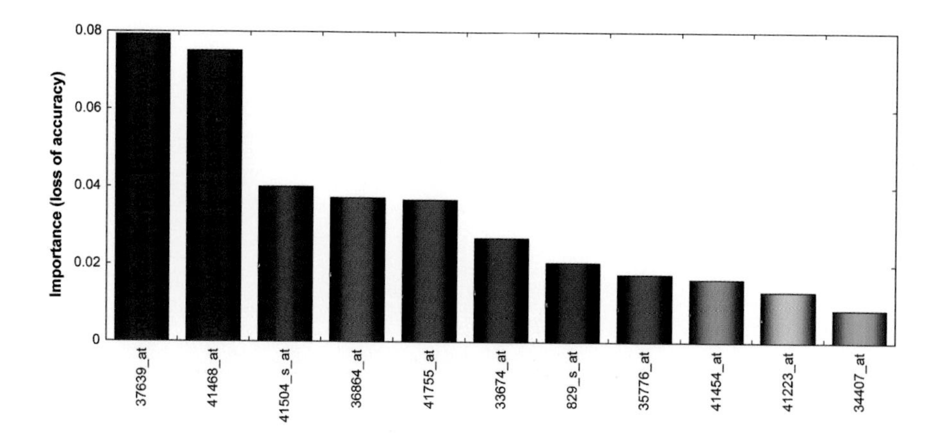

FIGURE 16.1 Importance scores of 11 genes selected with greedy PTA from the two-class prostate cancer dataset.

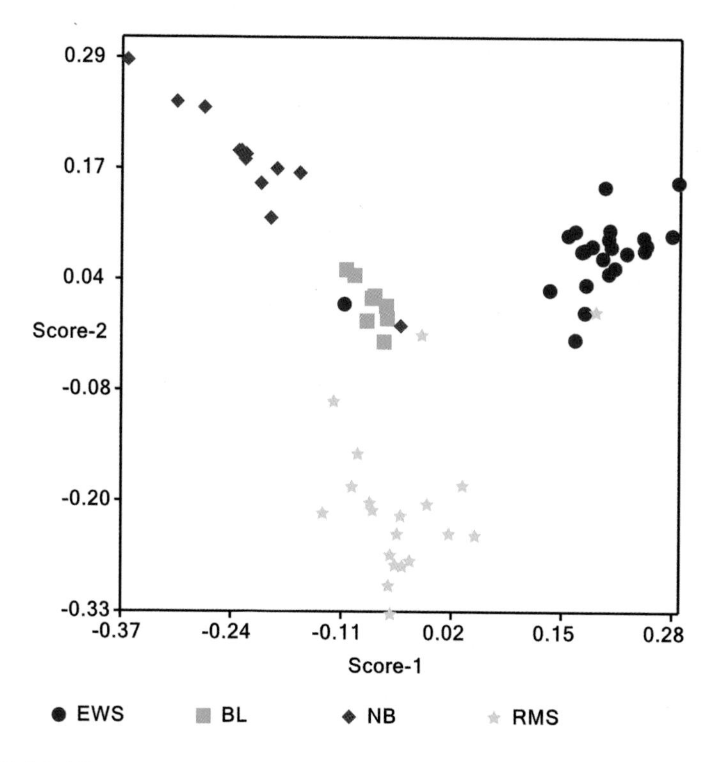

FIGURE 16.13 Supervised clustering results for the SRBCT four-class dataset based on log-transformed input feature values. Array-specific loadings (i.e., $\sqrt{\lambda_j} e_{ij}$) for the first and second greatest eigenvalues of matrix τ are shown.

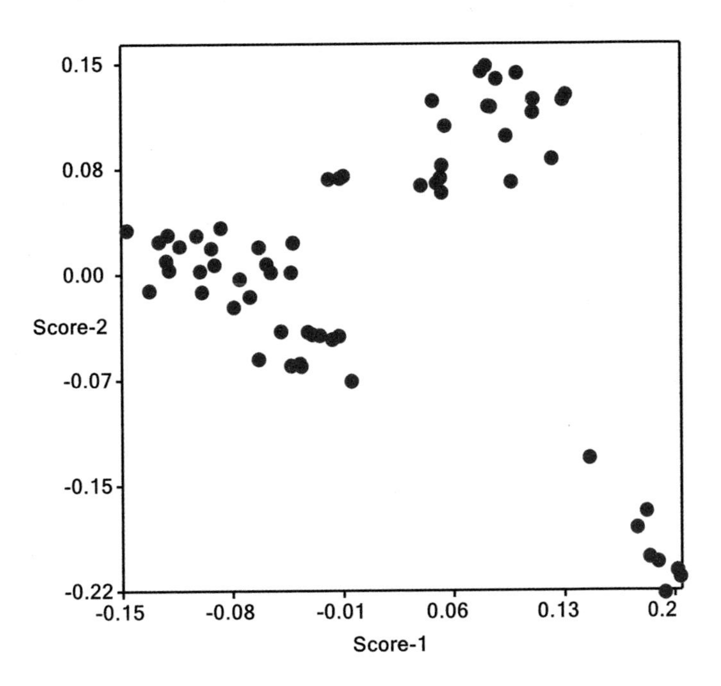

FIGURE 16.16 Unsupervised cluster analysis results for the SRBCT4 dataset.

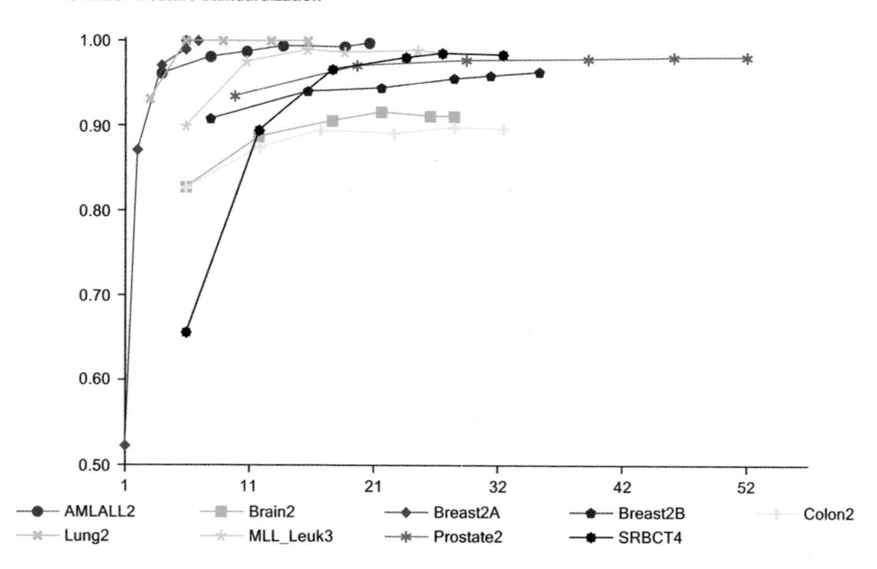

FIGURE 22.26 SVMLS accuracy with feature standardization as a function of bootstrap sample size used during training. The x-axis values represent the number of arrays randomly sampled $B = 40$ times at fractions $f = 0.1$, 0.2, 0.3, 0.4, 0.5, or 0.6 of the available arrays.

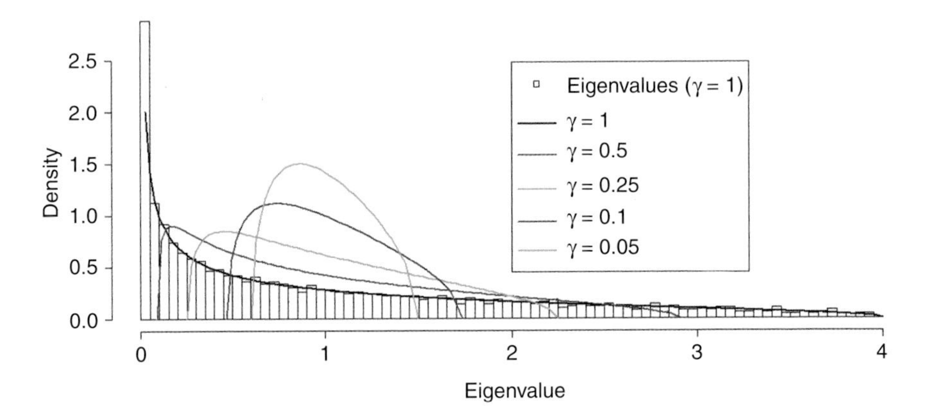

FIGURE 28.2 Marčenko-Pastur limiting eigenvalue densities for Wishart ensemble matrices $W_p(n, \mathbf{I})$ for $n = 1000$, $p = 1000(\gamma = 1)$, $p = 500(\gamma = 0.5)$, $p = 250(\gamma = 0.25)$, $p = 100(\gamma = 0.1)$, and $p = 50(\gamma = 0.05)$.

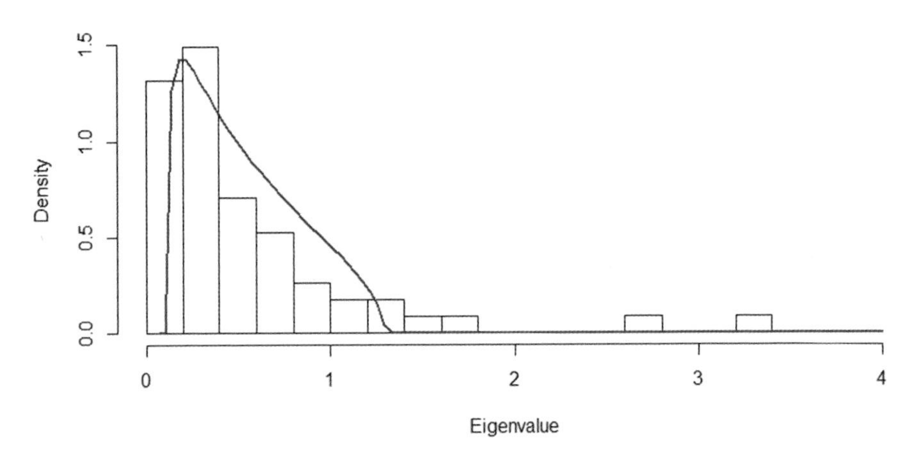

FIGURE 28.4 Histogram of eigenvalues from a 59×59 array-by-array covariance matrix $\mathbf{S}_{\mathrm{raw}}$ based on the mean-zero standardized NCI-60 expression dataset using the first 500 genes in the 1375 gene set. Fitted line is the MP density fitted to the empirical eigenvalue density, based on fitted values of $\gamma = 0.2079$, $\sigma = 0.998$, $\lambda^- = 0.1045$, and $\lambda^+ = 1.296$. Eigenvalues above $\lambda^+ = 1.296$ are considered informative, while those below are considered to be in the noise region.

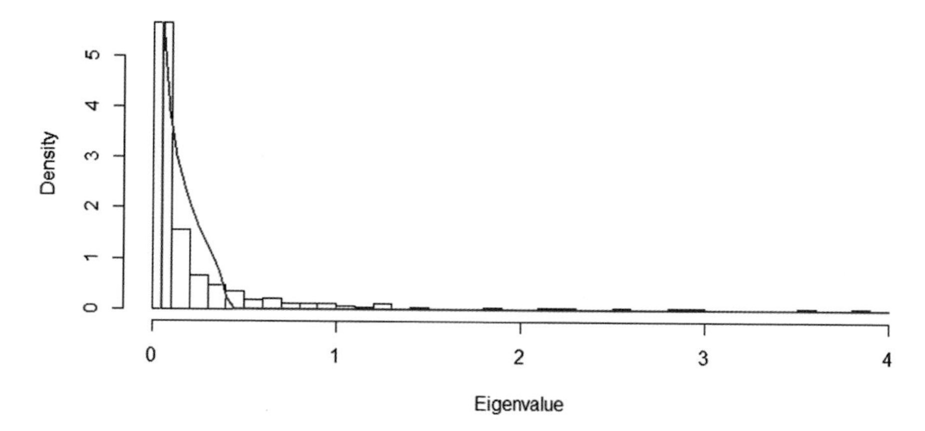

FIGURE 28.5 Histogram of eigenvalues from a 280 × 280 array-by-array covariance matrix $\mathbf{S}_{\text{raw}}$ based on the GCM expression dataset using the 999 informative genes with more than 80% present calls (log-transformed, mean-zero standardized). Fitted line is the MP density fitted to the empirical eigenvalue density, based on fitted values of $\gamma = 0.2392$, $\sigma = 1.2387$, $\lambda^- = 0$, and $\lambda^+ = 0.4107$. Eigenvalues above $\lambda^+ = 0.4107$ are considered informative, while those below are considered to be in the noise region.

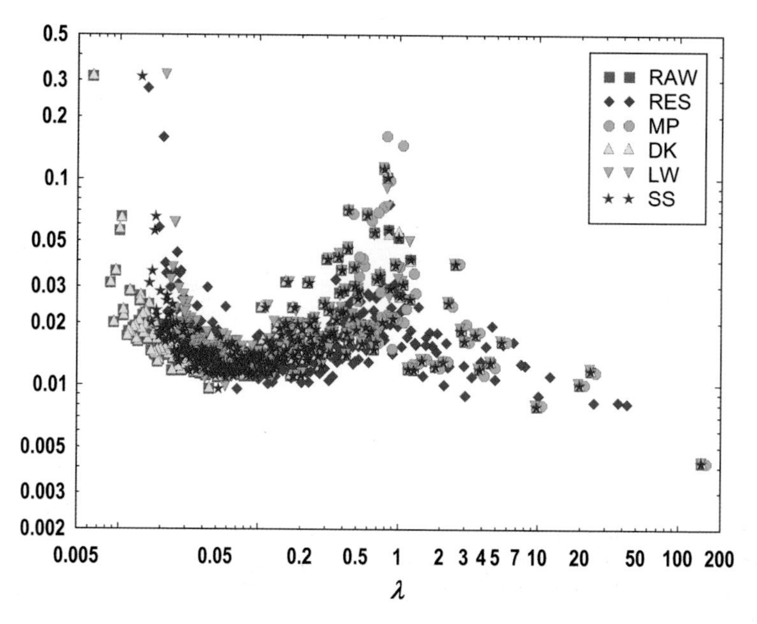

FIGURE 28.6 Inverse participation ratios for eigenvectors as a function of eigenvalue for the 280 × 280 (array-by-array) covariance matrix ($p = 280$) for the GCM expression dataset with all informative $n = 999$ genes ($\gamma = 0.28 = 280/999$) having more than 80% present calls. Covariance matrices evaluated include $\mathbf{S}_{\text{raw}}$, $\mathbf{S}_{\text{res}}$, $\mathbf{S}_{\text{MP}}$, $\mathbf{S}_{\text{DK}}$, $\mathbf{S}_{\text{LW}}$, and $\mathbf{S}_{\text{SS}}$.

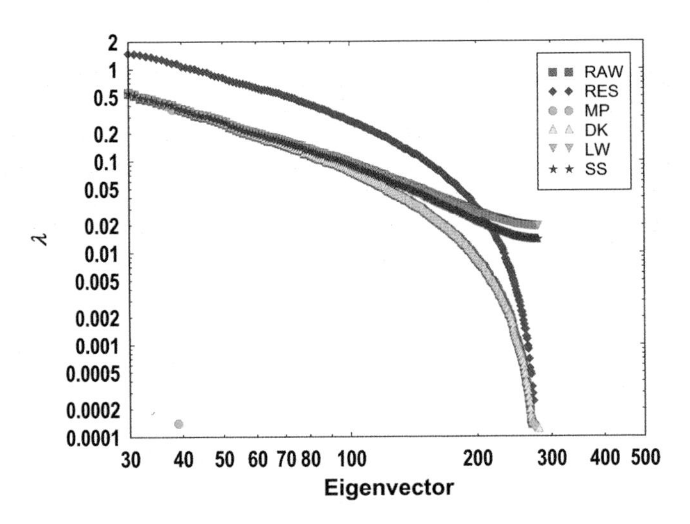

FIGURE 28.11 Empirical eigenvalue distribution for 280 × 280 (array-by-array) covariance matrix ($p = 280$) using the GCM expression dataset with $n = 280$ genes ($\gamma = 1 = 280/280$) having more than 80% present calls. Covariance matrices evaluated include $\mathbf{S}_{\text{raw}}$, $\mathbf{S}_{\text{res}}$, $\mathbf{S}_{\text{MP}}$, $\mathbf{S}_{\text{DK}}$, $\mathbf{S}_{\text{LW}}$, and $\mathbf{S}_{\text{SS}}$.

CHAPTER 8

TEXT MINING: DOCUMENT CLUSTERING

8.1 INTRODUCTION

Text mining is a language- and semantic-based technique that attempts to evaluate unstructured text documents for information retrieval. In its simplest form, text mining usually involves K-means cluster analysis of documents to agglomerate together documents having similar term frequencies. Understanding the cluster structure of documents can reveal the major groups of documents present and the concepts portrayed by each group. As cluster analysis is performed in text mining, other types of unsupervised methods can also be applied to the data generated. Examples include latent semantic indexing with singular value decomposition, principal component analysis, self-organizing maps, and neural gas, to name only a few.

While not exclusively used for DNA microarray analysis, text mining is very popular in business intelligence (BI) for credit risk determination, customer relations management, quality control, optimization in advertisement, and other forms of knowledge discovery.

8.2 DUO-MINING

Duo-mining involves simultaneous data mining of quantitative and qualitative data and text mining of textual data. The return on investment (ROI) from data mining alone can be increased by 10-20% when using duo-mining. Naturally, there are many ways in which data mining and text mining can be jointly accomplished, but the simplest fusion approach involves membership prediction for a particular document cluster using input text for a single document, and then applying this information in predictive analytics derived from data mining. For example, text data from customer complaints logged

Classification Analysis of DNA Microarrays, First Edition. Leif E. Peterson.

at call centers can be mined for predicting customers who may churn (go away) or return, followed by distribution of custom-tailored advertisement for increased profitability. Hospitals can use text data from customer satisfaction questionnaires to identify inefficiencies in patient care or correlations between worker satisfaction and customer care in specific work units.

The important point of duo-mining is the awareness of expanding the data capture to retrieve additional information for increasing the ROI, whether the approach involves molecular-genomic data or BI.

8.3 STREAMS AND DOCUMENTS

A term such as `microarray` is formed from a collection of characters or bytes such as `m`, `i`, and `c`. Below is an example stream of characters in the form of a PubMed abstract [1] identified with the search term `microarray`:

> As a branch of pharmacogenomics aimed at predicting drug safety concerns, toxicogenomics drew much excitement with the emergence of technologies such as gene expression microarrays. A few years down the line, the evidence is scant that current approaches to toxicogenomics are really making an impact in areas such as preclinical toxicology. It has been argued that there needs to be a re-focus of application toward high-throughput approaches which combine the best of tissue and genomic modelling. This commentary gives a brief introduction to in vitro toxicogenomics, drawn from the perspectives of the specialist toxicogenomics company, SimuGen. 2009 FRAME.

The raw stream of characters in the abstract in its present form is unusable for text mining until lexical analysis and stemming are performed.

8.4 LEXICAL ANALYSIS

The process of lexical analysis transforms a stream of characters into a stream of usable terms or words. The principal techniques used in lexical analysis are automatic indexing and stopword removal.

8.4.1 Automatic Indexing

There are several *automatic indexing* steps that need to be performed in order to preprocess a stream of characters. Firstly, standalone digits are removed. For example, if _ is assumed to be a blank, then digits such as _2_, _14_, or _-0.038_ are removed from the source stream. Note that digits next to alphabet characters, for example, `TP53`, are not removed since they bear special meaning and are likely an identifier or gene symbol. Once standalone digits are removed from the character stream, hyphens "(-)" are removed, and uppercase characters are transformed to lower case as in `Jaguar`→`jaguar`.

8.4.2 Removing Stopwords

The next preprocessing step in text mining is to remove individual occurrences of the 26 alphabet characters `a`, `b`, `c`, . . ., `z`, and other *stopwords*, such as `of`, `and`, and `the`. Stopwords provide no resemblance of the information contained in a document, and therefore are removed from all document strings. Frakes and Baeza-Yates have provided lists of stopwords [2]. Examples of stopwords include `a`, `about`, `above`, `across`, `after`, `again`, `against`, `all`, `almost`, `alone`, `along`, `already`, `also`, `although`, `always`, `among`, `an`, `and`, `another`,. . ., `year`, `years`, `yet`, `you`, `young`, `younger`, `youngest`, `your`, `yours`.

8.5 STEMMING

Another preprocessing step is stemming, which is a form of suffix stripping to reduce variants of terms to their common root word–or stem [3,4]. For example, the following words have similar meanings:

```
Class
Classes
Classified
Classification
Classifications
Classified
Classifier
Classifying
```

If the suffixes of these words are stripped off or *conflated*, the result will be the word `Class`, which forms the root word. Two popular algorithms for suffix stripping are the Porter stemmer [5,6] ands stemmer programs [2], and since the time of its introduction in 1980, the Porter stemmer has become one of the most widely used algorithms for stemming. Tables 8.1 and 8.2 list the suffix stripping steps in the Porter stemmer algorithm. Porter ran the stemming algorithm on an English vocabulary with 10,000 words, and the resulting vocabulary consisted of 6370 stems.

8.6 TERM WEIGHTING

Documents vary by size according to the number of words or terms they contain. The informativeness of a term is often revealed by its frequency

TABLE 8.1 Steps 1 and 2 in the Porter Stemming Algorithm

Suffix	Result	Input	Output
sses	ss	classes	class
ies	i	cookies	cooki
ss	ss	fitness	fitness
s	(drop)	iterations	iteration
eed	dd	speed	speed
ed	(drop)	filtered	filter
ing	(drop)	filtering	filter
ated	ate	operated	operate
bled	ble	doubled	double
ized	ize	standardized	standardize
y	i	commonly	commonli
ational	ate	relational	relate
tional	tion	conditional	condition
izer	ize	vectorizer	vectorize
abli	vable	probabli	probable
alli	al	specialli	special
entli	ent	patentli	patent
ousli	ous	obviousli	obvious
ization	ize	standardization	standardize
ation	ate	operation	operate
ator	ate	operator	operate
alism	al	formalism	formal
iveness	ive	receptiveness	receptive
fulness	ful	usefullness	usefull
ousness	ous	callousness	callous
aliti	al	formaliti	formal
iviti	ive	productivity	productive
biliti	ble	abiliti	able

of occurrence and its uniqueness to the information content. Most modern information retrieval systems are based on term weighting, which accounts for the within-document (intra-document) frequency of occurrence of each term, the number of documents each term appears in, and the length of each document [7].

Text mining results would be biased if term weighting were not used, since longer documents commonly have more terms and greater term frequencies [8]. Small documents will have more terms with zero term frequency, and large documents with more terms will therefore have more nonzero term frequencies, which are greater [9]. For this reason, term frequencies are penalized by using weights based on document length [10].

TABLE 8.2 Steps 3 and 4 in the Porter Stemming Algorithm

Suffix	Result	Input	Output
icate	ic	duplicate	duplic
ative	(drop)	derivative	deriv
alize	al	formalize	formal
iciti	ic	electriciti	electric
ical	ic	statistical	statistic
ness	(drop)	usefulness	useful
ful	(drop)	useful	use
al	(drop)	survival	surviv
ance	(drop)	impedance	imped
ence	(drop)	difference	differ
er	(drop)	finer	fin
ic	(drop)	microscopic	microscop
able	(drop)	adjustable	adjust
ible	(drop)	defensible	defens
ant	(drop)	irritant	irrit
ement	(drop)	replacement	replac
ment	(drop)	adjustment	adjust
ent	(drop)	dependent	depend
tion	(drop)	adoption	adopt
ue	(drop)	prologue	prolog
ism	(drop)	formalism	formal
ate	(drop)	discriminate	discrim
iti	(drop)	polariti	polar
ous	(drop)	homologous	homolog
ive	(drop)	effective	effect
ize	(drop)	specialize	special

Let f_{ij} be the frequency of word j in document i, and the number of documents containing word j as d_j. Let $\mathbf{x}_1, \mathbf{x}_2, \ldots, \mathbf{x}_n (1 \le i \le n)$ represent n document vectors of length $d (1 \le j \le d)$. Set the jth component of each document vector [8] to

$$x_{ji} = t_{ji} g_j s_i, \tag{8.1}$$

where $t_{ji} = f_{ij}$ is the *term frequency*, $g_j = \log(n/d_j)$ is the *inverse document frequency*, and

$$s_i = \frac{1}{\sqrt{\left(\sum_{j=1}^{d} (t_{ji} g_j)^2\right)}}. \tag{8.2}$$

is the *document length normalization*. Kolda [11] introduced 50 choices of different weighting schemes for t_{ji}, g_j, and s_i, and we have chosen two of the most popular schemes, called `tfn` and `txn`. The first scheme, `tfn`, is called *normalized term frequency-inverse document frequency* and uses $g_j = \log(n/d_j)$. The second scheme, `txn`, is called *normalized term frequency*, which uses $g_j = 1$.

Function Words. An additional preprocessing technique for reducing computation time is to remove terms with the lowest and highest term frequencies, and retain the most valuable *function words*. For example, terms whose frequencies range from $(0.002 < f_{ij} < 0.15)$ are kept in the training. Keeping vital function words removes little information but shortens computation time [9].

Now that the preprocessing (automatic indexing, stopwords, stemming), term weighting and normalization are complete, the next step is to determine the total number of unique terms d that were learned from all the documents. An $\mathbf{X}$ array of dimensions $d \times n$ can now be used to store the x_{ji} values from (8.1).

8.7 CONCEPT VECTORS

Spherical K-means cluster analysis [12] of the n document vectors can now be performed to partition $\mathbf{x}_1, \mathbf{x}_2, \ldots, \mathbf{x}_n$ into K disjoint clusters $\omega_1, \omega_2, \ldots, \omega_K$ using $K = \sqrt{n}$. This will result in K centers of the form $\mathbf{m}_1, \mathbf{m}_2, \ldots, \mathbf{m}_K$. The central component of spherical K-means cluster analysis is the concept vector for each center $\mathbf{m}_k$, whose functional form is

$$\mathbf{c}_k = \frac{\mathbf{m}_k}{||\mathbf{m}_k||}. \tag{8.3}$$

Dhillon and Modha [12] reported that for any unit vector $\mathbf{z} \in \mathbb{R}$, the concept vectors $\mathbf{c}_1, \mathbf{c}_2, \ldots, \mathbf{c}_k$ satisfy the Cauchy-Schwartz inequality

$$\sum_{\mathbf{x} \in \omega_k} \mathbf{x}^T \mathbf{z} \leq \sum_{\mathbf{x} \in \omega_k} \mathbf{x}^T \mathbf{c}_k. \tag{8.4}$$

Therefore, a concept vector is the vector that has the greatest correlation (cosine similarity) with all of the documents in a given cluster.

8.8 MAIN TERMS REPRESENTING CONCEPT VECTORS

The cluster of documents represented by each concept vector requires a set of labels to reflect the information content. The labels are derived from the most prevalent terms in each concept vector. This section describes an approach for identifying the top 10 terms for each cluster of documents.

Once K-means cluster analysis of the documents is complete, and concept vectors are determined to represent each cluster center, we must define a $d \times K$ matrix $\mathbf{C}$, and fill each column with a concept vector $\mathbf{c}_k$. Our next goal is to identify the top 10 terms in each concept vector by performing a descending sort on the c_{jk} values within each cluster. However, before doing this, we need to decide whether we want to allow the same term to represent multiple clusters, or whether we prefer each term to represent a single cluster only.

Allowing Term Redundancy. If we are to allow terms to be used to represent multiple clusters, that is, term redundancy, we simply perform a descending sort in each kth column of $\mathbf{C}$ and extract the terms associated with the top 10 values of c_{jk}. These are the primary or top terms that represent the main information behind each concept vector, or essentially each cluster of documents. When allowing term redundancy, it is not known a priori whether terms will show up more than once in the top 10 lists for multiple clusters, but there is no reason why they can't.

Preventing Term Redundancy. When term redundancy is not desired, we must first identify column $l = \arg\max_k\{c_{jk}\}$ in each row, for which the c_{jk} is the greatest, retain the value of c_{jl} in column l, and set all other column entries ($k \neq l$) to zero. When done, each row of $\mathbf{C}$ will have only one nonzero entry, which will occur at the concept vector with the greatest c_{jk}. Again, a descending sort is applied to kth column of $\mathbf{C}$, and the terms associated with the top 10 values of c_{jk} are extracted.

8.9 ALGORITHM

The text mining algorithm employs the use of hashtables $\mathcal{H}$ and $\mathcal{H}_g$, which perform the bookkeeping for all input words for each document and for all learned words in all documents, respectively. Regarding notation, n_D represents the total number of documents and n_W is the total number of learned words in all documents. The output for text mining is the set of document vectors $\mathbf{x}_i$ consisting of elements x_{ji} for each term. Spherical K-means cluster analysis is performed on document vectors, and the top terms of each cluster are then used as labels for representing the information content. Obviously, the distance or correlation between each cluster center and a new future document can be determined once an input document's term frequencies x_{ji} are estimated. The workflow for inverse document frequency text mining is presented in Algorithm 8.1.

Algorithm 8.1: Text Mining

```
Data: Documents in ASCII text format
Result: Clustered documents based on spherical K means
Determine number of documents n_D
Determine total number of learned words n_W
for i ← 1 to n_D do
    Read document i
    Tokenize document i
    Fill readwords()
    Run stemmer on document i
    Remove stop words in document i
    foreach word ∈ document i do
        if word ∉ H then
            Append H
        endif
        if word ∉ Hg then
            Append Hg
        endif
    endfch
    foreach token ∈ H do
        j = j + 1
        T_ij = token.Key
        TF_ij = token.Value
    endfch
endfor
foreach token ∈ Hg do
    j = j + 1
    for i ← 1 to n_D do
        for k ← 1 to n_W do
            if T_ik = token then
                d_j = d_j + 1
            endif
        endfor
    endfor
endfch
for j ← 1 to n_W do
    f_j = d_j/n_D
endfor
t_low = max(1/n_D, 0.002)
tupper = 0.15
for j ← 1 to n_W do
    if f_j ≤ t_low then
        drop word j
    endif
    if f_j ≥ tupper then
        drop word j
    endif
endfor
for j ← 1 to n_W do
    if tfn then
        g_j = log n_D/d_j
    endif
    if txn then
        g_j = 1
    endif
endfor
for i ← 1 to n_D do
    sum = 0
    for j ← 1 to n_W do
        sum = sum + TF_ij TF_ij
    endfor
    s_i = 1/√sum
endfor
for j ← 1 to n_W do
    for i ← 1 to n_D do
        x_ji = TF_ij g_j s_i
    endfor
endfor
```

8.10 PREPROCESSING

Reading Words in Each Document. The first step in text mining involves partitioning, or *tokenizing* all the characters in the input stream for each. Breaking up the stream of characters into words is performed by use of a blank or space delimiter. The code below lists a procedure for splitting up each line read from a delimited ASCII text file, splitting each line into the constituent words, and then "reading" words by storing them in array `readwords()`; at this point, we have only read the input words, but not yet "learned" the words that we want to keep for bookkeeping:

```
Dim readwords(), inbuff, buffer() As String
NumReadWords = 0
NumInputWords = 0
Do While Not EOF(1)
  inbuff = LineInput(1)
  dynamarray = inbuff.Split(DELIMITER)
  nrow = UBound(dynamarray) + 1
  ReDim buffer(nrow)
  buffer = inbuff.Split(DELIMITER)
  For j = LBound(buffer) To UBound(buffer)
    NumReadWords = NumReadWords + 1
    ReDim Preserve readwords(NumReadWords)
    readwords(NumReadWords) = buffer(j)
  Next j
Loop
```

Now that the words in the document are read, character case needs to be lowered, format characters removed, stopwords removed, and stemming performed.

Character Formatting and Removal with Regular Expressions. All read words are first set to lowercase using the `readwords(j).ToLower` command. Regular expression techniques are then used to strip out any undesired characters in words, standalone digits, and nonwords. The syntax definitions for these regular expression sequences are listed as follows:

```
Dim REDropFMTCHAR As New Regex("\t*\r*\v*\f*\n\d")
Dim REDropDIGIT As New Regex("\d*")
Dim REDropNONWORD As New Regex("\W")
```

In these definitions, zero or more matches of the tab character are removed with the regular expression sequence `\t*`, matches of carriage return are removed with `\r*`, matches of vertical tab characters with `\v*`, formfeed with `\f*`, newline character with `\n`, and the decimal digits $0, 1, 2, \ldots, 9$ using `\d`. Zero or more digits (numbers) that exist by themselves are dropped using the regular expression sequence `\d*`. Standalone nonword characters (0-9, a-z, A-Z) are removed using the sequence `\W`.

Removing Stopwords and Stemming. Stopwords are removed by comparing each of the read words with a hashtable called `TestStopWords` that contains stopwords. Any read word that equals a stopword is not used in further training. After stopwords are eliminated, stemming is performed:

```
Dim inputwords() As String = Nothing
Dim Learnedwords() As String
Dim wordfreq(,) As Integer = Nothing
Dim wordsindocument() As Integer = Nothing
For j = 1 To NumReadWords
   If readwords(j) = "" Then GoTo nextword
   If TestStopWords(readwords(j)) = True Then GoTo nextword
   If readwords(j) = " " Then GoTo nextword
   buffstr = readwords(j).ToLower
   buffstr = REDropFMTCHAR.Replace(buffstr, "")
   buffstr = REDropDIGIT.Replace(buffstr, "")
   buffstr = REDropNONWORD.Replace(buffstr, "")
   'begin stemming
   buffstr = Stemmer.stemTerm(buffstr)
   If buffstr = "" Then GoTo nextword
   If TestStopWords(buffstr) = True Then GoTo nextword
   NumInputWords = NumInputWords + 1
   ReDim Preserve inputwords(NumInputWords)
   inputwords(NumInputWords) = buffstr 'readwords(j)
nextword:
Next j
Array.Clear(readwords, 0, UBound(readwords))
wordsindocument(ThisDocument) = NumInputWords
If ThisDocument = 1 Then Learnedwords(1) = inputwords(1)
```

Learning Unique Words in All Documents. The next step is to construct a hashtable `ht` for learning words in each document and a hashtable `htglob` for learning the unique list and frequency of all words read from all documents. Hashtable `htglob` must be defined before any documents are read, and hashtable `ht` is redefined prior to reading words from each document. The two-dimensional arrays `htinputwords(,)` and `htinputwordfreq(,)` are instantiated to pick off the words and their frequencies using the document-specific hashtable `ht`, which contains learned words only for a given document. This is accomplished using the syntax listed below:

```
Dim ht As New Hashtable
For i = 1 To NumInputWords ' consider each input word
  'input word is not known, so learn it (increment known word
  'array by 1 and add word)
```

```
    If ht.ContainsKey(inputwords(i)) = True Then
      Cnt = ht(inputwords(i))
      ht(inputwords(i)) = Cnt + 1
    End If
    If ht.ContainsKey(inputwords(i)) = False Then
      ht.Add(inputwords(i), 1)
    End If
    If htglob.ContainsKey(inputwords(i)) = True Then
      Cnt = htglob(inputwords(i))
      htglob(inputwords(i)) = Cnt + 1
    End If
    If htglob.ContainsKey(inputwords(i)) = False Then
      maxCnt += 1
      htglob.Add(inputwords(i), 1)
    End If
Next i

ReDim Preserve htinputwords(NumDocuments, maxCnt)
ReDim Preserve htinputwordfreq(NumDocuments, maxCnt)
Dim de As DictionaryEntry
Cnt = 0
For Each de In ht
  Cnt += 1
  htinputwords(ThisDocument, Cnt) = de.Key
  htinputwordfreq(ThisDocument, Cnt) = de.Value
Next
```

Note that in this code we are really focusing on the frequency of each word; however, the word order in hashtables is unordered. Thus, we not only fill the array `htinputwordfreq(,)` with word frequencies for each document, but in addition fill the array `htinputwords(,)` with each word so that we can later pick off the word when constructing arrays for which frequencies represent the same word across the same row. We next fill in jagged arrays to minimize memory allocation for the large sparse matrices `htinputwords(,)` and `htinputwordfreq(,)`. Jagged arrays can store a varying number of vector elements with words and their frequencies for each document, so sparseness is eliminated:

```
Dim jaggedword(NumDocuments)() As String
Dim jaggedfreq(NumDocuments)() As Integer
Dim jaggedelements(NumDocuments) As Integer
For i = 1 To NumDocuments
  Cnt = 0
  For k = 1 To NumLearnedWords
    If htinputwordfreq(i, k) > 0 Then
      Cnt += 1
      ReDim Preserve jaggedword(i)(Cnt)
      jaggedword(i)(Cnt) = htinputwords(i, k)
```

```
      ReDim Preserve jaggedfreq(i)(Cnt)
      jaggedfreq(i)(Cnt) = htinputwordfreq(i, k)
    End If
  Next k
  jaggedelements(i) = Cnt
Next i
```

The next section provides code to determine d_j, the number of documents that each word in the global hashtable is in, and store results in the vector `NumDocsWithWord()`:

```
Dim refword As String = Nothing
Dim NumDocsWithWord(NumLearnedWords) As Integer

For j = 1 To NumLearnedWords
  refword = htglobwords(j)
  For i = 1 To NumDocuments
    For k = 1 To jaggedelements(i)
      If jaggedword(i)(k) = refword Then
        NumDocsWithWord(j) += 1
        Exit For
      End If
    Next k
  Next i
Next j
```

Next, divide the number of documents that a word appears in by the total number of documents to obtain the array `wordindocfreq()`:

```
Dim wordindocfreq(NumLearnedWords) As Double
For j = 1 To NumLearnedWords
  wordindocfreq(j) = 0
  If NumDocsWithWord(j) > 0 Then wordindocfreq(j) =
     NumDocsWithWord(j) / NumDocuments
Next j
```

Now, drop words that are either too infrequent or too frequent, and change `NumLearnedWords` to the number of new words remaining:

```
Dim dropword(NumLearnedWords) As Byte
Dim lowerthreshold, upperthreshold As Double
lowerthreshold = Math.Max(1 / NumDocuments, 0.002)
upperthreshold = 0.15
For j = 1 To NumLearnedWords
  If wordindocfreq(j) <= lowerthreshold Then dropword(j) = 1
Next j
For j = 1 To NumLearnedWords
  If wordindocfreq(j) >= upperthreshold Then dropword(j) = 1
Next j
```

```
'redefine htglobwords

Dim buffernewlearnedwords() As String = Nothing
Dim buffernewlearnedwordfreq() As String = Nothing
Cnt = 0
For j = 1 To NumLearnedWords
  If dropword(j) = 0 Then
    Cnt += 1
    ReDim Preserve buffernewlearnedwords(Cnt)
    buffernewlearnedwords(Cnt) = htglobwords(j)
    ReDim Preserve buffernewlearnedwordfreq(Cnt)
    buffernewlearnedwordfreq(Cnt) = htglobwordfreq(j)
  End If
Next j
NumLearnedWords = Cnt
ReDim htglobwords(NumLearnedWords)
ReDim htglobwordfreq(NumLearnedWords)
For j = 1 To NumLearnedWords
  htglobwords(j) = buffernewlearnedwords(j)
  htglobwordfreq(j) = buffernewlearnedwordfreq(j)
Next j
```

Recall that hashtables don't order or sort their contents, so we step through the unordered hashtable of global learned words $\mathcal{H}_g$ and fill in the frequency counts of each learned word one at a time, across all documents, filling up array `TF(,)`. This ensures that row values represent the same word across all documents. This is accomplished using elements from the memory-saving jagged arrays `jaggedfreq(,)`, using the following code:

```
Dim TF(NumDocuments, NumLearnedWords) As Double
Dim IDF(NumLearnedWords) As Double
For j = 1 To NumLearnedWords
  refword = htglobwords(j)
  For i = 1 To NumDocuments
    For k = 1 To jaggedelements(i)
      If jaggedword(i)(k) = refword Then
        TF(i, j) = jaggedfreq(i)(k)
        Exit For
      End If
    Next k
  Next i
Next j
```

Note that frequencies of terms across all documents in array `TF(,)` are ordered only according to the order of words as they appear in the global hashtable of learned words. However, the word order represented by rows of `TF(,)` is not sorted in any particular fashion. At this point, the goal is merely to ensure that the frequencies represent the same words across the documents, so that we can construct document vectors for spherical *K*-means cluster analysis.

We now calculate g_j, s_i, and x_{ji} as described in Equation (8.1), which form the document vector elements used as input into a spherical *K*-means algorithm:

```
Dim g_j(NumLearnedWords) As Double
For j = 1 To NumLearnedWords
  g_j(j) = 0
  If tfn specified AndAlso NumDocsWithWord(j) > 0 Then g_j(j) =
                        Math.Log(NumDocuments / NumDocsWithWord(j))
  If txn method Then g_j(j) = 1
Next j

Dim s_i(NumDocuments) As Double
For i = 1 To NumDocuments
  sum = 0
  For j = 1 To NumLearnedWords
    sum += TF(i, j) * TF(i, j)
  Next j
  s_i(i) = 0
  If sum > 0 Then s_i(i) = 1 / Math.Sqrt(sum)
Next i

'x_i
Dim x_ij(NumDocuments, NumLearnedWords) As Double
For j = 1 To NumLearnedWords
  For i = 1 To NumDocuments
    x_ij(i, j) = TF(i, j) * g_j(j) * s_i(i)
  Next i
Next j
```

Example 8.1 A PubMed search was conducted using the term *microarrays*, and 373 abstracts with titles were obtained. Abstracts were stored to disk in separate ASCII text files using the PMID number as the filename. Each abstract was assumed to be a document, and at runtime $K = 10$ clusters was specified. Text mining was performed with both the `tfn` and `txn` term weighting schemes, with and without allowance for term redundancy. Table 8.3 lists the top 10 terms for each of the 10 concept vectors extracted from the 373 abstracts, when `tfn` term weighting was used without allowing term redundancy. Table 8.4 lists the top 10 terms for `tfn` weighting, but when term redundancy was allowed. Table 8.5 lists the top terms for `txn` weighting without allowing term redundancy. Table 8.6 lists the top terms for `tfn` weighting with term redundancy allowed.

Looking at the resulting tables, it appears that the normalized term frequency weighting scheme `txn` provided better results concerning cluster labels when compared with the `tfn` normalized term frequency-inverse document frequency approach. In Tables 8.5 and 8.6, cluster 6 for breast cancer

TABLE 8.3 Top 10 Terms for 10 Concept Vectors Generated from 373 PubMed Abstracts Identified with Search Term *Microarrays*[a]

1	2	3	4	5	6	7	8	9	10
technolog	surviv	test	genet	sequenc	breast	cnv	carcinoma	infect	probe
system	prognost	sensit	region	speci	posit	copi	renal	strain	hybrid
pcb	cd	fc	loss	annot	er	antibodi	sa	immun	experi
vegfr	predict	type	map	network	gastric	featur	confirm	mice	acid
applic	tumour	splice	famili	inform	ihc	variant	histolog	isol	assai
advanc	independ	stress	pcr	transcriptom	neg	identif	hcc	induc	diagnost
field	egfr	plai	line	tool	growth	previous	report	pathogen	design
quantit	treatment	total	realtim	promot	prolifer	microrna	classif	indic	platform
collect	invas	cultur	chromosom	base	peptid	nedd	cluster	metabol	techniqu
biomark	prognosi	cellular	complex	bind	progenitor	delet	diagnosi	downregul	measur

[a] `tfn` term weighting scheme used, with term redundancy turned off.

TABLE 8.4 Top 10 Terms for 10 Concept Vectors with Term Redundancy Allowed[a]

1	2	3	4	5	6	7	8	9	10
technolog	surviv	test	genet	sequenc	breast	cnv	carcinoma	infect	probe
system	prognost	sensit	region	speci	line	copi	renal	strain	hybrid
pcb	cd	fc	loss	annot	posit	antibodi	sa	immun	experi
vegfr	predict	immun	map	network	er	region	confirm	mice	acid
applic	tumour	type	famili	inform	gastric	platform	histolog	isol	assai
advanc	independ	splice	pcr	predict	ihc	featur	hcc	induc	diagnost
field	egfr	assai	line	transcriptom	neg	chromosom	report	pathogen	test
quantit	treatment	stress	predict	experi	growth	gener	classif	indic	sequenc
collect	invas	predict	realtim	tool	bind	sequenc	cluster	metabol	design
biomark	prognosi	network	chromosom	promot	prolifer	variant	antibodi	downregul	platform

[a]Term weighting scheme with `tfn` used.

TABLE 8.5 txn without Redundancy

1	2	3	4	5	6	7	8	9	10
system	treatment	stress	genet	sequenc	breast	cnv	surviv	infect	probe
vegfr	line	sensit	region	network	mirna	antibodi	prognost	immun	platform
pcb	liver	splice	famili	annot	microrna	copi	tumour	strain	experi
technolog	proteom	plai	mrna	inform	estrogen	featur	cd	mice	hybrid
applic	mm	fc	map	speci	neg	domain	carcinoma	host	test
quantit	bone	cultur	affect	base	er	chromatin	stage	induc	oligonucleotid
biomark	chemic	sp	pcr	tool	progenitor	chromosom	grade	isol	amplif
toxicogenom	mitochondri	met	realtim	resist	overexpress	dynam	stain	pathogen	measur
field	growth	dai	loss	avail	type	ii	independ	indic	snp
advanc	melanoma	carcinogen	complex	promot	prolifer	size	predict	metabol	assai

TABLE 8.6 `txn` **with Redundancy**

1	2	3	4	5	6	7	8	9	10
system	treatment	stress	genet	sequenc	breast	cnv	surviv	infect	probe
vegfr	line	test	region	network	mirna	antibodi	prognost	immun	platform
pcb	liver	immun	famili	annot	microrna	copi	tumour	strain	experi
technolog	proteom	sensit	mrna	experi	estrogen	region	cd	mice	hybrid
applic	induc	splice	map	inform	neg	featur	carcinoma	host	test
quantit	mm	plai	affect	speci	er	domain	stage	induc	oligonucleotid
biomark	bone	integr	line	base	amplif	chromatin	grade	isol	amplif
toxicogenom	chemic	fc	stage	predict	marker	chromosom	stain	pathogen	measur
field	mitochondri	network	pcr	tool	posit	dynam	independ	indic	snp
advanc	mrna	assai	realtim	resist	grade	ii	predict	treatment	assai

seems to include terms for ER, estrogen, miRNA, microRNA, amplification, positive, and grade. Under the scheme `tfn` in Tables 8.3 and 8.4, the cluster with breast as the top term also included gastric, line, positive ER, immunohistochemistry, bind, and profiler. In Tables 8.5 and 8.6, there also appears to be a breast cancer cluster (column 8), which lists carcinoma, renal, sa, confirm, histologogy, hcc, report, classification, cluster, and antobodi.

8.11 SUMMARY

Inverse document frequency text mining is one of the earliest forms of document-based information retrieval. Many of today's text mining applications use most, if not all, the techniques described in this chapter. Document-based information retrieval, and specifically stemming [3], originated much earlier than many of the commonly used machine learning techniques that are widely employed today. The use of text mining in biology is steadily increasing as the capacity for more advanced information processing systems and workflow continues to grow [13,14].

REFERENCES

[1] Q. Wills, C. Mitchell. Toxicogenomics in drug discovery and development—making an impact. (PMID: 19807202). *Altern. Lab. Anim.* **37**(S1):33–37, 2009.

[2] W.B. Frakes, R. Baeza-Yates. *Information Retrieval: Data Structures and Algorithms.* Prenctice-Hall, Englewood Cliffs, NJ, 1992.

[3] J.B. Lovins. Development of a stemming algorithm. *Mech. Trans. Comput. Ling.* **11**:23–31, 1968.

[4] J.L. Dawson. Suffix removal and word conflation. *Assoc. Lit. Ling. Comput. Bull.* **2**(3):33–46, 1974.

[5] M.F. Porter. An algorithm for suffix stripping. *Program* **14**(3):130–137, 1980.

[6] K. Sparck-Jones, P. Willet. *Readings in Information Retrieval.* Morgan-Kaufmann, San Mateo, CA, 1997.

[7] C. Buckley. The importance of proper weighting methods. In M. Bates, ed., *Human Language Technology*, Morgan-Kaufmann, San Mateo, CA, 1993.

[8] G. Salton, C. Buckley. Term weighting approaches in automatic text retrieval. *Infor. Process. Manage.* **24**(5):513–523, 1988.

[9] G. Salton, M.J. McGill. *Introduction to Modern Information Retrieval.* McGraw-Hill, New York, 1983.

[10] A. Singhal, G. Salton, M. Mitra, C. Buckley. *Document Length Normalization.* Technical Report TR95-1529, Dept. Computer Science, Cornell Univ., Ithaca, NY, 1995.

[11] T.G. Kolda. *Limited Memory Matrix Methods with Applications*. PhD Thesis, Dept. Mathematics, Univ. Maryland, College Park, MD, 1997.

[12] I.S. Dhillon, D.S. Modha. Concept decompositions for large sparse text data using clustering. *Machine Learn.* **42**:143–175, 2001.

[13] S. Zaremba, M. Ramos-Santacruz, T. Hampton, P. Shetty, J. Fedorko, J. Whitmore, J.M. Greene, N.T. Perna, J.D. Glasner, G. Plunkett, M. Shaker, D. Pot. Text-mining of PubMed abstracts by natural language processing to create a public knowledge base on molecular mechanisms of bacterial enteropathogens. *BMC Bioinform.* **10**:177, 2009.

[14] M. Ongenaert, L. Van Neste, T. De Meyer, G. Menschaert, S. Bekaert, W. Van Criekinge. PubMeth: A cancer methylation database combining text-mining and expert annotation. *Nucleic Acids Res.* **36**(D):842–846, 2008.

CHAPTER 9

TEXT MINING: N-GRAM ANALYSIS

9.1 INTRODUCTION

In the previous chapter on text mining, we observed how useful text mining is for information retrieval from a corpus of words in many documents and subsequent document clustering based on inverse document frequency of words. In this chapter, we address N-gram text mining [1], which is another semantic technique based on breaking up words into smaller N-length *tokens* or *N-grams*, which are used for frequency analysis. Earlier work using N-gram analysis focused on machine-read Morse code [2], spelling error detection [3], string matching [4], speech recognition [5], and document clustering [6].

In the general domain of biomedical applications, N-gram analysis has been applied to PubMed user actions [7], fMRI-based brain function models [8], partitioning medical documents for experts and consumers [9], compound medical diagnoses [10], record linkage [11], automated coding of diagnoses [12], pathology reports [13], histometry [14], and machine vision in histology [15]. In genomic sciences and molecular biology, more recent applications of N-gram analysis include protein subfamily profiles [16], analyzing gene-disease relationships [17], extracting biological relationships [18], motif identification [19], protein localization [20], coding and noncoding DNA sequences [21], molecular sequence classification [22], protein classification [23], compression of nucleic acid and protein sequence data [24], protein secondary structure [25], and sequence analysis [26].

N-gram text mining is more tolerant to errors in spelling, grammar, and OCR, because only the specific N-gram that is erroneous will bias the results. For example, if `airplane` is misspelled and written as `aorplane`, then only the `o` that is misplaced for an `i` is in error, while the remaining characters `a_rplane` are still correct.

Classification Analysis of DNA Microarrays, First Edition. Leif E. Peterson.

The construction of N-grams starts by padding each word of length K with a leading blank "_" and $K-1$ trailing blanks. Hence, the word `airplane` of length $K = 8$ is partitioned into the various N-grams using the scheme provided in Table 9.1. Notice in Table 9.1 that in the 8-grams set the $K-1$ trailing blanks help distinguish that the last character `e` originated from a word of length $K = 8$, and not a shorter or longer word. Not counting the two blanks _ in the 1-gram set, there are 71 N-grams used for representing the single word `airplane`. Such expansion of information is not without its cost during evaluation of the frequency distribution of N-grams. Quite simply, N-gram tokenization can result in thousands of grams from a few documents.

9.2 ALGORITHM

N-gram analysis starts by splitting each document into all possible words, padding each word with leading and trailing blanks, and then tokenizing each padded word into its constituent tokens. As the tokens are obtained, a global hashtable is used to store each new token learned, and a document-by-token array is used to store a running sum of each token's frequency in each document. Let i be the ith document, j be the jth N-gram or token, T_{ij} is the actual character value of the N-gram, and TF_{ij} is its frequency. Algorithm 9.1 lists the source code for N-gram analysis.

TABLE 9.1 Schema for Generating Text-Based *N*-Grams[a]

N-Gram length		Result
Word		`airplane` ($K = 8$)
Padded		`_airplane_ _ _ _ _ _ _` (1 leading blank, $K - 1$ trailing blanks)
1-gram	1	`_, a, i, r, p, l, a, n, e, _,`
2-gram	2	`_a, ai, ir, rp, pl, la, an, ne, e_`
3-gram	3	`_ai, air, irp, rpl, pla, lan, ane, ne_, e_ _`
4-gram	4	`_air, airp, irpl, rpla, plan, lane, ane_, ne_ _, e_ _ _`
5-gram	5	`_airp, airpl, irpla, rplan, plane, lane_, ane_ _, ne_ _ _, e_ _ _ _`
6-gram	6	`_airpl, airpla, irplan, rplane, plane_, lane_ _, ane_ _ _, ne_ _ _ _, e_ _ _ _ _`
7-gram	7	`_airpla, airplan, irplane, rplane_, plane_ _, lane_ _ _, ane_ _ _ _, ne_ _ _ _ _, e_ _ _ _ _ _`
8-gram	8	`_airplan, airplane, irplane_, rplane_ _, plane_ _ _, lane_ _ _ _, ane_ _ _ _ _, ne_ _ _ _ _ _, e_ _ _ _ _ _ _`

[a]Each token is represented by the notation T_{ij}, where i represents the document and j represents the token, or N-gram.

Algorithm 9.1: *N*-Gram Text Mining

Data: Documents in ASCII text format
Result: Clustered documents based on spherical *K*-means
Determine number of documents n_D
Determine total number of learned words n_W
for $i \leftarrow 1$ **to** n_D **do**
  Read document *i*
  Fill `readwords()`
  foreach *word* ∈ *document i* **do**
    Pad leading and trailing edges with "_"
    Tokenize *word*
    foreach *token* ∈ *word* **do**
      if $token \notin \mathcal{H}$ **then**
        Append $\mathcal{H}$
      endif
      if $token \notin \mathcal{H}_g$ **then**
        Append $\mathcal{H}_g$
      endif
    endfch
    foreach $token \in \mathcal{H}$ **do**
      $j = j + 1$
      $T_{ij} = token.Key$
      $TF_{ij} = token.Value$
    endfch
  endfch
endfor

9.3 IMPLEMENTATION

Implementation of *N*-gram text mining is very similar to inverse document frequency text mining. The only difference is that in text mining, the tokenization is performed on the document to parse the words and append them to a hashtable, while in *N*-gram text mining we token a word and append the individual grams to a hashtable.

Processing each Document. Each document is in the form of a delimited ASCII text file and each line is parsed, and separate words are appended to the `readwords()` array. Unwanted characters are then stripped out of each word, and the filtered words are then appended to the `inputwords()` array.

```
'For the current document:
Do While Not EOF(1)
```

```
  inbuff = LineInput(1)
  dynamarray = inbuff.Split(DELIMITER)
  nrow = UBound(dynamarray) + 1
  ReDim buffer(nrow)
  buffer = inbuff.Split(DELIMITER)
  For j = LBound(buffer) To UBound(buffer)
    NumReadWords = NumReadWords + 1
    ReDim Preserve readwords(NumReadWords)
    readwords(NumReadWords) = buffer(j)
  Next j
Loop
For j = 1 To NumReadWords
  If readwords(j) = "" Then GoTo nextword
  If readwords(j) <> "" Then
    NumInputWords = NumInputWords + 1
    ReDim Preserve inputwords(NumInputWords)
    buffstr = readwords(j).ToLower
    buffstr = REDropFMTCHAR.Replace(buffstr, "")
    buffstr = REDropDIGIT.Replace(buffstr, "")
    buffstr = REDropNONWORD.Replace(buffstr, "")
    inputwords(NumInputWords) = buffstr 'readwords(j)
  End If
nextword:
Next j
```

Append Hashtables. Once all of the input words have been determined, the next step is to pad each word with a leading underscore "_" and then use the `Mid(,,)` string function for tokenizing the characters in a word before appending the hashtable:

```
'Still processing same document:
For i = 1 To NumInputWords
  numgrams = Len(inputwords(i)) + 1
  inputwords(i) = "_" & inputwords(i) & "_ _ _ _"
  'Learn and append 1 grams to hashtable
  For k = 2 To numgrams
    If ht(Mid(inputwords(i), k, 1)) > 0 Then
      Cnt = ht(Mid(inputwords(i), k, 1))
      ht(Mid(inputwords(i), k, 1)) = Cnt + 1
    End If
    If ht(Mid(inputwords(i), k, 1)) = 0 Then
      ht.Add(Mid(inputwords(i), k, 1), 1)
    End If
    If htlearnedngrams(Mid(inputwords(i), k, 1)) > 0 Then
      Cnt = htlearnedngrams(Mid(inputwords(i), k, 1))
      htlearnedngrams(Mid(inputwords(i), k, 1)) = Cnt + 1
    End If
```

```
    If htlearnedngrams(Mid(inputwords(i), k, 1)) = 0 Then
      htlearnedngrams.Add(Mid(inputwords(i), k, 1), 1)
    End If
  Next k
  'Learn and append 2-5 grams to hashtable
  For j = 2 To 5
    For k = 1 To numgrams
      If ht(Mid(inputwords(i), k, j)) > 0 Then
        Cnt = ht(Mid(inputwords(i), k, j))
        ht(Mid(inputwords(i), k, j)) = Cnt + 1
      End If
      If ht(Mid(inputwords(i), k, j)) = 0 Then
        ht.Add(Mid(inputwords(i), k, j), 1)
      End If
      If htlearnedngrams(Mid(inputwords(i), k, j)) > 0 Then
        Cnt = htlearnedngrams(Mid(inputwords(i), k, j))
        htlearnedngrams(Mid(inputwords(i), k, j)) = Cnt + 1
      End If
      If htlearnedngrams(Mid(inputwords(i), k, j)) = 0 Then
        htlearnedngrams.Add(Mid(inputwords(i), k, j), 1)
      End If
    Next k
  Next j
Next i
```

Fill N-Gram and Frequency Arrays. Next, we fill the string and frequency array for each of the *N*-grams in the current document. Again, recall that our focus is on the *N*-gram frequencies; however, the hashtables used do not order the *N*-grams:

```
'Still processing same document:
Dim de1 As New DictionaryEntry
Cnt = 0
For Each de1 In ht
  Cnt += 1
  If Cnt > maxCnt Then
    maxCnt = Cnt
    ReDim Preserve htinputtokens(NumDocuments, maxCnt)
    ReDim Preserve htinputtokenfreq(NumDocuments, maxCnt)
  End If
  htinputtokens(ThisDocument, Cnt) = de1.Key
  htinputtokenfreq(ThisDocument, Cnt) = de1.Value
Next
```

Sorting N-Grams and Frequencies within Documents. The *N*-gram frequency array filled using hashtables is not ordered in any particular fashion, so we next order the *N*-grams within each document on the basis of their frequency:

```
descending = True
For ThisDocument = 1 To NumDocuments
  For i = 1 To maxCnt
    arr(i) = htinputtokenfreq(ThisDocument, i)
    ndx(i) = i
  Next i
  sort(arr, ndx, maxCnt, descending)
  For i = 1 To maxCnt
    sortedtokens(ThisDocument, i) = htinputtokens(ThisDocument,
       ndx(i))
    sortedtokenfreq(ThisDocument, i) = arr(i)
  Next i
Next ThisDocument
```

Sorting Global List of Learned N-Grams. This step was carried out only for Example 9.1, for reasons of time complexity. For small vocabularies where the number of words is much less than 10,000, we can sort the global list of learned *N*-grams so that the row values can be used for picking off rows for given *N*-grams in the unsorted hashtable `htinputtokens`. The resulting array with sorted *N*-grams is `sortedlearnedngrams()`. Note that the rows of the sorted *N*-gram frequencies in the global hashtable are used later when assigning frequencies for the matrix `symmetriclearnedtokencounts()`:

```
Cnt = 0
ReDim arr(htlearnedngrams.Count)
ReDim ndx(htlearnedngrams.Count)
For Each de In htlearnedngrams
    Cnt += 1
    arr(Cnt) = de.Value
    ndx(Cnt) = Cnt
    unsortedlearnedngrams(Cnt) = de.Key
Next
sort(arr, ndx, NumLearnedWords, descending)
Cnt = 0
descending = True
ReDim WordNames(NumLearnedTokens)
For i = 1 To NumLearnedTokens
    If unsortedlearnedngrams(ndx(i)) <> Nothing Then
      Cnt += 1
      sortedlearnedngrams(Cnt) = unsortedlearnedngrams(ndx(i))
      sortedlearnedngramsrow(Cnt) = Cnt
      WordNames(Cnt) = unsortedlearnedngrams(ndx(i))
    End If
```

```
Next
NumLearnedTokens = Cnt
Dim htrowforthistoken As New Hashtable
For i = 1 To NumLearnedTokens
  If htrowforthistoken.ContainsKey(sortedlearnedngrams(i))
     = False Then
    htrowforthistoken.Add(sortedlearnedngrams(i), 1)
  End If
Next
For i = 1 To NumLearnedTokens
  htrowforthistoken(sortedlearnedngrams(i))
     = sortedlearnedngramsrow(i)
Next
ReDim symmetriclearnedtokencounts(NumDocuments,
   NumLearnedTokens)
For j = 1 To NumDocuments
  For i = 1 To NumLearnedTokens
    If sortedtokens(j, i) <> Nothing Then
     symmetriclearnedtokencounts(j, htrowforthistoken
        (sortedtokens(j, i)))_
       = sortedtokenfreq(j, i)
    End If
  Next i
Next j
```

A major point to be made here is that construction of a frequency matrix whose rows represent the same *N*-gram across all documents (SMILES strings) was performed for the example because of the smaller gram number of (1-gram to 5-gram range) for SMILES strings. With ordinary text documents, based on, say, a 10,000-word English vocabulary, we don't recommend construction of such a matrix because it is prohibitively expensive given the large number of *N*-grams in text documents.

Example 9.1 This example addresses the use of *N*-gram text mining to evaluate the similarity between SMILES strings,[1] which are a standard chemical nomenclature for molecules based on atoms and chemical bonds. SMILES strings were introduced by Weininger [27-29] in the late 1980s as a means of using a very comprehensible unique universal identifier for molecular structure. Several examples of SMILES string used to identify chemical structures are listed in Table 9.2.

For this example, we performed *N*-gram analysis on 200 SMILES obtained from the National Cancer Institute's Open Database, which is part of

[1]SMILES (simplified molecular input line entry system) strings are a trademark of Daylight Chemical Information Systems, Inc.

TABLE 9.2 Molecular SMILES Strings and Associated Structure

SMILES String	Structure
CC1=CC(=O)C=CC1=O	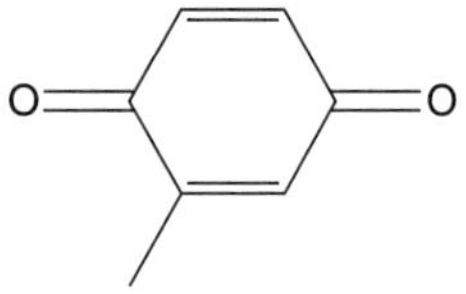
S(SC1=NC2=CC=CC=C2S1) C3=NC4=C(S3)C=CC=C4	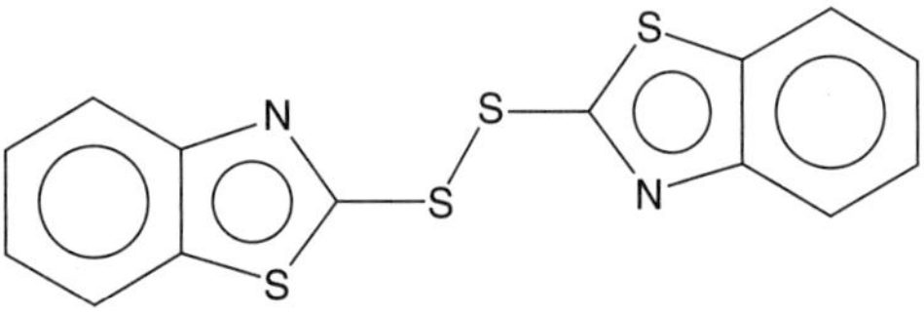
OC1=C(Cl)C=C(C=C1[N+] ([O-])=O)[N+]([O-])=O	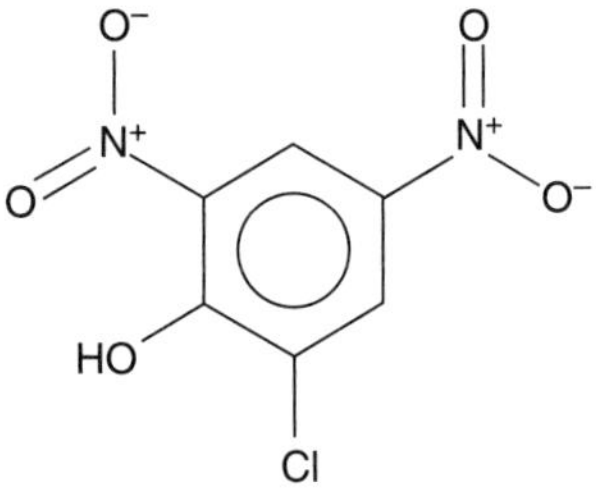
[O-][N+](=O)C1=CNC(=N)S1	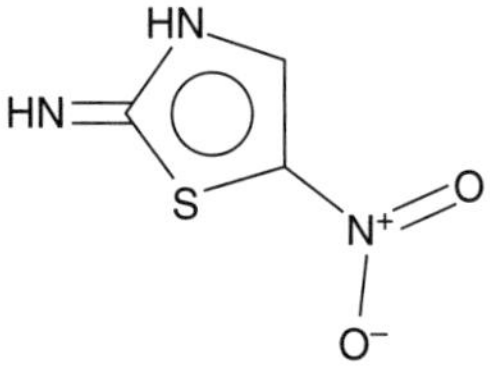
NC1=CC2=C(C=C1)C(=O) C3=C(C=CC=C3)C2=O	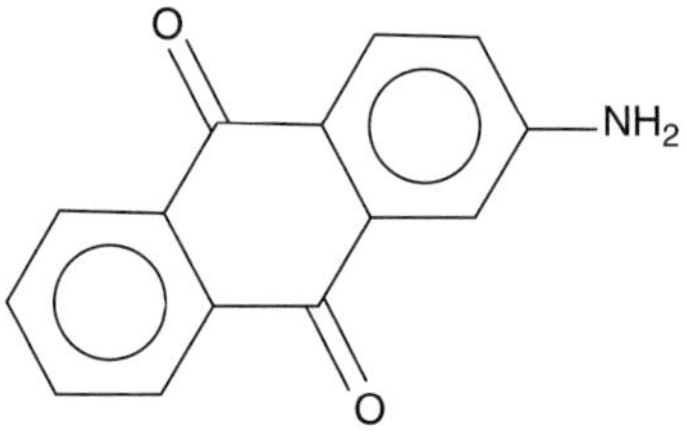
CN(C)C1=C(Cl)C(=O) C2=C(C=CC=C2)C1=O	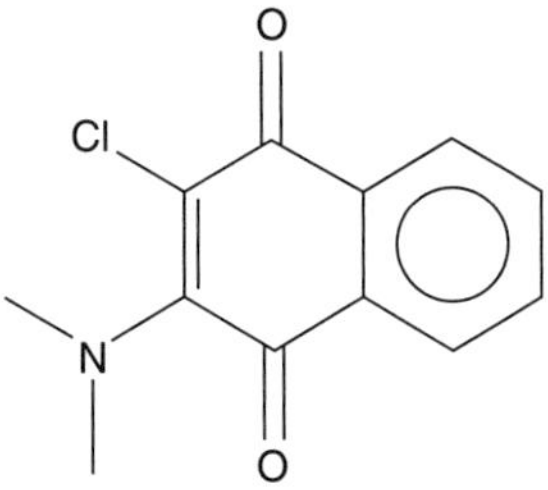

TABLE 9.3 Pairs of SMILES Strings that Clustered Together

NSC	SMILES	String
114	2,2,3,3,4,4,5,5-Octafluoro-1-pentanol	OCC(F)(F)C(F)(F)C(F)(F)C(F)F
115	2,2,3,3,4,4,5,5,6,6,7,7-Dodecafluoro-1-heptanol	OCC(F)(F)C(F)(F)C(F)(F)C(F)(F)C(F)(F)C(F)F
159	3-(3,5-Dibromo-2-hydroxyphenyl)-2-phenylacrylic acid	OC(=O)C(=CC1=C(O)C(=CC(=C1)Br)Br)C2=CC=CC=C2
160	3-(3,5-Dibromo-4-hydroxyphenyl)-2-phenylacrylic acid	OC(=O)C(=CC1=CC(=C(O)C(=C1)Br)Br)C2=CC=CC=C2
61	Triphenylacetic acid	OC(=O)C(C1=CC=CC=C1)(C2=CC=CC=C2)C3=CC=CC=C3
66	3,3,4-Triphenyl-2-butanone	CC(=O)C(CC1=CC=CC=C1)(C2=CC=CC=C2)C3=CC=CC=C3
58	*N*-Benzhydryl(diphenyl)methanamine	N(C(C1=CC=CC=C1)C2=CC=CC=C2)C(C3=CC=CC=C3)C4=CC=CC=C4
194	1,2,2,2-Tetraphenylethanone	O=C(C1=CC=CC=C1)C(C2=CC=CC=C2)(C3=CC=CC=C3)C4=CC=CC=C4

the Developmental Therapeutics Program (`http://dtp.nci.nih.gov/`) and generated each string a of 1-gram, 2-gram, 3-gram, 4-gram, 5-gram frequency for each string.

The resulting frequencies were employed in hierarchical cluster analysis using Euclidean distance for agglomeration during clustering. Table 9.3 lists examples of the closest pairs of SMILES strings observed during cluster analysis.

As one can notice in Figures 9.1-9.4, the N-gram technique enables the identification of SMILES strings for which the associated structures are similar. Table 9.4 lists resulting N-gram frequencies ($N < 6$) obtained from Algorithm 9.1 for 20 example SMILES strings.

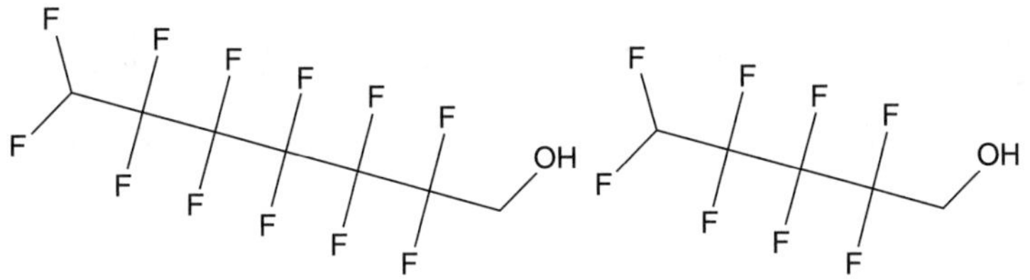

FIGURE 9.1 Structure of SMILES strings (NSC) 114 and 115.

TABLE 9.4 Example N-Gram ($N < 6$) Frequencies for 20 SMILES Strings

	SMILES string (NSC number)																			
N-Gram	118	27	60	63	145	18	140	134	104	54	34	166	58	151	152	87	198	40	129	15
C=	0	0	2	2	2	8	1	0	2	2	1	0	8	0	2	4	4	4	5	4
(=	0	2	0	3	1	0	2	0	0	1	1	0	0	3	3	0	1	0	0	0
1=	0	1	1	1	1	1	0	0	1	1	1	0	1	1	1	0	1	1	0	1
CC(	3	1	1	2	0	0	1	2	0	1	1	1	0	2	1	0	0	0	0	0
2=	0	0	0	1	0	1	0	0	0	1	0	0	1	0	1	1	1	1	1	1
C(=O)	0	0	0	1	0	0	2	0	0	1	0	0	0	0	1	0	1	0	0	0
)=	0	0	1	0	0	0	0	0	0	0	1	1	0	1	1	0	0	0	0	0
=O__	0	0	0	0	0	0	0	1	0	1	1	1	0	1	0	0	0	0	0	0
C2__	0	0	0	0	0	0	0	0	0	0	0	0	0	0	0	0	1	0	0	0
CH	0	0	0	0	0	0	0	0	1	1	0	0	0	0	0	1	2	0	0	0
OC(=O	0	0	0	0	0	0	1	0	0	0	0	0	0	0	1	0	0	0	0	0
(F)C	10	0	0	0	0	0	0	0	0	0	0	0	0	0	0	0	0	0	0	0
O)C1	0	0	1	0	1	0	0	0	0	0	0	0	0	0	0	0	1	0	0	0
)C(F)	10	0	0	0	0	0	0	0	0	0	0	0	0	0	0	0	0	0	0	0
(F)(	10	0	0	0	0	0	0	0	0	0	0	0	0	0	0	0	0	0	0	0
3=	0	0	0	0	0	1	0	0	0	0	0	0	1	0	0	1	0	0	1	1
C=C(	0	0	0	0	0	0	0	0	0	0	0	0	0	0	0	0	0	2	0	0
[O	0	0	0	0	0	0	0	0	0	0	1	0	0	0	0	0	0	0	0	0
=C2)	0	0	0	1	0	1	0	0	0	1	0	0	1	0	1	1	0	1	0	0
C=C3	0	0	0	0	0	1	0	0	0	0	0	0	1	0	0	1	0	0	1	1

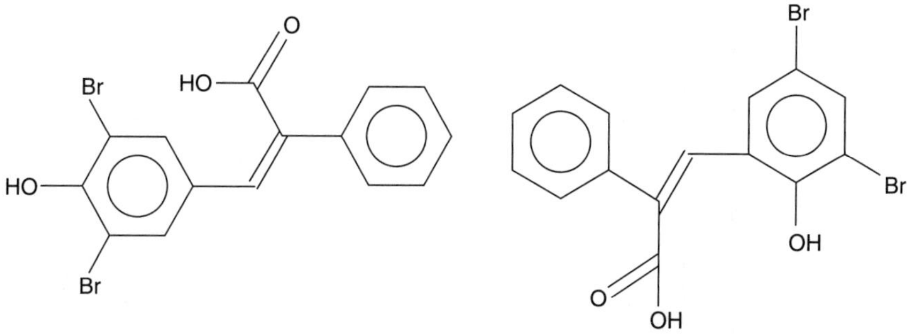

FIGURE 9.2 Strucure of SMILES strings (NSC) 59 and 60.

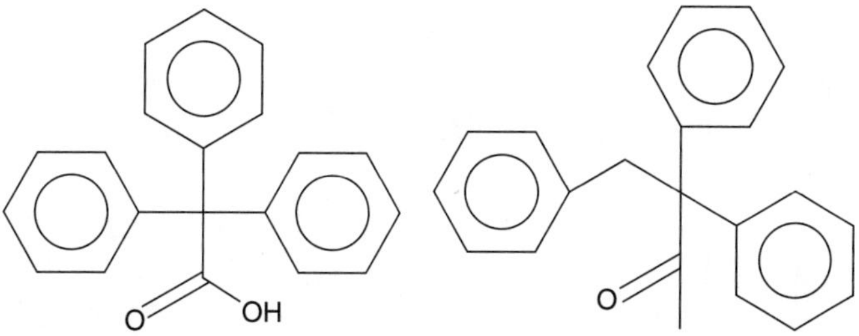

FIGURE 9.3 Structure of SMILES strings (NSC) 60 and 61.

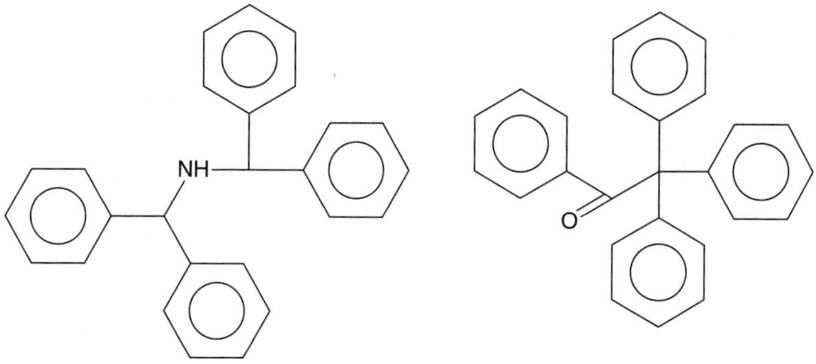

FIGURE 9.4 Structure of SMILES strings (NSC) 58 and 194.

Example 9.2 In this example, we employed 24 of the 373 abstracts used for the example in Chapter 8. N-gram analysis was used ($N < 6$) to derive all 1-, 2-, 3-, 4-, and 5-gram values from the words contained in the abstracts. Table 9.5 lists the most frequent N-grams within each abstract, while Table 9.6 lists the N-gram frequencies for N-grams as they appear in each abstract.

TABLE 9.5 Maximum *N*-Gram Frequencies ($N < 6$) for 10 Example PubMed Abstracts[a]

11140706	freq	12460888	freq	12542399	freq	14594711	freq	14752004	freq	15026786	freq	15117756	freq	15195116	freq	15474668	freq	15618527	freq
e	159	e	92	e	152	e	172	e	155	e	140	e	134	e	135	e	137	a	78
a	118	a	70	a	103	a	144	i	109	a	114	t	91	t	114	i	111	e	72
t	118	s	58	t	102	t	113	a	104	i	110	s	84	i	90	r	110	s	61
r	97	n	56	i	99	s	112	t	101	s	105	a	73	n	88	a	105	o	58
i	94	t	53	s	95	r	103	n	90	n	94	i	67	a	88	t	103	r	55
s	92	r	51	r	90	i	84	o	84	o	87	r	63	r	86	s	99	n	48
o	91	o	47	n	87	n	84	s	84	r	85	o	58	o	84	n	97	t	47
n	84	i	42	o	86	o	81	r	83	t	77	n	55	s	83	o	73	i	46
c	55	d	27	c	56	l	64	d	51	l	72	l	40	d	46	c	66	c	45
e__	53	c	26	h	52	c	50	c	50	c	70	e_	36	c	36	l	51	l	41
e_	53	m	25	l	45	m	48	l	49	e__	42	e___	36	e__	34	e_	43	_a	36
e___	53	l	24	p	44	p	47	e__	48	e____	42	e__	36	e____	34	e___	43	s____	28
_t	53	n____	21	s___	42	d	43	e___	48	e___	42	e____	36	e_	34	e__	43	s_	28
e____	53	n__	21	s__	42	s__	41	e____	48	e_	42	d	31	e___	34	e____	43	s___	28
h	50	n___	21	s_	42	s___	41	e_	48	m	40	c	31	l	31	d	40	s__	28
d	43	n_	21	s____	42	s____	41	m	45	h	40	s___	30	h	30	f	30	_c	24
m	35	e__	20	e__	41	s_	41	h	32	s____	39	s_	30	g	30	p	29	h	24
th	32	_a	20	m	41	er	33	u	32	s__	39	s__	30	in	30	g	29	d	23
u	31	e_	20	e___	41	g	33	p	30	s___	39	s____	30	n___	26	h	29	g	22
l	30	e___	20	e_	41	at	32	g	29	s_	39	m	29	n____	26	u	29	m	19
s_	30	e____	20	e____	41	h	30	s____	28	f	37	h	29	n__	26	s____	27	e__	19
s___	30	g	20	u	40	f	30	s___	28	u	37	p	28	n_	26	s__	27	e___	19
s____	30	d____	19	_t	39	u	27	s_	28	n__	36	f	24	u	26	s_	27	e_	19
s__	30	d__	19	d	37	e_	27	s__	28	n____	36	re	23	_a	25	s___	27	e____	19

TABLE 9.6 List of *N*-Gram Frequencies for *N*-Grams as They Appear in Each Pubmed for 2 Abstract Used

	11140706	12460888	12542399	14594711	14752004	15026786	15117756	15195116	15474668	15618527
s___	30	18	42	41	28	39	30	24	27	28
_a	29	20	24	26	21	34	20	25	14	36
th	32	11	31	16	25	24	21	24	20	10
n___	23	21	19	21	23	36	14	26	16	10
_th	29	10	27	14	16	19	18	15	18	7
ro	17	10	19	22	10	15	7	13	15	9
_o	19	5	17	13	11	18	10	17	12	12
t___	16	4	20	14	8	9	11	14	16	7
r___	12	9	19	17	9	15	12	16	19	12
y___	19	6	16	14	15	16	10	11	13	9
io	7	7	7	6	17	18	5	12	7	6
the	13	3	17	10	11	11	5	7	11	5
he_	13	3	17	10	11	11	5	7	11	5
he___	13	3	17	10	11	11	5	7	11	5
nt	12	2	14	8	11	7	6	11	13	2
ion	7	7	7	4	13	16	4	12	6	6
ed_	10	14	4	16	10	7	11	16	10	1
of	15	2	13	8	5	18	8	10	11	7
f___	15	3	13	8	4	13	7	10	11	7
a___	8	4	1	12	9	7	7	4	8	
es_	7	9	15	15	9	11	12	7	9	7
es___	7	9	15	15	9	11	12	7	9	7
of_	15	2	13	8	4	13	7	10	11	7
ca	9	7	9	15	8	15	5	3	12	8
co	7	3	4	4	3	5	4	11	6	4

(continued)

TABLE 9.6 (*Continued*)

	16646817	16849266	17069508	17074082	17342950	17459967	17463023	17948124	18372916	19179604
s___	23	31	30	28	19	62	24	20	26	31
_a	22	24	23	21	11	41	22	20	22	35
th	21	13	26	24	17	31	15	13	23	45
n___	11	14	28	18	14	29	15	16	24	41
_th	17	7	19	22	13	22	12	11	19	18
ro	8	11	10	23	17	21	9	7	14	45
_o	6	22	11	16	9	16	13	12	16	13
t___	15	11	9	16	6	13	17	10	15	14
r___	8	17	14	10	9	20	9	8	10	14
y___	10	5	7	13	9	16	16	13	7	14
io	15	8	24	7	12	17	11	4	15	18
the	10	5	10	12	12	17	6	5	12	13
he_	10	5	10	12	12	17	6	5	12	13
he___	10	5	10	12	12	17	6	5	12	13
nt	9	10	6	9	9	19	14	10	10	14
ion	14	8	24	6	11	12	8	4	12	15
ed_	4	5	9	14	7	19	7	11	13	11
of	6	12	8	10	6	11	8	10	9	11
f___	5	12	8	7	6	10	8	9	14	11
a___	10	9	7	12	8	16	14	7	9	6
es_	5	11	5	5	8	13	5	5	7	5
es___	5	11	5	5	8	13	5	5	7	5
of_	5	12	8	7	6	10	8	9	9	11
ca	5	5	8	5	8	11	9	11	7	14
co	8	5	3	8	8	15	5	4	0	4

[a]Frequencies sorted in descending order. Document identification is the PubMed PMID number.

When working with a 10,000-word English vocabulary, the total N-grams number is prohibitively expensive for straightforward class discovery or class prediction due to the immense number of N-grams (features) extracted. Recall that there can be thousands of 1-gram, 2-gram, 3-gram, 4-gram, and 5-gram values occurring in documents as small as an abstract, and when considering overlap between documents it is highly likely that many sparse feature values will be present because of nonshared N-gram values. To avoid sparseness in feature values and minimize the impact of the curse of dimensionality, we must resort to methods based on distance in the absence of available feature values.

An appropriate distance measure for N-gram values is the out-of-place distance function, described by Cavnar and Trenkle [6]. While they did not provided the mathematical notation for the out-of-place distance function [6], we believe a reliable formalism to be:

$$d(\mathbf{x}_i, \mathbf{x}_j) = \sum_{k=1}^{|\mathcal{H}|} |R_{ik} - R_{jk}|, \tag{9.1}$$

where $|\mathcal{H}|$ is the cardinality of the global hashtable containing all learned N-grams from all documents, R_{ik} is the rank of the kth N-gram frequency TF_{ik} in the ith document, R_{jk} is the rank of the kth N-gram frequency TF_{jk} in the jth document, and $|R_{ik} - R_{jk}|$ is the out-of-place distance. Any N-grams not shared between documents i and j are assigned the maximum value of $|R_{ik} - R_{jk}|$ that is observed between the two documents. The out-of-place distance metric can be easily incorporated into a distance matrix implemented in hierarchical cluster analysis for document class discovery or K nearest neighbor for class prediction when document class is known a priori.

Example 9.3 Out-of-place distance was calculated for 10 PubMed abstracts, 5 of which were identified using the PubMed search term "Alzheimer's diagnosis," and 5 which were identified with the search term "healthcare reform." Table 9.7 lists the between-document out-of-place distance matrix for the 10 abstracts. Note that, on average, the between-abstract out-of-place distance for abstracts on Alzheimer's diagnosis is lower when compared with the between-abstract distance among abstracts on healthcare reform. Certainly, on occasion there were lower distances between abstracts in the different topics; this was attributed to greater between-abstract similarity at finer levels of granularity. For example, if two abstracts focus on the same proteins but the disease domains are different (e.g., prostate cancer vs. arteriosclerosis), then the out-of-place distance might be lower when compared to the potentially greater out-place distances between abstracts for prostate cancer and abstracts for arteriosclerosis. Use of document titles would likely minimize the chance for such between-class homogeneity of

TABLE 9.7 Out-of-Place Distance Matrix for Five PubMed Abstracts Identified with Search Term *Alzheimers's Diagnosis*[a] and Five Abstracts Identified with Search Term *Healthcare Reform*[b]

	ALZ1	ALZ2	ALZ3	ALZ4	ALZ5	HCR1	HCR2	HCR3	HCR4	HCR5
ALZ1	0	746	762	531	920	2001	1161	729	1055	849
ALZ2	0	0	9082	1024	2933	4150	5135	1252	2215	3137
ALZ3	0	0	0	1035	1917	2878	1380	3445	2562	1429
ALZ4	0	0	0	0	3250	5936	3906	1331	1866	3084
ALZ5	0	0	0	0	0	861	4331	1228	3701	1496
HCR1	0	0	0	0	0	0	1507	899	2411	1788
HCR2	0	0	0	0	0	0	0	960	1013	1756
HCR3	0	0	0	0	0	0	0	0	859	1494
HCR4	0	0	0	0	0	0	0	0	0	2452

[a]PMID numbers for Alzheimer's-related abstracts ALZ1-ALZ5 are 20202073, 20204386, 20205671, 20205826, and 20207459.
[b]PMID numbers for healthcare-reform-related asbtracts HCR1-HCR5 are 20202507, 20207927, 20214086, 20215333, and 20220736.

out-of-place distance, but with such little text available from document titles, an approach involving "title mining" could hardly be couched as a text mining effort. Granted, there is merit in matching titles, but the typical goal of text mining is to perform information-retrieval-based document clustering and classification by drilling down into the deeper levels of text.

9.4 SUMMARY

There are many advantages of *N*-gram text mining over the inverse document frequency text mining introduced in Chapter 8. *N*-gram text mining is useful when there essentially is no vocabulary consisting of individual words or tokens, but rather when each document consists of a stream or string of characters joined together. Our examples on SMILES strings was one such case, where the characters represented atoms and molecular bonds. Proteins in the form of peptide strings of amino acids is another potential application of *N*-gram text mining.

We caution the reader to note that *N*-gram text mining has also been applied in the context of individual words serving as the "grams." A good example of this approach is Google's *N*-gram research program.[2] As of September 22, 2006, Google had constructed an *N*-gram set consisting of the following data counts:

[2]Further information is available at `http://googleresearch.blogspot.com/2006/08/all-our-n-gram-are-belong-to-you.html`.

Type	Number
Tokens	1,024,908,267,229
Sentences	95,119,665,584
1-gram	13,588,391
2-gram	314,843,401
3-gram	977,069,902
4-gram	1,313,818,354
5-gram	1,176,470,663

An example 3-grams listing (and frequency) in the Google database is as follows:

```
ceramics collectables collectibles 55
ceramics collectables fine 130
ceramics collected by 52
ceramics collectible pottery 50
ceramics collectibles cooking 45
ceramics collection , 144
ceramics collection . 247
ceramics collection <S> 120
ceramics collection and 43
ceramics collection at 52
ceramics collection is 68
ceramics collection of 76
```

Examples of 4-gram values (and frequency) in the Google set include:

```
serve as the incoming 92
serve as the incubator 99
serve as the independent 794
serve as the index 223
serve as the indication 72
serve as the indicator 120
serve as the indicators 45
serve as the indispensable 111
serve as the indispensible 40
serve as the individual 234
serve as the industrial 52
serve as the industry 607
```

Use of words as grams in *N*-gram text mining is useful for spell checking and grammar checking. Assume that you are developing a word processing software package, and a user misspelled the 4-gram term: `the qyick brown fox`. A comparison of the misspelled 4-gram term with all 4-gram terms in the Google database would indicate that `the quick brown fox` is much more common than the misspelled 4-gram. Hence, you would suggest to the user the word `quick` as a replacement for `qyick`. Grammar checking follows a similar theme, as several groups are augmenting their information retrieval and natural language processing research with information in the Google *N*-gram database [30-32]. The Linguistic Data Consortium (LDC) is a reliable online research resource site for readers who are interested in additional information on *N*-gram analysis with words as grams, as well as other text mining approaches.[3]

REFERENCES

[1] C.Y. Suen. N-gram statistics for natural language understanding and text processing. *IEEE Trans. Pattern Analy. Machine Intell.* **1**(2):164–172, 1979.

[2] C.K. McElwain, M.B. Evens. The Degarbler–a program for correcting machine-read Morse code. *Inform. Control* **5**:368–384, 1962.

[3] E.M. Zamora, J.J. Pollock, A. Zamora. The use of trigram analysis for spelling error detection. *Inform. Process. Manage.* **17**(6):305–316, 1981.

[4] J.-Y. Kim, J. Shawe-Taylor. Fast string matching using an n-gram algorithm. *Software Pract. Experience* **24**(1):79–88, 1994.

[5] R. Schwartz, Y.-L. Chow, The n-best algorithm: An efficient and exact procedure for finding the n most likely sentence hypotheses. *Proc. 1990 IEEE Int. Conf. Acoustics, Speech and Signal Processing*. IEEE Press, Piscataway NJ, 1990, vol. 1, pp. 81–84.

[6] W.B. Cavnar, J.M. Trenkle. N-gram-based text categorization. *Proc. SDAIR-94, 3rd Annual Symp. Document Analysis and Information Retrieval*, ISRI (Information Science Research Institute), University of Nevada, Las Vegas, 1994, pp. 161–175.

[7] J. Lin, W.J. Wilbur. Modeling actions of PubMed users with N-gram language models. *Inform. Retr.* (Boston) **12**:487–503, 2008.

[8] M.Y. Hsiao, D.Y. Chen, J.H. Chen. Constructing human brain-function association models from fMRI literature. *Conf. Proc. IEEE Eng. Med. Biol. Soc.* **2007**:1188–1191, 2007.

[9] M. Poprat, K. Marko, U. Hahn. A language classifier that automatically divides medical documents for experts and health care consumers. *Stud. Health Technol. Inform.* **124**:503–508, 2006.

[3]Further information is available at `http://www.ldc.upenn.edu/`.

[10] G. Heja, G. Surjan. Using n-gram method in the decomposition of compound medical diagnoses. *Stud. Health Technol. Inform.* **90**:455–459, 2002.

[11] T. Churches, P. Christen. Some methods for blindfolded record linkage. *BMC Med. Inform. Decis. Mak.* **4**:9, 2004.

[12] P. Franz, A. Zaiss, S. Schulz, U. Hahn, R. Klar. Automated coding of diagnoses-three methods compared. *Proc. AMIA Symp.* Los Angeles, California, 2000, pp. 250–254.

[13] L.M. de Bruijn, A. Hasman, J.W. Arends. Supporting the classification of pathology reports: Comparing two information retrieval methods. *Comput. Methods Programs Biomed.* **62**(2):109–113, 2000.

[14] P.W. Hamilton, P.H. Bartels, R. Montironi, N.H. Anderson, D. Thompson, J. Diamond, S. Trewin, H. Bharucha. Automated histometry in quantitative prostate pathology. *Anal. Quant. Cytol. Histol.* **20**(5):443–460, 1998.

[15] P.H. Bartels, H.G. Bartels, R. Montironi, P.W. Hamilton, D. Thompson. Machine vision in the detection of prostate lesions in histologic sections. *Anal. Quant. Cytol. Histol.* **20**(5):358–364, 1998.

[16] J.K. Vries, X. Liu. Subfamily specific conservation profiles for proteins based on n-gram patterns. *BMC Bioinform.* **9**:72, 2008.

[17] Y.T. Yen, B. Chen, H.W. Chiu, Y.C. Lee, Y.C. Li, C.Y. Hsu. Developing an NLP and IR-based algorithm for analyzing gene-disease relationships. *Methods Inform. Med.* **45**(3):321–329, 2006.

[18] M. Palakal, M. Stephens, S. Mukhopadhyay, R. Raje, S. Rhodes. A multi-level text mining method to extract biological relationships. *Proc. IEEE Comput. Soc. Bioinform. Conf.* **1**:97–108, 2002.

[19] C.H. Wu, S. Zhao, H.L. Chen, C.J. Lo, J. McLarty. Motif identification neural design for rapid and sensitive protein family search. *Comput. Appl. Biosci.* **12**(2):109–118, 1996.

[20] B.R. King, C. Guda. ngLOC: An n-gram-based Bayesian method for estimating the subcellular proteomes of eukaryotes. *Genome Biol.* **8**(5):R68, 2007.

[21] R.N. Mantegna, S.V. Buldyrev, A.L. Goldberger, S. Havlin, C.K. Peng, M. Simons, H.E. Stanley. Systematic analysis of coding and noncoding DNA sequences using methods of statistical linguistics. *Phys. Rev. E Stat. Phys. Plasmas Fluids Relat. Interdiscip. Top.*, **52**(3):2939–2950, 1995.

[22] C. Wu, M. Berry, Y.S. Fung, J. McLarty. Neural networks for molecular sequence classification. *Proc. First Int. Conf. Intelligent Systems for Molecular Biology (ISMB'93) Bethesda (MD)*, 1993, vol. 1, pp. 429–437

[23] C. Wu, G. Whitson, J. McLarty, A. Ermongkonchai, T.C. Chang. Protein classification artificial neural system. *Protein Sci.* **1**(5):667–677, 1992.

[24] J.R. Walker, P. Willett. Compression of nucleic acid and protein sequence data. *Comput. Appl. Biosci.* **2**(2):89–93, 1986.

[25] J.K. Vries, X. Liu, I. Bahar. The relationship between n-gram patterns and protein secondary structure. *Proteins* **68**(4):830–838, 2007.

[26] M. Ganapathiraju, V. Manoharan, J. Klein-Seetharaman. BLMT: Statistical sequence analysis using N-grams. *Appl. Bioinform.* **3**(2-3):193–200, 2004.

[27] D. Weininger. SMILES, a chemical language and information system. 1. Introduction to methodology and encoding rules. *J. Chem. Inform. Comput. Sci.* **28**:31–36, 1988.

[28] D. Weininger, A. Weininger, J.L. Weininger. SMILES, 2. Algorithm for generation of unique SMILES notation. *J. Chem. Inform. Comput. Sci.* **29**:97–101, 1989.

[29] D. Weininger. SMILES, 3. DEPICT. Graphical depiction of chemical structures. *J. Chem. Inform. Comput. Sci.* **30**(3):237–243, 1990.

[30] M. Federico, M. Cettolo. Efficient handling of N-gram language models for statistical machine translation. *Proc. 2nd Workshop on Statistical Machine Translation,* Prague, June 2003. ACL (Assocication for Computational Linguistics), Morristown, NJ, 2007, p. 8895.

[31] I. Bulyko, M. Ostendorf, A. Stolcke. Getting more mileage from web text sources for conversational speech language modeling using class-dependent mixtures. *Proc. 2003 Confe. North American Chapter of the Association for Computational Linguistics on Human Language Technology,* companion volume of *Proc. HLT-NAACL 2003,* Edmonton, Canada. ACL, Morristown, NJ, 2003.

[32] P. Nakov, M. Hearst. Search engine statistics beyond the n-gram: Application to noun compound bracketing. *Proc. 9th Conf. Computational Natural Language Learning (CoNLL),* Ann Arbor, MI, June 2005. ACL, Morristown, NJ, 2005, pp. 17–24.

PART II

DIMENSION REDUCTION

CHAPTER 10

PRINCIPAL COMPONENTS ANALYSIS

10.1 INTRODUCTION

Principal components analysis (PCA) is used primarily for reproducing the total variance among a large number of variables using a much smaller number of artificial variables called *latent factors* [1-5]. Other terms used for the artificial variables are *constructs* or *unobservable dimensions*. In machine learning, the use of PCA can help minimize overfitting, which arises when a system learns both the noise and nonrandom components of a dataset. PCA minimizes noise by collapsing the majority of variance in a dataset down to a smaller number of dimensions. Multicollinearity is another problem experienced during system learning, since redundant correlations can throw off future predictions. This chapter addresses numerical methods of PCA for gene expression datasets and provides the multivariate statistical theory behind PCA, including the principal factor solution to the factor model of the correlation matrix **R**, extraction of factors (components) from **R**, and identification of genes with similar component loadings to construct groups of genes with similar expression profiles.

10.2 MULTIVARIATE STATISTICAL THEORY

This section describes the multivariate statistical theory of principal component analysis for PCA of gene expression on microarrays. The correlation of expression between two genes (pairwise) is first described, followed by spectral decomposition of the square symmetric correlation matrix to extract eigenvalues and eigenvectors for component loadings. Section 10.4 describes the assembly of gene groups based on component loadings.

Classification Analysis of DNA Microarrays, First Edition. Leif E. Peterson.

10.2.1 Matrix Definitions

Let $\mathbf{M}_{(n,p)} \in \mathbb{R}$ be the set of all data variables over the field of real numbers and $\mathbf{M}_{(p)} \in \mathbb{R}$, the set of symmetric p-square matrices over the field of real numbers. Define the *data* matrix $\mathbf{Y} \in \mathbf{M}_{n,p}(\mathbb{R})$ with n rows and p columns as a function on the pairs of integers (i, j), $1 \leq i \leq n, 1 \leq j \leq p$, with values in $\mathbf{Y}$ in which y_{ij} designates the value of $\mathbf{Y}$ at the pair (i, j):

$$\mathbf{Y}_{n\times p} = \begin{bmatrix} y_{11} & y_{12} & \cdots & y_{1p} \\ y_{21} & y_{22} & \cdots & y_{2p} \\ \vdots & \vdots & \ddots & \vdots \\ y_{n1} & y_{n2} & \cdots & y_{np} \end{bmatrix}. \quad (10.1)$$

Each element of $\mathbf{Y}$, namely, y_{ij}, represents an expression value of gene j on array i. Thus, columns of $\mathbf{Y}$ represent expression of a single gene on multiple arrays, and rows represent arrays.

Let us consider performing PCA on a high-dimensional ($p \gg n$) set of genes in order to reduce the dimensionality. We will therefore need to develop the gene-by-gene ($p \times p$) correlation matrix $\mathbf{R}$. Standardization of $\mathbf{Y}$ is performed using the gene-specific mean vector $\bar{y}$, consisting of gene (column)-specific means and standard deviations:

$$s_j = \sqrt{\frac{\sum_{i=1}^{n}(y_{ij} - \bar{y}_j)^2}{n-1}}. \quad (10.2)$$

Elements of the $n \times p$ standardized data matrix $\mathbf{Z}$ are

$$z_{ij} = \frac{y_{ij} - \bar{y}_j}{s_j} \quad . \quad (10.3)$$

The pairwise correlation between genes k and l given is then calculated from the product of gene-specific Z scores over all of the arrays in the form

$$r_{kl} = \frac{\sum_{i=1}^{n} z_{ik} z_{il}}{n-1}, \quad (10.4)$$

with summation over arrays 1 to n. This results in a $p \times p$ (*gene-by-gene*) correlation matrix:

$$\underset{p\times p}{\mathbf{R}} = \begin{bmatrix} r_{11} & r_{12} & \cdots & r_{1p} \\ r_{21} & r_{22} & \cdots & r_{2p} \\ \vdots & \vdots & \ddots & \vdots \\ r_{p1} & r_{p2} & \cdots & r_{pp} \end{bmatrix}. \quad (10.5)$$

During calculation of pairwise gene correlations, the $p(p-1)/2$ elements in the upper triangle of $\mathbf{R}$ are stored in random access memory (RAM) as a vector (vech $\mathbf{R}$).

10.2.2 Principal Component Solution of R

The principal axis theorem states that, there exists a rotation matrix $\mathbf{E}$ and diagonal matrix $\mathbf{\Lambda}$ such that $\mathbf{ERE}' = \mathbf{\Lambda}$. Premultiplying both sides by $\mathbf{E}$ and post-multiplying by $\mathbf{E}'$ yield the principal form (or spectral decomposition) of $\mathbf{R}$ given as

$$\underset{p\times p}{\mathbf{R}} = \underset{p\times p}{\mathbf{E\Lambda E}'} = \begin{bmatrix} e_{11} & e_{12} & \dots & e_{1p} \\ e_{21} & e_{22} & \dots & e_{2p} \\ \vdots & \vdots & \ddots & \vdots \\ e_{p1} & e_{p2} & \dots & e_{pp} \end{bmatrix} \begin{bmatrix} \lambda_1 & 0 & \dots & 0 \\ 0 & \lambda_2 & \dots & 0 \\ \vdots & \vdots & \ddots & \vdots \\ 0 & 0 & \dots & \lambda_p \end{bmatrix} \begin{bmatrix} e_{11} & e_{21} & \dots & e_{p1} \\ e_{12} & e_{22} & \dots & e_{p2} \\ \vdots & \vdots & \ddots & \vdots \\ e_{1p} & e_{2p} & \dots & e_{pp} \end{bmatrix}, \tag{10.6}$$

where columns of $\mathbf{E}$ and $\mathbf{E}'$ are the *eigenvectors* and diagonal entries of $\mathbf{\Lambda}$ are the *eigenvalues*. Eigenvectors and associated eigenvalues are extracted from $\mathbf{E\Lambda E}'$ using the iterative Jacobi method.

The Jacobi method, which is a foolproof but time-consuming algorithm for finding the eigenvalues and eigenvectors of a symmetric square matrix, goes as follows. Before iterating (iter), set $\mathbf{\Lambda}_0 = \mathbf{R}$ and $\mathbf{E}_0 = \mathbf{I}$. Use Meglicki's [6] method to speed up the Jacobi transformation by looping 50 times, calculating a threshold each time as

$$\text{Threshold} = \begin{cases} \frac{1}{5}(1/p^2)\sum_{q=1}^{p-1}\sum_{r=q+1}^{p}\lambda_{qr} & 1 \leq \text{iter} \leq 3 \\ 0 & \text{iter} > 3 \end{cases} \tag{10.7}$$

where λ_{qr} $(q < r)$ are off-diagonal elements in the upper triangular of $\mathbf{\Lambda}_0$ and p is the number of genes. During each iteration, loop through all λ_{qr} off-diagonal elements of $\mathbf{\Lambda}_0$ (i.e., $q = 1$ to $p-1$; $r = q+1$ to p). Within the loops for q and r, if λ_{qr} exceeds the *threshold* for that iteration, then rotate the off-diagonal element λ_{qr} and calculate new corresponding eigenvalues λ_{qq} and λ_{rr} on the diagonal of $\mathbf{\Lambda}_0$. After the third iteration, off-diagonals λ_{qr} that do not strongly influence eigenvalues undergoing rotation (i.e., $|\lambda_{qq}| + 100|\lambda_{qr}| = |\lambda_{qq}|$ and $|\lambda_{rr}| + 100|\lambda_{qr}| = |\lambda_{rr}|$) are set to zero. During the rotations, let $\mu = (\lambda_{qq} - \lambda_{rr})/2$ and calculate ω as

$$\omega \equiv sgn(\mu)\frac{\lambda_{qr}}{\sqrt{\lambda_{qr}^2 + \mu^2}}. \tag{10.8}$$

Calculate the trigonometric relationships

$$\sin(\theta) = \frac{\omega}{\sqrt{2(1 + \sqrt{1 - \omega^2})}} \tag{10.9}$$

$$\cos(\theta) = \sqrt{1 - \sin^2(\theta)} \tag{10.10}$$

and substitute into the $p \times p$ rotation matrix $\mathbf{S}$ as

$$\mathrm{S} = \begin{bmatrix} \mathrm{I} & 0 & 0 \\ 0 & T & 0 \\ 0 & 0 & \mathrm{I} \end{bmatrix}, \tag{10.11}$$

where the top row partition has $q - 1$ rows, the bottom row partition has $p - r$ rows, and the $\mathbf{T}$ matrix with $r - q+1$ rows is

$$T = \begin{bmatrix} \cos(\theta) & 0 & 0 & \cdots & 0 & 0 & -\sin(\theta) \\ 0 & 1 & 0 & \cdots & 0 & 0 & 0 \\ & & & \vdots & & & \\ 0 & 0 & 0 & \cdots & 0 & 1 & 0 \\ \sin(\theta) & 0 & 0 & \cdots & 0 & 0 & \cos(\theta) \end{bmatrix}. \tag{10.12}$$

The iterative matrix operations $\mathbf{\Lambda}_{m+1} = \mathbf{S}\mathbf{\Lambda}_m\mathbf{S}'$ and $\mathbf{E}_m = \mathbf{S}\mathbf{E}_m$ are carried out during rotations within each iteration. When the sum of the off-diagonal elements of $\mathbf{\Lambda}_m$ is less than, say, 0.01, then terminate the program. Most runs are completed after about six to eight iterations. After m iterations, $\mathbf{\Lambda}_m$ will be a diagonal matrix with near-zero off-diagonals and diagonal elements equal to the p eigenvalues of $\mathbf{R}$, and $\mathbf{E}_m$ will be the matrix of eigenvectors of $\mathbf{R}$. One should check to ensure that $\mathbf{R} = \mathbf{E}\mathbf{\Lambda}\mathbf{E}'$.

10.2.3 Extraction of Principal Components

Now that the eigenvalues of $\mathbf{R}$ based on standardized data are known, they are sorted in descending order. Because the total variance of $\mathbf{R}$ is p and each eigenvalue contributes λ_{jj}/p to total variance, only components whose eigenvalues exceed unity ($\lambda_{jj} > 1$) are extracted from $\mathbf{\Lambda}$ and used in the quicksort. (At this point, notation for eigenvalues changes from λ_{pp} to λ_m.) Thus, m sorted eigenvalues are selected with values greater than unity so that $\lambda_1 \geq \lambda_2 \geq \cdots \geq \lambda_m$.

Since each eigenvalue represents a *component*, there will be m components selected. The correlation between expression of individual genes and the m components is represented by a matrix of *loadings* given as

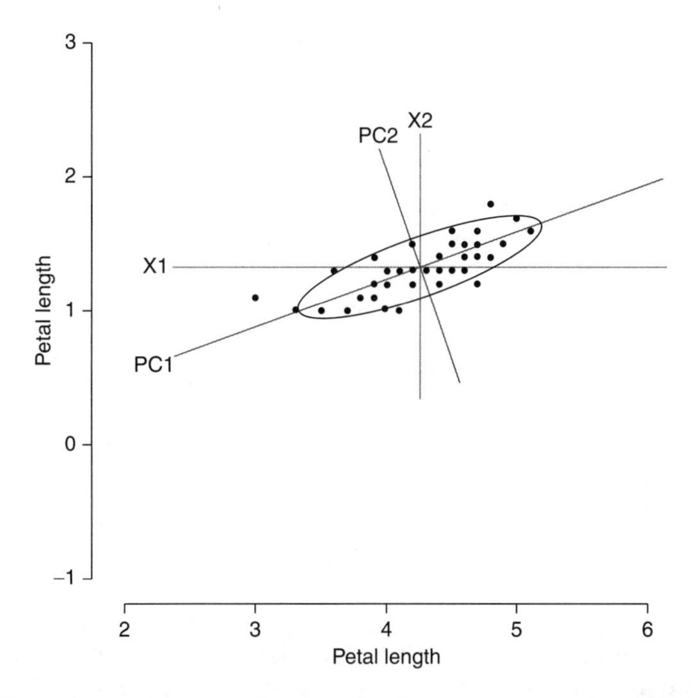

FIGURE 10.1 Geometric projection of the two principal components (eigenvectors) and eigenvalues extracted from the 2×2 correlation matrix $\mathbf{R}$ for the petal length and petal width features (Versicolor class) of the Fisher iris dataset. The length of the ellipse for PC1 is equal to $2\sqrt{\lambda_1}$, while the length of the ellipse for PC2 is $2\sqrt{\lambda_2}$.

$$\underset{p\times m}{\mathbf{L}} = \begin{bmatrix} \sqrt{\lambda_1}e_{11} & \sqrt{\lambda_2}e_{12} & \cdots & \sqrt{\lambda_m}e_{1m} \\ \sqrt{\lambda_1}e_{21} & \sqrt{\lambda_2}e_{22} & \cdots & \sqrt{\lambda_m}e_{2m} \\ \vdots & \vdots & \ddots & \vdots \\ \sqrt{\lambda_1}e_{p1} & \sqrt{\lambda_2}e_{p2} & \cdots & \sqrt{\lambda_m}e_{pm} \end{bmatrix}. \tag{10.13}$$

where rows represent genes and columns represent components, and, for example, $\sqrt{\lambda_1}e_{11}$ is the loading (correlation) between gene 1 and component 1.

It is common to construct a simple two-way scatterplot of x_1 versus x_2 and overlay the orthogonal lines representing the first and second orthogonal principal components (PC1 and PC2). Figure 10.1 illustrates the geometric relationship between two principal components, the corresponding eigenvalues and eigenvectors, for the petal length and petal width features (Versicolor class) from the Fisher iris dataset.

To begin, the value of PC-1 for each array is represented in the form

$$y = x_{i1}\, e_{11} + x_{i2}\, e_{21} \tag{10.14}$$

where x_{i1} is the expression value of gene 1 on the ith microarray, x_{i2} is the expression of gene 2 on the ith microarray, e_{11} is the eigenvector element for gene 1 and e_{21} is the eigenvector gene 2. First, set $y_i = 0$ to get

$$0 = x_{i1}e_{11} + x_{i2}e_{21}. \tag{10.15}$$

Next, subtract $x_{i2}e_{21}$ from both sides to get

$$x_{i2}e_{21} = x_{i1}e_{11}. \tag{10.16}$$

Solve for x_{i2} to get

$$x_{i2} = \frac{x_{i1}e_{11}}{e_{21}}. \tag{10.17}$$

To make the straight line for PC-1, simply loop over values of x_{i1} at fixed intervals (over the range of centered x_1 values) to get the x_2 values. Similarly, PC-2 is defined as

$$y = x_{i1}e_{12} + x_{i2}e_{2,2}, \text{ (notice column 2 is used for eigenvectors)} \tag{10.18}$$

and solving for x_{i1} gives

$$x_{i1} = \frac{x_{i2}e_{12}}{e_{22}}. \tag{10.19}$$

Next, loop over values of x_{i2} at fixed intervals over the range of the centered values. This will essentially overlay the fitted values of the PC-1 and PC-2 over the scatter plot of centered x_1 and x_2 values.

10.2.4 Varimax Orthogonal Rotation of Components

Varimax rotation [7] is performed on all m components to achieve parsimony in the loadings. The goal in Varimax rotation is to determine the rotation angle ϕ for each component pair r, s that maximizes $\sum(h_j^2 - \bar{h}_j^2)$, where h_j^2 is the *communality* for the jth variable (gene) ($j = 1, 2, ..., p$), equal to the rowwise sum of the squared loadings across all components $\sum_r L_{jr}$.

The final component matrix $\mathbf{L}^*$ is based on cross-multiplication of all $(m(m-1)/2)$ transformation matrices, given as

$$\mathbf{L}* = \mathbf{L}\mathbf{T}_{12}\mathbf{T}_{13}\cdots\mathbf{T}_{rs}\cdots\mathbf{T}_{(m-1),m}, \tag{10.20}$$

where $r = 1, 2, ..., m - 1$ and $s = r + 1, r + 2, ..., m$. The transformation matrix for component pair r, s is

$$\bar{\mathbf{T}}_{rs} = \begin{pmatrix} \cos\phi & -\sin\phi \\ \sin\phi & \cos\phi \end{pmatrix}. \tag{10.21}$$

For gene j, the rotation angle ϕ between components r and s is estimated by setting

$$\begin{aligned} x_j &= \frac{L_{jr}}{\sqrt{h_j^2}} \\ y_j &= \frac{L_{js}}{\sqrt{h_j^2}} \end{aligned} \tag{10.22}$$

and solving for ϕ in the relationship

$$\begin{aligned} \cos(4\phi) = & \left[\sum_{j=1}^{p} (x_j^2 - y_j^2) - \frac{\left(\sum_{j=1}^{p} (x_j^2 - y_j^2)\right)^2 - \left(\sum_{j=1}^{p} 2x_j y_j\right)^2}{p} \right] \\ & \times 2 \sum_{j=1}^{p} (x_j^2 - y_j^2) 2x_j y_j - \left\{ \left[\frac{2\left(\sum_{j=1}^{p} (x_j^2 - y_j^2)\right)\left(\sum_{j=1}^{p} 2x_j y_j\right)}{p} \right]^2 \right. \\ & \left. + \left[\sum_{j=1}^{p} (x_j^2 - y_j^2) - \frac{\left(\sum_{j=1}^{p} (x_j^2 - y_j^2)\right)^2 - \left(\sum_{j=1}^{p} 2x_j y_j\right)^2}{p} \right]^2 \right\}^{-1/2}, \end{aligned} \tag{10.23}$$

$$\cos(2\phi) = \left[\frac{1 + \cos(4\phi)}{2} \right]^{1/2}, \tag{10.24}$$

$$\cos\phi = \left[\frac{1 + \cos(2\phi)}{2} \right]^{1/2}, \tag{10.25}$$

$$|\sin\phi| = \left[\frac{1 - \cos(2\phi)}{2} \right]^{1/2}. \tag{10.26}$$

If the squared term in line 2 of (10.23) is < 0, then $\sin\phi = -|\sin\phi|$, whereas if it is ≥ 0, then $\sin\phi = |\sin\phi|$.

A transformation matrix is estimated for each $m(m-1)/2$ pair of components. The first step involves determination of $\mathbf{T}_{12}$ for components 1 and 2 and the new loadings for each jth gene ($j = 1, 2, ..., p$) as follows:

$$(L^*_{j1}, L^*_{j2}) = \left[x_{j1}\sqrt{h_j^2}, y_{j2}\sqrt{h_j^2}\right] \begin{pmatrix} \cos\phi_1 & -\sin\phi_1 \\ \sin\phi_1 & \cos\phi_1 \end{pmatrix}. \tag{10.27}$$

The second step sets

$$\begin{aligned} x_j &= \frac{L^*_{j1}}{\sqrt{h_j^2}} \\ y_j &= \frac{L_{j3}}{\sqrt{h_j^2}}, \end{aligned} \tag{10.28}$$

with which $\mathbf{T}_{13}$ is obtained. The new loadings for components 1 and 3 become

$$(L^*_{j1}, L^*_{j3}) = \left[x_{j1}\sqrt{h_j^2}, y_{j3}\sqrt{h_j^2}\right] \begin{pmatrix} \cos\phi_2 & -\sin\phi_2 \\ \sin\phi_2 & \cos\phi_2 \end{pmatrix}. \tag{10.29}$$

This is repeated until all $m(m-1)/2$ pairs of the p components have been rotated [8]. Figure 10.2 shows the workflow for which PCA can be employed for dimensional reduction of a dataset. The main conclusion from the workflow and formalisms above is that loadings are generated for each dimension of the input correlation matrix and PC scores are generated for each dimension of the vectors used for correlation. In short, if there are p n-dimensional vectors used for correlation, then there are p loadings and n PC scores.

10.2.5 Principal Component Score Coefficients

Calculation of the *principal component score coefficients* is required to obtain the principal component scores. PC score coefficients are not used in plotting and are seldom reported in publications, since they essentially map the genes back to the microarrays via the PCs, assuming, that eigenanalysis was performed with the *gene* × *gene* ($p \times p$) correlation matrix $\mathbf{R}$. The matrix of

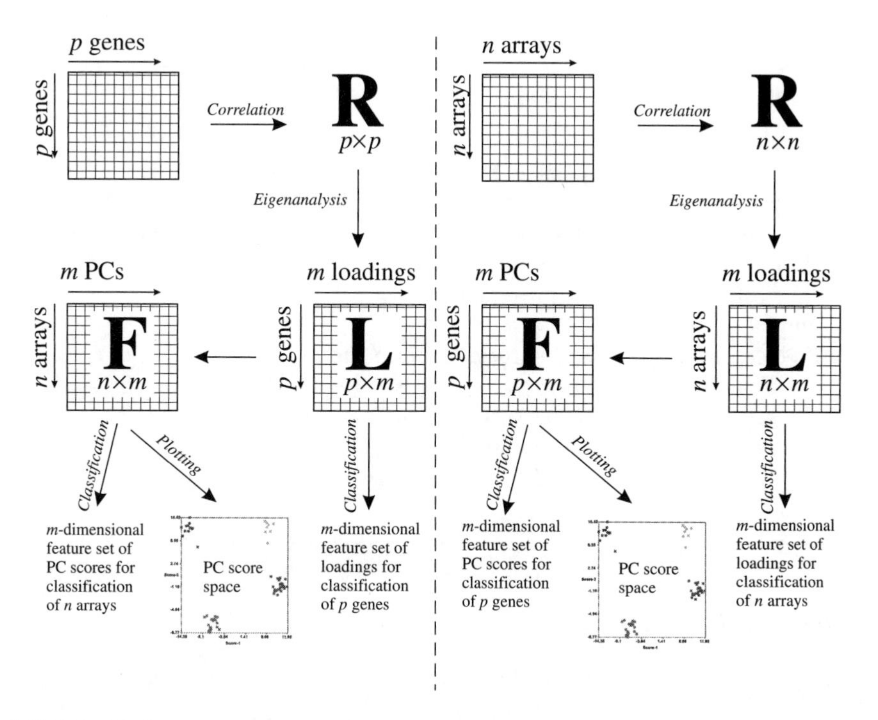

FIGURE 10.2 Workflow options for PCA when employed for dimensional reduction of gene expression datasets. **R** is the correlation matrix, **L** the loading matrix, and **F** is the principal component (PC) score matrix.

PC score coefficients is obtained using the matrix operation

$$\underset{p\times m}{\mathbf{W}} = \underset{p\times m}{\mathbf{L}} \; \underset{m\times m}{(\mathbf{L}'\mathbf{L})^{-1}}, \tag{10.30}$$

where **L** is the loading matrix reflecting the correlation between each gene expression profile and the extracted PC scores.

10.2.6 Principal Component Scores

Once the PC score coefficients are obtained, the PC scores are determined. The array-by-component ($n \times m$) **F** matrix of *PC scores* is determined with the matrix of standardized expression values (standardized with mean and SD over the genes) as follows:

$$\underset{n\times m}{\mathbf{F}} = \underset{n\times p}{\mathbf{Z}} \; \underset{p\times m}{\mathbf{W}}. \tag{10.31}$$

10.3 ALGORITHM

Algorithm 10.1: Principal Component Analysis (PCA)

```
Result: PCs, PC scores
if PC scores for n arrays desired then
    Generate p × p (gene × gene) correlation matrix, R
    Use Jacobi method to extract m × m E and Λ matrices from R
    Determine p × m loading matrix L
    Determine n × m F matrix of PC scores
endif
if PC scores for p genes (features) desired then
    Generate n × n (array × array) correlation matrix, R
    Use Jacobi method to extract m × m E and Λ matrices from R
    Determine n × m loading matrix L
    Determine p × m F matrix of PC scores
endif
```

10.4 WHEN TO USE LOADINGS AND PC SCORES

There are generally two reasons for performing PCA:

1. To reduce the dimensions of the genes or arrays. The reduced dimensions are represented by m orthogonal (uncorrelated) principal components, which can be further used for classification analysis or modeling.
2. To determine the loadings (correlation) between the original input dimensions and the principal components in order to partition the input dimensions into groups.

Principal component analysis can be performed on either genes or the arrays. The former analysis to reduce the dimensions in the number of genes is called *R-mode* analysis, whereas PCA on the arrays is called *Q-mode* analysis [9]. When the primary goal of an analysis is to identify novel groups of genes or arrays using correlation, then loadings should be used. If the goal is dimensional reduction, cluster analysis, or graphical display of results in PC score space, then the PCs should be used. Consider the loading of an array with a PC and the PC score for an array. Assume that n is the number of arrays, p is the number of genes (features), and m is the number of PCs. To obtain loadings for arrays, that is, the correlation between an array and a PC, PCA has to be run using $\mathbf{R}$ with dimensions $n \times n$. When the PC scores for each array are desired, $\mathbf{R}$ must have dimensions $p \times p$. In summary, loadings are correlation coefficients, and PC scores reflect the PCs in sample space, as shown in the following table, where n denotes the number of arrays (objects),

p denotes the number of genes (features), and m denotes the number of principal components (PCs).

Goal	Correlation based on:	R	L	F
Reflect correlation (loading) between arrays and PCs	Arrays	$n \times n$	$n \times m$	$p \times m$
Reflect correlation (loading) between genes and PCs	Genes	$p \times p$	$p \times m$	$n \times m$
Reflect geometric location (PC score) of arrays in PC-space	Genes	$p \times p$	$p \times m$	$n \times m$
Reflect geometric location (PC score) of genes in PC-space	Arrays	$n \times n$	$n \times m$	$p \times m$

10.5 IMPLEMENTATION

10.5.1 Correlation Matrix R

The computational goal of PCA is to extract eigenvectors and eigenvalues from either the covariance matrix **V** or correlation matrix **R**. When compared with covariance, correlation is safer to use for PCA because it is scale-independent, whereas covariance can be dominated by one or more features whose range and scale are much larger (smaller) than other features. The code below shows how the $p \times p$ (gene $\times$ gene) correlation matrix is determined from a `z(,)` matrix that contains mean-zero standardized values. Vectors `genemean` and `genesd` store the gene-specific means and standard deviations of the mean-zero standardized values calculated over all microarrays:

```
For j = 2 To NumGenes
  For k = 1 To j - 1
    ReDim Preserve dist(j)(k)
    dis = 0
    For i = 1 To NumObjects
      dis += (z(i, j) - genemean(j)) * (z(i, k) - genemean(k))
    Next i
    dis /= ((NumObjects - 1) * genesd(j) * genesd(k))
    dist(j)(k) = 1 - dis
  Next k
Next j
ReDim r(NumGenes, NumGenes)
For i = 2 To NumGenes
  For j = 1 To i - 1
    r(i, j) = dist(i)(j)
  Next j
Next i
```

```
For i = 1 To NumGenes
  r(i, i) = 1
Next i
For j = 1 To NumGenes
  For k = 1 To j
    r(k, j) = r(j, k)
  Next k
Next j
```

10.5.2 Eigenanalysis of Correlation Matrix R

An implementation of the Jacobi method using rotation equations given by Greenstadt [10] proceeds as follows. Let `p`,`r(,)`, `eigvec(,)`, and `eigval(,)` be the dimension, the $p \times p$ correlation matrix, eigenvector, and eigenvalue matrices representing **R**, **E**, and **Λ** described above. Pass `n` along with `r(,)` to the *eigen1* subroutine as follows:

```
Sub eigen1(n, r, eigval, eigvec, ifault)
Dim eps, cos, sine As Double
Dim y1 , z1, a1, b1, d1 As Double
Dim i, j, k, l, m, iter, q, r As Integer
Dim sm, thresh, g As Double
eps = 0.01
For i = 1 To p
 For j = 1 To p
   eigval(i, j) = r(i, j)
 Next j
Next i
For i = 1 To p
   eigvec(i, i) = 1
Next i
For iter = 1 To 50
  'sm = sum of upper triangular
  sm = 0
  For i = 1 To p - 1
    For j = i + 1 To p
      sm += Math.Abs(eigval(i, j))
    Next j
  Next i
  If sm < 0.01 Then Exit Sub
  If iter <= 3 Then thresh = 0.2 * sm / (p ^ 2)
  If iter > 3 Then thresh = 0
  For q = 1 To p - 1
    For r = q + 1 To p
      g = 100 * Math.Abs(eigval(q, r))
        If iter > 4 And Math.Abs(eigval(q,q))+g=Math.Abs(eigval(q,q)) _
        AndAlso Math.Abs(eigval(r,r))+g=Math.Abs(eigval(r, r)) Then
          eigval(q, r) = 0
          ElseIf Math.Abs(eigval(q, r)) > thresh Then
          Call SCHUR(eigval, q, r, cos, sine, ifault)
```

```
          For i = 1 To p
            If i <> q And i <> r Then
              y1 = eigval(i, q)
              z1 = eigval(i, r)
              eigval(i, q) = y1 * cos - z1 * sine
              eigval(i, r) = y1 * sine + z1 * cos
            End If
          Next i
          For k = 1 To p
            If k <> q And k <> r Then
              y1 = eigval(q, k)
              z1 = eigval(r, k)
              eigval(q, k) = y1 * cos - z1 * sine
              eigval(r, k) = y1 * sine + z1 * cos
            End If
          Next k
          For i = 1 To p
            y1 = eigvec(i, q)
            z1 = eigvec(i, r)
            eigvec(i, q) = y1 * cos - z1 * sine
            eigvec(i, r) = y1 * sine + z1 * cos
          Next i
          a1 = eigval(q, q)
          b1 = eigval(q, r)
          d1 = eigval(r, r)
          eigval(q, q) = (a1 * cos^2) + (d1 * sine ^2)-_
            (2 * b1 * sine * cos)
          eigval(r, r) = (a1 * sine^2) + (d1 * cos ^2)+_
            (2 * b1 * sine * cos)
          eigval(q, r) = (a1 - d1) * sine * cos + _
           (b1 * (cos^2 - sine^2))
          eigval(r, q) = (a1 - d1) * sine * cos + _
           (b1 * (cos^2 - sine^2))
        End If
    Next r
  Next q
Next iter
End Sub

Sub SCHUR(eigval, q, r, cos, sine, ifault)
Dim mu, lambda As Double
lambda = -eigval(q, r)
mu = 0.5 * (eigval(q, q) - eigval(r, r))
If mu = 0 Then omega = 1
If mu <> 0 Then omega = Math.Sgn(mu) * lambda / _
  Math.Sqrt(lambda^2 + mu^ 2)
sine = omega / Math.Sqrt(2 * (1 + Sqr(1 - omega ^ 2)))
cos = Math.Sqrt(1 - sine ^ 2)
End sub
```

On return from *eigen1*, one receives the `eigvec(,)` and `eigval(,)` arrays for **E** and Λ.

10.5.3 Determination of Loadings and Varimax Rotation

Varimax orthogonal rotation should always be employed to ensure that the vectors used for correlation load mostly on one PC. Without use of Varimax, loadings can be dispersed over several PCs, which is not an optimal result that is sought from PCA. The source code below lists the Varimax procedures, with the `loading(,)` array used for the rotated loadings, and the `comm(,)` array as an intermediate work array containing communalities:

```
'Extract eigenvectors, eigenvalues from R(ncol,ncol)
Call eigen1(ncol, s, ipca)
ReDim arr(ncol)
ReDim ndx(ncol)
descending = True
For j = 1 To ncol
  arr(j) = eigval(j, j)
  ndx(j) = j
Next j
Call sort(arr, ndx, ncol, descending)
NumComps = 0
sumvar = 0
totcomponents = ncol
For i = 1 To ncol
  sumvar = sumvar + arr(i)
  If arr(i) > 1 Then NumComps = NumComps + 1
Next i
ReDim loading(ncol, NumComps)
For i = 1 To ncol
  For j = 1 To NumComps
    loading(i, j) = Math.Sqrt(arr(j)) * eigvec(i, ndx(j))
  Next j
Next i
'Perform Varimax rotation on loadings
Dim comm() As Double
Dim x(,) As Double
Dim a, B, c, D1, e, F, g, v1, U As Double
Dim cos4phi, cos2phi, cosphi, abssinphi, sinphi As Double
Dim h, r, s1 As Integer
Dim vnew, vold,t(2,2) As Double
ReDim comm(ncol)
ReDim x(ncol, NumComps)
For iter = 1 To ncol
  For i = 1 To ncol
    For j = 1 To NumComps
      comm(i) = comm(i) + loading(i, j) * loading(i, j)
    Next j
  Next i
  For r = 1 To NumComps - 1
```

```
    For s1 = r + 1 To NumComps
      a = 0
      B = 0
      c = 0
      D1 = 0
      fac1 = 1 / ncol
      For h = 1 To ncol
        If comm(h) > 0 Then x(h, r) = loading(h, r) /
           Math.Sqrt(comm(h))
        If comm(h) > 0 Then x(h, s1) = loading(h, s1) /
           Math.Sqrt(comm(h))
        v1 = 2 * x(h, r) * x(h, s1)
        U = (x(h, r) * x(h, r)) - (x(h, s1) * x(h, s1))
        a += U
        B += v1
        c += ((U * U) - (v1 * v1))
        D1 += (U * v1) + (U * v1)
      Next h
      e = D1 - 2 * a * B * fac1
      F = c - (a * a - B * B) * fac1
      g = Math.Sqrt(e * e + F * F)
      If g <> 0 Then cos4phi = F / g
      cos2phi = Math.Sqrt((1 + cos4phi) * 0.5)
      cosphi = Math.Sqrt((1 + cos2phi) * 0.5)
      abssinphi = Math.Sqrt((1 - cos2phi) * 0.5)
      If e < 0 Then sinphi = abssinphi * -1
      If e >= 0 Then sinphi = abssinphi
      t(1, 1) = cosphi
      t(1, 2) = sinphi * -1
      t(2, 1) = sinphi
      t(2, 2) = cosphi
      For h = 1 To ncol
        loading(h, r) = (x(h, r) * t(1, 1)) + (x(h, s1) *
           t(2, 1))
        loading(h, s1) = (x(h, r) * t(1, 2)) + (x(h, s1) *
           t(2, 2))
        loading(h, r) = loading(h, r) * Math.Sqrt(comm(h))
        loading(h, s1) = loading(h, s1) * Math.Sqrt(comm(h))
      Next h
    Next s1
  Next r
  e = 0
  For r = 1 To NumComps
    For h = 1 To ncol
      diff = (loading(h, r) / Math.Sqrt(comm(h)))
      If comm(h) > 0 Then e += diff * diff * diff * diff
    Next h
  Next r
```

```
  g = 0
  For r = 1 To NumComps
    F = 0
    For h = 1 To ncol
      If comm(h) <> 0 Then F += (loading(h, r) *
         loading(h, r)) / comm(h)
    Next h
    g += F * F
  Next r
  vnew = ncol * e - g
  If iteration > 1 And Math.Round(vnew, 4) = Math.Round(vold, 4)
     Then Exit For
  vold = vnew
Next iter
```

10.5.4 Calculating Principal Component (PC) Scores

Loadings rotated with Varimax rotation are used for obtaining the PC scores via cross-multiplication of matrices. Several primitives are called for matrix transpose and multiplication, which is followed by the final cross-multiplication to derive the **F** matrix of PC scores. Programatically, the array `PC(,)` is used to represent **F**:

```
Dim LT(NumComps, ncol) As Double
Dim LTL(NumComps, NumComps) As Double
Dim w(ncol, NumComps) As Double
Dim PC(nrow, NumComps) As Double

Call Primitives.TRNPOS(loading, LT, ncol, NumComps)
Call Primitives.MATMUL(LT, loading, LTL, NumComps, ncol, NumComps)
Call eigen1(NumComps, LTL, 0), 'Provides UTU, the inverse of LTL
Call Primitives.MATMUL(loading, UTU, w, ncol, NumComps, NumComps)
If ClusterChoice = 1 Then 'PCA performed on microarrays (objects)
    For j = 1 To ncol
      For i = 1 To nrow
        If zmatrix(i, j) <> MissDataCode And arraysd(j) <> 0 Then_
          zmatrix(i, j) = (zmatrix(i, j) - arraymean(j)) / arraysd(j)
      Next i
    Next j
End If

If ClusterChoice = 2 Then 'PCA performed on genes (features)
    For j = 1 To ncol
      For i = 1 To nrow
        If zmatrix(i, j) <> MissDataCode And genesd(j) <> 0 Then _
          zmatrix(i, j) = (zmatrix(i, j) - genemean(j)) / genesd(j)
      Next i
    Next j
End If

Call Primitives.MATMUL(zmatrix, w, PC, nrow, ncol, NumComps)
```

Example 10.1 ***PCA Runs on Genes.*** The BRCA1/BRCA2 dataset with $n = 15$ arrays (objects) and $p = 9$ genes (features) was used to generate a 9×9 (gene-by-gene) correlation matrix **R**, which was used for eigenanalysis to generate loadings and PC scores. Table 10.1 lists the $p \times p$ **R** matrix, with pairwise correlation coefficients for all genes. Table 10.2 lists the $p \times m$ **E** matrix of eigenvectors for the top two eigenvalues that exceeded unity (i.e., $\lambda > 1$), listed in Table 10.3. Tables 10.4 and 10.5 list the loadings showing correlation between the two extracted PCs and each $n \times 1$ gene expression vector, with and without use of Varimax rotation, respectively. Table 10.6 shows the two PCs with the sample-specific PC scores.

The effect of Varimax rotation can be observed by comparing loadings in Table 10.4 with loadings in Table 10.5. For a given table row (gene), when Varimax is not used, loading values tend to be distributed across the PCs, whereas when Varimax is applied the loading tends to be greater (less) for a single PC. This is the preferred result for PCA because the intended purpose of PCA is to decorrelate the extracted PCs to render them orthogonal. Table 10.6 lists the PC scores for the 15 arrays based on the first and second major components, which could be plotted to reveal PCA-based array clustering.

Example 10.2 ***PCA Runs on Arrays.*** Another PCA run was performed using the BRCA1/BRCA2 dataset, but this time the 15×15 (microarray-by-microarray) correlation matrix **R** was used for eigenanalysis. Table 10.7 lists the $n \times n$ **R** matrix, with pairwise correlation coefficients for all microarrays. Table 10.8 lists the $n \times m$ **E** matrix of eigenvectors for the top two eigenvalues that exceeded unity (i.e., $\lambda > 1$), listed in Table 10.9. Tables 10.10 and 10.11 list the loadings showing correlation between the two extracted PCs and each $p \times 1$ gene expression vector, with and without use of Varimax rotation, respectively. Table 10.6 shows the two PCs with the gene-specific PC scores.

Again, one can note that the loadings after Varimax rotation are greater (less) for mostly one PC (Table 10.11) when compared with loadings that were not rotated (Table 10.10). Analogously, Table 10.12 lists the PC scores for the 9 genes based on the first and second major components, which when plotted could reveal PCA-based gene clustering.

Example 10.3 The array-by-array (63×63) correlation matrix for the four-class SRBCT dataset was used in PCA in order to determine loadings of the arrays on the extracted PCs. Figure 10.3 shows the microarray-specific loadings for the top 4 principal components extracted from the 63×63 (i.e., $n \times n$) correlation matrix for the SRBCT dataset, when Varimax rotation was not employed. Figure 10.4 illustrates the microarray-specific loadings for the top four principal components extracted from the 63×63 correlation matrix for the SRBCT dataset when Varimax rotation was used. It is apparent that when Varimax rotation is not used, the loading values are similar for multiple PCs, whereas when Varimax rotation is used, loading is greatest (least) for

TABLE 10.1 Correlation Matrix R for All Pairwise Gene-by-Gene ($p \times p$) Correlations Based on the Two-Class Breast Data for BRCA1 and BRCA2

	Interleukin	TP53BP1	ARP1	ESTs_293977	Catenin	KIAA0191	Cathepsin K	cAMPBP	MAD
Interleukin	1.000	0.896	−0.769	0.631	−0.516	0.684	−0.600	0.655	0.694
TP53BP1	0.896	1.000	−0.643	0.676	−0.555	0.740	−0.539	0.586	0.746
ARP1	−0.769	−0.643	1.000	−0.532	0.692	−0.530	0.580	−0.679	−0.614
ESTs_293977	0.631	0.676	−0.532	1.000	−0.484	0.527	−0.468	0.421	0.509
Catenin	−0.516	−0.555	0.692	−0.484	1.000	−0.532	0.610	−0.272	−0.606
KIAA0191	0.684	0.740	−0.530	0.527	−0.532	1.000	−0.549	0.760	0.466
Cathepsin K	−0.600	−0.539	0.580	−0.468	0.610	−0.549	1.000	−0.442	−0.142
cAMPBP	0.655	0.586	−0.679	0.421	−0.272	0.760	−0.442	1.000	0.254
MAD	0.694	0.746	−0.614	0.509	−0.606	0.466	−0.142	0.254	1.000

TABLE 10.2 Eigenvectors Extracted from Correlation Matrix in Table 10.1 for Which Eigenvalues Exceed Unity ($\lambda > 1$)

Gene	$\mathbf{e}_1$	$\mathbf{e}_2$
Interleukin	0.386	0.019
TP53BP1	0.382	0.119
ARP1	−0.359	−0.009
ESTs_293977	0.311	0.118
Catenin	−0.309	−0.265
KIAA0191	0.344	−0.283
Cathepsin K	−0.290	0.367
cAMPBP	0.302	−0.535
MAD	0.301	0.633

TABLE 10.3 Eigenvalues Extracted from Correlation Matrix in Table 10.1 Whose Values Exceed Unity ($\lambda > 1$)

	λ_1	λ_2
Eigenvalues (>1)	5.636	1.013
Variance explained	0.626	0.113
Cumulative variance explained (%)	0.626	0.739

TABLE 10.4 Loadings between Genes and the Two PCs Extracted from Correlation Matrix in Table 10.1 without Varimax Orthogonal Rotation

Gene	$\mathbf{l}_1$	$\mathbf{l}_2$
Interleukin	0.917	0.020
TP53BP1	0.907	0.119
ARP1	−0.851	−0.009
ESTs_293977	0.737	0.119
Catenin	−0.734	−0.267
KIAA0191	0.816	−0.285
Cathepsin K	−0.689	0.369
cAMPBP	0.718	−0.538
MAD	0.715	0.637

a single PC. The latter (when Varimax is used) is more advantageous for grouping objects based on the PCs they load on the most (least).

Example 10.4 The gene-by-gene correlation matrix for the four-class SRBCT dataset ($p \times p = 23 \times 23$) and 3-class MLL_Leukemia dataset ($p \times p = 20 \times 20$)

were used in PCA to determine array-specific PC scores. Figure 10.5 shows PC score plots for the two PCs having the greatest eigenvalues extracted from the gene-by-gene correlation matrix of the SRBCT ($n = 63$ arrays) and MLL_Leukemia ($n = 57$ arrays) datasets. The upper panels illustrate the PC scores for the SRBCT arrays determined without (a) and with (b) use of Varimax rotation, while the lower panels show PC scores for the MLL_Leukemia arrays derived without (a) and with (b) Varimax rotation. PC scores calculated after Varimax rotation of all PCs in each run degraded separability of arrays in the PC score plots.

TABLE 10.5 Loadings between Genes and the Two PCs Extracted from Correlation Matrix in Table 10.1 with Varimax Orthogonal Rotation

Gene	$\mathbf{l}_1$	$\mathbf{l}_2$
Interleukin	0.674	−0.622
TP53BP1	0.736	−0.543
ARP1	−0.619	0.584
ESTs_293977	0.614	−0.426
Catenin	−0.714	0.317
KIAA0191	0.390	−0.772
Cathepsin K	−0.240	0.743
cAMPBP	0.144	−0.885
MAD	0.957	−0.037

TABLE 10.6 PCs (Columns) Extracted from Correlation Matrix in Table 10.1[a]

Microarray	PC1	PC2
BRCA1_s1996_1	1.386	−0.146
BRCA1_s1822_5	0.794	−0.757
BRCA1_s1714_3	0.982	−0.308
BRCA1_s1224_7	1.547	0.010
BRCA1_s1252_2	−0.576	−2.277
BRCA1_s1510_4	0.540	−0.890
BRCA1_s1905_6	0.494	−0.474
BRCA2_s1900_10	−0.463	0.158
BRCA2_s1787_9	0.183	1.315
BRCA2_s1721_8	−0.591	0.461
BRCA2_s1486_10	−1.920	−0.689
BRCA2_s1816_13	−0.234	1.476
BRCA2_s1616_14	−1.548	−0.064
BRCA2_s1063_11	−0.692	0.873
BRCA2_s1936_12	0.097	1.313

[a]Each table cell value is a principal component (PC) score. Loadings used for PC scores were based on Varimax rotation.

TABLE 10.7 Correlation Matrix R for All Pairwise Microarray-by-Microarray ($n \times n$) Correlations Based on the Two-Class Breast Data for BRCA1 and BRCA2

	B1_ s1996_1	B1_ s1822_5	B1_ s1714_3	B1_ s1224_7	B1_ s1252_2	B1_ s1510_4	B1_ s1905_6	B2_ s1900_10	B2_ s1787_9	B2_ s1721_8	B2_ s1486_10	B2_ s1816_13	B2_ s1616_14
B1_s1996_1	1.000	0.895	0.855	0.846	0.687	0.919	0.864	0.306	0.064	0.199	0.153	0.007	0.062
B1_s1822_5	0.895	1.000	0.872	0.737	0.679	0.814	0.870	0.220	0.006	0.154	0.110	−0.030	0.068
B1_s1714_3	0.855	0.872	1.000	0.941	0.851	0.894	0.714	0.620	0.413	0.576	0.495	0.373	0.448
B1_s1224_7	0.846	0.737	0.941	1.000	0.822	0.915	0.647	0.669	0.484	0.575	0.526	0.430	0.468
B1_s1252_2	0.687	0.679	0.851	0.822	1.000	0.868	0.402	0.782	0.630	0.684	0.740	0.598	0.646
B1_s1510_4	0.919	0.814	0.894	0.915	0.868	1.000	0.694	0.600	0.400	0.453	0.494	0.347	0.398
B1_s1905_6	0.864	0.870	0.714	0.647	0.402	0.694	1.000	−0.037	−0.288	−0.063	−0.141	−0.333	−0.171
B2_s1900_10	0.306	0.220	0.620	0.669	0.782	0.600	−0.037	1.000	0.954	0.950	0.966	0.939	0.922
B2_s1787_9	0.064	0.006	0.413	0.484	0.630	0.400	−0.288	0.954	1.000	0.888	0.945	0.997	0.935
B2_s1721_8	0.199	0.154	0.576	0.575	0.684	0.453	−0.063	0.950	0.888	1.000	0.932	0.886	0.916
B2_s1486_10	0.153	0.110	0.495	0.526	0.740	0.494	−0.141	0.966	0.945	0.932	1.000	0.942	0.968
B2_s1816_13	0.007	−0.030	0.373	0.430	0.598	0.347	−0.333	0.939	0.997	0.886	0.942	1.000	0.941
B2_s1616_14	0.062	0.068	0.448	0.468	0.646	0.398	−0.171	0.922	0.935	0.916	0.968	0.941	1.000
B2_s1063_11	0.066	0.014	0.411	0.474	0.623	0.413	−0.267	0.952	0.993	0.882	0.952	0.992	0.938
B2_s1936_12	0.047	−0.002	0.412	0.478	0.619	0.381	−0.293	0.953	0.999	0.899	0.945	0.998	0.938

[a]Columns for the last two arrays are not shown because of space limitations.

TABLE 10.8 Eigenvectors Extracted from Correlation Matrix in Table 10.7 for Which Eigenvalues Exceed Unity ($\lambda > 1$)

Microarray	$\mathbf{e}_1$	$\mathbf{e}_2$
B1_s1996_1	0.149	−0.398
B1_s1822_5	0.132	−0.399
B1_s1714_3	0.245	−0.288
B1_s1224_7	0.253	−0.256
B1_s1252_2	0.283	−0.159
B1_s1510_4	0.236	−0.294
B1_s1905_6	0.047	−0.442
B2_s1900_10	0.317	0.085
B2_s1787_9	0.292	0.194
B2_s1721_8	0.296	0.111
B2_s1486_10	0.302	0.142
B2_s1816_13	0.285	0.215
B2_s1616_14	0.289	0.169
B2_s1063_11	0.291	0.191
B2_s1936_12	0.291	0.198

TABLE 10.9 Eigenvalues, Variance, and Cumulative Variance Explained by the Two Major PCs

	λ_1	λ_2
Eigenvalues (>1)	9.498	4.679
Variance explained	0.633	0.312
Cumulative variance explained (%)	0.633	0.945

10.6 RULES OF THUMB FOR PCA

The basic rules of thumb for PCA analysis are as follows:

- The most common application of PCA is to identify uncorrelated dimensions identified from the correlation matrix of a set of variables, and then observe which variables have loadings that are, for example, greater than 0.45 or less than −0.45. Variables that load on a particular principal component are likely to be correlated and therefore may share the same characteristics or same generative source. Principal components are not correlated, so variables that load on different principal components are greatly different in terms of variance explanation of the dataset.

TABLE 10.10 Loadings between Microarrays and the Two PCs Extracted from Correlation Matrix in Table 10.7 without Varimax Orthogonal Rotation

Microarray	l_1	l_2
B1_s1996_1	0.459	−0.861
B1_s1822_5	0.407	−0.863
B1_s1714_3	0.754	−0.623
B1_s1224_7	0.780	−0.553
B1_s1252_2	0.873	−0.343
B1_s1510_4	0.728	−0.636
B1_s1905_6	0.144	−0.955
B2_s1900_10	0.978	0.183
B2_s1787_9	0.899	0.419
B2_s1721_8	0.911	0.240
B2_s1486_10	0.932	0.306
B2_s1816_13	0.877	0.466
B2_s1616_14	0.891	0.365
B2_s1063_11	0.898	0.412
B2_s1936_12	0.895	0.429

TABLE 10.11 Loadings between Microarrays and the Two PCs Extracted from Correlation Matrix in Table 10.7 with Varimax Orthogonal Rotation

Microarray	l_1	l_2
B1_s1996_1	−0.006	−0.976
B1_s1822_5	−0.052	−0.953
B1_s1714_3	0.366	−0.907
B1_s1224_7	0.422	−0.858
B1_s1252_2	0.605	−0.718
B1_s1510_4	0.338	−0.906
B1_s1905_6	−0.328	−0.909
B2_s1900_10	0.947	−0.304
B2_s1787_9	0.990	−0.059
B2_s1721_8	0.916	−0.222
B2_s1486_10	0.965	−0.174
B2_s1816_13	0.993	−0.008
B2_s1616_14	0.957	−0.103
B2_s1063_11	0.986	−0.065
B2_s1936_12	0.992	−0.049

TABLE 10.12 PCs (Columns) Extracted from Correlation Matrix in Table 10.7[a]

Gene	PC_1	PC_2
Interleukin	−0.437	−1.531
TP53BP1	−0.253	−1.320
ARP1	0.051	0.797
ESTs_293977	−0.401	−0.088
Catenin	−0.476	1.673
KIAA0191	−0.458	0.455
Cathepsin K	2.622	0.020
cAMPBP	−0.146	−0.367
MAD	−0.502	0.361

[a]Each table cell value is a PC score. Loadings used for PC scores were based on Varimax rotation.

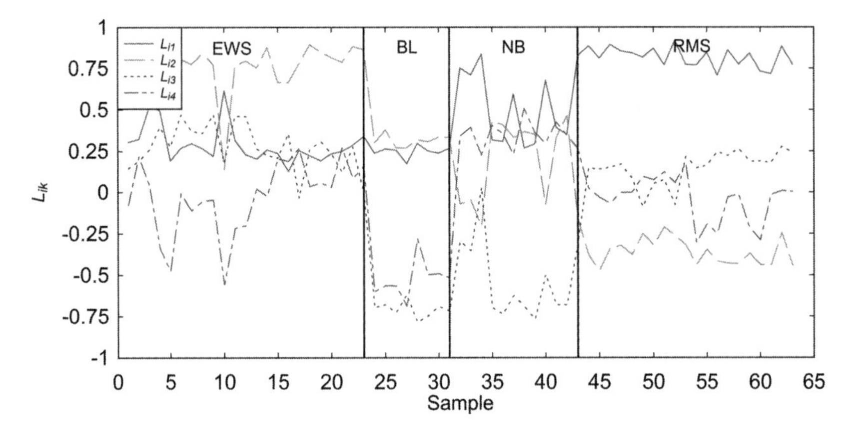

FIGURE 10.3 Microarray-specific loadings for the top four PCs extracted from the 63 × 63 (i.e., $n \times n$) correlation matrix for the SRBCT dataset. Varimax rotation not used.

- Assume the number of arrays to be n; the number of features for each microarray, p; and m, the number of principal components extracted from the correlation matrix **R**.
- It is common to extract m principal components from **R** whose eigenvalues are greater than unity (i.e., $\lambda > 1$), in order to explain a sufficient amount of the total variance. Otherwise, it is customary to use either the first or second principal component ($m = 1$, $m = 2$). When this approach is used, PC scores are calculated and used to replace the original dimensions that formed the correlation matrix.

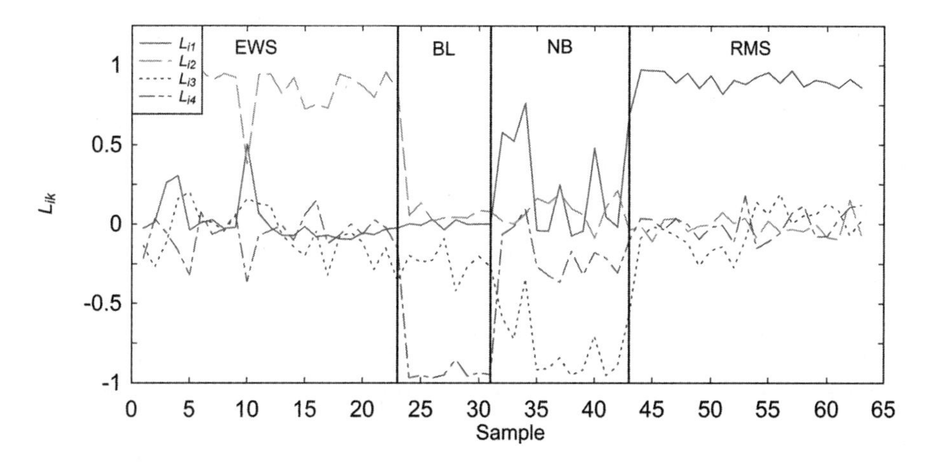

FIGURE 10.4 Microarray-specific loadings for the top four PC extracted from the 63 × 63 (i.e., $n \times n$) correlation matrix for the SRBCT dataset. Varimax rotation used.

- When a large number of genes are to be reduced to several vectors, **R** must have dimensions $p \times p$, and the resulting columns of the $n \times m$ PC score matrix **F** should be used as the new dimensions in further analysis. Conversely, if a large number of arrays are to be reduced to several vectors, **R** must have dimensions $n \times n$, and the resulting columns of the $p \times m$ PC score matrix **F** should be used as the new dimensions in further analysis.
- Correlation matrices are commonly used for input to PCA, although covariance matrices can be used as well. While correlation matrices are scale-independent, covariance matrices retain the original scale and range among variables, potentially causing detrimental problems if one or more features with a large scale and/or range strongly influences model results.
- The number of loadings is equal to the dimensions of **R**.
- The number of PC scores is equal to the length of the vectors used in **R**.
- The majority of x-y scatterplots of PC scores for microarrays reported in the literature are based on PCA runs using a gene-by-gene (feature-by-feature) correlation matrix.
- When the gene-by-gene ($p \times p$) **R** matrix is used for input to PCA, that is, when the $n \times 1$ vectors are used for correlation, there are p loadings. In this case, there will be n PC scores, which represent the microarrays.
- When the array-by-array ($n \times n$) **R** matrix is used for input (i.e., when the $p \times 1$ vectors are used for correlation), there are n loadings. Here, there will be p PC scores, which represent the genes.

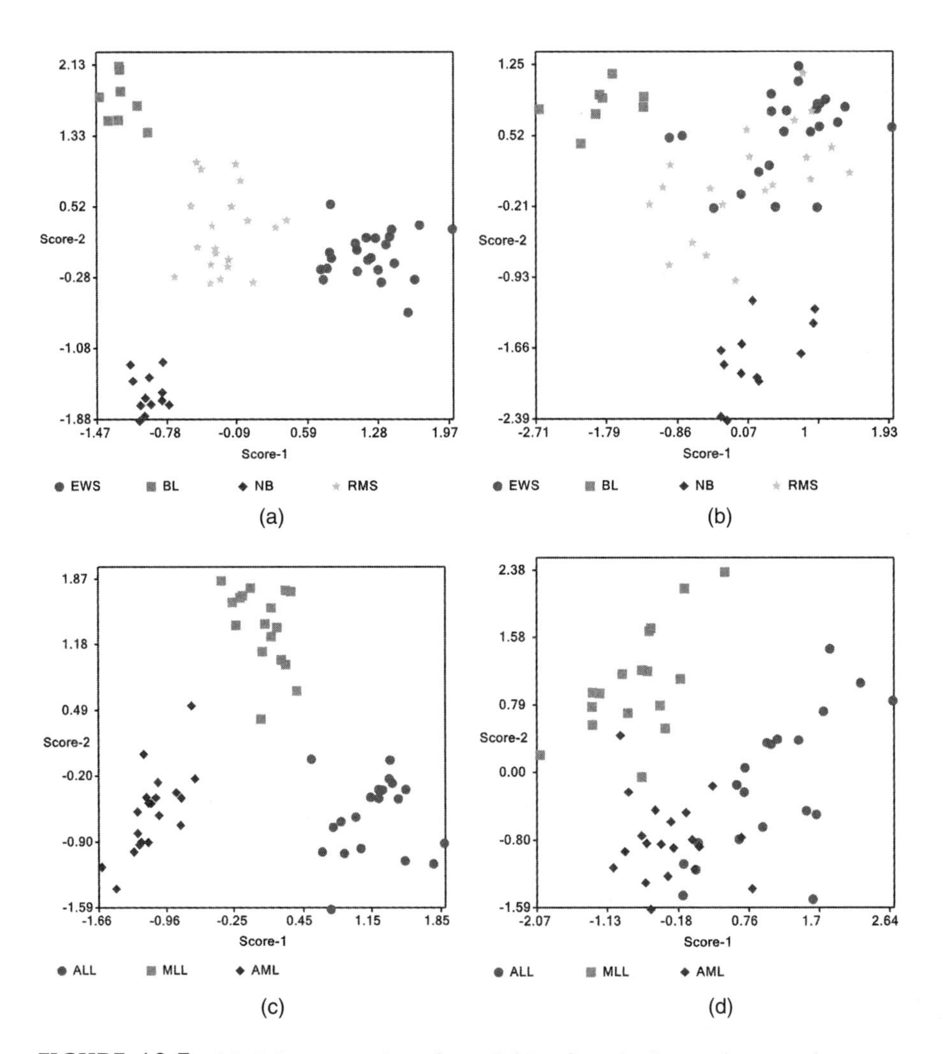

FIGURE 10.5 2D PC, score plots from PCA of each dataset's gene-by-gene correlation matrix. Upper panels contain PC scores for the four-class SRBCT arrays ($n = 63$) determined without (a) and with (b) use of Varimax rotation. Lower panels show PC scores for the three-class MLL_Leukemia arrays ($n = 57$) derived without (c) and with (d) Varimax rotation. (*See insert for color representation of the figure.*)

10.7 SUMMARY

Principal component analysis does not provide the *absolute* theory about groups of genes based on selection due to component loadings, but rather provides a *heuristic* for assembling genes based on similar correlation patterns within the components. In the case of bipolar components with positive and negative loadings, genes (arrays) with strong negative loadings (assumed <-0.45) contribute the same amount of information as do genes (arrays) that strongly load positively (>0.45), since their absolute values exceed 0.45. As the signs of their correlations (loadings) are different, however, they can be separated into two groups of data–one group whose standardized expression profiles are positively correlated with the component and the others that are negatively correlated with the component. Crossloading on multiple components is also not dealt with in any particular fashion, and such genes (arrays) will appear in more than one group.

Principal component analysis of expression data from microarrays can be used to construct groups of a limited number of genes (arrays) whose expression profiles have strong positive or negative correlations with components. Unlike PCA, cluster analysis provides results for all genes regardless of pairwise correlation between expression profiles. Self-organizing maps can provide results similar to those for PCA. However, since the nodes used in self-organizing maps start at random locations in the feature space, and training is a form of nonlinear neural adaptive learning, results can be drastically different.

REFERENCES

[1] J.C. Gower. Multivariate analysis and multidimensional geometry. *Statistician* **17**:13–28, 1967.

[2] E.J. Williams. Comparing means of correlated variates. *Biometrika* **57**:459–461, 1970.

[3] J.C. Gower. Some distance properties of latent root and vector methods used in multivariate analysis. *Biometrika* **53**:325–338, 1966.

[4] I.T. Jolliffe. Principal components analysis and exploratory factor analysis. *Stat. Methods Med. Res.* **1**:69–95, 1992.

[5] B.J.T. Morgan, A.P.J. Ray. Non-uniqueness and inversions in cluster analysis. *Appl. Stat.* **44**:117–134, 1995.

[6] Z. Meglicki. Lecture notes—Advanced Scientific Computing—B673 Univ. Indiana, Bloomington, 2001 (available at `http://beige.ucs.indiana.edu/B673/B673.html`).

[7] H.F. Kaiser. The varimax criterion for analytic rotation in factor analysis. *Psychometrika* **23**:187–200, 1958.

[8] H.H. Harman. Objective orthogonal multiple-factor solutions. In *Modern Factor Analysis*, Univ. Chicago Press, 1967, chap. 13.

[9] R.B. Cattell. The data box: Its ordering of total resources in terms of possible relational systems. In R.B. Cattell, ed., *Handbook of Multivariate Experimental Psychology*, Rand-McNally, Chicago, 1966, pp. 67–128.

[10] J. Greenstadt. The determination of the characteristic roots of a matrix by the Jacobi method. In A. Ralston and H. S. Wilf, eds., *Mathematical Methods for Digital Computers*, Wiley, New York, 1967, vol. I, chap 7.

CHAPTER 11

NONLINEAR MANIFOLD LEARNING

11.1 INTRODUCTION

High-dimensional datasets commonly suffer from the small-sample problem ($n \ll p$). The two primary reasons for reducing dimensions for gene expression data are to reduce (1) storage requirements and (2) redundancy in the form of multicollinearity. Reducing multicollinearity does not guarantee improved classifier performance in terms of accuracy; however, there is no compelling reason not to remove multicollinearity when it exists in a dataset, especially when parsimony is desired. Dimension reduction can be performed using linear or nonlinear manifold learning methods. Manifold learning attempts to retrieve the original structure of a high-dimensional dataset and can identify a surprisingly small number of dimensions to describe characteristics of the original data structure. A *manifold* is defined as the image of a low-dimensional space that is linearly or nonlinearly embedded in a high-dimensional domain. One of the most common linear manifold learning methods is principal component analysis (PCA), which maps a high-dimensional set of features to a low-dimensional space using the minimal number of orthogonal dimensions required to describe the majority of the variance in the high-dimensional space. Nonlinear manifold learning (NLML) can tackle problems when the manifold is not linearly embedded in the original domain. Data integration of high-dimensional molecular and genomic datasets can be accelerated by using NLML, since patterns in the original data are compacted and preserved so that they can be utilized for association mapping between datasets.

More recent implementations of NLML include kernel PCA [1], diffusion maps [2], Laplacian eigenmaps [3], local linear embedding [4], and locality

Classification Analysis of DNA Microarrays, First Edition. Leif E. Peterson.

preserving projections [5]. Earlier introduced methods include Kohonen self-organizing maps [6], unsupervised neural gas [7], and Sammon's projection [8]. The basic assumption of these algorithms is that the high-dimensional data lie on a low-dimensional manifold embedded in a high-dimensional structure.

11.2 CORRELATION-BASED PCA

Let $\mathbf{X}$ be an $n \times p$ data matrix with row vectors $\mathbf{x}_i$ of dimension $1 \times p$. In addition, we assume that data points in a manifold $\mathbf{Y}$ with dimensions $n \times p*$ ($p* << p$) are embedded and lie near data points in $\mathbf{X}$. Dimensionality reduction maps $\mathbf{X} \in \mathbb{R}^p \mapsto \mathbf{Y} \in \mathbb{R}^{p*}$, which has the same number of rows (objects) but much fewer columns, namely, two. For a general idea of manifold embedding, Figure 11.1 shows a 2D manifold (plane) embedded in 3D space.

Correlation PCA (CPCA) is the traditional form of PCA. For each microarray (object) we seek a two-dimensional (2D) vector $\mathbf{y}_i \in \mathbb{R}^2$ that is mapped from the higher-dimensional vector $\mathbf{x}_i \in \mathbb{R}^p$, where p is the number of input genes. Let $\mathbf{R}$ be the gene-by-gene ($p \times p$) correlation matrix. According to the principal axis theorem, there exists a rotation matrix $\mathbf{E}$ and diagonal matrix $\mathbf{\Lambda}$ such that $\mathbf{ERE}' = \mathbf{\Lambda}$. Columns of $\mathbf{E}$ and $\mathbf{E}'$ are the eigenvectors and diagonal entries of $\mathbf{\Lambda}$ are the eigenvalues $\lambda_1, \lambda_2, \ldots, \lambda_p$. For CPCA, the two principal eigenvalues and their accompanying eigenvectors are of interest, so $m = 2$, and we use only the corresponding two columns of $\mathbf{E}$ associated with the two

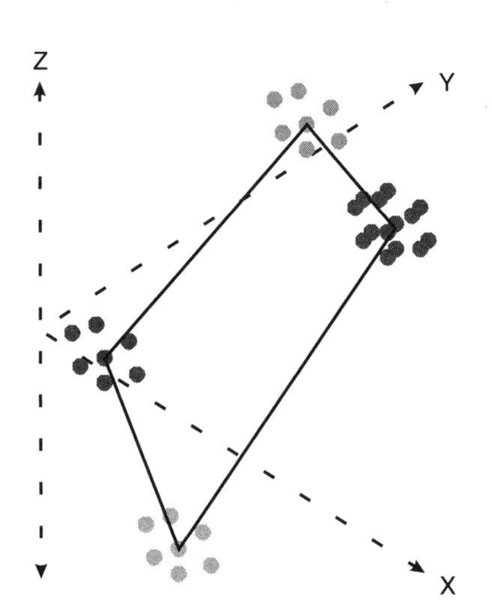

FIGURE 11.1 2D manifold (plane) embedded in a 3D space. (*See insert for color representation of the figure.*)

principal eigenvalues. The matrix of PC score coefficients is obtained using the matrix operation

$$\underset{p\times m}{\mathbf{W}} = \underset{p\times m}{\mathbf{L}} \underset{m\times m}{(\mathbf{L}'\mathbf{L})^{-1}}, \tag{11.1}$$

where $\mathbf{L}$ is the loading matrix reflecting the correlation between each gene expression profile and the extracted PC scores. The array-by-component $(n \times m)$ $\mathbf{Y}$ matrix of *PC scores* is determined with the matrix of standardized expression values (standardized with mean and standard deviation (s.d) over the genes) as follows

$$\underset{n\times m}{\mathbf{Y}} = \underset{n\times p}{\mathbf{Z}} \underset{p\times m}{\mathbf{W}}, \tag{11.2}$$

where $\mathbf{Z}$ is the mean-zero standardized data matrix. Since m is preset to 2, each row of $\mathbf{Y}$ represents a 2D vector $\mathbf{y}_i = (y_{i1}, y_{i2})$ in the low-dimensional feature space.

11.3 KERNEL PCA

Kernel-based PCA [1] using a distance-based kernel (KDPCA), Gaussian kernel (KGPCA), and Tanimoto distance-based kernel (KTPCA) are nonlinear methods that map the kernel $K(\mathbf{x}, \mathbf{x})$ onto the low-dimensional feature space. KDPCA uses the Euclidean distance between each pair of arrays as the kernel in the form $K(\mathbf{x}_i, \mathbf{x}_j) = \|\mathbf{x}_i - \mathbf{x}_j\|^2$, KGPCA employs the radial basis kernel $K(\mathbf{x}_i, \mathbf{x}_j) = \exp(-\|\mathbf{x}_i - \mathbf{x}_j\|^2/2\sigma^2)$, and KTPCA uses the continuous variant of the Tanimoto distance in the form

$$K(\mathbf{x}_i, \mathbf{x}_j) = \frac{\mathbf{x}_i'\mathbf{x}_j}{\|\mathbf{x}_i\| + \|\mathbf{x}_j\| - \mathbf{x}_i'\mathbf{x}_j}, \tag{11.3}$$

where $\mathbf{x}_i'\mathbf{x}_j$ is the dot product of vectors (arrays) $\mathbf{x}_i$ and $\mathbf{x}_j$ and $\|\cdot\|$ is the Euclidean norm. The goal of kernel PCA is to find the mapping eigenvector matrix $\mathbf{E}$ that minimizes the cost function

$$\varsigma(\mathbf{Y}) = \sum_{ij}(\|\mathbf{x}_i - \mathbf{x}_j\|^2 - \|\mathbf{y}_i - \mathbf{y}_j\|^2), \tag{11.4}$$

the solution of which is the eigenvalues obtained from the eigenanalysis $\mathbf{G} = \mathbf{K}\mathbf{K}^T = \mathbf{A}\Lambda\mathbf{A}'$, where $\mathbf{G}$ is the gram matrix. The eigenvectors $\mathbf{e}$ for the covariance matrix $\mathbf{K}^T\mathbf{K}$ are obtained from the eigenvectors $\mathbf{a}$ of $\mathbf{G}$ through the relationship $\mathbf{e} = 1/\sqrt{\lambda}\mathbf{a}$. Now let $\mathbf{k}_j$ be the column vector with kernel values for array j versus all other ith arrays and $\mathbf{e}_1$ and $\mathbf{e}_2$ the eigenvectors for the top two eigenvalues. The low-dimensional vector $\mathbf{y}_i \in \mathbb{R}^2$ for array j is equal to $\{\mathbf{e}_1'\mathbf{k}_j, \mathbf{e}_2'\mathbf{k}_j\}$.

11.4 DIFFUSION MAPS

Diffusion maps (DMs) are based on a random walk on the data to create a Markov matrix for which eigenvectors and eigenvalues are extracted for construction of a diffusion map [2]. The data are embedded in Euclidean space by the diffusion map in order to reflect connectivity. The construction of the Markov matrix $\mathbf{P}^t$ at timestep t begins with creation of Gaussian weights

$$w_{ij} = \exp\left(\frac{-\|\mathbf{x}_i - \mathbf{x}_j\|^2}{2\sigma^2}\right), \tag{11.5}$$

which are normalized to yield elements of $\mathbf{P}^{(t)}$ at step t in the form

$$p_{ij}^{(t)} = \frac{w_{ij}}{\sum_k w_{ik}}. \tag{11.6}$$

The probability $p_{ij}^{(t)}$ is determined as the power t of Markov matrix $\mathbf{P}$ as in $\mathbf{P}^{(t)}$ and represents the probability of transition from point x to point y in one timestep. At this point, the eigenanalysis of $\mathbf{P}$ could be performed to obtain a solution; however, we proceed further to develop the diffusion distance matrix $\mathbf{D}$ based on $\mathbf{X}$. The distance between two data points now becomes

$$d_{ij}^{(t)} = \sqrt{\sum_k \frac{(p_{ik}^{(t)} - p_{jk}^{(t)})^2}{\phi^{(0)}(\mathbf{x}_k)}}, \tag{11.7}$$

where $\phi^{(t)}(\mathbf{x}_k) = m_k / \sum_j m_j$ where $m_k = \sum_j p_{ij}$. The elements d_{ij} will be small if there are many short connections between point x and y, thus reflecting the concept of a cluster. In addition, d_{ij} is very robust to noise, and combines all information about connectivity between data points, thereby extending its usefulness for inference on data. The objective function for DM is

$$\zeta(\mathbf{Y}) = \sum_{ij} \|\mathbf{y}_i - \mathbf{y}_j\|^2 p_{ij}, \tag{11.8}$$

and is minimized by finding the smallest nonzero eigenvalues of

$$\mathbf{PE} = \mathbf{\Lambda E}. \tag{11.9}$$

The reduced form of $\mathbf{X} \in \mathbb{R}^p \mapsto \mathbf{Y} \in \mathbb{R}^2$ is represented by $\mathbf{Y} = \{\lambda_1\mathbf{e}_1, \lambda_2\mathbf{e}_2\}$.

11.5 LAPLACIAN EIGENMAPS

Laplacian eigenmaps (LEM) constructs an adjacency matrix of objects and then applies a K-nearest-neighbor approach to the distance matrix [3]. Let $\mathbf{W}$

be a sparse $n \times n$ adjacency matrix with elements $w_{ij} = \exp(-\|\mathbf{x}_i - \mathbf{x}_j\|^2)$ given array i and j among the K nearest neighbors; otherwise $w_{ij} = 0$. Next, let $\mathbf{D}$ be a diagonal weight matrix having elements $d_{ii} = \sum_j w_{ij}$, which are the row sums of $\mathbf{W}$. The Laplacian matrix is then $\mathbf{L} = \mathbf{D} - \mathbf{W}$, which is square symmetric. The objective function for LEM is

$$\zeta(\mathbf{Y}) = \sum_{ij} \|\mathbf{y}_i - \mathbf{y}_j\|^2 w_{ij} = \mathbf{Y}^T \mathbf{L} \mathbf{Y}, \tag{11.10}$$

and is minimized by finding the smallest nonzero eigenvalues of

$$\mathbf{LE} = \mathbf{\Lambda DE}. \tag{11.11}$$

The reduced form of $\mathbf{X} \in \mathbb{R}^p \mapsto \mathbf{Y} \in \mathbb{R}^2$ is represented by $\mathbf{Y} = \mathbf{E}$, where $\mathbf{E}$ has dimensions $n \times 2$ based on the two lowest eigenvalues of $\mathbf{\Lambda}$.

11.6 LOCAL LINEAR EMBEDDING

Local linear embedding (LLE) eliminates the need to estimate distance between distant objects and recovers global nonlinear structure by local linear fits [4]. LLE is advantageous because it involves no parameters such as learning rates or convergence criteria. LLE also scales well with the intrinsic dimensionality of $\mathbf{Y}$. The objective function for LLE is

$$\begin{aligned} \zeta(\mathbf{Y}) &= (\mathbf{Y} - \mathbf{WY})^2 \\ &= \mathbf{Y}^T(\mathbf{I} - \mathbf{W})^T(\mathbf{I} - \mathbf{W})\mathbf{Y}. \end{aligned} \tag{11.12}$$

The weight matrix $\mathbf{W}$ elements w_{ij} for arrays i and j are set to zero if j is not a nearest neighbor of i; otherwise, the weights for the K nearest neighbors of array i are determined via a least-squares fit of

$$\mathbf{U} = \mathbf{G}\boldsymbol{\beta}, \tag{11.13}$$

where the dependent variable $\mathbf{U}$ is a $K \times 1$ vector of ones, $\mathbf{G}$ is a $(K \times K)$-gram matrix for all nearest neighbors of array i, and $\boldsymbol{\beta}$ is a $K \times 1$ vector of weights that follow sum-to-unity constraints. Let $\mathbf{D}$ be a symmetric positive semidefinite $K \times K$ distance matrix for all pairs of the K nearest neighbors of p-dimensional array $\mathbf{x}_i$. It can be shown that $\mathbf{G}$ is equal to the doubly centered distance matrix $\boldsymbol{\tau}$ [9] with elements

$$\tau_{lm} = -\frac{1}{2}\left(d_{lm}^2 - \frac{1}{K}\sum_k^K d_{lk}^2 - \frac{1}{K}\sum_k^K d_{km}^2 + \frac{1}{K^2}\sum_g^K\sum_h^K d_{gh}^2\right). \tag{11.14}$$

Calculation of $\boldsymbol{\tau}$ from the distance matrix $\mathbf{D}$ requires $\mathcal{O}(n^2p)$ operations. The K regression coefficients are determined numerically using

$$\underset{K\times 1}{\boldsymbol{\beta}} = \underset{K\times K}{(\boldsymbol{\tau}^T\boldsymbol{\tau})^{-1}} \underset{K\times 1}{\boldsymbol{\tau}^T\mathbf{U}} \tag{11.15}$$

and are checked to confirm that they sum to unity. Values of $\boldsymbol{\beta}$ are embedded into row i of $\mathbf{W}$ at the various column positions corresponding to the K nearest neighbors of array i, as well as the transpose elements. This is repeated for each ith array in the dataset. It warrants noting that if the number of nearest neighbors K is too low, then $\mathbf{W}$ can be sparse, rendering eigenanalysis difficult. We observed that $K = 9$ nearest neighbors resulted in $\mathbf{W}$ matrices that did not contain pathologies during eigenanalysis. The objective function (11.12) is minimized by finding the smallest nonzero eigenvalues of

$$(\mathbf{I} - \mathbf{W})^T(\mathbf{I} - \mathbf{W})\mathbf{E} = \boldsymbol{\Lambda}\mathbf{D}\mathbf{E}. \tag{11.16}$$

The reduced form of $\mathbf{X}$ is represented by $\mathbf{Y} = \mathbf{E}$, where $\mathbf{E}$ has dimensions $n \times 2$ based on the two lowest eigenvalues of $\boldsymbol{\Lambda}$.

11.7 LOCALITY PRESERVING PROJECTIONS

Locality preserving projections (LPP) is a linear dimension reduction technique that constructs a graph with information about the neighborhoods within the data points [5]. A Laplacian of the graph is used to compute a transformation matrix that maps the data points to a subspace, thus preserving local neighborhood relationships. LPP is essentially a discrete approximation to a continuous map based on the manifold geometry. Preservation of the locality also renders LPP insensitive to noise and outliers as well. For LPP, we set up the data matrix as a feature-by-object matrix with dimensions $p \times n$. Let $\mathbf{W}$ be a sparse $n \times n$ adjacency matrix with elements $w_{ij} = \exp(-\|\mathbf{x}_i - \mathbf{x}_j\|^2)$, assuming that objects i and j are among the K nearest neighbors; otherwise $w_{ij} = 0$. Next, let $\mathbf{D}$ be a diagonal weight matrix having elements $d_{ii} = \sum_j w_{ij}$, which are the row sums of $\mathbf{W}$. The Laplacian matrix is then $\mathbf{L} = \mathbf{D} - \mathbf{W}$, which is square symmetric. The objective function for LPP is

$$\zeta(\mathbf{Y}) = \sum_{ij} |\mathbf{y}_i - \mathbf{y}_j\|^2 w_{ij} \tag{11.17}$$

and is minimized by finding the smallest nonzero eigenvalues of the $p \times p$ matrices $\mathbf{XLX}^T$ and $\mathbf{XMX}^T$:

$$\mathbf{XLX}^T\mathbf{E} = \boldsymbol{\Lambda}\mathbf{XMX}^T\mathbf{E}. \tag{11.18}$$

The reduced form of **X** is represented by $\mathbf{Y} = \mathbf{X}^T\mathbf{E}$, where **E** has dimensions $n \times 2$ based on the two lowest eigenvalues of $\mathbf{\Lambda}$.

11.8 SAMMON MAPPING

Sammon mapping [8] was one of the first nonlinear dimension reduction techniques introduced. The objective function to be minimized for Sammon mapping is

$$\zeta(\mathbf{Y}) = \frac{1}{\sum_{i<j} d_{ij}^*} \sum_{i<j}^{m} \frac{(d_{ij}^* - d_{ij})^2}{d_{ij}^*}, \tag{11.19}$$

where $d_{ij}^* = \|\mathbf{x}_i - \mathbf{x}_j\|^2$ and $d_{ij} = \|\mathbf{y}_i - \mathbf{y}_j\|^2$. Instead of using a steepest descent for function minimization, we used particle swarm optimization. First, assign a 2D **y** vector for each p-dimensional input data vector **x**. Let each particle form a vector of length $2 \times n$, where 2 is the reduced dimensional size and n is the number of input objects. In this fashion, there will be n 2D $\mathbf{y}_i$ vectors representing n p-dimensional $\mathbf{x}_i$ arrays. Let the position and velocity vectors for particle l be $\mathbf{r}_l$ and $\mathbf{v}_l$. The position of particle l is $\mathbf{r}_l = \{r_{1,l}, \ldots, r_{n\times 2,l}\}$, and the velocity of particle l is $\mathbf{v}_l = \{v_{1,l}, \ldots, v_{n\times 2,l}\}$. Particle positions are set equal to random uniform variates $U(0,1)$ at initialization, whereas velocity is initialized to zero. At each iteration, the reduced 2D vector for $\mathbf{y}_i$ is set equal to the corresponding two vector elements of the $(n \times 2)$-length particle position vector, d_{ij} and d_{ij}^* are determined, and then fitness for each particle is determined as $1/\zeta(\mathbf{Y})$, which increases as $\zeta(\mathbf{Y})$ decreases. The velocity update is

$$\begin{aligned} v_l(t+1) = wv_l(t) + c_1 U(0,1) \otimes (\mathbf{b}_l(t) \\ -\mathbf{r}_l(t)) + c_2 U(0,1) \otimes (\mathbf{b}_g(t) - \mathbf{r}_l(t)), \end{aligned} \tag{11.20}$$

where w is the *inertia factor*, c_1 is the cognitive parameter, and c_2 is the social parameter; $\mathbf{b}_l(t)$ is the best historical fitness for particle l, and $\mathbf{b}_g(t)$ is the global best particle. The inertia at iteration t is $w(t) = w_{\text{start}} - (w_{\text{start}} - w_{\text{end}})t/T_{\max}$. The particle position update is $\mathbf{r}_l(t+1) = \mathbf{r}_l(t) + \mathbf{v}_l(t+1)$. Parameter values for PSO were set to: #numparticles (number of particles) $= 50, T_{\max} = 200$, $v_{\min} = -0.7, v_{\max} = 0.7, c_1 = 2, c_2 = 2, w_{\min} = 0.4$, and $w_{\max} = 0.9$. After $T_{\max}$ training iterations, the $n \times 2$ matrix **Y** is based on particle position values of the most fit particle.

11.9 NLML PRIOR TO CLASSIFICATION ANALYSIS

The methods described above were used for dimension reduction of the nine datasets used throughout this book (see Table 12.9). We also employed self-organizing maps (SOM) and unsupervised neural gas (UNG). For SOM, a 900-node 30×30 grid was used such that for each array, the 2D results were based on grid row and grid column values for the best matching unit after training. The UNG approach used two prototypes exclusively for all datasets, and it is realized that the three- and four-class datasets may underperform when UNG is based on only two prototypes (instead of three and four, respectively). However, the intent is to reduce the higher dimension datasets down to 2D, irrespective of the number of classes. Specifically, the numbers of features for each dataset listed in Table 12.9 were reduced to 2D, where we assumed the higher-dimension data to be $\mathbf{X}$ and the 2D data to be $\mathbf{Y}$. The 2D data were then either normalized to range [0,1] or mean-zero standardized and then input into nine supervised classifications methods: K nearest neighbor (KNN), naïve Bayes classifier (NBC), linear discriminant analysis (LDA), Fisher's discriminant analysis (FDA), learning vector quantization (LVQ1), least squares support vector machines (SVMLS), artificial neural networks (ANN), constricted particle swarm optimization (CPSO), and polytomous logistic regression (PLOG). For KNN, K was set equal to 5 (5NN), and for the LVQ1 we used a single prototype per class. For SVMs, we used an L^2 soft-norm least-squares approach to SVM. A weighted exponentiated RBF kernel was employed to map microarrays in the original space into the dot-product space, given as $K(\mathbf{x}, \mathbf{x}^T) = \exp[-(\gamma/m)||\mathbf{x} - \mathbf{x}^T||]$, where m = #features. Such kernels are likely to yield the greatest class prediction accuracy providing that a suitable choice of γ is used. To determine an optimum value of γ for use with RBF kernels, a grid search was done using incremental values of γ from 2^{-15}, $2^{-13}, \ldots, 2^3$ in order to evaluate accuracy for all training microarrays. We also used a grid search in the range of $10^{-2}, 10^{-1}, \ldots, 10^4$ for the SVM margin parameter C. The optimal choice of C was based on the grid search for which classification accuracy is the greatest, resulting in the optimal value for the separating hyperplane and minimum norm $||\xi||$ of the slack variable vector. SVM tuning was performed by taking the median of parameters during grid search iterations when the test microarray misclassification rate was zero. For the ANN classifier, the logistic activation function was used with each hidden node, and the softmax function used to compute class membership probabilities for output node weight connections. We also used 500 sweeps for training, with a grid search for each ANN model in which the learning rate ϵ and momentum α ranged from $2^{-9}, 2^{-8}, \ldots, 2^{-1}$. The number of hidden nodes was fixed at 2, since the input vectors were always 2D. In cases when there were multiple values of grid search parameters for the same error rate, we used the median value. Two 10-fold CV was employed for all runs, where arrays were permuted and repartitioned into the 10 folds prior to each CV run.

Classifiers were trained with the same 2D feature sets, and then classifier votes were combined using the ensemble majority voting (EMV) and ensemble weighted majority voting (EWMV) ensemble combination techniques, described in Section 13.4. Results of ensemble methods are presented under the classifier names EMV and EWMV in tabular form with results of the individual classifiers. The silhouette index (Section 2.5.5) was also calculated for each 2D feature set (dataset) after NLML to determine the degree of cluster separation of classes.

11.10 CLASSIFICATION RESULTS

Boxplots for the silhouette index for each NLML method for all datasets for normalized and mean-zero standardized 2D feature sets are shown in Figure 11.2. First, it is readily apparent that when the 2D feature sets from NLML were normalized, KGPCA and KTPCA resulted in the greatest

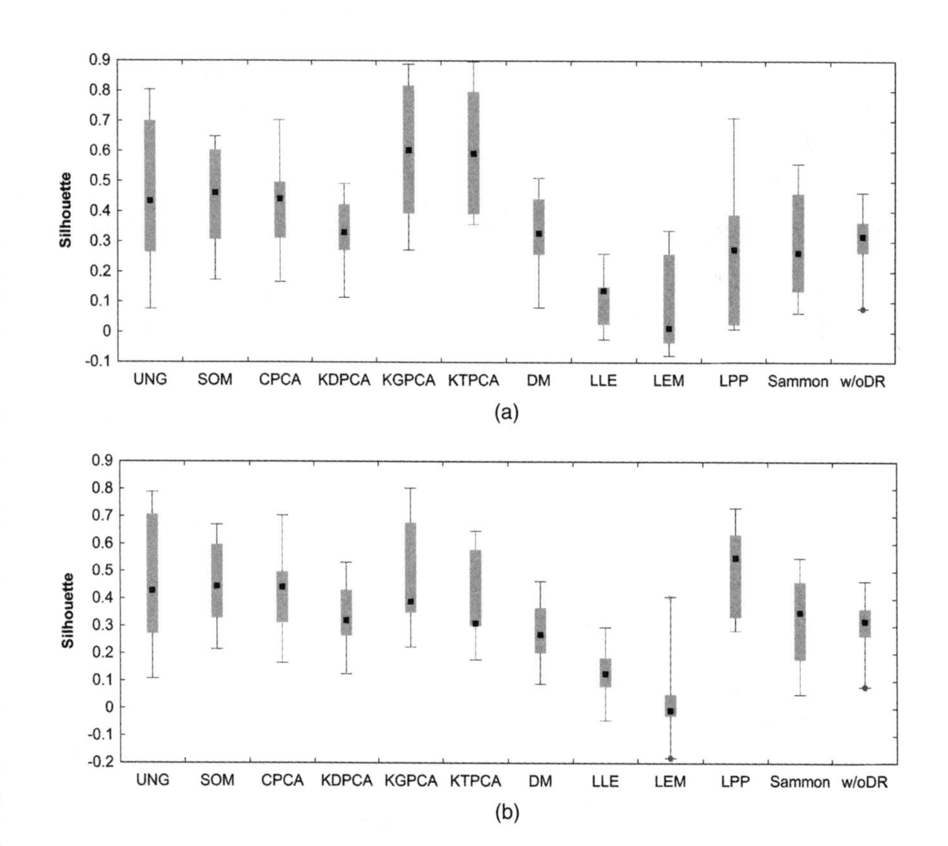

FIGURE 11.2 Boxplots of silhouette index for all datasets based on normalized (a) and mean-zero standardized (b) feature sets from NLML.

median silhouette index for all nine datasets, which was approximately 0.6. When 2D input feature sets were mean-zero-standardized, LPP resulted in the greatest median silhouette index (0.55). However, SOM resulted in greater silhouette values for both normalization and mean-zero standardization when compared with silhouette for higher-dimension data (i.e., "w/oDR" and silhouette values for the remaining NLML methods). An example output of NLML is shown specifically for the 2D four-class SRBCT dataset in Figure 11.3, which provides a graphical visualization of the cluster structure.

Boxplots of classification accuracy by classifier over all datasets and all NLML methods based on normalized and mean-zero-standardized input 2D features are shown in Figure 11.4. The majority of classifiers performed equally well for both feature transformations, with the exception of the NBC and SVMLS classifiers. The interquartile range of classification accuracy was smaller for mean-zero standardization than the interquartile range of accuracy based on normalized input 2D feature sets. Moreover, the classification accuracy of EMV and EWMV classifier fusion methods were consistently high and performed reliably for the ensemble of classifiers used. Classification accuracy with and without NLML over all datasets and all classifiers except for EMV and EWMV is shown for normalized and mean-zero standardized input 2D features in Figure 11.5. When considering both normalization and mean-zero standardization, we observed that CPCA, DM. KDPCA, KGPCA, KTPCA, SOM, and UNG resulted in 2D feature sets

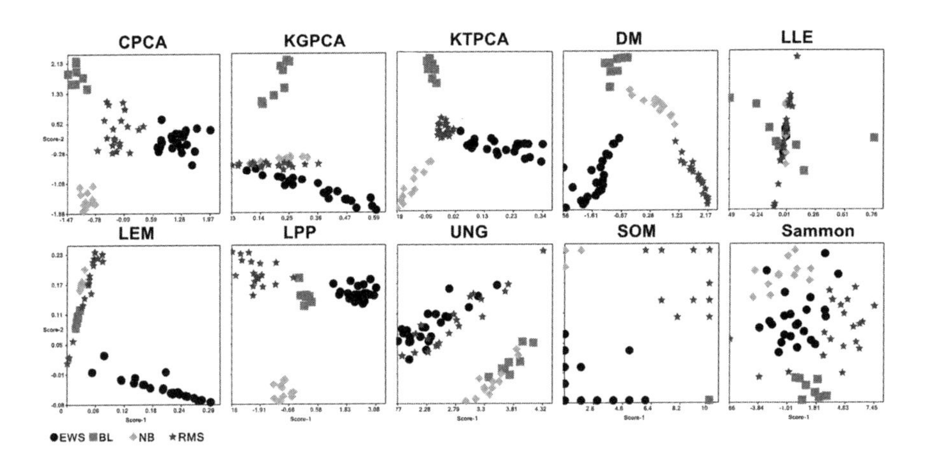

FIGURE 11.3 Cluster structure of four-class SRBCT dataset based on the 2D feature set from NLML. (*See insert for color representation of the figure.*)

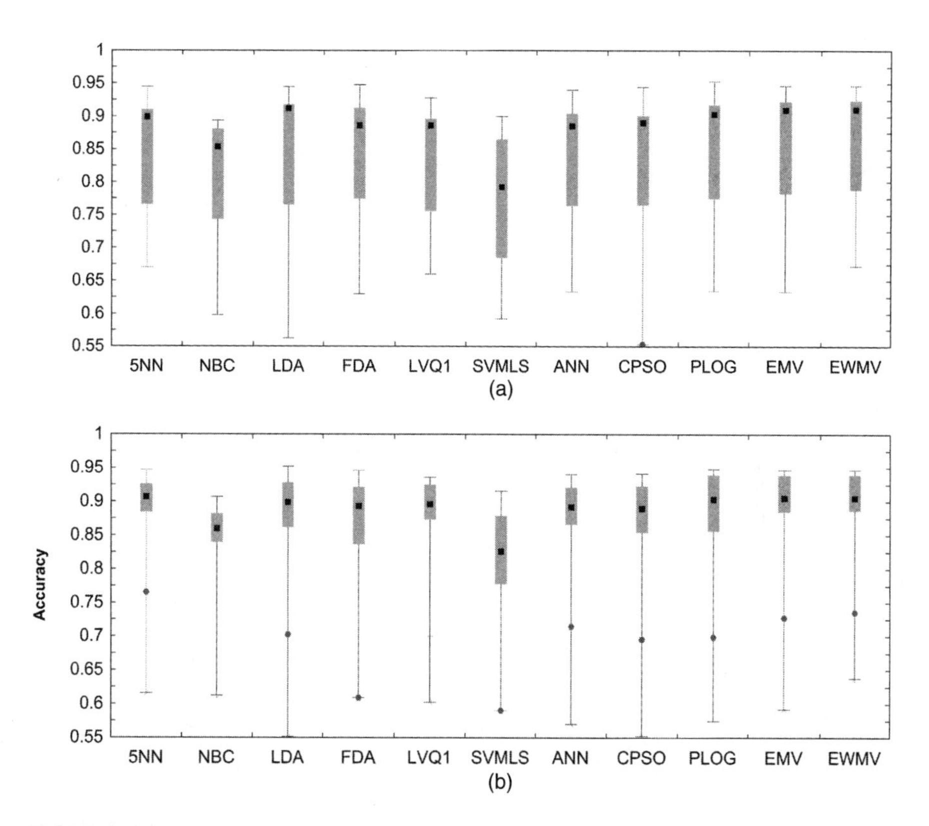

FIGURE 11.4 Boxplots of classification accuracy by classifier for all datasets and all NLML methods based on normalized (a) and mean-zero standardized (b) 2D feature sets from NLML.

yielding the greatest classification accuracy for the classifiers considered. Whereas LEM, LLE, LPP, and Sammon resulted in 2D feature sets whose classification accuracy yields were the lowest. Tables 11.1 and 11.2 list specific results for classification accuracy based on two 10-fold CV runs of each 2D dataset. Two-dimensional feature sets from CPCA and SOM resulted in the greatest EWMV accuracy when compared with other methods. Following this, 2D feature sets from the kernel PCA methods resulted in consistently high levels of EWMV accuracy. LPP feature sets resulted in high levels of EWMV accuracy when mean-zero standardized prior to classification, and low EWMV results when normalized. Sammon mapping, LLE, and LEM resulted in feature sets that yielded lower (60-80%) EWMV accuracy. In light of these findings, linear CPCA and nonlinear SOM yielded feature sets that resulted in the greatest ensemble classification accuracy.

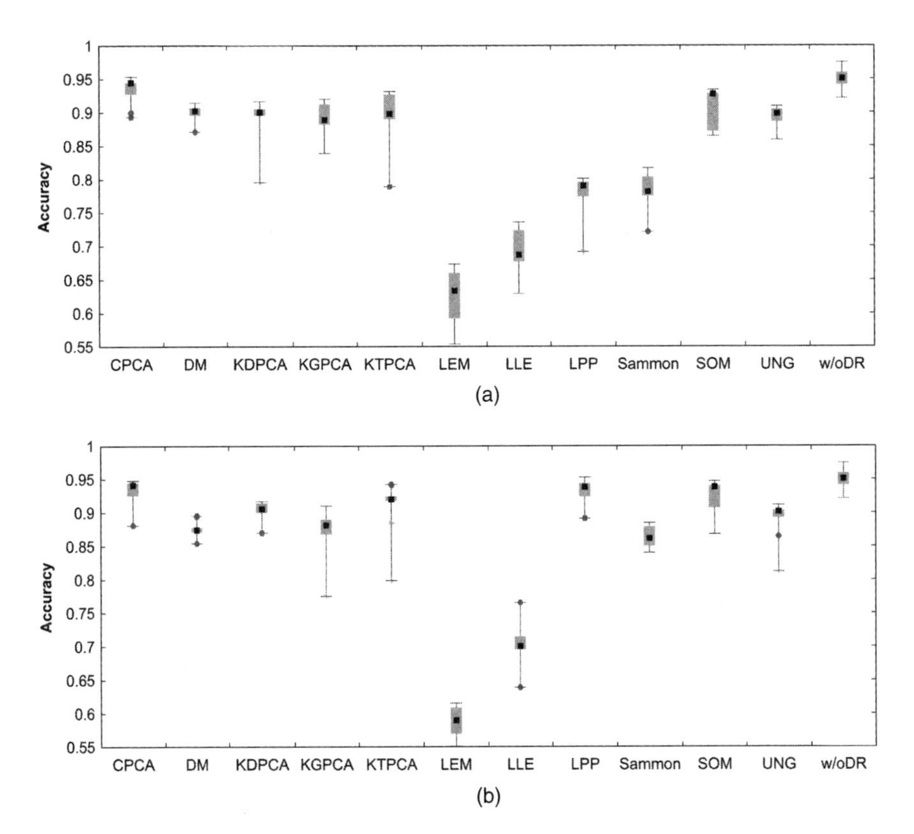

FIGURE 11.5 Boxplots of classification accuracy by NLML method for all datasets and all classifiers based on normalized (a) and mean-zero standardized (b) 2D feature sets from NLML. (w/oDR represents mean classifier accuracy without dimensional reduction of features via NLML prior to classification analysis).

11.11 SUMMARY

A central issue with NLML is whether the low-dimensional structure of the manifold is an original representation of the data up to some rotation, translation, or scaling factor. The underlying low-dimensional structure is seldom known, and therefore one is left with assuming the algorithm can asymptotically recover the original structure as the number of data points tends to infinity. Noise is another factor that is assumed to be avoided by most algorithms.

The major impetus for developing methods described in this chapter was to introduce nonlinear dimensional reduction techniques that can be used as an alternative to PCA, which is wholly linear. The original feature sets used as inputs into the NLML methods were selected using a greedy PTA approach

TABLE 11.1 Classification Accuracy by NLML Method and Classifier Based on Normalized 2D Feature Sets from NLML[a]

Method	5NN	NBC	LDA	FDA	LVQ1	SVMLS	ANN	CPSO	PLOG	EMV	EWMV
w/oDR	0.948	0.929	0.957	0.959	0.941	0.921	0.974	0.950	0.970	0.971	0.971
CPCA	0.944	0.893	0.944	0.948	0.928	0.900	0.940	0.944	0.953	0.947	0.947
SOM	0.927	0.871	0.928	0.864	0.904	0.870	0.933	0.930	0.927	0.934	0.934
KTPCA	0.929	0.880	0.917	0.927	0.898	0.789	0.890	0.892	0.931	0.928	0.928
KDPCA	0.900	0.884	0.917	0.914	0.896	0.796	0.906	0.900	0.904	0.921	0.923
KGPCA	0.904	0.848	0.920	0.912	0.887	0.839	0.882	0.889	0.914	0.918	0.920
UNG	0.898	0.859	0.909	0.908	0.886	0.863	0.889	0.904	0.902	0.912	0.913
DM	0.902	0.883	0.914	0.910	0.896	0.871	0.904	0.898	0.907	0.908	0.908
Sammon	0.780	0.774	0.817	0.791	0.764	0.721	0.781	0.803	0.814	0.817	0.820
LPP	0.776	0.764	0.798	0.796	0.773	0.691	0.790	0.794	0.801	0.806	0.804
LLE	0.736	0.679	0.667	0.723	0.730	0.629	0.712	0.677	0.687	0.713	0.739
LEM	0.670	0.673	0.562	0.630	0.660	0.592	0.633	0.553	0.634	0.633	0.672

[a]Sorted by decreasing accuracy from ensemble weighted majority voting (EWMV). Classification accuracy values reported are averages of two 10-fold CV runs made for each of the 9 datasets using the 2D feature set derived from each NLML method (w/oDR represents classifier accuracy without dimensional reduction of features via NLML prior to classification analysis).

TABLE 11.2 Classification Accuracy by NLML Method and Classifier Based on Mean-Zero Standardized 2D Feature Sets from NLML[a]

Method	5NN	NBC	LDA	FDA	LVQ1	SVMLS	ANN	CPSO	PLOG	EMV	EWMV
w/oDR	0.948	0.929	0.957	0.959	0.941	0.921	0.974	0.950	0.970	0.971	0.971
SOM	0.947	0.907	0.938	0.868	0.936	0.904	0.940	0.938	0.941	0.947	0.947
CPCA	0.944	0.881	0.942	0.944	0.926	0.912	0.933	0.941	0.948	0.943	0.944
LPP	0.943	0.891	0.952	0.946	0.931	0.916	0.923	0.938	0.939	0.944	0.944
KTPCA	0.920	0.884	0.924	0.941	0.924	0.799	0.920	0.918	0.942	0.938	0.938
KDPCA	0.917	0.879	0.914	0.914	0.900	0.870	0.911	0.903	0.906	0.922	0.922
UNG	0.911	0.864	0.907	0.903	0.892	0.812	0.903	0.900	0.901	0.912	0.912
DM	0.896	0.854	0.873	0.878	0.894	0.868	0.874	0.871	0.874	0.897	0.898
KGPCA	0.887	0.848	0.890	0.910	0.881	0.776	0.880	0.868	0.907	0.898	0.897
Sammon	0.876	0.848	0.884	0.882	0.850	0.840	0.860	0.879	0.861	0.888	0.888
LLE	0.766	0.687	0.702	0.741	0.700	0.639	0.714	0.696	0.699	0.728	0.736
LEM	0.616	0.612	0.551	0.609	0.602	0.590	0.570	0.552	0.574	0.592	0.638

[a]The sets are sorted by decreasing accuracy from ensemble weighted majority voting (EWMV). Classification accuracy values reported are averages of two 10-fold CV runs made for each of the nine datasets using the 2D feature set derived from each NLML method (w/oDR represents classifier accuracy without dimensional reduction of features via NLML prior to classification analysis).

that guarantees the best class separation with reduced bias created by the "nesting" problem. The EWMV results reported in this chapter based on nine microarray datasets and nine classifiers indicate surprisingly good ensemble accuracy when reducing larger feature sets down to 2D, and may be more robust than results of previous investigations of gene expression data [10,11]. A very small difference ($\delta = 0.024$) was observed between EWMV accuracy derived from SOM 2D feature sets (0.947) and EWMV accuracy obtained without using NLML (0.971).

REFERENCES

[1] B. Schölkopf, A.J. Smola. Nonlinear component analysis as a kernel eigenvalue problem. *Neural Comput.* **10**(5):1299–1319, 1998.

[2] S. Lafon, A.B. Lee. Diffusion maps and coarse graining: A unified framework for dimensionality reduction, graph partitioning, and dataset parameterization. *IEEE Trans. Pattern Anal. Machine Intell.* **28**(9):1393–1403, 2006.

[3] M. Belkin, P. Niyogi. Laplacian eigenmaps and spectral techniques for embedding and clustering. In *Advances in Neural Information Processing Systems*, MIT Press, Cambridge, MA, 2004, vol. 14, pp. 585–591.

[4] S.T. Roweis, L.K. Saul. Nonlinear dimensionality reduction by locally linear embedding. *Science* **290**(5500):2323–2326, 2000.

[5] X. He, P. Niyogi. Locality preserving projections. *Proc. Neural Information Processing Systems (NIPS'03)*, MIT Press, Boston, 2003.

[6] T. Kohonen. *Self-Organizing Maps.* Springer, Berlin, 2001.

[7] T. Martinetz, S. Berkovich, K. Schulten. Neural gas network for vector quantization and its application to time-series prediction. *IEEE Trans. Neural Networks* **4**(4):558–568, 1993.

[8] J.W. Sammon. A nonlinear mapping for data structure analysis. *IEEE Trans. Comput.* **18**(5):401–409, 1969.

[9] W.S. Torgerson. Multidimensional scaling: I. Theory and method. *Psychometrika* **17**:401–419, 1952.

[10] C. Bartenhagen, H.U. Klein, C. Ruckert, X. Jiang, M. Dugas. Comparative study of unsupervised dimension reduction techniques for the visualization of microarray gene expression data. *BMC Bioinform.* **11**:567, 2010.

[11] G. Lee, C. Rodriguez, A. Madabhushi. Investigating the efficacy of nonlinear dimensionality reduction schemes in classifying gene and protein expression studies. *IEEE/ACM Trans. Comput. Biol. Bioinform.* **5**(3):368–384, 2008.

PART III

CLASS PREDICTION

CHAPTER 12

FEATURE SELECTION

12.1 INTRODUCTION

DNA microarrays provide a considerable wealth of information on transcriptomic changes during disease and controlled experimentation. Large microarray data sets can easily contain expression values of ~50,000 transcripts for each array and are therefore amenable to feature selection.

In Part I, the goal was to explore the cluster structure of a dataset. When class structure is unknown, it makes sense to cluster-analyze the data using the entire feature set in order to understand genome-wide genetic regulatory and pathway influences. For class prediction, however, we start with a priori known phenotypic or diagnostic classes, and search for expression profiles that are predictive of class. The efficiency of this predictive search is greatly improved when redundant information in the features is reduced—essentially by reducing the number of features used. There may also be a considerably high level of noise among many gene expression profiles, rendering them useless for information retrieval purposes. If we are not careful during feature selection, noise can be piled on top of noise, and the resulting predictive model will be less informative. Feature selection is performed to increase parsimony, remove noise, and eliminate redundancy of information.

For class prediction, the objective for maximizing parsimony should be a priority. Parsimony entails the search for the smallest number of features to ensure optimal class prediction performance. Parsimony also follows the rule of Occam's razor, which generally states that it is better to have fewer complexities and assumptions involved in solving a particular problem. In statistics, parsimony is one of the most fundamental rules for evaluating the features in a model, since resource allocation will not be optimized when an overabundance of wasteful features are employed. The premise for this assumption is that resources are required for measuring both observations and variables, which don't contribute to the performance if a model is

Classification Analysis of DNA Microarrays, First Edition. Leif E. Peterson.

wasteful. Therefore, it is more parsimonious to identify a smaller model that can predict with the same performance as a larger model.

Feature selection is a two-sided process where, the goal is to both reduce the number of features and increase class prediction performance. It is important to realize that for a given set of data and features used, the reduction of correlation does not affect the resulting predictive accuracy what so ever. What matters most is that for the genes selected, the redundancy is minimized while classification accuracy is maximized. In this chapter, we will explore the effects of several gene selection methods on between-gene correlation and class prediction performance.

12.2 FILTERING VERSUS WRAPPING

Gene selection for class prediction can be performed as a preprocessing step prior to classification, or bundled directly into the classification process. The process of *filtering* identifies genes prior to and independently from the classification method into which the features are input. During filtering, features are ranked according to a criterion or statistical test result, and used as needed for classification. The separation between the filtering process and later classification minimizes the potential for selection bias caused by selecting features with the class prediction method. There are, nevertheless, potential drawbacks of filtering. One major disadvantage of feeding classifiers with genes significantly differentially expressed on the basis of filtering is that features may have an unusually high level of correlation within the classes. For example, genes having the most significant t-test results across two classes of arrays will undoubtedly have both the greatest between-class difference in mean expression and the smallest within-class variation (low values of standard deviation). The small within-class variation will ensure that expression values are consistent within each class. A classifier founded on this approach can break down during the initial presentation of an observation that has the opposite expression pattern, and will likely end up being misclassified. Another drawback of using genes with correlated within-class features is that the correlation may be due to shared upstream signaling molecules. If pathway heterogeneity plays a major role in disease causation, and the majority of genes selected are coregulated by the same pathway, then misclassification can be increased.

The process of *wrapping* embeds feature selection directly into the classifier and recursively identifies features using classifier performance. Features are ranked according to the performance of the wrapping classifier used. Wrapping increases the potential for selection bias because the deployed class prediction method itself is solely used as the feature selection criterion [1]. Selection bias can be minimized by augmenting the wrapping procedure with external validation or bootstrapping. Wrapping does tend to alleviate problems associated with within-class feature correlation, but this is not nearly as severe as the impact of selection bias on external validity and generalization of results.

The favored approach to feature selection in this book is to use filtering, rendering the feature selection process entirely independent of classifier performance. The remaining sections discuss data types, data format, and filtering methods for features having continuous and qualitative values.

12.3 DATA

One of the first issues addressed when approaching the data analysis phase of a study is how the data are distributed. This issue is multifactorial, as the types of numbers, responses, measurement scale, and variables used must be considered. Without understanding the details of the data to be analyzed, it is impossible to select the appropriate methods required.

12.3.1 Numbers

Integers. Integers are negative and positive whole numbers such as $-3, -2, -1, 0, 1, 2, 3, \ldots$ in the range $[-\infty, \infty]$. Integers are used for representing counts and ranks, and frequency of occurrence. Examples include the number of arrays, number of genes, number of children in a family, and number of times a telephone rings per day.

Rational Numbers. Rational numbers are defined as ratios of integers such as a/b, and result in terminating and repeating decimal fractions. For example, a rational number like 0.25 has a terminating decimal component that is based on the ratio of 1 to 4 (1 : 4). A rational number with a repeating decimal component is 0.33... based on the ratio 1 : 3. For this example, the decimal component 0.33333333... never ends.

Irrational Numbers. Irrational numbers cannot be defined as a ratio of integers and are usually derived from mathematical operations such as the square root, as in $\sqrt{2} = 1.4142$. Examples of irrational numbers are $\pi = 3.1416\ldots$ and $e = 2.7183$.

12.3.2 Responses

Discrete. Discrete responses are typically represented by integers or counts. Discrete responses may represent answers to yes/no, low/medium/high and similar mutually exclusive, questions in which an integer is used to represent the result. The number of patients in age groups 20-24, 25-29, 30-34, 35-39, 40-44, and 45-49 in a research project is based on discrete responses hinged to counting.

Continuous. Continuous responses typically involve rational numbers such as 2.314, 231.33, and 1.2×10^{-7}, which do not reflect counts or ordinal ranking. Fluorescent intensity, $\log(R/G)$ ratios, contrast, brightness, temperature, weight, and volume of sound on a radio, are examples of continuous responses. Continuous responses usually contain both an integer and decimal component to reflect the level of precision used in measurement. Most data used in statistical analyses are continuous.

12.3.3 Measurement Scales

Nominal. Data that are on the nominal scale represent groups or categories. Examples of data on the nominal measurement scale are groups $1, 2, \ldots, \Omega$; gender; race; ethnicity; or diseases of various classifications. Nominal data do not have a quantitative relationship between each group. For example, gender can be coded using two class labels ($\Omega = 2$), with 1—male and 2—female. However, the difference between class labels 1 and 2 cannot be expressed in a quantitative mathematical relationship, as gender lacks a mathematical definition.

Ordinal. Data arranged on the ordinal scale reflect classification with rank. For ordinally ranked data, there can be a greater or lesser degree of information at each level; however, the relationship between differences of each level does not have to be consistently linear, exponential, or mathematical. For example, in tumor tissue, expression of protein A can rank higher than protein B, higher than protein C, and so on, although rank $1, 2, 3, \ldots$ does not reflect exact percentile in expression (i.e., 38%, 31%, 24%).

Whenever the values of a variable reflect an outcome lower on one end of the scale and higher on the other, or an outcome worse on one end and better on the other, the data are ordinally ranked, and the variable is ordinal. An example of an ordinal variable is the measurement scale used for assessing worry about cancer. For example, the response to the question: "How much did you worry about getting cancer? 1—rarely, 2—a little, 3—somewhat, 4—a lot," reflects little worry on the low end of the scale and greater worry at the other end of the scale.

Interval. Data measured on the interval scale have the same unit distance between each successive point, but with no absolute zero. For example, if patients are categorized into groups on the basis of weight with cutpoints

120-129, 130-139, 140-149, 150-159, and so on, and the grouping variable for people with weight 120-129 is set to 1, 130-139 to 2, 140-149 to 3, and so forth, then the grouping variable $(1, 2, \ldots, \Omega)$ does not reflect the absolute weight of people in each group. The only information in the grouping variable is that (1) the midpoint of weight in each successive group differs by 10 lb and (2) there is more weight as the grouping variable increases. A single-unit change of the grouping variable does not indicate a 10-lb change in weight from subject to subject, but does indicate a category having a midpoint whose weight is 10 lb greater.

Ratio. Measurement on the ratio scale reflects the same ratio of information or amount between a pair of units as well as an absolute value from zero. In the previous example for interval scales, people weighing 120-129 lb represented only the lowest category. On the ratio scale, we would use the individual values of weight such as 120, 121, 122, and so on, where each weight reflects the absolute values from a common zero.

12.3.4 Variables

Qualitative. Qualitative variables are usually in the form of integers and represent count frequencies or groups of data measured on the nominal scale. When analyzed, they are usually evaluated for frequency histograms (i.e., $0, 12, 5, 7, 22, \ldots$) over a range of categories or to form grouping variables $(1, 2, \ldots, \Omega)$ to partition the data being used.

Quantitative. Quantitative variables represent a finer level of measured precision and usually contain an integer and decimal component. For example, body weight can be in the form of 153.29 lb, provided that the instrument used for weighing a patient or research subject had a level of precision out to the hundredths of a pound. Understanding the types of data described above is essential for creating a successful class prediction strategy. As we will see in later chapters, functional transformation of input variables and recoding are sometimes necessary for applying a given class prediction strategy. Figure 12.1 shows the hierarchy of data types and portrays how various types of data are subsumed by one another.

12.4 DATA ARRANGEMENT

A set of observations are often arranged into records and variables. A *record* represents an individual measurement or observation made from one unit of study, such as a patient, rat, or mouse. *Variables*, on the other hand, represent the collection of observations for all the records, such as measurements of

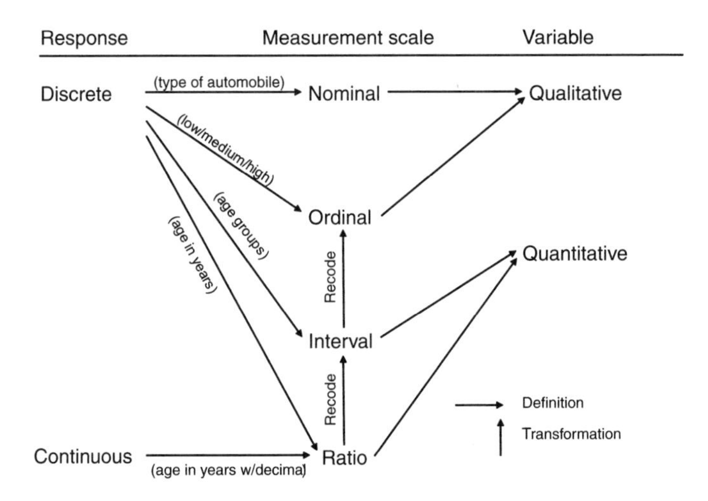

FIGURE 12.1 Schematic of the types of numbers, responses, measurement scales, and variables commonly used in statistical analysis.

mRNA or protein expression, gender, age, or weight. When arranging the data in tabular form, records are arranged in rows, where the index i is used to denote each record, with range $i = 1, 2, \ldots, n$. Variables are arranged in columns where the index j is used to denote each variable, with range $j = 1, 2, \ldots, p$. Let us consider the random variable x_{ij} to represent a single observation, where the subscript i denotes the ith record and subscript j denotes the jth variable. A collection of multiple x_{ij} values for many observations can be arranged in tabular form according to the arrangement in Table 12.1. The symmetric arrangement of data in Table 12.1 can also be represented by use of a *matrix*. For n records and p variables, we introduce the matrix $\mathbf{X}$ in the following form:

$$\underset{n \times p}{\mathbf{X}} = \begin{pmatrix} x_{11} & x_{12} & \cdots & x_{1j} & \cdots & x_{1p} \\ x_{21} & x_{22} & \cdots & x_{2j} & \cdots & x_{2p} \\ \vdots & \vdots & \ddots & \vdots & & \vdots \\ x_{i1} & x_{i2} & \cdots & x_{ij} & \cdots & x_{ip} \\ \vdots & \vdots & & \vdots & \ddots & \vdots \\ x_{n1} & x_{n2} & \cdots & x_{nj} & \cdots & x_{np} \end{pmatrix}. \tag{12.1}$$

Each row in $\mathbf{X}$ represents a record, and each column represents a variable. A matrix is characterized, among other things, by the *dimension*. The dimension of $\mathbf{X}$ for our example is $n \times p$ representing the n records and p variables. (Another example of a matrix is the correlation matrix $\mathbf{R}$, which is a square symmetric matrix with dimensions $p \times p$.) Each element of $\mathbf{X}$, denoted x_{ij},

TABLE 12.1 Arrangement of Records and Variables for a Set of Continuously Scaled Observations

	Variable					
Record	1	2	...	j	...	p
1	x_{11}	x_{12}	...	x_{1j}	...	x_{1p}
2	x_{21}	x_{22}	...	x_{2j}	...	x_{2p}
⋮	⋮	⋮	⋱	⋮	⋮	⋮
i	x_{i1}	x_{i2}	...	x_{ij}	...	⋮
⋮	⋮	⋮	...	⋮	⋱	⋮
n	x_{n1}	x_{n2}	...	...	...	x_{np}

represents the observation for the ith record and the jth variable in the dataset. Because the notation for matrix element x_{ij} can be used to represent any record or variable in the dataset, matrices occupy a very powerful area of statistics known as *matrix algebra*. As we shall see in the chapters throughout this book, matrix algebra is fundamental to many areas of class prediction. Appendix B covers the basic principles of matrix algebra.

However, before we address more advanced topics, let's stick to methods used for describing the *distribution* of a random variable such as x_{ij}. When considering a set of data, a fundamental issue among all analysts is how the data are distributed. With regard to scale of measurement, are the data categorical and measured on the nominal scale, ordinal scale, or are the data continuous and measured on the ratio scale? In terms of probability distributions, when the data are arranged into a histogram, are they normally distributed with a bell-shaped curve, distributed according to the binomial distribution, Poisson distribution, or multinomial distribution? A frequency histogram (distribution) of most data used in statistics can be characterized by a specific *probability distribution* that captures information related to the scale of measurement, observation response, and type of variable. In this chapter, we will be concerned primarily with the normal distribution, and apply parametric and nonparametric tests for feature selection.

12.5 FILTERING

12.5.1 Continuous Features

Normal Distribution. There are many kinds of continuous distributions. The one we will be most concerned with is the *normal* distribution. Normal

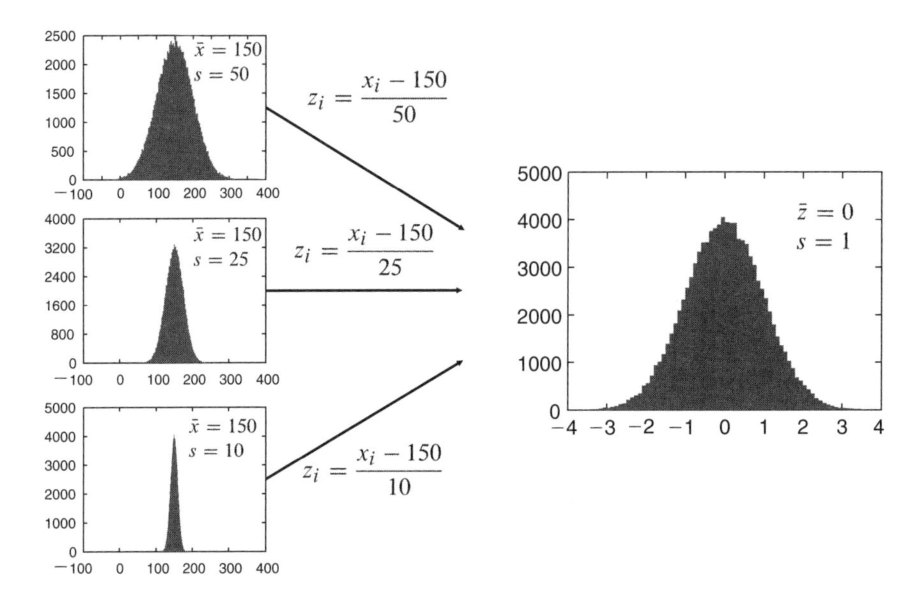

FIGURE 12.2 Three normal distributions with the same mean value $\bar{x} = 150$ but with different standard deviations (s) of 50, 25, and 10. Standardization of each normal distribution by use of the Z transformation results in the same standard normal distribution with $\bar{z} = 0$ and $s = 1$.

distributions are also known as *Gaussian* distributions, and consist of a sample of *normal variates*. A frequency histogram of normal variates is typically bell-shaped, with the bulk of data lying in the middle of the curve and smaller less frequent data residing in vanishing tails (see Figure 12.2). The particular shape of a histogram is characterized by its *central tendency* and *dispersion*.

Figure 12.2 shows several normal distributions based on 100,000 simulated normal variates. The panels each show a normal distribution with the same mean of $\bar{x} = 150$, but with varying values of standard deviation (σ_x) of $50, 25,$ and 10. Because the means are the same among the three distributions, their central location is the same. However, the spread of each distribution decreases with decreasing standard deviation. Greater values of standard deviation reflect more uncertainty about the mean.

Central Tendency and Dispersion. For normal distributions, two parameters used to distinguish central tendency and dispersion are the *mean*, or average, and *standard deviation*. Let $x_{1j}, x_{2j}, \ldots, x_{nj}$ be a random sample of size n from the distribution of random variable x_j. The *mean* of the n variates is denoted $\bar{x}_j$ and is a measure of central tendency that reflects the average value of x_{ij}:

$$\bar{x}_j = \frac{\sum_{i=1}^{n} x_{ij}}{n}. \tag{12.2}$$

Without loss of generality, we can say that the mean is also equal to the *expected value* of x, in the form

$$\mu_x = \mathrm{E}(x) = \sum_i p_i x_i, \tag{12.3}$$

where p_i is the percentage of counts in the ith histogram bin and x_i is the quantile value. Terminology used for the mean is μ for a population, and $\bar{x}$ for a sample. Typically $\bar{x}$ is used since one seldom knows the true population mean, but rather an estimate of the sample mean through some kind of sampling strategy.

The sample *variance* of variable j is a measure of dispersion reflecting the overall width of the distribution, and takes the form

$$s_j^2 = \frac{\sum_{i=1}^{n}(x_{ij} - \bar{x}_j)^2}{n-1}, \tag{12.4}$$

and the *standard deviation* of variable j is simply the square root of the variance, functionally expressed as

$$s_j = \sqrt{\frac{\sum_{i=1}^{n}(x_{ij} - \bar{x}_j)^2}{n-1}}. \tag{12.5}$$

The variance can also be determined by using the expected value of x^2. The expected values of a random variable x^2 are

$$\mathrm{E}(x^2) = \sum_i p_i x_i^2 \tag{12.6}$$

and the variance of x becomes

$$\begin{aligned} \sigma_x^2 = \mathrm{Var}(x) &= \mathrm{E}(x^2) - \mathrm{E}(x)^2 \\ &= \sum_i \frac{(x_i - \mu_x)^2}{n} \\ &= \sum_i p_i x_i^2 - \left(\sum_i p_i x_i\right)^2. \end{aligned} \tag{12.7}$$

Note that there was a shift in the notation used for several of the parameters introduced above. Various forms of notation are used interchangeably in practice; however, one set is called *parameters*, which represent these metrics in a population, and the other set is called *statistics*, which are obtained from a sample drawn from a population. These are described as follows.

$$
\begin{aligned}
&\text{Population notation (parameters):}\\
&\qquad \mu_x = \text{population mean}\\
&\qquad \sigma_x = \text{population standard deviation}\\
&\qquad \sigma_x^2 = \text{population variance}\\
&\qquad \sigma_{xy} = \text{population covariance between } x \text{ and } y\\
&\qquad \rho_{xy} = \text{population correlation between } x \text{ and } y\\
&\text{Sample notation (statistics):}\\
&\qquad \bar{x} = \text{sample mean}\\
&\qquad s_x = \text{sample standard deviation}\\
&\qquad s_x^2 = \text{sample variance}\\
&\qquad s_{xy} = \text{sample covariance between } x \text{ and } y\\
&\qquad r_{xy} = \text{sample correlation between } x \text{ and } y.
\end{aligned}
\tag{12.8}
$$

There are several other important parameters for the normal distribution: range, median, mode, skewness, and kurtosis. The *range* can be simply determined by use of the relationship

$$\Delta_j = \max(x_{ij}) - \min(x_{ij}). \tag{12.9}$$

The range Δ_j of a variable is informative about how far apart the most extreme data points are, which is useful when comparing variables with another. For example, if $\max(x_{i1})$ and $\min(x_{i1})$ for variable $j = 1$ are 2.53 and 0.4, respectively, the range is 2.13; if for variable $j = 2$, the values of $\max(x_{i2})$ and $\min(x_{i2})$ are 79.6 and 21.2, the range is 58.4. A large difference in range is also known as a large change in *scale*, affecting the mean difference and covariance between two variables. Another parameter of interest is the *median*, which reflects the value of a variable in the midpoint of the distribution. The median can be easily found by sorting the data in a distribution in ascending order and identifying the midpoint. Leaving the variable subscript j aside, let $x_1, x_2, \ldots, x_n$ be a set of normal variates and $x_{(1)} \leq x_{(2)} \leq \cdots \leq x_{(n)}$ be their rank-ordered counterpart. The median is equal to the rank-ordered variate $x_{(n/2)}$. The definition of the median introduces a new concept called the pth *quantile*. A quantile is any value among the rank-ordered variates $x_{(1)} \leq x_{(2)} \leq \cdots \leq x_{(n)}$. Consider 100 rank-ordered normal variates such that $n = 100$. The *lower quartile* is the 25th percentile, which is equal to $x_{(25)}$. The median is the 50th percentile, which for this example would be equal to $x_{(50)}$. The *upper quartile* is the 75th percentile and is equal to is $x_{(75)}$. In cases when the most frequent quantile is sought, the *mode* is determined, which is the most frequently occurring quantile. The mode occurs at the peak of the distribution, since the x value at this location is the most frequent.

We now introduce the rth moment about the mean of a distribution. The rth moment about the mean is defined as

$$\kappa_r = \frac{1}{n}\sum_{i=1}^{n}(x_i - \bar{x})^r. \tag{12.10}$$

The coefficient of skewness is

$$\gamma_1 = \frac{\kappa_3}{(\kappa_2)^{3/2}}, \tag{12.11}$$

and the coefficient of kurtosis is

$$\gamma_2 = \frac{\kappa_4}{(\kappa_2)^2} - 3. \tag{12.12}$$

The coefficient of variation is a unitless parameter for comparing features, and is defined as

$$\%\text{CV} = \frac{s}{\bar{x}} \times 100. \tag{12.13}$$

The geometric mean is

$$\text{GM} = \sqrt[n]{\prod_{i=1}^{n} x_i} = \exp\left(\frac{\sum_{i=1}^{n}\ln x_i}{n}\right), \tag{12.14}$$

and the harmonic mean is

$$H = \frac{n}{\sum_i^n(1/x_i)}. \tag{12.15}$$

The geometric mean is more appropriate when averaging lognormally distributed data or multiplicative processes such as the product of growth rates. For example, if something grew by a factor of 10% during the first period, -30% in the second period, and 20% during the third period, the geometric mean would be $\text{GM} = \exp(\log(1.1) + \log(0.7) + \log(1.2)) - 1 = -0.026$, while the arithmetic average would be $0 = ((1.1 + 0.7 + 1.2)/3) - 1$. In this example, the arithmetic mean was upwardly biased by 0.026. The harmonic mean is more appropriate when there are outliers among the data to be averaged. As an example, the arithmetic mean of 1, 4, 55, 3, and 6 is $13.8 = (1 + 4 + 55 + 3 + 6)/5$, while the harmonic mean is $2.83 = 5/(\frac{1}{1} + \frac{1}{4} + \frac{1}{55} + \frac{1}{3} + \frac{1}{6})$. Therefore, in the presence of an outlier, the arithmetic mean is upwardly biased because of the much greater value of the outlier value of 55.

Standard Normal Distribution. Normal distributions can vary by central location and dispersion on the basis of the mean and variance. A frequency histogram for a group of patients having a mean weight of 130 lb and standard deviation (s.d.) 12, will be centrally located 70 units (lb) to the left of a histogram for patients with mean weight 200 lb and s.d. 12, when plotted on the same graph. Distributions with varying means and standard deviations are called *scale-* and *range*-dependent, since their location and spread on the same histogram will be different.

A powerful transformation that can be applied to any normal distribution to remove the effects of scale and range is called the *Z transformation*. Let $\bar{x}$ and s be the mean and standard deviation of a normal distribution of sample size $n = 1000$. The *Z score* for each of the n normal variates is given as

$$Z_i = \frac{x_i - \bar{x}}{s}, \tag{12.16}$$

where x_i represents a normal variate, and Z_i is a newly derived *standard normal variate*. The process of subtracting the mean from each normal variate and dividing by the standard deviation is called *mean-zero standardization*, which results in a *standard normal distribution*. Looking back at Figure 12.2, we notice how the Z transformation converts each of the normal distributions with different standard deviations into the same standard normal distribution. The standard normal distribution always has a mean of zero and standard deviation of one, with variance equal to one ($1^2 = 1$). In addition, the Z score equates to the number of standard deviations that each normal variate lies away from the mean. Large values of $|Z_i| > 2$ are said to be 2-σ away from the mean and are considered extreme observations. Figure 12.3 shows the negative and positive tails of the standard normal distribution. For a tail probability of $\alpha = 0.05$, the value of $Z_{0.05}$ in the negative tail is -1.65, whereas in the positive tail the value of $Z_{0.95} = 1.65$. When the tail probability is divided by 2, that is, $\alpha/2$ then values of $Z_{0.025}$ in the negative tail is -1.96, while $Z_{0.975}$ in the positive tail is 1.96. A probabilistic definition of tail probabilities is that they represent the probability that a given value of Z is greater than or equal to the reference value of Z_α. For example, $P(Z \leq -1.96) = 0.025$, $P(Z \leq -1.65) = 0.05$. Analogously, $P(Z \geq 1.65) = 0.05$, $P(Z \geq 1.96) = 0.025$.

To test whether a single normal variate x_i among a sample of n normal variates is significantly different from a given x value (e.g., x_0), we would

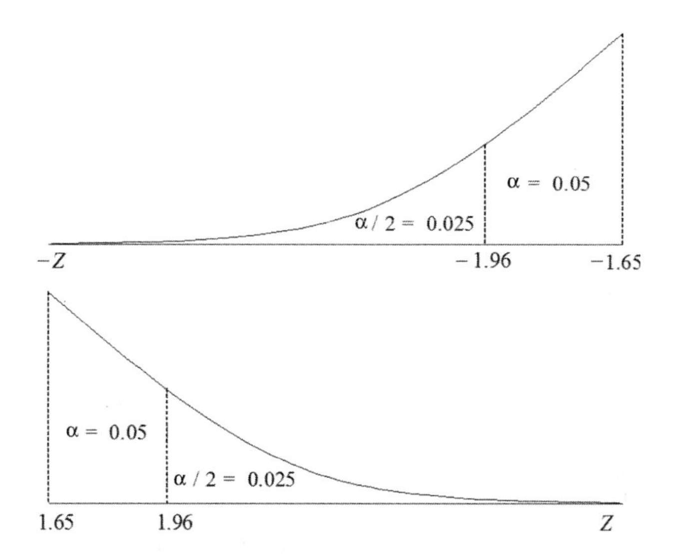

FIGURE 12.3 Regions of rejection for the standard normal distribution. Absolute values of Z for tail probabilities $\alpha = 0.05$ and $\alpha = 0.025$ are $Z = 1.65$ and $Z = 1.96$, respectively.

use the relationship

$$t = \frac{x - x_0}{s/\sqrt{n}}, \tag{12.17}$$

where the test statistic t is the number of standard deviations x_i is from x_0 and n is the sample size. If $n \to \infty$ and $|t| > 1.65$, then the probability that $|x_i|$ exceeds or equals $|t|$ is less than 0.05. If, for example, $n \to \infty$ and $|t| > 2$, then the probability that x_i is at least 2 s.d. from the mean is $P = 0.023$, where P is the *P value*. Standard normal distributions also form the basis of many areas in statistics,
including

- Determination of outliers for a normal distribution (i.e., $|Z_i| > 2$)
- Transformation of standard normal variates into a χ^2 distribution with ν degrees of freedom ($\chi^2_\nu = Z_1^2 + Z_2^2 + \cdots + Z_\nu^2$)
- Correlation between two random variables x and y as $\sum_i (Z_{i,x} Z_{i,y})/(n-1)$

12.5.2 Best Rank Filters

The simplest approach for continuously scaled feature selection is to select the "best ranked" features on the basis of selection criteria, and report classifier performance as a function of genes used. Popular metrics for continuously-scaled feature selection are two-class t tests and Mann-Whitney tests, and multiclass F tests and Kruskal-Wallis tests.

Two-Class t Tests. The t test is used to determine whether two means are the same. Naturally, since the t test is based on the average and standard deviation of a variable's values within each group, the test is of the parametric type. In addition, the t test assumes normality among the data being tested, so the data must be continuously scaled. Specific t tests have been developed for cases in which the variances among variables being tested are assumed to be equal or unequal, the latter of which is known as the *Behrens-Fisher problem*. The null and alternative hypotheses for the t test are

$$\boxed{\begin{aligned} H_0 &: \mu_1 = \mu_2 \\ H_a &: \mu_1 \neq \mu_2. \end{aligned}}$$

The belief (alternative hypothesis) is that the two means are significantly different, and that the null hypothesis will be discredited by sampling the data and performing the test.

Homogeneity of Variance Test for Two Samples. Before conducting a t test, it is instructive to know whether the variances for the two samples are equal or unequal. A simple set of hypotheses can be constructed for this scenario, where the null and the alternative hypotheses for *homoscedasticity* (equal variances) among Ω groups are

$$\boxed{\begin{aligned} H_0 &: \sigma_1^2 = \sigma_2^2 = \cdots = \sigma_\Omega^2 \\ H_a &: \sigma_k^2 \neq \sigma_l^2. \end{aligned}}$$

Tests for equal variance among groups are also known as *homogeneity of variance tests*. The Levene test [2] was originally introduced to test whether all variances are equal among Ω groups

$$W = \frac{(n - \Omega) \sum_{\omega=1}^{\Omega} n_\omega (\bar{Z}_{\cdot\omega} - \bar{Z}_{\cdot\cdot})^2}{(\Omega - 1) \sum_{\omega=1}^{\Omega} \sum_{i=1}^{n_\omega} (Z_{i\omega} - \bar{Z}_{\cdot\omega})^2}, \tag{12.18}$$

where n is the total sample size, n_ω is the sample size in each group, $Z_{i\omega} = |x_{i\omega} - \bar{x}_{\cdot\omega}|$ is the absolute value of an observation centered by its group average, $x_{i\omega}$ is an individual response, $\bar{x}_{\cdot\omega}$ is the mean response in group ω, $\bar{Z}_{\cdot\omega}$ is the mean centered response in group ω, and $\bar{Z}_{\cdot\cdot}$ is the grand mean of the centered responses over all n arrays. For the t test, with only two groups, $\Omega = 2$. If $W > F_{1-\alpha;\Omega-1;n-\Omega}$, then the test is rejected at the

α level of significance, implying that the $\Omega = 2$ independent samples have unequal variance. Thus, if the P value is less than $\alpha = 0.05$, the variances are not equal, whereas if the P value is not less than $\alpha = 0.05$, then the two variances are considered equal (not significantly different).

***t* Tests for Independent Samples.** Assume that two samples of size n_1 and n_2 were drawn from two infinitely large populations. Calculate the sample means $\bar{x}_1$ and $\bar{x}_2$, replace the sample data, resample, recalculate means, resample, and so on. The mean differences $d = \bar{x}_1 - \bar{x}_2$ obtained from multiple samplings would be normally distributed with variance σ^2. A typical question is: Is the difference ($d = \bar{x}_1 - \bar{x}_2$) between two sample means significant? The hypotheses (where δ is the hypothesized difference in means) are as follows:

$$\boxed{\begin{array}{l} H_0 : \delta = 0 \\ H_a : \delta \neq 0. \end{array}}$$

Because we are limited to samples from the infinitely large populations only, we obtain samples of size n_1 and n_2, which have averages $\bar{x}_1$ and $\bar{x}_2$, with sample variances σ_1^2 and σ_2^2. For the two independent samples, the test statistic is

$$\begin{aligned} t &= \frac{d - \delta}{\sqrt{(s_1^2/n_1) + (s_2^2/n_2)}} \\ &= \frac{(\bar{x}_1 - \bar{x}_2) - 0}{\sqrt{(s_1^2/n_1) + (s_2^2/n_2)}} \\ &= \frac{\bar{x}_1 - \bar{x}_2}{\sqrt{(s_1^2/n_1) + (s_2^2/n_2)}}, \end{aligned} \tag{12.19}$$

which follows a t distribution with $\nu = n_1 + n_2 - 2$ degrees of freedom. When sampling during a study or experiment, a priori values of the population variances can either be known or unknown. If they are known, then they can either be equal or unequal. If they are unknown, it is a common practice to assume that they are unequal, since this would be the most conservative approach. In the remaining sections, we address the formulation of the t test when the sample variances are equal and unequal:

Variances Assumed Equal ($\sigma_1^2 = \sigma_2^2$). When the population variances are assumed to be equal, it makes sense to combine together the variance from each sample. This is known as the *common pooled variance*. To calculate the common pooled variance, we must weight each sample variance by the sample size in the form

$$s_p^2 = \frac{(n_1 - 1)s_1^2 - (n_2 - 1)s_2^2}{n_1 + n_2 - 2}. \tag{12.20}$$

Returning to the general equation for the t test in (12.19), a substitution is made replacing the variance for each sample with the common pooled variance, so that the test statistic now becomes

$$\begin{aligned} t &= \frac{\bar{x}_1 - \bar{x}_2}{\sqrt{(s_p^2/n_1) + (s_p^2/n_2)}} \\ &= \frac{\bar{x}_1 - \bar{x}_2}{\sqrt{s_p^2\left[(1/n_1) + (1/n_2)\right]}}. \end{aligned} \tag{12.21}$$

Significance of the test statistic is compared with tail probabilities for Student's distribution at α for $\nu = n_1 + n_2 - 2$ degrees of freedom.

Variances Assumed Unequal ($\sigma_1^2 \neq \sigma_2^2$). When the sample variances for the populations are unequal, we keep the sample variances intact in the general equation for the t statistic given as

$$t = \frac{\bar{x}_1 - \bar{x}_2}{\sqrt{(s_1^2/n_1) + (s_2^2/n_2)}} \tag{12.22}$$

however, the value of t no longer represents the t distribution. This problem was first recognized by Behrens [3], who introduced one of the first solutions. Behrens' approach was later verified by Fisher [4]. We implement the approach introduced by Dixon and Massey [5], which involves a correction to the degrees of freedom, ν_{adj}, in the form

$$\nu_{adj} = \frac{\left[(s_1^2/n_1) + (s_2^2/n_2)\right]^2}{\dfrac{(s_1^2/n_1)^2}{n_1 - 1} + \dfrac{(s_2^2/n_2)^2}{n_2 - 1}}. \tag{12.23}$$

The computed value of ν_{adj} is rounded to the nearest integer. The significance of the test statistic is compared with tail probabilities for Student's t distribution at values of α and ν_{adj} degrees of freedom.

***t* Test for Paired Samples.** The tests discussed above applied for determining whether mean expression values of a single gene are significantly different between two independent groups of arrays. Paired t tests are used for determining whether the average expression of two genes varies within the same individuals. For a sample size of n arrays, the paired t test is

$$t = \frac{\bar{d} - \delta}{s_d/\sqrt{n}} = \frac{\bar{d}}{s_d/\sqrt{n}}, \tag{12.24}$$

where $\bar{d}$ is the mean of the n pairwise differences $d_i = x_{i1} - x_{i2}$ for each array, and s_d is the standard deviation of differences, given as

$$s_d = \sqrt{\frac{\sum_{i=1}^{n}(d_i - \bar{d})^2}{n-1}}. \tag{12.25}$$

In words, the paired t test is based on a total of n arrays, with each array represented by the subscript i. Within each ith array, the difference between expression values of two genes is determined as $d_i = x_{i1} - x_{i2}$. The n values of d_i are averaged to obtain $\bar{d}$, and the standard deviation of d_i over the n arrays is set to s_d. The null hypothesis is that the mean difference between the two genes over all the arrays does not vary (i.e., remains the same), which is denoted as $\delta = 0$. The null hypothesis is rejected if $|t|$ exceeds the tabled t based on $\nu = n - 1$ degrees of freedom. t tests are used less frequently than independent t tests mostly because genes whose expression values differ between two groups are usually sought in order to identify differential expression and possibly to infer disease causality for a given gene. Nevertheless, there are occasions for which differential expression between two genes is sought.

Two-Class Mann-Whitney Test. The Mann-Whitney U test is a *nonparametric* two-sample test to determine whether two independent samples are from the same distribution [6]. Unlike the t test, which assumes an underlying normal distribution of the data, the Mann-Whitney test requires that the data be at least ordinal and not necessarily normally distributed. The Mann-Whitney test should be applied when little is known about the distributional properties for both samples, or the assumptions of normality do not hold. The goal of the Mann-Whitney test is to compare two sets of data to determine whether they are from the same distribution. By "same distribution" we mean that values of one distribution are either less or greater than values of the other distribution; if this is the case, then the two distributions are different. Stated formally, the null (H_0) and alternative (H_a) hypotheses of the Mann-Whitney test are as follows:

H_0: Both samples from the same distribution (null hypothesis)
H_a: Both samples from different distributions (alternative hypothesis)

Given these hypotheses, the goal is to discredit or reject the null hypothesis, and disprove that values of the two distributions are not the same. We begin calculations for the Mann-Whitney test by first determining the U statistic for each group in the form

$$\begin{aligned} U_1 &= n_1 n_2 + \frac{n_1(n_1+1)}{2} - R_1 \\ U_2 &= n_1 n_2 + \frac{n_2(n_2+1)}{2} - R_2, \end{aligned} \tag{12.26}$$

where n_1 and n_2 are the sample sizes for each of the two groups and R_1 and R_2 are the sum of the ranks within each group. The U statistic is taken as the lowest value among U_1 and U_2:

$$U = \min(U_1, U_2). \tag{12.27}$$

The test statistic for the Mann-Whitney test is standard normal distributed as

$$\begin{aligned} Z &= \frac{U - \mu}{\sigma} \\ &= \frac{U - (n_1 n_2/2)}{\sqrt{[n_1 n_2(n_1 + n_2 + 1)]/12}}. \end{aligned} \tag{12.28}$$

Example 12.1 Let's consider an example that tests whether expression of a single mRNA in two independent groups of arrays are from the same distribution. A laboratory assay was carried out in which normalized fluorescent intensity levels were generated for expression of gene A in two groups of mice: wild-type and knockout. The data are shown in Table 12.2, which includes the normalized fluorescent intensity levels and ranks. Ranks are determined by first sorting the data for both samples in ascending order, and then assigning ranks. Tied expression values are handled by averaging their original rank assignments.

Values of U_1 and U_2 are found by substituting in the sample sizes and sum of ranks:

$$\begin{aligned} U_1 &= 140 = (10)(15) + \frac{10(11)}{2} - 65 \\ U_2 &= 10 = (10)(15) + \frac{15(16)}{2} - 260. \end{aligned} \tag{12.29}$$

TABLE 12.2 Tissue Expression of Gene A in Two Samples

Wild-type	Rank	Knockout	Rank
1.9	1	4.5	8.5
2.9	2	4.5	8.5
3.4	3	4.6	11
3.9	4.5	4.7	13
3.9	4.5	4.9	14
4.1	6.5	5.5	16.5
4.1	6.5	5.5	16.5
4.6	11	5.7	18
4.6	11	5.8	19
5.3	15	5.9	20
	$R_1 = 65$	6.2	21
	$n_1 = 10$	6.3	22
		6.5	23
		6.7	24
		7.1	25
			$R_2 = 260$
			$n_2 = 15$

The minimum value among U_1 and U_2 is 10, so the test statistic becomes

$$\begin{aligned} Z &= \frac{U - (n_1 n_2)/2}{\sqrt{[(n_1)(n_2)(n_1 + n_2 + 1)]/12}} \\ &= \frac{10 - [(10)(15)]/2}{\sqrt{[(10)(15)(10 + 15 + 1)]/12}} \\ &= \frac{10 - 75}{18.03} \\ &= -3.61. \end{aligned} \tag{12.30}$$

Because the calculated Z statistic of -3.61 is much less than $Z_{0.025} = -1.96$, we reject the null hypothesis that expression of gene A in the two groups is the same. Another way of looking at the results is to compare the left tail probability of $\Phi(-3.61) = 0.0003$, which is much less than $\alpha = 0.05$. Algorithm 12.1 lists the steps necessary for the two-class Mann-Whitney test.

Implementation. Coding the Mann-Whitney test is straightforward. Below we provide the regular and permutation forms of the test. One of the most important aspects of accurately calculating the Mann-Whitney test statistics

Algorithm 12.1: Two-Class Mann-Whitney Test

Data: 2 groups of arrays
Result: Mann-Whitney test statistic, Z score.

1. Collect data from two samples.
2. Determine the sample sizes n_1 and n_2.
3. Rank-order the data across both groups. For tied data values, average the ranks.
4. Calculate the sum of ranks in group 1, R_1 and group 2, R_2.
5. Calculate the U statistic for each group, i.e., U_1 and U_2 setting $U = \min(U_1, U_2)$.
6. Calculate the expected U statistic μ and standard deviation σ.
7. Calculate the test statistic Z.
8. Determine the P value or probability that the test statistic exceeds the tabled critical value, $P(|Z| > Z_\alpha)$.

is that the beginning and end of runs of ranks for tied observations need to be identified, and the ranks between averaged. Arrays `xvec(n)` and `cvec(n)` store the continuous x values and class membership for each observation, respectively, while `ranksum(2)` and `ranksort(n)` contain the sum of ranks for the two classes and individual ranks for observations; vector `run(n)` stores the runs of tied observations which are averaging ranks;

```
(NOTE: Uses function NORMCDF(x), where x is a standard
normal variate and NORMCDF(x) is the P-value.)

 Sub mannwhitney(ByRef vartested, ByRef rn1, ByRef rn2, _
    ByRef obsr1, ByRef obsr2, ByRef obsu1, ByRef obsu2, _
    ByRef obsmeanu, ByRef obsU, ByRef obssdU, ByRef obsz, _
    ByRef obstailprob, ByRef meanpermz, ByRef
       permtailprob, _
    ByRef iterations)

    Dim i, j, k, m As Integer
    Dim n1, n2, N As Integer
    Dim descending As Object
    Dim MWClasses As Byte
    Dim iter As Integer
    Dim xvec() As Double
    Dim cvec() As Integer
    Dim irow As Integer
    Dim x1() As Double
```

```
    Dim x2() As Double
    Dim meanU, sigmaU As Double
    Dim r1 As Double
    Dim r2 As Double
    Dim u, u1, u2 As Double
    Dim buffer As Integer
    Dim z As Double
    Dim numless As Integer
    Dim xsort(nrow) As Double
    Dim rankxsort(nrow) As Double
    Dim runstart As Integer
    Dim runfinish As Integer
    Dim arr() As Double
    Dim ndx() As Integer

    rn1 = 0
    rn2 = 0
   'rncol = # arrays
   'ncol = grouping variable
   'vartested = test feature

    'Fill arrays for each of the two groups
    For i = 1 To rncol
        If GroupingVar = 1 Then
            If xmatrix(i, ncol) = 1 Then
                rn1 = rn1 + 1
                ReDim Preserve x1(rn1)
                x1(rn1) = xmatrix(i, vartested)
            End If
            If xmatrix(i, ncol) = 2 Then
                rn2 = rn2 + 1
                ReDim Preserve x2(rn2)
                x2(rn2) = xmatrix(i, vartested)
            End If
        End If

    Next i

    ReDim xvec(rn1 + rn2)
    ReDim cvec(rn1 + rn2)
    For i = 1 To rn1
        xvec(i) = x1(i)
        cvec(i) = 1
    Next i

    For i = 1 To rn2
        xvec(rn1 + i) = x2(i)
        cvec(rn1 + i) = 2
```

```
Next i

nrow = rn1 + rn2

'Mann-Whitney test (with ties)

ReDim arr(nrow)
ReDim ndx(nrow)
For i = 1 To nrow
    ndx(i) = i
    arr(i) = xvec(i)
Next i

descending = False
Call sort(arr, ndx, nrow, descending)

For i = 1 To nrow
    xsort(i) = arr(i)
    rankxsort(i) = i
Next i

Dim run(nrow) As Byte

For i = 1 To nrow - 1
    If xsort(i) = xsort(i + 1) Then
        run(i) = 1
        run(i + 1) = 1
    End If
    If xsort(i) <> xsort(i + 1) AndAlso run(i) = 1
       Then run(i) = 2
Next i

If xsort(nrow - 1) = xsort(nrow) Then run(nrow) = 2

For i = 0 To nrow - 1 'start at zero since known 0
   then 1 in (1)
    If (run(i) = 0 AndAlso run(i + 1) = 1) Then
       runstart = i + 1
    If (run(i) = 2 AndAlso run(i + 1) = 1) Then
       runstart = i + 1
    If run(i) = 1 AndAlso run(i + 1) = 2 Then
        runfinish = i + 1
        For k = runstart To runfinish
            rankxsort(k) = (runstart + runfinish) / 2
        Next k
    End If
Next i
```

```
MWClasses = 2
Dim sampsize(MWClasses) As Integer
Dim ranksum() As Double
For k = 1 To MWClasses
    For i = 1 To nrow
        If cvec(ndx(i)) = k Then sampsize(k) += 1
    Next i
Next k
n1 = sampsize(1)
n2 = sampsize(2)

For iter = 1 To iterations

    If iter > 1 Then
        N = nrow
        Do While N >= 1
            m = System.Math.Round((N - 1) * Rnd()) + 1
            buffer = cvec(N)
            cvec(N) = cvec(m)
            cvec(m) = buffer
            N = N - 1
        Loop
    End If

    ReDim ranksum(MWClasses)
    For k = 1 To MWClasses
        For i = 1 To nrow
            If cvec(ndx(i)) = k Then ranksum(k) +=
               rankxsort(i)
        Next i
    Next k

    u1 = (n1 * n2) + (n1 * (n1 + 1) / 2) - ranksum(1)
    u2 = (n1 * n2) + (n2 * (n2 + 1) / 2) - ranksum(2)
    u = minimum(u1, u2)

    meanU = n1 * n2 / 2
    sigmaU = System.Math.Sqrt(n1 * n2 *
       (n1 + n2 + 1) / 12)
    z = (u - meanU) / sigmaU

    If iter = 1 Then
        FileClose(15)
        FileOpen(15, OutputDirectory & "MW_test.csv",
           OpenMode.Output)
        For k = 1 To 2
            irow = 0
```

```
                For i = 1 To nrow
                    If cvec(ndx(i)) = k Then
                        irow += 1
                        PrintLine(15, irow & "," &
                           cvec(ndx(i)) & "," & xsort(i)_
                                    & "," & rankxsort(i))
                    End If
                Next i
            Next k
            FileClose(15)

            obsr1 = ranksum(1)
            obsr2 = ranksum(2)
            obsu1 = u1
            obsu2 = u2
            obsmeanu = meanU
            obsU = u
            obssdU = sigmaU
            obsz = z
            obstailprob = normcdf(z)
            numless = 0
            meanpermz = 0
        End If

        If iter > 1 Then
            If z < obsz Then numless += 1
            meanpermz += z
        End If
    Next iter

    permtailprob = numless / iterations
    meanpermz = meanpermz / iterations

End Sub
```

Multiclass F Test. Analysis of variance (ANOVA) is used for testing whether the averages in three or more groups are equal. ANOVA is based on means and variances and therefore lies in the area of parametric statistics. Formally stated, the null hypothesis is that all group means being tested are equal, and the alternative hypothesis states that at least a single pair of any means are not equal. In statistical notation, the hypotheses are as follows:

$$\begin{aligned} H_0 &: \mu_1 = \mu_2 = \cdots = \mu_\Omega \\ H_a &: \mu_k \neq \mu_l. \end{aligned}$$

The only criterion needed for rejecting the null hypothesis is for at least two means to be significantly different. ANOVA requires one dependent variable y that is normally distributed, and homoscedasticity, or equal variances of the outcome response in the groups considered. The Levene test of equal variances was introduced in the section on two-class t tests. If the Levene test is significant, suggesting unequal variances, then the Kruskal-Wallis test (discussed later) should be used. In addition, there may be one or more independent variables, all of which are required to be categorical. Variants of ANOVA are one-way ANOVA, two-way ANOVA, analysis of covariance, and repeated measures ANOVA. The simplest type of ANOVA is called a *one-way* or *one-factor* ANOVA. Because it is always assumed that the dependent variable y for an ANOVA model is normally distributed and continuously scaled, the majority of the nomenclature for different ANOVA models rests with the structure of the independent variables. Thus, one-way ANOVA is based on one independent variable, which is categorical. Independent variables are more commonly called *factors*, having several *levels*. As an example, for `gender` there are two levels: males and females, whereas for `boxcolor` there may be three levels: red, green, and blue.

Data Format. Define $y_{i\omega}$ as a continuously scaled observation for the ith subject in group ω with sample size n_ω ($i = 1, 2, \ldots, n_\omega$; $\omega = 1, 2, \ldots, \Omega$). For data arrangement, Table 12.3 lists the responses ($y_{i\omega}$) for all subjects in a study. Columns in the table represent treatment groups, and because each group can have a different (i.e., its own) sample size n_ω, the number of rows only reflect the greatest sample size among all groups.

ANOVA Model. The ANOVA model states that each random observation can be represented by the model

$$y_{i\omega} = \mu + \alpha_\omega + \varepsilon_{i\omega}, \tag{12.31}$$

where μ is the grand mean, α_ω is the treatment mean for group ω, and $\varepsilon_{i\omega}$ is the random error. Another form of the grand mean is

$$\mu = \bar{y}_{\cdot\cdot} = \frac{\sum_{\omega=1}^{\Omega} \sum_{i=1}^{n_\omega} y_{i\omega}}{n}, \tag{12.32}$$

and treatment mean is

$$\alpha_\omega = \bar{y}_{\cdot\omega} = \frac{\sum_{i=1}^{n_\omega} y_{i\omega}}{n_\omega}. \tag{12.33}$$

Using the model effects based on y, we can rewrite (12.31) as

$$y_{i\omega} = \bar{y}_{\cdot\cdot} + (\bar{y}_{\cdot\omega} - \bar{y}_{\cdot\cdot}) + (y_{i\omega} - \bar{y}_{\cdot\omega}). \tag{12.34}$$

TABLE 12.3 Data Format for One-Way ANOVA

Group 1	Group 2	$\cdots$	Group ω	$\cdots$	Group Ω
y_{11}	y_{11}	$\cdots$	$y_{1\omega}$	$\cdots$	$y_{1\Omega}$
y_{21}	y_{22}	$\cdots$	$y_{2\omega}$	$\cdots$	$y_{2\Omega}$
y_{31}	y_{32}	$\cdots$	$y_{3\omega}$	$\cdots$	$y_{3\Omega}$
$\vdots$			$\vdots$		$\vdots$
y_{i1}	y_{i2}	$\cdots$	$y_{i\omega}$	$\cdots$	$y_{i\Omega}$
$\vdots$	$\vdots$		$\vdots$		$\vdots$
$y_{n_1 1}$	$y_{n_2 2}$	$\cdots$	$y_{n_\omega \omega}$	$\cdots$	$y_{n_\Omega \Omega}$

Subtracting $\bar{y}_{\cdot\cdot}$ from the right side of (12.34) yields

$$y_{i\omega} - \bar{y}_{\cdot\cdot} = (\bar{y}_{\cdot\omega} - \bar{y}_{\cdot\cdot}) + (y_{i\omega} - \bar{y}_{\cdot\omega}), \tag{12.35}$$

and squaring all terms and summing over i and ω gives us

$$\sum_{\omega=1}^{\Omega}\sum_{i=1}^{n_\omega}(y_{i\omega} - \bar{y}_{\cdot\cdot})^2 = \sum_{\omega=1}^{\Omega}\sum_{i=1}^{n_\omega}(\bar{y}_{\cdot\omega} - \bar{y}_{\cdot\cdot})^2 + \sum_{\omega=1}^{\Omega}\sum_{i=1}^{n_\omega}(y_{i\omega} - \bar{y}_{\cdot\omega})^2. \tag{12.36}$$

A more numerically efficient method for calculating the sum of squares is in the form

$$\sum_{\omega=1}^{\Omega}\sum_{i=1}^{n_\omega} y_{i\omega}^2 - \frac{y_{\cdot\cdot}^2}{n} = \sum_{\omega=1}^{\Omega}\frac{y_{\cdot\omega}^2}{n_\omega} - \frac{y_{\cdot\cdot}^2}{n} + \sum_{\omega=1}^{\Omega}\sum_{i=1}^{n_\omega} y_{i\omega}^2 - \sum_{\omega=1}^{\Omega}\frac{y_{\cdot\omega}^2}{n_\omega}. \tag{12.37}$$

In summary, the sum of squares total is equal to the *treatment sum of squares* (error between treatments) and the *error sum of squares* (within treatments)

$$\mathrm{SST}_{(\mathrm{total})} = \mathrm{SST}_{(\mathrm{between})} + \mathrm{SSE}_{(\mathrm{within})}. \tag{12.38}$$

Dividing each of these sums of squares by their individual degrees of freedom, one can obtain the mean-square errors

$$\mathrm{MST} = \frac{\mathrm{SST}}{\Omega - 1}, \tag{12.39}$$

and

$$\mathrm{MSE} = \frac{\mathrm{SSE}}{n - \Omega}. \tag{12.40}$$

The F statistic is calculated as the ratio of MST to MSE in the form

$$F = \frac{\text{MST}}{\text{MSE}} \sim F_{\Omega-1,n-\Omega}, \tag{12.41}$$

with numerator degrees of freedom $\Omega - 1$ and demominator degrees of freedom $n - \Omega$.

Example 12.2 Table 12.4 lists example data for normalized tissue expression of gene A for three treatment groups. Notice that the table is asymmetric because the sample sizes are not the same. Using the formulation described above, we now work through calculations for obtaining the sum of squares (SST, SSE), mean squares (MST, MSE), degrees of freedom for the F test, and the F statistic. The P value for the F test is based on looking up the tabled critical values of F for the specified numerator and denominator degrees of freedom. Now that the data are arranged appropriately in Table 12.4, we are ready to start calculating the F statistic for the observed data, to determine whether the means in the three groups are different. The first element

TABLE 12.4 Normalized Tissue Expression of Gene A in Three Treatment Groups: Control, Drug A, and Drug B

	Control	Drug A	Drug B	
	1.26	1.46	5.34	
	1.89	1.55	4.77	
	1.50	1.43	5.19	
	2.01	1.34	4.71	
	2.52	1.06	4.62	
	1.59	1.80	5.16	
	1.90	1.73	4.86	
	1.64	1.64	4.85	
	2.00	1.42	5.25	
	1.97	1.29	5.42	
	1.67	1.46	4.76	
	2.41		5.13	
			5.07	
			4.48	
			4.74	
$y_{\cdot\omega}$:	22.27	16.18	74.35	$y_{\cdot\cdot} = 112.8$
n_ω :	12	11	15	$n = 38$
$\bar{y}_\omega$:	1.856	1.471	4.957	
s :	0.167	0.003	0.058	
$\sum_{i=1}^{n_\omega} y_{i\omega}^2$:	42.7737	24.2328	369.6571	$\sum_{\omega=1}^{\Omega}\sum_{i=1}^{n_\omega} y_{i\omega}^2 = 436.6636$

to determine is SST, which captures the *between-treatment* error based on differences between the treatment means and the grand mean, weighted by sample size. If the treatments means are close to the grand mean, then there might be little difference due to the categorization of $y_{i\omega}$ responses in the treatment groups. This effect is commonly known as *effect size*. Substituting in the sums from the bottom of Table 12.4, we obtain

$$\begin{aligned} \text{SST} &= \sum_{\omega=1}^{\Omega} n_\omega (\bar{y}_{.\omega} - \bar{y}_{..})^2 \\ &= \sum_{\omega=1}^{\Omega} \frac{y_{\cdot\omega}^2}{n_\omega} - \frac{y_{..}^2}{n} \\ &= \frac{(22.27)^2}{12} + \frac{(16.18)^2}{11} + \frac{(74.35)^2}{15} - \frac{(112.8)^2}{38} \\ &= 98.82. \end{aligned} \tag{12.42}$$

The degrees of freedom (d.f.) for SST are determined next, based the number of groups:

$$\begin{aligned} \text{d.f.} &= \Omega - 1 \\ &= 3 - 1 \\ &= 2. \end{aligned} \tag{12.43}$$

The SSE or "residual error," which picks off the error between each observation and its treatment mean reveals the *within treatment* error. When SSE, is large, the residual error can outweigh the treatment effects resulting in a nonsignificant test result:

$$\begin{aligned} \text{SSE} &= \sum_{\omega=1}^{\Omega} \sum_{i=1}^{n_\omega} (y_{i\omega} - \bar{y}_{\cdot\omega})^2 \\ &= \sum_{\omega=1}^{\Omega} \sum_{i=1}^{n_\omega} y_{i\omega}^2 - \sum_{\omega=1}^{\Omega} \frac{y_{\cdot\omega}^2}{n_\omega} \\ &= \left\{ 436.66 - \left[\frac{(22.27)^2}{12} + \frac{(16.18)^2}{11} + \frac{(74.35)^2}{15} \right] \right\} \\ &= 3.01, \end{aligned} \tag{12.44}$$

$$\begin{aligned} \text{d.f.} &= n - \Omega \\ &= 38 - 3 \\ &= 35. \end{aligned} \tag{12.45}$$

The F statistic can now be calculated as the ratio of the two mean-square errors:

$$F = \frac{\text{MST}}{\text{MSE}} = \frac{98.82/2}{3.01/35} = \frac{49.41}{0.09} = 575.16. \tag{12.46}$$

The tabled $F_{0.95;2,35}$ is 3.27, suggesting that the means are not all equal, with a P value much less than 0.05.

Multiclass Kruskal-Wallis Test. The Kruskal-Wallis test is used to determine whether three or more samples are from the same distribution [7]. It is a nonparametric test based on ranks. The null hypothesis is $H_0 : y_{i1}, y_{i2}, \ldots, y_{i\omega}, \ldots, y_{i\Omega} \sim \mathcal{D}$, which states that variates in groups $\omega = 1, 2, \ldots, \Omega$ are from same distribution. The alternative hypothesis is $H_a : y_{i1}, y_{i2}, \ldots, y_{i\omega}, \ldots, y_{i\Omega} \nsim \mathcal{D}$, which implies that variates in each sample are not from the same distribution $\mathcal{D}$. The variables used in a Kruskal-Wallis test include one test variable that is at least measured on the ordinal scale, and one categorical variable that represents the groups.

Example 12.3 To begin a worked example, Table 12.5 lists total tissue expression values of gene A on $n = 38$ array partitioned into three treatment groups: control, drug A, and drug B. The 38 expression values were

TABLE 12.5 Tissue Expression of Gene A on n = 38 Arrays in Three Treatment Groups and the Assigned Ranks[a]

Control, y_{i1}	Rank	Drug A, y_{i2}	Rank	Drug B, y_{i3}	Rank
1.26	2	1.06	1	4.48	24
1.50	9	1.29	3	4.62	25
1.59	11	1.34	4	4.71	26
1.64	12.5	1.42	5	4.74	27
1.67	14	1.43	6	4.76	28
1.80	16.5	1.46	7.5	4.77	29
1.90	18	1.46	7.5	4.85	30
1.97	19	1.55	10	4.86	31
2.00	20	1.64	12.5	5.07	32
2.01	21	1.73	15	5.13	33
2.41	22	1.80	16.5	5.16	34
2.52	23		$R_2 = 88$	5.19	35
	$R_1 = 188$		$n_2 = 11$	5.25	36
	$n_1 = 12$			5.34	37
				5.42	38
					$R_3 = 465$
					$n_3 = 15$

first sorted in ascending order, and then assigned a rank from 1 to 38. Ranks of tied expression values were averaged, resulting in decimal-valued ranks.

The test statistic used for the Kruskal-Wallis test is the χ^2 distribution with $\Omega - 1$ degrees of freedom. Substituting in the calculated values in Table 12.5 into the equation for the test statistic, we have

$$\begin{aligned}\chi^2_{\Omega-1} &= \frac{12}{n(n+1)} \sum_{\omega=1}^{\Omega} \frac{R_\omega^2}{n_\omega} - 3(n+1) \\ &= \frac{12}{38(38+1)} \left[\frac{(188)^2}{12} + \frac{(88)^2}{11} + \frac{(465)^2}{15} + \right] - 3(38+1) \\ &= 29.27 \end{aligned} \tag{12.47}$$

The tabled critical χ^2 value χ is approximately distributed with 2d.f., and therefore the calculated test statistic value of 29.27 is very significant.

12.5.3 Randomization Tests

Randomization tests, also known as *permutation tests* or *empirical P-value tests*, are tests based on permutations of either the data or class labels. A stringent requirement for randomization tests is that the null hypothesis is enforced during the permutation process. This is easily achieved for pairwise correlation, where reshuffling one of the data vectors is sufficient to enforce the null hypothesis of no correlation. However, permutation tests with a null hypothesis based on distributional assumptions may be more challenging to develop. The following sections address various ways to employ randomization tests as an alternative to performing distribution-based inferential hypothesis testing.

Within-Gene Randomization Test. This test shuffles class labels across the arrays considered. It can be applied to two-class or multiclass problems, and here we show a worked example for a two-class problem. For each gene, calculate the observed t statistic t_j $(j = 1, 2, \ldots, p)$ based on the raw data for groups A and B. Next use B permutations, where the number of permutations required is

$$B = \binom{n_1 + n_2}{\min\{n_1, n_2\}}. \tag{12.48}$$

During each permutation, calculate the t statistic $t_j^{(b)}$ for permutation $(b = 1, 2, \ldots, B)$ after randomly shuffling either the feature values or array class labels. The empirical P value for gene j after B iterations is

$$p_j = \frac{\#\{b : |t_j^{(b)}| \geq |t_j|\}}{B}. \tag{12.49}$$

Table 12.6 shows results of 19 permutations of expression values for a single gene and the resulting t statistics for a two-class problem with $n = 16$ arrays. In this table, note the large difference between the observed t statistic when $b = 0$ and the test statistic based on permuted class labels for iterations $b = 1, 2, \ldots, 19$. For these example reshufflings, the P value would be equal to $P = 0/19$, since $|t|$ never exceeds $|-17.15|$ when $b > 0$.

Between-Gene Randomization Test. Another approach for reshuffling during each iteration involves a comparison between the observed t test for a given gene and the t tests for other genes after their values have been reshuffled. For each gene, calculate the observed t statistic t_j from the raw data for groups A and B. Next, permute the class labels across both groups (A and B) and calculate the t statistic $t_j^{(b)}$ for permutation b (b = 1, 2, . . ., B). During the bth iteration, count the number of times that the t statistics of other genes, based on permuted labels [i.e., $t_j^{(b)}$] exceed the observed t_j for the jth gene. After, say, $B = 100$ iterations, determine the P value for the jth gene as

$$P_j = \frac{\#\{j, b : |t_j^{(b)}| \geq |t_j|\}}{pB} \tag{12.50}$$

This method preserves the dependence structure between genes by accounting for test results of other genes rather than results obtained after reshuffling data within a single gene. The Westfall-Young method adopts this approach in a more complete fashion.

12.5.4 Multitesting Problem

Microarray datasets often suffer from the "small-sample problem," where the the number of features greatly outweighs the number of arrays, $n \ll p$. This exacerbates the multitesting problem, since the use a P-value criterion of $\alpha = 0.05$ to identify significant differential expression among 1000 genes will result in 50 false positives.

Bonferroni-Sidak One-Step Procedures. The *familywise error rate* (FWER), introduced by Westfall and Young [8] as a means for controlling the

TABLE 12.6 Within-Gene Permutation Test Results for Differential Expression of a Single Gene Across $n = 16$ Arrays and 2 Classes ($x_1, \ldots, x_8, y_1, \ldots y_8$)[a]

b	x_1	x_2	x_3	x_4	x_5	x_6	x_7	x_8	y_1	y_2	y_3	y_4	y_5	y_6	y_7	y_8	PSE[b]	t
0	1	2	3	4	5	6	7	8	22	23	24	25	26	27	28	29	1.22	−17.15
1	24	26	25	29	23	3	8	6	4	22	27	7	28	5	1	2	5.52	1.09
2	24	7	6	29	28	1	23	27	3	2	4	25	5	8	22	26	5.50	1.14
3	24	29	5	25	7	28	6	4	2	23	3	8	22	1	26	27	5.72	0.35
4	25	24	6	27	1	3	26	23	7	29	4	28	5	22	8	2	5.66	0.66
5	26	22	23	24	4	27	2	25	29	3	8	28	7	5	5	1	5.30	1.56
6	7	25	4	22	8	29	23	3	1	6	24	26	5	27	2	28	5.74	0.04
7	26	22	28	2	5	24	4	25	3	23	27	7	1	8	29	6	5.64	0.71
8	1	8	26	5	3	28	23	22	29	27	24	7	2	6	25	4	5.74	−0.17
9	3	4	22	27	5	25	7	1	24	23	6	28	29	8	2	26	5.48	−1.19
10	26	3	6	25	22	4	2	28	1	5	8	29	7	27	23	24	5.74	−0.17
11	23	28	2	26	27	5	7	1	22	6	4	8	24	29	25	3	5.74	−0.04
12	27	25	28	24	2	22	8	1	23	4	6	5	29	7	3	26	5.63	0.75
13	6	22	25	27	3	5	2	28	7	24	8	29	1	4	26	23	5.74	−0.09
14	22	8	26	23	28	7	4	27	29	5	2	3	6	25	1	24	5.50	1.14
15	2	27	6	5	29	8	25	1	22	3	4	7	23	28	24	26	5.63	−0.75
16	22	8	24	3	2	1	27	6	7	26	23	28	4	29	5	25	5.45	−1.24
17	23	26	4	22	7	8	5	24	1	25	29	27	6	3	2	28	5.74	−0.04
18	6	23	8	22	1	27	4	25	7	26	3	24	2	29	5	28	5.74	−0.17
19	5	26	24	29	27	23	7	28	8	25	3	22	4	2	6	1	4.72	2.60

[a]When $b = 0$ the unshuffled raw input data are used; when $b = 1, 2, \ldots, 19$, the class labels are randomly shuffled within each row to obtain t statistics.

[b]Pooled standard error.

false positive error rate, is defined as the probability of at least one Type I error. One-step methods are the simplest form of bounding the FWER and involve a straightforward single-step calculation. The Bonferroni one-step adjustment is the most popular correction for multiple testing because of its simplicity. The Bonferroni adjusted α level for Type I errors is determined as

$$\alpha^* = \frac{\alpha}{\#\text{tests}}. \quad (12.51)$$

Hence, for 100 tests, α^* is equal to $0.0005 = 0.05/100$. The Sidak adjustment is

$$\alpha^* = 1 - (1 - \alpha)^{1/\#\text{tests}}, \quad (12.52)$$

which approximates the Bonferroni adjustment. The disadvantages of the Bonferroni-Sidak methods are that they assume independence between the tests, and are highly conservative, resulting in reduced power and greater chances of missing an effect.

Benjamini-Hochberg False Discovery Rate (FDR). The false discovery rate (FDR) was introduced by Benjamini and Hochberg [9] as an alternative to FWER, and is defined as the expected proportion of Type I errors among all rejected hypotheses. The Benjamini-Hochberg method is a *stepdown* method and one of the first FDR methods introduced. Let $P_1, P_2, \ldots, P_m$ represent P values for m genes ($j = 1, 2, \ldots, m$), and let $P_{(1)} \leq P_{(2)} \leq \cdots \leq P_{(m)}$ represent their ranked counterparts. Define k as the greatest value of j when the statement $P_{(j)} \leq \alpha j/m$ is true, where α is the desired FDR (e.g., 5%, 10%, or 15%). Reject all $P_{(j)}$ for which $j \leq k$, and define all other P values for which $j > k$ as null.

Storey q Values for PFDR. Storey q values [10] represent a positive FDR method (pFDR). Let m_0 represent the number of truly null tests among m tests. Since it is known that null tests are distributed $\mathcal{U}(0, 1)$, the expected number of null tests in the interval $(\lambda, 1)$ is approximated simply as $W(\lambda) \approx m_0(1 - \lambda)$. For well-chosen values of λ, we can safely assume that P values greater than λ are also null, and thus $W(\lambda) = \#\{j : P_j > \lambda\}$. Combining the two relationships, we obtain an estimate of the number of null tests assuming λ as

$$\hat{m}_0(\lambda) = \frac{\#\{j : P_j > \lambda\}}{1 - \lambda}. \quad (12.53)$$

Let us assume that all P values greater than λ are null. An estimate of the number of null tests is then

$$\hat{m}_0 = \frac{\#\{j : P_j > 0.5\}}{0.5}. \quad (12.54)$$

The proportion of null tests out of all tests is

$$\hat{\pi}_0 = \frac{\hat{m}_0}{m} = \frac{\#\{j : P_j > 0.5\}}{m(0.5)}. \tag{12.55}$$

The false discovery rate controlled at level α is

$$\text{FDR}(\alpha) = \frac{\#\text{ false positive tests}}{\#\text{ significant tests}} = \frac{\hat{m}_0\alpha}{\#\{j : P_j < \alpha\}}, \tag{12.56}$$

where the numerator $\hat{m}_0\alpha$ represents the number of null tests times the probability of rejecting the null tests given that the null is true (Type I error), and the denominator is the number of P values less than α that are assumed to be significant. When sorting P values in descending order, it warrants noting that $\#\{j : P_j < \alpha\}$ is equal to j when looping through the descending rank-ordered P values from $P_{(m)}, P_{(m-1)}, P_{(m-1)}, \ldots, P_{(1)}$. The FDR at a given P value can now be expressed as

$$\text{FDR}(P_{(j)}) = \frac{\hat{m}_0 P_{(j)}}{j}. \tag{12.57}$$

Note that $\text{FDR}(P_{(j)})$ is not the false discovery rate for the specific gene having the jth rank-ordered P value, but rather for all genes assumed significant with P values less than $P_{(j)}$, that is, the j remaining genes when P values are ranked in descending order. The q value associated with each P value is the minimum FDR that can be attained assuming all P values less than $P_{(j)}$ to be significant:

$$\hat{q}(P_{(j)}) = \min_{P_{(j)}<t}\left(\frac{\hat{m}_0 t}{\#\{P_j < t\}}\right) = \min\left(\frac{\hat{m}_0 P_{(j)}}{j}, \hat{q}(P_{(j+1)})\right). \tag{12.58}$$

The q values are ranked in increasing order, and the number of genes below, say, 0.05, are significant, and 5% of them are false positives (i.e., the FDR is 5%). The FDR reflects the proportion of null statistics among all statistics considered significant, that is, the proportion of false positives among all significant statistics. P values, however, reflect the false positive rate, which is the probability that a null statistic is significant.

Statistical Analysis of Microarrays (SAM). The statistical analysis of microarrays (SAM) correction for multiple testing provides an estimate of the FDR as a function of a threshold parameter called Δ [11]. For gene $j, (i = 1, 2, \ldots, p)$ assume a sample of size n_1 of gene expression values among control arrays labeled A, and a sample size of n_2 for expression values due to treatment, all of which are labeled B. The hypothesis test applied depends on the choice of the sample pooled standard deviation in the denominator of the t statistic. For *t statistics*, the pooled standard deviation under the assumption of *equal variance* is

$$s_j = \sqrt{\frac{(n_1 - 1)s_1^2 - (n_2 - 1)s_2^2}{n_1 + n_2 - 2}}, \tag{12.59}$$

whereas for Welch statistics, where *unequal variance* is assumed, the standard deviation is

$$s_j = \sqrt{\frac{s_1^2}{n_1} + \frac{s_2^2}{n_2}}. \tag{12.60}$$

The statistic t_j for gene j is

$$t_j = \frac{\bar{X}_1 - \bar{X}_2}{s_0 + s_j}, \tag{12.61}$$

where s_0 is set equal to the 5th percentile of ascending sorted s_j used to make the pooled standard deviation independent of gene expression level. For each gene j, calculate

$$\Delta_j = |t_j| - \frac{\sum_{b=1}^{B} |t_j^{(b)}|}{B}, \tag{12.62}$$

where t_j is the observed t statistic based on the raw data and $t_j^{(b)}$ $(b = 1, 2, \ldots, B)$ is the t statistic for permutation b after randomly shuffling the group labels A and B across both groups. The number of permutations required for each gene is

$$B = \binom{n_1 + n_2}{\min\{n_1, n_2\}}. \tag{12.63}$$

Next, perform an ascending sort of all Δ_j and obtain 40 percentiles of sorted Δ_j, that is, $\Delta_{\text{cut},l} = \Delta(l * 0.025 * p)$, $l = 1, 2, \ldots, 40$. At each value of $\Delta_{\text{cut},l} (l = 1, 2, \ldots, 40)$, significant genes are those for which $\Delta_j \geq \Delta_{\text{cut},l}$. Among the significant genes identified, determine the lowest $|t_j|$ as $\min\{|t_j|\}$ and set as a cutoff, $t_{\min,l}$. The average number of false positives at each level of $\Delta_{cut,l}$ is

$$\overline{\#FP_l} = \frac{\sum_{j=1}^{k} \#\{b : |t_j^{(b)}| \geq t_{\min,l}\}}{p}, \tag{12.64}$$

and the false discovery rate (FDR) at $\Delta_{\text{cut},l}$ is based on the relationship

$$\text{FDR}_\Delta = \frac{\overline{\#FP_l}}{\#\{j : \Delta_j > \Delta_{\text{cut},l}\}}. \tag{12.65}$$

These steps are repeated by determining the next group of significant genes with $\Delta_j > \Delta_{\text{cut},l+1}$ and average number of false positives based on $\#\overline{\text{FP}}_{l+1} > t_{\min,l+1}$.

Westfall-Young Method. The Westfall-Young method [8,12] is a FWER method that preserves the dependence structure between features. For p genes, start by calculating the observed P values for testing whether each gene is differentially expressed between the classes considered. This will yield $P_1, P_2, \ldots, P_k$. Determine $c = \arg\min_i\{P_i\}$ as the index for the gene with the lowest P value. During the first iteration when $b = 1$ ($b = 1, 2, \ldots, B$), shuffle the class labels of all of the arrays being used and perform the same test on the p genes to obtain $P_1^{(1)}, P_2^{(1)}, \ldots, P_k^{(1)}$, and identify the lowest P value, namely, $P_{\min}^{(1)}$. If $P_{\min}^{(1)} < P_c$, add 1 to the running total `numsig`. For iterations $b = 2, 3, \ldots, B$, repeat the same steps and add a 1 to `numsig` whenever $P_{\min}^{(b)} < P_c$. The steps outlined above will result in a Westfall-Young P value of $P* =$ `numsig`$/B$ for gene c. Next, set `numsig`=0, set $p = p - 1$, remove gene c from the list of genes being tested, and apply the same procedures to the $p - 1$ genes again to identify gene c with the lowest observed P value, reshuffle, and determine the Westfall-Young P value for the new gene c. The steps described above are illustrated in the Table 12.7 for three genes.

The Westfall-Young FDR can identify more putatively significant hypotheses when compared with FDR methods. Because the FDR finds as many false null hypotheses as possible, new FDR methods seem more appropriate than FWER approaches. The FWER is likely to be too conservative for the exploratory phase of gene discovery, since it does not permit more than a single null hypothesis from being erroneously rejected.

12.5.5 Filtering Qualitative Features

Qualitative features are categorical in nature, and can be ordinal or nominal. *Information gain* (entropy) and *Gini diversity* are two metrics that can be used for qualitative feature values. When the original scale of feature values is ratio, that is, continuously-scaled, we can use discretization to transform the continuous feature values into ordinal categorical codes. Features that are already categorically coded, such as gender, race, or diagnostic class, are straightforwardly amenable to information gain and Gini diversity and can be directly fed into these algorithms.

Discretization and Entropy. Discretization is central to the use of information gain and Gini diversity metrics, which reflect the amount of information encapsulated by a feature. Discretization is used to recode continuous feature values into ordinally ranked categories reflecting the

TABLE 12.7 Westfall-Young Stepdown *P*-Value Calculations for Genes g_1, g_2, g_3[a]

b	g_1	g_2	g_3	$P_{\text{min}}^{(b)}$	P_2	$I(p_{\text{min}}^{(b)} < P_2)$
0	0.0185	0.0021	0.0244	—	—	—
1	0.0572	0.0789	0.0665	0.0572	0.0021	—
2	0.0284	0.0686	0.0011	0.0011	0.0021	1
3	0.0463	0.0366	0.0379	0.0366	0.0021	—
⋮	⋮	⋮	⋮	⋮	⋮	27
9999	0.1543	0.1000	0.1714	0.1000	0.0021	—
10000	0.1688	0.0351	0.0820	0.0351	0.0021	—
—	—	—	—	—	$P =$	28/10,000
—	—	—	—	—	—	—
b	g_1	g_3	—	$P_{min}^{(b)}$	P_1	$I(P_{\text{min}}^{(b)} < P_1)$
0	0.0185	0.0244	—	—	—	—
1	0.0789	0.0157	—	0.0157	0.0185	1
2	0.0668	0.0607	—	0.0607	0.0185	—
3	0.0589	0.0736	—	0.0589	0.0185	—
⋮	⋮	⋮	—	⋮	⋮	48
9999	0.1997	0.1749	—	0.1749	0.0185	—
10000	0.0442	0.1630	—	0.0442	0.0185	—
—	—	—	—	—	$P =$	49/10,000
—	—	—	—	—	—	—
b	g_3	—	—	$P_{min}^{(b)}$	P_1	$I(P_{\text{min}}^{(b)} < P_1)$
0	0.0244	—	—	—	—	—
1	0.0148	—	—	0.0148	0.0244	1
2	0.1602	—	—	0.1602	0.0244	—
3	0.0913	—	—	0.0913	0.0244	—
⋮	⋮	—	—	⋮	⋮	264
9999	0.0611	—	—	0.0611	0.0244	—
10000	0.1842	—	—	0.1842	0.0244	
—	—	—	—	—	$P =$	265/10,000

[a]Values obtained using $B = 10{,}000$ iterations per gene. Final P value for each gene is equal to the number of times $P_{\text{min}}^{(b)} < P_c$ divided by B.

quantile equivalent for each feature value. Let us define the *entropy* of a dataset $\mathcal{S}$ according to the information theoretic relationship

$$H(\mathcal{S}) = -\sum_{\omega}^{\Omega} P_{\omega} \log_2(P_{\omega}), \tag{12.66}$$

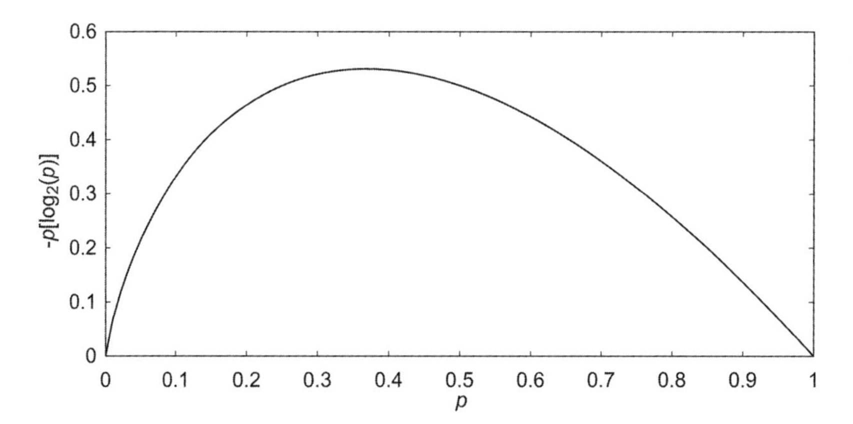

FIGURE 12.4 Plot of $-p[\log_2(p)]$ as a function of p. Maximum value of $-p[\log_2(p)]$ occurs when $p = 1/e$.

where P_ω is the class *prior* which is equal to n_ω/n and reflects the proportion of arrays in class ω. Assuming p is a class prior for a single class, Figure 12.4 illustrates that the maximum value of $-p[\log_2(p)]$ occurs when $p = 1/e$. Assume a two-class dataset $\mathcal{S}$ with class priors p and $q = 1 - p$. The entropy is determined as

$$H(\mathcal{S}) = -\left(p \log_2(p) + q \log_2(q)\right). \tag{12.67}$$

Examples of the entropy values for the two-class dataset are listed in tabular form as follows:

p	q	$H(\mathcal{S})$
0.0001	0.9999	0.001
0.2	0.8	0.722
0.3	0.7	0.881
0.4	0.6	0.971
0.5	0.5	1.000
0.6	0.4	0.971
0.7	0.3	0.881
0.8	0.2	0.722
0.9	0.1	0.469
0.9999	0.0001	0.001

One can note that the greatest value of $H(\mathcal{S})$ occurs when p and q are close to 0.5. This relationship is plotted over the range of p in Figure 12.5, which reflects maximum entropy $H(\mathcal{S})$ at $p = q = 0.5$.

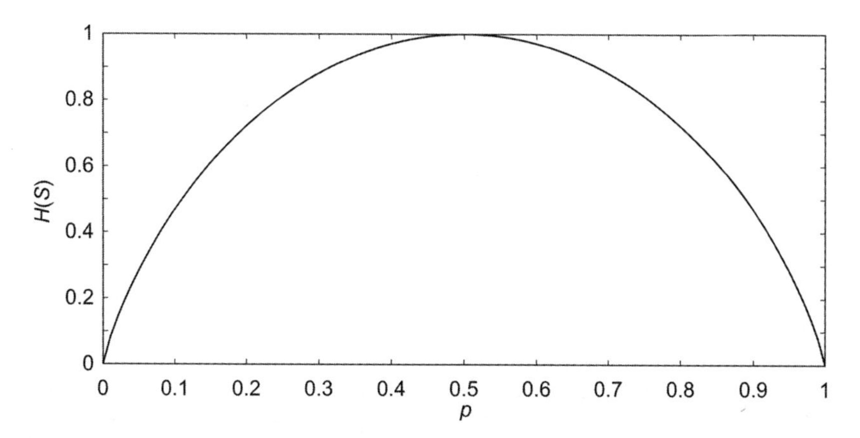

FIGURE 12.5 Plot of entropy $H(\mathcal{S})$ as a function of class prior p for a two-class dataset $\mathcal{S}$ having class priors p and $q = 1 - p$.

Multiclass Information Gain. Information gain is first determined for the entire dataset using the total number of training arrays and the number of arrays in each class, and then determined for each feature using discretized features values. Specifically, global information gain for the entire dataset is calculated in the form

$$I(G)_{\text{entropy}} = \sum_{\omega}^{\Omega} - \left(\frac{n_\omega}{n}\right) \log_2 \left(\frac{n_\omega}{n}\right), \tag{12.68}$$

where n_ω is the number of arrays in class ω ($\omega = 1, 2, \ldots, \Omega$) and n is the number of training arrays. We already noted that global information is at a maximum when the class priors are equal, and this should be considered during the design stage of a study or experiment. Next, for feature-specific information gain, cutpoint values for the 25th, 50th, and 75th percentiles (quartiles) are identified for each feature over all training arrays and all classes. Individual feature values are then assigned categorical codes from the set $\{1, 2, 3, 4\}$ according to which quartile their continuous value falls in. Information gain for each jth feature and quantile value $q = 1, 2, 3, 4$ is determined as

$$I(j, q) = \sum_{\omega} - \frac{n(j, q_j, \omega)}{n(q)} \log_2 \left(\frac{n(j, q_j, \omega)}{n(q)}\right), \tag{12.69}$$

where q is the quantile, ω is the class, $n(j, q_j, \omega)$ is the number of training arrays in class ω with a quantile value of q_j, and $n(q)$ is the number of training

TABLE 12.8 Discretization Method for a Single Continuously Scaled Feature ($j = 1$) Across 12 Arrays with Uniform Membership in Four Classes[a]

Class	Class 1			Class 2			Class 3			Class 4		
Value	2.1	1.9	3.7	−2.3	−0.5	−0.1	−0.3	−0.3	−0.8	−3.1	−0.7	−0.4
Quartile	4	4	4	1	2	2	2	2	1	1	1	2
$n(1, q_4, \omega)$	$n(1,4,1) = 4$			$n(1,4,2) = 1$			$n(1,4,3,) = 1$			$n(1,4,4) = 1$		
$n(1, q_3, \omega)$	$n(1,3,1) = 1$			$n(1,3,2) = 1$			$n(1,3,3,) = 1$			$n(1,3,4) = 1$		
$n(1, q_2, \omega)$	$n(1,2,1) = 1$			$n(1,2,2) = 3$			$n(1,2,3,) = 3$			$n(1,2,4) = 2$		
$n(1, q_1, \omega)$	$n(1,1,1) = 1$			$n(1,1,2) = 2$			$n(1,1,3,) = 2$			$n(1,1,4) = 3$		

[a]Feature values are recoded using the code set $\{1, 2, 3, 4\}$ according to the quartile (Q_1, Q_2, Q_3, Q_4) into which each array's value falls. All values of $n(j, q_j, \omega)$ are padded with a one before tabulating class and quartile counts. The 25th percentile value of expression is −0.675, while the median expression is zero, and 75th percentile 0.675.

arrays with quantile values of q over all training arrays. Finally, the total information gain for feature j is

$$I(j)_{\text{entropy}} = I(G)_{\text{entropy}} - \sum_{q} \frac{n(q)}{n} I(j, q). \tag{12.70}$$

Table 12.8 shows tabular notation for the discretization process for a single feature, including the feature's cutpoint values at the 25th, 50th, and 75th percentiles, and resulting totals for cell counts $n(j, q_j, \omega)$.

12.5.6 Multiclass Gini Diversity Index

For Gini diversity, the global Gini index is computed as

$$I(G)_{Gini} = 1 - \sum_{\omega}^{\Omega} \left(\frac{n_\omega}{n} \right)^2, \tag{12.71}$$

with the feature-quantile-specific Gini index as

$$I(j, q) = 1 - \sum_{\omega}^{\Omega} \left(\frac{n(j, q_j, \omega)}{n(q)} \right)^2. \tag{12.72}$$

Finally, the overall feature-specific Gini is computed as

$$I(j)_{\text{Gini}} = I(G)_{\text{Gini}} - \sum_{q} \frac{n(q)}{n} I(j, q). \tag{12.73}$$

Information gain and Gini diversity are determined for each feature, and features are then ranked.

12.5.7 Class Comparison Techniques

Implementing the abovementioned *suboptimal* best rank methods by simply ranking features and employing them in classification analysis is insufficient. Instead, a structured *modular* approach resulting in nonredundant selection of features is used. The paragraphs below outline steps necessary for each of the modular feature selection methods.

All-Possible-Pairs (APP). This method is based on identification of discriminating features using all-possible-pairs (APP) of class comparisons. We know that for Ω classes, there are $\Omega(\Omega-1)/2$ possible pairs of classes. Assume a four-class problem (i.e., $\Omega = 4$). Evaluation begins by first identifying the most differentially expressed genes for arrays in classes 1 and 2, and then ranking the genes according to their test results. This is repeated using the same genes for classes 1 and 3, 1 and 4, 2 and 3, 2 and 4, and 3 and 4. In total, there will be six class comparisons made, since $6 = 4(4-1)/2 = 12/2$. Algorithm 12.2 lists the procedures necessary for APP modular feature identification. In addition, the schematic in Figure 12.6 shows how increments of nonredundant top ranked genes are identified. Because pairs of classes were used for identifying genes, we will need to use the top set of nonredundant genes that best discriminate the classes considered; therefore, increments of $\Omega(\Omega-1)/2$ are applied to classification analysis. This means that for a four-class problem, classification performance should be assessed only using sets of 6,12,18,24,... nonredundant genes, based on increments of $\Omega(\Omega-1)/2$ genes. To elaborate further, let p be the total number of genes being filtered, and let $k = 1, 2, \ldots, \Omega(\Omega-1)/2$ be the class comparison. Next, perform testing for each class comparison and rank genes in ascending order according to the filter used, which for each kth class comparison will result in ranks $R_{1k} < R_{2k} < \cdots < R_{pk}$. For our first classification run, we select the top ranked gene from each class comparison, which is

$$g_k = \arg\min_j \{R_{jk}\}, \tag{12.74}$$

where g_k is the top ranked unselected gene for the kth class comparison, and R_{jk} is the rank of gene j for the kth class comparison. In our example with six class comparisons, we first select gene set $\{g_1, g_2, g_3, g_4, g_5, g_6\}$, remove it from the list of genes, and then employ this set in classification analysis. The idea

Algorithm 12.2: All-Possible-Pairs (APP) Modular Feature Selection

Data: p informative genes across all arrays $\mathbf{x}_i$
Result: Increments of $\Omega(\Omega - 1)/2$ genes
Use t-test, Mann-Whitney, information gain, Gini
Let t_j be a test result for the jth gene
Let k be a pairwise class comparison
for $l \leftarrow 1$ ***to*** $\Omega - 1$ **do**
  for $m \leftarrow l + 1$ ***to*** Ω **do**
    $k = k + 1$
    for $j \leftarrow 1$ ***to*** p **do**
      Determine t_j for $\mathbf{x}_i \in \omega_l$ and $\mathbf{x}_i \in \omega_m$;
    endfor
    Rank genes $R_{1k} < R_{2k} < \cdots < R_{jk} \cdots < R_{pk}$ based on $|t_j|$
  endfor
endfor
Apply increments of $\Omega(\Omega - 1)/2$ nonredundant genes up to ~150 genes and input these ~150 genes into greedy PTA (next section)

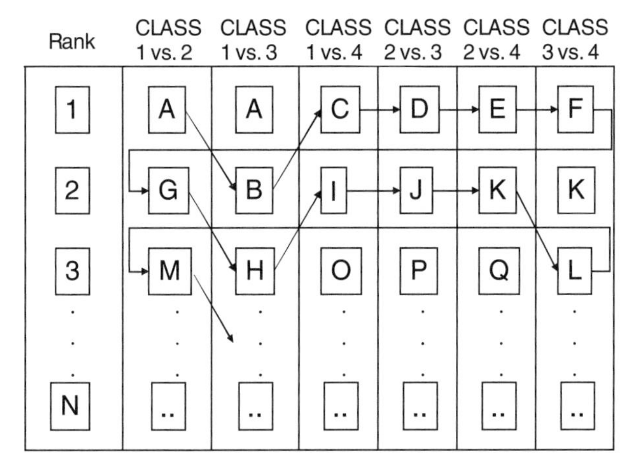

FIGURE 12.6 All-possible-pairs (APP) modular feature selection. Genes are tested for differential expression between all possible pairs of classes, and ranked for each comparison. This schematic shows $\Omega = 4$ classes, and $6 = 4(4 - 1)/2 = 12/2$ pairwise class comparisons, based on $\Omega(\Omega - 1)/2$. Increments of six nonredundant genes are used in increments of 6 for classification analysis. Even though gene A best discriminates between classes 1 and 2 and classes 1 and 3, we select gene B for the class 1-class 3 comparison for nonredundancy. The first set of genes used for classification is {A,B,C,D,E,F}, whereas the second set is {G,H,I,J,K,L}.

Algorithm 12.3: One-Against-All (OAA) Modular Feature Selection

Data: p informative genes across all arrays $\mathbf{x}_i$
Result: Increments of Ω genes
Use t test, Mann-Whitney, information gain, Gini
Let t_j be a test result for the jth gene
for $l \leftarrow 1$ ***to*** Ω **do**
  for $j \leftarrow 1$ ***to*** p **do**
    Determine t_j for $\mathbf{x}_i \in \omega_l$ vs. $\mathbf{x}_i \notin \omega_l$
  endfor
  Rank genes $R_{1l} < R_{2l} \cdots < R_{jl} \cdots < R_{pl}$ based on $|t_j|$
endfor
Apply increments of Ω nonredundant genes for classification up to ~150 genes and input these ~150 genes into greedy PTA (next section)

is that for each classification run, whenever we increment the genes used, we would like to include the set of $\Omega(\Omega - 1)/2$ genes that discriminate between all of the classes simultaneously. If more genes are sought for classification, the next set of 6 $\{g_1, \ldots, g_6\}$ is selected from the remaining genes and combined with the original 6 to make a set of 12 genes. The total list of genes used for classification is incremented by 6 each time more genes are desired for classification.

One-Against-All (OAA). This method identifies genes that best discriminate each class singly against all remaining classes. Expression values of genes for a particular class are tested against expression of the same genes in arrays in all other classes, and genes with the greatest test value, information gain, or Gini coefficient are guaranteed to exhibit properties that are unique for the single class or remaining classes. Again, consider $\Omega = 4$ classes for a particular classification problem. For OAA feature selection, there will be four OAA comparisons, so the genes identified will be employed in classification analyses in increments of size Ω. The OAA workflow is carried out methodologically in Algorithm 12.3.

Using our scheme developed in APP feature selection, the first set of genes used for classification contains only Ω genes, representing the Ω OAA class comparisons. For our example four-class problem, Figure 12.7 shows the gene ranks for each Ω OAA class comparison, and selects set $\{g_1, g_2, g_3, g_4\}$, removes these genes from the list, and applies them to classification analysis. Another set of four genes is selected if more are needed for classification, such that the total number of genes used for any classification runs is incremented as 4, 8, 12, and so on.

All-at-Once (AAO). The all-at-once (AAO) criterion for feature selection applies a test that can accommodate all classes in a single run. Unlike the APP and OAA methods, this test involves no class comparisons for AAO, and therefore the incremental addition of genes is on a single basis. Recall that the F test and Kruskal-Wallis test are multiclass tests, and the multiclass versions of information gain and Gini coefficient are a variation on a theme. Algorithm 12.4 lists the workflow for AAO modular feature selection.

During multiclass classification runs, the highest ranking gene is applied first, followed by genes with the next greatest level of significance or information.

Example 12.4 Feature selection was carried out using gene expression values for the three-class MLL_Leukemia3 and four-class SRBCT4 dataset described in Table 1.3. Pairwise class tests involved t tests, Mann-Whitney, entropy, and Gini, while three- and four-class tests involved F tests, Kruskal-Wallis, entropy, and Gini. K-nearest neighbor (KNN) classification results for ($K = 5$) (5NN) are provided as a function of the number of best ranked features used during classification. The classification accuracy for the three-class MLL_Leukemia3 dataset shown in Figure 12.8 indicates that the OAA approach [panel (b)] resulted in better performance when compared with APP results [panel (a)] and AAO results [panel (c)]. Specifically, looking at panel (b) for feature selection using the OAA approach, the Mann-Whitney test, as well as Gini coefficient performed the best. Different results were obtained for the four-class SRBCT dataset, shown in Figure 12.9, where APP feature selection [panel (a)] resulted in the best performance for 5NN. OAA feature selection and 5NN performed almost as well as APP feature selection; however, AAO feature selection before 5NN classification did not perform nearly as well. In conclusion, classification results for modular nonredundant feature selection depend on the data as well as the classification technique used. We did not exhaustively evaluate the effect of dataset, feature selection method, and various classification methods on performance, but do suggest that a full evaluation be considered whenever deploying a solution.

12.5.8 Generation of Nonredundant Gene List

We now describe methods for generating a unique nonredundant list of filtered genes for multiclass problems. Recall that $\mathbf{x}_i = (x_{i1}, x_{i2}, \ldots, x_{ip})$ is an object (array) with $p(j = 1, 2, \ldots, p)$ features (genes), $n(i = 1, 2, \ldots, n)$ is the total number of arrays, and $\Omega(\omega = 1, 2, \ldots, \Omega)$ the total number of classes. In addition, let $M = \Omega(\Omega - 1)/2$ be the possible pairs of class comparisons and $M = \Omega$ be all possible one-against-all (OAA) class comparisons ($m = 1, 2, \ldots, M$). For each mth class comparison, the top N_m biomarkers with the greatest informativeness are identified. Informativeness can be based

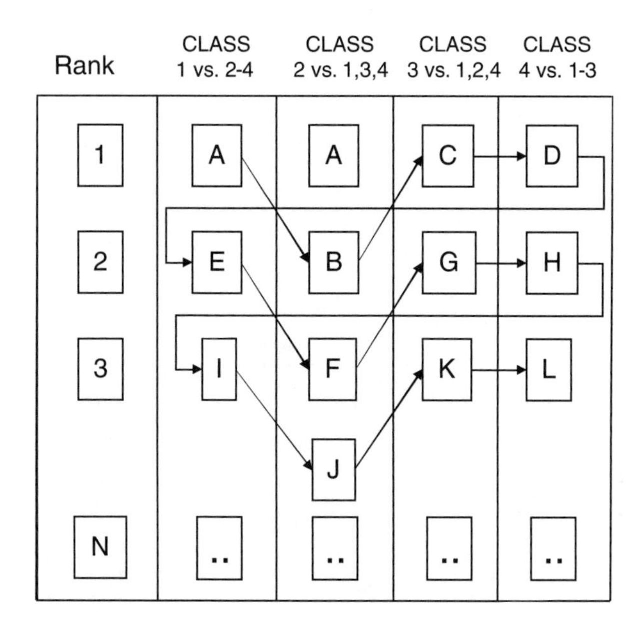

FIGURE 12.7 Nonredundant modular feature selection for one-against-all (OAA). The top genes are identified for each of the Ω comparisons between a class and the remaining classes, and increments of Ω genes are used.

Algorithm 12.4: All-at-Once (AAO) Modular Feature Selection

Data: p informative genes across all arrays $\mathbf{x}_i$
Result: Increments of single genes
Use F test, Kruskal-Wallis, information gain, Gini
Let t_j be a test result for the jth gene
for $j \leftarrow 1$ ***to*** p **do**
| Determine t_j for all $\mathbf{x}_i$
endfor
Rank genes $R_1 < R_2 < \cdots < R_j \cdots < R_p$ based on $|t_j|$
Identify ~150 genes and input these ~150 genes into greedy PTA (next section)

on the t test, Mann-Whitney test, F test, Kruskal-Wallis test, Gini index, and entropy in the form of information gain. For statistical tests, N_m is equal to the number of biomarkers for which the P value $P_j \leq 0.05$ during the mth class comparison. A list of nonredundant biomarkers among the $N = N_1 + N_2 + \cdots + N_m + \cdots + N_M$ biomarkers is then constructed. For large gene lists, we commonly identify 150 unique biomarkers from the N

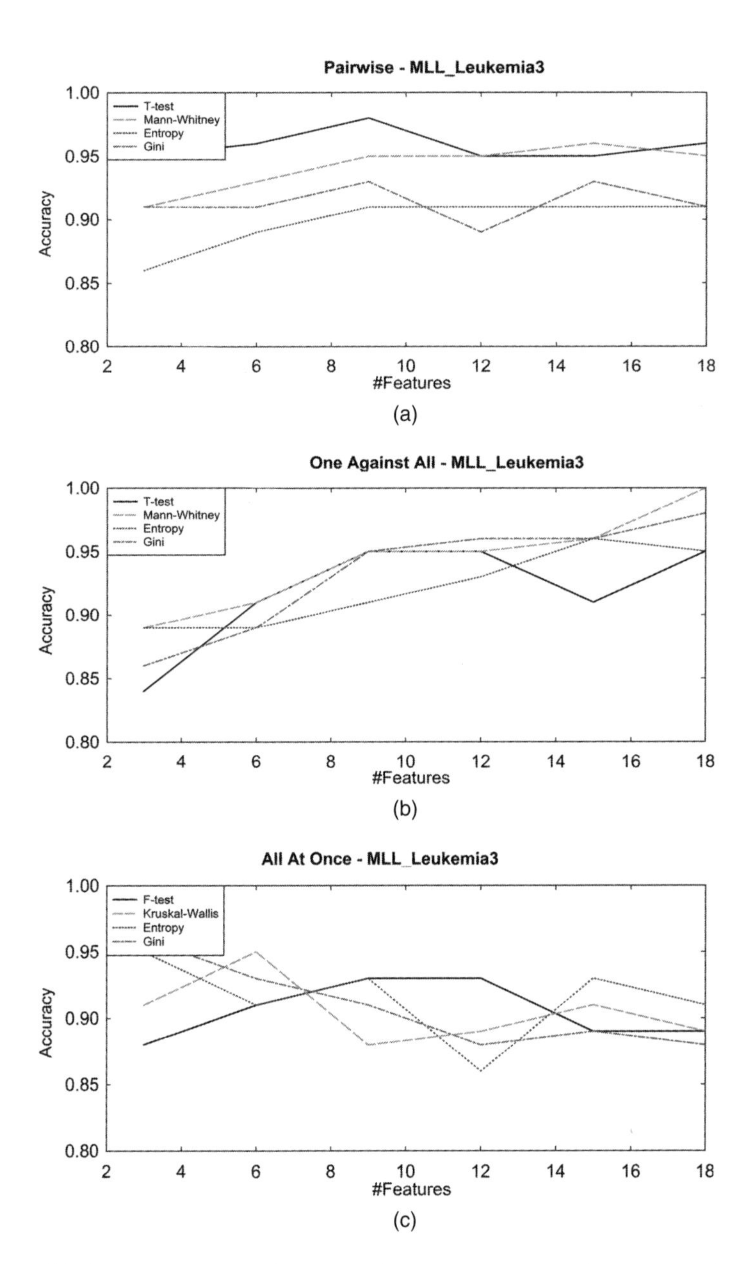

FIGURE 12.8 Best ranked feature selection and classification performance results for 5NN classification. Input dataset is the MLL_Leukemia3 dataset with 57 arrays, 12,582 genes, and three classes (20 ALL, 17 MLL, 20 AML): (a) all-possible-pairs (APP) results; (b) three one-against-all (OAA) results; and (c) all-at-once (AAO) results.

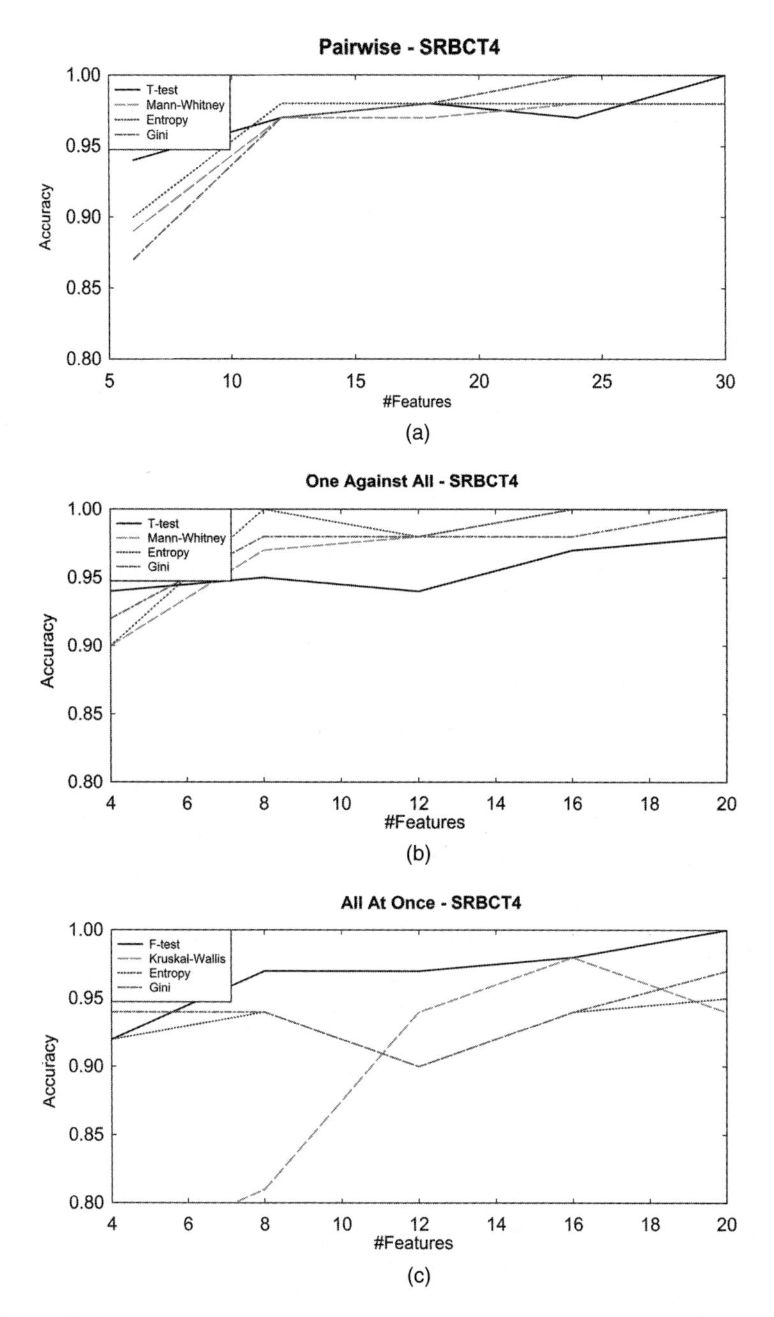

FIGURE 12.9 Best ranked classification performance for 5NN for the four-class SRBCT dataset with 63 arrays (23 EWS, 8 BL, 12 NB, 20 RMS) and 2308 genes: (a) all-possible-pairs (APP) results; (b) one-against-all (OAA) results; (c) all-at-once (AAO) results.

TABLE 12.9 Number of Class Comparisons M and Number of Genes N_m Filtered for Each Comparison

Test	All-possible-pairs ($M = \Omega(\Omega - 1)/2$)	One-against-all ($M = \Omega$)	All-at-once $M = 1$
t test	$N_m = \#\{j : P_j \leq 0.05\}$	$N_m = \#\{j : P_j \leq 0.05\}$	-
Mann-Whitney	$N_m = \#\{j : P_j \leq 0.05\}$	$N_m = \#\{j : P_j \leq 0.05\}$	-
F test	-	-	$N = \#\{j : P_j \leq 0.05\}$
Kruskal-Wallis	-	-	$N = \#\{j : P_j \leq 0.05\}$
Gini index	$N_m = p/M$	$N_m = p/M$	$N = p$
Entropy	$N_m = p/M$	$N_m = p/M$	$N = p$

total genes. Table 12.9 lists the number of genes filtered for the M possible class comparisons. It warrants noting that the genes identified from various class comparisons can be redundant, so a unique list is obtained from the M comparisons before performing gene selection (see next section).

12.6 SELECTION METHODS

12.6.1 Greedy Plus Takeaway (Greedy PTA)

Once a list of unique genes is generated, it is still not clear how many genes are required in order to optimally separate arrays in the various classes. Moreover, the suboptimal feature filtering methods discussed earlier in this chapter do not result in lists of genes that simultaneously discriminate classes from one another. Instead, they add genes independently with total disregard for how the final set of genes cooperate collectively to discern decision boundaries between classes. This situation tends to result in the *nesting* effect, where the models hold onto genes that were important only during their initial selection. Discriminatory effects of genes are not evaluated once they are selected. Another term used to represent this problem is called the *marginal effect* of the gene set, meaning that the final discriminative ability of the genes is really based on the marginals over all genes.

Forward-reverse sequential floating feature selection methods can be employed to maximize separation between arrays in multiple classes. We call genes selected using sequential floating methods an *optimal set* of genes. Through a hill-climbing approach, genes are added (removed) to (from) the model during each iteration [13-15]. Example PTA methods include sequential forward floating, sequential forward-backward floating, plus-k-takeaway k, and plus-1-takeaway 1. We favor the greedy plus-takeaway (greedy PTA) approach, which typically requires more computation, but the return on the investment (ROI) can be very high [16].

In this section, we focus exclusively on selection using greedy PTA via a stepwise plus 1-takeaway 1 heuristic. The process starts by first normalizing, transforming, and standardizing the input expression data. Next, select the top 150 best ranked genes using APP, OAA, or AAO filtering. Let n be the total number of arrays in all classes and n_ω be the number of arrays in class ω. Let p be the number of genes in a full model, $p-1$ be the number of genes in a reduced model without the pth gene, $y_{ij\omega}$ the expression value of gene j for the ith array in class ω, and $\bar{y}_{j\omega}$ the mean value of feature j for all training arrays in class ω. For each class ω, the centered matrix is

$$\underset{(n_\omega \times p)}{\bar{\mathbf{Y}}_\omega} = \begin{bmatrix} y_{11\omega}-\bar{y}_{1\omega} & y_{12\omega}-\bar{y}_{2\omega} & \cdots & y_{1j\omega}-\bar{y}_{j\omega} & \cdots & y_{1p\omega}-\bar{y}_{p\omega} \\ y_{21\omega}-\bar{y}_{1\omega} & y_{22\omega}-\bar{y}_{2\omega} & \cdots & y_{2j\omega}-\bar{y}_{j\omega} & \cdots & y_{2p\omega}-\bar{y}_{p\omega} \\ \vdots & & \vdots & & \vdots & \\ y_{i1\omega}-\bar{y}_{1\omega} & y_{i2\omega}-\bar{y}_{2\omega} & \cdots & y_{ij\omega}-\bar{y}_{j\omega} & \cdots & y_{ip\omega}-\bar{y}_{p\omega} \\ \vdots & & \vdots & \vdots & & \\ y_{n_\omega 1\omega}-\bar{y}_{1\omega} & y_{n_\omega 2\omega}-\bar{y}_{2\omega} & \cdots & y_{n_\omega j\omega}-\bar{y}_{j\omega} & \cdots & y_{n_\omega p\omega}-\bar{y}_{p\omega}. \end{bmatrix}, \tag{12.75}$$

where n_ω is the number of arrays in class ω. The $p \times p$ covariance matrix for class ω is

$$(n_\omega - 1) \underset{(p\times p)}{\mathbf{S}_\omega} = \bar{\mathbf{Y}}'_\omega \bar{\mathbf{Y}}_\omega, \tag{12.76}$$

and when summed over all classes, forms the pooled covariance matrix

$$\underset{(p\times p)}{\mathbf{S}_{pl}} = \frac{1}{n-\Omega} \sum (n_\omega - 1)\mathbf{S}_\omega. \tag{12.77}$$

Next, the mean feature values within each class are arranged into a column vector as follows:

$$\underset{(p\times 1)}{\bar{\mathbf{y}}_\omega} = \begin{bmatrix} \bar{y}_{1\omega} \\ \bar{y}_{2\omega} \\ \vdots \\ \bar{y}_{j\omega} \\ \vdots \\ \bar{y}_{p\omega}. \end{bmatrix} \tag{12.78}$$

Now we assume a *reduced model* with $p-1$ genes. Each unselected gene is evaluated singly by adding it to the reduced model in order to obtain a *full model* with p genes. If there are K unselected remaining genes, then there will be K full models to evaluate, each having p genes. For each full model

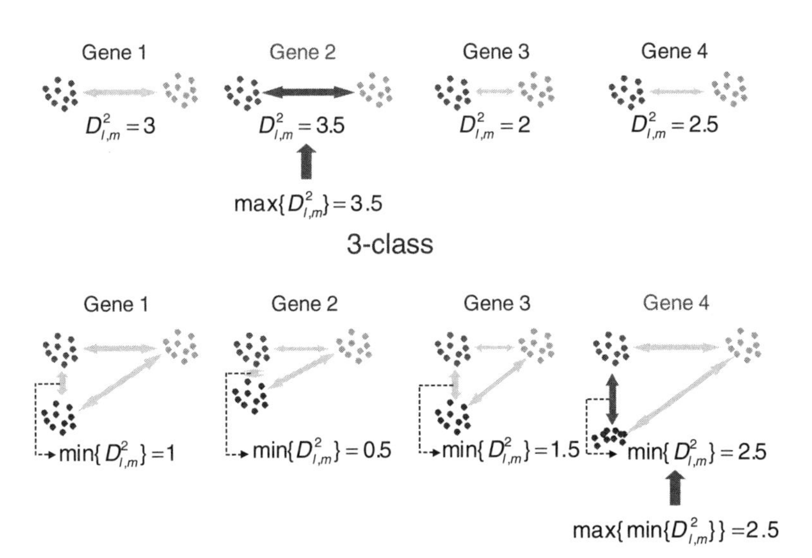

FIGURE 12.10 Schematic showing the Mahalanobis distance D criterion for greedy PTA gene selection. Dots represent arrays, and the distance between groups of arrays represents the Mahalanobis distance. During each forward step in the model for two-class problems the gene resulting in the greatest between-class Mahalanobis distance is selected, whereas in three-class problems the gene resulting in the greatest minimum between-class Mahalanobis distance is added to the model. The process of selecting only genes that progressively widen the between-cluster distance is called "greedy" hill climbing. (*See insert for color representation of the figure.*)

with unselected gene k added ($k = 1, 2, \ldots, K$), the minimum Mahalanobis distance is calculated for all possible pair of classes

$$d_k^2 = \min_{\ell \neq m} \left\{ (\bar{\mathbf{y}}_l - \bar{\mathbf{y}}_m) \mathbf{S}_{pl}^{-1} (\bar{\mathbf{y}}_l - \bar{\mathbf{y}}_m) \right\}, \tag{12.79}$$

where $\bar{\mathbf{y}}_l$ and $\bar{\mathbf{y}}_m$ are the mean expression vectors for p genes in the full model with unselected gene k for arrays in class l and arrays in class m. The full model with the greatest value of d_k^2 is identified as

$$g = \arg\max_k \{d_k^2\}, \tag{12.80}$$

and gene g is the top choice for the final full model. Figure 12.10 shows an example of gene selection for two- and three-class problems based on the greedy PTA method.

The methods described above were used to determine *which* gene should be added to the existing reduced model, but do not provide a means for quantifying the statistical significance for adding gene g to the model. Significance is determined by using F tests for hypothesis testing. Let $\mathbf{B}$ be the $p \times p$ between-class sum-of-squares matrix and $\mathbf{W}$ be the $p \times p$ within-class sum-of-squares matrix [17]. The jkth elements of these matrices are found as

$$\begin{aligned} b_{jk} &= \sum_{\omega=1}^{\Omega} \frac{1}{n_\omega} T_{j\omega} T_{k\omega} - \frac{1}{n} G_j G_k \\ w_{jk} &= \sum_{\omega=1}^{\Omega} \sum_{i=1}^{n_\omega} y_{ij\omega} y_{ik\omega} - \sum_{\omega=1}^{\Omega} \frac{1}{n_\omega} T_{j\omega} T_{k\omega}, \end{aligned} \tag{12.81}$$

where $y_{ij\omega}$ is the ith value of training feature j in class ω, $T_{j\omega} = \sum_i^{n_\omega} y_{ij\omega}$ is the sum of all values of training feature j in class ω, $G_j = \sum_\omega^\Omega T_{j\omega}$ is the grand total of all values of training feature j over the training arrays, and $n = n_1 + n_2 + \cdots + n_\Omega$.

An equivalent form of the F test ($H_0 : \boldsymbol{\mu}_1 = \boldsymbol{\mu}_2 = \cdots = \boldsymbol{\mu}_p$) is based, in part, on *Wilks' lambda*, given as

$$\Lambda(y_1, y_2, \ldots, y_p) = \frac{|\mathbf{W}|}{|\mathbf{B}| + |\mathbf{W}|} = \prod_{j=1}^{p} \left(\frac{1}{1 + \lambda_j} \right), \tag{12.82}$$

where $\lambda_1, \lambda_2, \ldots, \lambda_p$ are the eigenvalues of $\mathbf{W}^{-1}\mathbf{B}$ [18]. Although the matrix $\mathbf{W}^{-1}\mathbf{B}$ is square, it is nevertheless nonsymmetric such that elements in the upper triangular are not a mirror reflection of elements in the lower triangular. Since the matrix $\mathbf{W}^{-1}\mathbf{B}$ is not symmetric, we must resort to using, for example, the *generalized eigenvalue problem*, as described in Section B.10. Once the generalized eigenvalue problem is solved, the hypothesis at the pth step for determining the F-to-enter statistic for the additional variable y_p is determined by

$$\Lambda^* = \Lambda(y_p | y_1, y_2, \ldots, y_{p-1}) = \frac{\Lambda(y_1, y_2, \ldots, y_p)}{\Lambda(y_1, y_2, \ldots, y_{p-1})}, \tag{12.83}$$

where $\Lambda(y_1, y_2, \ldots, y_p)$ in the numerator is for the *full model* with variables $y_1, y_2, \ldots, y_p$, and $\Lambda(y_1, y_2, \ldots, y_{p-1})$ in the denominator is for the *reduced model* without variable y_p. When $p = 1$, the denominator $\Lambda(y_1, y_2, \ldots, y_{p-1})$ is set to unity. The F-to-enter statistic is

$$F_{\Omega-1, n-\Omega-p-1} = \frac{n - \Omega - p - 1}{\Omega - 1} \left(\frac{1 - \Lambda^*}{\Lambda^*} \right). \tag{12.84}$$

Forward Stepping. Stepwise selection of genes is used to add genes if they result in the most class separability. During forward stepping, genes are eligible for entry if their F-to-enter statistic exceeds 3.84, which is usually a default F-to-enter criteria. Among genes whose F-to-enter statistic > 3.84, the specific gene selected is the gene whose standardized expression results in the greatest squared Mahalanobis distance between the closest groups. The p variables in the final model are then used for backward stepping.

Backward Stepping. For each full model containing p variables, $\Lambda(y_1, y_2, \ldots, y_p)$ is determined. Next, each variable is removed singly from the full model, after which $\Lambda(y_1, y_2, \ldots, y_{p-1})$ and the F-to-remove statistic are determined. The first variable for which the F-to-remove statistic is less than the F-to-remove criterion of 2.71 is removed. If a variable is removed, forward stepping is resumed without further testing of F-to-remove criteria for other variables. If no variables are removed, then all variables in the full model remain intact, and forward stepping is resumed. Stepping stops when the F-to-enter criterion is not exceeded by any of the genes. At this point the number of genes selected by greedy PTA is denoted as p.

Algorithm for Greedy PTA. Let the indicator for a gene be denoted as $i, j, k = 1, 2, \ldots, p$, and let $D_{l,m}$ be the Mahalanobis distance between classes l and m for p genes, and D_i be the smallest between-class Mahalanobis distance when gene i is added to the set of genes for the current stepwise model. Let $\Lambda_i(y_1, y_2, \ldots, y_p)$ be Wilks' lambda when gene i is included in the model and $\Lambda(y_1, y_2, \ldots, y_{p-1})$ be Wilks' lambda without gene i in the model. If $p = 1$, $\Lambda(y_1, y_2, \ldots, y_{p-1}) = 1$. Algorithm 12.5 lists the computational steps for greedy PTA gene selection.

12.6.2 Best Ranked Genes

Once greedy PTA is run and the final number of genes is identified, the same number of best ranked genes are selected using the identical filtering methods. This best ranked gene selection alternative provides a suitable way for determining whether greedy PTA or best ranked results in greater performance.

Example 12.5 Greedy PTA gene selection was performed on each of the nine datasets listed in Table 1.3. Expression values of all genes in the nine datasets were $\log_e$-transformed before being employed in filtering and selection. For each dataset, 150 genes were identified using APP with filtering and pairwise t tests assuming unknown variance. The 150 genes were then used for greedy PTA. Results are listed in Table 12.10. The sets of filtered genes generated in this example form the basis of the main datasets used throughout this book.

Algorithm 12.5: Greedy Plus 1-Takeaway 1 (PTA) Stepwise Gene Selection of Optimal Gene Set

Data: List of ~150 filtered (top ranked) genes obtained by filtering with APP, OAA, AAO. Data are in the form or matrix with dimensions $p \times n$, or genes-by-microarrays. Expression values standardized by use of column(array)-specific mean and s.d. over the ~150 genes.
Result: p selected genes for classification

Set $\Lambda(y_1, y_2, \ldots, y_{p-1}) = 1$
while $F_{\text{enter}} > 3.84$ **do**
  Set $p \leftarrow p + 1$
  foreach *unselected gene i* **do**
    $D_i = \min\{D_{l,m}\}$
  endfch
  $j = \arg\max_i\{D_i\}$
  Determine $\Lambda_j(y_1, y_2, \ldots, y_p)$
  Determine F_{enter} using
  $\Lambda^* = \Lambda_j(y_1, y_2, \ldots, y_p)/\Lambda(y_1, y_2, \ldots, y_{p-1})$
  If $F_{\text{enter}} < 3.84$ **Return**
  Add gene j to the base model
  $\Lambda(y_1, y_2, \ldots, y_p) = \Lambda_j(y_1, y_2, \ldots, y_p)$
  for $k \leftarrow 1$ ***to*** p **do**
    Determine $\Lambda_k(y_1, y_2, \ldots, y_{p-1})$ with gene k removed from model
    Determine F_{remove} using
    $\Lambda^* = \Lambda(y_1, y_2, \ldots, y_p)/\Lambda_k(y_1, y_2, \ldots, y_{p-1})$
    if $F_{\text{remove}} < 2.71$ **then**
      remove gene k from model
      $p \leftarrow p - 1$
      exit for
    endif
  endfor
  Set $\Lambda(y_1, y_2, \ldots, y_{p-1}) = \Lambda(y_1, y_2, \ldots, y_p)$
endw

12.7 MULTICOLLINEARITY

It is desirable to reduce the amount of between-feature correlation, or *multicollinearity*, to the maximum extent possible [19,20]. However, it is important to realize that a lower degree of multicollinearity does not always guarantee greater classification performance. Below we define seven measures of multicollinearity studied by Heo [21], which can be used to gauge

TABLE 12.10 Greedy PTA Gene Selection of Optimal Gene Sets for the Nine Datasets[a] Used throughout This Book

Dataset	Original genes	Selected
AMLALL2	7,129	15
Brain2	7,129	11
Prostate2	12,600	11
Breast2	3,170	9
Breast2	24,481	13
Colon2	2,000	3
Lung2	12,533	29
MLL_Leukemia3	12,582	13
SRBCT4	2,308	23

[a]These datasets are described and listed in Table 1.3. Initially, 150 genes were identified via APP filtering with pairwise *t* tests. Greedy PTA was then applied to the sets of 150 genes.

the degree of between-feature correlation

$$q_1 = \sqrt{\frac{\left(\sum_{j=1}^{p} \lambda_j^2\right) - p}{p(p-1)}} \qquad q_2 = \left(1 - \frac{\min\{\lambda_j\}}{\max\{\lambda_j\}}\right)^{p+2}, \tag{12.85}$$

$$q_3 = 1 - \frac{p}{\sum_{j=1}^{p}(1/\lambda_j)} \qquad q_4 = 1 - \sqrt{|\mathbf{R}|}, \tag{12.86}$$

$$q_5 = \left(\frac{\max\{\lambda_j\}}{p}\right)^{3/2}, \qquad q_6 = \left(1 - \frac{\min\{\lambda_j\}}{p}\right)^{5}, \tag{12.87}$$

$$q_7 = \sum_{j=1}^{p} \frac{1 - 1/r^{jj}}{p}. \tag{12.88}$$

where $\lambda_1, \lambda_2, \ldots, \lambda_j, \ldots, \lambda_p$ are eigenvalues of the correlation matrix $\mathbf{R}$ and r^{jj} is the jth diagonal element of the matrix $\mathbf{R}^{-1}$. Further information on multicollinearity among features can be obtained by performing Bartlett's test of sphericity, using the null hypothesis $H_0 : \mathbf{R} = \mathbf{I}$, defined by

$$\chi^2_{p(p-1)/2} = -\left[n - 1 - \frac{(2p+5)}{6}\right] \log_e(|\mathbf{R}|), \tag{12.89}$$

where $n - 1$ is the degrees of freedom of the correlation matrix $\mathbf{R}$, p is the number of features, $|\mathbf{R}|$ is the determinant of $\mathbf{R}$, and $\chi^2_{p(p-1)/2}$ is a χ^2

test statistic that is distributed with $p(p-1)/2$ degrees of freedom [22]. If the null hypothesis $H_0 : \mathbf{R} = \mathbf{I}$ is rejected, then the correlation matrix **R** being tested is not an identity matrix, and the covariances in the off-diagonal elements are nonzero. Hence, for a fixed number of degrees of freedom, a larger value of the χ^2 test statistic implies a greater degree of multicollinearity.

Example 12.6 Modular gene filtering using APP, OAA, and AAO was performed on the full 2308-gene four-class SRBCT dataset. After selecting ~150 of the top ranked genes using modular filtering, a greedy PTA was employed whose results are compared with the same number of best ranked genes. Table 12.11 lists the modules used, filters, number of genes selected from the full set of 2308 genes, Bartlett's sphericity test, and multicollinearity parameters $q_1 - q_7$ derived from the correlation matrix for the genes selected. All of the Bartlett's χ^2 statistics to test for zero correlation were highly significant, suggesting that between-gene correlation is present for all of the feature sets. Table 12.12 lists the classifier accuracy for the same runs. Plots of discriminant scores for the 63 arrays were obtained with Fisher discriminant analysis (FDA), which is described in Chapter 19, and are shown in Figure 12.11 for the modular APP, OAA, and AAO techniques. Figure 12.12 shows the Fisher discriminant scores for the same number of best ranked genes. Figure 12.13 shows the workflow for analyses performed in this example. In order to characterize parameters q_1-q_7 and gain a more in-depth understanding of the information that each parameter portrays, we determined the correlation between $q_1 - q_7$ and the number of features selected, Bartlett's sphericity test and maximum eigenvalue of **R** (see Table 12.13). Both q_1 and q_5 correlated the greatest with the maximum eigenvalue of **R**, while q_2, q_3, q_6, and q_7 were correlated the most with the χ^2 test statistic for Bartlett's sphericity test. We did not expect any of the parameters to correlate with the number of features of **R**, since feature size is independent of correlation.

Table 12.14 shows the correlation between classifier accuracy and the multicollinearity parameters $q_1 - q_7$. Results indicate that 5NN, FDA, LVQ1, SVMGD, SVMLS, KREG, CPSO, and PLOG correlated the most with q_5, which is dependent mostly on $\lambda_{\max}$. NBC was weakly correlated with q_4 ($r = 0.241$), which is moderately correlated with Bartlett's test statistic. LDA, on the other hand, is strongly correlated with q_1, which is dependent mostly on $\lambda_{\max}$. Finally, the advantage of ANN is reflected with its very weak correlation with all of the multicollinearity parameters $q_1 - q_7$, and its low negative correlation ($r = -0.225$) with the number of features. This observation suggests that ANN is an optimal classifier to employ because of its lack of dependence on feature correlation.

TABLE 12.11 Comparison of Multicollinearity of Feature Correlation Matrix R for Final Sets of Genes Selected with Greedy PTA and Best Ranked Method

Method[a]	Filter[b]	Test[c]	#Features	$\chi^2_{p(p-1)/2}$ [d]	λ_1 [e]	λ_2	λ_3	q_1 [f]	q_2	q_3	q_4	q_5	q_6	q_7
Greedy PTA	APP	*t* test	23	745.47	0.106	4.524	3.837	0.275	0.570	0.310	1.000	0.087	0.978	0.670
		MW test	21	749.10	5.670	0.273	0.108	0.312	0.778	0.271	1.000	0.140	0.985	0.684
		Gini	21	691.46	0.101	0.368	0.244	0.279	0.607	0.297	0.999	0.092	0.978	0.672
		Entropy	22	748.79	5.293	3.686	0.063	0.306	0.749	0.295	1.000	0.118	0.986	0.661
	OAA	*t* test	22	736.45	5.733	0.134	0.091	0.298	0.702	0.296	0.999	0.133	0.981	0.678
		MW test	24	944.09	5.446	0.057	3.571	0.293	0.782	0.217	1.000	0.108	0.989	0.742
		Gini	29	1172.57	5.541	0.228	6.092	0.284	0.913	0.169	1.000	0.096	0.997	0.800
		Entropy	20	680.91	4.402	0.143	0.454	0.289	0.733	0.278	0.999	0.103	0.985	0.691
	AAO	*F* test	20	555.39	4.598	0.077	3.255	0.284	0.690	0.362	0.996	0.110	0.981	0.606
		KW test	23	780.28	3.542	5.184	0.109	0.290	0.687	0.294	1.000	0.107	0.983	0.686
		Gini	18	480.21	4.295	0.110	3.269	0.279	0.596	0.384	0.992	0.117	0.970	0.585
		Entropy	18	560.32	3.494	4.657	3.026	0.310	0.629	0.342	0.996	0.132	0.971	0.638
Best ranked	APP	*t* test	23	1072.38	7.405	0.237	0.039	0.371	0.877	0.192	1.000	0.183	0.992	0.783
		MW test	21	751.19	5.969	0.175	0.311	0.323	0.745	0.290	1.000	0.152	0.982	0.683
		Gini	21	710.01	6.044	2.910	3.235	0.316	0.718	0.305	0.999	0.154	0.980	0.662
		Entropy	22	824.83	0.087	5.339	0.220	0.298	0.828	0.229	1.000	0.120	0.991	0.746

OAA	*t* test	23	986.23	0.108	4.117	6.189	0.336	0.805	0.219	1.000	0.140	0.988	0.756
	MW test	21	775.13	6.017	3.924	0.066	0.331	0.775	0.273	1.000	0.153	0.984	0.687
	Gini	21	847.04	6.360	0.127	3.118	0.351	0.780	0.255	1.000	0.167	0.984	0.726
	Entropy	22	987.38	6.316	0.235	4.996	0.360	0.828	0.214	1.000	0.154	0.989	0.761
AAO	*F* test	23	1270.56	8.885	0.165	5.158	0.432	0.909	0.158	1.000	0.240	0.993	0.820
	KW test	21	1037.95	9.175	3.273	0.200	0.440	0.915	0.177	1.000	0.289	0.992	0.799
	Gini	21	997.91	8.700	3.994	1.406	0.428	0.942	0.170	1.000	0.267	0.995	0.771
	Entropy	22	995.67	8.260	3.791	2.512	0.394	0.841	0.210	1.000	0.230	0.987	0.758

[a] Greedy PTA and best ranked were used for final gene selection based on inputs of 150 genes identified by APP, OAA, and AAO filtering methods.

[b] APP—All-possible-pairs; OAA—one-against-all; AAO—all-at-once. These filtering methods were first used to filter 150 genes from the original larger datasets.

[c] MW denotes Mann–Whitney test, KW denotes Kruskal–Wallis test.

[d] Bartlett's sphericity test for $H_0 : \mathbf{R} = \mathbf{I}$; that is, there is no correlation among features (**R** generated for each row).

[e] $\lambda_1, \lambda_2, \lambda_3$ are the three greatest eigenvalues of **R**.

[f] $q_1 - q_7$ multicollinearity parameters described in (12.88).

Notation: Values were obtained using input sets of 150 genes identified with APP, OAA, and AAO filtering. Dataset used was the four-class SRBCT dataset with $n = 63$ arrays and $p = 2,308$ genes. Each row represents a different selection (greedy PTA or best ranked) and filtering (APP, OAA, AAO) combination. A correlation matrix **R** was generated for the final set of features, which was used for performing the Bartlett's sphericity test and generating multicollinearity parameters q_1–q_7.

TABLE 12.12 Classification Performance (Accuracy) of 11 Classifiers for Modular Selection and Filtering Methods Listed in Table 12.11[a]

Method	Filter	Test	5NN	NBC	LDA	FDA	LVQ1	SVMGD	SVMLS	KREG	CPSO	PLOG	ANN	EMV	EWMV
Greedy PTA	APP	*t* test	1	0.98	1	1	1	1	1	1	1	1	1	1	1
		MW test	1	0.96	1	1	1	1	1	1	1	1	1	1	1
		Gini	1	0.97	1	1	1	1	1	1	0.99	1	1	1	1
		Entropy	1	0.95	1	1	0.99	0.99	1	1	0.99	1	1	1	1
	OAA	*t* test	1	0.99	1	1	1	1	1	1	0.99	1	1	1	1
		MW test	1	0.97	0.99	0.99	1	0.99	0.99	1	0.99	1	0.99	1	1
		Gini	1	0.97	1	1	1	1	1	1	1	1	1	1	1
		Entropy	1	0.98	1	1	1	0.99	1	1	0.99	1	1	1	1
	AAO	*F* test	1	0.97	1	1	1	1	1	0.99	1	1	1	1	1
		KW test	1	0.97	0.99	0.98	1	0.99	1	0.98	1	1	1	1	1
		Gini	0.98	0.94	1	1	0.98	0.96	1	0.98	0.98	0.99	1	0.99	0.99
		Entropy	1	0.96	1	1	1	1	1	0.99	1	1	1	1	1
Best ranked	APP	*t* test	0.97	0.95	0.98	0.98	0.99	1	1	1	0.98	1	1	1	1
		MW test	0.99	0.99	1	0.98	1	1	1	1	0.99	1	1	1	1
		Gini	0.98	0.95	1	0.97	0.99	0.99	0.99	0.99	0.98	1	1	1	1
		Entropy	0.96	0.92	0.98	0.96	0.97	0.98	0.97	0.98	0.98	1	1	1	1
	OAA	*t* test	0.99	0.93	0.98	0.98	1	1	0.99	1	0.98	0.99	1	1	1
		MW test	1	0.97	1	1	1	1	1	1	1	1	1	1	1
		Gini	0.99	1	0.98	0.97	1	1	1	1	1	1	1	1	1
		Entropy	1	0.98	1	0.97	1	1	1	1	1	1	1	1	1
	AAO	*F* test	1	0.97	1	0.98	1	1	1	1	1	1	1	1	1
		KW test	0.98	0.98	0.94	0.95	0.98	0.95	0.93	0.98	0.95	0.97	1	0.98	0.99
		Gini	0.97	0.95	0.91	0.92	0.96	0.96	0.91	0.96	0.86	0.91	1	0.97	0.99
		Entropy	0.97	0.99	0.97	0.95	0.98	0.99	0.94	0.98	0.95	0.98	1	0.99	1

[a]The columns denoted with headers EMV and EWMV represent the ensemble majority vote and ensemble weighted majority vote results for the 11 classifiers

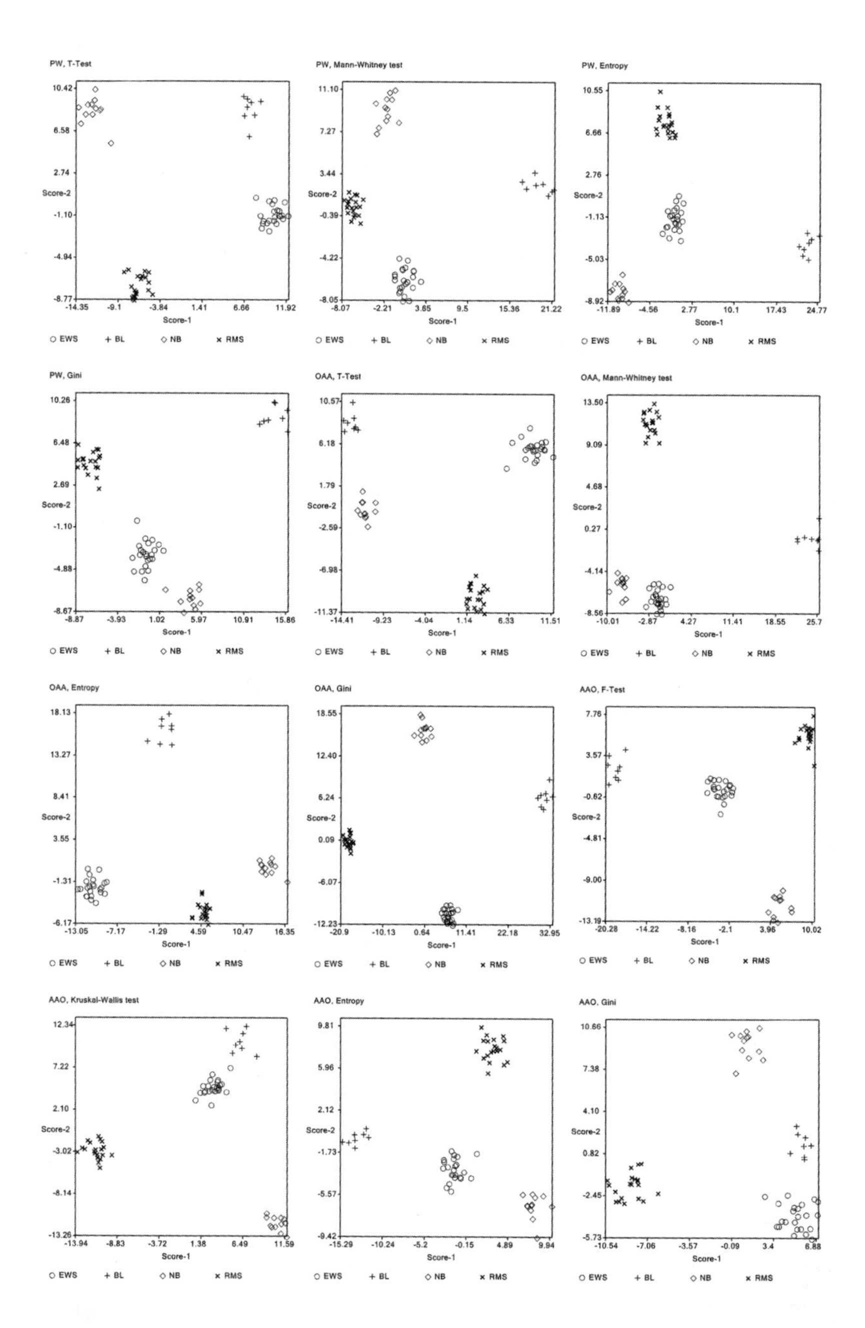

FIGURE 12.11 Discriminant scores from Fisher discriminant analysis (FDA) of the four-class SRBCT dataset (2308 genes) based on initial selection of 150 genes using all-possible-pairs (APP), one-against-all (OAA), and all-at-once (AAO), and final feature selection based on greedy PTA. The size of each gene set is listed in rows 1–12 of Table 12.11.

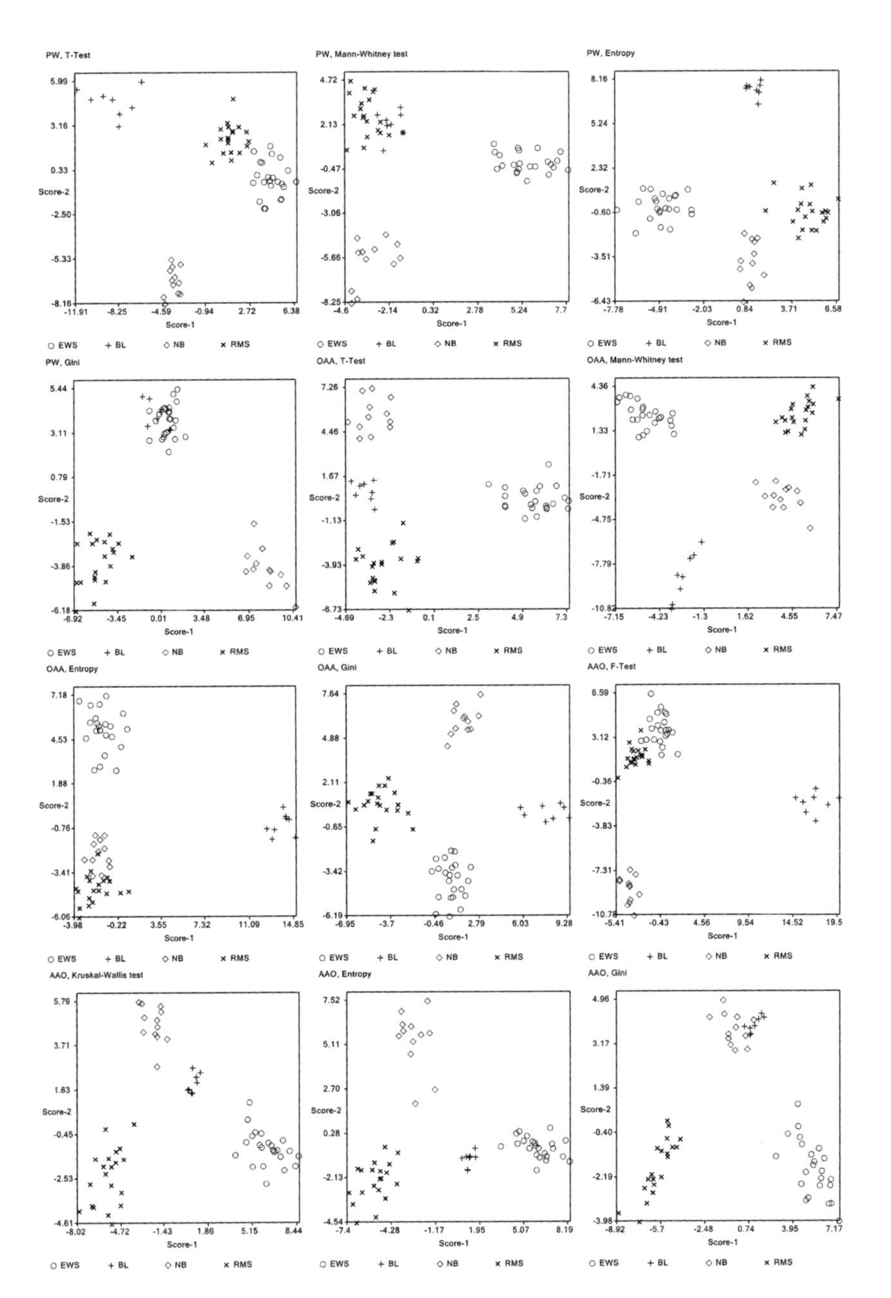

FIGURE 12.12 Discriminant scores from Fisher discriminant analysis (FDA) of the four-class SRBCT dataset (2308 genes) based on initial selection of 150 genes using all-possible-pairs (APP), one-against-all (OAA), and all-at-once (AAO) approaches, and final feature selection based on the best rank genes. The size of each gene set is listed in rows 13-24 of Table 12.11.

TABLE 12.13 Correlation between Number of Features (# Features), Bartlett's Test Statistic, and Multicollinearity Parameters $q_1 - q_7$ in Table 12.11 and Performance of 11 Classifiers Listed in Table 12.12[a]

Criterion	5NN	NBC	LDA	FDA	LVQ1	SVMGD	SVMLS	KREG	CPSO	PLOG	ANN
#Features	0.119	0.059	0.028	0.005	0.181	0.277	0.057	0.278	0.116	0.116	**−0.225**
Bartlett's test	0.191	0.069	−0.393	−0.472	−0.122	−0.003	0.340	0.037	−0.246	−0.246	−0.115
q_1	−0.417	0.115	−0.687	−0.749	−0.431	−0.362	−0.658	−0.346	−0.574	−0.584	0.148
q_2	0.391	0.056	0.578	0.627	0.378	0.225	0.535	0.193	0.452	0.438	0.032
q_3	0.303	0.015	0.540	0.586	0.268	0.130	0.491	0.076	0.393	0.378	0.141
q_4	0.005	**0.241**	0.253	0.357	0.083	0.277	0.219	0.225	0.109	0.071	0.099
q_5	**−0.492**	0.115	**−0.739**	**−0.774**	**−0.528**	**−0.484**	**−0.743**	**−0.467**	**−0.651**	**−0.649**	0.160
q_6	0.239	−0.053	−0.477	−0.493	−0.248	−0.097	−0.414	−0.054	−0.340	−0.329	−0.139
q_7	−0.289	0.053	−0.488	−0.560	−0.206	−0.085	−0.433	−0.032	−0.308	−0.294	−0.104

[a] Maximum correlation values are shown in bold typeface.

2308 Genes
APP ▸ | T-test ▸ | 150 genes ▸ | Greedy PTA ▸ | q1-q7
OAA | MW Test | Best ranked | Bartlett's sphericity test
APP | Information gain | Classification accuracy
Gini

FIGURE 12.13 Telescoping workflow for Example 12.6 analyses of the four-class 2308-gene SRBCT dataset, showing available filtering methods within APP, OAA, and AAO modular approaches.

TABLE 12.14 Correlation between Multicollinearity Coefficients $q_1 - q_7$ and Number of Features (# Features), Bartlett's Sphericity Test, and Maximum Eigenvalue $\lambda_{\max}$ for Columns of Table 12.11[a]

Criterion	q_1	q_2	q_3	q_4	q_5	q_6	q_7
#Features	−0.013	0.465	−0.609	0.594	−0.106	0.686	0.619
Bartlett's test	0.672	**0.855**	**−0.960**	**0.666**	0.557	**0.877**	**0.968**
$\lambda_{\max}$	**0.849**	0.806	−0.692	0.373	**0.848**	0.623	0.681

[a]Maximum correlation values are shown in bold typeface.

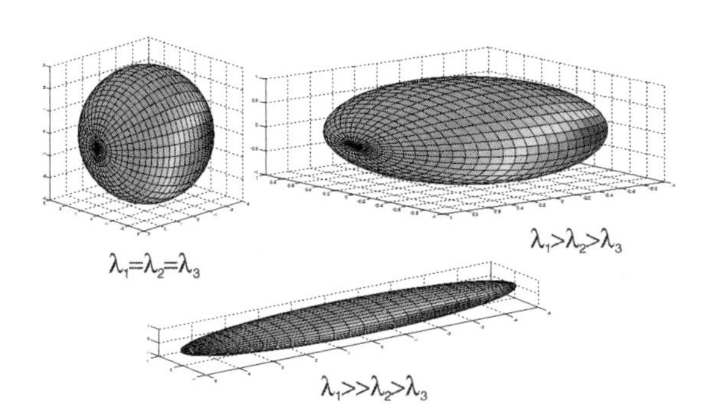

FIGURE 12.14 Contours of equal density for a 3D vector showing the 3-ellipse in 3-space. (*See insert for color representation of the figure.*)

Eigenspheres. The degree of correlation among features can be visually checked by viewing a 3D image of the three greatest eigenvalues, $\lambda_1, \lambda_2, \lambda_3$, of the correlation matrix **R**. When there is zero interfeature correlation, all eigenvalues will be equal, that is, $\lambda_1 = \lambda_2 = \lambda_3$, and an *eigensphere* will be a perfect sphere. Figure 12.14 shows contours of equal density for a 3D vector illustrating 3-ellipses in 3-space. A 3D eigensphere has principal radii equal to the square root of the greatest eigenvalues of **R**. If $\lambda_1 > \lambda_2 > \lambda_3$, the eigensphere will form more of an ellipse in 2D, and if the greatest eigenvalue

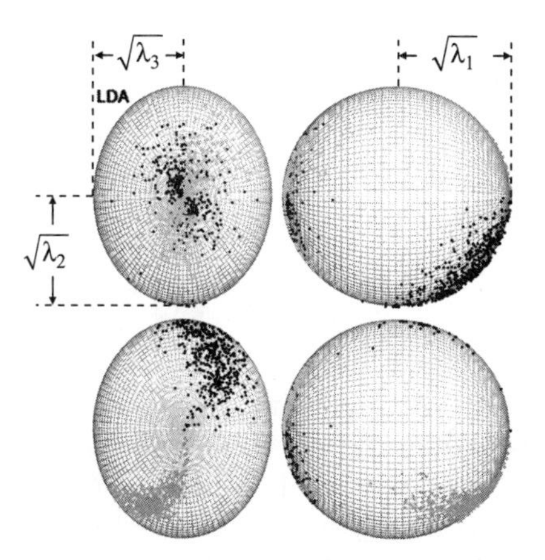

FIGURE 12.15 Rotated eigenspheres with radii equal to the eigenvalues of the first three principal components. The more spherical the eigensphere, the greater the equality among eigenvalues—and the lower the between-feature correlation.

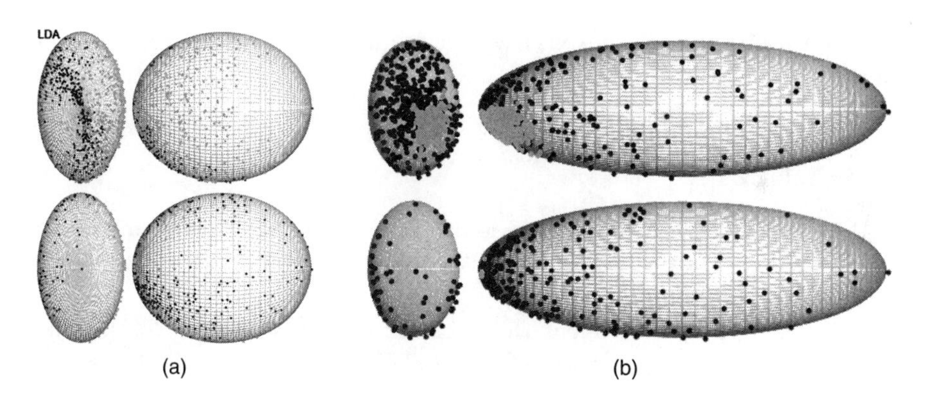

FIGURE 12.16 (a) Eigenspheres for three genes from colon cancer dataset selected with greedy PTA; (b) eigenspheres for three genes selected using the best ranked approach from the colon cancer data.

is much greater than the other two (i.e., $\lambda_1 \gg \lambda_2 > \lambda_3$), then the eigensphere will assume a cigar shape, as shown in Figure 12.14. Figure 12.15 illustrates an eigensphere and the radii and eigenvalue equivalents.

Example 12.7 Eigenanalysis of the correlation matrix **R** for the two-class colon cancer dataset was performed for the correlation matrix based on the three genes selected first using greedy PTA, and then by using the top ranked three genes. Figure 12.16a shows the 3D eigensphere of **R** and its rotation for the three genes selected with greedy PTA (APP, *t* test); Figure 12.16b shows the eigenspheres for the best ranked three genes. Note that the eigenstructure of greedy PTA is far more spherical, indicating less multicollinearity among genes in comparison to the best rank selection.

12.8 SUMMARY

Numerous comparisons can be made for the various feature selection methods presented in this chapter. Dimensional reduction potentially reduces redundancy of information and noise, thereby improving the efficiency of a discriminative model. Greedy PTA filtering can reduce multicollinearity and increase orthogonality, but there is no guarantee that the reduced between-feature correlation always results in greater performance. Observations made during analysis for Example 12.7 suggest that ANNs are less biased by within-feature correlation, which is a beneficial characteristic of classifiers. It can be argued that one should not prioritize use of a classifier known to be less dependent on correlation, since the use of other classifiers more dependent on correlation would require de-correlation. However, use of such a classifier (i.e., ANN) would minimize the risk of results being influenced by correlation.

REFERENCES

[1] C. Ambroise, G.J. McLachlan. Selection bias in gene extraction on the basis of microarray gene-expression data. *Proc. Natl. Acad. Sci. USA* **99**(10):6562–6566, 2002.

[2] H. Levene. Contributions to probability and statistics. In I. Olkin et al., eds., *Essays in Honor of Harold Hotelling*, Stanford Univ. Press, 1960, pp. 278–292.

[3] W. Behrens. Journal of Agriculture Scientific Archives of the Royal Prussian State College-Economy, Berlin, Prussian Ministry of Agriculture, Forests and Domains, **68**:807–837, 1929.

[4] R. Fisher. The comparison of samples with possibly unequal variances. *Ann. Eugen.* **9**:174–180, 1932.

[5] W. Dixon, F. Massey. *Introduction to Statistical Analysis*. McGraw-Hill, New York, 1969.

[6] H. Mann, D. Whitney. On a test of whether one of two random variables is stochastically larger than the other. *Ann. Math. Stat.* **18**:50–60, 1947.

[7] W. Kruskal, W. Wallis. Use of ranks on one-criterion variance analysis. *Am. Stat. Soc.* **47**:583–621, 1952.

[8] P.H. Westfall, S.S. Young. *Resampling-based Multiple Testing: Examples and Methods for p-value Adjustment*. Wiley, New York, 1993.

[9] Y. Benjamini, Y. Hochberg. Controlling the false discovery rate: A practical and powerful approach to multiple testing. *J. Roy. Stat. Soc. B* **57**:289–300, 1995.

[10] J.D. Storey. A direct approach to false discovery rates. *J. Roy. Stat. Soc. B* **64**:479–498, 2002.

[11] V. Tusher, R. Tibshirani, G. Chu. Significance analysis of microarrays applied to the ionizing radiation response. *Proc. Natl. Acad. Sci. USA* **98**:5116–5121, 2001.

[12] Y. Ge, S. Dudoit, T. Speed. *Resampling-Based Multiple Testing for Microarray Data Analysis*, Technical Report 633, Dept. Statistics, Univ. California, Berkeley, Jan. 2003.

[13] T.B. Farver, O.J. Dunn. Stepwise variable selection in classification problems. *Biometr. J.* **21**:145–153, 1979.

[14] K.N. Berk. Forward and backward stepping in variable selection. *J. Stat. Comput. Simul.* **10**:177–185, 1980.

[15] S. Ganeshanandam, W.J. Krzanowski. On selecting variables and assessing their performance in linear discriminant analysis. *Austral. J. Stat.* **31**:433–447, 1989.

[16] P. Somol, P. Pudil, J. Nonovicova, J. Paclik. Adaptive floating search methods in feature selection. *Pattern Recogn. Lett.* **20**:1157–1163, 1999.

[17] D.F. Morrison. The multivariate analysis of variance. In *Multivariate Statistical Methods*, 3rd ed., McGraw-Hill, New York, 1990, chap. 5.

[18] R.A. Johnson, D.W. Wichern. Comparision of several multivariate means. In *Applied Multivariate Statistical Methods*, 4th ed., Prenctice-Hall, Upper Saddle River, NJ, 1998, chap. 6.

[19] Y. Haitovsky. Multicollinearity in regression analysis: Comment. *Rev. Econ. Stat.* **50**:486–489, 1969.

[20] R.F. Gunst. Regression analysis with multicollinearity predictor variables: Definition, detection, and effects. *Commun. in Stat.—Series A, Theory and Methods*, **12**:2217–2260, 1983.

[21] T.Y. Heo. *The Comparison of Eigensystem Techniques for Measuring Multicollinearity in Multivariate Normal Data*, Master's Thesis, Dept. Statistics, Brigham Young Univ., Provo, UT, 1987.

[22] M.S. Bartlett. Tests of significance in factor analysis. *Br. J. Psychiatr.* **3**:77–85, 1950.

CHAPTER 13

CLASSIFIER PERFORMANCE

13.1 INTRODUCTION

Performance assessment is the primary outcome of developing a classification algorithm [1]. Knowing how well a classification algorithm (classifier) performs depends on characteristics of the sample and feature space, the range and scale of the data, and the fundamental characteristics of the classifier. After gaining some experience in classification, and learning the characteristics of different classification methods, we can begin to develop some criteria and rules of thumb about various methods. Criteria such as speed, robustness, and generalization are key factors in selection of a classification scheme. We now briefly discuss these criteria and provide the reader with general recommendations [2].

13.2 INPUT-OUTPUT, SPEED, AND EFFICIENCY

Data Input. The size, format, and method of inputting data can have a profound effect on performance of an algorithm. For large networked systems, various network connection technologies such as ODBC, OLEDB, ADO, ADOX are commonly used to transmit data between an algorithm and disk storage. SOAP/XML (simple object access protocol/extensive markup language) can also be used to send and receive data requests between servers, and it is not uncommon to wrap algorithms into a workflow so that input and output data are transmitted over networks automatically. Smaller applications running on standalone workstations, desktops, and notebooks can also connect with databases stored on the same machine; however, there are many applications that use flatfiles in the form of tab- or comma-delimited text, or spreadsheet formatted files such as Microsoft (MS) Excel.

Classification Analysis of DNA Microarrays, First Edition. Leif E. Peterson.

In this chapter we are interested in the algorithm performance of a stand-alone system on a single machine that processes single or multiple datasets during a classification run. One of the most expedient ways to process large datasets of raw expression values on a single machine is to input the data directly into the algorithm using a connection technology such as HDF, OLEDB or ADOX with the data originating from a database based on the technology available in Oracle, MySQL, MS SQL, or MS Access. Thus, there are two steps to be concerned with: connection technology and database technology. Depending on the operating system used, database storage can offer tremendous advantages for moving and processing large datasets with an algorithm. A word of caution here for MS users is that there is a cottage industry for repairing corrupted Access files, while for MS SQL there is no such industry. In addition, MS Access has an upper storage limit of ~2 GB (gigabytes), so its use may be limited if large datasets are needed. If you plan to develop large systems in the workplace, then it is suggested that you don't incorporate system designs based on MS Access, but implement more advanced database technologies. Some IT departments don't have a choice with respect to Oracle, MySQL, or MS SQL, so you will need to look into this during the initial planning stages.

The last choice of data input that we discuss is text in the form of tab- or comma-delimited flatfiles, or spreadsheets. If the raw expression values are needed for initial feature selection, then this method will be much slower than connecting to and reading directly from a database. Our experience is that once initial feature selection is performed, the use of small text files containing feature sets will increase speed by an appreciable extent. Small files containing preselected genes also eliminates the need to program database connection and retrieval calls. In summary, if you need to operationally use large expression sets repeatedly to develop feature selection methods or perform cluster analysis with all features, then database storage is recommended. However, if repeated classification runs are required using reduced sets of preselected features, then performance will be improved if data are retrieved using flatfiles in text format. The rule of thumb is that, when using large datasets repeatedly it is better to input data from a database, whereas small datasets used repeatedly can be input directly from disk storage in text format.

Data Output. Processing results are commonly written to disk in the form of text or spreadsheet, or sent directly to a grid, table, or list that's visible in the graphical user interface (GUI)—that is, the user's monitor. If the goal is to simply publish results of classification runs produced in textual form, then it may suffice to write directly to disk and import text into slide presentation or word processing software. The generation of image files with charts, plots, or heatmaps is considerably more time-consuming, especially if pixel-specific bitmaps are generated (vs. vector images based on line art and text). A picture is worth a thousand words, so textual information listed in tables can't nearly achieve the levels of informativeness offered by plots and charts generated on the fly. Obviously, the tradeoff is how far you want to go with generation of output images versus output in the form of text. Another important factor that impacts output speed is where and when, in an algorithm, the data are sent to output. Lines of code that write output in the middle of a loop commonly will adversely affect performance, so keeping track of when and where output occurs in an algorithm is vitally important. We have typically incorporated output options for users to enable them to write the internal computational results of scalars, vectors, and matrices to disk for review and confirmation of accuracy. It is also helpful to give users a lot of output choices, enabling them to access most of the intermediate computational results for their own use. After all, methodological development is based to a large extent on novel methods and interpretations made with the intermediate data. A remaining caveat with regard to providing different output options in an algorithm is that too much logical syntax (i.e., if-then-else) can present a burden to the algorithm, since it has to interpret each if-then statement. If logical syntax is not placed too deeply within loops, then the degradation in performance may be acceptable.

Speed. The speed at which classification operations are carried out within an algorithm depends on the intrinsic structure and information flow in procedures related to reading input data, writing output data, graphical image generation, GUI activities, interpreting logical statements, memory allocation, garbage collection, and numerical methods related to the CPU instruction set. The overall size of an algorithm can degrade performance by virtue of the demands on memory allocation or time required for numerous calculations. There are many ways to degrade the efficiency of a classification algorithm. Unnecessarily making too many procedure calls at the wrong time and wrong place can reduce the speed. Drawing excessive system resources when only small amounts are needed will reduce the rate at which operations are carried out. A successful algorithm can benefit the most from good programming, which incorporates time-saving measures to the extent possible.

Only programming experience combined with running an optimizer allow you to fully optimize an algorithm. The majority of experience one has is devoted to looking ahead in the calculations to determine when and where a procedure or method has to be invoked as well as the minimum number of occurrences. As an example, consider the k nearest neighbor (KNN) classifier when implemented with object repartitioning and cross-validation. Here, the between-microarray Euclidean distance for all possible pairs of microarrays needs to be determined only once during program initiation, because the distance between almost every microarray pair will need to be known as the microarrays used in training and testing are re-determined during each repartitioning and cross-validation fold. One can't assume that the inexperienced programmer does or does not have the insight to perform the calculations once, or to erroneously embed the between-microarray distance calculation deep within a loop and unnecessarily recalculate the same value. Effective programming techniques for optimizing runtimes is based extensively on knowing when and where to place a procedure or method the minimum number of times.

Some known truths about optimizing program performance are reflected in the following guidelines:

- Run an optimizer to understand where the bottlenecks are.
- Know the machine error (ϵ) for the processor on which the software is being developed, and know how to exploit this to your advantage using single- and double-precision variables.
- Avoid use of mixed math to the extent possible; that is, avoid using a mixture of byte, integer, and single- and double-precision variables together during division, multiplication, addition, and subtraction.
- When powers such as $x^2, x^3, \ldots, x^n$ are needed, use $x * x$ for x^2, $x * x * x$ for x^3, and so on. Using the syntax ^{2} or $**$ for squaring a term commonly results in lower performance.
- When generating loops within loops, it is good practice to place the loop with more steps on the inside, and leave the smaller steps on the outer loops. For example, consider the slow syntax followed by fast syntax given below:

```
Slow:

   .
For i = 1 to 1000
 For j = 1 to 5
     .
     .
 Next j
Next i

Fast:

For j = 1 to 5
 For i = 1 to 1000
     .
     .
 Next i
Next j
```

- When possible, use a constant such as $a = x(i)$ for $x(i)$ if i is not referred to inside a loop; determining the value of $x(i)$ unnecessarily inside a loop requires a memory call, which can be expensive:

```
Slow:

For i = 1 to 20
 For j = 1 to 1000
  z(j)=3*x(i)
 Next j
Next i

Fast:

For i = 1 to 20
 a=x(i)
 For j = 1 to 1000
  z(j)=3*a
 Next j
Next i
```

- Depending on the compiler-processor combination, when dividing by a scalar value x such as $y(i,j) = z(i,j)/x$ that won't change inside a loop, it may be faster to determine the factor $fac = 1/x$ before the loop and then multiply by $y(i,j) = z(i,j) * fac$. You would need to be using a processor compiler-for which division is slower than multiplication, which may not be necessary. Currently used processors can parallelize floating-point operations (ADD/SUB/MUL/DIV), and provide performance levels that are not substantially different.

13.3 TRAINING, TESTING, AND VALIDATION

Rule evaluation entails calculation of the accuracy, or the probability of correctly predicting the true class membership of a test object $\mathbf{x}$. Another definition of accuracy is the proportion of objects correctly assigned to their original classes. Classification accuracy results are routinely presented as a function of the classifier used, percentage of training microarrays used during training, and accuracy during training and testing. Using such results, one can determine the most accurate classification method, best number of features, and the optimal feature selection method and number of features.

Objects and Attributes. Machine learning terminology often uses the term *object* to denote each microarray. Each object has a collection of attribute or feature values, which represent gene expression values.

Training. Training is the process of using a classifier to iteratively learn the patterns in training data. Microarrays for which class membership is being predicted are omitted from training in order to prevent learning bias. If microarrays are used concurrently in both training and testing, then the informativeness contributed by each microarray will be directly applied for predicting class membership of the same microarrays used in training–resulting in serious bias. Various methods are proposed for omitting microarrays from training when predicting class membership of test microarrays and are discussed in the following paragraphs.

Confusion Matrix and Accuracy. The basis of determining classification *accuracy* is to compare for each test microarray $\mathbf{x}$ left out of training the true class label ω with the predicted class label $\hat{\omega}$. The *confusion matrix* used for tabulating test microarray class prediction during testing is denoted as $\mathbf{C}$ and is a square symmetric matrix with dimensions $\Omega \times \Omega$, expressed as

$$\underset{\Omega\times\Omega}{\mathbf{C}} = \begin{pmatrix} c_{11} & c_{12} & \cdots & c_{1c} \\ c_{21} & c_{22} & \vdots & \vdots \\ \vdots & \vdots & \ddots & \vdots \\ c_{c1} & c_{c2} & \cdots & c_{cc} \end{pmatrix}, \tag{13.1}$$

where Ω is the total number of classes. The rows of $\mathbf{C}$ represent true class ω, while columns represent predicted class $\hat{\omega}$. During testing, if the predicted class of test microarray $\mathbf{x}$ is correct (i.e., $\hat{\omega} = \omega$), then the diagonal element $c_{\omega\omega}$ of the confusion matrix is incremented by 1. However, if the predicted class is incorrect (i.e., $\hat{\omega} \neq \omega$), then the off-diagonal element $c_{\omega\hat{\omega}}$ is incremented by 1. After the predicted class membership of microarrays has been assigned, and construction of the confusion matrix is completed, we can determine the class-specific accuracy and overall accuracy. Class-specific accuracy (Acc) for class ω is calculated using the relationship

$$\text{Acc}_\omega = \frac{c_{\omega\omega}}{c_{\omega\cdot}}, \tag{13.2}$$

where $c_{\omega\omega}$ is the diagonal element of $\mathbf{C}$ for which microarrays were correctly classified and $c_{\omega\cdot} = \sum_j c_{\omega j}$ is the row sum of $\mathbf{C}$ for microarrays having true class label ω. Analogously, overall accuracy is determined as

$$\text{Acc} = \frac{\sum_\omega^\Omega c_{\omega\omega}}{n}, \tag{13.3}$$

where $n = \sum_\omega c_{\omega\cdot}$ is the total number of microarrays tested (i.e., the sum of row sums). When Acc is equal to 1, all of the microarrays will be correctly assigned to their true class.

***k*-Fold Cross-Validation.** One of the most efficient methods for building the confusion matrix **C** is to use *k-fold cross-validation*. Let $\mathcal{D}$ represent all of the microarrays in a dataset. In k-fold cross-validation, we split $\mathcal{D}$ into k mutually exclusive partitions $\mathcal{D}_1, \mathcal{D}_2, \ldots, \mathcal{D}_k$ and train and test the classifier k times. In 10-fold cross-validation (CV10), we set $k = 10$. We then perform stratified 10-fold cross-validation using 10 *repartitionings* for each classification run [3]. During a repartitioning, we first shuffle the order of the microarrays and then assign the microarrays to the 10 ordered partitions (folds), while ensuring that at least one microarray from each class is in each partition. Using the 10 partitions $\mathcal{D}_1, \mathcal{D}_2, \ldots, \mathcal{D}_{10}$, we first train the classifier with microarrays in partitions $\mathcal{D}_2, \ldots, \mathcal{D}_{10}$, and then use microarrays in partition $\mathcal{D}_1$ for testing (class prediction). Next, we train with partitions $\mathcal{D}_1, \mathcal{D}_3, \ldots, \mathcal{D}_{10}$ and test microarrays in partition $\mathcal{D}_2$. We repeat this procedure until training with partitions $\mathcal{D}_1, \mathcal{D}_2, \ldots, \mathcal{D}_9$ and test microarrays in partition $\mathcal{D}_{10}$ is completed. We repeat this scheme 10 times, each time randomly assigning all microarrays into 10 partitions and then performing training and testing on microarrays in partitions $\mathcal{D}_1, \mathcal{D}_2, \ldots, \mathcal{D}_{10}$ sequentially. An example of fold partitions and training and testing steps for 5CV is shown in Figure 13.1.

Leave-One-Out Cross-Validation. Another method for constructing the confusion matrix involves *leave-one-out cross-validation* (LOOCV), where $k = 1$. In LOOCV, class prediction is made on each test microarray when is it held out from training. LOOCV is known to be a pessimistically unbiased estimate of the true error rate [3], but like other approaches, it does not guarantee the same accuracy for future microarrays [4]. Figure 13.2 illustrates example workflow for LOOCV.

0.632 Bootstrap. This technique determines classifier accuracy using cross-validation based on the 0.632 bootstrapping method [5]. For the bth iteration ($b = 1, 2, \ldots, B$), microarrays are randomly selected with replacement and denoted as sample $\mathcal{D}^b$. The probability that a microarray is not selected during bootstrapping is $(1 - 1/n)^n = \exp(-1) = 0.368$, and we

n/5 arrays	n/5 arrays	n/5 arrays	n/5arrray	n/5 arrays
Test	Train	Train	Train	Train
Train	Test	Train	Train	Train
Train	Train	Test	Train	Train
Train	Train	Train	Test	Train
Train	Train	Train	Train	Test

FIGURE 13.1 Workflow for 5-fold cross-validation (5CV).

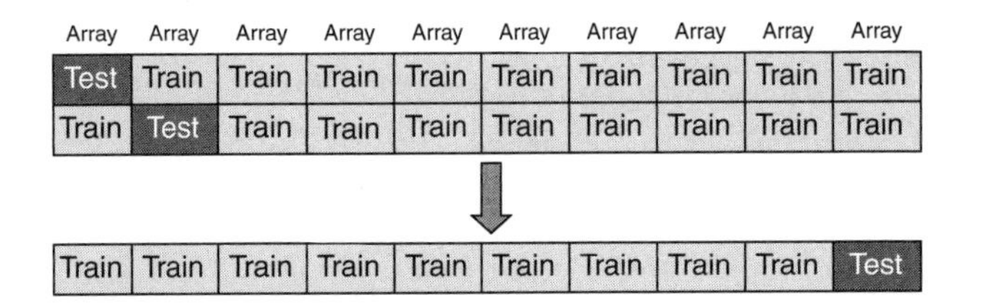

FIGURE 13.2 Workflow for LOOCV. (*See insert for color representation of the figure.*)

denote this sample of arrays as $\mathcal{D}^0$. Next, the classifier is trained using $\mathcal{D}^b$. The classification accuracy of microarrays in $\mathcal{D}^b$ is known as the apparent accuracy and is denoted as Acc_b. The classification accuracy of microarrays in $\mathcal{D}^0$ based on training with $\mathcal{D}^b$ is denoted as Acc_0.

1. Select n microarrays with replacement, and call this the *bootstrap* sample $\mathcal{D}^b$. Denote the set of microarrays not selected as sample $\mathcal{D}^0$.
2. Train the classifier using microarrays in $\mathcal{D}^b$.
3. Obtain Acc_0, by predicting class membership of arrays in $\mathcal{D}^0$.
4. Obtain Acc_b, by predicting class membership of arrays in $\mathcal{D}^b$.
5. The 0.632 cross-validation accuracy is determined by the relationship

$$\text{Acc}_{0.632} = 0.368\text{Acc}_b + 0.632\text{Acc}_0. \tag{13.4}$$

These steps are carried out, for example, $B = 40$ times, and the average of $\text{Acc}_{0.632}$ over the B iterations is used as the final estimate of accuracy.

Holdout Method. The *holdout* method for constructing the confusion matrix partitions the dataset $\mathcal{D}$ so that, for example, two-thirds of the microarrays are used for training and the remaining one-third, for testing. Random subsampling is often repeated k times, in which microarrays are randomly selected followed by accuracy determination. A drawback of the holdout method is that it makes inefficient use of $\mathcal{D}$, since a third of the microarrays are not used for training the classifier.

13.4 ENSEMBLE CLASSIFIER FUSION

Ensemble techniques involve the use of combination rules to fuse results from multiple classifiers. Studies have shown that the results of ensemble techniques often result in a more robust solution with improved generalization performance when compared with use of a single classifier [6,7]. Ensemble techniques can be applied to investigate whether complementary information can be extracted from the gene sets used for classification, and whether classifier fusion will result in improved performance.

Majority Vote. Under this approach, classifiers are trained with the same feature sets, and then classifier votes are combined using the *ensemble majority voting* (EMV) and *ensemble weighted majority voting* (EWMV) ensemble combination techniques [8]. Let $d_{l,\omega}(\mathbf{x}) \in \{0,1\}$ be the decision rule for an array by the lth classifier ($l = 1, 2, \ldots, L$) for class ω ($\omega = 1, 2, \ldots, \Omega$). The support for EMV and EWMV, respectively, is functionally expressed as

$$\boldsymbol{\mu}_{\omega}(\mathbf{x}) = \sum_{l=1}^{L} d_{l,\omega}(\mathbf{x}), \qquad d_{l,\omega}(\mathbf{x}) \in \{0,1\}, \tag{13.5}$$

$$\boldsymbol{\mu}_{\omega}(\mathbf{x}) = \sum_{l=1}^{L} w_l d_{l,\omega}(\mathbf{x}), \qquad d_{l,\omega}(\mathbf{x}) \in \{0,1\}, \tag{13.6}$$

where w_l is the normalized weight reflecting the accuracy of the lth classifier. Here, accuracy is based on the proportion of classified test arrays assigned to the diagonal of the confusion matrix divided by the number of test arrays. Let the set of class labels be $\omega = 1, 2, \ldots, \Omega$ and the ensemble decision for array $\mathbf{x}$ be $E(\mathbf{x} \rightsquigarrow \omega)$. The decision rule for test array $\mathbf{x}$ is

$$E(\mathbf{x} \rightsquigarrow \omega) \equiv \arg\max_{c}\{\boldsymbol{\mu}_c(\mathbf{x})\}. \tag{13.7}$$

Random Oracles. Random oracles can also be used for ensemble classifier fusion, to partition the training and testing of microarrays (arrays) into two disjoint partitions, or "oracles" formed by use of miniclassifiers. The advantages of random oracles are that (1) partitioning microarrays for training and testing can be easier than jointly performing the same operations on the combined microarrays and (2) more diversity is introduced into training and testing because errors are not shared between subclassifiers [9]. Ensemble diversity has been investigated for boosting, bagging, and the sum–product rule for majority voting [10-12]. Ensemble diversity can be measured using the kappa statistic κ, which reflects both the accuracy and diversity of an ensemble of classifiers [13]. Kappa is defined as

$$\kappa = \frac{\Theta_1 - \Theta_2}{1 - \Theta_2}, \tag{13.8}$$

where

$$\Theta_1 = \frac{\sum_{i=1}^{L} C_{ii}}{n}, \tag{13.9}$$

and

$$\Theta_2 = \sum_{i=1}^{L} \left(\frac{\sum_{j=1}^{L} C_{ij}}{n} \right) \left(\frac{\sum_{j=1}^{L} C_{ji}}{n} \right). \tag{13.10}$$

In this formulation, C_{ii} is the number of arrays that classifier i voted for and C_{ij} is the number of arrays for which classifiers i and j had the same vote. Smaller κ values indicate a larger diversity; when $\kappa = 0$, the agreement between two classifiers is based on chance, whereas when $\kappa = 1$, the two classifiers agree on every example. An ensemble kappa-error plot [14] of average classifier pairwise error versus κ with $L(L-1)/2$ points is typically generated to reflect average error versus diversity for each possible pair of classifiers in an ensemble for a given dataset. In a previous study [15], we assessed diversity of an ensemble of 11 classifiers when they were employed as miniclassifiers of a principal direction linear oracle (PDLO) system. Principal direction hyperplane splits were used for assigning arrays from the nine datasets used in this book to a pair of the same type of miniclassifiers. Figure 13.3 shows a kappa-error plot of error and diversity for the 11-classifier ensemble with and without use of PDLO. Each plot contains 495 points representing all possible pairs of the 11 classifiers [$55 = 11(10)/2$] times nine datasets. The x axis represents values of κ for each pair of classifiers, and the y axis is the average error rate of a pair of classifiers. The left Figure 13.3a illustrates more ensemble diversity when PDLO was used, since there are more classifier pairs with κ

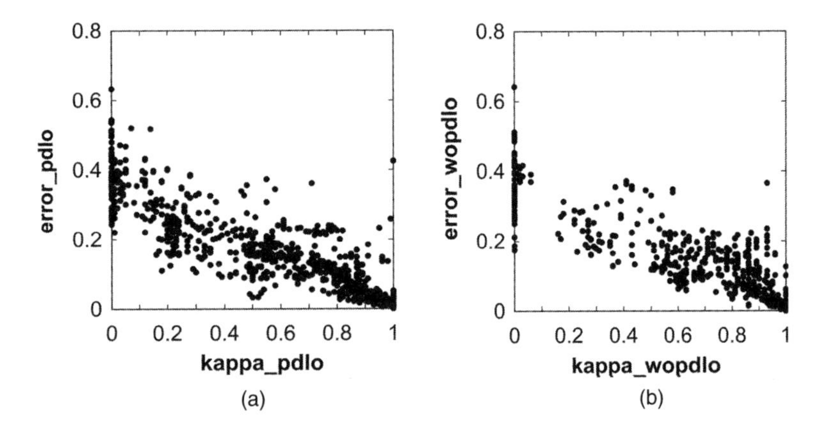

FIGURE 13.3 Kappa-error plot of error and diversity of an ensemble of 11 classifiers based on use of a principal direction linear oracle (PDLO) to randomly perform hyperplane splits of arrays before assigning them to a pair of miniclassifiers. There are 495 data points in each panel, representing all possible pairs of the 11 classifiers [$55 = 11(10)/2$] times nine datasets. Panel (a), based on use of PDLO, reflects more diversity, since there are more classifier pairs with κ in the range 0–0.5 when compared with results when PDLO was not used (b).

in the range 0-0.5 when compared with results in Figure 13.3b, for which PDLO was not used. We reported in later research that a random spherical linear oracle resulted in greater performance when compared with PDLO [16].

13.5 SENSITIVITY AND SPECIFICITY

Another method for evaluating classifier performance involves determination of *screening diagnostics*, originally developed for evaluating the validity of a diagnostic test used in clinical screening for disease.

Table 13.1 lists counts of tested microarrays for which a are the true positives (TP), b are the false positives (FP), c are the false negatives (FN), and d are true negatives (TN). Under these definitions, *sensitivity* is the proportion of diseased subjects who have a positive test, which is functionally composed as

$$\text{Sensitivity} = \frac{a}{a+c} = \frac{\text{TP}}{\text{TP}+\text{FN}}. \tag{13.11}$$

On the other hand, the *specificity* is the proportion of disease-free subjects with a negative test

$$\text{Specificity} = \frac{d}{d+b} = \frac{\text{TN}}{\text{TN}+\text{FP}}. \tag{13.12}$$

Generally speaking, if sensitivity and specificity are not both above 95%, then the test will likely not be approved for use in clinical settings. Typically, tests in the initial stage of development have a sensitivity close to or above 95% and a lower specificity on the order of 80-90%. Development of a new test often employs a *gold standard* test against which results are compared.

Sensitivity and specificity are based on test subjects for which the true disease state is known; however, in the real clinical setting the true disease state of patients is unknown. Therefore, the merit of a test in clinical applications is contingent on the predictive values. The *positive predictive value* PV^+ is defined as the proportion of subjects with a positive test who

TABLE 13.1 Truth Table and Predicted Class of Microarrays and Corresponding Sensitivity and Specificity

	Disease (Truth)		
Test	Present	Absent	
Positive	a—true positives	b—false positives	$a+b$
Negative	c—false negatives	d—true negatives	$c+d$
	$a+c$	$b+d$	n

have the disease

$$PV^+ = \frac{a}{a+b} = \frac{TP}{TP+FP}, \tag{13.13}$$

and the *negative predictive value*, PV^- is defined as the proportion of subjects with a negative test who are disease-free, which is functionally expressed as

$$PV^- = \frac{d}{d+c} = \frac{TN}{TN+FN}. \tag{13.14}$$

If the proportion of truly diseased subjects is much lower than the proportion of subjects truly disease-free, then the positive predictive value will be low ($\leq 50\%$). Likewise, if the true proportion of diseased is much greater than the proportion of disease-free, then the negative predictive value will be low ($\leq 50\%$).

For class prediction, we shall be concerned with comparing the true and predicted fractions of microarrays assigned to a given class. Table 13.2 is an example two-way table used for determining sensitivity and specificity for microarrays assigned to class ω versus microarrays not in class ω. The sensitivity is equivalent to $a/(a+c)$ and is the probability of correctly predicting that a microarray is in class ω given that the microarray is truly in class ω. The specificity is equivalent to the ratio $d/(b+d)$, and reflects the probability of correctly assigning a microarray to a class other than ω given that the microarray is truly not in class ω. The PV^+, which is equivalent to $a/(a+b)$, is defined as the probability that a microarray predicted to belong to class ω is truly in class ω. Lastly, the PV^-, is equivalent to the ratio $d/(c+d)$, and is the probability that a microarray not predicted to belong to class ω is truly not class ω. Table 13.3 lists example class-specific sensitivity, specificity, PV^+, and PV^- for the two-class prostate cancer dataset.

13.6 BIAS

The bias of a classifier reflects how classification accuracy depends on the data used, and is assessed by comparing mean accuracy results for varying degrees of k-fold cross-validation. A classifier that is not biased will not

TABLE 13.2 Truth Table and Predicted Class of Microarrays and Corresponding Sensitivity and Specificity

	Truth		
Prediction	Class ω	Not Class ω	
Class ω	a—true positives	b—false positives	$a+b$
Not class ω	c—false negatives	d—true negatives	$c+d$
	$a+c$	$b+d$	n

TABLE 13.3 Class-Specific Sensitivity, Specificity, Predictive Value Positive, and Predictive Value Negative for Various Classifiers Applied to the Two-Class Prostate Cancer Dataset

Classifier	Class	Accuracy	Sensitivity	Specificity	PV_+	PV_-
5NN	Tumor	0.949	1.000	0.906	0.900	1.000
	Normal	0.949	0.906	1.000	1.000	0.900
NBC	Tumor	0.927	0.979	0.884	0.877	0.980
	Normal	0.927	0.884	0.979	0.980	0.877
LDA	Tumor	0.981	1.000	0.963	0.963	1.000
	Normal	0.981	0.963	1.000	1.000	0.963
SVMGD	Tumor	0.961	0.988	0.936	0.935	0.988
	Normal	0.961	0.936	0.988	0.988	0.935
LOG	Tumor	0.968	0.975	0.961	0.962	0.974
	Normal	0.968	0.961	0.975	0.974	0.962
EMV	Tumor	0.971	1.000	0.943	0.942	1.000
	Normal	0.971	0.943	1.000	1.000	0.942
EWMV	Tumor	0.971	1.000	0.943	0.942	1.000
	Normal	0.971	0.943	1.000	1.000	0.942

display varying results across the different cross-validation runs. Figure 13.4 illustrates mean accuracy for 10 k-fold cross-validations runs using a 5-nearest-neighbor (5NN) and naïve Bayes classifier (NBC) on the nine datasets listed in Table 12.10. The 5NN classifier (Figure 13.4a) is less biased than the NBC classifier, which was more biased for two-fold cross-validation.

13.7 VARIANCE

Variance reflects stability of a classifier and is measured as the standard deviation of accuracy over various k-fold cross-validation runs. Using the

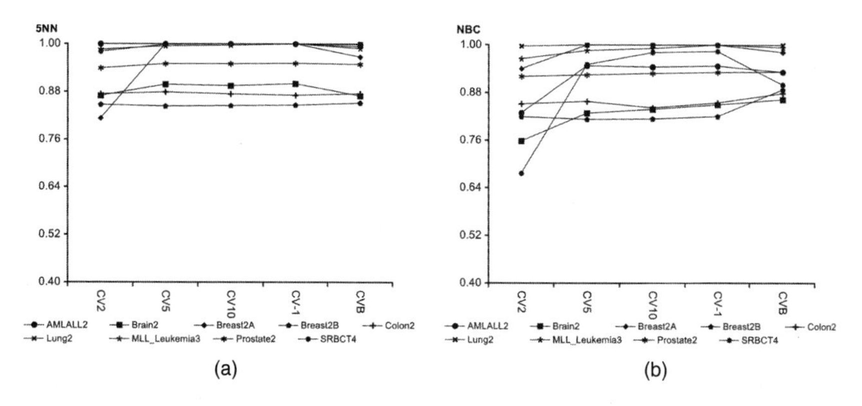

FIGURE 13.4 Bias of 5NN and NBC for various cross validation analyses. The NBC classifier is more biased when compared with the 5NN classifier. Points on the plot are mean accuracy for 10 k-fold cross-validation runs of nine datasets.

nine datasets listed in Table 12.10, Figure 13.5 portrays much greater variance of the NBC classifier [panel (b)] when compared with standard deviation of accuracy for the 5NN classifier. After evaluation of bias and variance of both the 5NN and NBC classifier in Figures 13.4 and 13.5, it is apparent that 5NN is lessed biased and has lower variance for the nine datasets considered. Therefore, the choice of a classifier should be based on a full evaluation of both the bias and variance for multiple datasets, as this helps to establish the robustness needed for deploying a classifier in a different environment with new data.

13.8 RECEIVER-OPERATOR CHARACTERISTIC (ROC) CURVES

Threshold-Based TPR and FPR. Receiver-operator characteristic (ROC) curves originated in radar research as a means of discriminating between the radar return of an actual airborne target and the noise created by electronics, the ground, atmosphere, or other phenomena. A typical plot of an ROC curve is shown in Figure 13.6. The x axis is represented by the false positive rate (FPR) equal to 1-specificity, while the y axis is represented by the true positive rate (TPR), which is equal to the sensitivity. The ROC curve itself forms a line reflecting the TPR as a function of the FPR. If a discrimination method is uninformative, the ROC curve will form a straight line spanning from the origin to the upper right corner of the plot where both the TPR and FPR are equal to 1, and the underlying *area under the curve* (AUC) will be equal to 0.5. Otherwise, the curve will lie above the uninformative line and preferably extend into the upper left corner where TPR = 1 and FPR = 0.

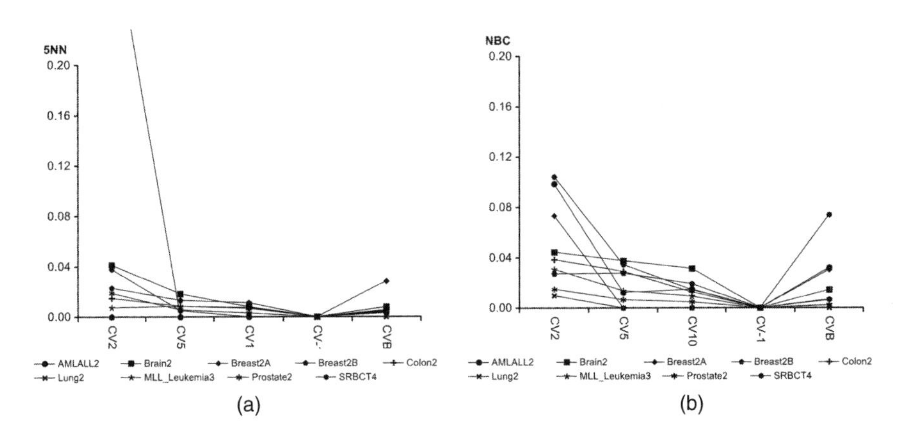

FIGURE 13.5 Standard deviation (SD) of accuracy for 10 k-fold cross-validation runs using 5NN and NBC classifiers for nine datasets. The 5NN classifier (a) is more stable than the the NBC classifier (b), which has a higher degree of s.d. variation.

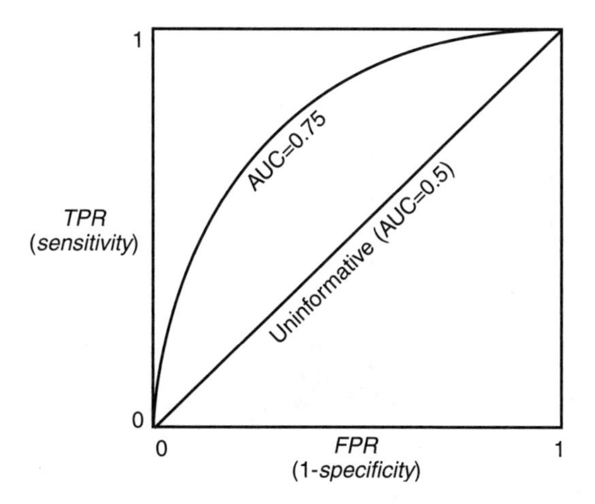

FIGURE 13.6 Receiver-operator characteristic (ROC) curve.

Threshold for Two-Class Problems. ROC curve construction is based on use of a *threshold*, which can be implemented two ways. In the first approach, assume a two-class problem with true outcome values of microarrays $y = 0$ and $y = 1$, for normal and diseased, respectively. Let the predicted outcome of a test microarray $\mathbf{x}$ be $\hat{y} = P(y = 1|\mathbf{x})$. Whenever $P(y = 1|\mathbf{x}) >$threshold, we set $\hat{y} = 1$; otherwise $\hat{y} = 0$. This is one of the more common methods implemented in computer software packages for logistic regression.

Threshold for Multiclass Problems. The approach we favor involves forming smoothed probability density functions of classifier accuracy based on the null and alternative distributions. In this sense, the alternative distribution of accuracy represents the signal, while the null accuracy distribution represents the noise. The sequential steps for constructing a multiclass ROC curve is completely developed in the following paragraphs.

Step 1: Assemble distributions for signal and noise. For this worked example, assume that the signal and noise distributions are normally distributed with means of 7 and 3, respectively. Figure 13.7 shows the signal and noise distributions as well as the frame of the ROC curve. Assume that the single threshold value is called x_m^* and represents a bin value (x-value) of the histogram for the null accuracy distribution divided into M ($m = 1, 2, \ldots, M$) equally spaced nonoverlapping bins.

Step 2: Increase threshold values to obtain FPR and TPR. This step involves increasing the threshold value x_m^* over only the range of the noise distribution to obtain FPR and TPR values for plotting. At each step, the threshold value is increased and FPR is determined as the integral

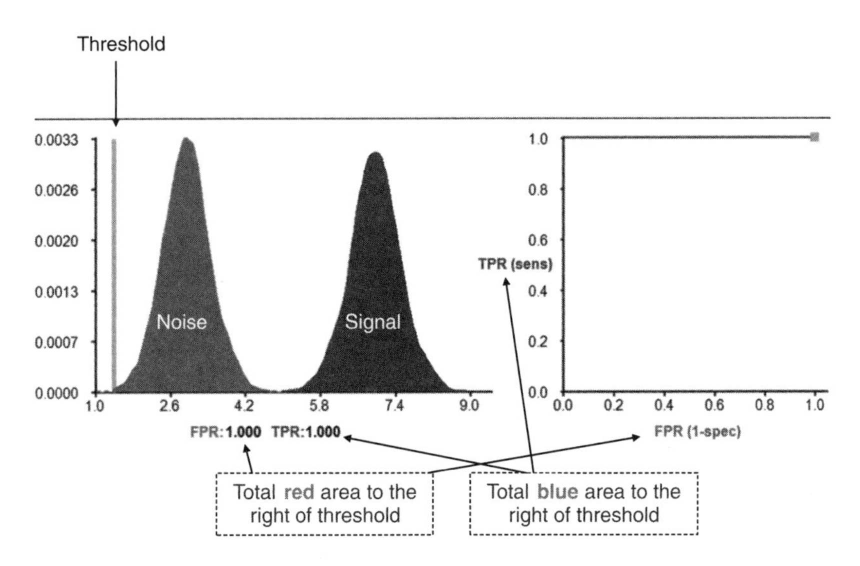

FIGURE 13.7 FPR and TPR of ROC curve when threshold value of x_m^* is small. The noise (null) distribution of classifier accuracy (a) is based on training and testing with permuted class labels, while the signal (alternative) distribution of accuracy (b) is based on training and testing with the true class labels. (*See insert for color representation of the figure.*)

probability of the remainder of the noise distribution that is greater than the threshold value. Likewise, the TPR is the integral probability of the signal distribution that exceeds the threshold value. For each threshold value, plot the FPR and TPR obtained from integrating the PDFs. Figure 13.8 illustrates the plotting of FPR and TPR values for the noise and signal distributions as the threshold value is increased. When the difference between the noise and the signal distributions is large, the entire span of the noise distribution is traversed by the increasing threshold value, to the extent that the entire range of FPR is covered with no change in TPR. Figure 13.9 shows the resulting ROC curve for the 2 distributions, which results in an AUC of unity. By the time the threshold value x_m^* approaches a value of 4 (Figure 13.9), the FPR, or integral probability of the noise distribution, is approaching zero and we essentially do not need to advance the threshold to greater values.

Figure 13.10 shows an ROC curve for the case when the distribution of the signal overlaps with the noise distribution. As the threshold value x_m^* is increased (Figure 13.11), the integral of the signal distribution to the right of the threshold value starts decreasing along with the decrease in integral value of the noise distribution, resulting in a decrease in FPR and TPR simultaneously. Figure 13.12 illustrates the final ROC curve for the

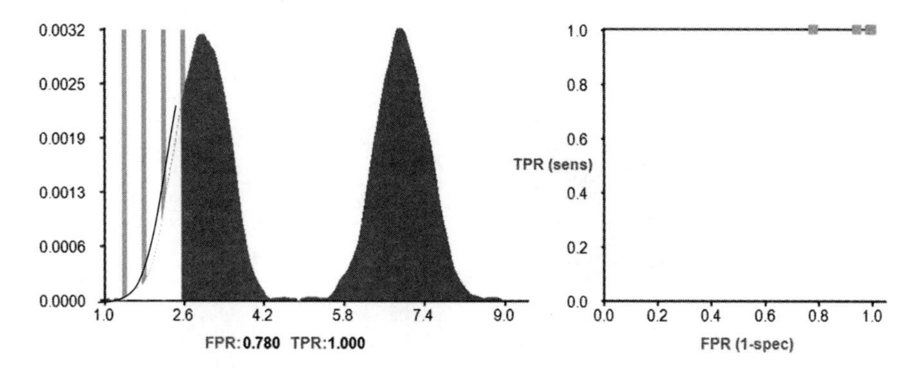

FIGURE 13.8 ROC curve for intermediate threshold value of x_m^*.

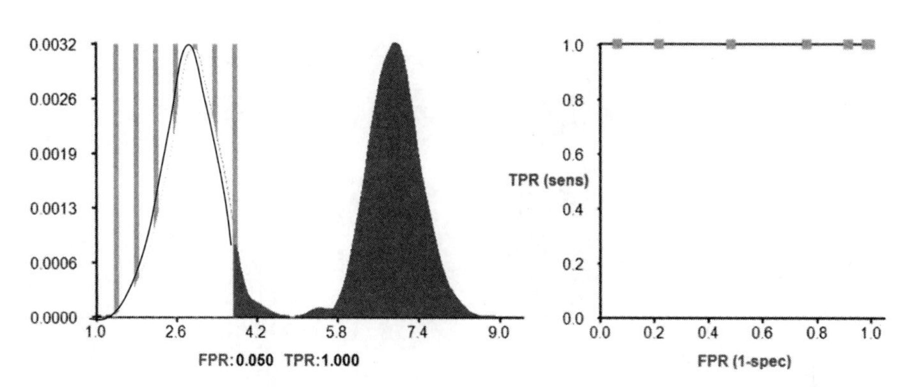

FIGURE 13.9 Final ROC curve results when null and alternative accuracy distributions differ greatly.

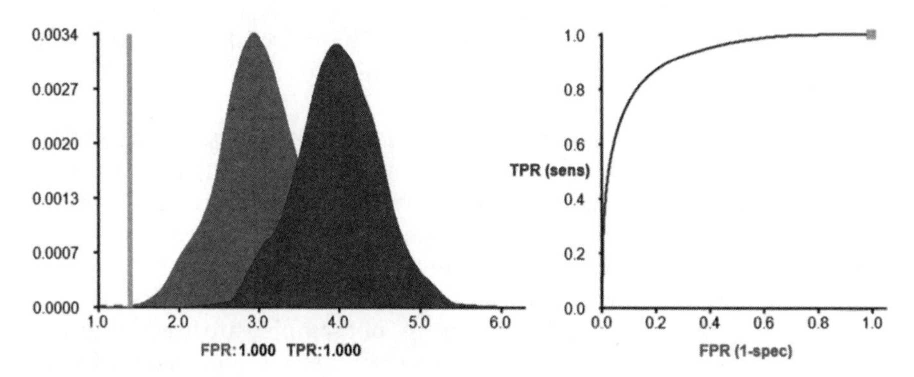

FIGURE 13.10 Early development of FPR and TPR elements of ROC curve when null and alternative distributions of accuracy are similar.

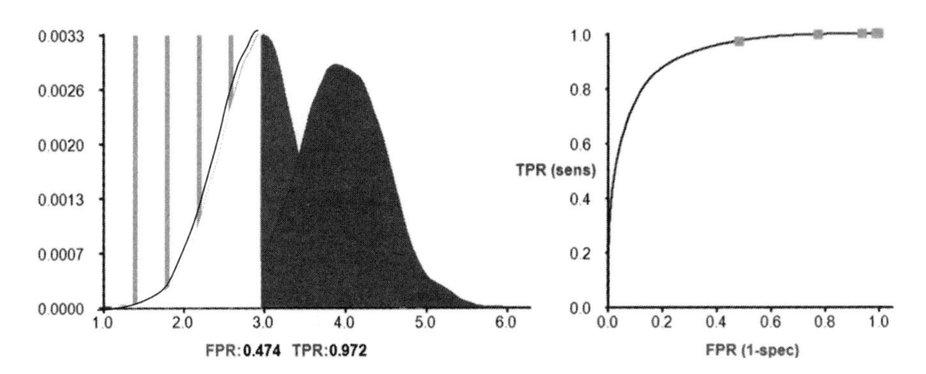

FIGURE 13.11 Resulting ROC curve when null and alternative accuracy distributions are similar.

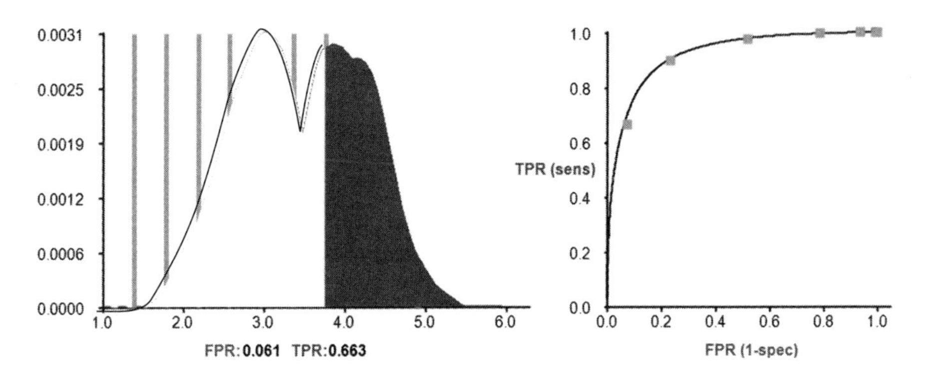

FIGURE 13.12 Final ROC curve when null and alternative accuracy distributions are alike.

two distributions, which is not impressive. The value of AUC needs to be determined using methods described in the next section.

Bootstrapping Microarrays for Null and Alternative Distributions of Accuracy. Data used to create the alternative (i.e., signal) distribution are based on a bootstrap sample of microarrays selected randomly with replacement whose sample size is equal to the number of total microarrays [17,18]. For example, if $n = 100$ microarrays are available, then $n = 100$ microarrays are randomly selected with replacement. The microarrays sampled are used for training the classifier, while microarrays not sampled are used for testing, and classifier accuracy is determined for only the test microarrays. During each bootstrap, accuracy for the alternative distribution (signal) is calculated using the true class labels of training and test microarrays to obtain Acc_b ($b = 1, 2, \ldots, B$). Accuracy values for the null distribution (noise) are determined in the same way, but after shuffling the class labels of all microarrays prior to training and testing -resulting in Acc_b^*. Bootstrapping is repeated $B = 40$ times, yielding 40 realizations of accuracy for the alternative and null distributions. Mukherjee et al. used $B = 50$ iterations during randomization tests to obtain SVM error rates for linear kernels based on 150 or all features for some of the same datasets [19]. Therefore, it is our belief that 40 iterations should provide an adequate number of accuracy realizations for PDF generation.

Kernel Density Estimation (KDE). Determination of area under the curve (AUC) and ROC curve construction with FPR and TPR requires smoothing the B values of Acc_b and Acc_b^* separately using kernel density estimation (KDE). We begin by estimating the mean, μ, and standard deviation, σ, of Acc_b and Acc_b^*. Two histograms are then constructed: one for the null accuracy distribution and one for the alternative distribution. Each histogram is constructed using $M = 1000$ equally spaced bins over the range $\Delta_x = \min\{\mu + 4\sigma, 1\} - \max\{\mu - 4\sigma, 0\}$. The histogram bin counts are functionally composed as

$$f(m) = \frac{1}{Nh} \sum_{i=1}^{N} K\left(\frac{a_i - x_m}{h}\right), \tag{13.15}$$

where $f(m)$ is the smoothed bin count for the mth histogram bin, $a_i = \mu + N(0, 1)\sigma$ is a simulated realization ($i = 1, 2, \ldots, N$) of accuracy using the mean and standard deviation of either Acc_b or Acc_b^*, x_m is the lower wall of the mth bin, $h = 2 \times 1.06\sigma N^{-0.2}$ is the bandwidth [20], and K is the Gaussian kernel function, defined as

$$K(u) = \frac{1}{\sqrt{2\pi}} \exp\left(-\frac{u^2}{2}\right), \tag{13.16}$$

where $u = (a_i - x_m)/h$. We have learned through experience that $N = 1000$ realizations of a_i is reasonable. The smoothed histogram bin heights $f(m)$ are then normalized so that their integral is unity, resulting in a PDF of $P(x_m)$ for the alternative accuracy distribution and $P(x_m^*)$ for the null accuracy distribution. For later use, we also denote the quantile value x_m as the lower bin walls of the histogram for the alternative accuracy distribution, and the quantile x_m^* as the lower bin walls of the histogram for null accuracy.

Bootstrap Bias. The bias-variance tradeoff of a classifier can be evaluated using the bootstrap, where $B = 40$ iterations are used with fractions of $f = 0.1$, 0.2, 0.3, 0.4, 0.5, or 0.6 of the available microarrays. Microarrays selected in the bootstrap sample are used for classifier training, while objects not selected that are omitted from training are used for testing and accuracy determination. Figure 13.13a shows the smooth null and alternative distributions obtained from KDE for various sample sizes based on the sampling

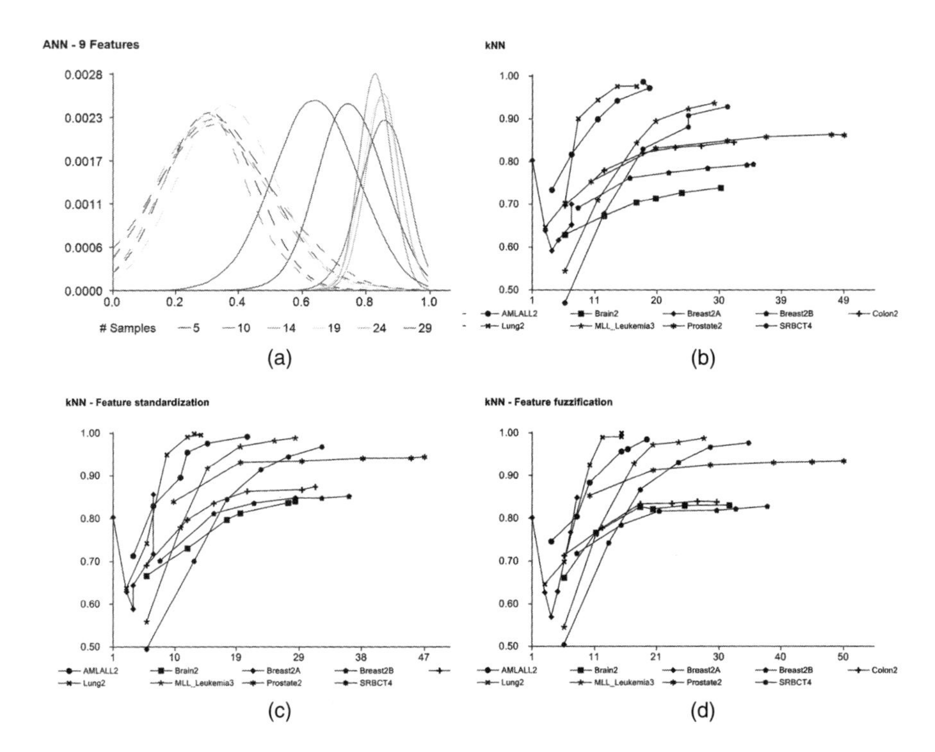

FIGURE 13.13 (a) Kernel density estimates of ANN classifier accuracy for $B = 40$ bootstrap samples based on sampling fractions of $f = 0.1$, 0.2, 0.3, 0.4, 0.5, or 0.6. PDFS to the right (solid lines) form the alternative distributions of x_m based on original true class labels, while PDFS on the left side of the plot (dashed lines) form the null distributions of x_m^* based on shuffled class labels. (b-d) Bootstrap bias of 5NN accuracy for nine datasets as a function of bootstrapped sample size based on f—without feature transforms, with feature standardization, and with feature fuzzification.

fraction f during bootstrapping. Figure 13.13b-d show the boostrap-based accuracy of objects not sampled as a function of the sampling fraction f.

Plotting the ROC Curve. Plotting the ROC curve is based on the integral areas of the null and alternative accuracy distributions (obtained from KDE) beyond a specified threshold null accuracy x_m^* $(m = 1, 2, \ldots, M)$. For a given threshold value of null accuracy x_m^*, the FPR and TPR for plotting are based on the relationships

$$\mathrm{FPR}(x_m^*) = \sum_{j=m}^{M} P(x_j^*) \qquad m = 1, 2, \ldots, M, \tag{13.17}$$

and

$$\mathrm{TPR}(x_m^*) = \sum_{j=1}^{M} P(x_j) I(x_j \geq x_m^*) \qquad m = 1, 2, \ldots, M. \tag{13.18}$$

Values of FPR and TPR are plotted using M equally spaced null accuracy values over the range of x^* based on histogram bin widths of $\Delta x^* = (x_{\max}^* - x_{\min}^*)/M$. Figure 13.14 shows an ROC curve for 5NN classifier results using the four-class SRBCT data based on 23 features. Algorithm 13.1 lists the steps for obtaining FPR and TPR for an ROC curve for a fraction f of the total sample size (the ROC curve itself is based on a plot of TPR vs. FPR).

Area Under the Curve (AUC) as a Function of Sample Size. The alternative and null PDFS from KDE are also employed to estimate the area under the curve (AUC) of the ROC curve. AUC is determined with the relationship

$$\mathrm{AUC} = 1/M \sum_{j=1}^{M} \sum_{i=1}^{M} P(x_i) I\{x_i \geq x_j^*\}, \tag{13.19}$$

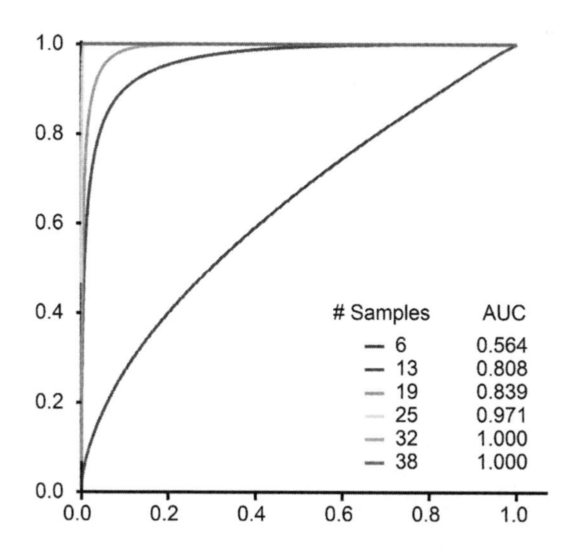

FIGURE 13.14 ROC curves as a function of bootstrapped sample sizes for 5NN classifier run on the four-class SRBCT data with 23 features selected using greedy PTA. The legend key "# Samples" represents the number of arrays randomly selected. (*See insert for color representation of the figure.*)

Algorithm 13.1: Two-Class and Multiclass ROC Curves for Sampling Fraction f

Data: Input dataset $\mathcal{D}$ of size n
Result: ROC curve, AUC as a function of sampling fraction f

```
for b ← 1 to B do
    Get bootstrap sample D_b of size nf from D
    Train classifier using microarrays in D_b
    Determine accuracy Acc_b of microarrays left out of D_b
endfor
for b ← 1 to B do
    Shuffle class labels of all microarrays
    Get bootstrap sample D_b of size nf from D
    Train classifier using microarrays in D_b
    Determine accuracy Acc*_b of microarrays left out of D_b
endfor
Obtain x_m and P(x_m) for PDF of Acc_b from KDE using M bins
Obtain x*_m and P(x*_m) for PDF of Acc*_b from KDE using M bins
'Determine FPR and TPR for M intervals
for j ← 1 to M do
    xcutoff = x*_j
    FPR_j = 0
    TPR_j = 0
    for m ← j to M do
        FPR_j = FPR_j + P(x*_m)
    endfor
    for m ← 1 to M do
        if x_m > xcutoff then
            TPR_j = TPR_j + P(x_m)
        endif
    endfor
endfor
'Determine AUC
sum = 0
for j ← 1 to M do
    xcutoff = x*_j
    TPR = 0
    for m ← 1 to M do
        if x_m > xcutoff then
            TPR = TPR + P(x_m)
        endif
    endfor
    sum = sum + TPR
endfor
AUC = sum/M
```

where $I(\cdot)$ is the indicator function set to one when true and zero when false. For each classifier, we generate ROC curves using randomly selected proportions of $f = 0.1, 0.2, 0.3, 0.4, 0.5$, or 0.6 of the total number of available microarrays. Algorithm 13.1 lists the steps for obtaining AUC for an ROC curve as a function of the bootstrap sampling fraction f. Statistical power is the determined as the value of TPR when FPR is 5%, which can be found by identifying the value of TPR when the FPR $= 0.05$.

For a given sample size, we also can vary the number of features used based on the best ranked features chosen. The number of steps used for varying the feature count is set equal to $20/(\Omega(\Omega - 1)/2)$, to ensure that no more than 20 features are used for any run. For each fixed set of randomly selected training microarrays and best ranked features, the proportion of input microarrays is resampled $B = 40$ times in order to generate 40 realizations of observed and null accuracy. Examples of null and alternative PDFS as a function of sample size are shown in Figure 13.13.

It is important to note that the minimum and maximum quantile values of accuracy for the null and alternative distributions are different. Thus, in the case of highly significant classification, the lowest quantile (accuracy) value of the alternative distribution will be greater than the greatest accuracy value in the null distribution. This is a common observation with the use of randomization tests for highly significant effect measures, where test statistics based on permuted labels are much lower than test statistics for the observed data.

REFERENCES

[1] G.K. Gupta. *Teaching Computer Science as the Science of Information*. Technical Report JCU-CS-96/10, Dept. Computer Science, James Cook Univ., Townsville, Australia, 1996.

[2] J. Horst, E. Messina, T. Kramer, H.-M. Huang. Precise definition of software component specifications, 1997. *Proc. IFAC Computer-Aided Control System Design Conf.*, Gent, Belgium, 1997.

[3] R. Kohavi. A study of cross-validation and bootstrap for accuracy estimation and model selection. *Proc. Int. Joint Conf. Artificial Intelligence (IJCAI)*, 1995, pp. 1137–1145.

[4] B. Efron. Estimating the error rate of a prediction rule: Improvement on cross-validation. *J. Am. Stat. Soc.* **78**:316–331, 1983.

[5] B. Efron, R. Tibshirani. Improvements on cross-validation: The .632+ bootstrap method. *J. Amer. Stat. Assoc.*, **92**:548–560, 1997.

[6] L.I. Kuncheva. *Combining Pattern Classifiers: Methods and Algorithms*. Wiley, New York, 2005.

[7] R. Polikar. Ensemble based systems in decision making. *IEEE Circuits Sys. Mag.* **6**(3):21–45, 2006.

[8] M. van Erp, L. Vuupijl, L. Shomaker. An overview and comparison of voting methods for pattern recognition. *Proc. 8th Int. Workshop on Frontiers in Handwriting Recognition (WFHR02)*, IEEE Press, Piscataway, NJ, 2002.

[9] A. Sharkey, N. Sharkey. Combining diverse neural net. *Knowl. Eng. Rev.* **12**(3):231–247, 1997.

[10] C. Domeniconi, B. Yan. On error correlation and accuracy of nearest neighbor ensemble classifiers. *Proc. 2005 SIAM Int. Data Mining Conf.*, Society for Industrial and Applied Mathematics Press, Philadelphia, 2005.

[11] L.I. Kuncheva, C.J. Whitaker. Measures of diversity in classifier ensembles. *Machine Lear.* **51**:181–207, 2003.

[12] L.I. Kuncheva. That elusive diversity in classifier ensembles. *Lecture Notes Comput. Sci.* **2652**:1126–1138, 2003.

[13] G. Brown, J. Wyatt, R. Harris, X. Yao. Diversity creation methods: A survey and categorisation. *J. Inform. Fusion* **6**(1):5–20, 2005.

[14] D. Margineantu, T. Dietterich. Pruning adaptive boosting. *Proc. 14th Int. Conf. Machine Learning*. Morgan Kaufmann Press, San Francisco, 1997.

[15] L.E. Peterson, M.A. Coleman. Principal direction linear oracle for gene expression ensemble classification. *Proc. Workshop on Computational Intelligence Approaches for the Analysis of Bioinformatics Data (CIBIO07); 2007 International Joint Conference on Neural Networks (IJCNN07)*, IEEE Press, Piscataway, NJ, 2007.

[16] L.E. Peterson, M.A. Coleman. Logistic ensembles of random spherical linear oracles for microarray classification. *Int. J. Data Mining Bioinform.*, **3**(4):382–397, 2009.

[17] L.E. Peterson, R.C. Hoogeveen, H.J. Pownall, J.D. Morrisett. Classification analysis of surface-enhanced laser desorption/ionization mass spectral serum profiles for prostate cancer. *Proc. 2006 IEEE World Congress on Computational Intelligence (WCCI 2006)*, IEEE Press, Piscataway, NJ, 2006.

[18] L.E. Peterson, M.A. Coleman. Machine learning-based receiver operating characteristic (ROC) curves for crisp and fuzzy classification of DNA microarrays in cancer research. *Int. J. Approx. Reason.* **47**:17–36, 2008.

[19] S. Mukherjee, P. Tamayo, S. Rogers, R. Rifkin, A. Engle, C. Campbell, T.R. Golub, J.P. Mesirov. Estimating dataset size requirements for classifying DNA microarray data. *J. Comput. Biol.* **10**(2):119–142, 2003.

[20] D. Fadda, E. Slezak, A. Bijaoui. Density estimation with non-parametric methods. *Astron. Astrophys. Suppl. Ser.* **127**(2):335–352, 1998.

CHAPTER 14

LINEAR REGRESSION

14.1 INTRODUCTION

Linear regression (LREG) is concerned with prediction based on a wholly linear model that uses *independent* variables or predictors to estimate the expected value of some *dependent* or response (outcome) variable [1,2]. Linear regression has found widespread use in classification analysis [3-5] and has become an essential first-step routine for modeling and predictive analytics when data are *linearly separable*. New datasets involving mixtures of distributions from molecular, genomic, and clinical sources are naturally less linearly separable, and therefore, require more advanced flexible methods. Nevertheless, there is merit in learning, early in the analysis stage, whether linear methods break down, because if the data structure and feature characteristics are not robust on a linear scale, then the chance of overfitting with flexible methods becomes a greater concern during real-world application on future datasets.

Although regression is similar to correlation, correlation is essentially the tightness of data points around a straight line irrespective of slope, whereas regression focuses on the dependence of the y variable(s) on one or more predictor x variables. The goal of regression is to determine the amount of variance explanation due to the independent variables and the slope measured by *regression coefficients* due to each independent variable. The degree of slope reflects the linear relationship between the dependent and independent variables.

We now introduce a linear multivariate regression model for which there are multiple dependent variables and multiple independent variables [6]. Let $\mathbf{x}_i = (x_{i1}, x_{i2}, \ldots, x_{ij}, \ldots, x_{ip})$ be a vector of expression values for p genes for the ith microarray (array) and $\mathbf{y}_i = (y_{i1}, y_{i2}, \ldots, y_{i\omega}, \ldots, y_{i\Omega})$ be the vector of class indicators for the same array. We set $y_{i\omega}$ to $+1$ for the true class membership and -1 for the remaining classes. For example, if there a four classes and a given array is in class 2, the dependent variable vector for this array would

Classification Analysis of DNA Microarrays, First Edition. Leif E. Peterson.

be $\mathbf{y}_i = (-1, +1, -1, -1)$. The regression equation for each array and each class is

$$y_{i\omega} = \beta_{0\omega} + \beta_{1\omega}x_{i1} + \beta_{2\omega}x_{i2} + \cdots + \beta_{p\omega}x_{ip} + \varepsilon_{i\omega}, \tag{14.1}$$

where $\beta_{0\omega}$ is the intercept, or *constant* term, $\beta_{j\omega}$ is the regression *coefficient* for gene x_j, and $\varepsilon_{i\omega}$ is the random error term or *residual*. Regression coefficients are used for predicting the y values for each array, using the relationship

$$\hat{y}_{i\omega} = \beta_{0\omega} + \sum_{j=1}^{p} \beta_{j\omega}x_{ij} \,. \tag{14.2}$$

Coefficients are determined directly from the matrix operation

$$\mathbf{Y} = \mathbf{XB} + \mathbf{E}. \tag{14.3}$$

The matrix form of the dependent variables is given as

$$\underset{n\times\Omega}{\mathbf{Y}} = \begin{bmatrix} y_{11} & y_{12} & \cdots & y_{1\Omega} \\ y_{21} & y_{22} & \cdots & y_{2\Omega} \\ \vdots & \vdots & \ddots & \vdots \\ y_{n1} & y_{n2} & \cdots & y_{n\Omega} \end{bmatrix}. \tag{14.4}$$

The typical x-value matrix is

$$\underset{n\times p+1}{\mathbf{X}} = \begin{bmatrix} 1 & x_{11} & x_{12} & \cdots & x_{1p} \\ 1 & x_{21} & x_{22} & \cdots & x_{2p} \\ 1 & \vdots & \vdots & \ddots & \vdots \\ 1 & x_{n1} & x_{n2} & \cdots & x_{np} \end{bmatrix}. \tag{14.5}$$

Regression coefficients are formed using the following matrix notation

$$\underset{p+1\times\Omega}{\mathbf{B}} = \begin{bmatrix} \beta_{01} & \beta_{02} & \cdots & \beta_{0\Omega} \\ \beta_{11} & \beta_{12} & \cdots & \beta_{1\Omega} \\ \vdots & \vdots & \ddots & \vdots \\ \beta_{p1} & \beta_{p2} & \cdots & \beta_{p\Omega} \end{bmatrix}, \tag{14.6}$$

and the matrix of random error terms is in the form

$$\underset{n\times\Omega}{\mathbf{E}} = \begin{bmatrix} \varepsilon_{11} & \varepsilon_{12} & \cdots & \varepsilon_{1\Omega} \\ \varepsilon_{21} & \varepsilon_{22} & \cdots & \varepsilon_{2\Omega} \\ \vdots & \vdots & \ddots & \vdots \\ \varepsilon_{n1} & \varepsilon_{n2} & \cdots & \varepsilon_{n\Omega} \end{bmatrix}. \tag{14.7}$$

Rearranging the matrices, we can solve for the regression coefficients using the following one-step operation:

$$\mathbf{B} = (\mathbf{X}^T\mathbf{X})^{-1}\mathbf{X}^T\mathbf{Y}. \quad (14.8)$$

The inverse of the dispersion matrix $\mathbf{X}^T\mathbf{X}$ can be obtained using the Jacobi method, singular value decomposition, or some other matrix inversion technique. After obtaining the coefficient matrix $\mathbf{B}$, we obtain the predicted y values by applying $\mathbf{B}$ to the data matrix $\mathbf{X}$ using the matrix operation $\hat{\mathbf{Y}} = \mathbf{XB}$. The decision rule for assigning each ith test microarray $\mathbf{x}_i$ to a class is

$$D(\mathbf{x}_i \rightsquigarrow \omega) = \underset{\omega \in \Omega}{\arg\max}\{\hat{y}_{i\omega}\}, \quad (14.9)$$

which follows the concept of assigning the test object to the class for which the predicted $\hat{y}_{i\omega}$ is the most positive.

14.2 ALGORITHM

Algorithm 14.1 lists the computational steps required for implementing LREG classification analysis.

14.3 IMPLEMENTATION

Linear regression is primarily a matrix algebra algorithm that simply carries out several matrix cross-multiplications and one matrix inversion. The $\mathbf{X}$ matrix has dimensions $n \times p$, so the transposed matrix $\mathbf{X}^T$ when multiplied by $\mathbf{X}$ yields the square symmetric matrix $\mathbf{X}^T\mathbf{X}$ whose dimensions are $p \times p$ or gene-by-gene. Between-gene correlation, or multicollinearity, can influence the singularity of the $\mathbf{X}^T\mathbf{X}$ covariance matrix, causing one or more of its eigenvalues to be near zero or zero. Recall that the covariance matrix must remain positive definite, such that its eigenvalues are all positive (nonzero). To prevent such pathology from degrading the solution, we recommend removing correlated genes to the extent possible. You should also mean-zero-standardize the gene expression values prior to matrix operations to minimize scale effects on the $\mathbf{X}^T\mathbf{X}$ matrix. After all, gene expression values of 4559.2, 234.8, and 17,103.1 for three subjects will be squared and summed in elements of $\mathbf{X}^T\mathbf{X}$. Performing eigenanalysis on $\mathbf{X}^T\mathbf{X}$ with such large matrix elements will severely degrade the ability to solve the symmetric eigenvalue problem on $\mathbf{X}^T\mathbf{X}$. Log-transforming and mean-zero standardizing gene expression values when appropriate will result in expression values that are standard normal distributed with mean zero and variance unity, essentially in the range from -3 to 3, causing little scale effect on eigenanalysis and inversion of $\mathbf{X}^T\mathbf{X}$.

Algorithm 14.1: LREG

Data: $n \times p$ data matrix **X** with n arrays and p genes, $n \times \Omega$ class membership output matrix **Y**
Result: $p + 1 \times \Omega$ matrix **B** of regression coefficients
$m \leftarrow 10$, $\text{Acc}_{\text{total}} = 0$
foreach Repartition $= 1 \leftarrow 10$ **do**
  Set $\mathbf{C} \leftarrow \mathbf{0}$, $n_{\text{test}} \leftarrow 0$
  Randomly partition input arrays into $\mathcal{D}_1, \mathcal{D}_2, \ldots, \mathcal{D}_m$ folds
  for *each cross-validation fold* $\mathcal{D}_m$ **do**
    Select all arrays in $\mathcal{D}_m$ for testing
    Use all remaining arrays not in $\mathcal{D}_m$ for training
    Determine $\mathbf{X}^T$ matrix using training arrays
    Cross-multiply $\mathbf{X}^T$ and $\mathbf{X}$ matrices to get $\mathbf{X}^T\mathbf{X}$
    Cross-multiply $\mathbf{X}^T$ and $\mathbf{Y}$ matrices to get $\mathbf{X}^T\mathbf{Y}$
    Invert $\mathbf{X}^T\mathbf{X}$ matrix to obtain $(\mathbf{X}^T\mathbf{X})^{-1}$ matrix
    Obtain coefficient matrix using $\mathbf{B} = (\mathbf{X}^T\mathbf{X})^{-1}\mathbf{X}^T\mathbf{Y}$
    for *each test array* $\mathbf{x}_i \in \mathcal{D}_m$ **do**
      $n_{\text{test}} += 1$
      for *each class* $\omega \in \Omega$ **do**
        Determine $\hat{y}_{i\omega} = \beta_{0\omega} + \sum_{j=1}^{p} \beta_{j\omega} x_{ij}$
      endfor
      Predict class $\hat{\omega} = \arg\max_{\omega \in \Omega}\{\hat{y}_{i\omega}\}$
      Increment confusion matrix **C** by 1 in element $c_{\omega,\hat{\omega}}$
    endfor
  endfor
  $\text{Acc}_{\text{total}} += (\sum_i c_{ii})/n_{\text{test}}$
endfch
$\text{Acc} = \text{Acc}_{\text{total}}/10$

14.4 CROSS-VALIDATION RESULTS

Both K-fold and bootstrap cross-validation (CV) were used to assess the bias and variance of LREG as a function of input feature transformations. Bootstrap accuracy (''CVB'') was based on average 0.632 bootstrap accuracy (13.4) for 10 bootstraps. Table 14.1 lists the average accuracy based on 10 repartitions of the input microarrays and subsequent cross-validation, with no input feature transformations applied. Table 14.2 lists the average accuracy based on 10 repartitions of the input microarrays and subsequent cross-validation, using mean-zero standardized input feature values. Table 14.3 lists the average accuracy based on 10 repartitions of the input microarrays and subsequent cross-validation, using fuzzified input feature values. Surprisingly, the performance measure by classification accuracy of LREG for the nine datasets is quite high without input feature transformation. Mean-zero standardization did not increase average performance, and feature fuzzification

TABLE 14.1 Performance without Feature Transformation

Dataset	Cross-validation method CV2	CV5	CV10	CV−1	CVB
AMLALL2	98.68	100	100	100	100
Brain2	91.17	94.67	95.83	96.67	96.73
Breast2A	100	100	100	100	100
Breast2B	92.82	95.38	94.62	94.87	97.51
Colon2	87.58	88.55	88.55	90.32	89.69
Lung2	100	100	100	100	100
MLL_Leukemia3	97.89	100	100	100	99.93
Prostate2	97.25	97.84	98.04	98.04	98.53
SRBCT4	97.94	100	100	100	99.82
Average	0.96	0.97	0.97	0.98	0.98

TABLE 14.2 Performance with Feature Standardization

Dataset	Cross-validation method CV2	CV5	CV10	CV−1	CVB
AMLALL2	98.16	100	100	100	99.98
Brain2	92.17	95.50	96.50	96.67	96.85
Breast2A	100	100	100	100	100
Breast2B	91.92	93.72	95.13	94.87	97.63
Colon2	87.74	88.06	88.87	90.32	89.40
Lung2	100	100	100	100	100
MLL_Leukemia3	97.89	100	100	100	99.93
Prostate2	97.55	98.14	98.04	98.04	98.52
SRBCT4	97.78	100	100	100	99.84
Average	0.96	0.97	0.98	0.98	0.98

reduced performance levels. With mean performance values above 95%, there is significant merit for using LREG as a robust classifier for future datasets.

Figure 14.1 illustrates accuracy as a function of cross-validation method for each dataset without the use of feature transformations. Performance is based on CV accuracy for 10 repartitions of microarrays. Figure 14.2 shows the standard deviation of accuracy as a function of the CV method for each dataset (without feature transformations). Standard deviation is based on CV accuracy for 10 repartitions of the training arrays. Figure 14.3 reveals average accuracy for all datasets as a function of feature transformation and CV method. The bias of LREG shown in Figure 14.1 suggests that LREG performance did not change considerably as more arrays were used for training. As the number of folds increase during cross-validation, there is an increase in the fraction of objects used for training. For example, leave-one-out cross-validation (CV − 1) trains with $n-1$ objects and predicts and

TABLE 14.3 Performance with Feature Fuzzification

Dataset	Cross-validation method				
	CV2	CV5	CV10	CV−1	CVB
AMLALL2	99.74	96.84	89.74	81.58	99.65
Brain2	65.50	74.83	80.67	75.00	87.02
Breast2A	100	100	100	100	100
Breast2B	58.72	80.38	81.79	83.33	89.35
Colon2	80.81	82.42	81.94	80.65	87.00
Lung2	100	100	100	100	100
MLL_Leukemia3	96.84	97.54	98.25	100	98.93
Prostate2	87.94	93.43	94.22	94.12	96.30
SRBCT4	99.37	99.21	98.57	95.24	99.70
Average	0.88	0.92	0.92	0.90	0.95

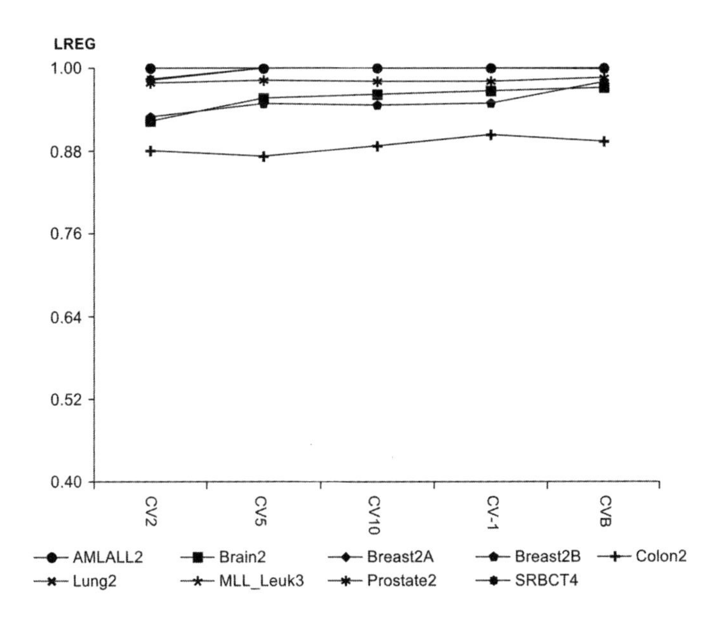

FIGURE 14.1 LREG accuracy as a function of cross-validation (CV) method for each dataset. Average is based on CV accuracy for 10 repartitions of arrays. No feature transformations applied.

tests accuracy of the class membership of the single objects when they are left out of training, whereas CV2 splits the dataset in half, and uses only 50% of the objects for training and 50% for testing. The decrease in the standard deviation of accuracy with increasing number of folds for cross-validation

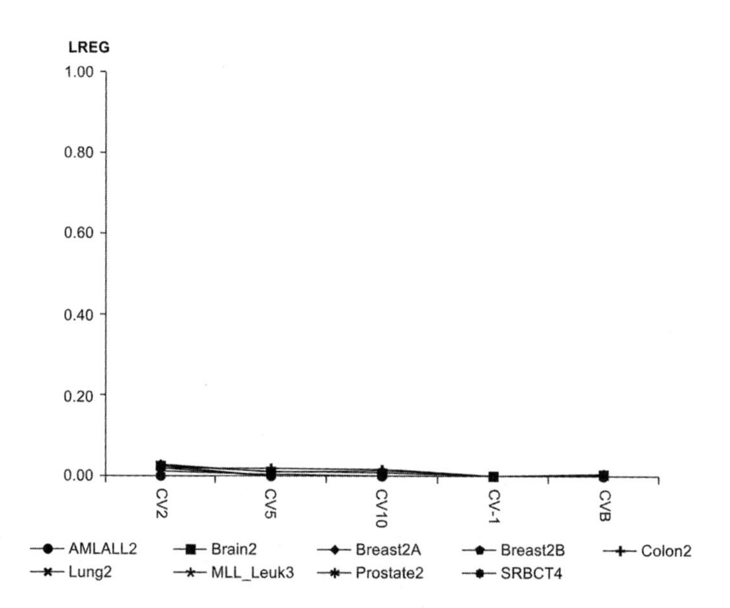

FIGURE 14.2 LREG standard deviation of accuracy as a function of cross-validation (CV) method for each dataset. Standard deviation based on CV accuracy for 10 repartitions of arrays. No feature transformations applied.

is expected, since the class predictions become more accurate, resulting in fewer deviations.

14.5 BOOTSTRAP BIAS

Bootstrap bias for LREG was also determined as a function of bootstrapped sample size, where $B = 40$ iterations were used with fractions of $f = 0.1$, 0.2, 0.3, 0.4, 0.5, or 0.6 of the available arrays. Bootstrap bias runs were based on determining the 0.632 bootstrap accuracy (see Eq. 13.4) for each sampling fraction. Acc_0 in 13.4 was based on test accuracy of unsampled arrays, while Acc_b in 13.4 was based on test accuracy for sampled arrays. Figure 14.4 shows LREG classification accuracy without transformations on features as a function of bootstrap sample size used during training, based on 10 resamplings from the input arrays. Figure 14.5 illustrates LREG accuracy with feature standardization as a function of bootstrap sample size based on 10 resamplings from the input arrays. Figure 14.6 shows LREG accuracy with feature fuzzification as a function of bootstrap sample size based on 10 resamplings from the input arrays. Figures 14.4, 14.5, and 14.6 show that LREG had little bias as there was a small change in performance with increasing sample size during bootstrapping for the nine microarray datasets. Datasets that are difficult for accurate class prediction remained difficult for the various feature transformations applied.

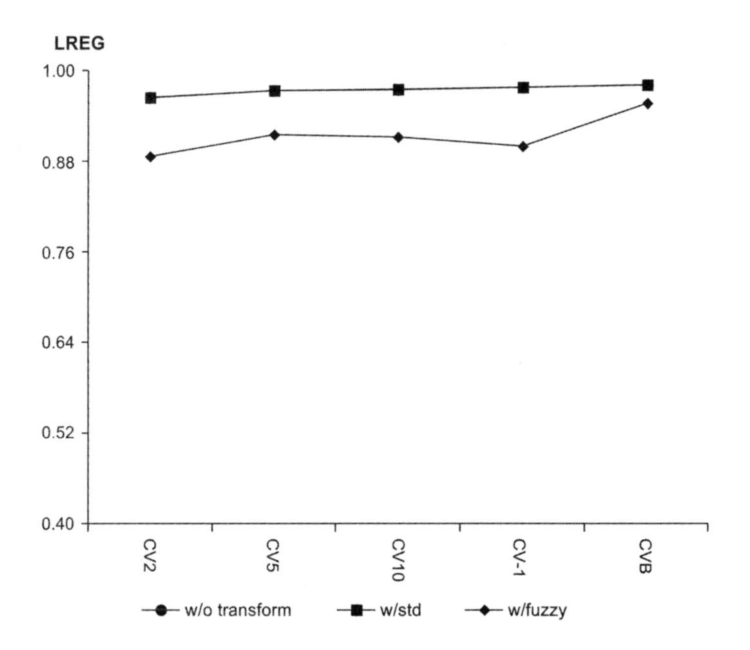

FIGURE 14.3 LREG average accuracy for all datasets as a function of feature transformation and cross-validation (CV) method.

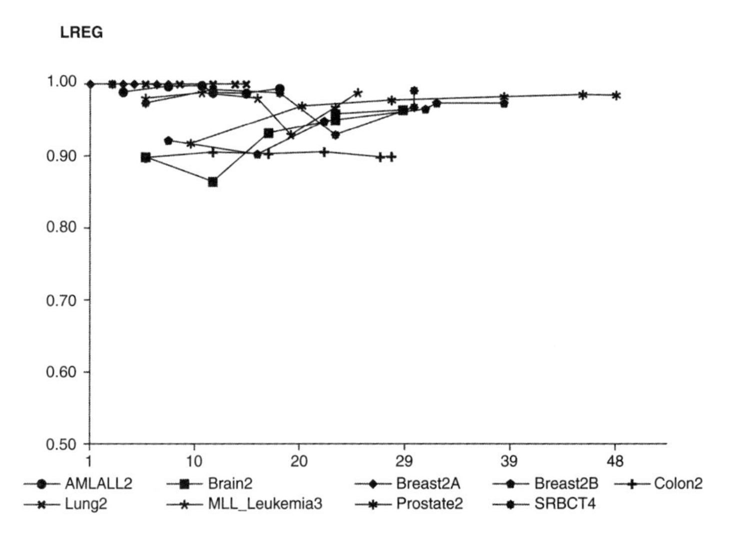

FIGURE 14.4 LREG accuracy without transformations on features as a function of bootstrap sample size used during training. The x-axis values represent the number of arrays randomly sampled $B = 40$ times at fractions $f = 0.1$, 0.2, 0.3, 0.4, 0.5, or 0.6 of the available arrays.

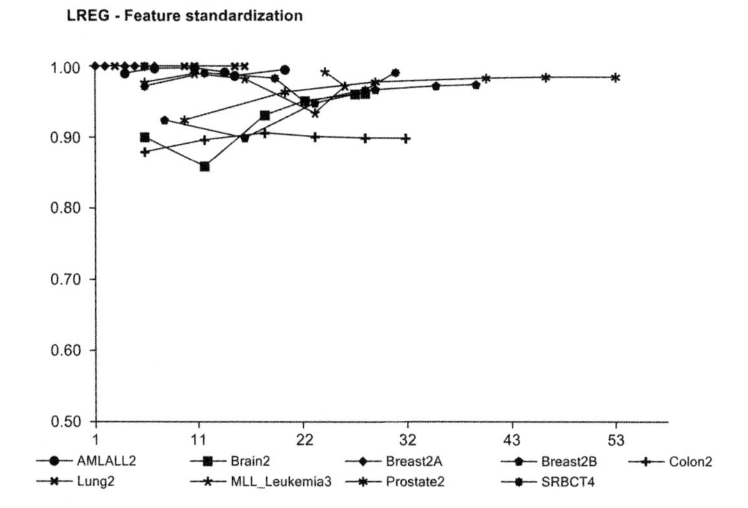

FIGURE 14.5 LREG accuracy with feature standardization as a function of bootstrap sample size used during training. The x-axis values represent the number of arrays randomly sampled $B = 40$ times at fractions $f = 0.1$, 0.2, 0.3, 0.4, 0.5, or 0.6 of the available arrays.

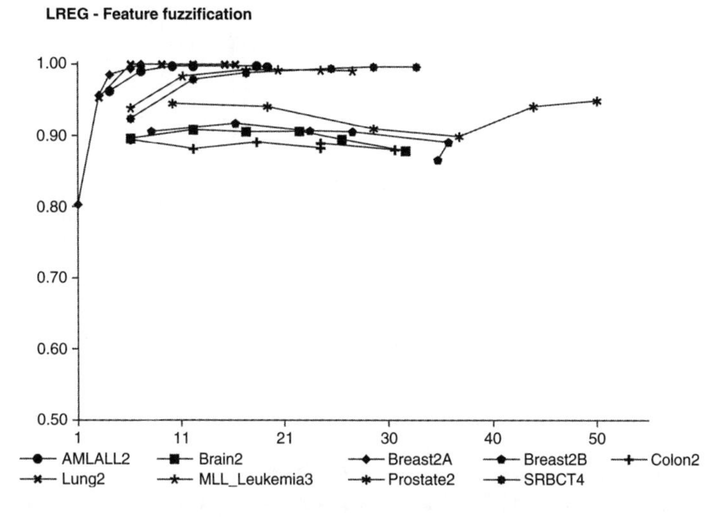

FIGURE 14.6 LREG accuracy with feature fuzzification as a function of bootstrap sample size used during training. The x-axis values represent the number of arrays randomly sampled $B = 40$ times at fractions $f = 0.1$, 0.2, 0.3, 0.4, 0.5, or 0.6 of the available arrays.

14.6 MULTICLASS ROC CURVES

Linear regression was also used to evaluate multiclass ROC curves with bootstrapping arrays $B = 40$ times at fractions $f = 0.1$, 0.2, 0.3, 0.4, 0.5, or 0.6 of the available arrays. Figure 14.7 illustrates ROC curves for LREG applied to the AMLALL2, Brain2, Breast2A, Breast2B, Colon2, Lung2,

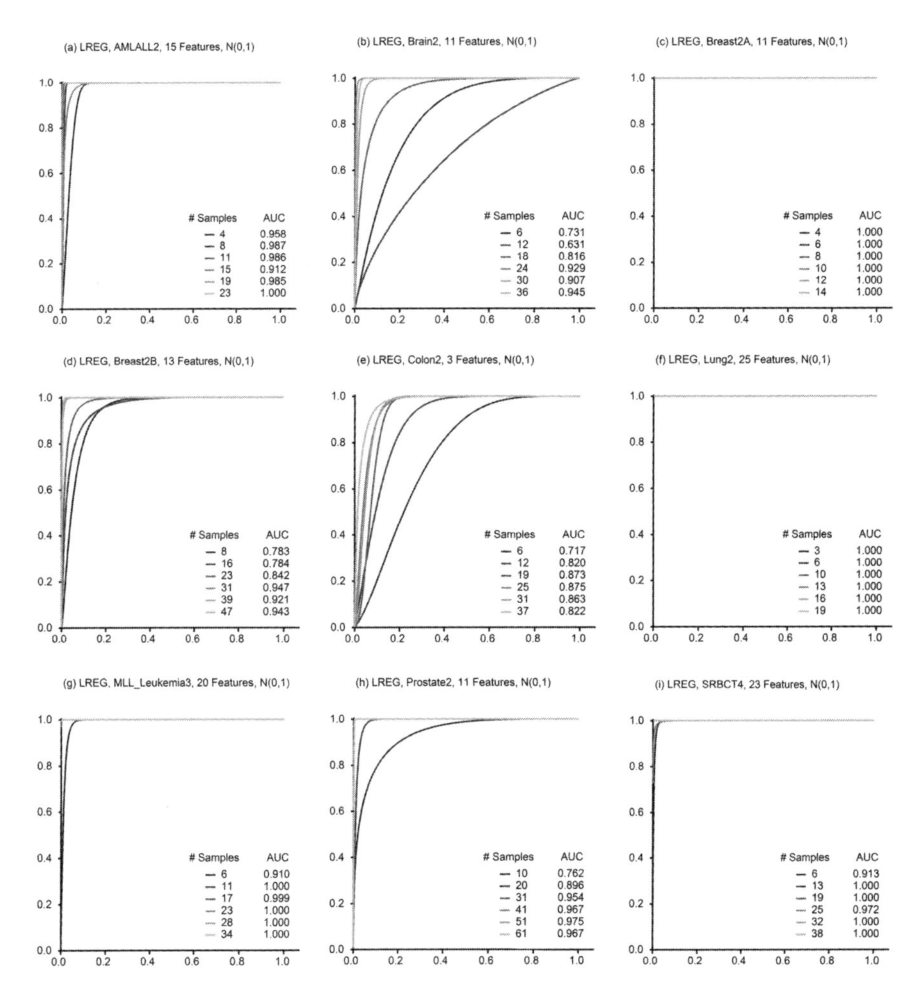

FIGURE 14.7 ROC curves for the LREG classifier applied to the AMLALL2, Brain2, Breast2A, Breast2B, Colon2, Lung2, MLL_Leukemia3, Prostate2, and SRBCT4 datasets based on mean-zero standardized input features: (a) AMLALL2; (b) Brain2; (c) Breast2A; (d) Breast2B; (e) Colon2; (f) Lung2; (g) MLL_Leukemia3; (h) Prostate2; (i) SRBCT4. The numbers of genes used for each dataset are listed in Table 12.10. Legend values represent the number of arrays randomly sampled $B = 40$ times at fractions $f = 0.1$, 0.2, 0.3, 0.4, 0.5, or 0.6 of the available arrays.

MLL_Leukemia3, Prostate2, and SRBCT4 datasets using mean-zero standardized input features. Figure 14.8 illustrates ROC curves for LREG based on the same datasets but with fuzzified input features. The Brain2 and Colon2 datasets had rather poor AUC values with increasing bootstrapped sample size, while the remaining datasets showed much higher AUC values for smaller sample sizes. Fuzzification of input feature values (Figure 14.8) seemed to reduce the AUCs for most of the datasets, especially for the smallest sampling fractions during bootstrapping.

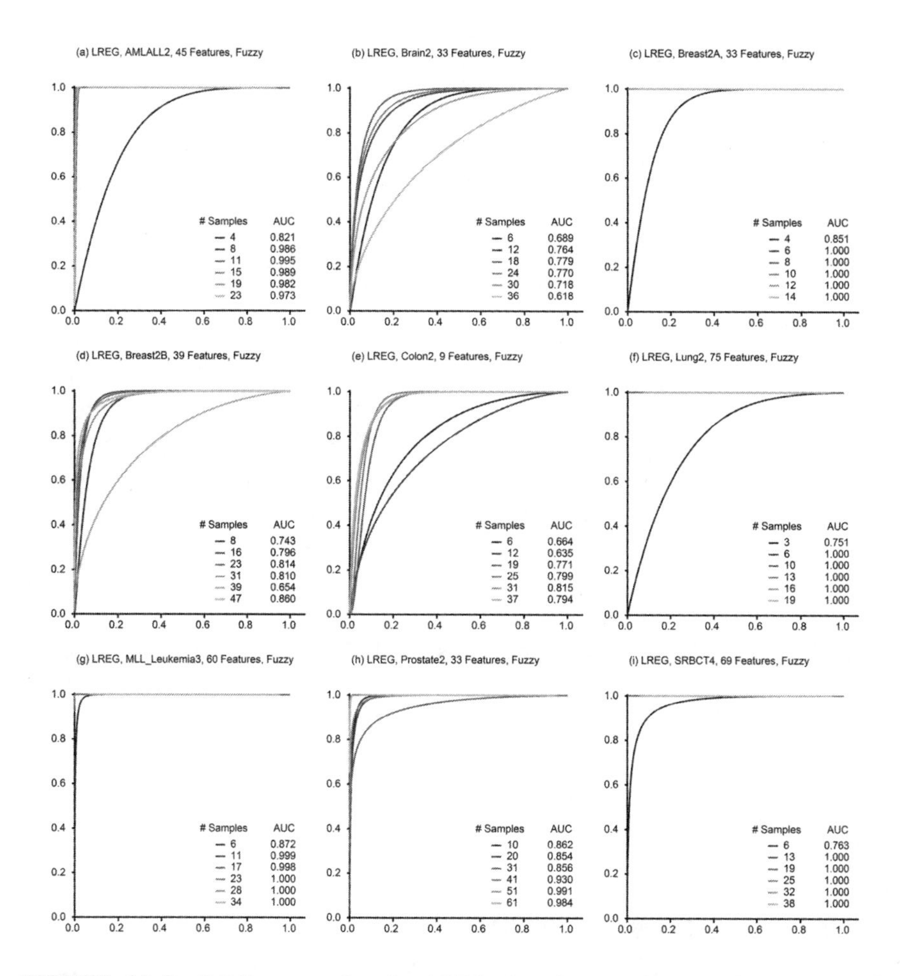

FIGURE 14.8 ROC curves for the LREG classifier applied to the AMLALL2, Brain2, Breast2A, Breast2B, Colon2, Lung2, MLL_Leukemia3, Prostate2, and SRBCT4 datasets based on fuzzified input features: (a) AMLALL2; (b) Brain2; (c) Breast2A; (d) Breast2B; (e) Colon2; (f) Lung2; (g) MLL_Leukemia3; (h) Prostate2; (i) SRBCT4. The numbers of genes used for each dataset are listed in Table 12.10.

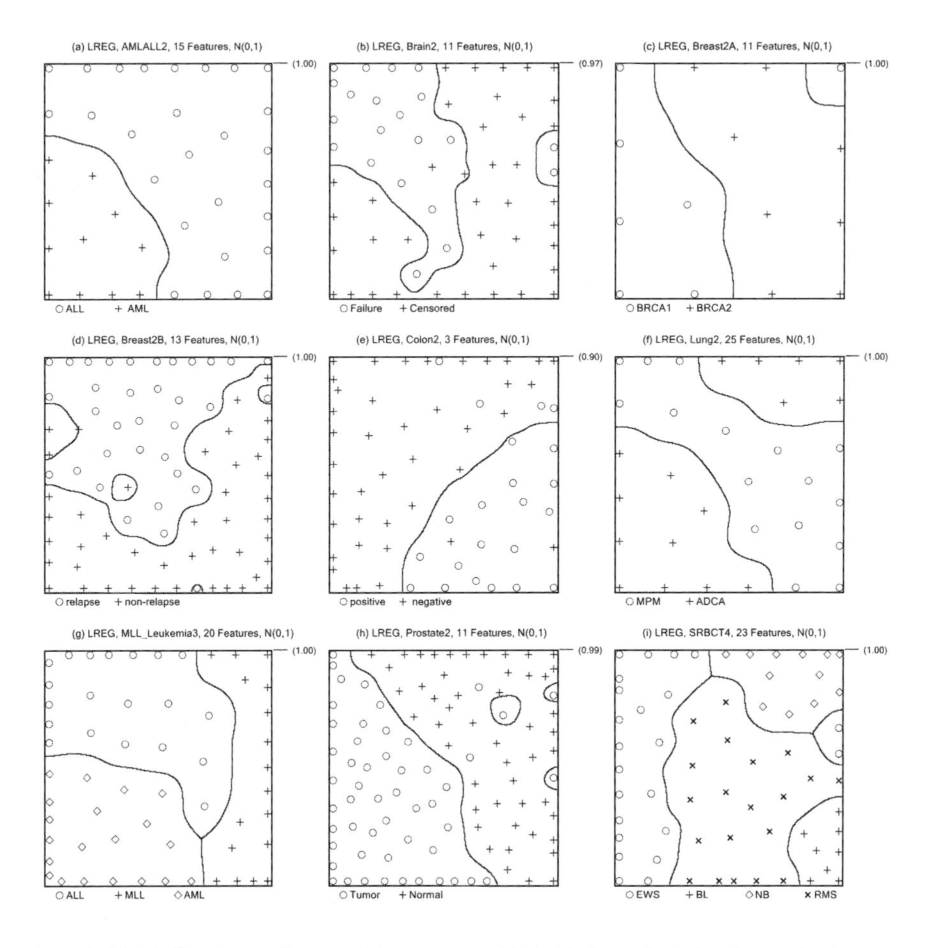

FIGURE 14.9 2D self-organizing maps of LREG class decision boundaries and classification accuracy (upper right of each panel) for the AMLALL2, Brain2, Breast2A, Breast2B, Colon2, Lung2, MLL_Leukemia3, Prostate2, and SRBCT4 datasets based on mean-zero standardized input features. Panel (a) AMLALL2, (b) Brain2, (c) Breast2A, (d) Breast2B, (e) Colon2, (f) Lung2, (g) MLL_Leukemia3, (h) Prostate2, and (i) SRBCT4. The number of genes used for each dataset are listed in Table 12.10.

14.7 DECISION BOUNDARIES

Linear regression was also used for classification runs of all available arrays using the reduced gene sets enumerated in Table 12.10. Class decision boundaries were identified by first running LREG on the arrays to extract the decision rules, which were applied to each array as well as each reference vector of a 300 × 300 2D SOM in order to predict class membership. The

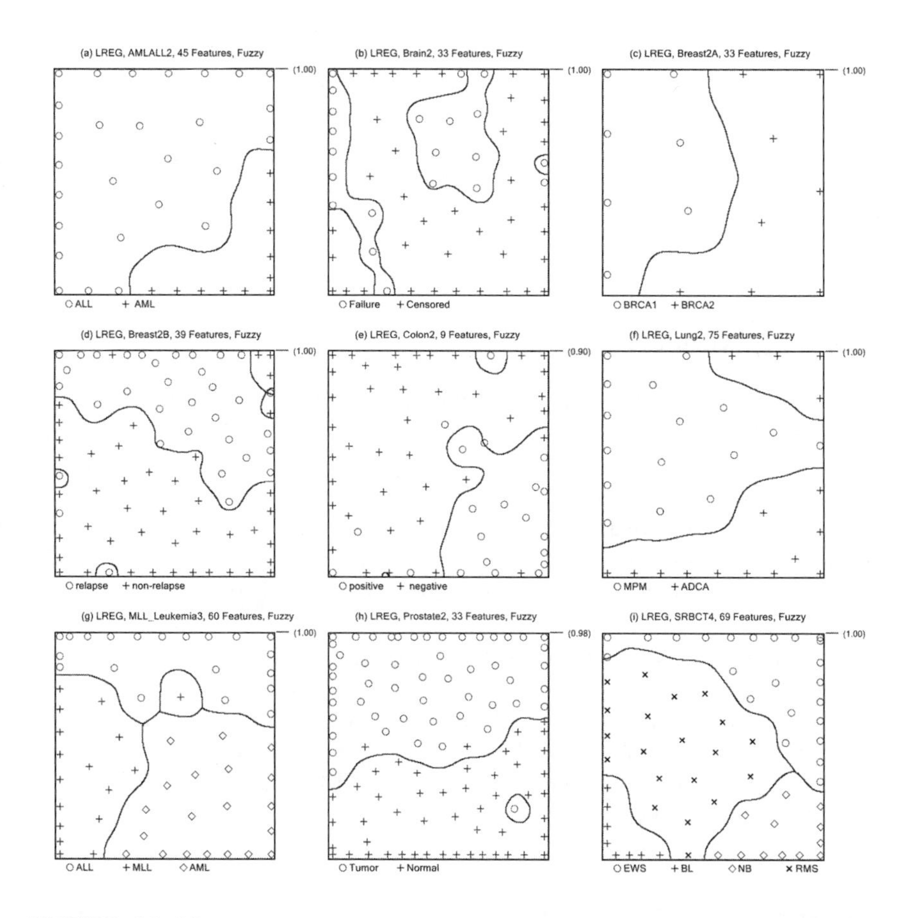

FIGURE 14.10 2D self-organizing maps of LREG class decision boundaries and classification accuracy (upper right of each panel) for the AMLALL2, Brain2, Breast2A, Breast2B, Colon2, Lung2, MLL_Leukemia3, Prostate2, and SRBCT4 datasets based on fuzzified input features: (a) AMLALL2; (b) Brain2; (c) Breast2A; (d) Breast2B; (e) Colon2; (f) Lung2; (g) MLL_Leukemia3; (h) Prostate2; (i) SRBCT4. The number of genes used for each dataset are listed in Table 12.10.

class decision boundaries based on class predictions of original reference vector values were then simultaneously displayed on the SOM along with the arrays using class-specific symbols for their true class labels. Figure 14.9 illustrates 2D self-organizing maps (SOMs) of LREG class decision boundaries based on the AMLALL2, Brain2, Breast2A, Breast2B, Colon2, Lung2, MLL_Leukemia3, Prostate2, and SRBCT4 datasets using mean-zero standardized input features. Figure 14.10 illustrates 2D self-organizing maps of LREG class decision boundaries for the same datasets but with fuzzified input features. The decision boundary results indicate that the Colon2 dataset in panel (e) of Figures 14.9 and 14.10 was one of the more challenging datasets to analyze, based on the lower performance results.

Linear regression produced relatively stable decision boundaries for the majority of datasets analyzed, which is a remarkable finding in terms of linear separability. The greedy PTA gene filtering used to select the genes in the datasets analyzed (see Table 12.10) improved class separability, which is, in fact, entirely linear.

14.8 SUMMARY

Determination of linear separability of objects (arrays) in a dataset should be the initial goal of any classification analysis. Certainly, the feature set should be reduced to the minimal set of genes, which best separates the arrays into their particular class membership. If use of the optimal feature set in LREG classification analysis can separate arrays into their respective classes with reasonably high accuracy, then nonlinear methods may not be needed. LREG also revealed high levels of accuracy for most of the CV approaches with varying fold size, and low values of standard deviation, which decreased when the number of CV folds increased. This was expected because as the number of folds increases, more data are used for training. Another major advantage of LREG as a classifier is that the decision boundaries are smoothly varying; therefore, LREG would be less likely to overfit the data as a result of learning the data too closely. With a low level of overfitting, the generalizability of LREG results for future objects would therefore be quite reliable.

REFERENCES

[1] F. Galton. Regression towards mediocrity in hereditary stature. *J. Anthropol. Inst.* **15**:246–263, 1886.

[2] M.S. Bartlett. Further aspects on the theory of multiple regression. *Proc. Cambridge Phil. Soc.* **34**:33–40, 1938.

[3] R.C. Holte. Very simple classification rules perform well on most commonly used datasets. *Machine Learn.* **11**:63–91, 1993.

[4] T.-S. Lim, W.-Y. Loh, Y.-S. Shih. A comparison of prediction accuracy, complexity, and training time of thirty-three old and new classification algorithms. *Machine Learn.* **40**:203–229, 2000.

[5] R. Tibshirani. Regression shrinkage and selection via the Lasso. *J. Roy. Stat. Soc. B* **58**:267–288, 1994.

[6] M.S. Bartlett. Multivariate analysis. *J. Royal Stat. Soc.* **9**:176–197, 1947.

CHAPTER 15

DECISION TREE CLASSIFICATION

15.1 INTRODUCTION

The historical reflection of decision tree induction begins with the automatic interaction detection (AID) algorithm, developed in 1963 by Morgan and Sonquist [1]. AID was a regression tree method in which objects of a parent node were split using the feature and cutpoint that yielded the least within-node sum of squares about the mean of the dependent variable. Soon after the introduction of AID for continuous variables, Hartigan [2] and Kass [3] introduced the CHAID algorithm, which was developed for categorical dependent variables and used within-node χ^2 values for determining optimal splits.

One pitfall of using statistical tests to determine tree nodes is that the trees became unnecessarily large, to the extent that some of the smaller nodes did not contribute to the overall fit of the model, suggesting increased risk for *overfitting*. Breiman et al. [4] reported that such nodes could undergo cost-complexity *pruning* via their classification and regression trees (CART) algorithm. More recent evaluations of pruning in decision tree analysis suggest that the main advantage is a factor of 10 reduction of tree size [5]. Overall, the effects of pruning depend on the method considered, the data domain, and the cost or loss functions and matrices used.

At about the same time that CART was introduced, Quinlan independently introduced the ID3 algorithm [6], which was based on an information-theoretic approach incorporating information entropy. Its successor algorithm C4.5 [7] and the commercial successor called C5.0 [8] are widely used in industry and are marketed by Rulequest Research [8], whereas CART is marketed by Salford Systems [9]. There are many other commercial

Classification Analysis of DNA Microarrays, First Edition. Leif E. Peterson.

implementations of decision tree algorithms—the ones listed above are used mainly for information purposes. An open-source version of C4.5 issued under a GNU general public license is available in the form of the J48 algorithm in the WEKA package [10].

Decision tree classification (DTC) is a greedy rule induction method that recursively splits the full training set of microarrays into smaller and smaller tuples (sets of microarrays) with the ultimate goal of creating tuples with class purity. DTC has proved to be one the better methods for identifying the most important features for a classification or regression problem. Applications with DNA microarrrays include pattern matching algorithms [11], toxicogenomics, [12], and probe quality evaluation [13]. Like most supervised classifiers, DTC can be applied as either a wrapper to select features or as a classifier after feature filtering. Recall that in wrapping, feature selection is performed by using the classifier to identify which features result in the best class prediction, whereas filtering is based on a separate technique applied prior to class prediction and is wholly independent of the classifier.

First, we define a *node* as a collection of microarrays. The DTC method builds a tree consisting of nodes, each of which identifies the best features and cutpoints to recursively split arrays into daughter nodes whose array class membership is preferably in a single class. The degree of node-specific array membership in a single class is called *node purity*. The goal of DTC is to generate a trained tree onto which test arrays are "dropped down" for class prediction resulting in minimal error.

Decision tree classification begins by assigning all arrays to a *parent* node. All possible pairs of classes are used for identifying the gene providing the best split. When node splitting is based on information gain, global information gain based on class priors is first calculated using the relationship

$$I(G)_{\text{entropy}} = -\frac{n_l}{n}\log_2\left(\frac{n_l}{n}\right) - \frac{n_m}{n}\log_2\left(\frac{n_m}{n}\right), \tag{15.1}$$

where n_l and n_m are the number of arrays in the current node from classes l and m, respectively, and n is the total number of arrays in the current node. If Gini index is used, the global Gini is calculated using the form

$$I(G)_{\text{Gini}} = 1 - \left[\left(\frac{n_l}{n}\right)^2 + \left(\frac{n_m}{n}\right)^2\right]. \tag{15.2}$$

Now let x_{ij} represent the expression value of gene j for the ith array and c_{jk} the kth cutpoint ($k = 1, 2, \ldots, 10$) of gene j. Arrays whose gene expression values $x_{ij} \leq c_{jk}$ are assigned a quantile value $q = 1$, whereas arrays whose $x_{ij} > c_{jk}$

are assigned quantile value $q = 2$. For each cutpoint value, information gain for each jth gene and quantile value $q = 1, 2$ is determined as

$$I(j,q) = -\frac{n(j,q_j,l)}{n(q)} \log_2 \left(\frac{n(j,q_j,l)}{n(q)} \right) - \frac{n(j,q_j,m)}{n(q)} \log_2 \left(\frac{n(j,q_j,m)}{n(q)} \right), \tag{15.3}$$

where q is the quantile, $n(j, q_j, l)$ is the number of arrays in class l with a quantile value of q_j and $n(q)$ is the number of arrays with quantile values of q over all arrays. If the Gini index is used as the splitting criterion, it is determined with the equation

$$I(j,q) = 1 - \left[\left(\frac{n(j,q_j,l)}{n(q)} \right)^2 + \left(\frac{n(j,q_j,m)}{n(q)} \right)^2 \right]. \tag{15.4}$$

Total entropy and Gini index are finally determined with the following forms:

$$I(j)_{\text{entropy}} = I(G)_{\text{entropy}} - \sum_{q=1}^{2} \frac{n(q)}{n} I(j,q), \tag{15.5}$$

$$I(j)_{\text{Gini}} = I(G)_{\text{Gini}} - \sum_{q=1}^{2} \frac{n(q)}{n} I(j,q). \tag{15.6}$$

These calculations are determined for each cutpoint over the range of expression for all genes considered, and the gene and its cutpoint value resulting in the greatest gain or Gini index for the pair of classes l and m are used to split arrays in the current parent node into two *child* nodes. This split will maximize the purity of class membership in the two child nodes. For a continuous feature, it is customary to send parent node microarrays to child node 2 if $x_{ij} \leq c_{jk}$ and the remaining microarrays for which $x_{ij} > c_{jk}$ to child node 3, as shown in Figure 15.1. Analogously, if a categorical feature results in the greatest information gain of Gini index, then the parent node microarrays are assigned to the K daughter nodes on the basis of their categorical feature values (Figure 15.2). After parent node microarrays are assigned to child nodes, the class purity in each child node is evaluated. A child node with one or more microarrays in the same class is considered pure, and is termed a *terminal* node. Each child node with class impurity becomes a parent node and is split in the same way using the cutpoints or categories of new features or the same feature already used. As we will see in the remaining sections, there are many variations of this theme for constructing a classification tree.

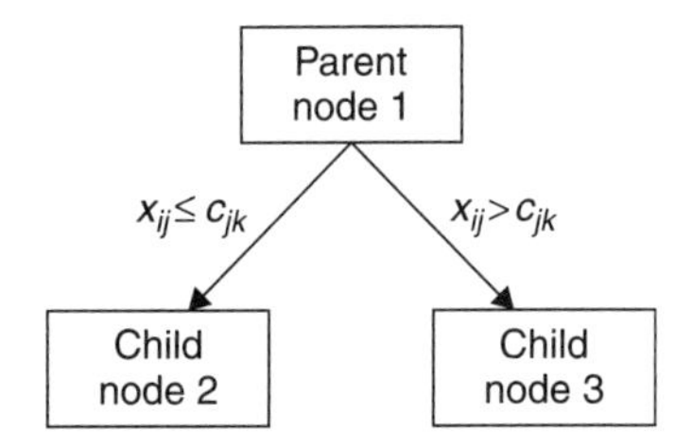

FIGURE 15.1 Splitting of parent node microarrays for a continuously scaled feature. Splitting at feature j cutpoint value c_{jk} results in greatest information gain or Gini index. Microarrays in parent node 1 are assigned to child node 2 if feature values are less than the selected cutpoint ($x_{ij} \leq c_{jk}$) and child node 3 otherwise.

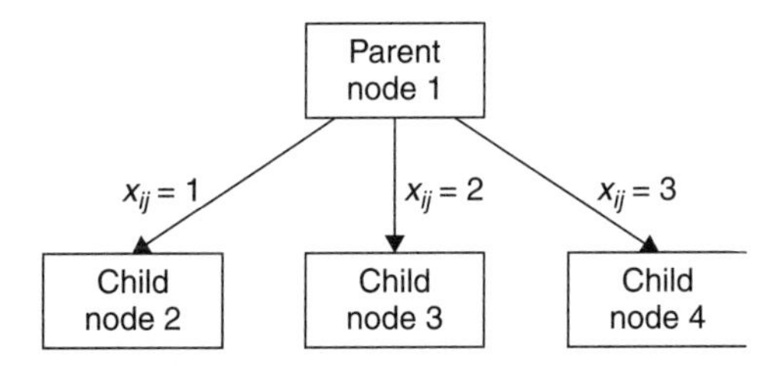

FIGURE 15.2 Splitting of parent node microarrays for a categorical feature with $K = 3$ values. Splitting with categorical feature j results in greatest information gain or Gini index. Microarrays are assigned to child node 2 if feature value $x_{ij} = 1$, child node 3 if $x_{ij} = 2$, and child node 4 if $x_{ij} = 3$.

15.2 FEATURES USED

Features used in DTC analysis can be either preselected or filtered using a heuristic or statistical test, or employed in the context of wrapping to identify informative features from a larger high-dimensional collection of features. Most of the time, DTC feature selection is employed for wrapping in order to identify features from a larger set. Features identified through filtering via an external technique prior to DTC are usually employed to reduce selection bias or for comparing DTC performance with other classifiers based on the same set of pre-selected features. Selection bias is an important issue with wrapping, since the classifier is used for identifying features that maximize its own performance. Filtering features prior to DTC analysis avoids the selection bias problem because feature selection is independent of classifier performance.

15.3 TERMINAL NODES AND STOPPING CRITERIA

Several stopping criteria can be imposed during training. Stopping in DTC is based on terminal node criteria. The most common criterion for a terminal node is that it has no class impurity and has one or more microarrays assigned to it. This criteria tends to work well when the features are very discriminating for class. Another criterion used is a minimum node size of 5 when there is class impurity. Tree growth can also be terminated when the node count reaches 100 or more or the level of child nodes in the tree exceeds 10. Depending on the data used, DTC does not always guarantee node purity in all terminal nodes, and therefore the tree size or child-level criteria can be used.

15.4 ALGORITHM

The algorithm for DTC is listed in AL 15.1, where arrays in each parent node $\mathcal{N}_p$ are split into child nodes $\mathcal{N}_c$, which can become either new parent nodes for additional splitting or terminal nodes $\mathcal{N}_t$.

15.5 IMPLEMENTATION

Implementation of DTC is straightforward, and the results of two example DTC runs are provided below. We did not use leave-one-out cross-validation or K-fold cross-validation, but rather used all microarrays from which optimal features and cutpoint values were extracted. Two example runs for DTC are reviewed below.

Example 15.1 A DTC run was performed using the $\log_e$-transformed two-class prostate dataset listed in Table 12.10. This dataset consists of 102 arrays with expression values for 11 genes selected using greedy PTA from the original 12,600. The Gini index was used for splits, and each of the continuously scaled features was evaluated at 100 nonoverlapping equally spaced cutpoints. Figure 15.3 shows the DTC results. There are five levels and eight terminal nodes with pure class membership, that is, final terminating nodes containing arrays from only one class. The root node at the top of the decision tree lists 52 tumor and 50 normal arrays, resulting in class priors of 51% and 49%, respectively. The first node split was made using an expression value cutpoint of −0.449 for gene 37639_at. In the left child node, there were 43 arrays, of which 41 (95%) had normal class membership. In the right child node, there were 59 arrays; 50 (85%) of these were in the tumor class. This is a

Algorithm 15.1: DTC

Data: p-dimensional input microarrays $\mathbf{x}_i$ $(i = 1, 2, \ldots, n)$
Result: Class discriminating features and cutpoints

```
foreach Parent node N_p do
    foreach Feature, j do
        if continuous then
            Determine K = 10 cutpoints c_jk
            for k = 1 to 10 do
                foreach Microarray x_i ∈ N_p do
                    if x_ij ≤ c_jk then
                        assign microarray to child node 1
                    endif
                    if x_ij > c_jk then
                        assign microarray to child node 2
                    endif
                endfch
                Determine Gini_k for split into 2 child nodes
            endfor
            c_jk = max{Gini_k}
        endif
        if categorical then
            for k = 1 to K do
                foreach Microarray x_i ∈ N_p do
                    if x_ij = k then
                        assign microarray to child node k
                    endif
                endfch
            endfor
            Determine Gini for split into K child nodes
        endif
    endfch
    Identify feature split with greatest Gini, j_best
    Split parent node into child nodes
    foreach child node, N_c do
        if N_c class impurity = true then
            child node becomes a terminal node, N_c → N_t
        endif
        if N_c class impurity = false then
            child node becomes new parent node, N_c → N_{p+1}
        endif
    endfch
    Add to bookkeeping the best split values j_best, c_jk for N_p
endfch
```

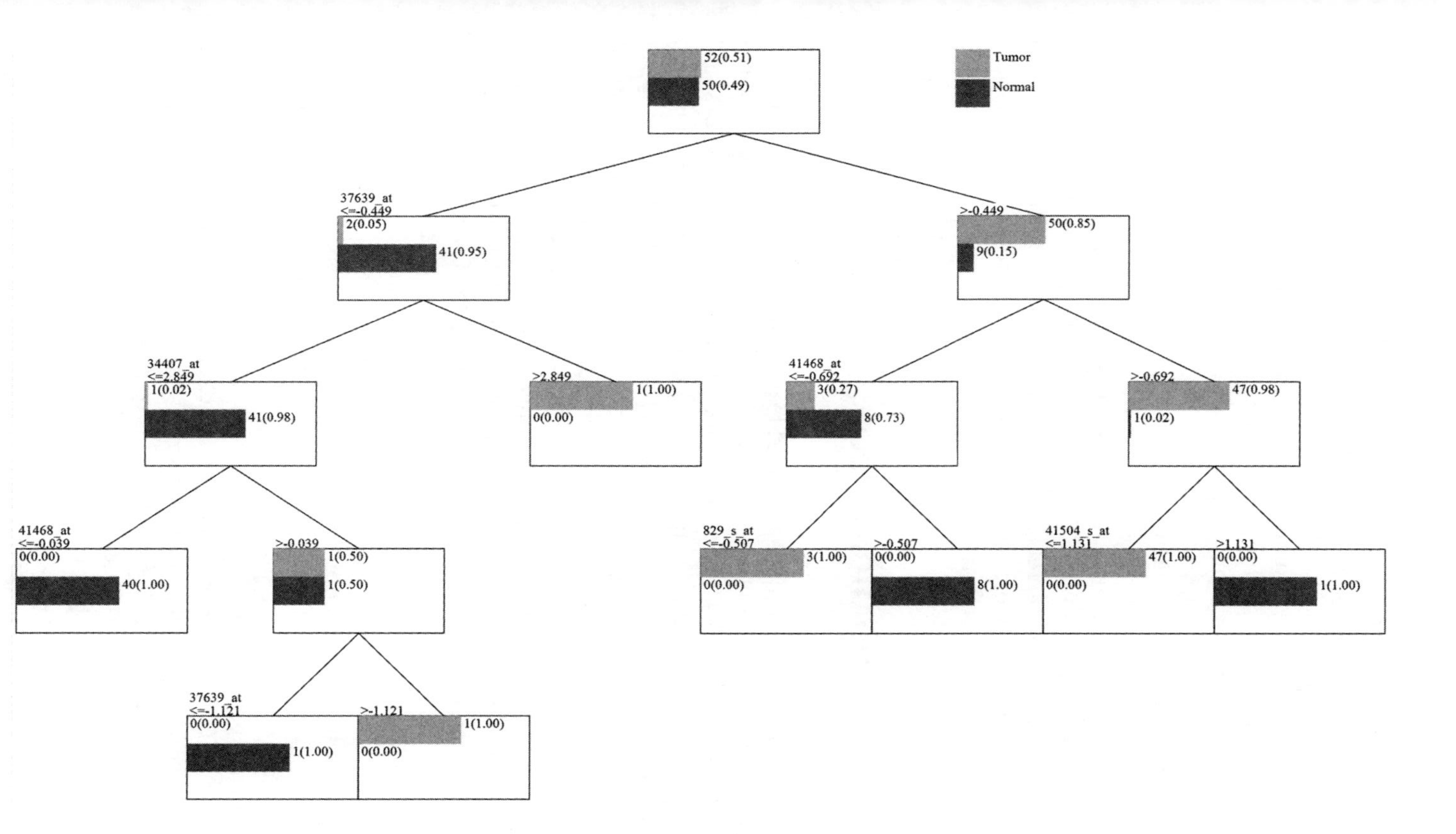

FIGURE 15.3 Decision tree classification (DCT) analysis of two-class prostate cancer dataset with 102 total arrays, partitioned into 52 (51%) tumor arrays and 50 (49%) normal arrays. Results reveal a five-level tree having eight terminal nodes with 100% class purity. (*See insert for color representation of the figure.*)

remarkably good split based on the basis of a highly informative gene, since the child nodes are populated with arrays predominantly from a single class. The remaining node splits occurred with relative ease and high quality due to genes whose optimal cutpoints resulted in high class purity in succeeding child nodes. The lack of such genes in a dataset will often force numerous less informative splits, resulting in potentially large trees.

Example 15.2 In this example, we ran DTC using the two-class breast cancer dataset with 78 arrays, with 34 relapse (44%) and 44 (56%) nonrelapse arrays. Figure 15.4 illustrates that the first cut with gene NM_013360 made with cutpoint 0.199 was able to partition 27 (87%) of the nonrelapse arrays into the right child node, and 30 relapse (64%) arrays into the left child node. All subsequent node splits were able to quickly partition the arrays into pure nodes with array membership in a single class.

15.6 CROSS-VALIDATION RESULTS

Both K-fold and bootstrap cross-validation (CV) were used to assess the bias and variance of DTC as a function of input feature transformations. Bootstrap accuracy ("CVB") was based on average 0.632 bootstrap accuracy (13.4) for 10 bootstraps. Table 15.1 lists the average accuracy of the input arrays and subsequent cross-validation, with no input feature transformations applied. Table 15.2 lists the average accuracy using mean-zero standardized input feature values. Table 15.3 lists the average accuracy based on fuzzified input feature values. It can be noticed that feature fuzzification slightly improved the average classification accuracy for all of the datasets used.

Figure 15.5 illustrates accuracy as a function of the CV method for each dataset when no feature transformations were applied. Figure 15.6 shows the standard deviation of accuracy as a function of the CV method for each dataset when no feature transformations were applied. In comparison to the LREG classifier, DTC resulted in unfavorably lower values of accuracy and quite high values of standard deviation. Figure 15.7 reveals average accuracy for all datasets as a function of feature transformation and CV method. Figure 15.8 reveals the DTC accuracy without transformations on expression values as a function of the number of equally spaced cutpoints within the range of each gene's expression value over all the microarrays. Figure 15.9 shows the DTC accuracy with feature standardization as a function of the number of cutpoints. Figure 15.10 illustrates the DTC accuracy with feature fuzzification as a function of the number of cutpoints. Figure 15.11 shows average accuracy for all datasets as a function of the number of cutpoints. It is apparent that feature (gene) transformations did not alter accuracy values for the majority of datasets. Also, the number of cutpoints used did not strongly vary the classification accuracy.

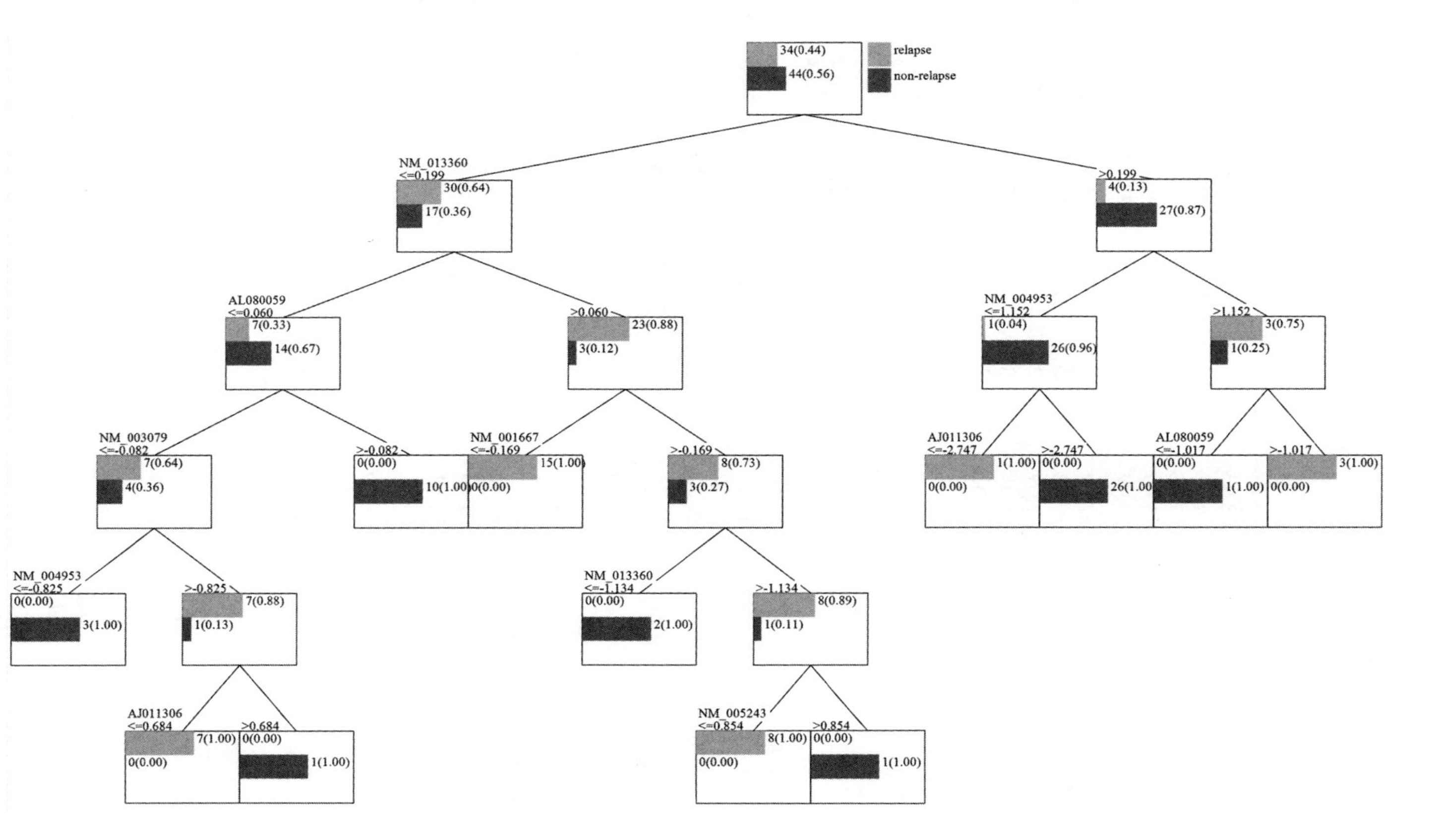

FIGURE 15.4 Decision tree classification (DCT) analysis of 2-class breast cancer dataset with 78 arrays, partitioned into 34 (44%) relapse arrays and 44 (56%) non-relapse arrays. Results reveal a 6-level tree having 12 terminal nodes with 100% class purity.

TABLE 15.1 Performance without Feature Transformation[a]

		Cross-validation method				
Dataset	#Cutpoints	CV2	CV5	CV10	CV − 1	CVB
AMLALL2	10	88.95	87.63	85.79	81.58	94.75
	20	84.21	89.21	87.63	89.47	95.17
	30	83.95	89.74	90.00	89.47	94.26
	40	84.47	88.42	89.47	92.11	94.19
	50	83.42	86.58	88.16	84.21	95.19
Brain2	10	67.17	70.67	70.67	73.33	89.32
	20	69.33	70.50	68.50	61.67	88.13
	30	67.67	67.67	66.50	65.00	88.16
	40	68.33	67.83	66.17	53.33	87.93
	50	71.33	63.33	62.17	61.67	88.02
Breast2A	10	91.33	93.33	95.33	93.33	97.13
	20	89.33	91.33	92.00	93.33	96.19
	30	91.33	92.00	93.33	93.33	97.21
	40	86.00	91.33	93.33	93.33	96.27
	50	83.33	93.33	93.33	93.33	96.79
Breast2B	10	66.67	68.46	66.92	67.95	88.79
	20	70.13	68.08	70.13	71.79	88.46
	30	65.77	66.92	67.05	65.38	88.14
	40	68.08	69.74	68.85	71.79	88.04
	50	68.97	67.95	67.82	64.10	88.65
Lung2	10	91.56	96.25	94.38	96.88	96.86
	20	93.44	92.19	94.38	93.75	96.81
	30	91.88	91.25	92.19	93.75	96.48
	40	93.75	93.44	93.44	93.75	97.52
	50	93.75	95.94	96.56	96.88	97.98
MLL_Leukemia3	10	81.40	79.30	81.40	78.95	92.75
	20	81.58	77.89	75.44	71.93	92.62
	30	82.98	78.60	83.51	82.46	92.65
	40	82.11	81.23	82.46	87.72	92.02
	50	78.95	76.49	75.96	73.68	92.73
Prostate2	10	87.16	86.96	87.16	89.22	95.10
	20	86.47	87.16	86.27	85.29	94.97
	30	86.27	87.94	84.61	83.33	95.05
	40	85.98	87.55	86.67	88.24	94.97
	50	85.39	87.84	89.22	91.18	95.05
SRBCT4	10	84.76	87.62	86.98	87.30	95.05
	20	83.49	89.68	91.11	93.65	94.74
	30	80.79	85.71	87.46	90.48	94.38
	40	82.22	86.51	87.62	88.89	93.92
	50	80.48	86.67	89.37	88.89	94.36
Average		0.73	0.74	0.74	0.74	0.83

[a]Total number of cutpoints (#Cutpoints) represent the number of equally spaced nonoverlapping bins into which the range of each feature was divided to determine the best cutpoint to partition microarrays into daughter nodes.

TABLE 15.2 Performance with Feature Standardization[a]

		Cross-validation method				
Dataset	#Cutpoints	CV2	CV5	CV10	CV − 1	CVB
AMLALL2	10	87.37	88.68	81.32	81.58	95.71
	20	86.84	88.68	88.16	89.47	94.90
	30	85.79	88.68	91.05	89.47	95.20
	40	82.63	85.79	92.89	92.11	94.60
	50	84.47	87.89	86.05	84.21	95.27
Brain2	10	64.00	69.83	70.50	73.33	88.92
	20	71.50	70.83	68.33	61.67	88.50
	30	68.50	65.67	63.83	65.00	88.22
	40	61.00	68.50	64.17	53.33	88.13
	50	71.67	67.33	64.67	61.67	87.95
Breast2A	10	91.33	94.67	94.67	93.33	97.73
	20	86.67	92.67	93.33	93.33	96.19
	30	89.33	91.33	92.67	93.33	96.41
	40	87.33	92.00	92.67	93.33	95.62
	50	86.00	92.00	92.67	93.33	96.15
Breast2B	10	70.38	67.95	70.64	67.95	88.30
	20	71.15	70.00	67.44	71.79	89.21
	30	71.79	68.97	65.38	65.38	88.78
	40	68.08	68.08	69.23	71.79	88.32
	50	68.46	70.13	67.82	64.10	88.20
Lung2	10	95.00	95.31	95.00	96.88	97.38
	20	90.94	92.19	93.75	93.75	97.29
	30	91.56	91.56	92.81	93.75	96.87
	40	93.75	91.88	93.13	93.75	97.68
	50	95.31	96.25	96.88	96.88	98.24
MLL_Leukemia3	10	80.88	80.88	78.42	77.19	93.20
	20	80.18	78.42	75.26	70.18	92.56
	30	80.35	78.77	81.93	82.46	92.67
	40	81.93	82.98	82.98	87.72	92.42
	50	77.02	80.53	76.14	73.68	92.17
Prostate2	10	85.88	86.86	85.69	86.27	95.00
	20	85.10	84.90	84.31	82.35	94.73
	30	85.98	85.78	84.61	82.35	94.65
	40	86.27	84.71	84.90	84.31	95.09
	50	85.98	87.25	90.78	91.18	94.98
SRBCT4	10	86.19	87.14	89.37	87.30	95.62
	20	86.83	87.94	91.90	93.65	94.70
	30	80.16	85.71	89.21	90.48	94.29
	40	81.27	86.83	86.03	88.89	94.31
	50	80.63	86.51	86.03	88.89	94.13
Average	Avg	0.73	0.74	0.74	0.73	0.83

[a]Total number of cutpoints (#Cutpoints) represents the number of equally spaced nonoverlapping bins into which the range of each feature was divided to determine the best cutpoint to partition microarrays into daughter nodes.

TABLE 15.3 Performance with Feature Fuzzification[a]

		Cross-validation method				
Dataset	#Cutpoints	CV2	CV5	CV10	CV − 1	CVB
AMLALL2	10	89.47	95.00	98.16	97.37	96.66
	20	84.74	90.53	93.42	94.74	96.55
	30	87.37	90.53	95.26	92.11	96.15
	40	88.16	90.26	93.16	92.11	96.08
	50	87.63	93.16	95.26	92.11	96.23
Brain2	10	66.33	69.83	71.83	66.67	88.58
	20	67.67	70.00	69.17	68.33	88.69
	30	69.67	71.00	70.17	66.67	88.74
	40	67.33	68.67	68.33	66.67	88.29
	50	66.33	68.33	69.83	66.67	88.33
Breast2A	10	100	100	100	100	100
	20	100	100	100	100	100
	30	98.67	99.33	100	100	99.82
	40	93.33	93.33	93.33	93.33	97.75
	50	93.33	93.33	93.33	93.33	98.05
Breast2B	10	68.85	69.49	70.13	69.23	88.44
	20	68.97	66.79	67.05	67.95	88.73
	30	69.10	68.72	67.05	57.69	88.45
	40	69.36	69.10	69.74	62.82	88.54
	50	66.79	66.54	67.44	56.41	88.16
Lung2	10	93.13	90.94	90.63	90.63	97.42
	20	95.31	96.25	100	100	99.55
	30	93.75	90.63	90.63	90.63	96.86
	40	95.31	96.25	96.88	96.88	97.92
	50	93.44	96.88	96.88	96.88	98.57
MLL_Leukemia3	10	82.81	78.25	81.75	82.46	92.81
	20	80.35	81.05	80.35	82.46	93.39
	30	83.16	84.21	80.88	80.70	93.61
	40	83.16	82.46	83.33	82.46	93.89
	50	83.51	81.75	84.91	84.21	93.43
Prostate2	10	88.14	87.25	89.12	89.22	95.07
	20	85.39	88.43	86.86	88.24	95.20
	30	86.08	85.49	86.27	88.24	95.12
	40	86.37	86.18	87.16	87.25	94.98
	50	84.51	85.98	86.76	88.24	95.19
SRBCT4	10	88.41	88.57	90.79	90.48	95.00
	20	85.56	87.30	88.89	90.48	95.39
	30	86.35	89.84	87.78	88.89	95.17
	40	87.14	89.52	87.78	90.48	95.18
	50	89.05	88.57	88.89	88.89	95.26
Average		0.75	0.75	0.76	0.75	0.84

[a]Total number of cutpoints (#Cutpoints) represents the number of equally spaced nonoverlapping bins into which the range of each feature was divided to determine the best cutpoint to partition microarrays into daughter nodes.

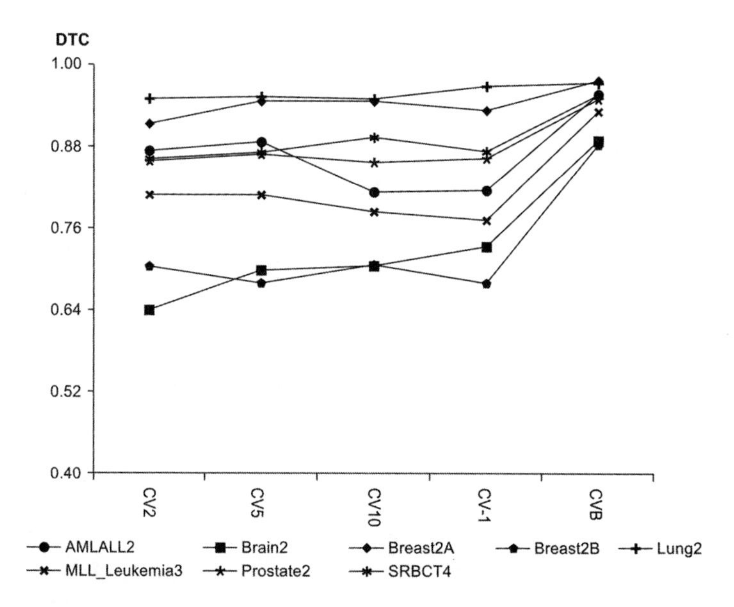

FIGURE 15.5 DTC accuracy as a function of CV method for each dataset. No feature transformations applied.

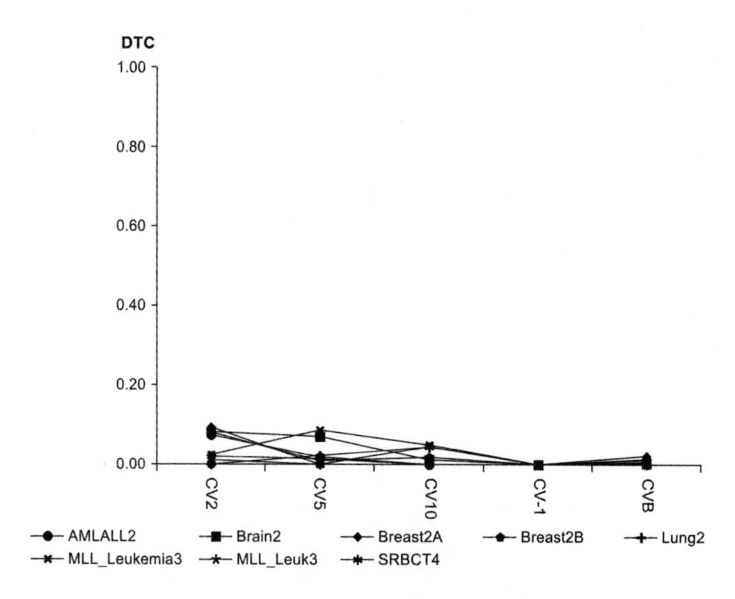

FIGURE 15.6 DTC standard deviation of accuracy as a function of CV method for each dataset. No feature transformations applied.

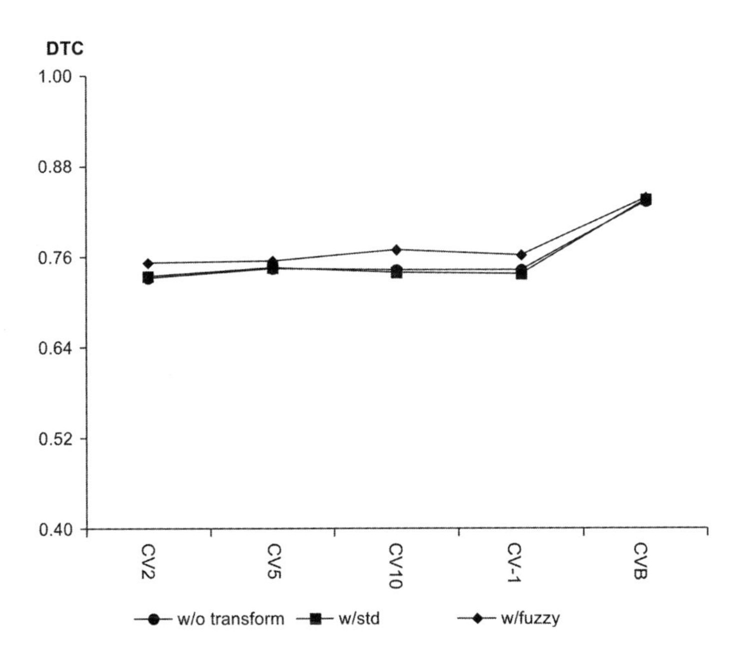

FIGURE 15.7 DTC average accuracy for all datasets as a function of feature transformation and CV method.

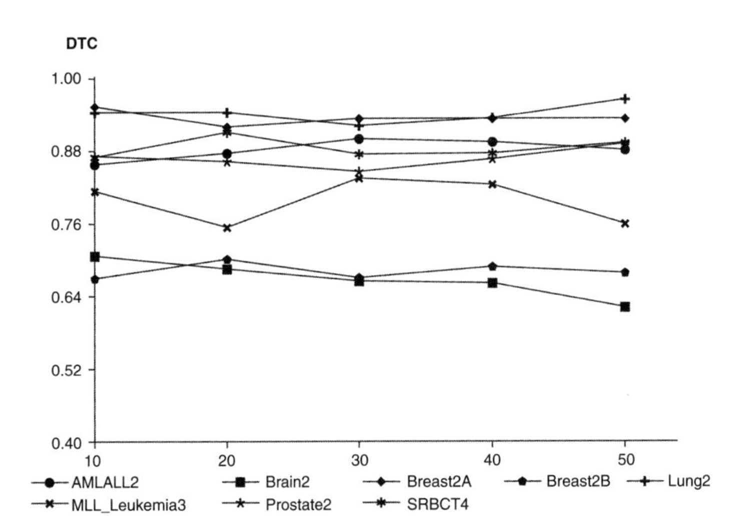

FIGURE 15.8 DTC accuracy without transformations on features as a function of the number of equally spaced cutpoints within the range of each gene's expression value over all the microarrays; 10-fold CV used.

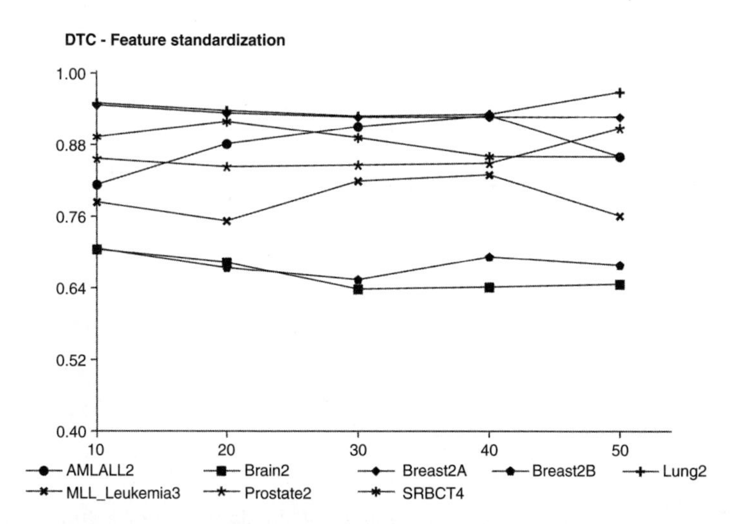

FIGURE 15.9 DTC accuracy with feature standardization as a function of the number of equally spaced cutpoints within the range of each gene's expression value over all the microarrays; 10-fold CV used.

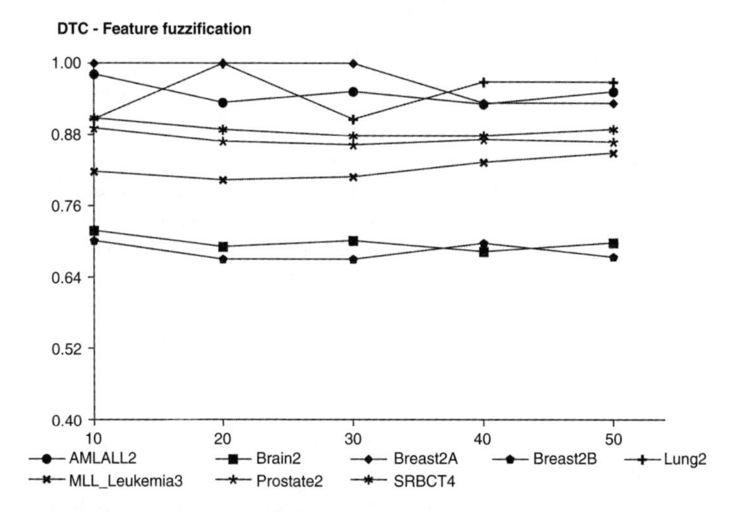

FIGURE 15.10 DTC accuracy with feature fuzzification as a function of the number of equally spaced cutpoints within the range of each gene's expression value over all the microarrays; 10-fold CV used.

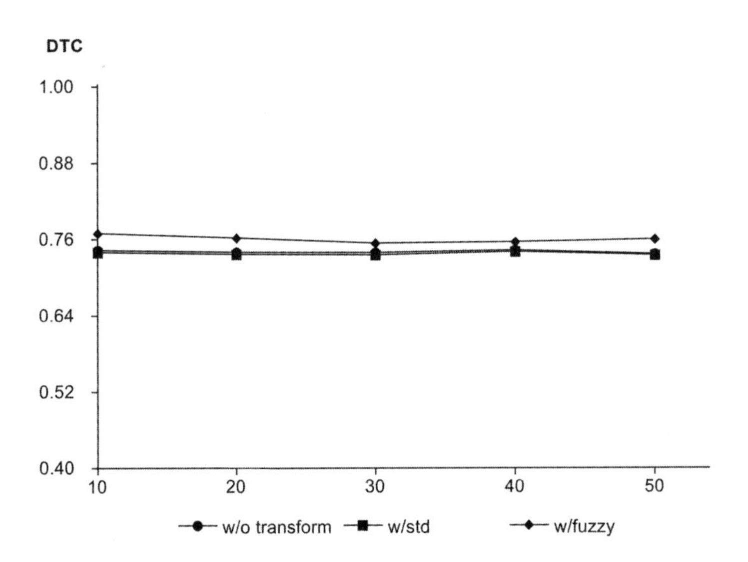

FIGURE 15.11 DTC average accuracy for all datasets as a function of the number of equally spaced cutpoints within the range of each gene's expression value over all the microarrays; 10-fold CV used.

15.7 DECISION BOUNDARIES

Classification analysis and construction of 2D self-organizing maps (SOM) was performed using all available arrays and the reduced gene sets enumerated in Table 12.10. Recall that 2D SOMs reduce the multimarker feature sets down to two dimensions so that the arrays and the class decision boundaries could be viewed simultaneously. Figure 15.12 illustrates a 2D self-organizing map of class decision boundaries for the DTC based on the AMLALL2, Brain2, Breast2A, Breast2B, Lung2, MLL_Leukemia3, Prostate2, and SRBCT4 datasets using mean-zero standardized input features. Figure 15.13 illustrates 2D self-organizing maps of class decision boundaries for the DTC based on the AMLALL2, Brain2, Breast2A, Breast2B, Lung2, MLL_Leukemia3, Prostate2, and SRBCT4 datasets using fuzzified input features.

The decision boundaries for the two-class Brain2, Breast2B, and Prostate2 datasets are highly variable and tend to be unnecessarily complex. The genes used for classification runs in this chapter were selected by *t* tests through a filtering approach. Had genes been selected directly with DTC, the decision boundaries in Figures 15.12 and 15.13 would likely have been different. However, it is important to realize that use of a wrapping approach to select genes during the actual classification process can increase selection bias.

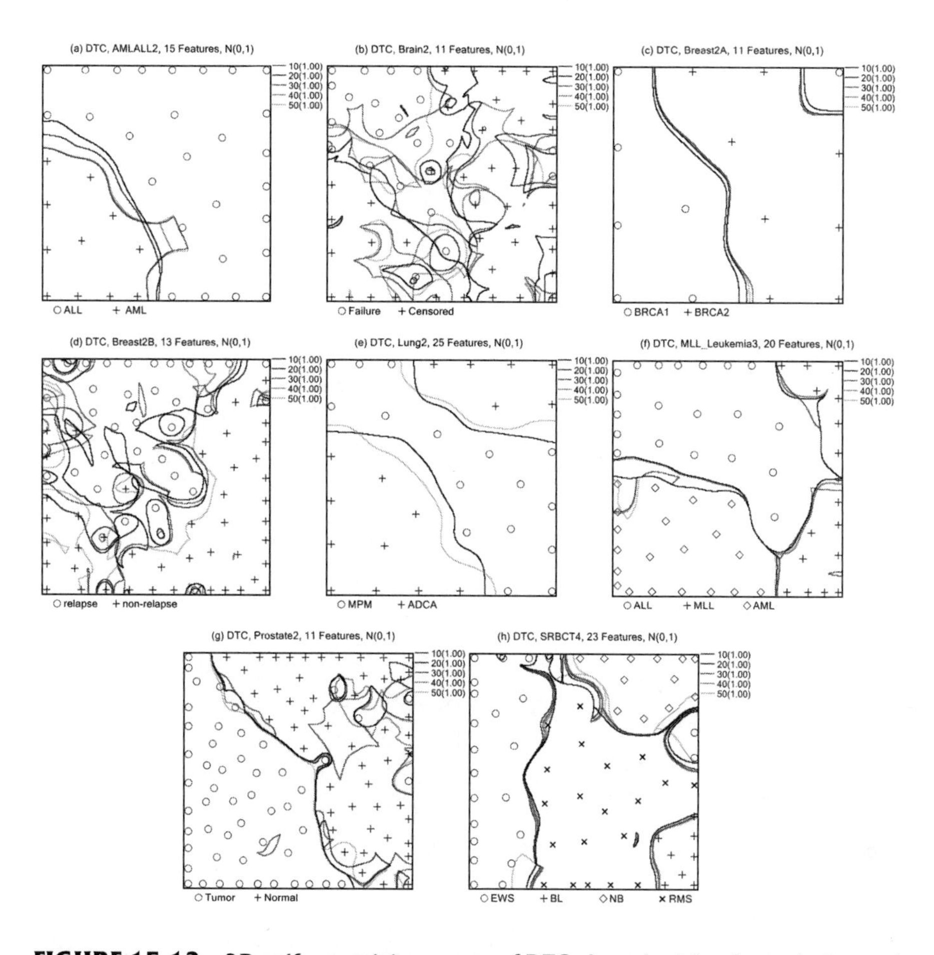

FIGURE 15.12 2D self-organizing maps of DTC class decision boundaries and classification accuracy (upper right of each panel) for the AMLALL2, Brain2, Breast2A, Breast2B, Lung2, MLL_Leukemia3, Prostate2, and SRBCT4 datasets based on mean-zero standardized input features: (a) AMLALL2; (b) Brain2; (c) Breast2A; (d) Breast2B; (e) Lung2; (f) MLL_Leukemia3; (g) Prostate2; (h) SRBCT4. The number of genes used for each dataset are listed in Table 12.10.

15.8 SUMMARY

Decision tree classification resulted in lower values of classification accuracy for the gene expression datasets used. DTC is likely to perform better when features (genes) are selected directly with DTC during classification analysis instead of preselecting features with filtering prior to classification. The advantages of DTC are that decision rules in the form of cutpoint values for continuously scaled features can be rapidly extracted from results, categorical

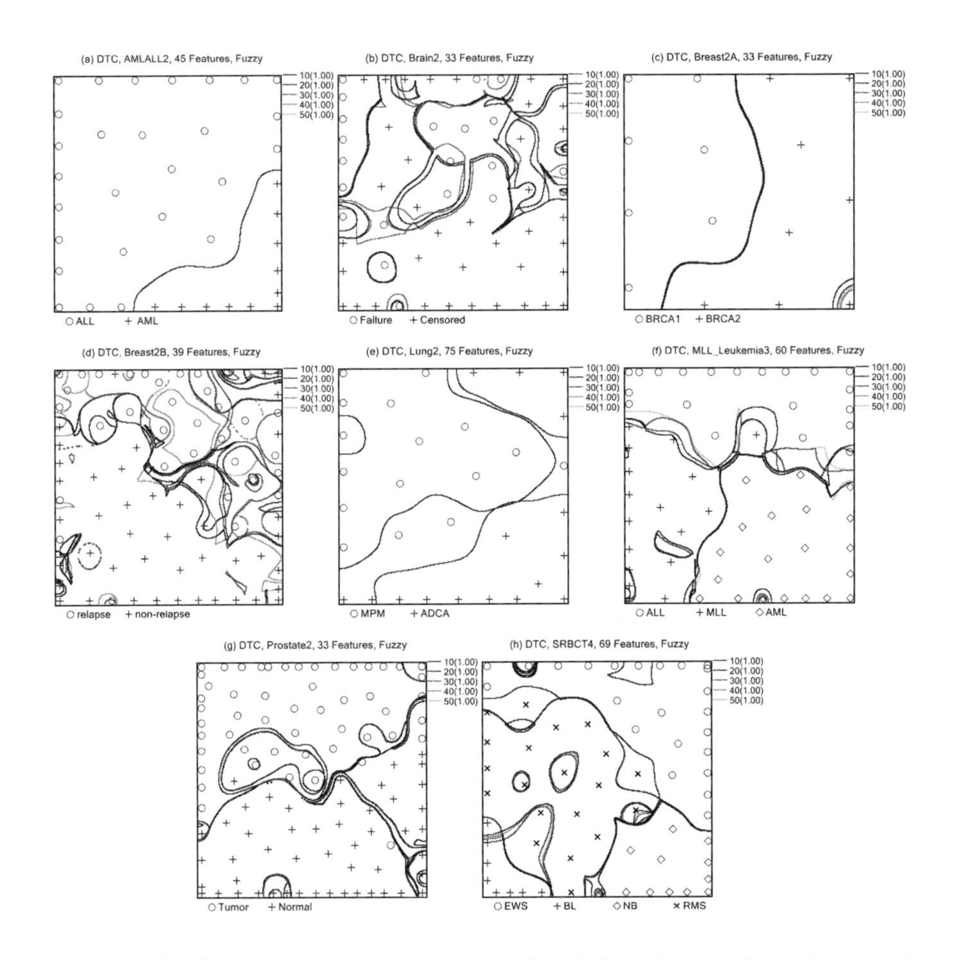

FIGURE 15.13 2D self-organizing maps of DTC class decision boundaries and classification accuracy (upper right of each panel) for the AMLALL2, Brain2, Breast2A, Breast2B, Lung2, MLL_Leukemia3, Prostate2, and SRBCT4 datasets based on fuzzified input features: (a) AMLALL2; (b) Brain2; (c) Breast2A; (d) Breast2B; (e) Lung2; (f) MLL_Leukemia3; (g) Prostate2; (h) SRBCT4. The numbers of genes used for each dataset are listed in Table 12.10.

features can be used, and feature selection is performed on an information theoretic basis (entropy and Gini index). Pruning was not discussed in this chapter, but it can be easily implemented to reduce overfitting. The disadvantages of DTC are that trees with many nodes may result in complex rule sets that cannot be easily understood, decision boundaries may be complex and not smoothly varying, and features selected prior to DTC may not result in the optimal best DTC classification performance.

REFERENCES

[1] J.N. Morgan, J.A. Sonquist. Problems in the analysis of survey data, and a proposal. *J. Am. Stat. Assoc.* **58**: 415-434, 1963.

[2] J.A. Hartigan. *Clustering Algorithms*. Wiley, New York, 1975.

[3] G.V. Kass. An exploratory technique for investigating large quantities of categorical data. *Appl. Stat.* **29**: 119-127, 1980.

[4] L. Breiman, J.H. Friedman, R.A. Olshen, C.J. Stone. *Classification and Regression Trees*. Wadsworth, Belmont, CA, 1984.

[5] J.P. Bradford, C. Kunz, R. Kohavi, C. Brunk, C.E. Brodley. Pruning decision trees with misclassification costs. *Lecture Notes Comput. Sci.* **1398**: 131-136, 1998.

[6] J.R. Quinlan. Induction of decision trees. *Machine Learn.* **1**: 81-106, 1986.

[7] J.R. Quinlan. *C4.5: Programs for Machine Learning*. Morgan-Kaufmann, San Mateo, CA, 1993.

[8] J.R. Quinlan. *C5.0 User Guide*, Rulequest Research, 2009 (available at `http://www.rulequest.com`).

[9] *CART(R) User's Guide*, Salford Systems, 2009 (available at `http://salford-systems.com`).

[10] M. Hall, E. Frank, G. Holmes, B. Pfahringer, P. Reutemann, I.H. Witten. The WEKA data mining software: An update. *ACM SIGKDD Explorations Newsletter* **11**(1): 10-18, 2009.

[11] R. Hulshizer, E.M. Blalock. Post hoc pattern matching: Assigning significance to statistically defined expression patterns in single channel microarray data. *BMC Bioinform.* **8**: 240, 2007.

[12] R. Martin, D. Rose, K. Yu, S. Barros. Toxicogenomics strategies for predicting drug toxicity. *Pharmacogenomics* **7**(7): 1003-1016, 2006.

[13] J.B. Tobler, M.N. Molla, E.F. Nuwaysir, R.D. Green, J.W. Shavlik. Evaluating machine learning approaches for aiding probe selection for gene-expression arrays. *Bioinformatics* **18**(S1): S164-S171, 2002.

CHAPTER 16

RANDOM FORESTS

16.1 INTRODUCTION

Random forests (RF) for classification and regression were developed by Breiman and Cutler in the form of an ensemble of decision tree classifiers [1,2]. RF performs stochastic discrimination [3] by simultaneously using bootstrapping and the random subspace method for selecting features randomly during tree training. Selection of features randomly was proposed earlier by Ho [4,5] and Amit and Geman [6], and RF combines these ideas with the *bagging* methodology introduced by Breiman.

The RF method has found widespread use in microarray analysis [7-13], genomewide association studies [14,15], genetic linkage [16] and disequilibrium [17], traffic flow prediction [18], short-term mortality and outcome classification [19], and classification in ecology [20]. In short, the RF classification approach has proved to be one the best performing classification algorithms in terms of accuracy and improved generalizability due to the reduced chance of overfitting.

Random forests employs bootstrapping of array data to determine classification accuracy of test data and *importance* of each input feature, or gene. Recall that the probability that an array will not be selected during bootstrapping is $(1-1/n)^n = \exp(-1) = 0.368$. In RF, arrays not selected during bootstrapping are termed *out-of-bag* (OOB) objects, and are used for testing. Thus, the arrays selected during bootstrapping form the training data, while OOB arrays form the test data.

To begin, draw a bootstrap sample of n arrays from the total set of n available arrays. Call arrays that are selected the *in-bag* arrays and arrays not selected, the OOB arrays. Let $\mathbf{x}_i (i = 1, 2, \ldots, n)$ represent an array with p genes, such that the elements of $\mathbf{x}_i$ are $(x_{i1}, x_{i2}, \ldots, x_{ij}, \ldots, x_{ip})$ and call in-bag arrays $\mathbf{x}_i$ and OOB arrays $\mathbf{x}_i^{\text{OOB}}$. In addition, let $\hat{\omega}$ be the predicted class, and ω the true class of an in-bag or OOB array $(\omega, \hat{\omega} = 1, 2, \ldots, \Omega)$. RF is initiated

Classification Analysis of DNA Microarrays, First Edition. Leif E. Peterson.

by deploying B decision trees, each of which is trained using the bth set of in-bag arrays selected during bootstrapping ($b = 1, 2, \ldots, B$). For each tree, a bootstrap sample of n arrays is randomly selected with replacement from the n available arrays. The tree is then trained using the in-bag arrays that were selected during the bootstrapping procedure, while OOB arrays that were not selected during bootstrapping are used for testing once the tree is trained. During testing, run each OOB array $\mathbf{x}_i$ down the tree and increment the $\Omega \times \Omega$ confusion matrix $\mathbf{C}_0$ by 1 in the element $c_0(\omega, \hat{\omega})$, where the row is equal to the true class ω and the column is equal to $\hat{\omega}$. Next, permute the expression values of gene $j = 1$ over all the OOB arrays, and run each OOB array down the same trained tree, obtaining a predicted class $\hat{\omega}$ for the array after it arrives in a terminal node. Increment the confusion matrix $\mathbf{C}_1$ by 1 in element $c_1(\omega, \hat{\omega})$ for each OOB array in a pure terminal node. Next, permute gene expression values for genes $j = 2, 3, \ldots, p$ in the OOB arrays, each time running the OOB arrays down the same tree to increment counts in the confusion matrix $\mathbf{C}_j$ based on the predicted and true class when each OOB array arrives at a terminal node with pure class membership. Repeat these steps for each bth tree used, incrementing elements of each confusion matrix as OOB samples arrive at terminal nodes. The importance score $I(j)$ for the jth gene is based on the difference between accuracy from testing OOB arrays without permuted expression values and accuracy of OOB arrays with expression for the jth gene permuted in only the OOB arrays, which is termed the *reduction in accuracy*. When you are finished, there will be $p + 1$ values of classification accuracy for all of the trees used, the first of which is based on OOB samples when expression values are not permuted, and the remainder will be used when expression for each gene is permuted over the OOB samples.

The key point to remember when developing the methodology for RF is that, for each tree, one bootstrap sample of n arrays is drawn from the pool of total arrays available. For each tree (bootstrap), the in-bag arrays selected during bootstrapping are used once to train the decision tree, and the OOB arrays are used for testing to establish (1) the baseline class prediction accuracy for OOB arrays without permutation of gene expression values and (2) the testing accuracy for the same OOB arrays when expression of each gene is permuted. Therefore, each tree is used $p + 1$ times based on the in-bag and OOB arrays drawn during a single bootstrapping procedure.

For the p genes, there are $p + 1$ confusion matrices used with subscript $\mathbf{C}_j$. After training the B trees and filling in the confusion matrices, the baseline classification accuracy Acc_0 is determined by dividing the sum of diagonal elements of $\mathbf{C}_0$ (which represent correctly classified OOB arrays) by the sum of all elements of $\mathbf{C}_0$. Analogously, Acc_j is determined by dividing the sum of diagonal elements of $\mathbf{C}_j$ by the sum of all elements of $\mathbf{C}_j$. It is important to note that once the arrays selected during bootstrapping are used for training the tree (to identify genes and the specific cutpoints that optimize class purity during splits), only the OOB arrays are used for testing to obtain baseline accuracy Acc_0 and accuracy Acc_j for each gene. Each tree is used a total of $p+1$ times, first to increment counts in $\mathbf{C}_0$ for determining Acc_0, and then once for each gene to increment counts in $\mathbf{C}_j$ when expression values of gene j are permuted in the same OOB arrays.

16.2 ALGORITHM

Number of Genes Used for Training `jtry`. A unique characteristic of RF is that only a small number of genes are randomly selected for training each tree. The parameter that specifies the number of genes used is `jtry`, and a typical value is `jtry`=$\sqrt{p}$. Small values of `jtry` of 1 and 2 have shown to yield high performance values. If there are a large number of genes to filter, then larger values of `jtry` will be required. During tree node splitting, `jtry` genes are randomly selected from the p genes, and used to create node splits.

Number of Trees `ntree`. The number of trees used is specified by `ntree`. It has been shown that as few as 50 trees can result in reliable results [20], but the majority of applications generally use 500, 1000, 2000, or more trees. Breiman has recommended using at least 1000 trees [1]; however, it is common to use 500 trees, unless, of course, more trees result in greater performance.

Number of Arrays in Terminal Nodes `nodesize`. The parameter `nodesize` controls the size of terminal nodes during node splitting while training a tree. Nodes with fewer than `nodesize` arrays are not split, and therefore become terminal nodes. The ability of a tree to generate terminal nodes with class purity depends on the quality of the `jtry` features randomly selected for making each split as well as the value of `nodesize`. If `nodesize`=1, however, then every terminal node will have class purity.

Number of Total Nodes `nnode`. The number of total nodes grown in a tree can also be limited by use of the parameter `nnode`. Genes that are less informative for splitting arrays according to class labels will result in trees with greater size, and therefore setting the upper bound to 500 or 1000 is certainly not unreasonable if there are hundreds of samples. Algorithm 16.1 lists the steps used for RF classification and importance scores for each gene.

Algorithm 16.1: RF

Data: n arrays, p genes
Result: Classification accuracy Acc_0, importance scores for each gene
Initialize $B = \texttt{ntree}$, $\mathbf{C}_j = \mathbf{0}$ $(j = 0, 1, 2, \ldots, p)$
for $b = 1$ **to** B **do**
  Generate $\mathcal{D}_b$, a bootstrap sample of n arrays
  Designate set of out-of-bag (OOB) arrays not selected as $\mathcal{D}_b^{\text{OOB}}$
  Call in-bag arrays $\mathbf{x}_i$, and OOB arrays $\mathbf{x}_i^{\text{OOB}}$
  Randomly select `jtry` genes from the available p genes $\rightarrow \mathcal{G}_b$
  for $j = 0$ **to** p **do**
    if $j = 0$ **then**
      Train this bth tree using in-bag arrays $\mathbf{x}_i$ in $\mathcal{D}_b$ and genes $\mathcal{G}_b$
      Run each OOB array $\mathbf{x}_i^{\text{OOB}}$ with genes $\mathcal{G}_b$ down the trained tree to predict class $\hat{\omega}$
      Increment $c_0(\omega, \hat{\omega})$ of $\mathbf{C}_0$ by 1 for each $\mathbf{x}_i^{\text{OOB}}$
    endif
    if $j > 0 \wedge j \in \mathcal{G}_b$ **then**
      Reset $\mathcal{D}_b^{\text{OOB}(j)} \leftarrow \mathcal{D}_b^{\text{OOB}}$
      Permute gene j expression values over all $\mathbf{x}_i^{\text{OOB}}$ in $\mathcal{D}_b^{\text{OOB}(j)}$
      Run each OOB array $\mathbf{x}_i^{\text{OOB}}$ with genes $\mathcal{G}_b$ down the tree trained when $j = 0$
      Predict class $\hat{\omega}$ for each array $\mathbf{x}_i^{\text{OOB}}$ in $\mathcal{D}_b^{\text{OOB}(j)}$
      Increment $c_j(\omega, \hat{\omega})$ of $\mathbf{C}_j$ by 1 for each $\mathbf{x}_i^{\text{OOB}}$ in $\mathcal{D}_b^{\text{OOB}(j)}$
    endif
  endfor
endfor
Determine $Acc_0 = \left[\sum_k c_0(k,k) / \sum_k \sum_l c_0(k,l)\right]$
Determine $Acc_j = \left[\sum_k c_j(k,k) / \sum_k \sum_l c_j(k,l)\right] \forall j$
Determine importance of gene j as $I(j) = Acc_0 - Acc_j$
Sort $I(j)$ in descending order
Note: For each bth tree, the fixed set of `jtry` genes are used in $\mathcal{D}_b$ and $\mathcal{D}_b^{\text{OOB}}$

16.3 IMPORTANCE SCORES

Importance scores reveal the drop in accuracy when each feature's values are permuted within the OOB arrays used for testing to determine accuracy prediction. The importance $I(j)$ for gene j is based on $Acc_0 - Acc_j$, the difference between baseline accuracy of OOB arrays and accuracy among OOB arrays when the jth gene's expression values are permuted.

Example 16.1 RF was used for classification analysis of the two-class prostate dataset listed in Table 12.10. The dataset consists of 102 arrays with $\log_e$-transformed expression values for 11 genes selected using greedy PTA from the original 12,600. A total of $B = 500$ trees were trained using bootstrap samples for each. The Gini index was used for splits, and each of the continuously scaled features was evaluated at 10 nonoverlapping equally spaced cutpoints. Moreover, we set `jtry=` $\sqrt{p}$, `ntree`=500, `nodesize`=1, and `nnode`=1000. Figure 16.1 shows the importance scores for the 11 genes selected during greedy PTA from the two-class prostate cancer dataset. Indeed, the first two genes, 37639_at and 41468_at, yielded the greatest importance and were actually used for the first node splits during DTC as shown in Figure 15.3. Several of the other genes in this figure were used for splits in the DTC classification. Naturally, there is no reason to assume that the order of appearance of genes employed for splits during DTC tracks with the level of importance identified from RF. Rather, the importance reflects the loss of classification accuracy of each gene based on 500 RF trees when genes are randomly selected for splitting during tree training. Next, we ran RF using the two-class breast cancer dataset with 78 arrays, with 34 relapse (44%) and 44 (56%) nonrelapse arrays and 13 genes selected using greedy PTA. Figure 16.2 illustrates that gene NM_013360 has the greatest importance value. This gene is employed for the first node split in Figure 15.4 of the DTC analysis of the same dataset. Other genes with varying levels of importance can be noticed at various stages of node splitting during the DTC run.

Importance scores also vary with the number of genes randomly selected (i.e., `jtry`) during tree training. Table 16.1 lists the between-gene correlation

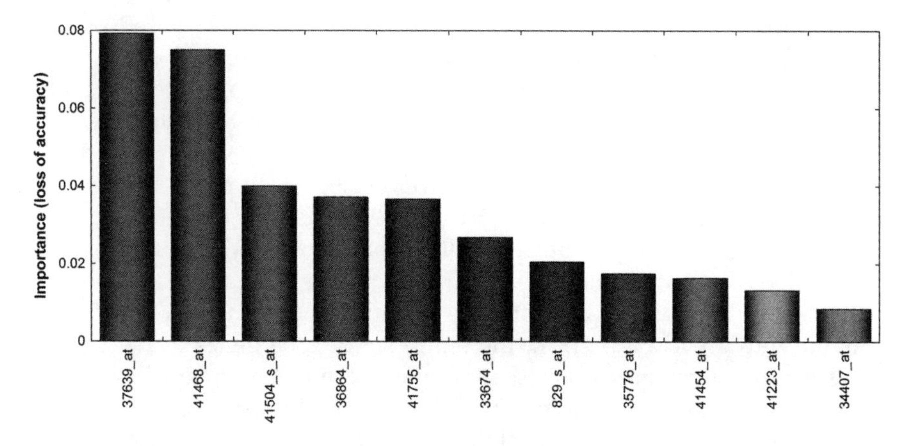

FIGURE 16.1 Importance scores of 11 genes selected with greedy PTA from the two-class prostate cancer dataset. (*See insert for color representation of the figure.*)

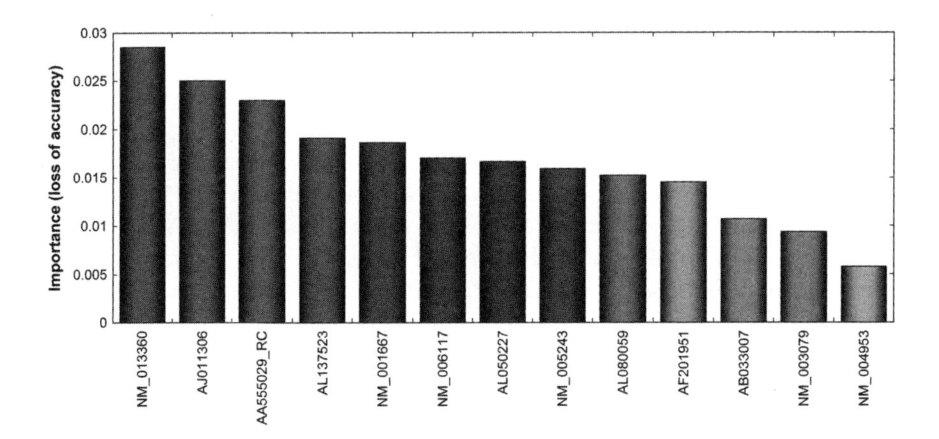

FIGURE 16.2 Importance scores for genes selected using greedy PTA from the two-class breast cancer dataset.

among the $p = 23$ genes selected in the SRBCT4 dataset. Note that there are numerous significant correlation coefficients, suggesting a moderate degree of multicollinearity. Figure 16.3 shows importance scores generated from RF runs (100 trees, Gini index node splits) for the SRBCT4 dataset using values of `jtry`=1,5,10 and $p = 23$. When `jtry`=1, only the marginal effects of each gene on importance are measured because there are no other variables in the model with which each gene can correlate. However, when `jtry`> 1, multiple genes are in the model, and as `jtry` increases, the importance begins to overtake correlation as less informative genes are not selected.

Figure 16.4 illustrates the frequency of selection of each of the 23 genes in the SRBCT4 dataset for values of `jtry`=1,5,10 and p=23 based on their Gini index during the split. Univariate results when `jtry`=1 indicate that each gene was selected relatively frequently and approximately the same number of times during splits. When `jtry`=5, there was a greater frequency of selection of the first six genes. At `jtry`=10, the same same phenomenon occurred in which the first six genes were selected more for the first and all splits. For first splits and `jtry`=10, gene 4 was selected less frequently when compared with all splits. Finally, when `jtry`=23, the first six genes were selected more frequently than the remaining genes. These results are testament to the unique value of RF as a feature selection tool. The 23 genes filtered with greedy PTA from the entire set of 2308 genes in the SRBCT4 dataset provide the greatest class separation based on Mahalanobis distance and are also statistically significantly differentially expressed in terms of the F-to-enter and F-to-remove statistics used during greedy PTA. That RF feature selection indicates that six genes were selected more frequently during all splits at `jtry`=10 and `jtry`=23 may suggest that the remaining genes were selected due to noise-related issues specific to the arrays in the dataset.

TABLE 16.1 Between-Gene Correlation for the $p = 23$ Genes Selected from SRBCT4 Dataset Using Filtering by Greedy PTA

	1	2	3	4	5	6	7	8	9	10	11	12	13	14	15	16	17	18	19	20	21	22
2	0.55^a	—	—	—	—	—	—	—	—	—	—	—	—	—	—	—	—	—	—	—	—	—
3	−0.4^a	0.1	—	—	—	—	—	—	—	—	—	—	—	—	—	—	—	—	—	—	—	—
4	0.8^a	0.6^a	−0.3^b	—	—	—	—	—	—	—	—	—	—	—	—	—	—	—	—	—	—	—
5	0.0	−0.1	−0.3^b	−0.1	—	—	—	—	—	—	—	—	—	—	—	—	—	—	—	—	—	—
6	−0.4^a	0.2	0.8^a	−0.3^b	−0.2	—	—	—	—	—	—	—	—	—	—	—	—	—	—	—	—	—
7	−0.1	−0.1	−0.1	−0.2	−0.2	−0.1	—	—	—	—	—	—	—	—	—	—	—	—	—	—	—	—
8	−0.1	−0.1	−0.1	0.0	−0.4^a	0.0	0.2	—	—	—	—	—	—	—	—	—	—	—	—	—	—	—
9	−0.3^b	−0.2	0.0	−0.2	−0.3^b	−0.2	0.4^a	0.6^a	—	—	—	—	—	—	—	—	—	—	—	—	—	—
10	0.1	0.2^b	0.5^a	0.1	−0.1	0.4^a	−0.3^b	0.1	−0.2	—	—	—	—	—	—	—	—	—	—	—	—	—
11	−0.5^a	−0.2	0.1	−0.4^a	0.4^a	0.1	0.0	−0.3^a	−0.1	−0.3^b	—	—	—	—	—	—	—	—	—	—	—	—
12	0.3^a	0.4^a	−0.1	0.3^b	0.3^b	−0.1	0.0	−0.4^a	−0.3^b	0.1	0.0	—	—	—	—	—	—	—	—	—	—	—
13	0.5^a	0.2	−0.2	0.4^a	0.2	−0.2	−0.1	−0.4^a	−0.5^a	0.1	0.0	0.2	—	—	—	—	—	—	—	—	—	—
14	0.3^b	0.4^a	0.1	0.4^a	−0.2	0.1	0.1	−0.1	0.0	−0.1	−0.3^b	0.4^a	0.1	—	—	—	—	—	—	—	—	—
15	−0.3	0.1	0.6^a	−0.2	−0.3^a	0.6^a	0.3	−0.1	−0.1	0.3^b	0.0	−0.1	−0.2	0.1	—	—	—	—	—	—	—	—
16	−0.4^a	−0.4^a	−0.1	−0.4^a	0.3	−0.2	0.3^b	0.1	0.4^a	−0.57^a	0.5^a	−0.3^b	−0.1	−0.4^a	−0.2	—	—	—	—	—	—	—
17	−0.4^a	−0.2	0.1	−0.3^b	0.6^a	0.1	−0.2	−0.2	−0.2	0.1	0.5^a	0.0	0.0	−0.2	−0.1	0.2	—	—	—	—	—	—
18	−0.3^b	0.0	0.4^a	−0.3^b	−0.2	0.4^a	−0.3^b	−0.1	−0.3^b	0.2	0.0	0.0	−0.2	0.0	0.3^a	−0.3^a	0.0	—	—	—	—	—
19	0.2	0.2	0.1	0.4^a	0.2	0.1	−0.3^b	−0.2	−0.3^b	0.1	0.0	0.0	0.2	0.3^b	0.0	−0.1	0.2	0.1	—	—	—	—
20	−0.3^b	−0.2	0.0	−0.3^b	0.6^a	0.0	−0.3^b	−0.3^b	−0.3^b	0.2	0.4^a	0.2	0.0	−0.2	−0.1	0.1	0.6^a	0.1	0.1	—	—	—
21	−0.2	0.1	0.5^a	−0.2	−0.2	0.4^a	−0.1	−0.1	−0.1	0.3^b	0.2	−0.1	0.0	0.0	0.3^b	−0.2	0.2	0.5^a	0.3^b	0.1	—	—
22	−0.4^a	−0.3^b	0.1	−0.3^a	−0.4^a	0.2	0.3^a	0.5^a	0.6^a	−0.1	0.0	−0.4^a	−0.5^a	−0.1	0.2	0.4^a	−0.2	−0.2	−0.3^b	−0.2	−0.1	—
23	−0.3^a	−0.3^b	−0.2	−0.3^b	0.5^a	−0.1	−0.2	−0.3^b	−0.2	−0.1	0.6^a	0.0	0.0	−0.3^b	−0.2	0.2	0.5^a	0.1	−0.1	0.6^a	0.0	−0.1

Note: A moderate level of multicollinearity is observed, based on significant Pearson correlation coefficients: $p < 0.05$,^{a} $p < 0.01$.^{b}

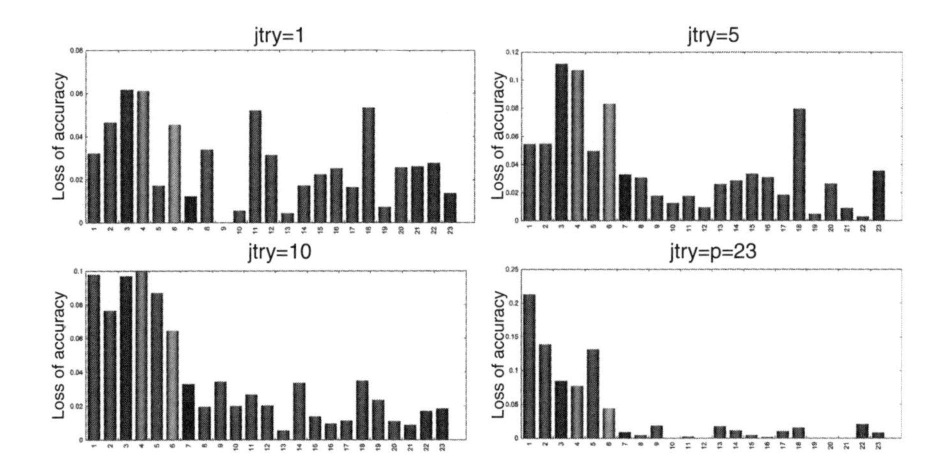

FIGURE 16.3 Effects of gene expression correlation on importance scores. When `jtry=1`, only the marginal effect of each gene is assessed singly, when there is no correlation with other genes. When `jtry>` 1, however, importance scores reveal the conditional effects when other genes are in the model.

The combination of bootstrapping and randomly selecting genes for node splitting during tree training is a more powerful way to evaluate feature robustness across multiple random realizations of a dataset.

16.4 STRENGTH AND CORRELATION

Generalization error of an RF classifier depends on *strength* and *correlation* [1]. Strength reflects accuracy of the trees in a forest, and correlation reflects the dependence between the trees. Let $P(\mathbf{x}_i, \omega)$ be the proportion of trained decision tree votes for the correct class label of OOB array $\mathbf{x}_i$, and let $P(\mathbf{x}_i, \bar{\omega})$ represent the greatest proportion of votes for incorrect predicted class label $\bar{\omega}$ of $\mathbf{x}_i$. The margin for OOB array $\mathbf{x}_i$ is

$$mr_i = P(\mathbf{x}_i, \omega) - P(\mathbf{x}_i, \bar{\omega}), \tag{16.1}$$

and the *strength* is

$$s = \mathrm{E}(mr) = \frac{1}{n}\sum_{i}^{n} mr_i. \tag{16.2}$$

Another estimate we will need is the mean square of the margin, in the form

$$\mathrm{E}(mr^2) = \frac{1}{n}\sum_{i}^{n} mr_i^2. \tag{16.3}$$

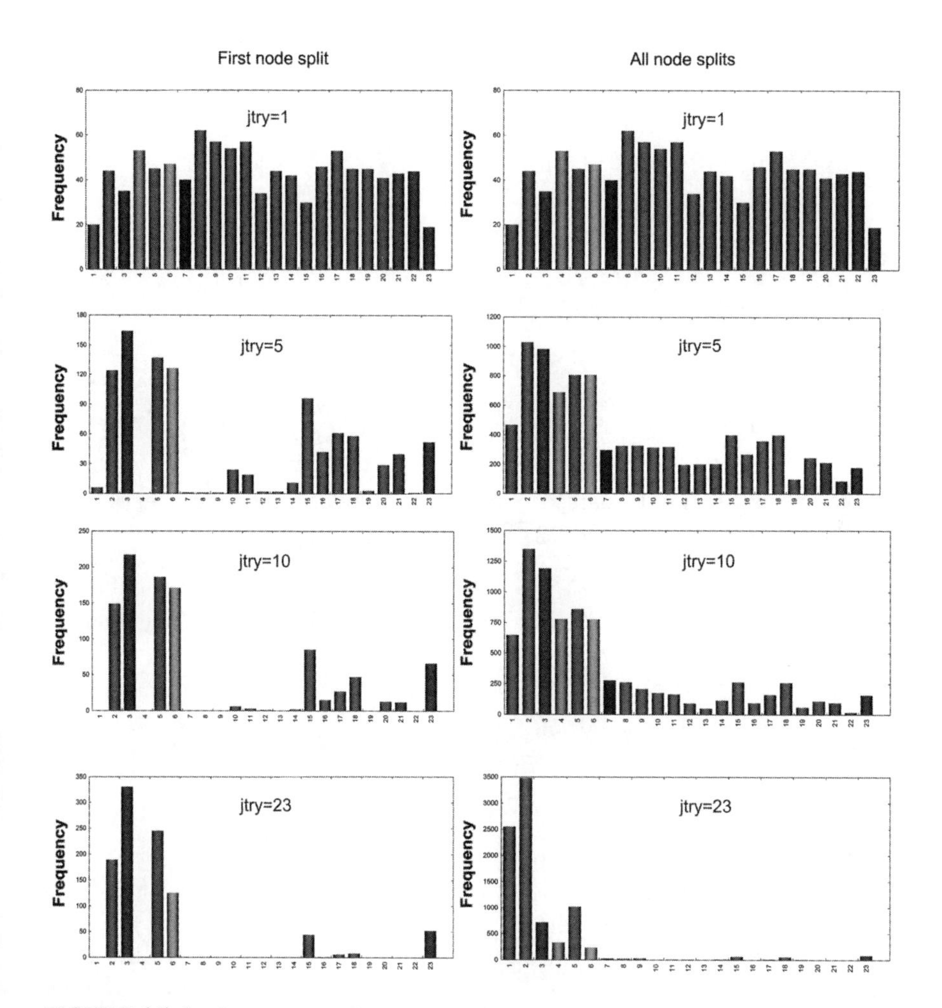

FIGURE 16.4 Frequency of selection of each of the $p = 23$ genes in the SBRCT4 dataset based on Gini index as a function of `jtry`.

Let $P(b, \omega)$ be the proportion of correct OOB array votes for tree b, and let $P(b, \bar{\omega})$ represent the greatest proportion of votes for the incorrect predicted class label from tree b.

The standard deviation for all trees in the forest is

$$\sigma = \frac{1}{B} \sum_{b}^{B} \sqrt{P(b,\omega) + P(b,\bar{\omega}) + (P(b,\omega) - P(b,\bar{\omega}))^2}. \tag{16.4}$$

The average *correlation* of the forest is the ratio of the variance of the margin to the square of the standard deviation of the forest [1]. Using the fact that

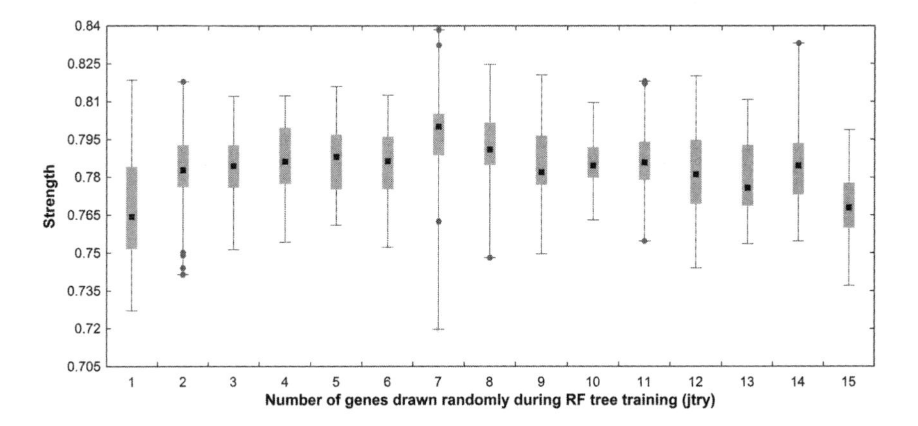

FIGURE 16.5 AMLALL2 dataset: strength as a function of the number of genes randomly drawn (i.e., `jtry`) during RF tree training.

$\mathrm{Var}(x) = \mathrm{E}(x^2) - \mathrm{E}(x)^2$, substituting in Equations (16.2) and (16.3), we have

$$\begin{aligned}\bar{\rho} &= \frac{\mathrm{E}(mr^2) - \mathrm{E}(mr)^2}{\sigma^2} \\ &= \frac{\mathrm{E}(mr^2) - s^2}{\sigma^2}. \end{aligned} \tag{16.5}$$

Example 16.2 Strength and correlation were determined for the AMLALL2, Breast2B, and SRBCT4 datasets by using 100 trees and varying `jtry`, the number of genes selected randomly during tree training. The Gini index was used by default for the node splitting criteria. The number of available genes used for each dataset are listed in Table 12.10. Figure 16.5 shows the strength as a function of `jtry` for the AMLALL2 dataset. A gradual increase in strength was observed for increasing values of `jtry`, which peaked at `jtry`=3. Finally, at the greatest value of `jtry`, strength decreased. Figure 16.6 illustrates correlation as a function of `jtry` for the AMLALL2 dataset. Correlation appeared to increase steadily over the range `jtry`. When `jtry` was equal to the number of genes available, correlation increased rapidly because the trees could no longer avoid less informative genes. Figure 16.7 shows that strength increased continually over the range of `jtry` for the Breast2B dataset. Correlation (Figure 16.8) increased rapidly until `jtry`=3, and then increased gradually for the remainder of `jtry` values. Results for the SRBCT4 dataset indicate that a substantial increase in strength occurred up to `jtry`=5 (Figure 16.9), and correlation (Figure 16.10) increased gradually.

The results for strength and correlation as a function of `jtry` varied over the datasets used. Within each run, the results were sometimes more jumpy than others. Strength and correlation plots reported by Breiman [1] appear more smooth than our results; however, the datasets used had hundreds of training objects and up to 60 input features. Also, we used only 100 trees for training, while Breiman commonly reported the use of no less than 1000 trees for each forest.

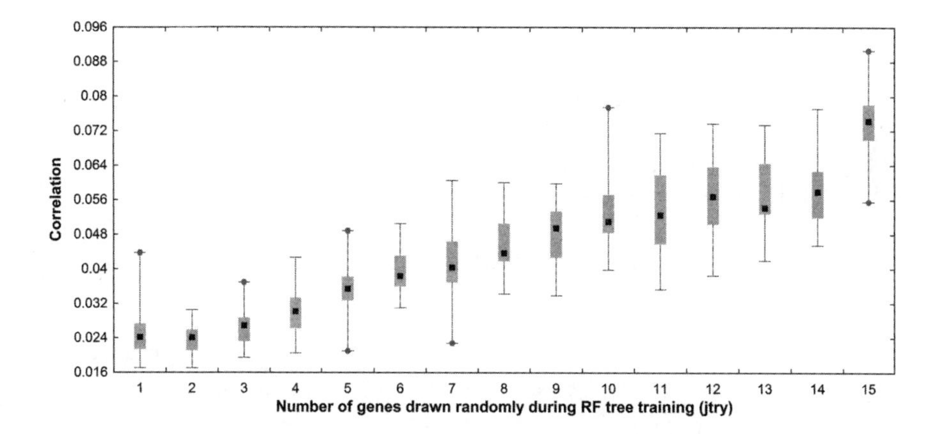

FIGURE 16.6 AMLALL2 dataset: correlation as a function of the number of genes randomly drawn (i.e., `jtry`) during RF tree training.

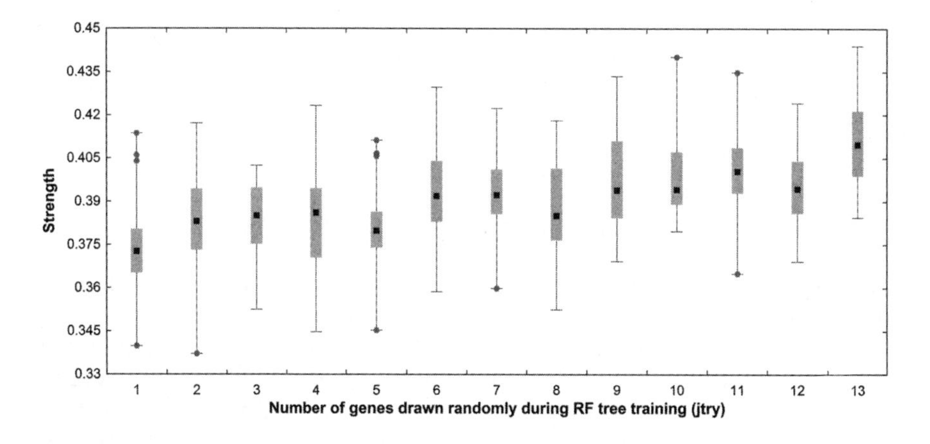

FIGURE 16.7 Breast2B dataset: strength as a function of the number of genes randomly drawn (i.e., `jtry`) during RF tree training.

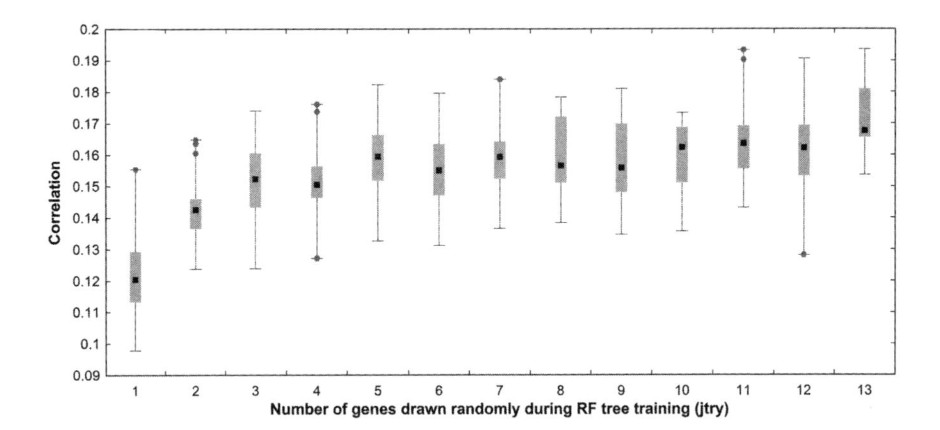

FIGURE 16.8 Breast2B dataset: correlation as a function of the number of genes randomly drawn (i.e., `jtry`) during RF tree training.

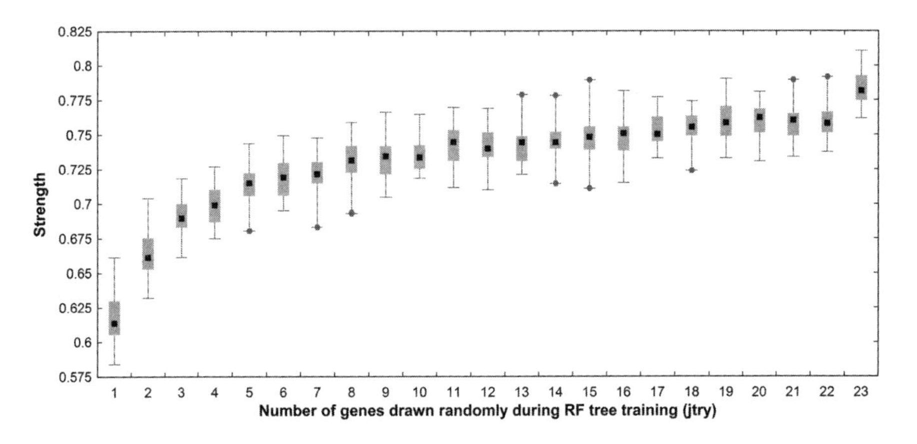

FIGURE 16.9 SRBCT4 dataset: strength as a function of the number of genes randomly drawn (i.e., `jtry`) during RF tree training.

16.5 PROXIMITY AND SUPERVISED CLUSTERING

Supervised cluster analysis using class labels can be carried out with RF by use of an array-by-array proximity matrix, **D**. Let f_{lm} $(l, m = 1, 2, \ldots, n)$ be the frequency of times that OOB arrays $\mathbf{x}_l$ and $\mathbf{x}_m$ happen to end up conjointly in the same terminal node in all the trees, and let f_l and f_m be the total number of times that OOB arrays $\mathbf{x}_l$ and $\mathbf{x}_m$ end up in terminal nodes conjointly or separately in all trees. Obviously, this will not occur in every tree, mainly because (1) these arrays can end up in different terminal nodes and (2) a pair of arrays will not always be OOB after each bootstrap sample is generated.

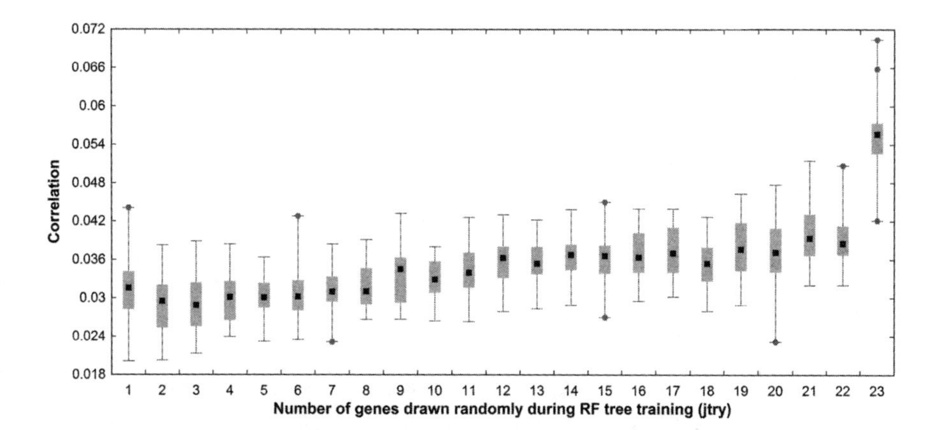

FIGURE 16.10 SRBCT4 dataset: correlation as a function of the number of genes randomly drawn (i.e., `jtry`), during RF tree training.

After training all the trees and running OOB arrays down their respective tree, calculate proximity as

$$d_{lm} = \frac{f_{lm}}{f_l + f_m}, \qquad f_l, f_m > 0 \tag{16.6}$$

and set all diagonal elements of **D** to one: $d_{ll} = 1$. Be sure to invoke the constraints $f_l > 0$ and $f_m > 0$, because if one of these counts is zero, then the particular array either was never an OOB array or never ended up in a terminal node. This will prevent calculation of a proximity value if an array was never observed as OOB during bootstrapping. The covariance matrix of **D** is equal to the doubly centered distance matrix $\boldsymbol{\tau}$ [21] with elements

$$\tau_{lm} = -\frac{1}{2}\left(d_{lm}^2 - \frac{1}{n}\sum_k^n d_{lk}^2 - \frac{1}{n}\sum_k^n d_{km}^2 + \frac{1}{n^2}\sum_g^n\sum_h^n d_{gh}^2\right). \tag{16.7}$$

Knowing that $\boldsymbol{\tau}$ is square symmetric and positive definite, perform eigenanalysis on $\boldsymbol{\tau}$ and find the two greatest eigenvalues after sorting eigenvalues in the order $\lambda_1 \geq \lambda_2, \geq \cdots \geq \lambda_n$. For the ith array, determine the 2 scores ($j = 1, 2$) defined as

$$l_{ij} = \sqrt{\lambda_j} e_{ij} \qquad j = 1, 2, \tag{16.8}$$

where λ_j is the first or second greatest eigenvalue and $e_{i,j}$ is the eigenvector element for the ith array and jth eigenvector. When completed, the result will be an $n \times 2$ matrix **L** of loadings, which are equal to the scores. Note that the loadings are the correlation between the arrays and the two major eigenvectors extracted from τ, which explains the majority of variance in the "array" space. The difference between loadings and PC scores described in section 10.2.6 is that PC scores require the additional multiplication with standardized expression values.

Supervised clustering is performed by calculating τ and generating an x-y scatterplot of all arrays based on the two scores. The cluster structure of **X** is reflected by the score plot of the n arrays. If the p genes used during an RF run to determine τ and **L** were already filtered or preselected for class prediction, then set `jtry=`p. Otherwise, it is appropriate to use `jtry=`$\sqrt{p}$ when a greater number of unfiltered genes are used.

Example 16.3 Proximity matrix determination and supervised clustering was performed for the AMLALL2, Breast2B, and SRBCT4 datasets, using 100 trees, `jtry=`p, and Gini index for the node splitting criterion. The numbers of available genes used for each dataset are listed in Table 12.10. Figures 16.11, 16.12, and 16.13 show the plots of array-specific scores for the two major eigenvectors, illustrating the within-class between-array distances for each dataset.

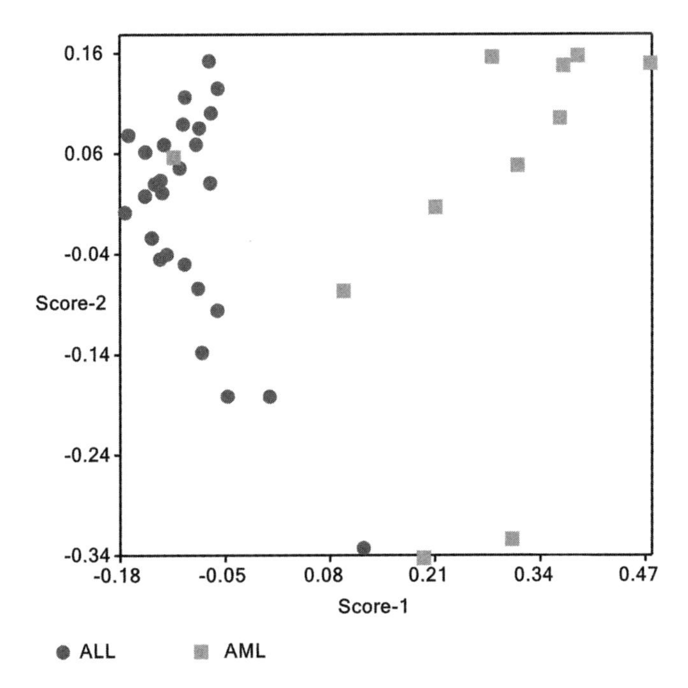

FIGURE 16.11 Supervised clustering results for the AMLALL2 two-class dataset based on log-transformed input feature values. Array-specific loadings (i.e., $\sqrt{\lambda_j}e_{ij}$) for the first and second greatest eigenvalues of matrix τ are shown.

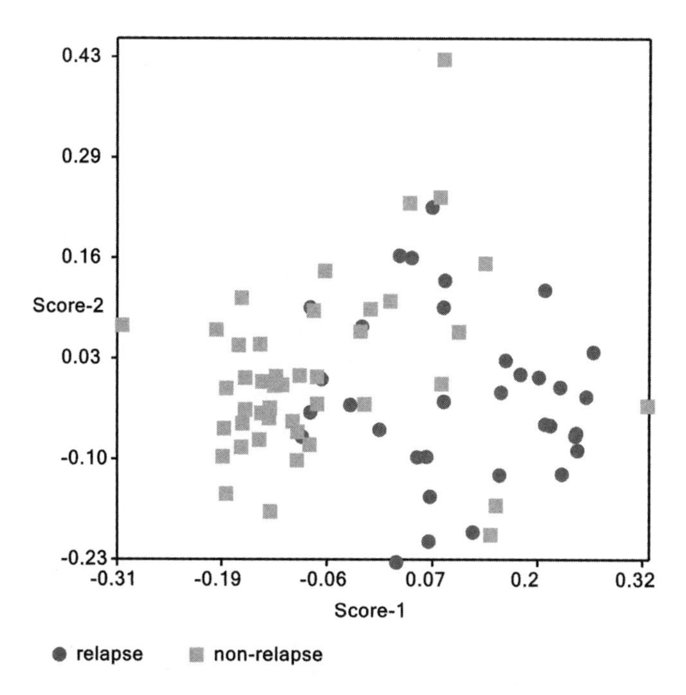

FIGURE 16.12 Supervised clustering results for the Breast2B two-class dataset based on log-transformed input feature values. Array-specific loadings (i.e., $\sqrt{\lambda_j}e_{ij}$) for the first and second greatest eigenvalues of matrix $\boldsymbol{\tau}$ are shown.

Supervised cluster results (Figures 16.11, 16.12, and 16.13) suggest that reasonably good separation is discernible among the clusters in each dataset. The SRBCT4 dataset had better separation between clusters when compared with the AMLALL2 and Breast2B datasets.

16.6 UNSUPERVISED CLUSTERING

Unsupervised clustering with RF involves setting the class label for all of the observed arrays to $\omega = 1$, and simulating an additional set of n arrays to form a new array-by-gene $(n \times p)$ matrix $\mathbf{X}^*$. The simulated expression values for gene j are based on randomly sampling expression values from the original observed arrays, so that $x^*_{ij} = x((n-1) * U(0,1) + 1, j)$, where $i = n+1, n+2, \ldots, 2n$. Class labels for the n simulated arrays are set equal to $\omega = 2$. The two arrays $\mathbf{X}$ and $\mathbf{X}^*$ are then stacked on top of one another, sample size is reset to $n \leftarrow 2n$, and an RF run is performed to determine $\boldsymbol{\tau}$ and $\mathbf{L}$. A score plot is constructed for only the original n arrays in class 1, that is, using the first n of the $2n$ arrays. Similarly, cluster structure is determined

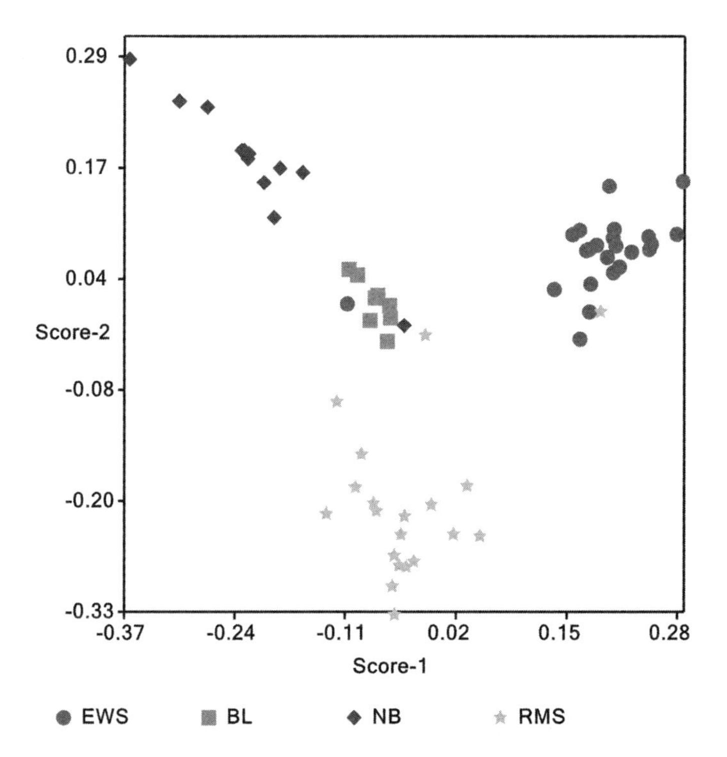

FIGURE 16.13 Supervised clustering results for the SRBCT four-class dataset based on log-transformed input feature values. Array-specific loadings (i.e., $\sqrt{\lambda_j}e_{ij}$) for the first and second greatest eigenvalues of matrix τ are shown. (*See insert for color representation of the figure.*)

by visualizing separation between the original n arrays in the score plot. If the classification accuracy value during an unsupervised RF run is close to 0.5, it suggests that the set of genes input into the RF run are much less informative for revealing the cluster structure of the data.

Example 16.4 Unsupervised cluster analysis was performed for the AMLALL2, Breast2B, and SRBCT4 datasets, using 10,000 trees, `jtry`=p, and Gini index as the node splitting criterion. The numbers of available genes used for each dataset are listed in Table 12.10. Figures 16.14, 16.15, and 16.16 show the plots of array-specific scores for the datasets. Better cluster separation was observed for the AMLALL2 and SRBCT4 datasets when compared with the Breast2B dataset, since several clusters of arrays are visible for the AMLALL2 and SRBCT4 datasets. Supervised cluster analysis results for the Breast2B dataset reflected class overlap among the arrays, so the lack of a discernible separation between the two classes during unsupervised classification was not surprising.

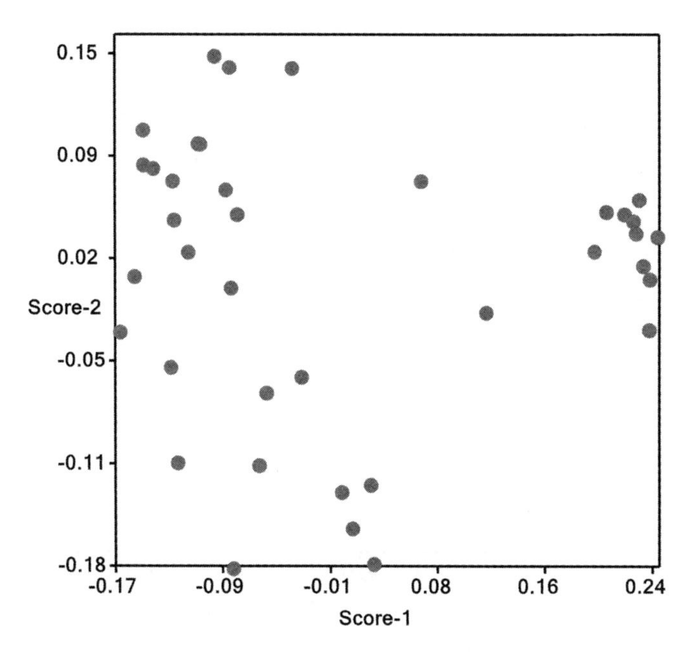

FIGURE 16.14 Unsupervised cluster analysis results for the AMLALL2 dataset.

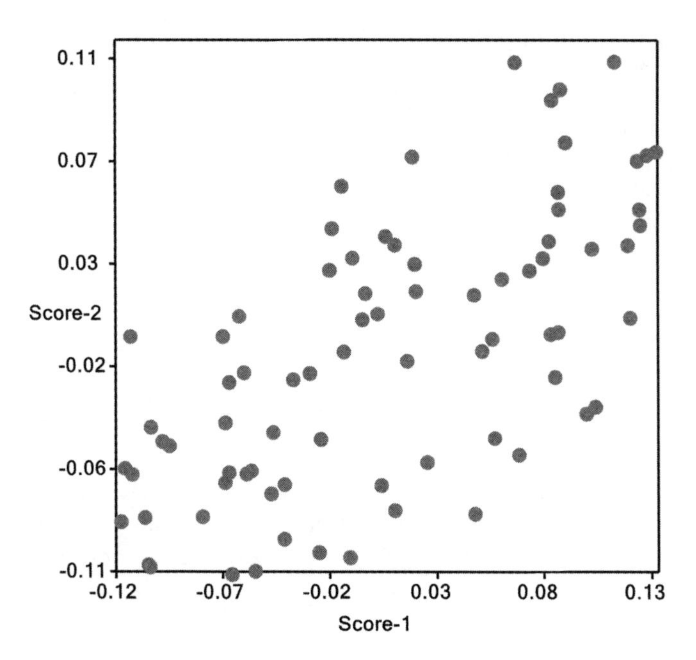

FIGURE 16.15 Unsupervised cluster analysis results for the Breast2B dataset.

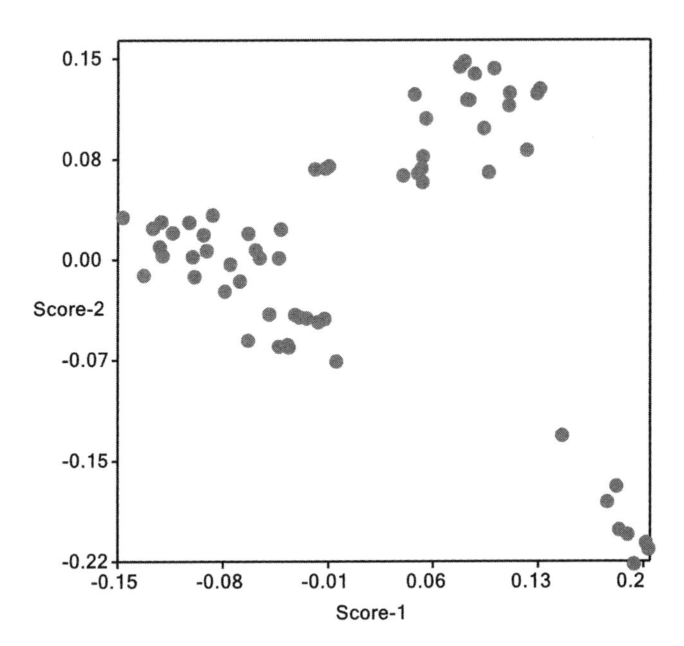

FIGURE 16.16 Unsupervised cluster analysis results for the SRBCT4 dataset. (*See insert for color representation of the figure.*)

16.7 CLASS OUTLIER DETECTION

Outliers among the n arrays in **X** can also be detected using RF. To begin, perform a supervised RF run using **X** and calculate the proximity matrix **D**. For array $\mathbf{x}_l$, calculate the inverse sum of squared values of proximity shared with other arrays (i.e., $\mathbf{x}_m$) in the same class ω, with the relationship

$$r_l = \frac{1}{\sum_{\mathbf{x}_l, \mathbf{x}_m \in \omega} d_{lm}^2}, \qquad l \neq m \tag{16.9}$$

Next, determine the mean absolute deviation, given in the form

$$\text{MAD} = \frac{1}{n} \sum_l |r_l - \text{med}(r)|, \tag{16.10}$$

where med(r) is the median of all r_l values. The residual representing "outlyingness" of each array is based on standardizing the inverse sum of squared proximity with MAD, determined with

$$sr_l = \frac{r_l - \text{med}(r)}{\text{MAD}}. \tag{16.11}$$

Residual values sr_l in excess of 10 are considered class outliers.

Example 16.5 Outlier analysis was performed for the AMLALL2, Breast2B, and SRBCT4 datasets, using 100 trees and `jtry`=p. The numbers of available genes used for each dataset are listed in Table 12.10. Figures 16.17, 16.18, and 16.19 show the plots of outlier residuals for the arrays in each dataset. There was one outlier identified ($sr_l > 10$) for the AMLALL2 dataset and one outlier observed for the SRBCT4 dataset. Cluster quality can also be discerned from the outlier residuals by judging the amount of scatter with a class. In Figure 16.17 there is more scatter among residuals in class AML when compared with ALL. For the Breast2B dataset, there is greater variation among residuals in the relapse class. Arrays having extreme values of sr_l can be excluded from an RF run to determine model improvement when outliers are removed.

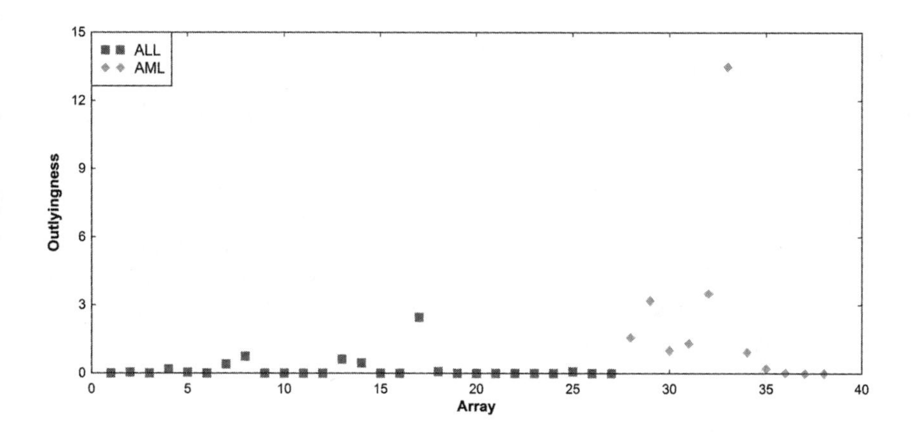

FIGURE 16.17 Outlier residuals sr_l for AMLALL2 arrays.

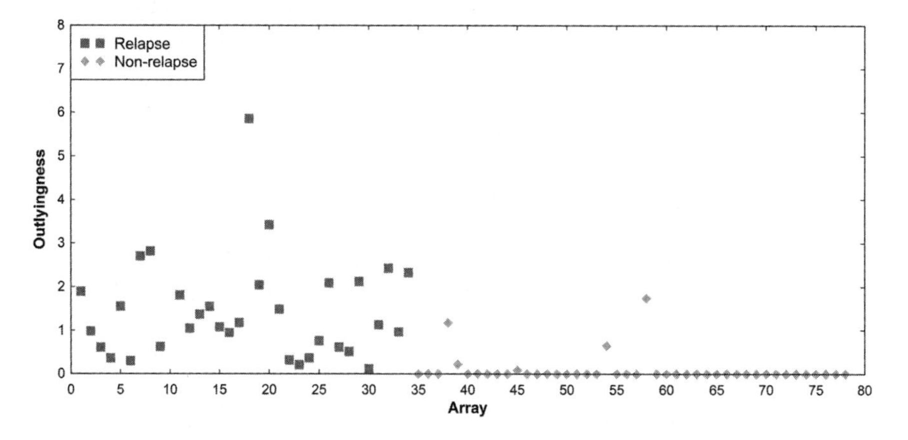

FIGURE 16.18 Outlier residuals sr_l for Breast2B arrays.

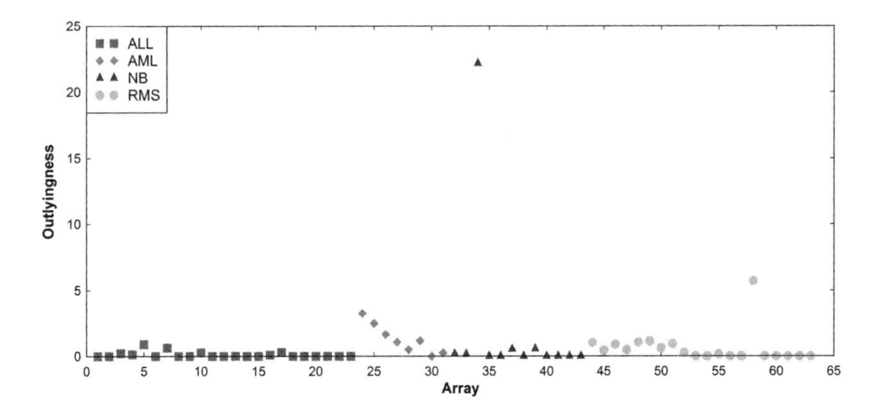

FIGURE 16.19 Outlier residuals sr_l for SRBCT4 arrays.

16.8 IMPLEMENTATION

Implementation of RF is a challenging endeavor that requires substantial bookkeeping when running OOB arrays down each trained decision tree. The $p + 1$ confusion matrices can be filled in as the OOB arrays are sent down all of the trees after training with in-bag arrays. After this, the accuracies are determined for each $p + 1$ confusion matrix, and importance scores for the genes are calculated and ranked in descending order. Plots of importance scores can be constructed to show the relative impact of each gene on overall classification accuracy for all of the trees.

16.9 PARAMETER EFFECTS

Random forest runs were made using the nine datasets and available numbers of genes selected by greedy PTA, which are listed in Table 12.10. Parameter values for `jtry` varied over the range $1, 2, \ldots, 5$, while the the number of trees used (`ntree`) was set at 25, 50, 100, 250, and 500. Table 16.2 lists the RF accuracy as a function of the number of genes randomly drawn during training specified with `jtry` as well as the number of trees employed specified with `ntree` when no feature transformations were applied. Interestingly, the variation of accuracy decreased with increasing number of trees used, but did not change appreciably with increasing values of the number of genes drawn randomly for tree training. The average accuracy value of 81% listed at the bottom of the table suggests that accuracy did not vary with the number of trees used. Table 16.3 lists the RF accuracy as a function of the number of genes randomly drawn and number of trees when mean-zero standardized features were used. Mean-zero standardization did not seem to drastically change the performance values. Table 16.4 lists the the RF

TABLE 16.2 RF Classification Performance without Feature Transformation[a]

		Number of trees (ntree)				
Dataset	jtry	25	50	100	250	500
AMLALL2	1	86.35	88.76	87.90	87.79	89.11
	2	92.40	90.81	87.39	89.41	88.90
	3	89.20	87.89	89.01	89.02	88.46
	4	90.49	89.29	88.90	88.76	88.79
	5	88.64	88.24	89.30	90.08	88.96
Brain2	1	71.02	70.04	68.37	68.41	68.91
	2	70.89	69.48	70.91	69.60	69.47
	3	66.31	67.26	67.01	69.52	69.41
	4	68.74	70.07	69.44	69.34	69.24
	5	70.02	66.52	67.83	68.82	70.07
Breast2A	1	83.66	88.46	86.12	83.00	83.10
	2	79.17	84.01	85.14	84.73	84.41
	3	86.52	85.86	86.30	88.23	87.33
	4	91.60	89.39	87.04	87.46	87.30
	5	84.38	92.06	88.91	89.85	88.92
Breast2B	1	70.55	66.53	69.86	68.85	68.04
	2	69.66	67.75	69.04	68.95	68.70
	3	72.22	70.78	69.80	70.30	69.74
	4	68.02	68.03	70.20	69.50	69.29
	5	67.45	70.42	69.23	70.37	69.68
Colon2	1	77.05	77.49	76.62	77.41	76.96
	2	76.61	74.40	75.33	76.76	76.48
	3	77.48	77.81	77.58	74.94	76.08
	4	75.00	78.43	75.79	77.05	76.58
	5	77.86	76.61	76.69	75.82	76.04
Lung2	1	94.01	91.95	91.72	92.19	92.12
	2	92.16	92.15	91.51	91.99	91.19
	3	93.13	88.93	92.13	90.58	91.79
	4	91.13	91.07	93.13	92.10	92.26
	5	93.38	91.36	91.27	91.54	92.25
MLL_Leukemia3	1	84.08	81.02	78.39	79.74	79.88
	2	81.23	81.60	82.17	81.20	80.66
	3	83.63	82.43	80.72	82.67	82.65
	4	82.18	82.08	81.95	83.17	82.28
	5	78.20	82.31	82.68	83.07	82.35
Prostate2	1	79.39	81.09	82.69	82.31	82.07
	2	82.11	84.15	83.62	83.59	83.08
	3	83.58	84.11	83.73	82.96	83.86
	4	84.86	83.61	84.79	84.44	84.09
	5	83.87	84.66	83.91	84.97	84.15
SRBCT4	1	74.21	75.84	75.14	74.44	74.15
	2	77.66	75.80	79.05	77.74	77.20
	3	77.70	79.52	77.92	80.28	78.98
	4	81.53	81.70	80.44	80.18	80.71
	5	80.20	80.03	80.54	81.98	81.28
Average		0.81	0.81	0.81	0.81	0.81

[a]The number of genes randomly drawn during training is represented by table row values of jtry and the number of trees employed by ntree, which is listed in columns.

TABLE 16.3 RF Classification Performance with Feature Standardization[a]

Dataset	jtry	Number of trees (ntree) 25	50	100	250	500
AMLALL2	1	83.62	88.86	89.69	88.70	88.54
	2	91.62	88.42	89.09	88.85	88.98
	3	86.44	86.69	87.55	88.96	89.11
	4	89.12	89.32	89.00	88.89	88.94
	5	86.00	89.06	87.72	88.64	87.93
Brain2	1	65.67	70.04	67.78	68.13	68.91
	2	68.29	68.96	70.00	68.71	70.01
	3	67.97	68.27	69.68	69.83	69.41
	4	66.08	69.83	71.07	69.99	69.11
	5	68.51	71.00	68.60	69.74	68.72
Breast2A	1	87.05	87.19	86.10	82.81	84.86
	2	85.83	85.97	83.52	83.95	85.03
	3	86.52	88.59	88.79	85.53	85.68
	4	89.13	87.33	87.25	88.28	85.56
	5	92.54	90.07	91.04	90.06	90.64
Breast2B	1	68.69	69.38	69.49	69.18	68.63
	2	69.86	69.72	69.77	69.23	69.17
	3	67.21	69.16	68.98	68.27	69.42
	4	70.41	68.25	69.72	69.50	69.21
	5	69.19	70.79	68.61	69.87	69.10
Colon2	1	74.24	77.86	77.90	77.27	75.99
	2	73.69	74.78	76.44	76.01	76.12
	3	74.43	74.67	77.69	75.88	75.97
	4	74.44	76.75	75.96	76.13	76.60
	5	73.63	74.66	74.88	76.14	76.15
Lung2	1	92.96	92.62	90.78	92.22	90.60
	2	91.00	89.92	91.16	91.95	91.78
	3	91.59	90.46	91.15	92.83	92.25
	4	92.63	90.42	92.68	92.00	92.29
	5	92.07	93.66	93.09	92.73	91.78
MLL_Leukemia3	1	81.31	81.02	80.55	78.97	79.82
	2	81.66	81.31	79.99	81.23	81.52
	3	79.32	82.25	81.54	81.72	81.93
	4	84.78	83.00	81.56	81.87	82.30
	5	83.08	81.47	80.24	82.76	83.21
Prostate2	1	81.22	81.04	82.64	81.13	82.08
	2	86.10	82.12	82.64	83.25	82.91
	3	84.13	84.99	83.79	83.47	83.15
	4	83.66	84.27	83.99	84.08	83.90
	5	85.94	85.67	85.64	84.58	84.36
SRBCT4	1	70.95	76.65	74.73	75.81	74.92
	2	79.08	76.95	77.89	77.98	77.57
	3	76.64	81.89	76.95	79.52	79.37
	4	78.27	80.02	79.48	81.02	80.07
	5	81.51	81.35	80.27	80.24	80.60
Average		0.80	0.81	0.81	0.81	0.81

[a]The number of genes randomly drawn during training is represented by table row values of jtry and the number of trees employed by ntree, which is listed in columns.

TABLE 16.4 RF Classification Performance with Feature Fuzzification[a]

		Number of trees (`ntree`)				
Dataset	`jtry`	25	50	100	250	500
AMLALL2	1	89.28	90.50	89.62	90.20	90.21
	2	92.74	92.50	92.12	91.59	92.81
	3	94.20	92.66	93.05	92.18	93.36
	4	92.88	90.27	93.39	93.29	92.89
	5	92.44	93.85	91.15	93.65	92.54
Brain2	1	64.63	65.76	65.56	66.65	66.24
	2	67.08	66.03	67.84	67.25	68.31
	3	67.87	69.25	66.24	69.70	68.83
	4	66.78	70.77	69.10	69.80	68.81
	5	71.51	68.98	69.49	70.26	68.41
Breast2A	1	91.11	87.78	92.34	89.19	89.79
	2	93.28	92.31	90.74	92.35	91.24
	3	92.20	94.12	91.08	91.38	92.37
	4	93.18	95.14	94.36	93.90	93.00
	5	89.71	94.51	94.04	94.03	92.75
Breast2B	1	68.36	66.50	66.01	66.19	67.46
	2	65.65	68.65	67.21	67.61	67.61
	3	68.92	69.18	66.68	68.31	67.99
	4	72.09	66.62	68.52	68.26	68.64
	5	69.19	69.23	69.46	68.30	68.40
Colon2	1	73.80	75.03	74.77	74.86	76.01
	2	74.00	73.45	75.43	75.17	74.68
	3	73.21	76.34	75.65	75.73	75.78
	4	72.41	74.83	74.25	74.99	75.64
	5	75.13	76.82	75.67	75.96	75.76
Lung2	1	93.75	92.43	89.97	91.32	90.27
	2	92.58	92.86	92.22	93.48	92.74
	3	94.18	95.66	92.86	92.70	93.36
	4	95.27	93.99	93.20	94.17	93.32
	5	92.63	94.56	93.97	93.74	93.77
MLL_Leukemia3	1	78.08	77.45	78.75	78.11	78.14
	2	84.05	80.52	81.94	79.73	81.40
	3	81.02	84.06	81.57	80.89	81.76
	4	81.70	81.51	83.74	83.41	82.17
	5	83.75	81.92	81.70	82.26	82.73
Prostate2	1	81.30	79.96	83.29	82.67	83.01
	2	86.02	85.01	84.34	85.44	84.55
	3	86.92	84.85	85.83	85.32	86.24
	4	86.61	85.05	86.30	86.52	85.92
	5	87.03	86.86	86.15	86.81	86.84
SRBCT4	1	79.68	75.04	77.00	75.94	77.06
	2	82.52	80.12	82.91	80.59	80.96
	3	82.75	84.60	81.49	82.25	83.38
	4	83.36	82.67	83.46	82.82	83.35
	5	85.46	83.69	83.86	84.33	84.42
Average		0.82	0.82	0.82	0.82	0.82

[a]The number of genes randomly drawn during training is represented by table row values of `jtry` and the number of trees employed by `ntree`, which is listed in columns.

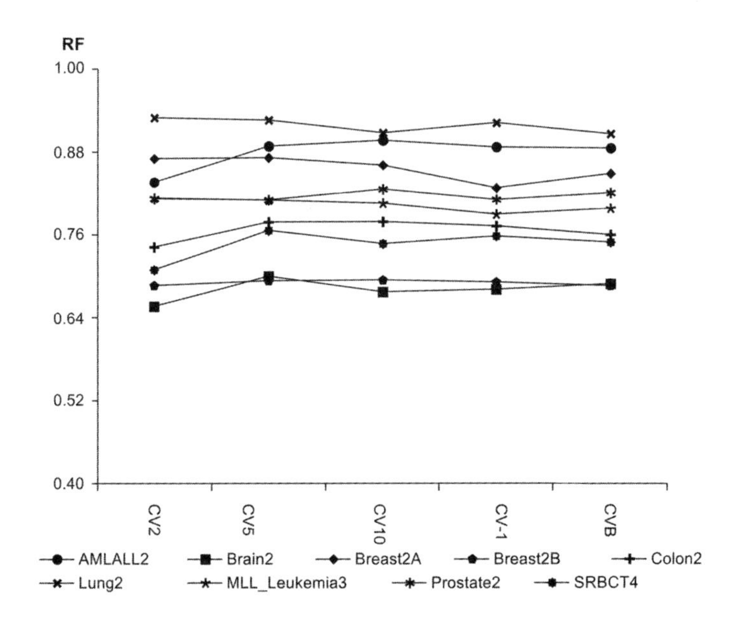

FIGURE 16.20 RF accuracy as a function of the number of arrays used for each dataset when no feature transformations were applied. (As the number of folds increases, the amount of data used for training increases). For each dataset, values shown are based on averaging over row and column values in Table 16.2.

accuracy as a function of the number of genes randomly drawn and number of trees when features were fuzzified. Input feature fuzzification improved performance several percentage points for the two-class AMLALL2 and the Breast2A datasets. Figure 16.20 illustrates RF accuracy for each dataset when no feature transformations were applied values based on dataset-specific average over rows and columns of Table 16.2. Figure 16.21 reveals average RF accuracy for all datasets as a function of feature transformation based on averaging over rows and columns within each dataset in the respective table (Tables 16.2, 16.3, 16.4). (The greater the number of folds used, the greater the amount of data used for training). Figure 16.22 reveals the RF accuracy without transformations on features as a function of the number of genes randomly selected for tree training, based on averaging over columns within each dataset in the Table 16.2. Figure 16.23 shows the RF accuracy with feature standardization as a function of the number of genes selected randomly during tree training, based on averaging over columns within each dataset in the Table 16.3. Figure 16.24 illustrates the RF accuracy with feature fuzzification as a function of number of genes randomly selected during tree training, based on averaging over columns within each dataset in the Table 16.4. Figure 16.25 shows average accuracy values for all datasets as a function of the number of genes selected randomly during tree training.

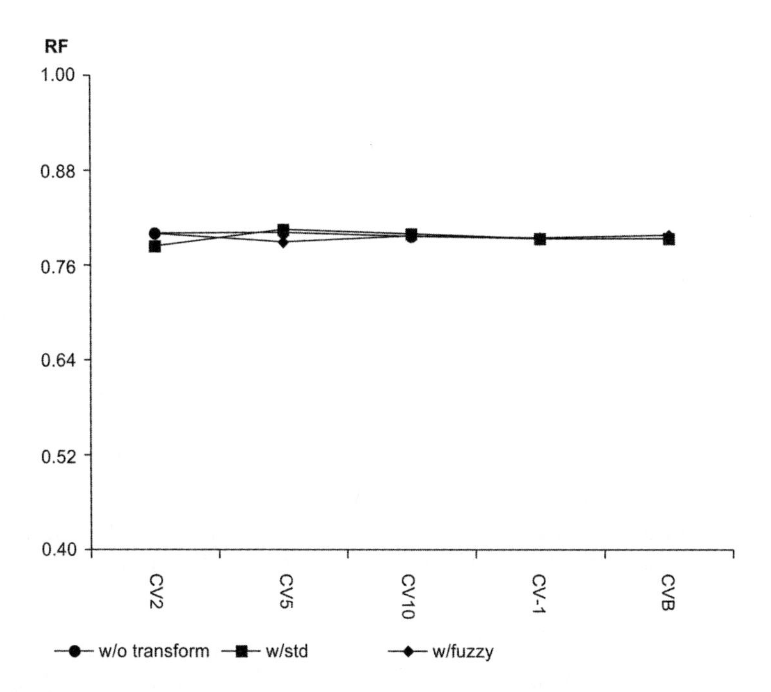

FIGURE 16.21 RF average accuracy for all datasets as a function of feature transformation and number of arrays used for training. Results shown are based on averaging over all rows and columns in Tables 16.2, 16.3, and 16.4.

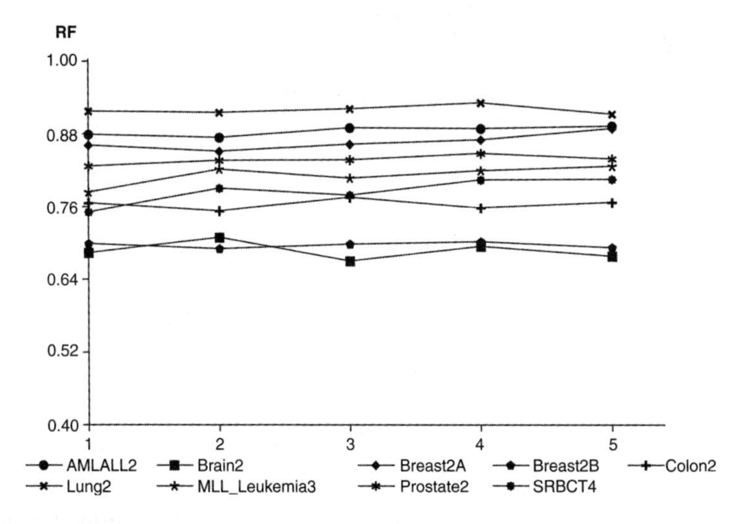

FIGURE 16.22 RF accuracy without transformations on features as a function of the number of genes randomly selected during tree training. Results shown are based on averaging over columns in Table 16.2.

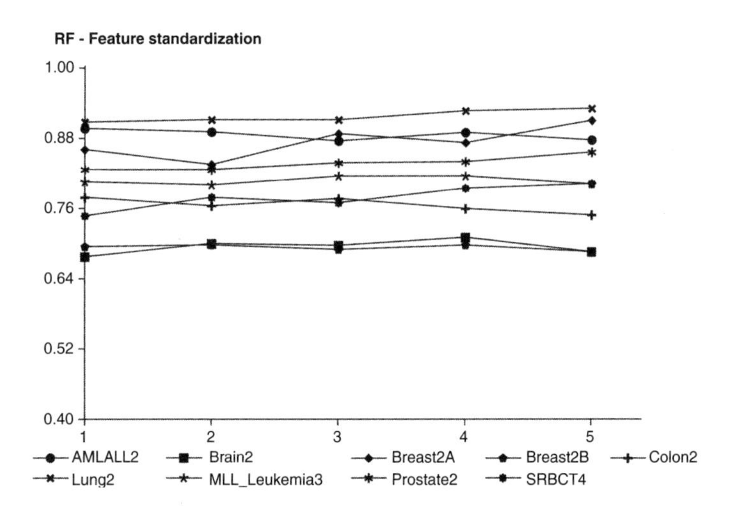

FIGURE 16.23 RF accuracy with feature standardization as a function of the number of genes randomly selected during tree training. Results shown are based on averaging over columns in Table 16.2.

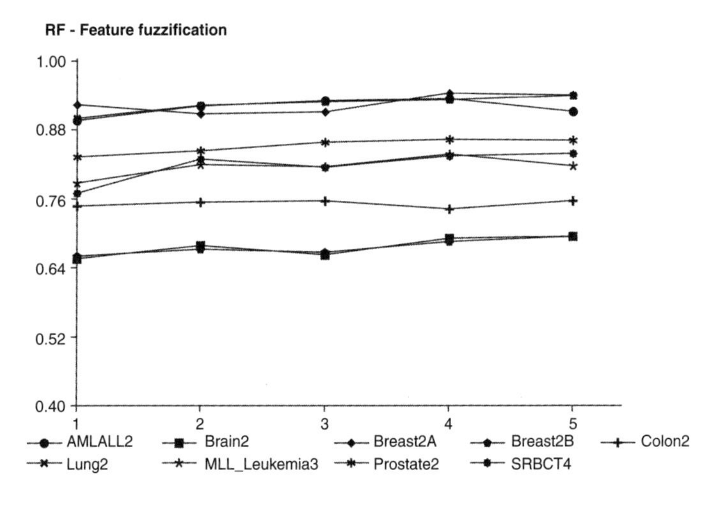

FIGURE 16.24 RF accuracy with feature fuzzification as a function of the number of genes randomly selected during tree training. Results shown are based on averaging over columns in Table 16.2.

Results in Figures 16.20-16.25 indicate that RF is quite stable over changing values of the number of genes randomly drawn during tree training as well as a varying number of trees. For cross-validation results in Figures 16.20 and 16.21, bootstrap accuracy ("CVB") was based on average 0.632 bootstrap

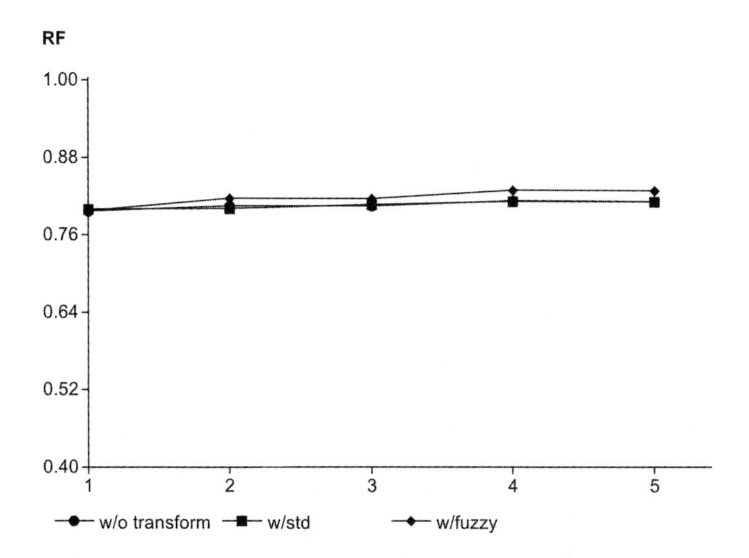

FIGURE 16.25 RF average accuracy for all datasets as a function of the number of genes randomly selected during tree training. Results shown are based on averaging over columns in Tables 16.2, 16.3, and 16.4.

accuracy (13.4) for 10 bootstraps. We did not evaluate the effect of `nodesize` or `nnode`.

16.10 SUMMARY

Random forests is a stable classifier that is not biased toward the amount of training data used, and offers little chance of overfitting the data. Generalization error is, for the most part, lower than that for other classifiers. RF classification accuracy values tend to be lower than other classifiers because, unlike other classifiers, RF randomly samples arrays and features (genes) used for each decision tree in the forest. This chapter introduced the reader to class discovery with unsupervised classification via RF, class prediction with supervised RF classification, feature importance scores, and object outlier determination. RF offers many advantages over other approaches, including

- Reduced generalization error
- Class discovery through unsupervised cluster prediction
- Class prediction via supervised learning
- Simultaneous interaction effects of multiple features
- Better handling of missing data
- Rapid learning times

One downturn of RF is that it is not necessary to perform cross-validation (e.g., 10-fold or leave-one-out CV), because of the advantage of using bootstrapping. Bootstrapping has been shown to result in low variance or less variation across datasets, and potentially greater levels of bias due to its dependence on the proportion of data used for training [22].

REFERENCES

[1] L. Breiman. Random forests. *Machine Learn.* **45**:5-32, 2001.

[2] L. Breiman, A. Cutler. *Random Forests User's Manual*, 2005 (available at `http://www.stat.berkeley.edu/~breiman/RandomForests/cc_manual.htm`).

[3] E. Kleinberg. An overtraining-resistant stochastic modeling method for pattern recognition. *Ann. Stat.* **24**(6):2319-2349, 1996.

[4] T. Ho. Random decision forest. *Proc. 3rd Int. Conf. Document Analysis and Recognition (ICDAR 1995)*, 1995, vol 1, pp. 278-282.

[5] T. Ho. The random subspace method for constructing decision forests. *IEEE Trans. Pattern Anal. Machine Intell.* **20**(8):832-844, 1998.

[6] Y. Amit, D. Geman. Shape quantization and recognition with randomized trees. *Neural Comput.* **9**(7):1545-1588, 1997.

[7] A. Cutler, J.R. Stevens. Random forests for microarrays. *Methods Enzymol.* **411**:422-432, 2006.

[8] R. Díaz-Uriarte, S. Alvarez de Andrés. Gene selection and classification of microarray data using random forest. *BMC Bioinform.* **7**:3, 2006.

[9] A. Statnikov, C.F. Aliferis. Are random forests better than support vector machines for microarray-based cancer classification? *Proc. AMIA Annu. Symp.*, 2007, pp. 686-690.

[10] A. Statnikov, L. Wang, C.F. Aliferis. A comprehensive comparison of random forests and support vector machines for microarray-based cancer classification. *BMC Bioinform.* **9**:319, 2008.

[11] H. Pang, H. Zhao. Building pathway clusters from random forests classification using class votes. *BMC Bioinform.* **9**:87, 2008.

[12] Y. Ma, Y. Qian, L. Wei, J. Abraham, X. Shi, V. Castranova, E.J. Harner, D.C. Flynn, L. Guo. Population-based molecular prognosis of breast cancer by transcriptional profiling. *Clin. Cancer Res.* **13**(7):2014-2022, 2007.

[13] H. Jiang, Y. Deng, H.S. Chen, L. Tao, Q. Sha, J. Chen, C.J. Tsai, S. Zhang. Joint analysis of two microarray gene-expression data sets to select lung adenocarcinoma marker genes. *BMC Bioinform.* **5**:81, 2004.

[14] B.A. Goldstein, A.E. Hubbard, A. Cutler, L.F. Barcellos. An application of random forests to a genome-wide association dataset: Methodological considerations and new findings. *BMC Genet.* **11**:49, 2010.

[15] K.L. Lunettta, L.B. Hayward, J. Segal, P. Van Eerdwwegh. Screening large-scale association study data: Exploiting interactions using random forests. *BMC Genet.* **5**:32, 2004.

[16] A.R. Torres, T.L. Sweeten, A. Cutler, B.J. Bedke, M. Fillmore, E.G. Stubbs, D. Odell. The association and linkage of the HLA-A2 class I allele with autism. *Human Immunol.* **67**(4-5):346-351, 2006.

[17] A.R. Torres, A. Maciulis, E.G. Stubbs, A. Cutler, D. Odell. The transmission disequilibrium test suggests that HLA-DR4 and DR13 are linked to autism spectrum disorder. *Human Immunol.* **63**(4):311-316, 2002.

[18] G. Leshem, Y. Ritov. Traffic flow prediction using Adaboost algorithm with random forests as a weak learner. *Int. J. Intell. Technol.* **2**(2):111-116, 2007.

[19] M.M. Ward, S. Pajevic, J. Dreyfuss, J.D. Malley. Short-term prediction of mortality in patients with systemic lupus erythematosus: Classification of outcomes using random forests. *Arthritis Rheum.* **55**(1):74-80, 2006.

[20] D.R. Cutler, T.C. Edwards, Jr., K.H. Beard, A. Cutler, K.T. Hess, J. Gibson, J.J. Lawler. Random forests for classification in ecology. *Ecology* **88**(11):2783-2792, 2007.

[21] W.S. Torgerson. Multidimensional scaling: I. Theory and method. *Psychometrika* **17**:401-419, 1952.

[22] R. Kohavi. A study of cross-validation and bootstrap for accuracy estimation and model selection. *Proc. Int. Joint Conf. Artificial Intelligence (IJCAI)*, 1995, pp. 1137-1145.

CHAPTER 17

K NEAREST NEIGHBOR

17.1 INTRODUCTION

The K-nearest-neighbor classification is an "instance-based" method developed from the need to perform discriminant analysis when reliable parametric estimates of Bayes probabilities densities are unknown or difficult to determine for a given classification problem. In 1951, Fix and Hodges introduced in an unpublished US Air Force School of Aviation Medicine report a nonparametric method for pattern classification that has since become known the K-nearest-neighbor (KNN) rule [1]. In 1967 the formal properties of the KNN rule were worked out, and it was shown that as $K \to 1$ the KNN error approaches the Bayes error rate [2]. Once the formal properties of KNN were established, a long line of investigation ensued, including new rejection approaches [3], refinements with respect to Bayes error rate [4], distance weighted approaches [5,6], soft methods [7], and fuzzy methods [8,9]. By the mid-1980s, the majority of present-day applications of KNN were known [10,11].

All KNN classification is based fundamentally on the same distance matrix used in hierarchical cluster analysis. The distance between all possible pairs of objects is first determined, and used for agglomerating together objects that are the closest until the entire hierarchy is constructed. In KNN, however, the K training objects closest to a test object are identified, and the most frequent class label among them is used for assigning the predicted class membership for the test object. An advantage of KNN classification is that the distance between all possible pairs of input objects (training and testing) can be precalculated and disk-stored. Figure 17.1 illustrates class prediction for a test object using the most frequent class label among the four nearest neighbors ($K = 4$) serving as training objects. Looking at the four objects that are the closest to the test object, three are members of class A (circles), while

Classification Analysis of DNA Microarrays, First Edition. Leif E. Peterson.

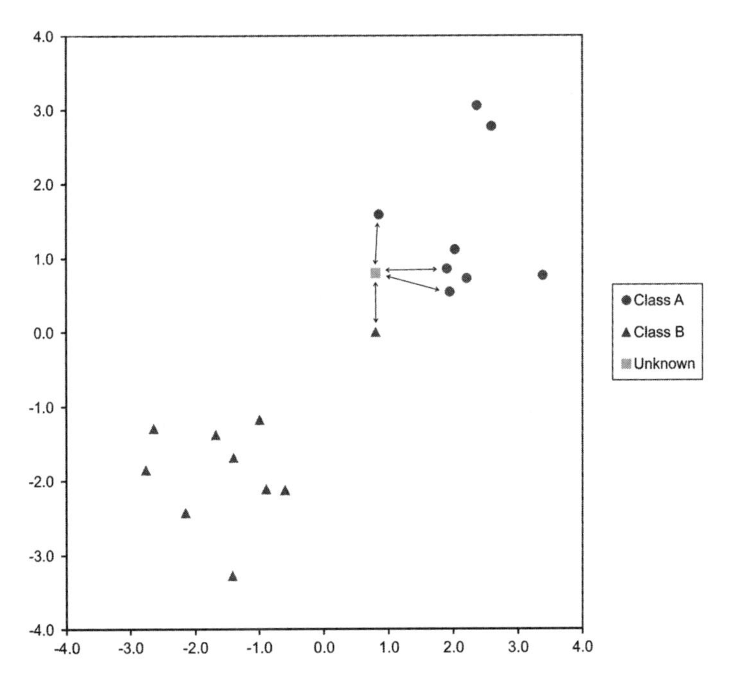

FIGURE 17.1 Class prediction for a test object using the most frequent class labels among the 4 closest ($K = 4$) neighbors. Each object depicted has an x value and a y value.

the remaining closest object is in class B (triangles). Because class A is the most frequent class label among the four closest objects, the class prediction for the test object is membership in class A. The KNN classifier is based on the Euclidean distance between a test object and the specified training objects. The nearest neighbor of a test object $\mathbf{x}$ is $\mathbf{x}_i$, based on the the relationship

$$D(\mathbf{x}_i, \mathbf{x}) = \min_{l}\{D(\mathbf{x}_l, \mathbf{x})\}. \tag{17.1}$$

The K nearest neighbors of $\mathbf{x}$ are identified, and the decision rule $D(\mathbf{x} \rightsquigarrow \hat{\omega})$ is to assign object $\mathbf{x}$ to the predicted class $\hat{\omega}$, which is the most frequent class label among the K nearest training objects.

17.2 ALGORITHM

Let $\mathbf{x}_i$ be an input array; n, be the total number of input arrays $(i, l = 1, 2, \ldots, n)$; and p, the total number of features $(j = 1, 2, \ldots, p)$. The Euclidean distance

Algorithm 17.1: *K* **Nearest Neighbor (KNN)**

Data: Initialize the $n \times n$ distance matrix **D**
$m \leftarrow 10$, $\text{Acc}_{\text{total}} = 0$
foreach *Repartition* $= 1 \leftarrow 10$ **do**
  Set $\mathbf{C} \leftarrow \mathbf{0}$, $n_{\text{test}} \leftarrow 0$
  Randomly partition input arrays into $\mathcal{D}_1, \mathcal{D}_2, \ldots, \mathcal{D}_m$ folds
  for *each cross-validation fold* $\mathcal{D}_m$ **do**
    Select all arrays in $\mathcal{D}_m$ for testing
    Use all remaining arrays not in $\mathcal{D}_m$ for training
    for *each test array* $\mathbf{x}_i \in \mathcal{D}_m$ **do**
      $n_{\text{test}} += 1$
      Determine the K training arrays closest to $\mathbf{x}_i$ based on **D**
      Determine $\hat{\omega}$, the most frequent class label among
        the K closest training arrays
      Increment confusion matrix **C** by 1 in element $c_{\omega,\hat{\omega}}$
    endfor
  endfor
  $\text{Acc}_{\text{total}} += (\sum_i c_{ii})/n_{\text{test}}$
endfch
$\text{Acc} = \text{Acc}_{\text{total}}/10$

between arrays $\mathbf{x}_i$ and $\mathbf{x}_l$ is defined as

$$d(\mathbf{x}_i, \mathbf{x}_l) = \sqrt{(x_{i1} - x_{l1})^2 + (x_{i2} - x_{l2})^2 + \cdots + (x_{ip} - x_{lp})^2}. \tag{17.2}$$

All possible values $d(\mathbf{x}_i, \mathbf{x}_l)$ are arranged in a square $n \times n$ matrix **D**. In addition, let ω represent the true class for array $\mathbf{x}_i$ ($\omega = 1, 2, \ldots, \Omega$). The confusion matrix used for tabulating the predictions of class membership during testing is denoted as **C** and has dimensions $\Omega \times \Omega$. During training, the term $\hat{\omega}$ is used to denote the predicted class label for test array **x**. Algorithm 17.1 lists the computational steps for KNN.

Using Algorithm 17.1, several measures of performance were evaluated for KNN, which included cross-validation, bias or accuracy as a function of training sample size, variance or standard deviation as a function of training sample size, and performance as a function of K.

17.3 IMPLEMENTATION

The concept of training with an input dataset during a KNN run is slightly different from the training used by other methods. Training for KNN amounts to determination of all possible pairwise between-object (array) Euclidean distances, which are stored in memory once at initialization. As objects are tested during various cross-validations (i.e., CV2, CV5, CV10, CV − 1, CVB) to predict class membership, their distance to the *K* nearest neighbors is already precalculated and fetched from the distance matrix stored in memory. Thus, when an object is tested in KNN, its feature values are not input into a trained parametric model but rather used for determining distance from training objects.

It is possible to use future test objects from an external validation analysis, which were not on hand earlier, and then determine *K* nearest neighbors for each in order to make a class prediction. Performance results of KNN will be greatest when the between-class distance of objects is wide, whereas overlap of the cluster structure among objects in different classes will degrade performance. KNN has been observed to outperform other linear methods, such as LREG and LDA, when a particular class of objects forms a cluster within a cluster, or a cluster of objects from a single class is contaminated with objects from another class. The advantage of KNN in these situations is due mostly to the use of Euclidean distance.

17.4 CROSS-VALIDATION RESULTS

Table 17.1 lists the average accuracy based on 10 repartitions of the input arrays and subsequent cross-validation (CV) without any input feature transformations applied. Bootstrap accuracy ("CVB") was based on average 0.632 bootstrap accuracy (13.4) for 10 bootstraps. Average accuracy over all datasets was 90-94%. Accuracy for the Brain2 and Breast2A datasets varied over values of *K* and CV method. In fact, the performance improved for the Breast2A dataset as more training data were used, that is, leave-one-out (CV − 1) accuracy was 100% and CV2 accuracy was lower. Table 17.2 lists the average accuracy based on 10 repartitions of the input arrays and subsequent cross-validation, using mean-zero standardized input feature values. Average accuracy over all datasets did not improve after feature mean-zero standardization. Previous results for the Brain2 and Breast2A datasets were not changed after mean-zero standardization was applied to input gene expression values, and average accuracy for most datasets did not improve. Table 17.3 lists the average accuracy based on 10 repartitions of the input arrays and subsequent cross-validation, using fuzzified input feature values. Feature fuzzification, on average, actually reduced performance over all datasets for the CV5, CV10, and CVB cross-validation methods.

Figure 17.2 illustrates average accuracy as a function of cross-validation method for each dataset based on 10 repartitions of arrays when no feature transformations are applied. Figure 17.3 shows standard deviation of accuracy as a function of CV method for each dataset, also based on 10 repartitions and no feature transformations. Figure 17.4 reveals average accuracy for all

TABLE 17.1 Performance without Feature Transformation

		Cross-validation method				
Dataset	*K*	CV2	CV5	CV10	CV − 1	CVB
AMLALL2	1	100	100	100	100	100
	3	100	100	100	100	99.92
	5	100	100	100	100	100
	7	97.63	100	100	100	98.62
	9	86.58	100	100	100	96.32
Brain2	1	77.00	79.50	79.67	78.33	77.92
	3	82.83	86.33	85.67	86.67	85.19
	5	87.33	88.50	88.50	90.00	86.67
	7	80.17	88.17	88.50	90.00	85.50
	9	78.83	86.50	86.83	86.67	85.13
Breast2A	1	99.33	100	100	100	99.93
	3	100	100	100	100	99.33
	5	82.00	100	100	100	97.77
	7	51.33	100	100	100	87.66
	9	36.67	92.00	100	100	64.21
Breast2B	1	84.23	81.28	82.18	82.05	83.28
	3	84.10	83.72	84.62	84.62	84.60
	5	83.85	84.23	84.49	84.62	85.25
	7	84.74	84.49	84.10	83.33	85.93
	9	86.03	85.90	87.05	85.90	87.14
Colon2	1	77.42	76.13	77.90	77.42	79.15
	3	86.61	86.61	85.32	85.48	86.79
	5	88.06	86.45	87.26	87.10	87.84
	7	86.94	88.39	87.74	87.10	87.50
	9	85.97	89.03	88.55	88.71	87.42
Lung2	1	100	100	100	100	100
	3	100	100	100	100	100
	5	100	100	100	100	100
	7	100	100	100	100	100
	9	100	100	100	100	99.96
MLL_Leukemia3	1	99.47	99.65	99.82	100	99.42
	3	99.12	99.82	100	100	99.31
	5	99.12	99.12	99.30	100	98.77
	7	98.60	98.25	98.25	98.25	98.68
	9	96.14	99.12	98.25	98.25	98.64
Prostate2	1	93.63	94.02	94.12	94.12	94.20
	3	94.71	94.90	95.10	95.10	94.74
	5	94.31	95.00	94.90	95.10	94.66
	7	93.92	94.61	94.90	95.10	94.34
	9	94.51	94.51	94.71	95.10	94.18
SRBCT4	1	99.52	100	100	100	99.71
	3	98.25	100	100	100	99.79
	5	96.03	99.84	100	100	99.49
	7	94.29	100	100	100	98.01
	9	91.43	99.05	100	100	96.20
Average		0.90	0.94	0.94	0.94	0.93

TABLE 17.2 Performance with Feature Standardization

		Cross-validation method				
Dataset	K	CV2	CV5	CV10	CV − 1	CVB
AMLALL2	1	100	100	100	100	100
	3	100	100	100	100	99.97
	5	100	100	100	100	99.86
	7	95.00	100	100	100	99.28
	9	88.16	99.74	100	100	98.25
Brain2	1	75.83	78.83	79.17	78.33	78.69
	3	85.50	85.17	87.00	86.67	85.54
	5	80.83	88.00	89.33	90.00	87.41
	7	82.17	89.00	87.50	90.00	85.37
	9	79.67	85.83	87.50	86.67	85.30
Breast2A	1	98.67	100	100	100	99.86
	3	96.00	100	100	100	99.47
	5	89.33	100	100	100	96.18
	7	38.00	100	100	100	87.37
	9	39.33	86.67	100	100	68.15
Breast2B	1	82.18	82.18	82.82	82.05	83.15
	3	85.26	84.10	84.74	84.62	84.55
	5	84.74	84.36	84.62	84.62	85.26
	7	86.54	85.51	84.62	83.33	86.18
	9	86.67	87.18	86.92	85.90	86.82
Colon2	1	78.23	78.23	77.26	77.42	79.23
	3	85.16	85.81	86.13	85.48	86.68
	5	87.74	87.42	87.42	87.10	87.66
	7	86.94	87.58	87.10	87.10	88.06
	9	85.81	88.06	88.55	88.71	87.45
Lung2	1	100	100	100	100	100
	3	100	100	100	100	100
	5	100	100	100	100	100
	7	99.69	100	100	100	100
	9	100	100	100	100	100
MLL_Leukemia3	1	98.95	99.47	100	100	99.52
	3	99.30	99.30	99.65	100	99.27
	5	98.25	99.30	99.82	100	98.86
	7	98.77	98.42	98.25	98.25	98.64
	9	98.07	98.60	98.25	98.25	98.93
Prostate2	1	93.24	94.31	94.61	94.12	94.07
	3	94.71	94.90	94.90	95.10	94.66
	5	94.02	95.00	94.80	95.10	94.54
	7	94.31	95.00	95.10	95.10	94.17
	9	94.41	94.41	94.71	95.10	94.26
SRBCT4	1	99.52	100	100	100	99.80
	3	99.52	100	100	100	99.90
	5	98.73	100	100	100	99.29
	7	95.24	100	100	100	98.26
	9	95.56	99.37	100	100	96.54
Average		0.90	0.94	0.94	0.94	0.93

TABLE 17.3 Performance with Feature Fuzzification

Dataset	K	Cross-validation method				
		CV2	CV5	CV10	CV − 1	CVB
AMLALL2	1	99.74	99.74	100	100	99.87
	3	97.11	99.74	100	100	99.78
	5	100	100	100	100	99.23
	7	91.58	100	100	100	99.30
	9	86.05	100	100	100	97.43
Brain2	1	72.00	73.67	75.17	76.67	74.47
	3	77.83	78.83	78.50	78.33	79.11
	5	81.50	80.67	80.50	81.67	81.80
	7	82.00	83.50	82.17	81.67	84.07
	9	83.83	82.33	83.83	83.33	84.04
Breast2A	1	100	100	100	100	100
	3	96.00	100	100	100	99.60
	5	90.00	100	100	100	97.30
	7	56.00	97.33	100	100	86.83
	9	42.00	81.33	100	100	65.26
Breast2B	1	81.41	80.13	79.10	78.21	79.33
	3	83.59	84.49	85.38	85.90	83.93
	5	82.82	85.51	86.15	87.18	84.37
	7	82.18	83.85	84.36	84.62	83.05
	9	83.08	83.72	84.49	84.62	82.83
Colon2	1	81.61	81.45	82.74	82.26	82.33
	3	83.06	83.39	84.03	83.87	84.08
	5	83.39	85.81	86.77	87.10	84.69
	7	84.19	85.32	85.48	87.10	84.23
	9	85.00	84.68	84.68	83.87	85.62
Lung2	1	100	100	100	100	100
	3	100	100	100	100	100
	5	100	100	100	100	100
	7	100	100	100	100	100
	9	100	100	100	100	100
MLL_Leukemia3	1	98.95	98.60	98.42	98.25	98.71
	3	98.25	99.65	99.82	100	99.27
	5	99.30	98.95	98.25	98.25	99.02
	7	99.12	99.47	98.77	98.25	99.24
	9	97.72	99.82	100	100	99.39
Prostate2	1	92.16	92.94	92.55	93.14	92.45
	3	93.24	93.82	93.73	94.12	93.27
	5	93.53	93.24	93.14	93.14	93.45
	7	93.33	93.63	93.24	93.14	93.48
	9	93.14	93.33	93.14	93.14	93.22
SRBCT4	1	98.73	99.05	99.84	100	98.86
	3	99.37	99.37	99.68	100	99.21
	5	97.62	100	100	100	99.59
	7	94.76	100	100	100	98.23
	9	92.54	100	100	100	97.85
Average		0.90	0.93	0.93	0.94	0.92

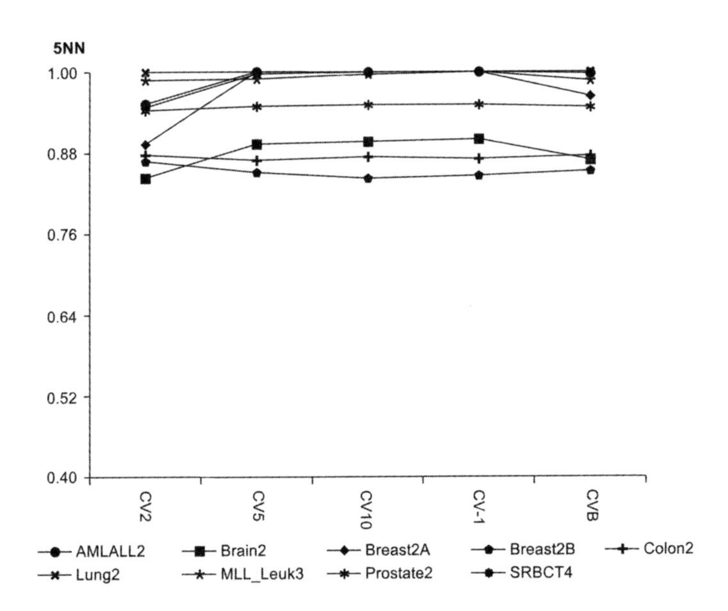

FIGURE 17.2 KNN accuracy as a function of cross-validation (CV) method for each dataset. Average based on CV accuracy for 10 repartitions of arrays. No feature transformations applied.

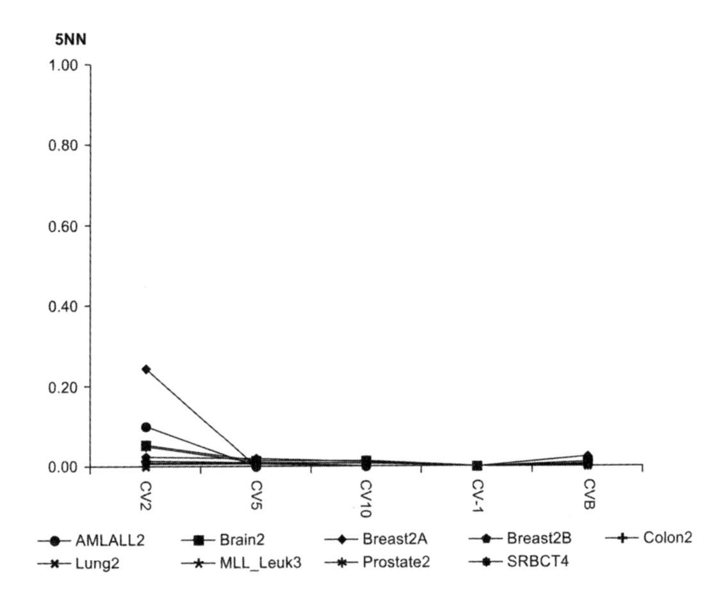

FIGURE 17.3 KNN standard deviation of accuracy as a function of CV method for each dataset. Standard deviation based on CV accuracy for 10 repartitions of arrays. No feature transformations applied.

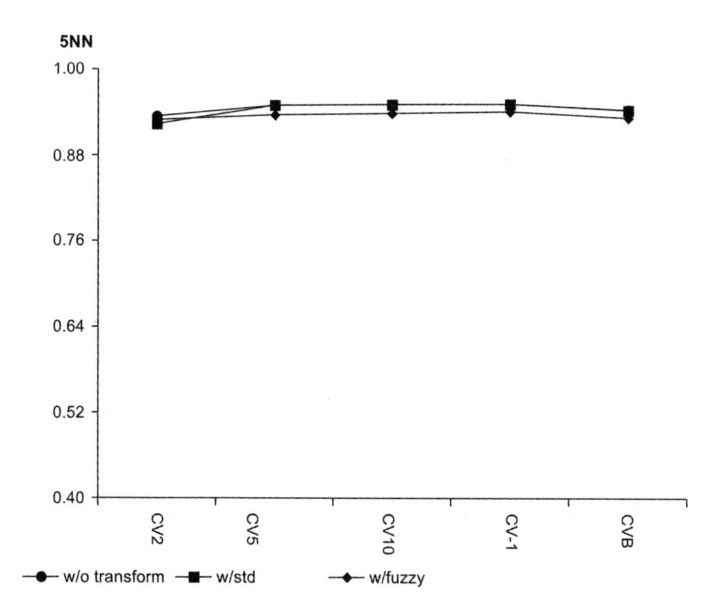

FIGURE 17.4 KNN average accuracy for all datasets as a function of feature transformation and CV method.

datasets as a function of feature transformation and CV method. The results presented in the previous figures indicate that KNN was not biased to an appreciable extent, since only the Brain2 and Breast2A datasets seemed to have altered performance when more or less data were used for training. There was a nominal degree of variance of KNN, due to the varying levels of performance observed over the datasets, and this was influenced by feature mean-zero standardization or fuzzification. Figure 17.5 reveals the KNN accuracy without transformations on features as a function of *K*. Figure 17.6 shows the KNN accuracy with feature standardization as a function of *K*. Figure 17.7 illustrates the KNN accuracy with feature fuzzification as a function of *K*. Figure 17.8 shows average accuracy for all datasets as a function of *K* when ten 10-fold CV was used. Performance values for KNN as a function of *K* were observed to change slightly for the Brain2, Breast2B, and Colon2 datasets, but not for the remainder of datasets. Mean-zero standardization and fuzzification also did not result in large changes observed to occur over the datasets.

17.5 BOOTSTRAP BIAS

Bias was also evaluated using bootstrapping, to reveal the change in performance as a function of the fraction of randomly sampled input arrays

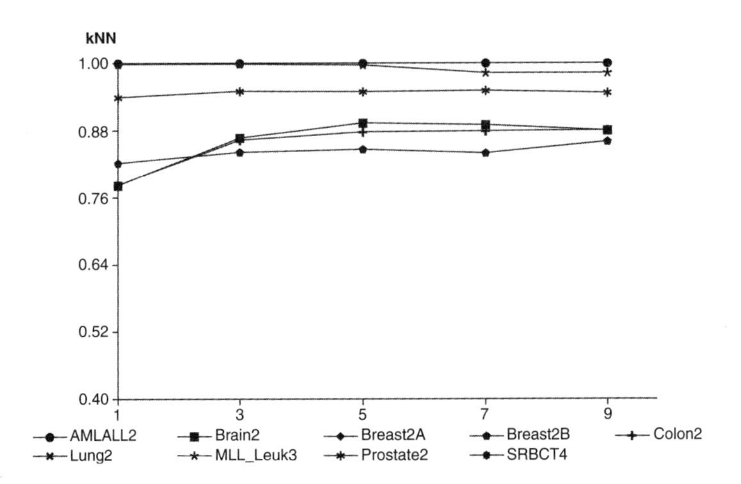

FIGURE 17.5 KNN accuracy without transformations on features as a function of *K*.

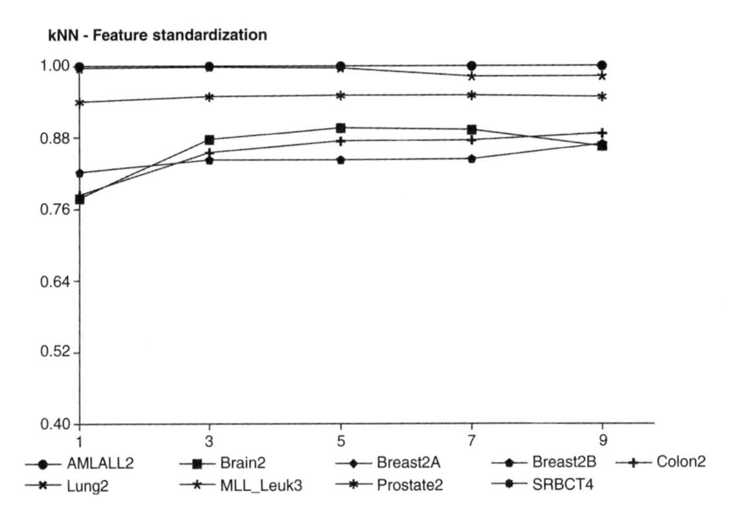

FIGURE 17.6 KNN accuracy with feature standardization as a function of *K*.

out of the total number of arrays. Bootstrap bias runs were based on determining the 0.632 bootstrap accuracy (see Eq. 13.4) for each sampling fraction. Acc_0 in 13.4 was based on test accuracy of unsampled arrays, while Acc_b in 13.4 was based on test accuracy for sampled arrays. Figure 17.9 shows the 5NN accuracy without transformations on features as a function of bootstrap sample size used during training. Figure 17.10 illustrates the 5NN accuracy

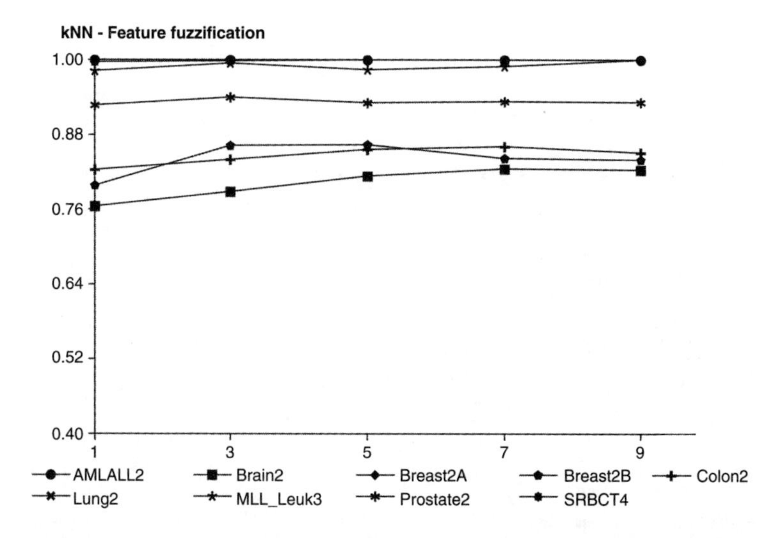

FIGURE 17.7 KNN accuracy with feature fuzzification as a function of K.

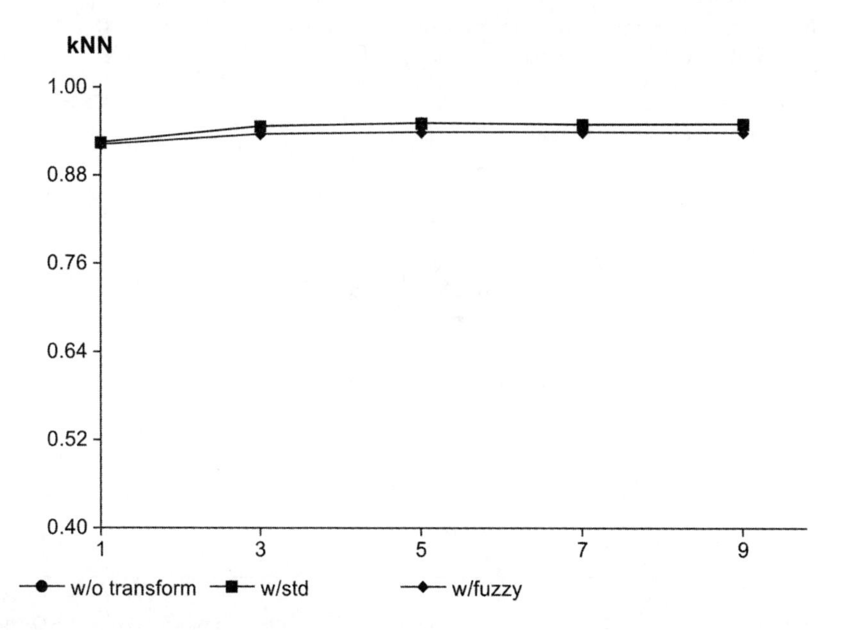

FIGURE 17.8 KNN average accuracy for all datasets as a function of K. Ten 10-fold cross-validation used.

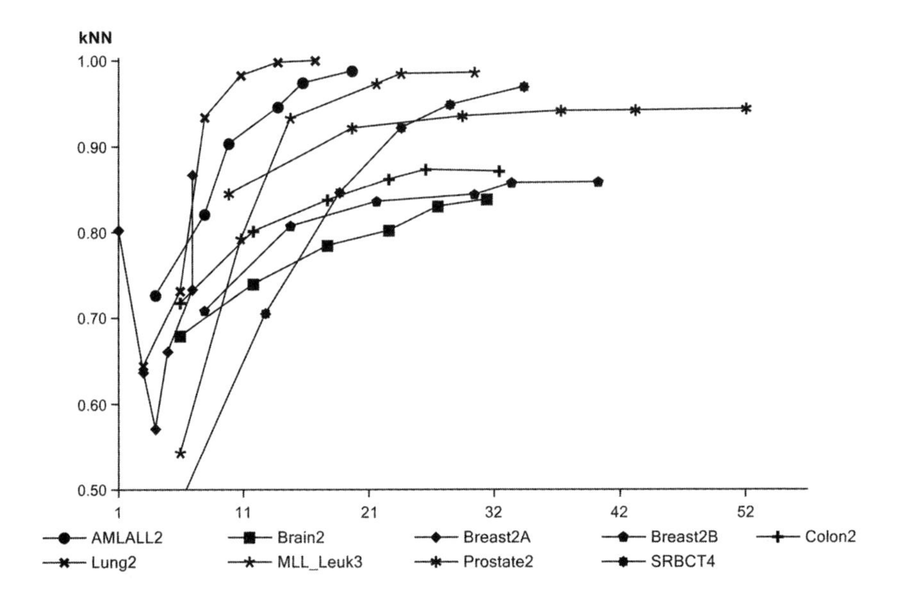

FIGURE 17.9 5NN accuracy without transformations on features as a function of bootstrap sample size used during training; x-axis values represent the number of arrays randomly sampled $B = 40$ times at fractions $f = 0.1, 0.2, 0.3, 0.4, 0.5$, or 0.6 of the available arrays.

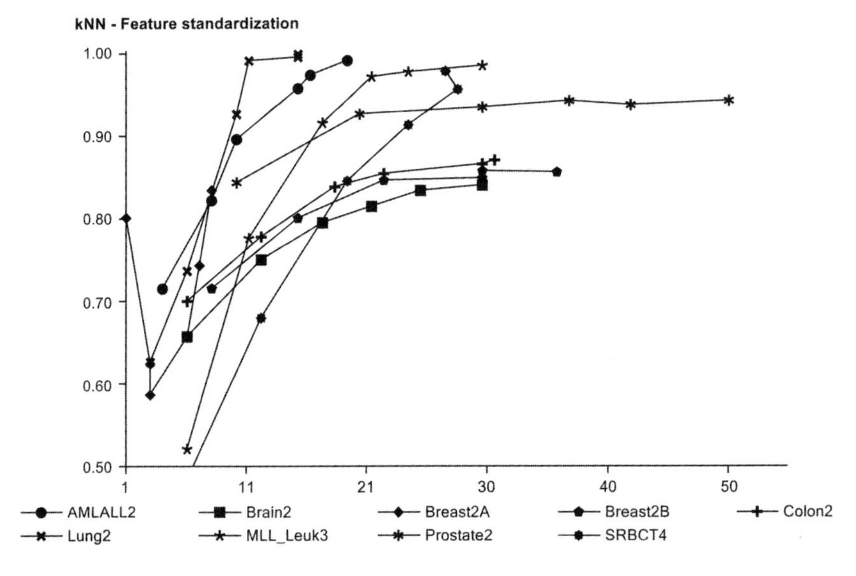

FIGURE 17.10 5NN accuracy with feature standardization as a function of bootstrap sample size used during training; x-axis values represent the number of arrays randomly sampled $B = 40$ times at fractions $f = 0.1, 0.2, 0.3, 0.4, 0.5$, or 0.6 of the available arrays.

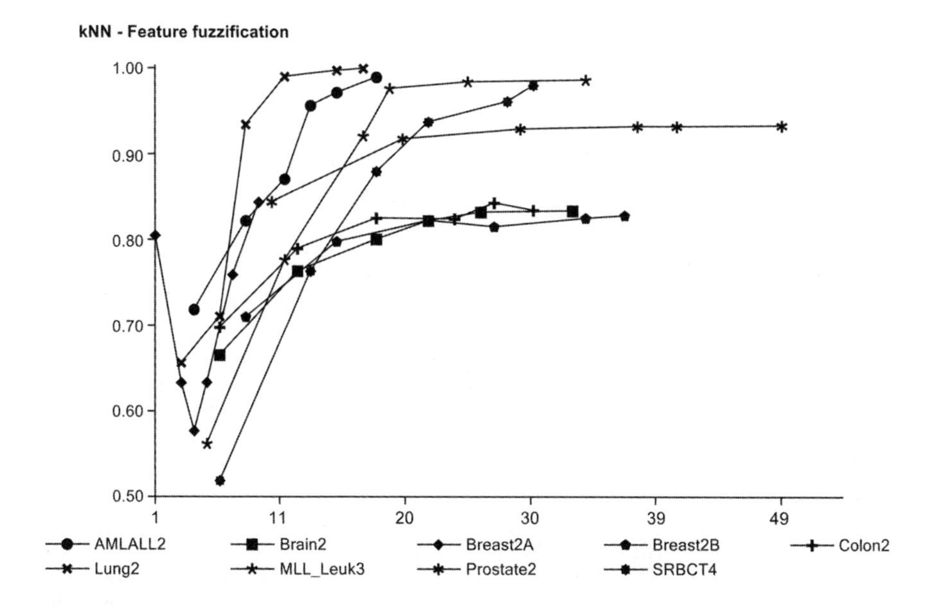

FIGURE 17.11 5NN accuracy with feature fuzzification as a function of bootstrap sample size used during training; x-axis values represent the number of arrays randomly sampled $B = 40$ times at fractions $f = 0.1, 0.2, 0.3, 0.4, 0.5$, or 0.6 of the available arrays.

with feature standardization as a function of bootstrap sample size used during training. Figure 17.11 shows the 5NN accuracy with feature fuzzification as a function of bootstrap sample size used during training. The results indicate that below 20 arrays, the performance dropped substantially, revealing a high degree of bootstrap bias of KNN. In addition, features transformations such as mean-zero standardizing and fuzzification did not appreciably change the bootstrap bias that was observed.

17.6 MULTICLASS ROC CURVES

Receiver-operator characteristic curves were also generated for the nine datasets using bootstrapping $B = 40$ times at fractions $f = 0.1, 0.2, 0.3, 0.4, 0.5$, or 0.6 of the available arrays. Figure 17.12 illustrates ROC curves for the 5NN applied to the AMLALL2, Brain2, Breast2A, Breast2B, Colon2, Lung2, MLL_Leukemia3, Prostate2, and SRBCT4 datasets using mean-zero standardized input features. Figure 17.13 illustrates ROC curves for the 5NN based on the AMLALL2, Brain2, Breast2A, Breast2B, Colon2, Lung2, MLL_Leukemia3, Prostate2, and SRBCT4 datasets using fuzzified input features. Results shown in the ROC curves suggest that more than 30% of the available arrays in each dataset are required in order to attain AUC levels on the order of 80-90% or greater.

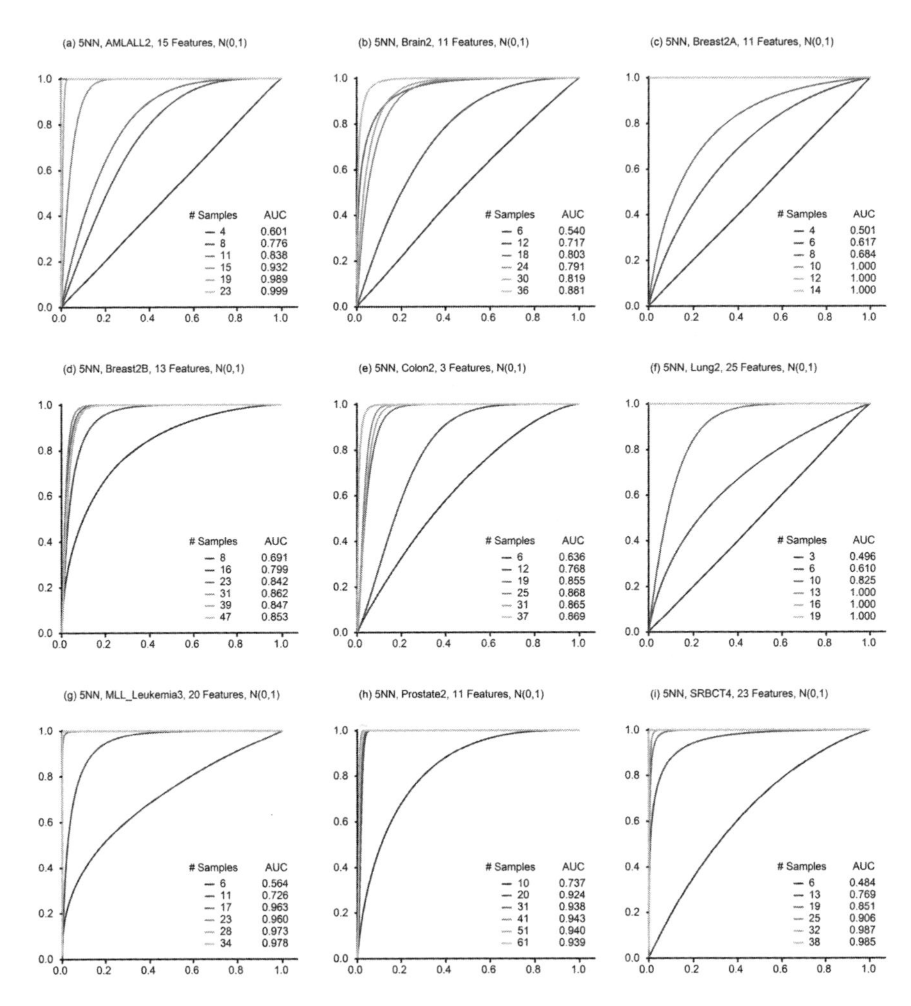

FIGURE 17.12 ROC curves for the 5NN classifier applied to the AMLALL2, Brain2, Breast2A, Breast2B, Colon2, Lung2, MLL_Leukemia3, Prostate2, and SRBCT4 datasets based on mean-zero standardized input features: (a) AMLALL2; (b) Brain2; (c) Breast2A; (d) Breast2B; (e) Colon2; (f) Lung2; (g) MLL_Leukemia3; (h) Prostate2; (i) SRBCT4. The numbers of genes used for each dataset are listed in Table 12.10. Legend values represent the number of arrays randomly sampled $B = 40$ times at fractions $f = 0.1, 0.2, 0.3, 0.4, 0.5$, or 0.6 of the available arrays.

17.7 DECISION BOUNDARIES

Figure 17.14 illustrates a 2D self-organizing map of class decision boundaries for the 9NN based on the AMLALL2, Brain2, Breast2A, Breast2B, Colon2, Lung2, MLL_Leukemia3, Prostate2, and SRBCT4 datasets using mean-zero

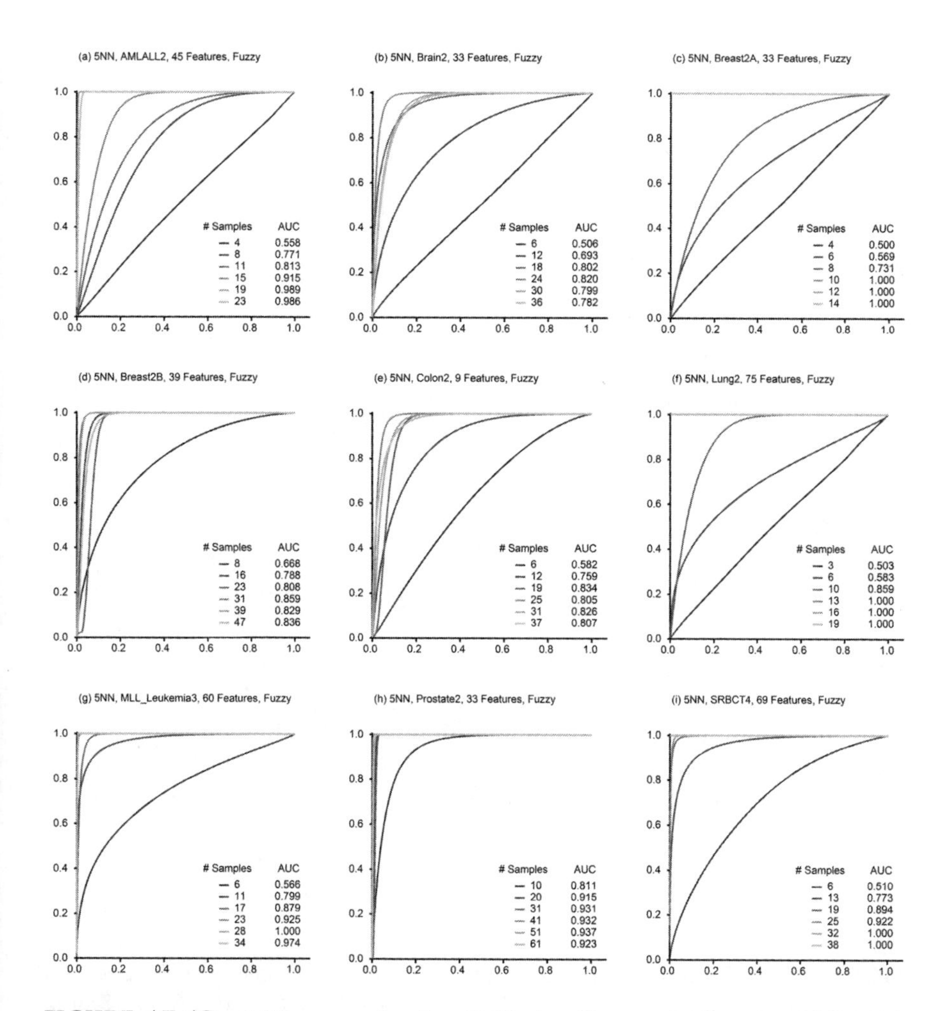

FIGURE 17.13 ROC curves for the 5NN classifier applied to the AMLALL2, Brain2, Breast2A, Breast2B, Colon2, Lung2, MLL_Leukemia3, Prostate2, and SRBCT4 datasets based on fuzzified input features: (a) AMLALL2; (b) Brain2; (c) Breast2A; (d) Breast2B; (e) Colon2; (f) Lung2; (g) MLL_Leukemia3; (h) Prostate2; (i) SRBCT4. The numbers of genes used for each dataset are listed in Table 12.10. Legend values represent the number of arrays randomly sampled $B = 40$ times at fractions $f = 0.1, 0.2, 0.3, 0.4, 0.5$, or 0.6 of the available arrays.

standardized input features. The legend entries in the plot represent varying values of K, followed by performance (accuracy) listed in parentheses. KNN class decision boundaries were more complex for the Brain2, Breast2B, Colon2, and Prostate2 datasets, whereas the classes in the remaining datasets were more easily separable. Figure 17.15 illustrates a 2D self-organizing map

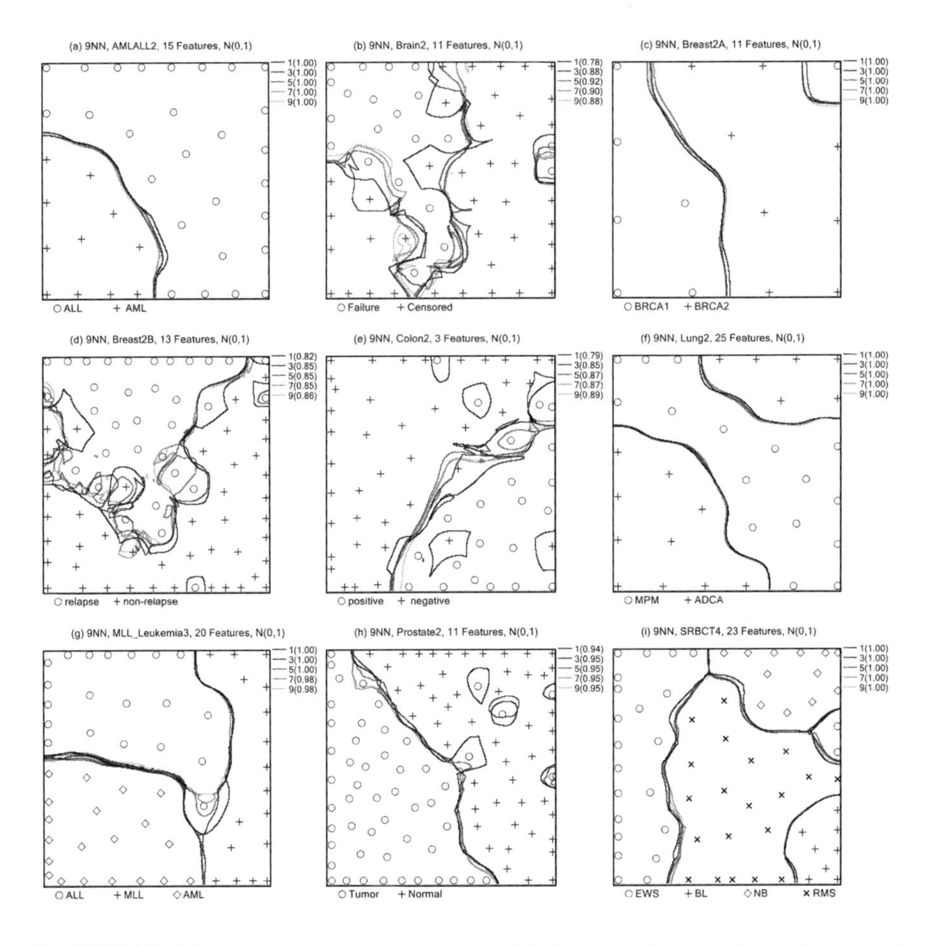

FIGURE 17.14 2D self-organizing maps of 9NN class decision boundaries and classification accuracy (upper right of each panel) for the AMLALL2, Brain2, Breast2A, Breast2B, Colon2, Lung2, MLL_Leukemia3, Prostate2, and SRBCT4 datasets based on mean-zero standardized input features: (a) AMLALL2; (b) Brain2; (c) Breast2A; (d) Breast2B; (e) Colon2; (f) Lung2; (g) MLL_Leukemia3; (h) Prostate2; (i) SRBCT4. The numbers of genes used for each dataset are listed in Table 12.10. Legend entries represent varying values of *K*, followed by performance (accuracy) listed between parentheses.

of class decision boundaries for the 9NN based on the AMLALL2, Brain2, Breast2A, Breast2B, Colon2, Lung2, MLL_Leukemia3, Prostate2, and SRBCT4 datasets using fuzzified input features. Fuzzification removed the complexity of the decision boundaries for the Brain2, Breast2B, Colon2, and Prostate2 datasets, but did not result in greater performance.

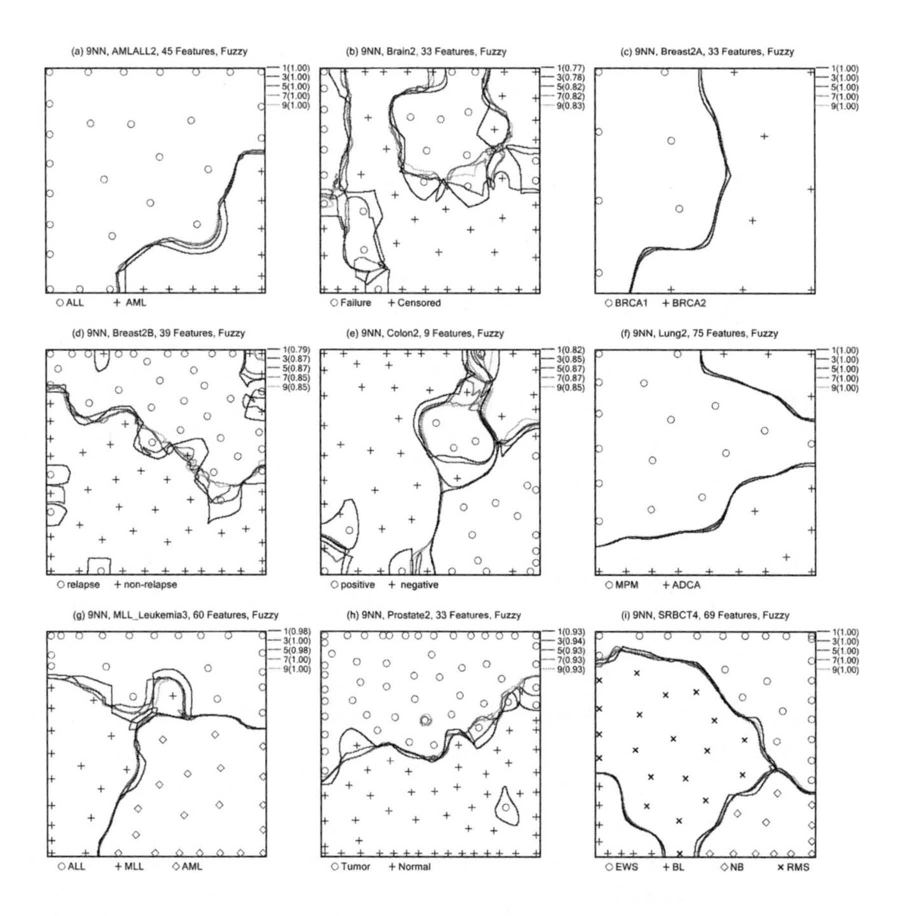

FIGURE 17.15 2D self-organizing maps of 9NN class decision boundaries and classification accuracy (upper right of each panel) for the AMLALL2, Brain2, Breast2A, Breast2B, Colon2, Lung2, MLL_Leukemia3, Prostate2, and SRBCT4 datasets based on fuzzified input features: (a) AMLALL2; (b) Brain2; (c) Breast2A; (d) Breast2B; (e) Colon2; (f) Lung2; (g) MLL_Leukemia3; (h) Prostate2; (i) SRBCT4. The numbers of genes used for each dataset are listed in Table 12.10. Legend entries represent varying values of K, followed by performance (accuracy) listed between parentheses.

17.8 SUMMARY

Because it retains all input feature data for making class predictions of test objects, KNN is a "lazy learner" algorithm. KNN was observed to be quite biased during bootstrap bias runs (e.g., see Figure 17.9) as well as during AUC determination in ROC curves when bootstrapping was also employed. The variance of KNN (Figure 17.9) was also more pronounced than, for example,

LREG, since there was a greater variation in performance over the various CV methods. One important advantage of KNN is that it can salvage a failed attempt when using another classifier whose performance becomes degraded if a class of objects is within another class. This ultimately presents itself as class contamination or impurity, which, for example, linear discriminant analysis fails to discern. Therefore, if other non-instance-based classifiers fail on a particular dataset, then reduce the dimensionality of the dataset to 2D or 3D to identify visually if clusters are present within clusters; if so, it may be possible to apply KNN to yield greater levels of performance. This is possible because of the Euclidean distance basis of KNN, which avoids many of the assumptions required of other classifiers. In summary, the KNN classification method should become a fundamental tool in the repertoire of classifiers used for workflow. We have observed KNN to often outperform linear discriminant analysis, which is one of the most commonly used linear methods.

REFERENCES

[1] E. Fix, J.L. Hodges. *Discriminatory Analysis, Non-parametric Discrimination: Consistency Properties*. Technical Report 4, USAF School of Aviation Medicine, Randolph Field, TX, 1951.

[2] T.M. Cover, P.E. Hart. Nearest neighbor pattern classification. *IEEE Trans. Inform. Theory* **IT-13**(1):21–27, 1967.

[3] M.E. Hellman. The nearest neighbor classification rule with a reject option. *IEEE Trans. Syst. Man Cybernet.* **3**:179–185, 1970.

[4] K. Fukunaga, L. Hostetler. k-Nearest-neighbor bayes-risk estimation. *IEEE Trans. Inform. Theory* **21**(3):285–293, 1975.

[5] S.A. Dudani. The distance-weighted k-nearest-neighbor rule. *IEEE Trans. Syst. Man Cybernet.* **SMC-6**:325–327, 1976.

[6] T. Bailey, A. Jain. A note on distance-weighted k-nearest neighbor rules. *IEEE Trans. Syst. Man Cybernet.* **8**:311–313, 1978.

[7] S. Bermejo, J. Cabestany. Adaptive soft k-nearest-neighbour classifiers. *Pattern Recogn.* **33**:1999–2005, 2000.

[8] A. Jozwik. A learning scheme for a fuzzy k-nn rule. *Pattern Recogn. Lett.* **1**:287–289, 1983.

[9] J.M. Keller, M.R. Gray, J.A. Givens. A fuzzy k-nn neighbor algorithm. *IEEE Trans. Syst. Man Cybern.* **SMC-15**(4):580–585, 1985.

[10] R.L. Morin, D.E. Raeside. A reappraisal of distance-weighted k-nearest neighbor classification for pattern recognition with missing data. *IEEE Trans. Syst. Man Cybern.* **SMC-11**(3):241–243, 1981.

[11] J.E. Macleod, A. Luk, D.M. Titterington. A re-examination of the distance weighted k-nearest neighbor classification rule. *IEEE Trans. Syst. Man Cybern.* **SMC-17**(4):689–696, 1987.

CHAPTER 18

NAÏVE BAYES CLASSIFIER

18.1 INTRODUCTION

Naïve Bayes classifiers (NBCs) were developed from probability-based rules derived from Bayes' rule, and therefore, are able to perform efficiently with minimum error rate [1]. NBCs have been widely used for protein function prediction [2], mining housekeeping genes [3], protein structure and fold prediction [4], multidrug resistance activity [5], nomograms for cancer recurrence [6], patient medical histories [7], athletic injuries [8], and MALDI-MS data [9]. NBC has also been applied for classification of microarrays in heart disease [10], cancer [11], and tissue microarrays [12].

The version of NBC presented in this chapter is based entirely on discretizing gene-specific expression values across arrays into categorical codes for quantiles. Training for NBC first requires calculation of the three cutpoints of quartiles (Q_{1j}, Q_{2j}, Q_{3j}) for the jth gene ($j = 1, 2, \ldots, p$) over all training arrays independent of class. Assuming the number of genes to be p, a matrix of size $p \times 3$ is used to store the three quartile cutpoints for all genes. Using the cutpoints for quantiles of each gene, we next transform continuous expression values into categorical quantile codes, and tabulate cell counts $n(j, q_j, \omega)$, which is the number of arrays having a quantile value of q_j ($q_j = 1, 2, 3, 4$) for the jth feature in class ω. At program start, each cell count is padded with $n(j, q_j, \omega) = 1$ in order to prevent multiplication by zero when probabilities are determined during testing. This is performed for all training arrays.

During testing, the array of quantile cutpoints is used to transform each test array's continuous feature values into categorical quantile values of $q_j = 1, 2, 3$, or 4. The assignment of a test array $\mathbf{x}$ to a specific class is based on the posterior probability of class ω, given as

$$P(\omega|\mathbf{x}) = P(\omega)P(\mathbf{x}|\omega), \qquad (18.1)$$

Classification Analysis of DNA Microarrays, First Edition. Leif E. Peterson.

where $P(\omega)$ is the class prior and $P(\mathbf{x}|\omega)$ is the conditional probability density. Note that $P(\omega)$ is constant and therefore only $P(\mathbf{x}|\omega)$ requires maximization. Thus, we compute

$$P(\mathbf{x}|\omega) = \prod_{j}^{p} \frac{n(j, q_j, \omega)}{n(\omega)}, \tag{18.2}$$

where $n(\omega)$ is the number of arrays in class ω used for training and p is the number of genes used for training. Clearly we are using the categorical quantile values of q_j for each feature of the test array to obtain the probability $n(j, q_j, \omega)/n(\omega)$, which is multiplied together for all features. The decision rule for the test array $\mathbf{x}$ is expressed as

$$D(\mathbf{x} \rightsquigarrow \omega) \equiv \arg\max_{\omega \in \Omega}\{P(\omega|\mathbf{x})\} = \arg\max_{\omega \in \Omega}\{P(\omega)P(\mathbf{x}|\omega)\}. \tag{18.3}$$

18.2 ALGORITHM

Algorithm 18.1 lists the steps used for NBC classification. The first step is to permute the order of all input arrays and partition them into m folds ($\mathcal{D}_1, \mathcal{D}_2, \ldots, \mathcal{D}_m$). Arrays not in a given fold are used for training, while arrays in a given fold are used for testing. All arrays, training or testing, have a true class ω, and when tested, their predicted class $\hat{\omega}$ is determined in order to increment elements of the confusion matrix $\mathbf{C}$ for accuracy estimation. Because the class priors are already known, computation of NBC involves only the determination of the posterior probability $P(\omega|\mathbf{x})$.

18.3 CROSS-VALIDATION RESULTS

Naïve Bayes classification using cross-validation (CV) was run for the nine datasets and genes selected using greedy PTA (see Table 12.10). Bootstrap accuracy ("CVB") was based on average 0.632 bootstrap accuracy (13.4) for 10 bootstraps. Table 18.1 lists the average accuracy based on 10 repartitions of the input arrays and subsequent CV, without any input feature transformations applied. Table 18.2 lists the average accuracy based on 10 repartitions of the input arrays and subsequent CV, using mean-zero standardized input feature values. Table 18.3 lists the average accuracy based on 10 repartitions of the input arrays and subsequent CV, using fuzzified input feature values.

Figure 18.1 illustrates accuracy as a function of the CV method for each dataset, based on 10 repartitions of arrays without feature transformations applied. Figure 18.2 shows the standard deviation of accuracy as a function of CV method for each dataset, where the standard deviation is based on CV accuracy for 10 repartitions of arrays and no feature

Algorithm 18.1: NBC

```
Data: n arrays, p genes, training arrays x_i, test arrays x
Result: 10-fold cross-validation (CV10) accuracy
m ← 10, Acc_total = 0
foreach Repartition = 1 ← 10 do
    Set C ← 0, n_test ← 0
    Randomly partition input arrays into D_1, D_2, ..., D_m folds
    for each cross-validation fold D_m do
        Select all arrays in D_m for testing
        Use all remaining arrays not in D_m for training
        Set n(j, q_j^0, ω) = 1   ∀j, ω, q_j^0
        for j = 1 to p do
            Determine Q_1j, Q_2j, Q_3j   ∀x_i ∉ D_m
        endfor
        for j = 1 to p do
            for each training array x_i ∉ D_m do
                q_j^0 = { 1  x_ij ≤ Q_1j
                        { 2  Q_1j < x_ij ≤ Q_2j
                        { 3  Q_2j < x_ij ≤ Q_3j
                        { 4  x_ij > Q_3j
                Increment n(j, q_j^0, ω)+ = 1        (x_i ∈ ω)
            endfor
        endfor
        for each test array x ∈ D_m do
            n_test+ = 1
            for j = 1 to p do
                q_j = { 1  x_j ≤ Q_1j
                      { 2  Q_1j < x_j ≤ Q_2j
                      { 3  Q_2j < x_j ≤ Q_3j
                      { 4  x_j > Q_3j
                for ω = 1 to Ω do
                    P(j|ω) = n(j, q_j, ω)/n(ω)
                endfor
            endfor
            for ω = 1 to Ω do
                P(x|ω) = ∏_{j=1}^{p} P(j|ω)
            endfor
            ω̂ = arg max_{ω∈Ω} {P(x|ω)}
            Increment confusion matrix C by 1 in element c_{ω,ω̂}
        endfor
    endfor
    Acc_total+ = (Σ_l c_ll)/n_test
endfch
Acc = Acc_total / 10
```

TABLE 18.1 Performance without Feature Transformation

Dataset	Cross-validation method				
	CV2	CV5	CV10	CV^{-1}	CVB
AMLALL2	83.68	93.95	94.74	94.74	93.65
Brain2	74.17	82.67	83.83	85.00	86.70
Breast2A	90.00	100	100	100	96.78
Breast2B	81.15	82.82	82.44	82.05	88.78
Colon2	81.77	84.84	85.48	85.48	87.75
Lung2	100	100	100	100	99.97
MLL_Leukemia3	98.07	98.25	99.12	100	99.23
Prostate2	92.65	92.94	93.04	93.14	93.20
SRBCT4	69.37	97.14	98.41	98.41	87.17
Average	0.86	0.93	0.93	0.93	0.93

TABLE 18.2 Performance with Feature Standardization

Dataset	Cross-validation method				
	CV2	CV5	CV10	CV−1	CVB
AMLALL2	84.47	93.68	95.26	94.74	92.31
Brain2	75.33	79.83	82.83	85.00	87.93
Breast2A	92.67	100	100	100	97.32
Breast2B	81.15	80.90	82.69	82.05	88.81
Colon2	82.42	84.35	85.65	85.48	88.68
Lung2	99.69	100	100	100	100
MLL_Leukemia3	97.37	98.25	99.47	100	99.09
Prostate2	92.75	92.94	92.65	93.14	93.27
SRBCT4	72.38	97.14	98.25	98.41	87.26
Average	0.86	0.92	0.93	0.93	0.93

TABLE 18.3 Performance with Feature Fuzzification

Dataset	Cross-validation method				
	CV2	CV5	CV10	CV−1	CVB
AMLALL2	78.16	91.05	94.21	94.74	90.56
Brain2	63.33	73.67	77.67	75.00	82.96
Breast2A	96.67	100	100	100	97.52
Breast2B	81.54	81.03	80.64	80.77	87.98
Colon2	82.10	83.87	82.74	82.26	86.39
Lung2	100	100	100	100	99.94
MLL_Leukemia3	96.14	97.37	97.37	98.25	98.82
Prostate2	92.55	92.84	92.75	93.14	93.06
SRBCT4	59.37	86.03	92.86	96.83	80.95
Average	0.83	0.90	0.91	0.91	0.91

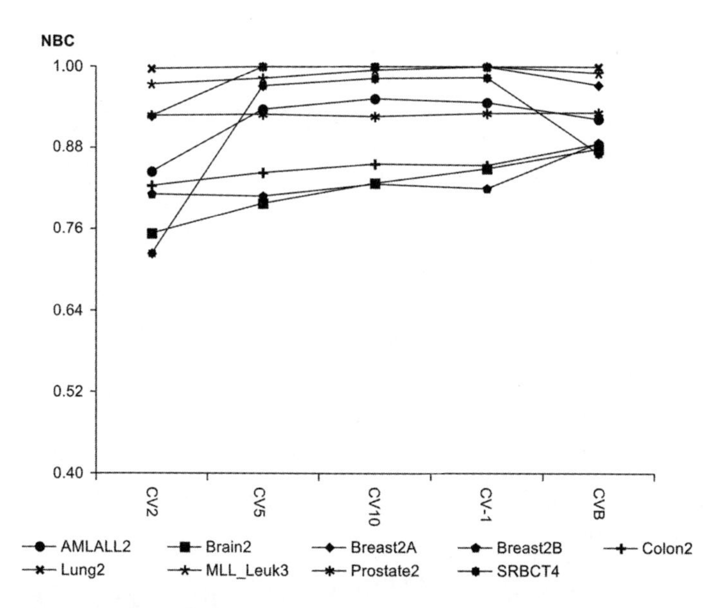

FIGURE 18.1 NBC accuracy as a function of cross-validation (CV) method for each dataset. Average based on CV accuracy for 10 repartitions of arrays. No feature transformations applied.

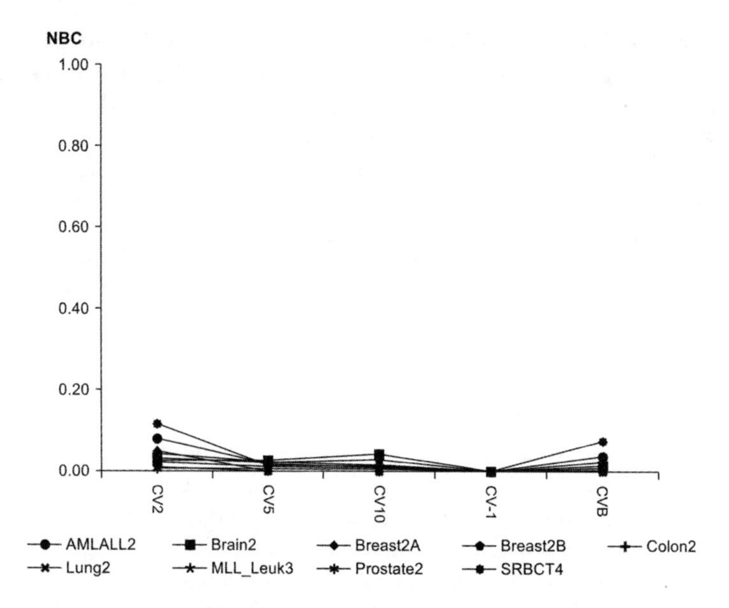

FIGURE 18.2 NBC standard deviation of accuracy as a function of CV method for each dataset. Standard deviation based on CV accuracy for 10 repartitions of arrays. No feature transformations applied.

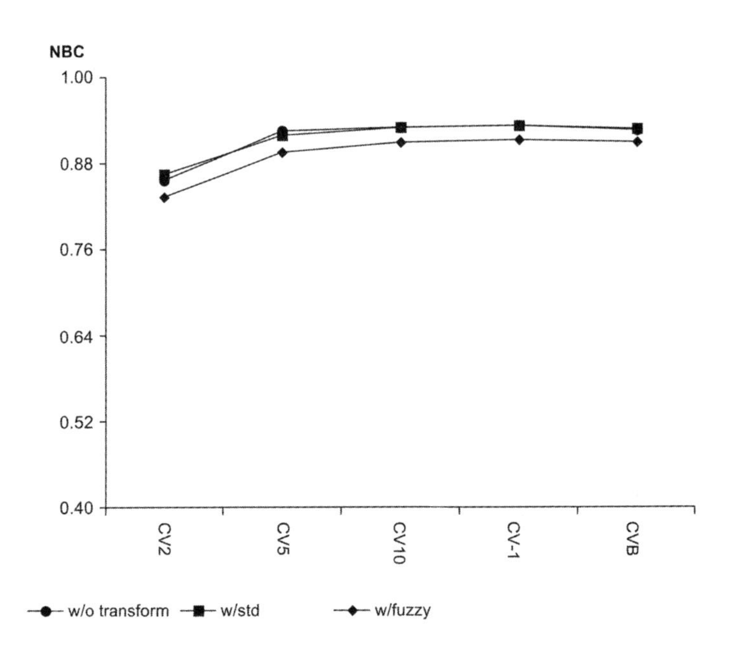

FIGURE 18.3 NBC average accuracy for all datasets as a function of feature transformation and CV method.

transformations. Figure 18.3 reveals average accuracy for all datasets as a function of feature transformation and CV method. Results of the CV runs made with the nine datasets indicate that NBC is biased and has a moderate level of variance. The bias can be observed in the change in accuracy as the number of fold change, since more data are used when the number of folds increases. With regard to the variance observed, accuracy levels varied over the datasets, and were not similar. NBC resulted in bias and variance that was similar to KNN, but dissimilar to LREG, which had less bias.

18.4 BOOTSTRAP BIAS

Figure 18.4 shows the NBC accuracy without transformations on features as a function of bootstrap sample size used during training, based on 10 resamplings from the input arrays. Bootstrap bias runs were based on determining the 0.632 bootstrap accuracy (see Eq. 13.4) for each sampling fraction. Acc_0 in 13.4 was based on test accuracy of unsampled arrays, while Acc_b in 13.4 was based on test accuracy for sampled arrays. The x-axis values represent the number of arrays randomly sampled $B = 40$ times at fractions $f = 0.1, 0.2, 0.3, 0.4, 0.5,$ or 0.6 of the available arrays. Figure 18.5 illustrates the NBC accuracy with feature standardization as a function of bootstrap sample size used during training, based on 10 resamplings from the input arrays. Figure 18.6 shows the NBC accuracy with feature fuzzification as a function

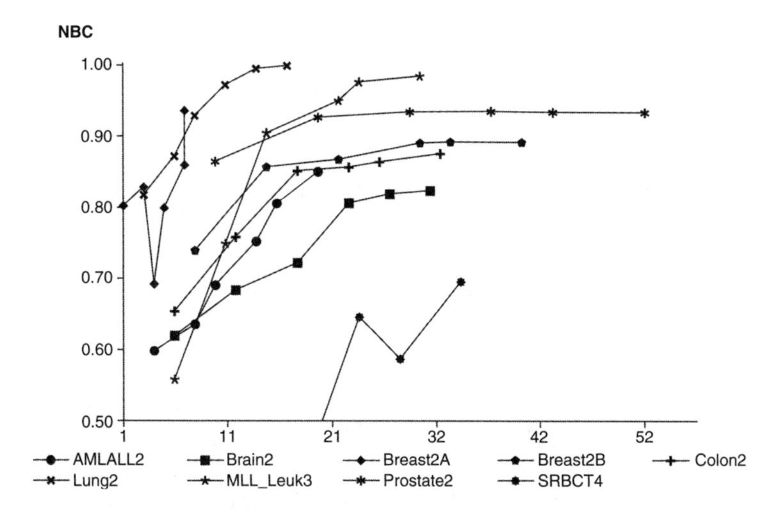

FIGURE 18.4 NBC accuracy without transformations on features as a function of bootstrap sample size used during training. The x-axis values represent the number of arrays randomly sampled $B = 40$ times at fractions $f = 0.1$, 0.2, 0.3, 0.4, 0.5, or 0.6 of the available arrays.

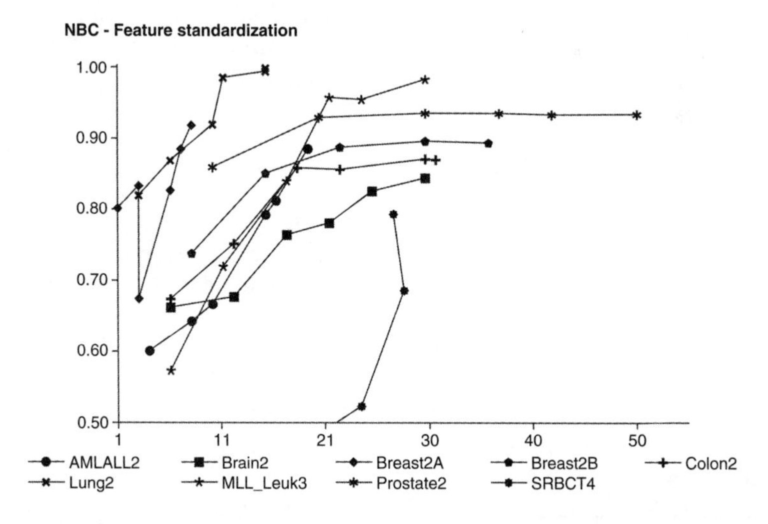

FIGURE 18.5 NBC accuracy with feature standardization as a function of bootstrap sample size used during training. The x-axis values represent the number of arrays randomly sampled $B = 40$ times at fractions $f = 0.1$, 0.2, 0.3, 0.4, 0.5, or 0.6 of the available arrays.

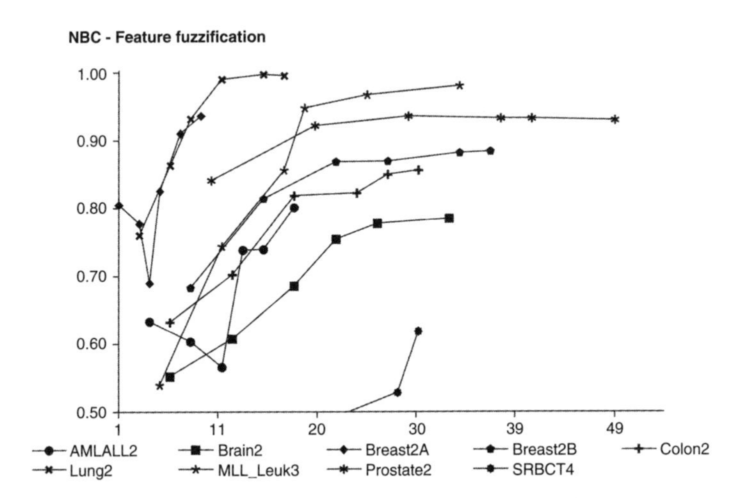

FIGURE 18.6 NBC accuracy with feature fuzzification as a function of bootstrap sample size used during training. The x-axis values represent the number of arrays randomly sampled $B = 40$ times at fractions $f = 0.1, 0.2, 0.3, 0.4, 0.5$, or 0.6 of the available arrays.

of bootstrap sample size used during training, based on 10 resamplings from the input arrays. Bootstrap results also suggest that NBC was biased because of the sharp rise in accuracy as the sample size of the randomly selected arrays increased. Mean-zero standardization and fuzzification did not seem to result in a noticeable change in bootstrap results.

18.5 MULTICLASS ROC CURVES

The higher degree of bias in NBC can also be observed in Figures 18.7 and 18.8. The AUC for NBC does not reach levels greater than 90% unless more than 50-60% of the available arrays are used for training. Input feature fuzzification also did not appear to increase the performance of NBC. Figure 18.7 illustrates ROC curves for the NBC applied to the AMLALL2, Brain2, Breast2A, Breast2B, Colon2, Lung2, MLL_Leukemia3, Prostate2, and SRBCT4 datasets using mean-zero standardized input features. The legend values represent the number of arrays randomly sampled $B = 40$ times at fractions $f = 0.1, 0.2, 0.3, 0.4, 0.5$, or 0.6 of the available arrays. Figure 18.8 illustrates ROC curves for the NBC based on fuzzified input features.

18.6 DECISION BOUNDARIES

Figure 18.9 illustrates a 2D self-organizing map of class decision boundaries for the NBC based on the AMLALL2, Brain2, Breast2A, Breast2B, Colon2,

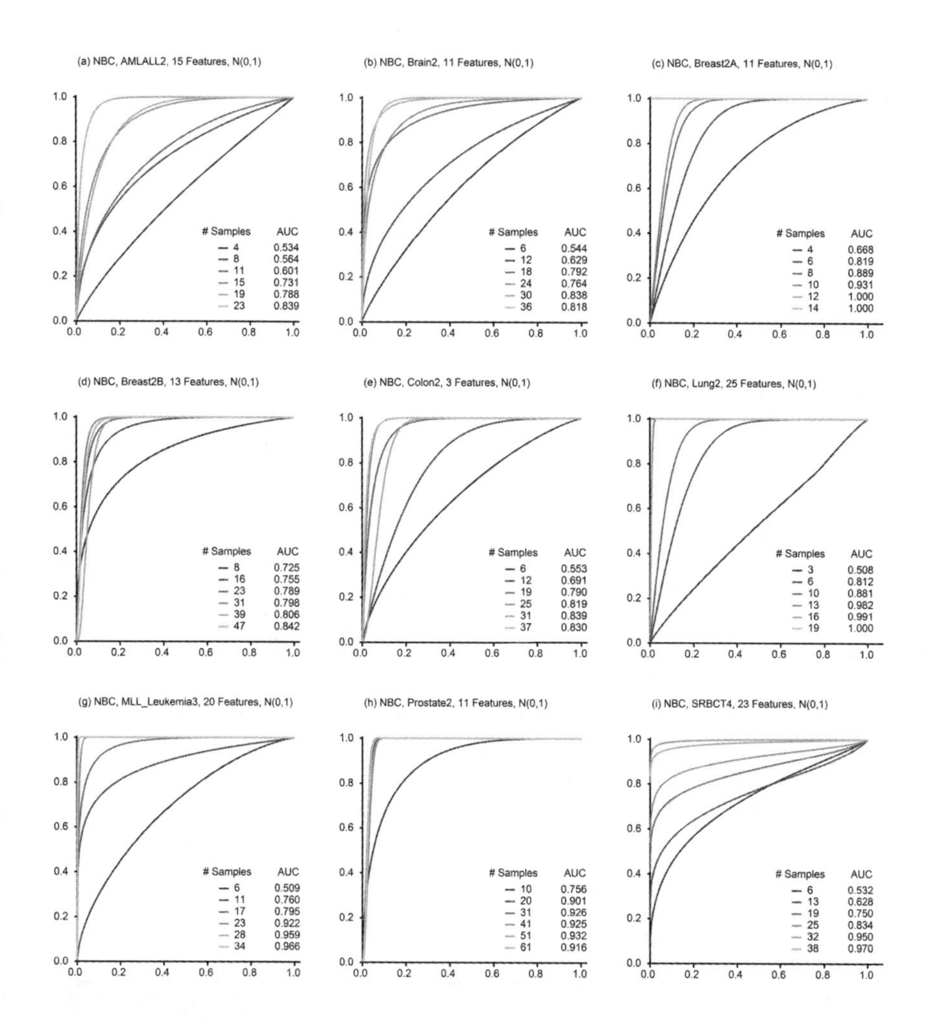

FIGURE 18.7 ROC curves for the NBC classifier applied to the AMLALL2, Brain2, Breast2A, Breast2B, Colon2, Lung2, MLL_Leukemia3, Prostate2, and SRBCT4 datasets based on mean-zero standardized input features; legend values represent the number of arrays randomly sampled $B = 40$ times at fractions $f = 0.1$, 0.2, 0.3, 0.4, 0.5, or 0.6 of the available arrays: (a) AMLALL2; (b) Brain2; (c) Breast2A; (d) Breast2B; (e) Colon2; (f) Lung2; (g) MLL_Leukemia3; (h) Prostate2; (i) SRBCT4. The numbers of genes used for each dataset are listed in Table 12.10.

Lung2, MLL_Leukemia3, Prostate2, and SRBCT4 datasets using mean-zero standardized input features. Figure 18.10 illustrates a 2D self-organizing map of class decision boundaries for NBC based on the AMLALL2, Brain2, Breast2A, Breast2B, Colon2, Lung2, MLL_Leukemia3, Prostate2, and SRBCT4 datasets using fuzzified input features. The Brain2, Breast2B, and

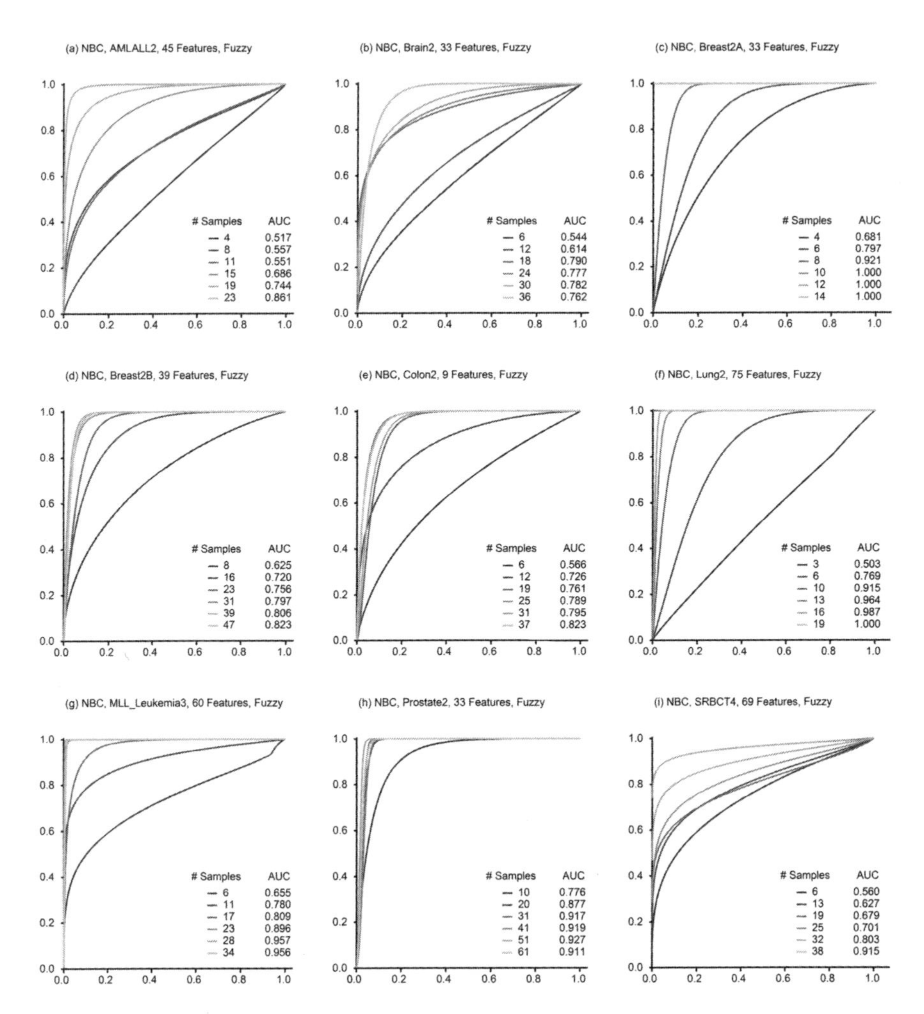

FIGURE 18.8 ROC curves for the NBC classifier applied to the AMLALL2, Brain2, Breast2A, Breast2B, Colon2, Lung2, MLL_Leukemia3, Prostate2, and SRBCT4 datasets based on fuzzified input features; legend values represent the number of arrays randomly sampled $B = 40$ times at fractions $f = 0.1$, 0.2, 0.3, 0.4, 0.5, or 0.6 of the available arrays: (a) AMLALL2; (b) Brain2; (c) Breast2A; (d) Breast2B; (e) Colon2; (f) Lung2; (g) MLL_Leukemia3; (h) Prostate2; (i) SRBCT4. The numbers of genes used for each dataset are listed in Table 12.10.

MLL_Leukemia3 datasets had decision boundaries that were more complex than those of the other datasets, and when input feature fuzzification was used, the decision boundaries for the Brain2 datasets became more complex with reduced accuracy. In fact, fuzzification did not result in greater classification performance when compared with mean-zero standardization.

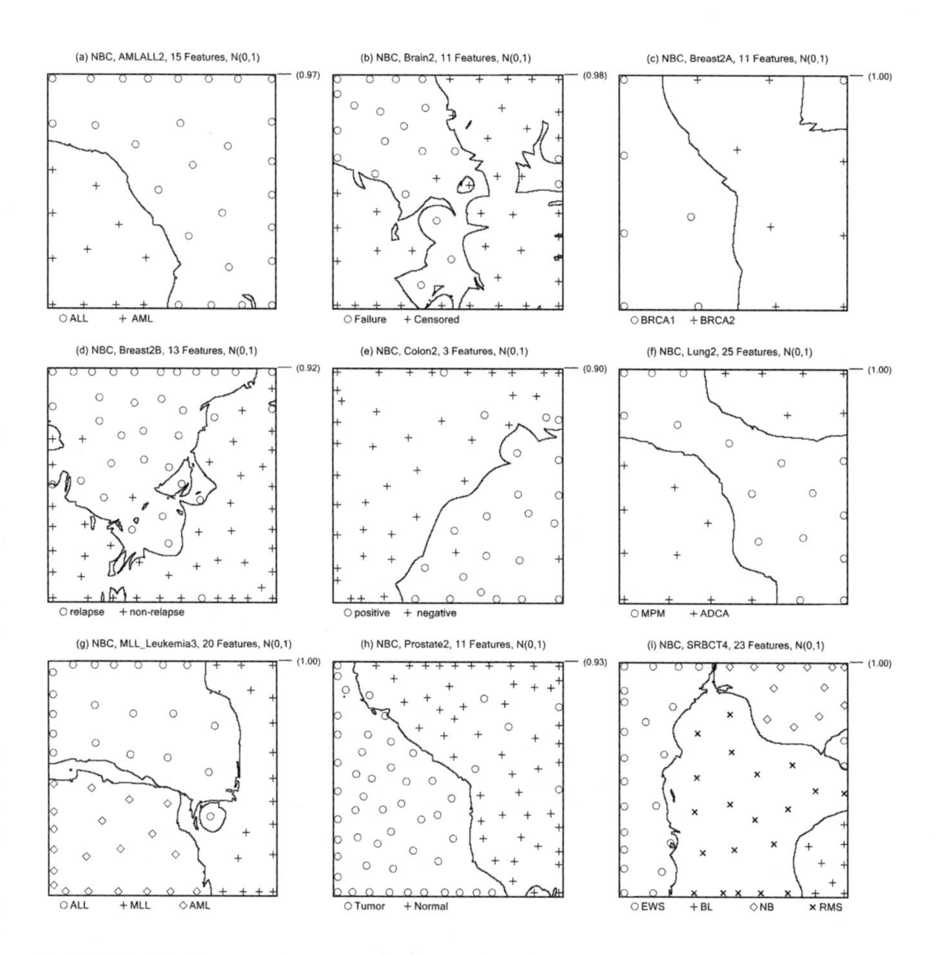

FIGURE 18.9 2D self-organizing maps of NBC class decision boundaries and classification accuracy (upper right of each panel) for the AMLALL2, Brain2, Breast2A, Breast2B, Colon2, Lung2, MLL_Leukemia3, Prostate2, and SRBCT4 datasets based on mean-zero standardized input features: (a) AMLALL2; (b) Brain2; (c) Breast2A; (d) Breast2B; (e) Colon2; (f) Lung2; (g) MLL_Leukemia3; (h) Prostate2; (i) SRBCT4. The numbers of genes used for each dataset are listed in Table 12.10.

18.7 SUMMARY

The NBC is a fast, simple classifier based on prior, conditional, and posterior probabilities and does not require optimization of an objective or cost function. It is also based on a generative model that can be stored in memory and used for class prediction of future unknown test cases. The major disadvantage of NBC is that it assumes independence between input features;

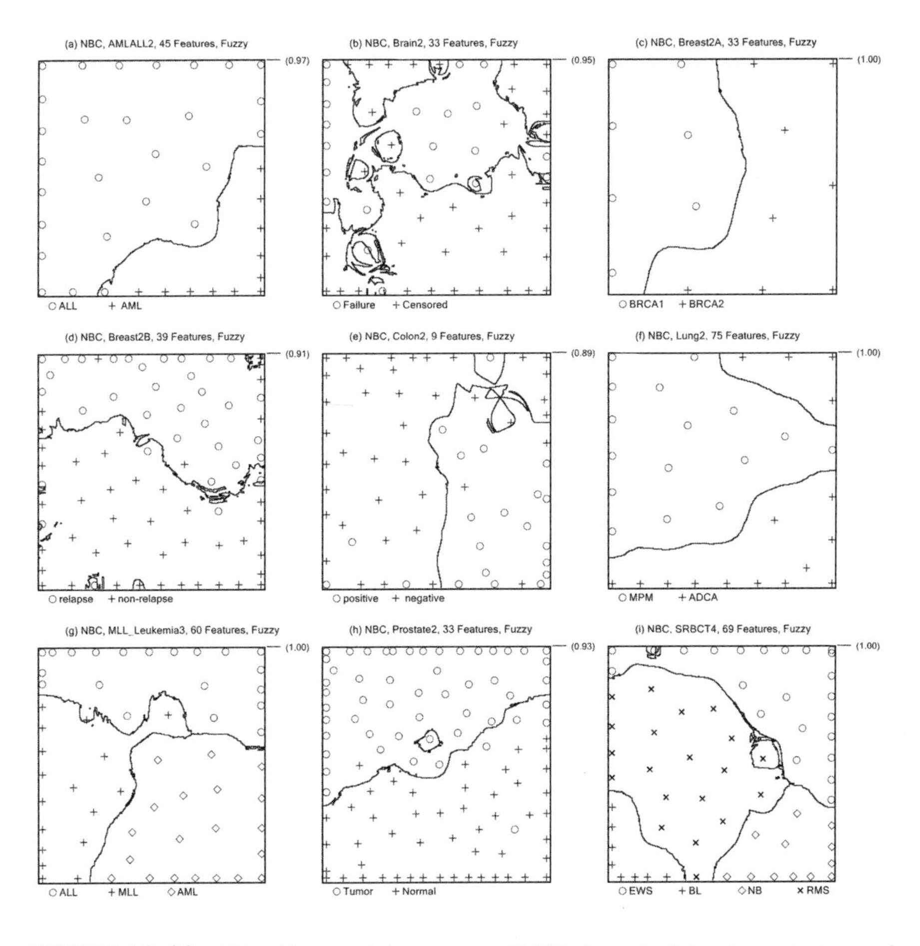

FIGURE 18.10 2D self-organizing maps of NBC class decision boundaries and classification accuracy (upper right of each panel) for the AMLALL2, Brain2, Breast2A, Breast2B, Colon2, Lung2, MLL_Leukemia3, Prostate2, and SRBCT4 datasets based on fuzzified input features; (a) AMLALL2; (b) Brain2; (c) Breast2A; (d) Breast2B; (e) Colon2; (f) Lung2; (g) MLL_Leukemia3; (h) Prostate2; (i) SRBCT4. The numbers of genes used for each dataset are listed in Table 12.10.

that is, it is "naïve," which translates into an assumption of zero correlation and no multicollinearity. NBC is incapable of evaluating simultaneous joint effects to two or more features on class prediction error rates, and therefore should not be employed for complex challenging problems with complex decision boundaries.

REFERENCES

[1] L.I. Kuncheva, Z.S. Hoare. Error-dependency relationships for the naive Bayes classifier with binary features. *IEEE Trans. Pattern Anal. Machine Intell.* **30**(4):735-740, 2008.

[2] J. Kohonen, S. Talikota, J. Corander, P. Auvinen, E. Arjas. A naive Bayes classifier for protein function prediction. *In Silico Biol.* **9**(1-2):23-34, 2009.

[3] L. De Ferrari, S. Aitken. Mining housekeeping genes with a naive Bayes classifier. *BMC Genomics* **7**:277, 2006.

[4] A. Chinnasamy, W.K. Sung, A. Mittal. Protein structure and fold prediction using tree-augmented naive Bayesian classifier. *J. Bioinform. Comput. Biol.* **3**(4):803-819, 2005.

[5] H. Sun. A naive Bayes classifier for prediction of multidrug resistance reversal activity on the basis of atom typing. *J. Med. Chem.* **48**(12):4031-4039, 2005.

[6] J. Demsar, B. Zupan, M.W. Kattan, J.R. Beck, I. Bratko. Naive Bayesian-based nomogram for prediction of prostate cancer recurrence. *Stud. Health Technol. Inform.* **68**:436-441, 1999.

[7] S. Monti, G.F. Cooper. The impact of modeling the dependencies among patient findings on classification accuracy and calibration, *Proc AMIA Symp.*, Orlando, Florida, 1998, pp. 592-596.

[8] I. Zelic, I. Kononenko, N. Lavrac, V. Vuga. Induction of decision trees and Bayesian classification applied to diagnosis of sport injuries. *J. Med. Syst.* **21**(6):429-444, 1997.

[9] Q. Liu, A.H. Sung, M. Qiao, Z. Chen, J.Y. Yang, M.Q. Yang, X. Huang, Y. Deng. Comparison of feature selection and classification for MALDI-MS data. *BMC Genomics* **10**(Suppl. 1):S3, 2009.

[10] C.G. Danko, A.M. Pertsov. Identification of gene co-regulatory modules and associated cis-elements involved in degenerative heart disease. *BMC Med. Genomics* **2**:31, 2009.

[11] L.E. Peterson, M.A. Coleman. Machine learning-based receiver operating characteristic (ROC) curves for crisp and fuzzy classification of DNA microarrays in cancer research. *Int. J. Approx. Reason.* **47**(1):17-36, 2008.

[12] F. Demichelis, P. Magni, P. Piergiorgi, M.A. Rubin, R. Bellazzi. A hierarchical naive Bayes model for handling sample heterogeneity in classification problems: An application to tissue microarrays. *BMC Bioinform.* **7**:514, 2006.

CHAPTER 19

LINEAR DISCRIMINANT ANALYSIS

19.1 INTRODUCTION

Linear discriminant analysis (LDA) was introduced by Fisher in 1936 [1]. Since then, LDA has been used for discriminant analysis of pulmonary flow volume [2], classification of wines [3], studies of the brain-computer-interface [4], classification of root canal microorganisms [5], classification of jet fuels [6], and classification of criminals using anthropometric measurements [7]. LDA has also been employed for DNA microarray classification of lung cancer [8], classification of small B-cell non-Hodgkin's lymphoma [9], expression value normalization [10], classification of colon biopsies [11], reclassification of subtypes of breast cancer [12], carbohydrate degradation in bacteriology [13], sample size estimation [14], and acute leukemia subtypes [15]. LDA should be a primary tool used for classification of DNA microarray data, due to its longstanding history.

In NBC, we discretized spectral values so that the posterior probabilities were based on the proportion of objects having a categorical (quantile) feature value in a given class. Linear discriminant analysis employs NBC but is hinged to use of the multivariate normal posterior, based on the quadratic form of χ^2 and each class-specific feature covariance matrix. The posterior probability that microarray $\mathbf{x}$ is assigned to class ω is the same as (18.1); however, the conditional probability density is

$$P(\mathbf{x}|\omega) = \frac{P(\omega)}{(2\pi)^{n/2}\,|\Sigma_\omega|^{1/2}} \exp\left[-\frac{1}{2}(\mathbf{x}_j - \bar{\mathbf{x}}_\omega)^T \Sigma_\omega^{-1} (\mathbf{x}_j - \bar{\mathbf{x}}_\omega)\right]. \qquad (19.1)$$

A major assumption in this approach is that the feature values across objects within each class are independent and, therefore, not correlated. Strongly

Classification Analysis of DNA Microarrays, First Edition. Leif E. Peterson.

correlated features within a class result in a singular covariance matrix for which the determinant is degenerate. This is due to essentially one feature describing all the variation, and thus the variance of that feature becomes the single principal component. In terms of the spectral decomposition of a singular covariance matrix, all except one of the eigenvalues is usually zero, and since the determinant is the product of all the eigenvalues, the determinant for a singular matrix is undefined. Furthermore, the quadratic form is assumed to be a spherical parallelepiped; however, when Σ_ω is singular, the projection is usually a flat surface or line, so there is no volume determined by the exp( ) term. In this case, one can determine the *content*; however, it would be much easier to simply avoid correlated features altogether.

There are other reasons for not selecting correlated features, such as, for example, if the feature values are equally large in only one class. The first time a classifier is introduced to a microarray with low values for the same features and same class there will be a significant drop in performance. Therefore, good classifiers build on features that are not correlated, in order to prevent large increases in misclassification error. Use of features with the greatest information gain is a good way to avoid correlated features, when compared with F tests, which enrich on invariant within-class feature variance (low group-specific standard deviations). In light of these shortcomings, we have chosen to use information gain as the basis for feature selection in all of our classifiers.

19.2 MULTIVARIATE MATRIX DEFINITIONS

Let $M_{n,p}(\mathbb{R})$ be the set of all data variables over the field of real numbers, with $M_p(\mathbb{R})$ as the set of symmetric p-square matrices over the field of real numbers. Define the *data* matrix $\mathbf{Y} \in M_{n,p}(\mathbb{R})$ with n rows and p columns as a function on the pairs of integers $(i,j), 1 \leq i \leq n, 1 \leq j \leq p$, with values in $\mathbf{Y}$ in which y_{ij} designates the value of $\mathbf{Y}$ at the pair (i,j):

$$\underset{n\times p}{\mathbf{Y}} = \begin{pmatrix} y_{11} & y_{12} & \cdots & y_{1p} \\ y_{21} & y_{22} & \cdot & y_{2p} \\ \vdots & \vdots & \ddots & \vdots \\ y_{n1} & y_{n2} & \cdots & y_{np} \end{pmatrix}. \tag{19.2}$$

The *sample mean* for the *jth* column of $\mathbf{Y}$ is

$$\bar{y}_j = \frac{1}{n} \sum_{i=1}^{n} y_{ij}, \tag{19.3}$$

and thus the $p \times 1$ *sample mean vector* is given by

$$\bar{\mathbf{y}} = \begin{pmatrix} \bar{y}_1 \\ \bar{y}_2 \\ \vdots \\ \bar{y}_p \end{pmatrix}. \tag{19.4}$$

The *sample covariance matrix* of **Y** is

$$\underset{p \times p}{\mathbf{S}} = \begin{pmatrix} s_{11} & s_{12} & \cdots & s_{1p} \\ s_{21} & s_{22} & \cdot & s_{2p} \\ \vdots & \vdots & \ddots & \vdots \\ s_{p1} & s_{p2} & \cdots & s_{pp} \end{pmatrix}, \tag{19.5}$$

with *diagonal* elements determined with the form

$$s_{jj} = \frac{1}{n-1} \sum_{i=1}^{n} (y_{ij} - \bar{y}_j)^2, \tag{19.6}$$

and *off-diagonal* elements as

$$s_{jk} = \frac{1}{n-1} \sum_{i=1}^{n} (y_{ij} - \bar{y}_j)(y_{ik} - \bar{y}_k). \tag{19.7}$$

The p-symmetric *sample correlation matrix* of **Y** is

$$\underset{p \times p}{\mathbf{R}} = \begin{pmatrix} r_{11} & r_{12} & \cdots & r_{1p} \\ r_{21} & r_{22} & \cdot & r_{2p} \\ \vdots & \vdots & \ddots & \vdots \\ r_{p1} & r_{p2} & \cdots & r_{pp} \end{pmatrix}, \tag{19.8}$$

with elements

$$r_{jk} = \frac{s_{jk}}{\sqrt{s_{jj} s_{kk}}}. \tag{19.9}$$

The $n \times p$ *standardized data matrix* of **Y** is

$$\underset{n \times p}{\mathbf{Z}} = \begin{pmatrix} z_{11} & z_{12} & \cdots & z_{1p} \\ z_{21} & z_{22} & \cdot & z_{2p} \\ \vdots & \vdots & \ddots & \vdots \\ z_{n1} & z_{n2} & \cdots & z_{np} \end{pmatrix}, \tag{19.10}$$

with elements

$$z_{ij} = \frac{y_{ij} - \bar{y}_j}{s_{jj}}. \tag{19.11}$$

The following sections describe the theory and implementation of linear discriminant analysis.

19.3 LINEAR DISCRIMINANT ANALYSIS

Linear discriminant analysis uses the class-specific variance-covariance matrices $\mathbf{S}_\omega$. For a given set of p features, calculation of $\mathbf{S}_\omega$ is based on arrays having class label ω. Therefore, (19.8) and (19.9) for the diagonal and off-diagonal elements of the original matrix $\mathbf{S}$ are rewritten in the form

$$s_{jj}^{\omega} = \frac{1}{n-1} \sum_{i=1}^{n} (y_{ij} - \bar{y}_j)^2 \qquad y_{ij} \in \omega \tag{19.12}$$

and *off-diagonal* elements as

$$s_{jk}^{\omega} = \frac{1}{n-1} \sum_{i=1}^{n} (y_{ij} - \bar{y}_j)(y_{ik} - \bar{y}_k) \qquad y_{ij} \in \omega \tag{19.13}$$

where s_{jj}^{ω} is the variance for gene j among arrays in class ω, s_{jk} is the covariance between genes j and k among arrays in class ω, and $\bar{y}_j$ is the mean of gene j for arrays in class ω. The major assumption for LDA is that the variance-covariance matrices are all equal; that is, $\mathbf{S}_1 = \mathbf{S}_2 = \cdots = \mathbf{S}_\Omega$. Using the class-specific variance-covariance matrices, we calculate the *pooled covariance matrix* as

$$\underset{(p\times p)}{\mathbf{S}_{pl}} = \frac{1}{n-\Omega} \sum (n_\omega - 1) \underset{(p\times p)}{\mathbf{S}_\omega} . \tag{19.14}$$

For a given microarray $\mathbf{y}$ represented by a $p \times 1$ vector of feature values, the Mahalanobis distance from the microarray to the centroid of class ω is defined as

$$\underset{(1\times 1)}{D_\omega(\mathbf{y})} = (\mathbf{y} - \bar{\mathbf{y}}_\omega)' \underset{(p\times p)}{\mathbf{S}_{pl}^{-1}} \underset{(p\times 1)}{(\mathbf{y} - \bar{\mathbf{y}}_\omega)}, \tag{19.15}$$

where $\bar{\mathbf{y}}_\omega$ is a $p \times 1$ vector of mean feature values for arrays in class ω, for which the individual elements are

$$\bar{y}_j = \frac{1}{n_\omega} \sum_{i=1}^{n_\omega} y_{ij} \qquad j = 1, 2, \ldots p \quad y_{ij} \in \omega \tag{19.16}$$

The decision rule $D(\mathbf{y} \rightsquigarrow \omega)$ is to assign microarray $\mathbf{y}$ to the class for which $D_\omega(\mathbf{y})$ is the smallest.

19.3.1 Algorithm

Linear discriminant analysis relies heavily on matrix algebra involving correlation and covariance matrices, and pooled covariance matrices, which are inverted and sandwiched between vector pairs. For large problems involving more than, for example, ~1000 genes, the matrix inversion could become expensive and time-consuming if a slower compiler-core combination is employed for finding the solution. We have always recommended matrix inversion for covariance matrices with dimensions not greater than 500×500, due to the lesser level of computational complexity. The procedural methods used for LDA are listed in Algorithm 19.1.

19.3.2 Cross-Validation Results

Cross-validation (CV) was performed on the nine datasets and genes selected using greedy PTA (see Table 12.10). Bootstrap accuracy ("CVB") was based on average 0.632 bootstrap accuracy (13.4) for 10 bootstraps. Table 19.1 lists the average accuracy based on 10 repartitions of the input arrays

Algorithm 19.1: LDA

Data: n arrays, p genes, training arrays $\mathbf{y}_i$, test arrays $\mathbf{y}$
Result: 10-fold cross-validation (CV10) accuracy
$m \leftarrow 10$, $Acc_{total} = 0$
foreach *Repartition* $= 1 \leftarrow 10$ **do**
  Set $\mathbf{C} \leftarrow \mathbf{0}$, $n_{test} \leftarrow 0$
  Randomly partition input arrays into $\mathcal{D}_1, \mathcal{D}_2, \ldots, \mathcal{D}_m$ folds
  for *each cross-validation fold* $\mathcal{D}_m$ **do**
    Select all arrays in $\mathcal{D}_m$ for testing
    Use all remaining arrays not in $\mathcal{D}_m$ for training
    for $\omega = 1$ *to* Ω **do**
      Determine pooled covariance matrix $\mathbf{S}_{pl}$
      Determine $\bar{\mathbf{y}}_\omega$ for training arrays in class ω
    endfor
    for *each test array* $\mathbf{y} \in \mathcal{D}_m$ **do**
      $n_{test} += 1$
      Determine $\hat{\omega} = \arg\min_{\omega \in \Omega} \{(\mathbf{y} - \bar{\mathbf{y}}_\omega)' \mathbf{S}_{pl}^{-1} (\mathbf{y} - \bar{\mathbf{y}}_\omega)\}$
      Increment confusion matrix C by 1 in element $c_{\omega,\hat{\omega}}$
    endfor
  endfor
  $Acc_{total} += \sum_\omega c_{\omega\omega} / n_{test}$
endfch
$Acc = Acc_{total}/10$

TABLE 19.1 Performance without Feature Transformation

	Cross-validation method				
Dataset	CV2	CV5	CV10	CV−1	CVB
AMLALL2	100	100	100	100	99.96
Brain2	92.33	95.33	95.50	95.00	96.47
Breast2A	94.00	86.67	96.00	100	95.32
Breast2B	93.33	94.74	95.00	94.87	97.52
Colon2	87.90	88.55	88.71	88.71	89.85
Lung2	97.19	90.63	90.00	90.63	98.09
MLL_Leukemia3	98.95	100	100	100	99.89
Prostate2	97.45	98.14	98.04	98.04	98.49
SRBCT4	96.19	99.84	100	100	99.88
Average	0.95	0.95	0.96	0.96	0.97

TABLE 19.2 Performance with Feature Standardization

	Cross-validation method				
Dataset	CV2	CV5	CV10	CV−1	CVB
AMLALL2	98.16	100	100	100	100
Brain2	92.17	95.17	95.17	95.00	96.58
Breast2A	94.67	90.67	93.33	100	96.00
Breast2B	93.21	94.87	95.00	94.87	97.54
Colon2	88.23	88.55	88.71	88.71	89.66
Lung2	99.69	84.69	91.56	90.63	98.55
MLL_Leukemia3	98.25	100	100	100	99.95
Prostate2	96.76	97.94	97.94	98.04	98.49
SRBCT4	98.10	100	100	100	99.83
Average	0.95	0.95	0.96	0.96	0.97

and subsequent CV, without any input feature transformations applied. Table 19.2 lists the average accuracy based on 10 repartitions of the input arrays and subsequent CV, using mean-zero standardized input feature values. Table 19.3 lists the average accuracy based on 10 repartitions of the input arrays and subsequent CV, using fuzzified input feature values. Figure 19.1 illustrates accuracy as a function of CV method for each dataset based on CV accuracy for 10 repartitions of arrays with no feature transformations applied. Figure 19.2 shows standard deviation of accuracy as a function of CV method for each dataset, where the standard deviation is based on CV accuracy for 10 repartitions of arrays with no feature transformations applied. Figure 19.3 reveals average accuracy for all datasets as a function of feature transformation and CV method. With the exception of the Breast2A

TABLE 19.3 Performance with Feature Fuzzification

	Cross-validation method				
Dataset	CV2	CV5	CV10	CV-1	CVB
AMLALL2	96.84	98.68	99.47	100	99.48
Brain2	75.67	75.83	75.67	75.00	88.87
Breast2A	91.33	98.00	99.33	100	95.62
Breast2B	73.33	79.62	80.90	82.05	90.25
Colon2	80.16	81.13	82.10	80.65	87.34
Lung2	98.75	97.19	97.50	96.88	99.45
MLL_Leukemia3	95.09	92.63	90.70	96.49	97.74
Prostate2	88.14	93.04	94.02	94.12	96.65
SRBCT4	88.89	89.84	89.05	96.83	96.70
Average	0.88	0.90	0.90	0.91	0.95

and Lung2 datasets, it was observed in Figure 19.1 that LDA has little bias, as the performance did not change substantially for all datasets as more arrays were used for training. The variance of LDA over the nine datasets, however, was greater than the variance of, for example, the LREG classifier shown in Figure 14.1. The standard deviation of performance was quite low (Figure 19.2), and input feature fuzzification seemed to increase the bias, as shown in Figure 19.3.

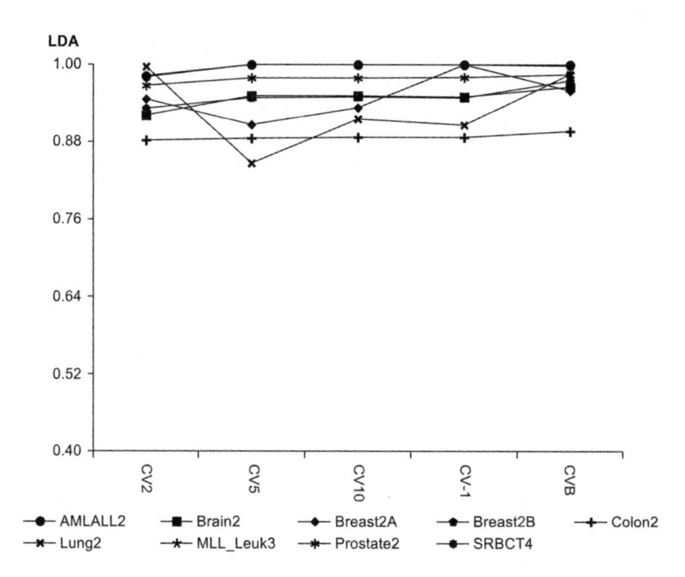

FIGURE 19.1 LDA accuracy as a function of cross-validation (CV) method for each dataset. Average based on CV accuracy for 10 repartitions of arrays. No feature transformations applied.

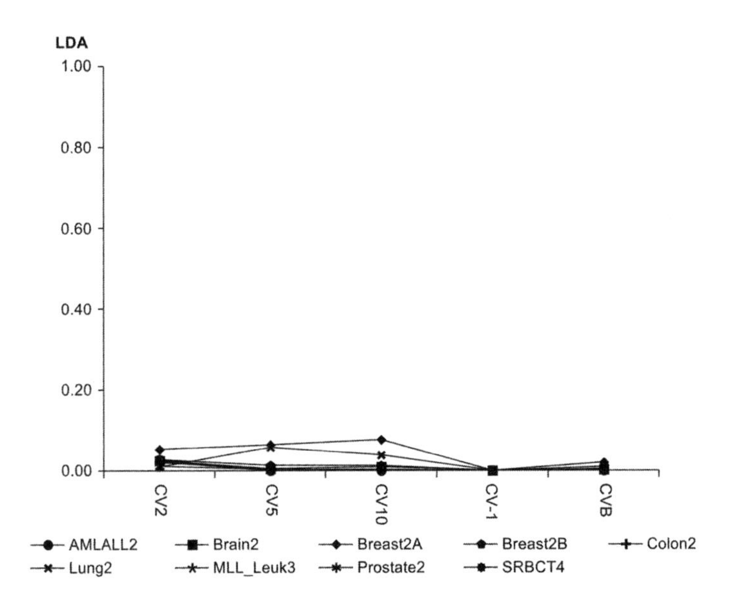

FIGURE 19.2 LDA standard deviation of accuracy as a function of CV method for each dataset. Standard deviation based on CV accuracy for 10 repartitions of arrays. No feature transformations applied.

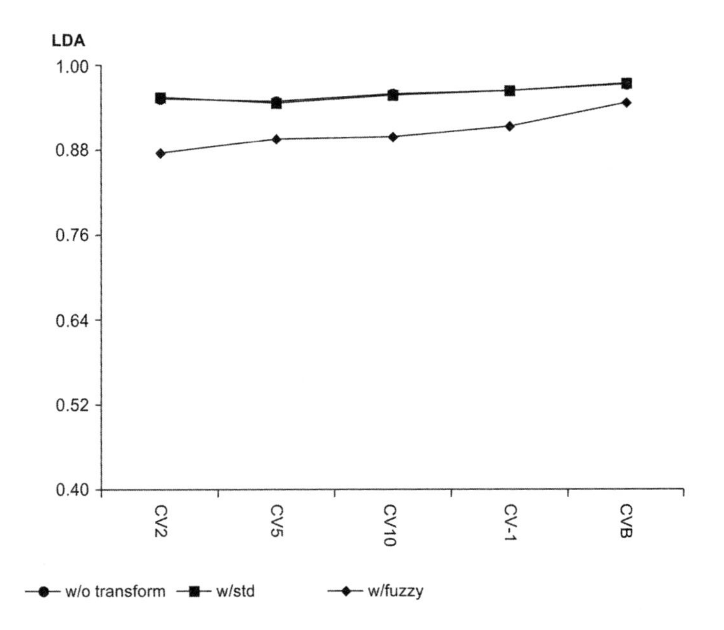

FIGURE 19.3 LDA average accuracy for all datasets as a function of feature transformation and CV method.

19.3.3 Bootstrap Bias

Bootstrap bias results were obtained for the nine datasets used in the previous section on CV. Bootstrap bias runs were based on determining the 0.632 bootstrap accuracy (see Eq. 13.4) for each sampling fraction. Acc_0 in 13.4 was based on test accuracy of unsampled arrays, while Acc_b in 13.4 was based on test accuracy for sampled arrays. Figure 19.4 illustrates the LDA accuracy without transformations on features as a function of bootstrap sample size used during training, based on 10 resamplings from the input arrays. The x-axis values represent the number of arrays randomly sampled $B = 40$ times at fractions $f = 0.1, 0.2, 0.3, 0.4, 0.5$, or 0.6 of the available arrays. Figure 19.5 illustrates the LDA accuracy with feature standardization as a function of bootstrap sample size used during training, based on 10 resamplings from the input arrays. Figure 19.6 depicts the LDA accuracy with feature fuzzification as a function of bootstrap sample size used during training, based on 10 resamplings from the input arrays. Bootstrap results indicate a strong dependence of LDA on the amount of training arrays used, since classification accuracy increased with increasing sample size of arrays used for training. Feature transformations such as mean-zero standardization and fuzzification did not seem to increase performance or decrease the bias observed.

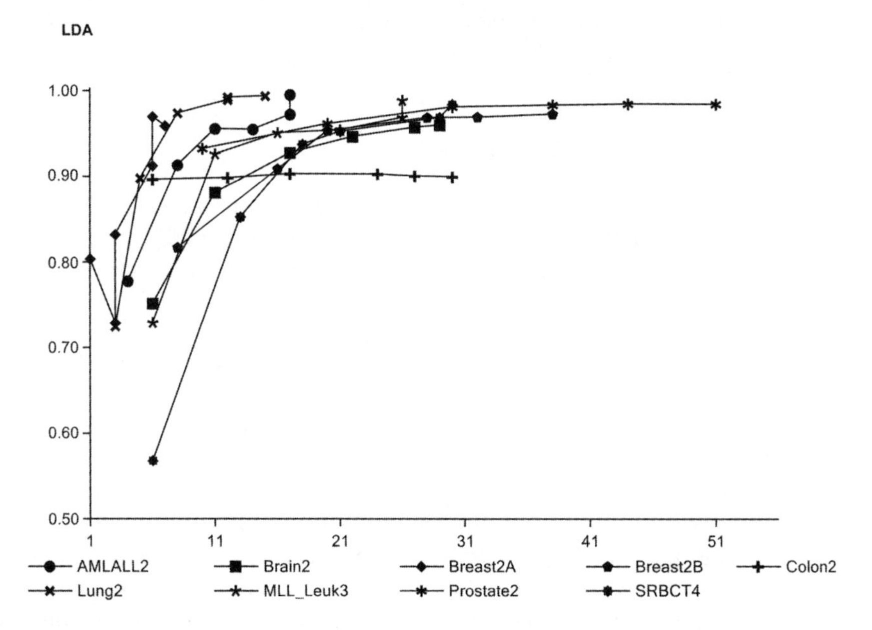

FIGURE 19.4 LDA accuracy without transformations on features as a function of bootstrap sample size used during training. The x-axis values represent the number of arrays randomly sampled $B = 40$ times at fractions $f = 0.1, 0.2, 0.3, 0.4, 0.5$, or 0.6 of the available arrays.

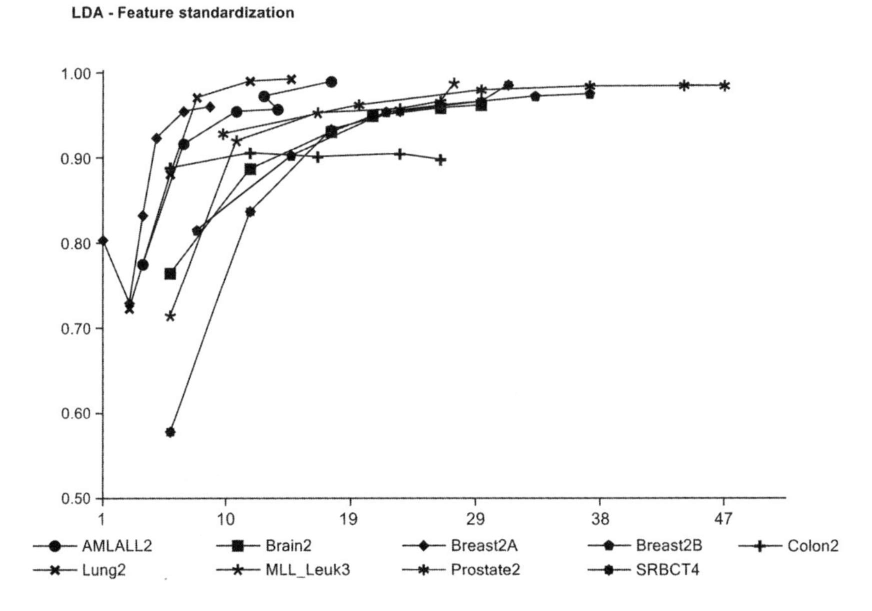

FIGURE 19.5 LDA accuracy with feature standardization as a function of bootstrap sample size used during training. The x-axis values represent the number of arrays randomly sampled $B = 40$ times at fractions $f = 0.1$, 0.2, 0.3, 0.4, 0.5, or 0.6 of the available arrays.

19.3.4 Multiclass ROC Curves

Receiver-operator characteristic (ROC) curves were run for each of the nine datasets for mean standardized input features and fuzzified input features. Figure 19.7 illustrates ROC curves for the LDA classifier applied to the AMLALL2, Brain2, Breast2A, Breast2B, Colon2, Lung2, MLL_Leukemia3, Prostate2, and SRBCT4 datasets using mean-zero standardized input features. The legend values represent the number of arrays randomly sampled $B = 40$ times at fractions $f = 0.1, 0.2, 0.3, 0.4, 0.5,$ or 0.6 of the available arrays. Figure 19.8 illustrates ROC curves for the LDA based on the AMLALL2, Brain2, Breast2A, Breast2B, Colon2, Lung2, MLL_Leukemia3, Prostate2, and SRBCT4 datasets using fuzzified input features. The bootstrap bias of LDA (see Section 19.3.3), is apparent in the multiclass ROC curves, since bootstrapping was used for ROC curve generation. Among the nine datasets, the MLL_Leukemia3, Prostate2, and SRBCT4 showed the greatest levels of AUC for smaller bootstrapped sampling fractions when feature mean-zero standardization was used. When feature fuzzification was used, the AUCs improved for mostly the AMLALL2 dataset, followed by the Prostate2, MLL_Leukemia3, and SRBCT4 datasets.

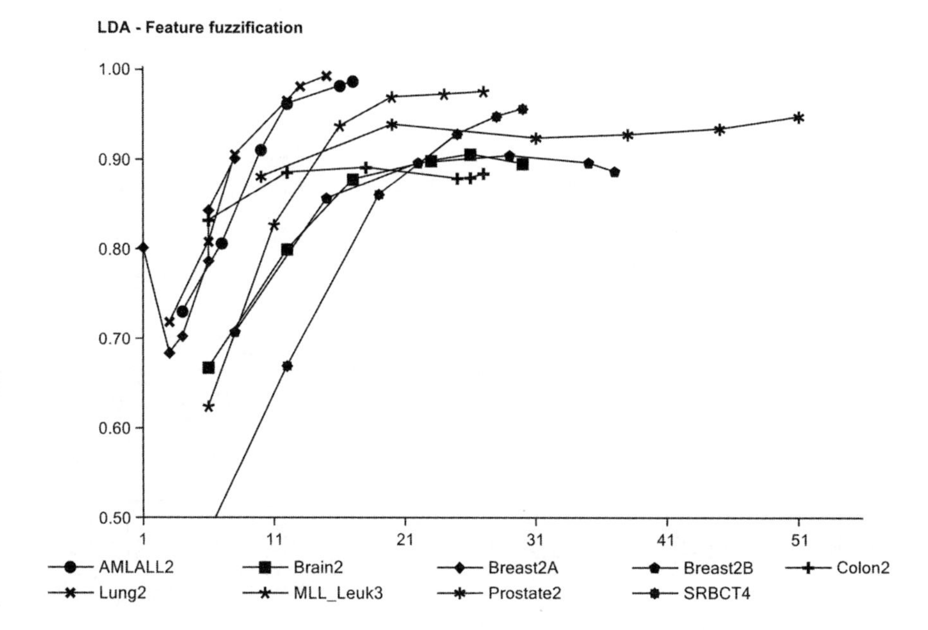

FIGURE 19.6 LDA accuracy with feature fuzzification as a function of bootstrap sample size used during training. The x-axis values represent the number of arrays randomly sampled $B = 40$ times at fractions $f = 0.1, 0.2, 0.3, 0.4, 0.5$, or 0.6 of the available arrays.

19.3.5 Decision Boundaries

Predicted class decision boundaries were generated for the nine datasets using input feature mean-zero standardization and fuzzification. Figure 19.9 illustrates 2D self-organizing maps of class decision boundaries for the LDA based on the AMLALL2, Brain2, Breast2A, Breast2B, Colon2, Lung2, MLL_Leukemia3, Prostate2, and SRBCT4 datasets using mean-zero standardized input features. Figure 19.10 illustrates 2D self-organizing maps of class decision boundaries for the LDA based on the AMLALL2, Brain2, Breast2A, Breast2B, Colon2, Lung2, MLL_Leukemia3, Prostate2, and SRBCT4 datasets using fuzzified input features. The decision boundaries were quite smooth for most datasets, and not as jumpy as the decision boundaries generated by the NBC and KNN classifiers.

19.4 QUADRATIC DISCRIMINANT ANALYSIS

When the covariance matrices are not equal, the distance from each microarray to class centroids is biased by large variance values on the matrix

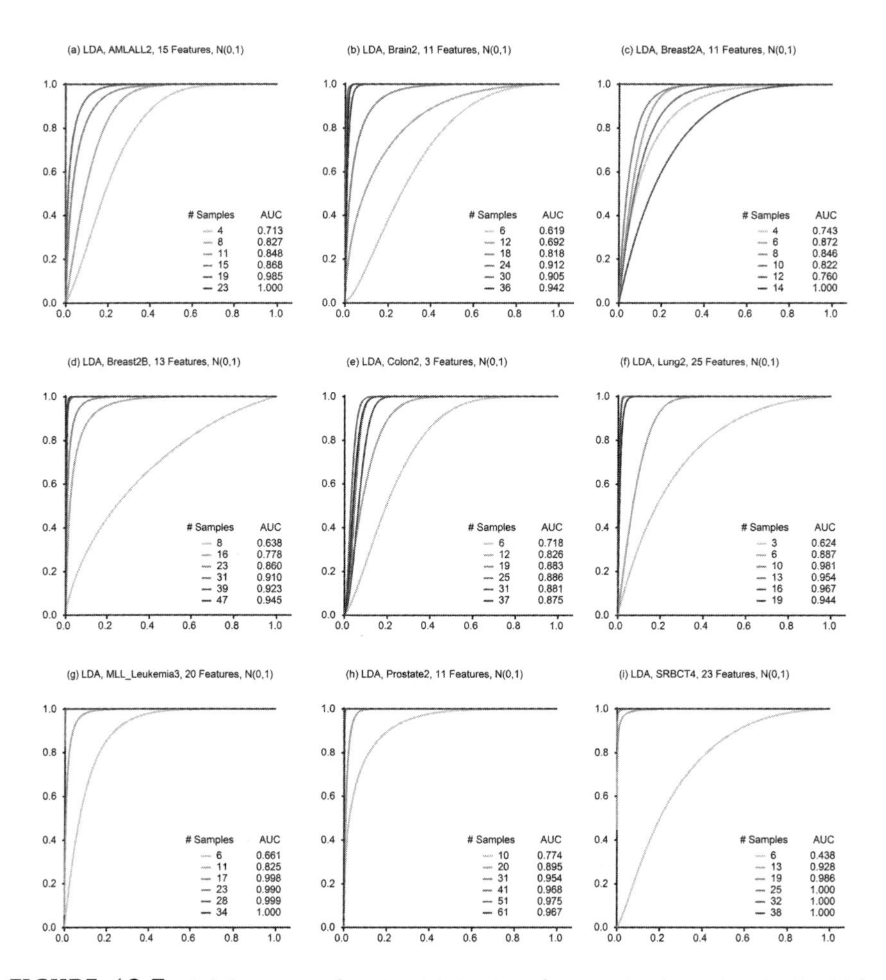

FIGURE 19.7 ROC curves for the LDA classifier applied to the AMLALL2, Brain2, Breast2A, Breast2B, Colon2, Lung2, MLL_Leukemia3, Prostate2, and SRBCT4 datasets based on mean-zero standardized input features; legend values represent the number of arrays randomly sampled $B = 40$ times at fractions $f = 0.1$, 0.2, 0.3, 0.4, 0.5, or 0.6 of the available arrays; (a) AMLALL2; (b) Brain2; (c) Breast2A; (d) Breast2B; (e) Colon2; (f) Lung2; (g) MLL_Leukemia3; (h) Prostate2; (i) SRBCT4. The numbers of genes used for each dataset are listed in Table 12.10.

diagonals. This is the major assumption of quadratic discriminant analysis (QDA). To minimize this bias among unequal covariance matrices, we replace the pooled covariance matrix $\mathbf{S}_{pl}$ in (19.15) with the class specific covariance matrices in the form

$$D_\omega(\mathbf{y}) = (\mathbf{y} - \bar{\mathbf{y}}_\omega)'\mathbf{S}_\omega^{-1}(\mathbf{y} - \bar{\mathbf{y}}_\omega). \tag{19.17}$$

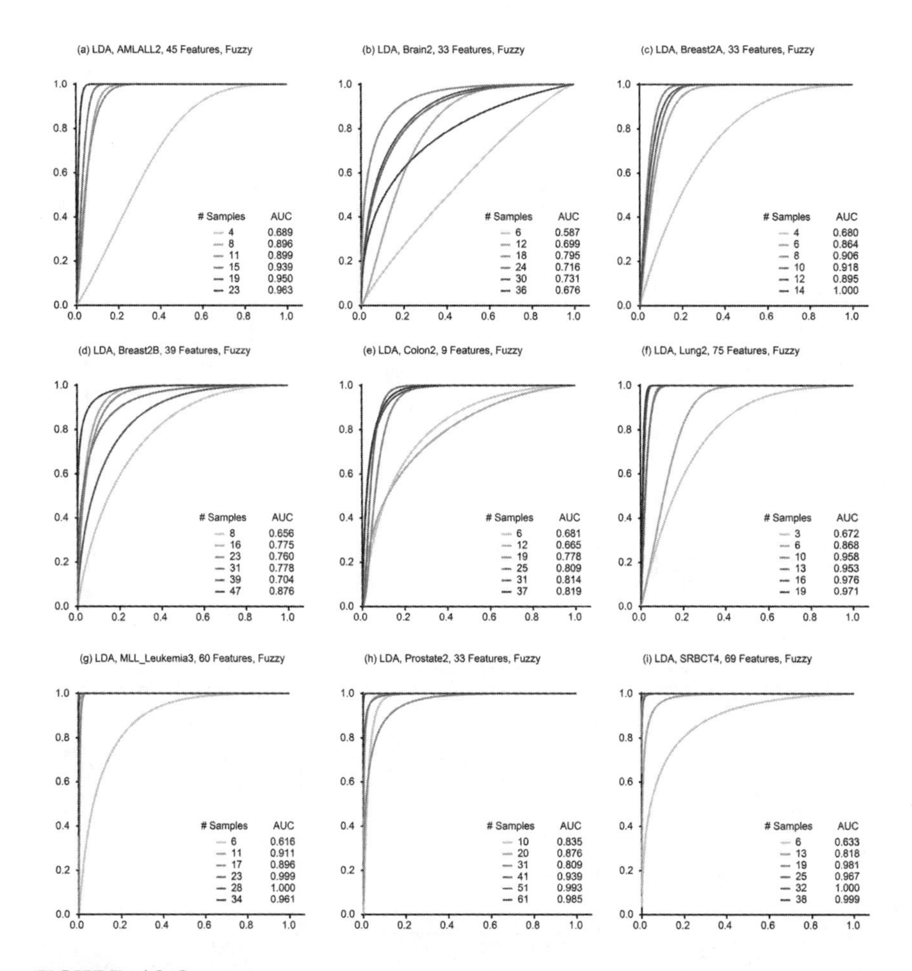

FIGURE 19.8 ROC curves for the LDA applied to the AMLALL2, Brain2, Breast2A, Breast2B, Colon2, Lung2, MLL_Leukemia3, Prostate2, and SRBCT4 datasets based on fuzzified input features; legend values represent the number of arrays randomly sampled $B = 40$ times at fractions $f = 0.1, 0.2, 0.3, 0.4, 0.5,$ or 0.6 of the available arrays: (a) AMLALL2; (b) Brain2; (c) Breast2A; (d) Breast2B; (e) Colon2; (f) Lung2; (g) MLL_Leukemia3; (h) Prostate2; (i) SRBCT4. The numbers of genes used for each dataset are listed in Table 12.10.

We use the same decision rule as before, in which microarray $\mathbf{y}$ is assigned to the class for which $D_\omega(\mathbf{y})$ is the smallest.

Cross-validation runs were made for the nine datasets using QDA, and Figure 19.11 reveals average accuracy for all datasets as a function of feature transformation and CV method. QDA is very biased and actually has poor performance levels; thus the assumption for class-specific covariance matrices does not increase performance for these data.

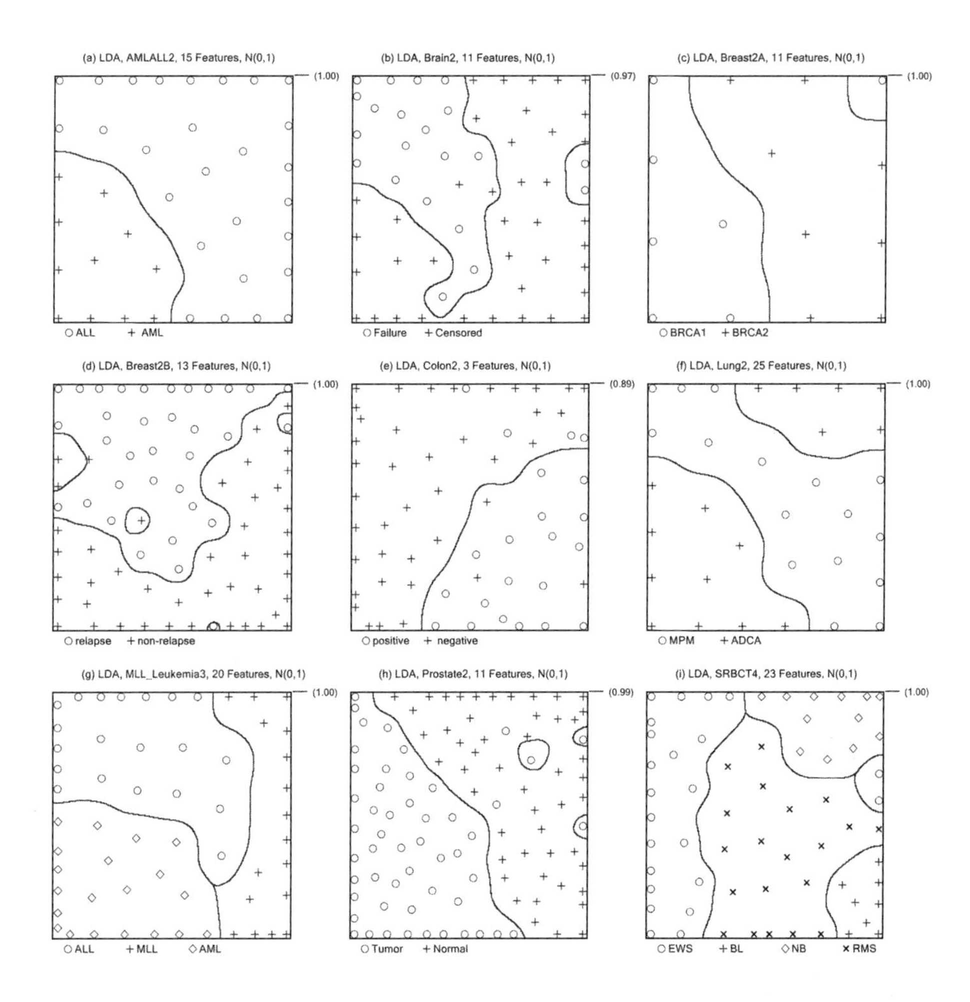

FIGURE 19.9 2D self-organizing maps of LDA class decision boundaries and classification accuracy (upper right of each panel) for the AMLALL2, Brain2, Breast2A, Breast2B, Colon2, Lung2, MLL_Leukemia3, Prostate2, and SRBCT4 datasets based on mean-zero standardized input features; (a) AMLALL2, (b) Brain2; (c) Breast2A; (d) Breast2B; (e) Colon2; (f) Lung2; (g) MLL_Leukemia3; (h) Prostate2; (i) SRBCT4. The numbers of genes used for each dataset are listed in Table 12.10.

19.5 FISHER'S DISCRIMINANT ANALYSIS

In this section we introduce Fisher's discriminant analysis (FDA). Let p be the number of features selected for training and n_ω the number of arrays from each class. Let $\mathbf{B}$ be the $p \times p$ between-class sum of squares matrix and $\mathbf{W}$ be the $p \times p$ within-class sum of squares matrix. The jkth element of these

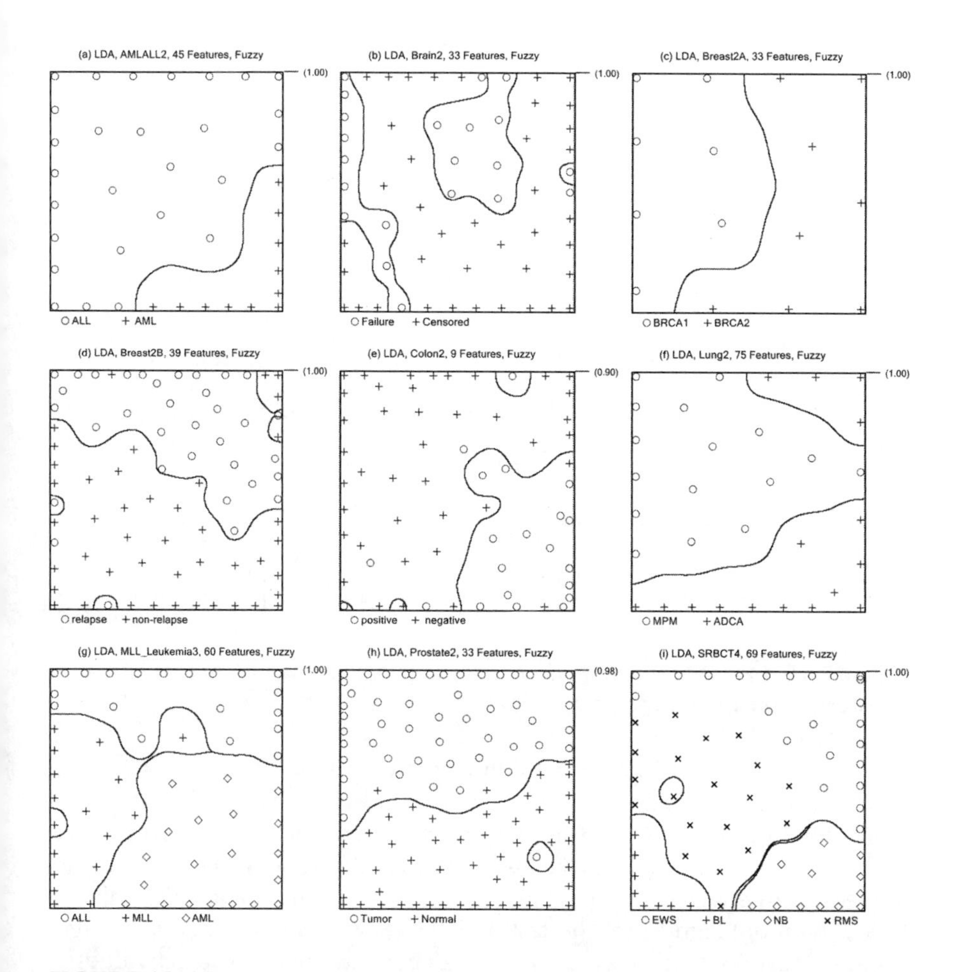

FIGURE 19.10 2D self-organizing maps of LDA class decision boundaries and classification accuracy (upper right of each panel) for the AMLALL2, Brain2, Breast2A, Breast2B, Colon2, Lung2, MLL_Leukemia3, Prostate2, and SRBCT4 datasets based on fuzzified input features; (a) AMLALL2; (b) Brain2; (c) Breast2A; (d) Breast2B; (e) Colon2; (f) Lung2; (g) MLL_Leukemia3; (h) Prostate2; (i) SRBCT4. The numbers of genes used for each dataset are listed in Table 12.10.

matrices are found as

$$
\begin{aligned}
b_{jk} &= \sum_{\omega=1}^{\Omega} \frac{1}{n_\omega} T_{j\omega} T_{kc} - \frac{1}{n} G_j G_k \\
w_{jk} &= \sum_{\omega=1}^{\Omega} \sum_{i=1}^{n_\omega} x_{ij\omega} x_{ikc} - \sum_{\omega=1}^{\Omega} \frac{1}{n_\omega} T_{j\omega} T_{k\omega}
\end{aligned}
, \qquad (19.18)
$$

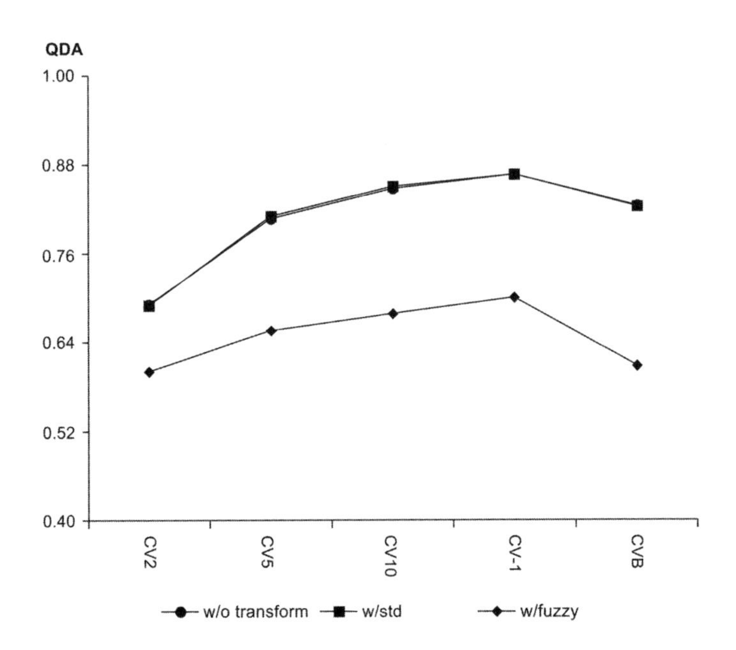

FIGURE 19.11 QDA average accuracy for all datasets as a function of feature transformation and CV method.

where $x_{ij\omega}$ is the ith value of training feature j in class ω, $T_{j\omega} = \sum_i^{n_\omega} x_{ij\omega}$ is the sum of all values of training feature j in class ω, $G_j = \sum_\omega^\Omega T_{j\omega}$ is the grand total of all values of training feature j over the training arrays, and $n = n_1 + n_2 + \cdots + n_\Omega$. It is well known that the ratio $\mathbf{B}/\mathbf{W}$ adjusted for degrees of freedom is the basis of the F-ratio test. The greater the value of $\mathbf{B}/\mathbf{W}$, the more significant the test and the greater the separation between classes of arrays. Let $\lambda_1, \lambda_2, \ldots, \lambda_m > 0$ denote the nonzero eigenvalues of the square *feature* $\times$ *feature* $\mathbf{W}^{-1}\mathbf{B}$ matrix and $\mathbf{e}_1$, $\mathbf{e}_2$, ..., $\mathbf{e}_m$ be the associated eigenvectors. The vector of coefficients $\mathbf{a}$ that maximize the ratio

$$\frac{\mathbf{a}^T\mathbf{B}\mathbf{a}}{\mathbf{a}^T\mathbf{W}\mathbf{a}} \tag{19.19}$$

is given by $\mathbf{a}_1 = \mathbf{e}_1$, which is the first vector of *unstandardized canonical discriminant functions*. The vector $\mathbf{a}_2 = \mathbf{e}_2$ is the second vector of unstandardized canonical discriminant functions, up to $\mathbf{a}_m = \mathbf{e}_m$. As we can see, the number of vectors of unstandardized canonical discriminant functions is equal to the number of nonzero eigenvalues extracted from the matrix $\mathbf{W}^{-1}\mathbf{B}$. Note that the matrix $\mathbf{W}^{-1}\mathbf{B}$ is not symmetric, so we must resort to using, for example, the *generalized eigenvalue problem*, as described in Section B.10. The unstandardized canonical discriminant functions are determined as $\mathbf{a} = \sqrt{n - \Omega}(\mathbf{W}^{1/2})^{-1}\mathbf{e}$, which are equal to the eigenvectors of $\mathbf{W}^{-1}\mathbf{B}$.

For the sake of efficiency, we store the m vectors of unstandardized discriminant functions $\mathbf{a}$ in the $p \times m$ $\mathbf{A}$ matrix. A constant is calculated for each vector $\mathbf{a}$ in the form

$$a_0^k = -\sum_{j=1}^{p} A_{jk}\bar{x}_j, \tag{19.20}$$

where $k = 1, 2, \ldots, m$, and $\bar{x}_j$ is the grand average for feature j over all training arrays. Class centroids of $\mathbf{a}$ are also determined in the form

$$f_{k\omega} = a_0^k + \sum_{j=1}^{p} A_{jk}\bar{x}_{j\omega}, \tag{19.21}$$

where $\bar{x}_{j\omega}$ is the average for feature j in class ω.

Formulation of the decision rule for each $p \times 1$ test microarray $\mathbf{x}$ starts with calculation of the *discriminant scores* given as

$$\mathbf{f} = \mathbf{x}^{\mathrm{T}}\mathbf{A} + \mathbf{a}_0. \tag{19.22}$$

The χ^2 distributed Mahalanobis distance to each centroid is

$$\chi_\omega^2 = (\mathbf{f} - \bar{\mathbf{f}}_\omega)\mathbf{D}_\omega^{-1}(\mathbf{f} - \bar{\mathbf{f}}_\omega)^T, \tag{19.23}$$

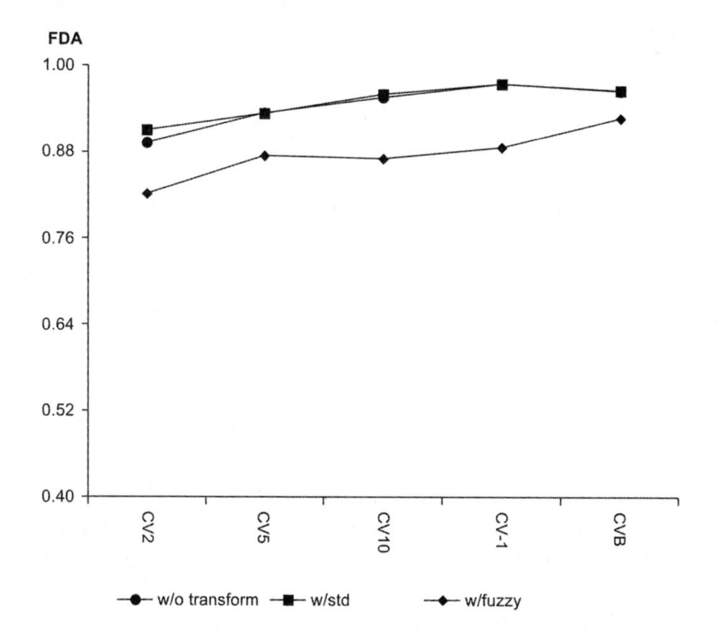

FIGURE 19.12 FDA average accuracy for all datasets as a function of feature transformation and cross-validation (CV) method.

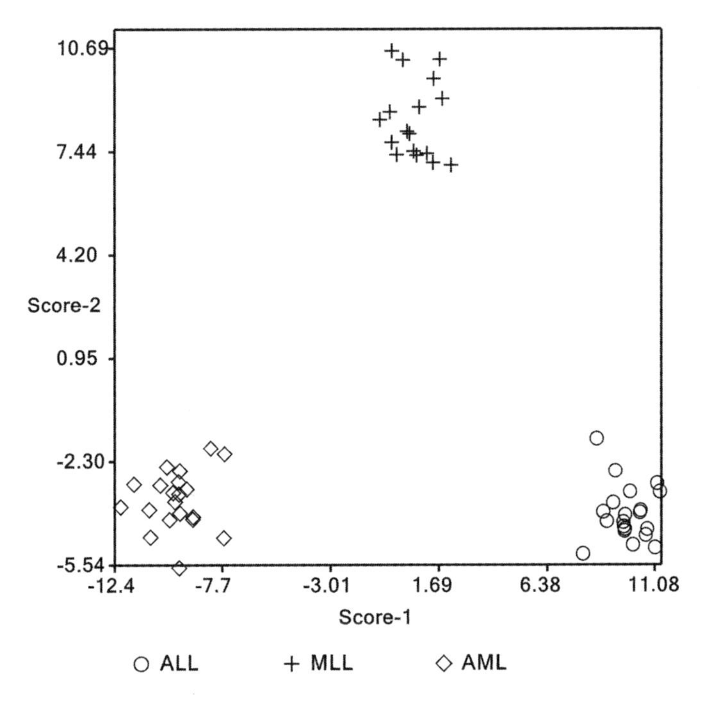

FIGURE 19.13 Plot of array-specific canonical scores from FDA for the MLL_Leukemia3 dataset.

where

$$\mathbf{D}_\omega = \mathbf{A}^T \mathbf{S}_\omega \mathbf{A}, \tag{19.24}$$

and $\mathbf{S}_\omega$ is the covariance matrix for class ω. A posterior probability of classification is

$$P(\omega|\mathbf{x}) = \frac{P_\omega \sqrt{|\mathbf{D}_\omega|} \exp(-\chi^2_\omega/2)}{\sum_{\omega=1}^{\Omega} P_\omega \sqrt{|\mathbf{D}_\omega|} \exp(-\chi^2_\omega/2)}. \tag{19.25}$$

The decision rule is

$$D(\mathbf{x} \leadsto \omega) \equiv \arg\max_{\mathbf{x} \in \omega} P(\omega|\mathbf{x}). \tag{19.26}$$

We ran the various cross-validation methods for FDA using the nine datasets, and results indicate that the FDA bias was lower than that of QDA, but greater than that of LDA, which had greater mean values of performance across all of the CV methods. Figure 19.12 reveals average accuracy for all datasets as a function of feature transformation and CV method.

The array-specific canonical scores for the MLL_Leukemia3 and SRBCT4 datasets are shown in Figures 19.13 and 19.14, respectively. In both the plots,

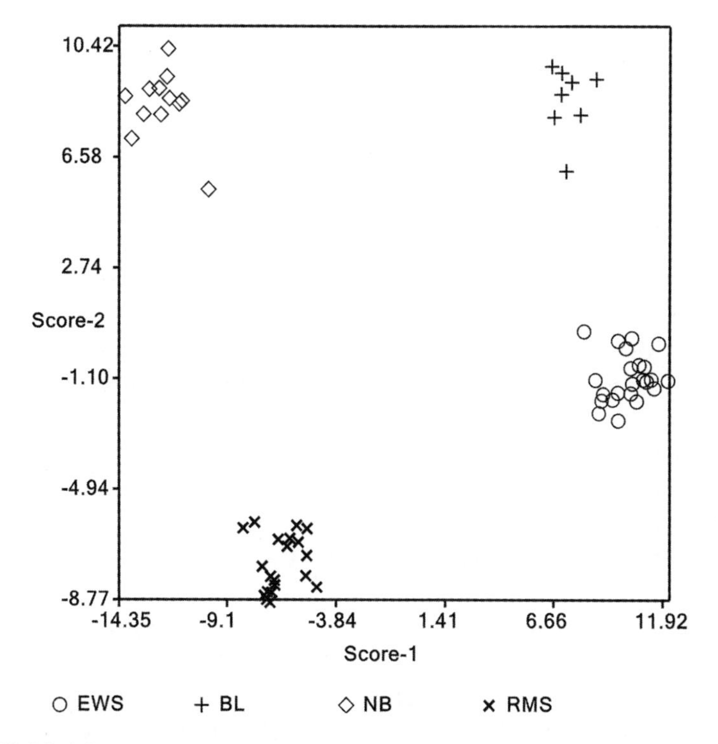

FIGURE 19.14 Plot of array-specific canonical scores from FDA for the SRBCT4 dataset.

there is clear separation between arrays in each class, suggesting a high degree of class separation.

19.6 SUMMARY

Linear discriminant analysis performed better than QDA and FDA; however, the canonical approach of FDA can occasionally outperform LDA. These algorithms are more challenging to implement, especially FDA, which requires considerable time and effort to develop. The features used with the various datasets were originally filtered using the Mahalanobis distance-based greedy PTA approach (see Table 12.10) to maximize class separation, so it was assumed that LDA and FDA would yield reasonably high levels of performance. LDA and FDA should be a first choice for classification methods employed in projects because of their long history of application and widespread use. If LDA or FDA do not perform well for the classes and genes identified, then the classes might not be linearly separable and the genes used are not as informative as expected. KNN can be used as a first alternative to LDA or FDA, since it can sometimes unravel arrays internal to a dispersed cluster or class of arrays that are problematic, and perform more accurate class membership predictions. Otherwise, NBC can be used as a third alternative.

REFERENCES

[1] R.A. Fisher. The use of multiple measurements in taxonomic problems. *Ann. Eugenics* **7**:179–188, 1936.

[2] H. Arabalibeik, S. Jafari, K. Agin. Classification of pulmonary system diseases patterns using flow-volume curve. *Stud. Health Technol. Inform.* **163**:25–30, 2011.

[3] C. Saenz, T. Cedron, S. Cabredo. Classification of wines from five Spanish origin denominations by aromatic compound analysis. *J. AOAC (Association of Official Analytical Chemists Int.* **93**(6):1916–1922, 2010.

[4] P. Xu, P. Yang, X. Lei, D. Yao. An enhanced probabilistic LDA for multi-class brain computer interface. *PLoS One* **6**(1):e14634, 2011.

[5] B.H. Aksebzeci, M.H. Asyali, Y. Kahraman, O. Er, E. Kaya, H. Ozbilge, S. Kara. Classification of root canal microorganisms using electronic-nose and discriminant analysis. *Biomed. Eng. Online* **9**:77, 2010.

[6] Z. Xu, C.E. Bunker, B. Harrington Pde. Classification of jet fuel properties by near-infrared spectroscopy using fuzzy rule-building expert systems and support vector machines. *Appl. Spectrosc.* **64**(11):1251–1258, 2010.

[7] G.G. Agrawal, A. Asthana, A. Maurya. Classification of hard core and petty criminals using anthropometric measurements. *Biosci. Trends* **4**(5):239–243, 2010.

[8] P. Guan, D. Huang, M. He, B. Zhou. Lung cancer gene expression database analysis incorporating prior knowledge with support vector machine-based classification method. *J. Exp. Clin. Cancer Res.* **28**:103, 2009.

[9] J.P. Gillet, T.J. Molina, J. Jamart, P. Gaulard, K. Leroy, J. Briere, I. Theate, C. Thieblemont, A. Bosly, M. Herin, J. Hamels, J. Remacle. Evaluation of a low density DNA microarray for small B-cell non-Hodgkin lymphoma differential diagnosis. *Leuk. Lymphoma* **50**(3):410–418, 2009.

[10] J. Hua, Y. Balagurunathan, Y. Chen, J. Lowey, M.L. Bittner, Z. Xiong, E. Suh, E.R. Dougherty. Normalization benefits microarray-based classification. *EURASIP J. Bioinform. Syst. Biol.* **v2006**:43056, 2006.

[11] O. Galamb, F. Sipos, E. Dinya, S. Spisak, Z. Tulassay, B. Molnar. mRNA expression, functional profiling and multivariate classification of colon biopsy specimen by cDNA overall glass microarray. *World J. Gastroenterol.* **12**(43):6998–7006, 2006.

[12] A.V. Ivshina, J. George, O. Senko, B. Mow, T.C. Putti, J. Smeds, T. Lindahl, Y. Pawitan, P. Hall, H. Nordgren, J.E. Wong, E.T. Liu, J. Bergh, V.A. Kuznetsov, L.D. Miller. Genetic reclassification of histologic grade delineates new clinical subtypes of breast cancer. *Cancer Res.* **66**(21):10292–10301, 2006.

[13] M.J. van der Werf, B. Pieterse, N. van Luijk, F. Schuren, B. van der Werff-van der Vat, K. Overkamp, R.H. Jellema. Multivariate analysis of microarray data by principal component discriminant analysis: Prioritizing relevant transcripts linked to the degradation of different carbohydrates in Pseudomonas putida S12. *Microbiology* **152**(1):257-272, 2006.

[14] D. Hwang, W.A. Schmitt, G. Stephanopoulos, G. Stephanopoulos. Determination of minimum sample size and discriminatory expression patterns in microarray data. *Bioinformatics* **18**(9):1184-1193, 2002.

[15] J.H. Cho, D. Lee, J.H. Park, K. Kim, I.B. Lee. Optimal approach for classification of acute leukemia subtypes based on gene expression data. *Biotechnol. Prog.* **18**(4):847-854, 2002.

CHAPTER 20

LEARNING VECTOR QUANTIZATION

20.1 INTRODUCTION

Learning vector quantization 1 (LVQ1) is a prototype learning method that uses a *punishment-reward* method for moving prototypes toward arrays with the same class labels and away from arrays with different class labels. LVQ1 has been widely employed in biomedicine for applications such as recognition of white blood cells [1], urinary tumor markers [2], investigation of thyroid lesions [3], diagnosis of acute appendicitis [4], ultrasound image processing [5], sleep classification in infants [6], and EEG classification [7]. DNA microarray classification with LVQ has not been widely used [8,9].

To begin our description of LVQ, we must first specify the number of *prototypes* per class. This can be done arbitrarily or through a grid search over the specified number of prototypes. Some authors recommend setting the number of prototypes the same in each class; however, this may be unnecessary since there may be more or fewer prototypes than are needed for class separability. One of the most parsimonious methods for finding the optimal number of prototypes per class is to perform a separate K-means cluster analysis on the arrays in each class ω ($\omega = 1, 2, \ldots, \Omega$). The prototypes representing a given class are set equal to the centers identified and are assigned class label ω. In this manner, the number of prototypes assigned to each class is optimized and is more parsimonious than arbitrarily using the same number of cluster per class.

Let $\mathbf{x}_i$ be the ith array ($i = 1, 2, \ldots, n$) and $\mathbf{v}_m$ ($m = 1, 2, \ldots, M$) be a prototype. In addition, let Ω be the number of classes with individual class labels, ω_i be the class label for array $\mathbf{x}_i$, and ω_m be the class label for prototype

Classification Analysis of DNA Microarrays, First Edition. Leif E. Peterson.

$\mathbf{v}_m$. Here, the M prototypes are based on the use of K-means cluster analysis of training arrays within each class, yielding K centers per class, times the number of classes, specifically, $M = K\Omega$.

During the initial iteration LVQ1 selects the first array $\mathbf{x}_i$ and derives the distance to each prototype $\mathbf{v}_m$ among all M prototypes in the form

$$d(\mathbf{x}_i) = \|\mathbf{x}_i - \mathbf{v}_m\|. \tag{20.1}$$

The closest prototype is then updated according to the rule

$$\begin{aligned} \mathbf{v}_m(t+1) &= \mathbf{v}_m(t) + \alpha(t)(\mathbf{x}_i(t) - \mathbf{v}_m(t)) \qquad \omega_m = \omega_i \\ \mathbf{v}_m(t+1) &= \mathbf{v}_m(t) - \alpha(t)(\mathbf{x}_i(t) - \mathbf{v}_m(t)) \qquad \text{otherwise} \end{aligned} \tag{20.2}$$

where $\mathbf{v}_m$ is the prototype closest to array $\mathbf{x}_i$ and $\alpha(t)$ is the learning rate at the tth iteration ($t = 1, 2, \ldots, T$). According to this updating rule, the closest prototype is rewarded if its class label is the same as the class label of the array and punished if different. A suitable choice for the learning rate $\alpha(t)$ is

$$\alpha(t) = \alpha_0 \left(1 - \frac{t-1}{T}\right) \tag{20.3}$$

Reliable results for LVQ1 are commonly obtained using an initial value of $\alpha_0 = 0.1$ and $T = 100$ iterations. The preceding calculations at iteration t are repeated for the remaining arrays, each time looping over all M prototypes to find the closest prototype. This calculation is then carried out over T total iterations. During testing, the decision rule is to assign array $\mathbf{x}$ to the class of which the closest prototype belongs, shown as

$$D(\mathbf{x} \rightsquigarrow \omega) = \min_m\{D(\mathbf{v}_m, \mathbf{x})\}. \tag{20.4}$$

It is important that the initial locations of prototypes representing each class be proximal to the cluster of arrays within the same class. Otherwise, if a prototype initially resides on the periphery of the array space outside a cluster of arrays with opposite class labels, then during the learning process the closest prototype to those arrays may always be the given prototype with the opposite class label. A prototype close to a cluster of arrays with opposite class labels that is far removed from arrays of like class will become repeatedly punished and repelled away from the global array space. New arrays having outlier characteristics may be proximal to the rejected prototype and therefore, may be misclassified. For this reason, it is important to select initial prototype locations within the geometric space of arrays with the same class label. As pointed out above, the use of K-means cluster analysis performed separately on arrays within each class will usually guarantee that the prototypes lie within or among the cluster of arrays of the same class.

20.2 CROSS-VALIDATION RESULTS

Table 20.1 lists the average accuracy as a function of the number of prototypes per class (M), based on 10 repartitions of the input arrays and subsequent cross-validation (CV), without any input feature transformations applied. Regarding bootstrap accuracy ("CVB"), we based results on average 0.632 bootstrap accuracy (13.4) for 10 bootstraps. Table 20.2 lists the average accuracy based on 10 repartitions of the input arrays and subsequent cross-validation, using mean-zero standardized input feature values. Table 20.3 lists the average accuracy based on 10 repartitions of the input arrays and subsequent cross-validation, using fuzzified input feature values. LVQ1 was the most biased on the Brain2, Breast2, and Colon2 datasets, while the remaining resulted datasets in very little bias. Mean-zero feature standardization and fuzzification of input features appeared to slightly stabilize the bias observed for these three datasets. Use of the bootstrap CV method resulted in a small increase in mean performance.

Figure 20.1 illustrates accuracy as a function of CV method for each dataset. The average is based on CV accuracy for 10 repartitions of arrays, and no feature transformations are applied. Figure 20.2 shows standard deviation of accuracy as a function of CV method for each dataset. Standard deviation is based on CV accuracy for 10 repartitions of arrays and no, feature transformations are applied. Figure 20.3 reveals average accuracy for all datasets as a function of feature transformation and CV method. Figure 20.4 reveals the LVQ1 accuracy without transformations on features as a function of M. Figure 20.5 shows the LVQ1 accuracy with feature standardization as a function of M. Figure 20.6 illustrates the LVQ1 accuracy with feature fuzzification as a function of M. Figure 20.7 shows average accuracy for all datasets as a function of M.

Varying the number of learning prototypes (M) used during training did not result in a noticeable effect on performance. This is a very positive observation, and suggests that for the material covered in previous chapters, LVQ1 is one of the more stable classifiers that is less biased from the amount of training data used.

20.3 BOOTSTRAP BIAS

Bootstrapping was also performed to determine the bootstrap bias of LVQ1. Bootstrap bias runs were based on determining the 0.632 bootstrap accuracy (see Eq. 13.4) for each sampling fraction. Acc_0 in 13.4 was based on test accuracy of unsampled arrays, while Acc_b in 13.4 was based on test accuracy for sampled arrays. Figure 20.8 shows the LVQ1 accuracy without transformations on features as a function of bootstrap sample size used during training, based on 10 resamplings from the input arrays. The x-axis values represent the number of arrays randomly sampled $B = 40$ times at

TABLE 20.1 Performance without Feature Transformation

Dataset	M	Cross-validation method				
		CV2	CV5	CV10	CV-1	CVB
AMLALL2	1	100	100	100	100	100
	3	100	100	100	100	100
	5	100	100	100	100	100
	7	100	100	100	100	100
	9	100	100	100	100	100
Brain2	1	81.33	83.67	84.67	78.33	92.39
	3	83.67	85.00	83.67	86.67	92.46
	5	84.83	85.33	83.33	83.33	91.69
	7	85.83	85.67	84.33	88.33	92.59
	9	84.00	82.33	83.83	80.00	92.18
Breast2A	1	100	100	100	100	100
	3	100	100	100	100	100
	5	100	100	100	100	99.94
	7	100	100	100	100	100
	9	100	100	100	100	100
Breast2B	1	85.38	86.67	86.79	84.62	93.02
	3	84.49	87.05	87.44	84.62	93.07
	5	85.51	87.18	86.54	85.90	93.59
	7	86.15	86.28	85.77	84.62	92.91
	9	85.64	87.05	86.67	85.90	93.24
Colon2	1	84.03	84.35	83.55	83.87	89.92
	3	83.55	83.23	84.84	82.26	89.71
	5	83.23	82.74	82.26	87.10	90.21
	7	83.71	84.03	83.87	88.71	89.86
	9	84.84	84.35	83.06	80.65	89.98
Lung2	1	100	100	100	100	100
	3	100	100	100	100	100
	5	100	100	100	100	100
	7	100	100	100	100	100
	9	100	100	100	100	100
MLL_Leukemia3	1	99.12	99.12	99.12	98.25	99.63
	3	98.95	99.12	99.12	98.25	99.63
	5	98.60	99.12	98.77	98.25	99.56
	7	98.25	98.77	98.95	98.25	99.67
	9	98.60	99.12	98.60	98.25	99.61
Prostate2	1	94.41	94.51	94.51	94.12	96.85
	3	94.51	94.22	94.80	92.16	96.76
	5	94.51	94.22	94.71	91.18	96.98
	7	94.51	95.00	94.41	94.12	96.59
	9	93.82	93.92	94.12	96.08	96.83
SRBCT4	1	99.84	100	100	100	99.92
	3	98.89	99.84	100	100	99.91
	5	99.68	99.84	100	100	99.96
	7	99.68	100	100	100	99.97
	9	99.84	100	100	100	99.99
Average		0.94	0.94	0.94	0.94	0.97

TABLE 20.2 Performance with Feature Standardization

Dataset	M	Cross-validation method				
		CV2	CV5	CV10	CV-1	CVB
AMLALL2	1	100	100	100	100	100
	3	100	100	100	100	100
	5	100	100	100	100	100
	7	100	100	100	100	100
	9	100	100	100	100	100
Brain2	1	83.67	83.00	83.17	80.00	92.22
	3	82.50	85.50	85.50	80.00	92.13
	5	81.17	84.50	83.33	83.33	92.32
	7	84.67	81.83	86.00	86.67	91.86
	9	84.50	84.50	82.00	83.33	91.45
Breast2A	1	100	100	100	100	100
	3	100	100	100	100	100
	5	100	100	100	100	100
	7	100	100	100	100	100
	9	99.33	100	100	100	100
Breast2B	1	85.64	85.77	87.69	84.62	92.89
	3	83.72	85.64	84.36	84.62	92.79
	5	84.62	86.28	85.38	82.05	92.55
	7	85.00	85.77	85.13	89.74	93.00
	9	86.79	84.62	85.77	88.46	93.45
Colon2	1	85.65	83.71	83.39	85.48	89.64
	3	83.23	83.87	83.87	82.26	89.74
	5	83.87	85.65	84.68	83.87	89.99
	7	84.19	85.32	83.23	83.87	89.80
	9	84.68	84.52	84.52	85.48	89.75
Lung2	1	100	100	100	100	100
	3	100	100	100	100	100
	5	100	100	100	100	100
	7	100	100	100	100	100
	9	100	100	100	100	100
MLL_Leukemia3	1	99.12	98.95	98.95	100	99.58
	3	98.77	98.77	98.07	98.25	99.63
	5	98.77	98.95	99.47	98.25	99.72
	7	98.95	98.95	98.77	100	99.74
	9	98.77	98.95	99.12	98.25	99.60
Prostate2	1	94.41	95.29	94.02	94.12	96.55
	3	93.92	94.61	95.20	93.14	96.54
	5	94.12	94.31	93.82	95.10	96.89
	7	93.53	93.82	94.41	93.14	96.61
	9	94.41	94.41	94.61	93.14	96.67
SRBCT4	1	99.68	100	100	100	99.97
	3	100	100	99.84	100	99.99
	5	99.52	100	100	100	99.94
	7	99.84	100	99.84	100	99.91
	9	99.37	100	100	100	99.96
Average		0.94	0.94	0.94	0.94	0.97

TABLE 20.3 Performance with Feature Fuzzification

		Cross-validation method				
Dataset	*M*	CV2	CV5	CV10	CV-1	CVB
AMLALL2	1	100	100	100	100	100
	3	100	100	100	100	100
	5	100	100	100	100	100
	7	100	100	100	100	100
	9	100	100	100	100	100
Brain2	1	82.33	80.33	82.17	83.33	91.39
	3	80.33	80.17	81.00	80.00	91.76
	5	82.17	83.33	81.67	80.00	91.60
	7	82.83	80.83	81.17	78.33	91.24
	9	81.50	82.33	82.00	76.67	91.65
Breast2A	1	100	100	100	100	100
	3	100	100	100	100	100
	5	100	100	100	100	100
	7	100	100	100	100	100
	9	100	100	100	100	100
Breast2B	1	80.00	84.36	84.36	79.49	91.45
	3	80.26	83.08	82.31	87.18	91.75
	5	81.92	82.82	83.21	79.49	91.60
	7	81.79	83.46	82.44	88.46	91.79
	9	82.18	83.21	83.85	80.77	92.09
Colon2	1	81.29	83.06	86.45	87.10	89.19
	3	80.48	84.52	82.90	82.26	89.57
	5	84.19	82.74	83.55	85.48	89.24
	7	81.94	84.19	85.48	85.48	89.43
	9	83.06	83.55	85.00	80.65	89.48
Lung2	1	100	100	100	100	100
	3	100	100	100	100	100
	5	100	100	100	100	100
	7	100	100	100	100	100
	9	100	100	100	100	100
MLL_Leukemia3	1	98.95	98.25	98.25	98.25	99.29
	3	98.25	98.42	98.25	98.25	99.45
	5	98.25	98.25	98.25	98.25	99.57
	7	98.60	98.25	98.25	98.25	99.64
	9	98.25	98.25	98.25	98.25	99.44
Prostate2	1	92.94	93.43	93.14	94.12	95.61
	3	92.45	92.65	93.33	94.12	95.47
	5	93.24	93.73	93.82	93.14	95.76
	7	92.65	93.82	93.73	93.14	95.65
	9	93.24	93.53	93.24	93.14	95.55
SRBCT4	1	99.68	99.84	99.84	100	99.86
	3	99.52	100	100	100	99.92
	5	99.84	100	99.84	100	99.86
	7	99.84	99.84	99.68	100	99.85
	9	99.68	99.68	99.68	100	99.99
Average		0.93	0.93	0.93	0.93	0.96

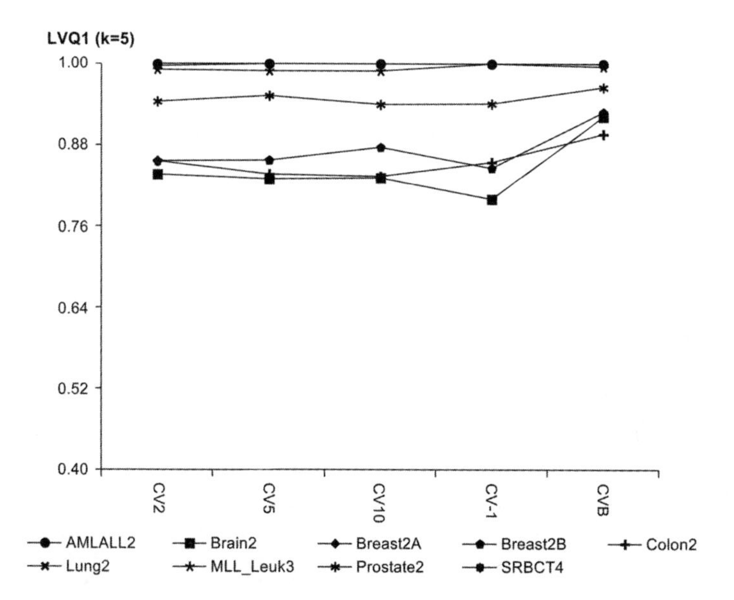

FIGURE 20.1 LVQ1 accuracy as a function of cross-validation (CV) method for each dataset. Average based on CV accuracy for 10 repartitions of arrays. No feature transformations applied.

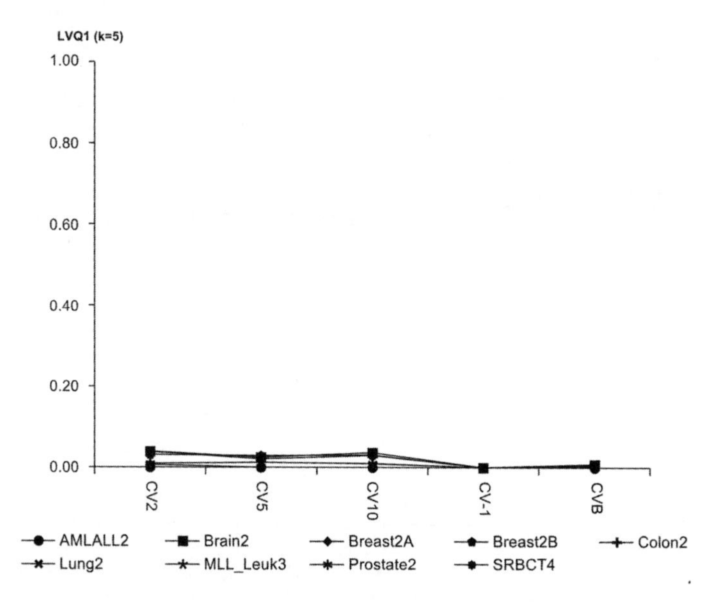

FIGURE 20.2 LVQ1 standard deviation of accuracy as a function of CV method for each dataset. Standard deviation based on CV accuracy for 10 repartitions of arrays. No feature transformations applied.

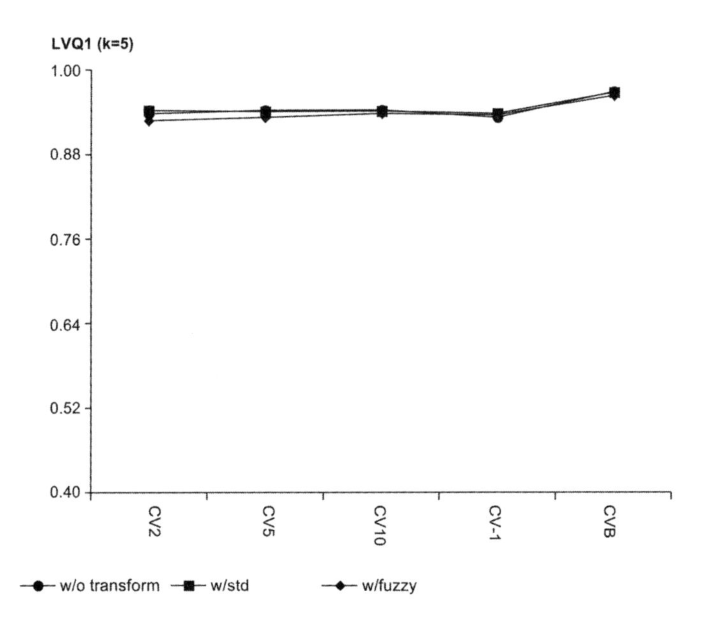

FIGURE 20.3 LVQ1 average accuracy for all datasets as a function of feature transformation and CV method.

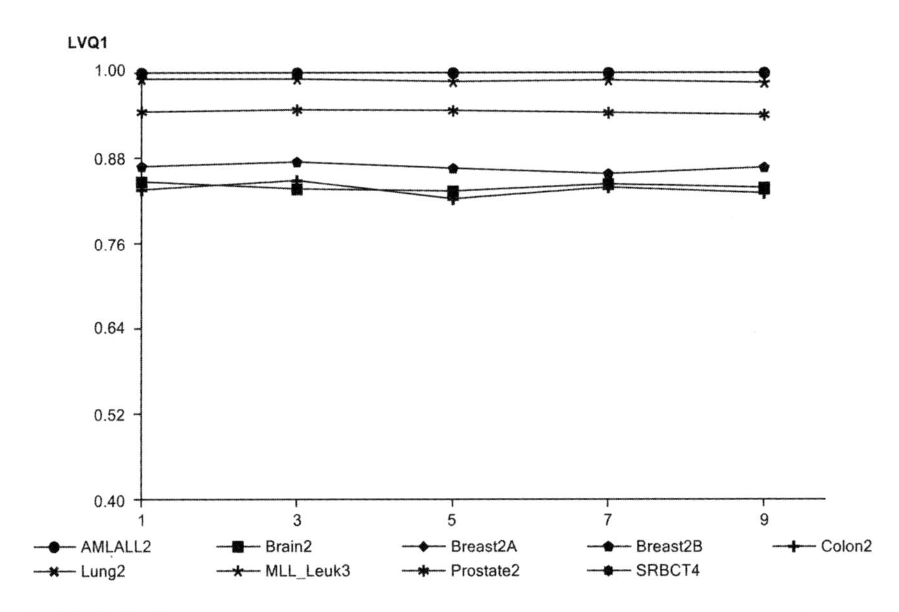

FIGURE 20.4 LVQ1 accuracy without transformations on features as a function of M.

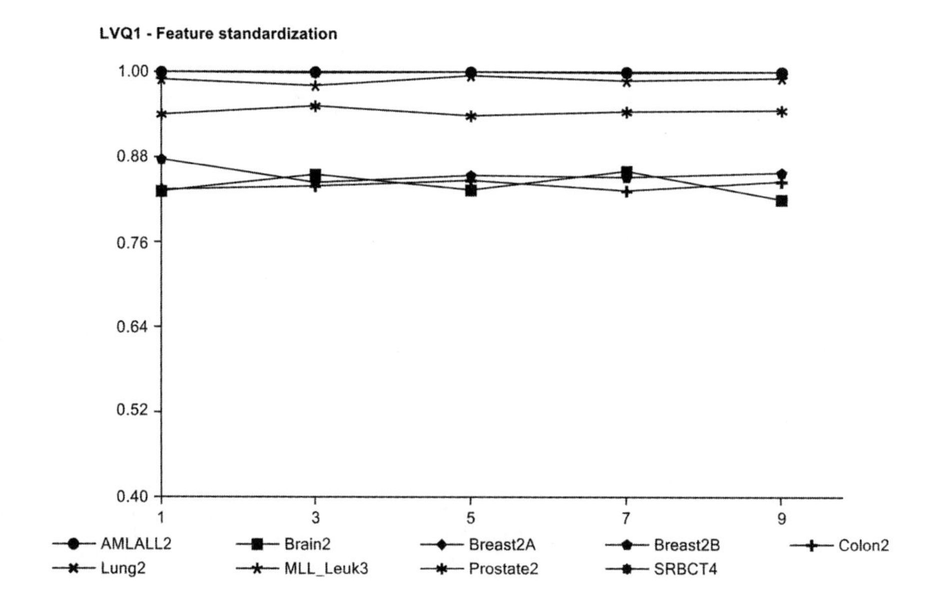

FIGURE 20.5 LVQ1 accuracy with feature standardization as a function of *M*.

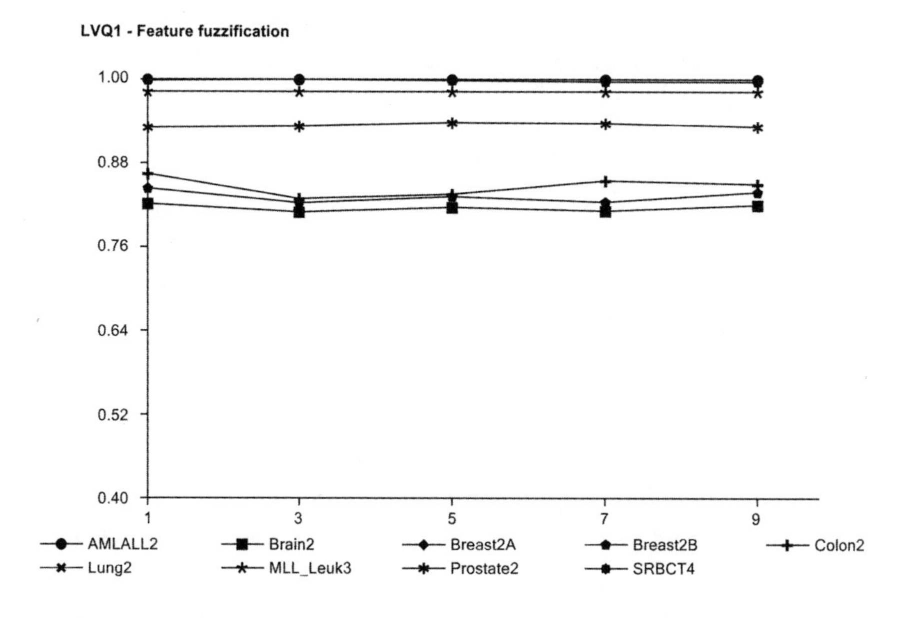

FIGURE 20.6 LVQ1 accuracy with feature fuzzification as a function of *M*.

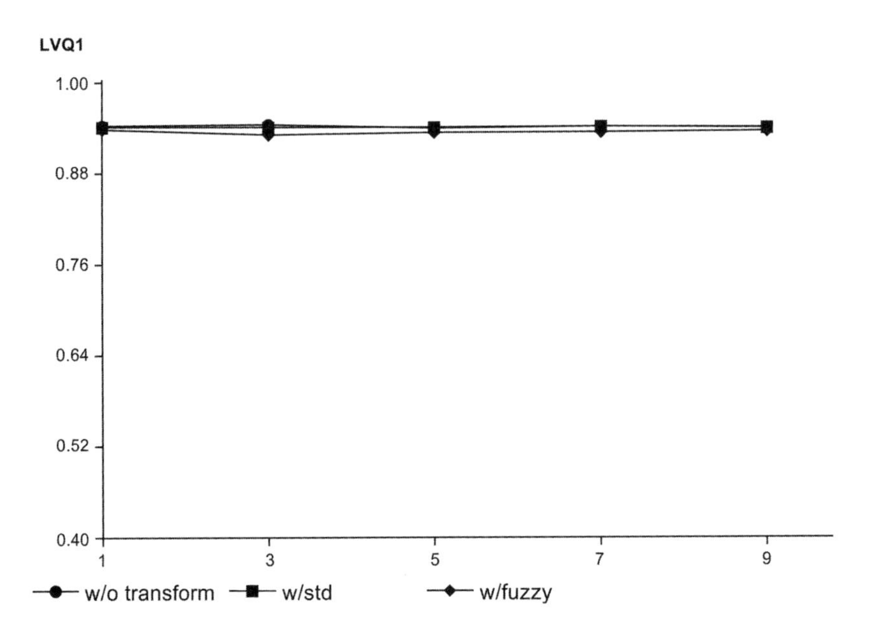

FIGURE 20.7 LVQ1 average accuracy for all datasets as a function of M. Ten 10-fold cross-validation used.

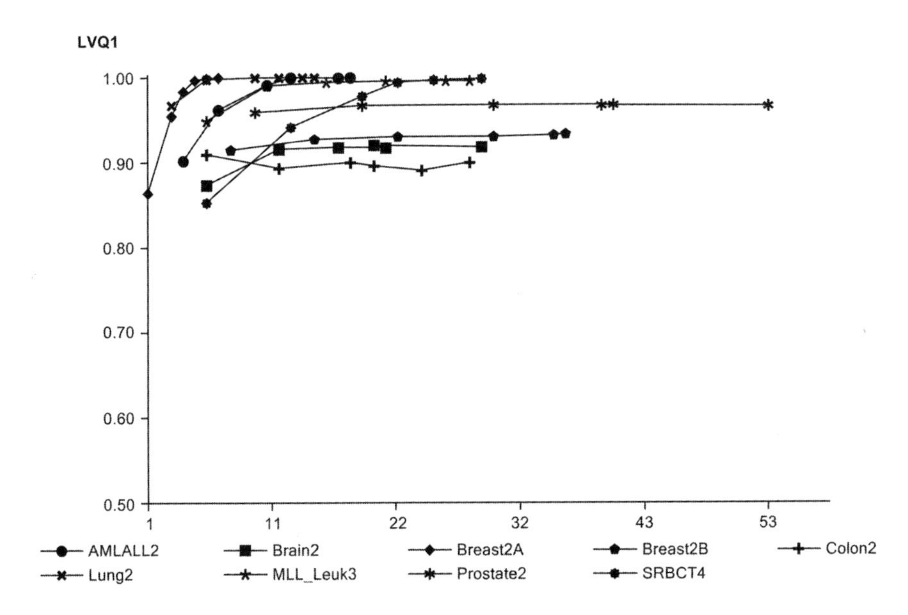

FIGURE 20.8 LVQ1 accuracy without transformations on features as a function of bootstrap sample size used during training. The x-axis values represent the number of arrays randomly sampled $B = 40$ times at fractions $f = 0.1$, 0.2, 0.3, 0.4, 0.5, or 0.6 of the available arrays.

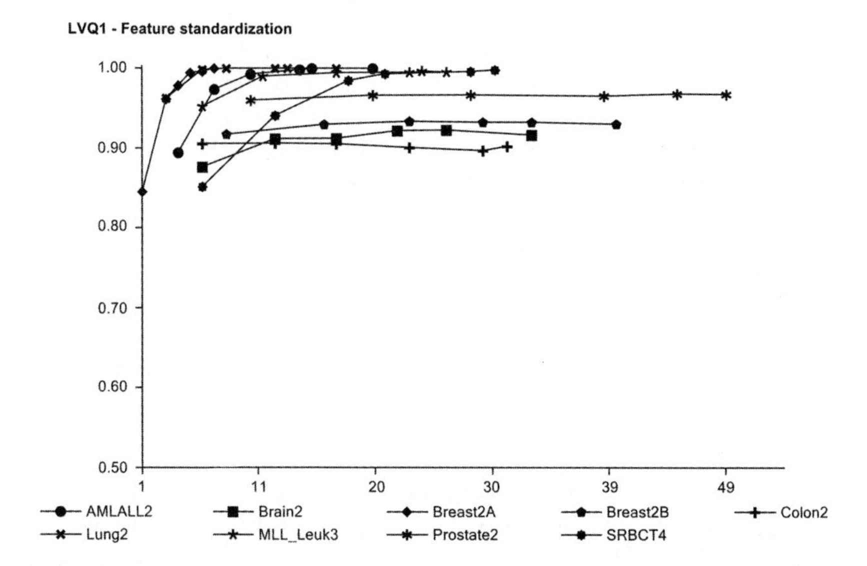

FIGURE 20.9 LVQ1 accuracy with feature standardization as a function of bootstrap sample size used during training. The x-axis values represent the number of arrays randomly sampled $B = 40$ times at fractions $f = 0.1, 0.2, 0.3, 0.4, 0.5$, or 0.6 of the available arrays.

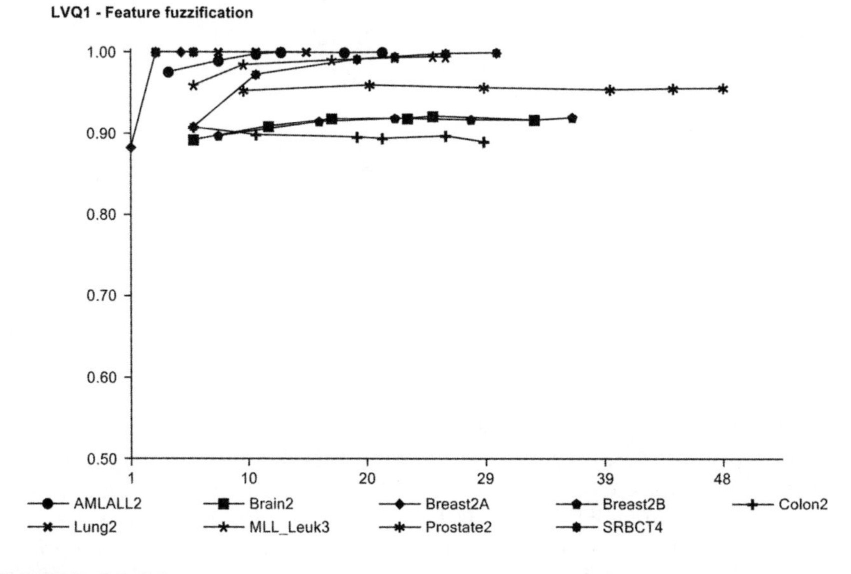

FIGURE 20.10 LVQ1 accuracy with feature fuzzification as a function of bootstrap sample size used during training. The x-axis values represent the number of arrays randomly sampled $B = 40$ times at fractions $f = 0.1, 0.2, 0.3, 0.4, 0.5$, or 0.6 of the available arrays.

fractions $f = 0.1$, 0.2, 0.3, 0.4, 0.5, or 0.6 of the available arrays. Figure 20.9 illustrates the LVQ1 accuracy with feature standardization as a function of bootstrap sample size used during training, based on 10 resamplings from the input arrays. Figure 20.10 shows the LVQ1 accuracy with feature fuzzification as a function of bootstrap sample size used during training, based on 10 resamplings from the input arrays. Results of bootstrapping suggest a small degree of bias early on, where class prediction by the learning algorithm was degraded for smaller sampling fractions and then increased dramatically with increasing bootstrap sample size. Input feature mean-zero standardization and fuzzification did not appreciably improve performance to any degree.

20.4 MULTICLASS ROC CURVES

Multiclass ROC curves were also generated for LVQ1 using bootstrapping. Figure 20.11 illustrates ROC curves for the LVQ1 classifier applied to the

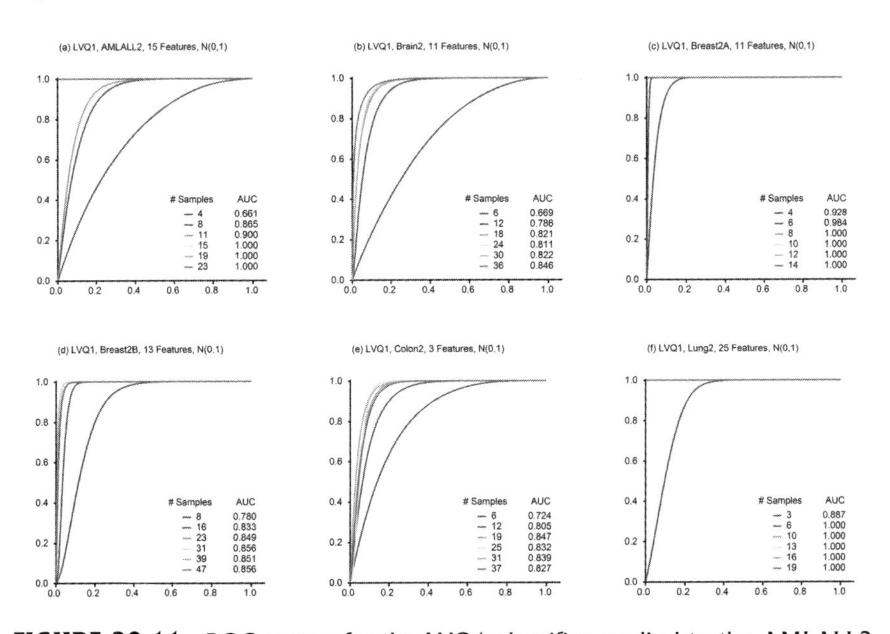

FIGURE 20.11 ROC curves for the LVQ1 classifier applied to the AMLALL2, Brain2, Breast2A, Breast2B, Colon2, Lung2, MLL_Leukemia3, Prostate2, and SRBCT4 datasets based on mean-zero standardized input features; legend values represent the number of arrays randomly sampled $B = 40$ times at fractions $f = 0.1$, 0.2, 0.3, 0.4, 0.5, or 0.6 of the available arrays: (a) AMLALL2; (b) Brain2; (c) Breast2A; (d) Breast2B; (e) Colon2; (f) Lung2; (g) MLL_Leukemia3, (h) Prostate2; (i) SRBCT4. The numbers of genes used for each dataset are listed in Table 12.10.

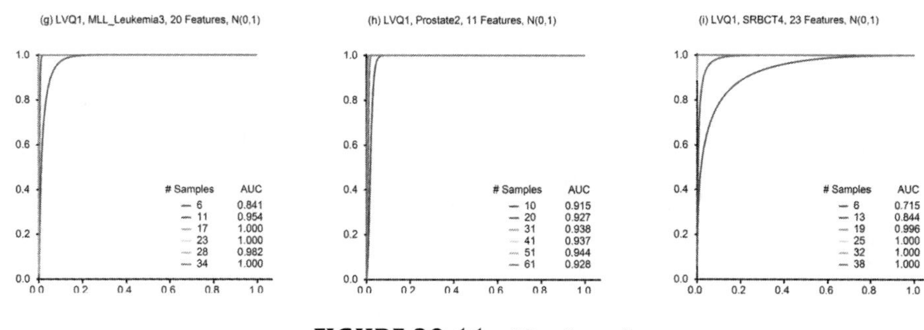

FIGURE 20.11 *(Continued)*

AMLALL2, Brain2, Breast2A, Breast2B, Colon2, Lung2, MLL_Leukemia3, Prostate2, and SRBCT4 datasets using mean-zero standardized input features. The legend values represent the number of arrays randomly sampled $B = 40$ times at fractions $f = 0.1, 0.2, 0.3, 0.4, 0.5$, or 0.6 of the available arrays. Figure 20.12 illustrates a ROC curves for the LVQ1 based on the AMLALL2, Brain2, Breast2A, Breast2B, Colon2, Lung2, MLL_Leukemia3, Prostate2, and SRBCT4 datasets using fuzzified input features. Surprisingly, input feature fuzzification increased AUC remarkably for most of the datasets. This characteristic had not been observed for any of the class prediction algorithms covered prior to this point.

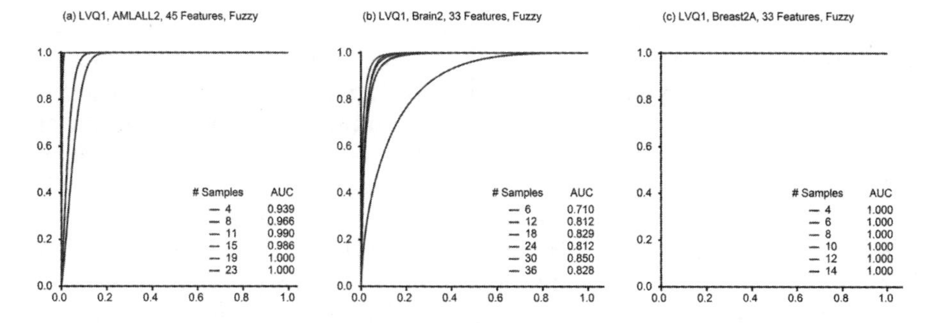

FIGURE 20.12 ROC curves for the LVQ1 classifier applied to the AMLALL2, Brain2, Breast2A, Breast2B, Colon2, Lung2, MLL_Leukemia3, Prostate2, and SRBCT4 datasets based on fuzzified input features; legend values represent the number of arrays randomly sampled $B = 40$ times at fractions $f = 0.1$, 0.2, 0.3, 0.4, 0.5, or 0.6 of the available arrays: (a) AMLALL2; (b) Brain2; (c) Breast2A; (d) Breast2B; (e) Colon2; (f) Lung2; (g) MLL_Leukemia3; (h) Prostate2; (i) SRBCT4. The numbers of genes used for each dataset are listed in Table 12.10.

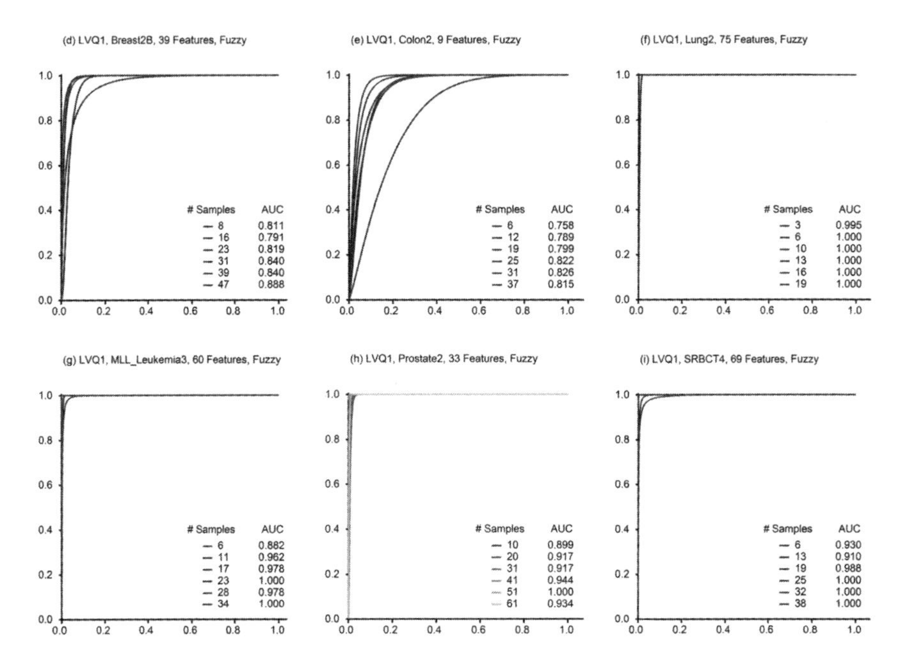

FIGURE 20.12 (*Continued*)

20.5 DECISION BOUNDARIES

The decision boundaries for class prediction of LVQ1 were not too complex and were rather smooth. Input feature fuzzification did not result in a noticeable improvement in performance or reduction in the complexity of class prediction boundaries. Figure 20.13 illustrates a 2D self-organizing map of class decision boundaries for the LVQ1 based on the AMLALL2, Brain2, Breast2A, Breast2B, Colon2, Lung2, MLL_Leukemia3, Prostate2, and SRBCT4 datasets using mean-zero standardized input features. Figure 20.14 illustrates a 2D self-organizing map of class decision boundaries for the LVQ1 based on the AMLALL2, Brain2, Breast2A, Breast2B, Colon2, Lung2, MLL_Leukemia3, Prostate2, and SRBCT4 datasets using fuzzified input features.

20.6 SUMMARY

The LVQ1 is a prototype learning algorithm that is closely related to Kohonen networks and neural gas. Prototypes are introduced into the object (feature) space, and a punishment-reward approach to learning is used to attract

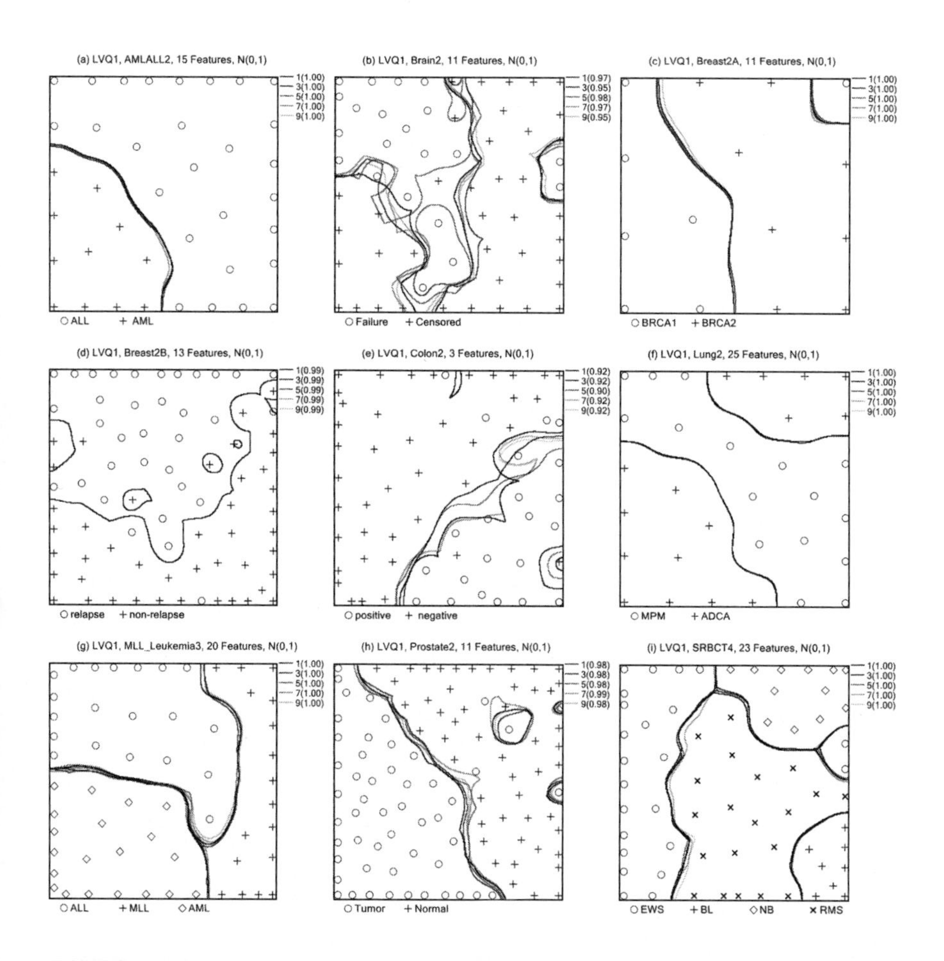

FIGURE 20.13 2D self-organizing maps of LVQ1 class decision boundaries and classification accuracy (upper right of each panel) for the AMLALL2, Brain2, Breast2A, Breast2B, Colon2, Lung2, MLL_Leukemia3, Prostate2, and SRBCT4 datasets based on mean-zero standardized input features: (a) AMLALL2; (b) Brain2; (c) Breast2A; (d) Breast2B; (e) Colon2; (f) Lung2; (g) MLL_Leukemia3; (h) Prostate2; (i) SRBCT4. The numbers of genes used for each dataset are listed in Table 12.10.

objects with the same class labels together and different class labels away. LVQ1 is an entirely distance-based method that is fundamentally similar to CKM, in which cluster centroids are initialized and objects with similar attributes are attracted to and different attributes repelled from each center. LVQ1 is not computationally expensive and will yield relatively good levels of performance.

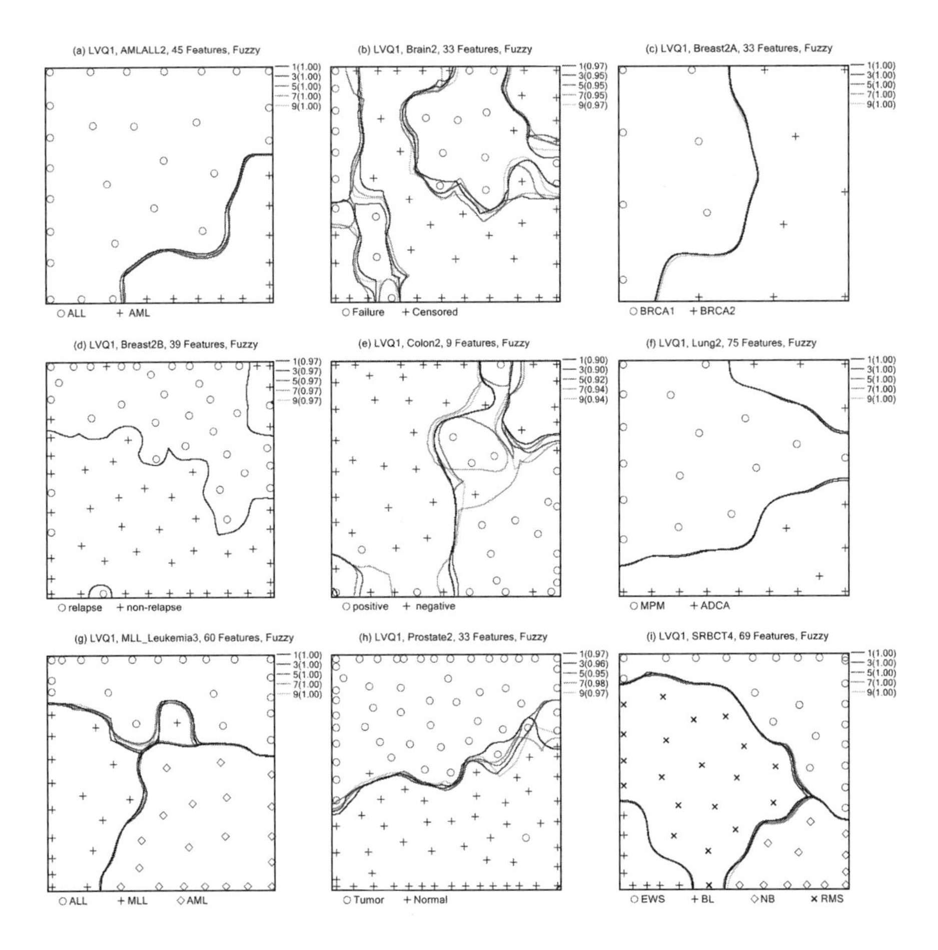

FIGURE 20.14 2D self-organizing maps of LVQ1 class decision boundaries and classification accuracy (upper right of each panel) for the AMLALL2, Brain2, Breast2A, Breast2B, Colon2, Lung2, MLL_Leukemia3, Prostate2, and SRBCT4 datasets based on fuzzified input features: (a) AMLALL2; (b) Brain2; (c) Breast2A; (d) Breast2B; (e) Colon2; (f) Lung2; (g) MLL_Leukemia3; (h) Prostate2; (i) SRBCT4. The numbers of genes used for each dataset are listed in Table 12.10.

REFERENCES

[1] P.R. Tabrizi, S.H. Rezatofighi, M.J. Yazdanpanah. Using PCA and LVQ neural network for automatic recognition of five types of white blood cells. *Conf. Proc. IEEE Eng. Med. Biol. Soc.* **2010**:5593-5596, 2010.

[2] F. Dieterle, S. Muller-Hagedorn, H.M. Liebich, G. Gauglitz. Urinary nucleosides as potential tumor markers evaluated by learning vector quantization. *Artif. Intell. Med.* **28**(3):265-279, 2003.

[3] P. Karakitsos, B. Cochand-Priollet, A. Pouliakis, P.J. Guillausseau, A. Ioakim-Liossi. Learning vector quantizer in the investigation of thyroid lesions. *Anal. Quant. Cytol. Histol.* **21**(3):201-208, 1999.

[4] E. Pesonen, C. Ohmann, M. Eskelinen, M. Juhola. Diagnosis of acute appendicitis in two databases. Evaluation of different neighborhoods with an LVQ neural network. *Methods Inform. Med.* **37**(1):59-63, 1998.

[5] C. Kotropoulos, X. Magnisalis, I. Pitas, M.G. Strintzis. Nonlinear ultrasonic image processing based on signal-adaptive filters and self-organizing neural networks. *IEEE Trans. Image Process.* **3**(1):65-77, 1994.

[6] M. Kubat, D. Flotzinger, G. Pfurtscheller. Towards automated sleep classification in infants using symbolic and subsymbolic approaches. *Biomed. Tech.* (Berlin) **38**(4):73-80, 1993.

[7] D. Flotzinger, J. Kalcher, G. Pfurtscheller. EEG classification by learning vector quantization. *Biomed. Tech.* (Berlin) **37**(12):303-309, 1992.

[8] L.E. Peterson, M.A. Coleman. Machine learning-based receiver operating characteristic (ROC) curves for crisp and fuzzy classification of DNA microarrays in cancer research. *Int. J. Approx. Reason.* **47**(1):17-36, 2008.

[9] J. Li, H. Zha. Simultaneous classification and feature clustering using discriminant vector quantization with applications to microarray data analysis. *Proc. IEEE Computer Society of Bioinformatics Conf.* **1**:246-255, 2002.

CHAPTER 21

LOGISTIC REGRESSION

21.1 INTRODUCTION

Logistic regression is one of the most commonly used class prediction methods for binary or multinomial outcomes [1,2]. This popularity stems from the fundamental assumption of the logistic model that there are no distributional requirements for independent variables, or features. As such, the features used can be nominally, ordinally, or continuously scaled. Logistic regression is one of the best-performing classifiers for class prediction problems. Logistic regression can be used to fit a model when the input features are already *filtered* prior to modeling, or for *wrapping*, whereby features are selected simultaneously during model fitting.

Logistic regression has been a workhorse for biological and clinical applications long before the advent of DNA microarray technology. Binary logistic regression has been popular in two-class clinical investigations focusing on normal and disease classes. Multinomial logistic regression has been popular with biological issues related to multiple outcomes or classes. Ordinal logistic regression is applied to ordinally ranked outcome categories for which, for example, there is either an increasing level of severity with increasing category or decreasing severity with increasing category.

The major difference between logistic regression and other classifiers used in class prediction is that it uses *maximum likelihood* as the optimization algorithm. The method of maximum likelihood uses the following steps for optimization:

1. Define the likelihood equation, which represents the product of individual (microarray-specific) class probabilities.
2. Apply the log transform to both sides of the likelihood equation.
3. Obtain the first and second partial derivatives of the loglikelihood with respect to the individual regression coefficients (weights).

Classification Analysis of DNA Microarrays, First Edition. Leif E. Peterson.

4. Set the derivatives equal to zero, and solve for the parameters, which are the maximum likelihood estimators, also known as the best asymptotic normal estimates.
5. Use the Newton-Raphson procedure to iteratively solve for the maximum likelihood estimates.

Logistic regression has been applied to DNA microarray data in various forms, such as finding markers of various cancers [3-6], antibody microarrays [7], data integration with clinical features [8], penalized logistic regression [9], and logistic discrimination [10]. This chapter discusses only the binomial and polytomous (nominal or multinomial) forms of logistic regression. Readers interested in ordinal logistic regression are referred to the text by Hosmer and Lemeshow [1].

21.2 BINARY LOGISTIC REGRESSION

Binary logistic regression (BLOG) assumes that there are two outcome classes to which microarrays are assigned. Let $y_i = 1$ represent disease and $y_i = 0$ nondisease for subject i having covariate vector $\mathbf{x}$. The probability of disease for subject i is

$$\pi_{i1} = P(y_i = 1|\mathbf{x}_i) = \frac{e^{g_1(\mathbf{x}_i)}}{e^{g_0(\mathbf{x}_i)} + e^{g_1(\mathbf{x}_i)}}, \tag{21.1}$$

and the probability of not having disease is

$$\pi_{i0} = 1 - \pi_{i1} = P(y_i = 0|\mathbf{x}_i) = \frac{e^{g_0(\mathbf{x}_i)}}{e^{g_0(\mathbf{x}_i)} + e^{g_1(\mathbf{x}_i)}}, \tag{21.2}$$

where $g_\omega(\mathbf{x})$ is the *logit* for response category $\omega = 0, 1$. The general definition for the logit is given as

$$\begin{aligned} g_\omega(\mathbf{x}_i) &= \log\left(\frac{\pi_{i\omega}}{\pi_{i0}}\right) \\ &= \log\left(\frac{P(y_i = \omega|\mathbf{x}_i)}{P(y_i = 0|\mathbf{x}_i)}\right) \\ &= \mathbf{x}_i^T \boldsymbol{\beta} \\ &= \beta_0 + \beta_1 x_{i1} + \beta_2 x_{i2} + \cdots + \beta_p x_{ip}. \end{aligned} \tag{21.3}$$

Using the general definition, the logit for nondisease is

$$g_0(\mathbf{x}_i) = \log\left(\frac{\pi_{i0}}{\pi_{i0}}\right)$$
$$= \log(1)$$
$$= 0. \tag{21.4}$$

and thus, $e^{g_0(\mathbf{x}_i)} = 1$. The logit for disease is

$$g_1(\mathbf{x}_i) = \log\left(\frac{\pi_{i1}}{\pi_{i0}}\right)$$
$$= \mathbf{x}_i^T\boldsymbol{\beta}$$
$$= \beta_0 + \beta_1 x_{i1} + \beta_2 x_{i2} + \cdots + \beta_p x_{ip}. \tag{21.5}$$

Through simple substitution of the logit, we obtain the conditional probabilities

$$\pi_{i1} = \frac{e^{g_1(\mathbf{x}_i)}}{1 + e^{g_1(\mathbf{x}_i)}}$$
$$= \frac{1}{1 + e^{-g_1(\mathbf{x}_i)}}, \tag{21.6}$$

and

$$\pi_{i0} = \frac{1}{1 + e^{g_1(\mathbf{x}_i)}}. \tag{21.7}$$

Development of the likelihood equation for logistic regression is hinged to the binomial probability distribution, where interest is in the probability of x successes and $n - x$ failures out of a total of n Bernoulli trials. Let p be the probability of success and $q = 1 - p$ be the probability of failure. A sequence of x successes within n trials occurs with probability

$$\Pr(x) = p^x(1 - p)^{n-x}. \tag{21.8}$$

The number of ways to assign x successes to n trials is determined as

$$\binom{n}{x} = \frac{n!}{x!(n - x)!}, \tag{21.9}$$

and finally, the probability of $X = x$ successes ($x = 1, 2, \ldots, n$) out of n independent trials is

$$\Pr(X = x) = \binom{n}{x} p^x (1-p)^{n-x}. \tag{21.10}$$

When dealing with an individual microarray for the ith patient, we find that the number of trials is one ($n = 1$) and the binomial coefficient $\binom{n}{x}$ drops out of the equation. The probability of disease for the ith patient then becomes

$$\Pr(D) = p_i^{y_i} (q_i)^{1-y_i}. \tag{21.11}$$

Rearranging the above equation, we obtain

$$\Pr(D) = (q_i)^{1-y_i} p_i^{y}, \tag{21.12}$$

and then setting the probability of disease p_i to π_{i1}, and the probability of no disease q_i equal to π_{i0}, the likelihood for n microarrays (patients) is

$$\begin{aligned} L(\boldsymbol{\beta}) &= \prod_{i=1}^{n} (\pi_{i0})^{1-y_i} \pi_{i1}^{y_i} \\ &= \prod_{i=1}^{n} (1 - \pi_{i1})^{1-y_i} \pi_{i1}^{y_i}. \end{aligned} \tag{21.13}$$

Taking the natural logarithm gives the loglikelihood in the form

$$\begin{aligned} \log(L(\boldsymbol{\beta})) &= \sum_{i=1}^{n} \log(1 - \pi_{i1}) - y_i \log(1 - \pi_{i1}) + y_i \log(\pi_{i1}) \\ &= \sum_{i=1}^{n} \log(1 - \pi_{i1}) + y_i \left\{ \log(\pi_{i1}) - \log(1 - \pi_{i1}) \right\} \\ &= \sum_{i=1}^{n} y_i \log\left(\frac{\pi_{i1}}{1 - \pi_{i1}} \right) + \log(1 - \pi_{i1}) \\ &= \sum_{i=1}^{n} y_i \log\left(\frac{\pi_{i1}}{\pi_{i0}} \right) + \log(\pi_{i0}) \\ &= \sum_{i=1}^{n} y_i g_1(\mathbf{x}_i) + \log\left(\frac{1}{1 + e^{g_1(\mathbf{x}_i)}} \right) \end{aligned}$$

$$= \sum_{i=1}^{n} y_i g_1(\mathbf{x}_i) - \log(1 + e^{g_1(\mathbf{x}_i)})$$

$$= \sum_{i=1}^{n} y_i(\mathbf{x}_i^T \boldsymbol{\beta}) - \log(1 + e^{(\mathbf{x}_i^T \boldsymbol{\beta})}). \tag{21.14}$$

The log form of the likelihood equation is linear, and therefore can be solved as a system of linear equations. Next, we solve for the vector of scores, and the information matrix, which will be used in Newton-Raphson iterations to solve for the maximum likelihood estimates. The score is

$$s_j(\boldsymbol{\beta}) = \frac{\partial \log L(\boldsymbol{\beta})}{\partial \beta_j} = \sum_{i=1}^{n} x_i \left(y_i - \frac{e^{(\mathbf{x}_i^T \boldsymbol{\beta})}}{1 + e^{(\mathbf{x}_i^T \boldsymbol{\beta})}} \right), \tag{21.15}$$

and element (j, k) of the information matrix is

$$I_{j,k}(\boldsymbol{\beta}) = \frac{-\partial^2 \log L(\boldsymbol{\beta})}{\partial \beta_j \partial \beta_k} = \sum_{i=1}^{n} x_{ij} x_{ik} \frac{e^{(\mathbf{x}_i^T \boldsymbol{\beta})}}{(1 + e^{(\mathbf{x}_i^T \boldsymbol{\beta})})^2}. \tag{21.16}$$

The Newton-Raphson procedure begins by setting the initial values of vector elements in $\boldsymbol{\beta}$ to low values (e.g., 0.1) for the 0th iteration, which are substituted into equations for s_j and $I_{j,k}$. Next, the update vector is determined with the matrix operation

$$\boldsymbol{\delta}_i = \mathbf{I}(\boldsymbol{\beta})^{-1} \mathbf{s}(\boldsymbol{\beta}), \tag{21.17}$$

which is added to the previous estimate of $\boldsymbol{\beta}_i$ in the form

$$\boldsymbol{\beta}_{i+1} = \boldsymbol{\beta}_i + \boldsymbol{\delta}_i. \tag{21.18}$$

At each iteration, newly calculated values of $\boldsymbol{\beta}_{i+1}$ are substituted into Equations (21.15) and (21.16), leading to new estimates of $\boldsymbol{\delta}_i$ and $\boldsymbol{\beta}_{i+1}$. Iterations are performed until convergence is reached when $||\boldsymbol{\delta}_i|| < \epsilon$. Values of ϵ are typically specified in the range $10^{-8} \leq \epsilon \leq 10^{-4}$. Depending on the sample space and features used, the loglikelihood will increase rapidly with the increasing number of iterations. Most software applications set the number of iterations to 20-25, since typical problems with hundreds of records and, for example, 10 features result in very rapid decreases of $||\boldsymbol{\delta}_i||$ within several iterations.

The importance of each feature is determined during model construction using the loglikelihood ratio statistic G. The G for a feature is equal to minus twice the ratio of the loglikelihood of a *reduced* model without the feature and the log-likelihood of the *full* model with the feature, given as

$$\begin{aligned} G &= -2\left(\frac{\mathrm{LL}_{\text{reduced}}}{\mathrm{LL}_{\text{full}}}\right) \\ &= -2(\mathrm{LL}_{\text{reduced}} - \mathrm{LL}_{\text{full}}), \end{aligned} \tag{21.19}$$

where $\mathrm{LL} = \log(L(\boldsymbol{\beta}))$ given in (21.14), representing the sum of loglikelihood values for individual microarrays. G is χ^2 distributed with one degree of freedom (1 d.f.). Therefore, values of G exceeding $\chi^2_{(1)} = 3.84$ one suggest that the additional feature in the full model is statistically significant in terms of contribution to the likelihood.

Assessing Model Goodness of Fit. A full examination of the appropriateness of a logistic model includes an evaluation of how well the assumed model fits the data; this criterion is called *goodness of fit* (GOF). GOF is largely a combination of the effects of the features selected for describing class membership of the microarrays. GOF is always applied after a model has been fitted. Recall that the residual for a microarray is essentially $(y_i - \hat{y}_i)$. The fitted regression value for the ith subject in class ω is $\hat{y}_{i\omega}$, and is determined as

$$\hat{\pi}_{i\omega} = P(y = \omega|\mathbf{x}_i) = \frac{e^{g_\omega(\mathbf{x}_i)}}{1 + \sum_{l=1}^{\Omega-1} e^{g_l(\mathbf{x}_i)}}. \tag{21.20}$$

The *Pearson residual* is

$$r(y_{i\omega}, \hat{\pi}_{i\omega}) = \frac{y_{i\omega} - \hat{\pi}_{i\omega}}{\sqrt{\hat{\pi}_{i\omega}(1 - \hat{\pi}_{i\omega})}}, \tag{21.21}$$

and the Pearson GOF statistic is

$$\chi^2_{n-p+1} = \sum_{i=1}^{n} r(y_{i\omega}, \hat{\pi}_{i\omega})^2. \tag{21.22}$$

Another commonly used residual is the *deviance residual*, which is calculated as

$$d(y_{i\omega}, \hat{\pi}_{i\omega}) = \sqrt{2|\log(\hat{\pi}_{i\omega})|}. \tag{21.23}$$

The deviance GOF statistic is

$$D = \sum_{i=1}^{n} d(y_{i\omega}, \hat{\pi}_{i\omega})^2. \tag{21.24}$$

Assuming that the model fits, the distribution of Pearson and deviance residuals is χ^2 distributed with $n-p+1$ degrees of freedom. For deviance, the GOF statistic D is the likelihood ratio of a model with n parameters and one with $p-1$ parameters. GOF values of χ^2 or D that are less than the d.f. $(n-p+1)$ indicate a significant fit; that is, the model used and data appropriately describe microarray membership in the outcome classes.

Classification Procedures. Binary logistic regression (BLOG) employs a maximum likelihood optimization approach to model all classification problems in the form of multiple two-class problems. For example, a four-class problem equates to $6=[4(4-1)/2]$ two-class problems. Assume a training scenario comparing training microarrays from class l and class m $(l \neq m, l, m = 1, 2, \ldots, \Omega)$. Set $y_i = 0$ if the ith microarray is from class l and $y_i = 1$ if the ith microarray is from class m. Logistic regression first requires computation of the logit $g_1(\mathbf{x}_i) = \beta_0 + \beta_1 x_{i1} + \beta_2 x_{i2} + \cdots + \beta_p x_{ip}$, which is hinged to regression coefficients modeled during the fitting procedure. Maximum likelihood modeling is performed, and after convergence, the probability that a test microarray $\mathbf{x}$ is in class l is $P(l|\mathbf{x}) = P(y=0|\mathbf{x}) = 1/[1+e^{g_1(\mathbf{x})}]$ and the probability of class m membership is $1-P(l|\mathbf{x})$. Hence, the two decision rules for test microarray $\mathbf{x}$ are $D_{lm} = P(l|\mathbf{x})$ and $D_{ml} = 1 - D_{lm}$. The intermediate decision rule for class l is

$$D_l(\mathbf{x}) = \sum_{m=1, m\neq l}^{\Omega} D_{lm}, \tag{21.25}$$

and the final decision rule for test microarray $\mathbf{x}$ is

$$D(\mathbf{x} \rightsquigarrow \omega) = \arg\max_{l\in\Omega}\{D_l(\mathbf{x})\}. \tag{21.26}$$

21.3 POLYTOMOUS LOGISTIC REGRESSION

Multinomial logistic regression was fundamentally developed for multiple class problems where $\Omega > 2$, unlike the dichotomous two-class nature of binary logistic regression described in the previous section. Specifically, a multinomial logistic model can be either *ordinal*, with ranked outcome categories, or *nominal*, with outcome categories having no mathematical relationship between them. In this chapter, we strictly limit discussion to the nominal type of multinomial regression, and call this *polytomous* logistic regression (PLOG). To begin, assume a multiclass problem in which the dependent variable $y_{\omega i}$ for each ith microarray is coded 0 or 1 to represent membership in class ω, where $\omega = 0, 1, \ldots, \Omega - 1$, and Ω is the number of classes. Let's assume that the reference class (corner point) is the first class for which $\omega = 0$. The general definition of the logit for multinomial logistic regression remains the same, as in

$$\begin{aligned} g_\omega(\mathbf{x}_i) &= \log\left(\frac{\pi_{i\omega}}{\pi_{i0}}\right) \\ &= \log\left(\frac{P(y_i = \omega|\mathbf{x})}{P(y_i = 0|\mathbf{x}_i)}\right) \\ &= \mathbf{x}_i^T \mathbf{b}_\omega \\ &= b_{\omega 0} + b_{\omega 1}x_{i1} + b_{\omega 2}x_{i2} + \cdots + b_{\omega p}x_{ip}, \end{aligned} \tag{21.27}$$

where $\mathbf{b}_\omega$ is the column of coefficients in the $p \times (\Omega - 1)$ matrix of coefficients $\mathbf{B}$. However, the conditional probability adds together the exponentiated logits in the denominator, shown in functional form as

$$\pi_{i\omega} = P(y_{i\omega} = \omega|\mathbf{x}_i) = \frac{e^{g_\omega(\mathbf{x}_i)}}{\sum_{l=0}^{\Omega-1} e^{g_l(\mathbf{x}_i)}} = \frac{e^{g_\omega(\mathbf{x}_i)}}{1 + \sum_{l=1}^{\Omega-1} e^{g_l(\mathbf{x}_i)}}. \tag{21.28}$$

For a three-class example ($\omega = 0, 1, 2$) the probability that a microarray $\mathbf{x}$ has membership in each of the three classes is

$$\begin{aligned} \pi_{i0} &= \frac{e^{g_0(\mathbf{x}_i)}}{e^{g_0(\mathbf{x}_i)} + e^{g_1(\mathbf{x}_i)} + e^{g_2(\mathbf{x}_i)}} \\ &= \frac{1}{1 + e^{g_1(\mathbf{x}_i)} + e^{g_2(\mathbf{x}_i)}} \\ \pi_{i1} &= \frac{e^{g_1(\mathbf{x}_i)}}{1 + e^{g_1(\mathbf{x}_i)} + e^{g_2(\mathbf{x}_i)}} \\ \pi_{i2} &= \frac{e^{g_2(\mathbf{x}_i)}}{1 + e^{g_1(\mathbf{x}_i)} + e^{g_2(\mathbf{x}_i)}}. \end{aligned} \tag{21.29}$$

The likelihood for PLOG is

$$\begin{aligned} L(\mathbf{B}) &= \prod_{i=1}^{n} (\pi_{i0})^{1-\sum_{\omega=1}^{\Omega-1} y_{i\omega}} \prod_{\omega=1}^{\Omega-1} \pi_{i\omega}^{y_{i\omega}} \\ &= \prod_{i=1}^{n} \left(1 - \sum_{\omega=1}^{\Omega-1} \pi_{i\omega}\right)^{1-\sum_{\omega=1}^{\Omega-1} y_{i\omega}} \prod_{\omega=1}^{\Omega-1} \pi_{i\omega}^{y_{i\omega}}. \end{aligned} \tag{21.30}$$

Taking the natural logarithm gives the loglikelihood in the following form:

$$
\begin{aligned}
\log(L(\mathbf{B})) &= \sum_{i=1}^{n} \log\left(1 - \sum_{\omega=1}^{\Omega-1} \pi_{i\omega}\right) - \log\left(1 - \sum_{\omega=1}^{\Omega-1} \pi_{i\omega}\right) \\
&\quad \times \sum_{\omega=1}^{\Omega-1} y_{i\omega} + \sum_{\omega=1}^{\Omega-1} \log(\pi_{i\omega}^{y_{i\omega}}) \\
&= \sum_{i=1}^{n} \log\left(1 - \sum_{\omega=1}^{\Omega-1} \pi_{i\omega}\right) - \log\left(1 - \sum_{\omega=1}^{\Omega-1} \pi_{i\omega}\right) \\
&\quad \times \sum_{\omega=1}^{\Omega-1} y_{i\omega} + \sum_{\omega=1}^{\Omega-1} y_{i\omega} \log(\pi_{i\omega}) \\
&= \sum_{i=1}^{n} \log\left(1 - \sum_{\omega=1}^{\Omega-1} \pi_{i\omega}\right) + \sum_{\omega=1}^{\Omega-1} y_{i\omega} \left\{ \log(\pi_{i\omega}) - \log\left(1 - \sum_{\omega=1}^{\Omega-1} \pi_{i\omega}\right)\right\} \\
&= \sum_{i=1}^{n} \log\left(1 - \sum_{\omega=1}^{\Omega-1} \pi_{i\omega}\right) + \sum_{\omega=1}^{\Omega-1} y_{i\omega} \left\{ \log\left(\frac{\pi_{i\omega}}{(1 - \sum_{\omega=1}^{\Omega-1} \pi_{i\omega})}\right)\right\} \\
&= \sum_{i=1}^{n} \sum_{\omega=1}^{\Omega-1} y_{i\omega} \log\left(\frac{\pi_{i\omega}}{\pi_{i0}}\right) + \log\left(1 - \sum_{\omega=1}^{\Omega-1} \pi_{i\omega}\right) \\
&= \sum_{i=1}^{n} \sum_{\omega=1}^{\Omega-1} y_{i\omega} g_{\omega}(\mathbf{x}_i) + \log(\pi_{i0}) \\
&= \sum_{i=1}^{n} \sum_{\omega=1}^{\Omega-1} y_{i\omega} g_{\omega}(\mathbf{x}_i) + \log\left(\frac{1}{1 + \sum_{\omega=1}^{\Omega-1} e^{g_{\omega}(\mathbf{x}_i)}}\right) \\
&= \sum_{i=1}^{n} \sum_{\omega=1}^{\Omega-1} y_{i\omega} g_{\omega}(\mathbf{x}_i) - \log\left(1 + \sum_{\omega=1}^{\Omega-1} e^{g_{\omega}(\mathbf{x}_i)}\right) \\
&= \sum_{i=1}^{n} \sum_{\omega=1}^{\Omega-1} y_{i\omega} (\mathbf{x}_i^T \mathbf{b}_{\omega}) - \log\left(1 + \sum_{\omega=1}^{\Omega-1} e^{(\mathbf{x}_i^T \mathbf{b}_{\omega})}\right). \qquad (21.31)
\end{aligned}
$$

In binary logistic regression the coefficients were obtained in the form of a $p \times 1$ column vector $\boldsymbol{\beta}$. In PLOG, however, there is a total of $p(\Omega - 1)$ coefficients, resulting in a $(p(\Omega - 1) \times p(\Omega - 1)$ matrix $\mathbf{B}$. The score value for logit ω ($\omega = 1, 2, \ldots, \Omega - 1$) and the jth variable ($j = 1, 2, \ldots, p$) will occupy the $s_{p\omega,j}$ element in the score matrix **S**, given as

$$\begin{aligned} s_{p(\omega-1),j}(\mathbf{B}) = \frac{\partial \log L(\mathbf{B})}{\partial b_{\omega j}} &= \sum_{i=1}^{n} x_{ij} \left(y_{i\omega} - \frac{e^{(\mathbf{x}_i^T \mathbf{b}_\omega)}}{1 + \sum_{l=1}^{\Omega-1} e^{(\mathbf{x}_i^T \mathbf{b}_l)}} \right) \\ &= \sum_{i=1}^{n} x_{ij} \left(y_{i\omega} - \pi_{i\omega} \right). \end{aligned} \tag{21.32}$$

The information matrix **I** is a partitioned matrix of size $p(\Omega - 1) \times p(\Omega - 1)$. Thus, if there are three classes ($J = 3$), there will be two logit functions, each of which is represented with a $p \times p$ matrix $\mathbf{I}_{\omega\omega}$:

$$\mathbf{I}(\mathbf{B}) = \begin{pmatrix} \mathbf{I}_{11}(\mathbf{B}) & \mathbf{I}_{12}(\mathbf{B}) \\ \mathbf{I}_{21}(\mathbf{B}) & \mathbf{I}_{22}(\mathbf{B}) \end{pmatrix}. \tag{21.33}$$

By definition, the second partial derivative for different variables (j, j') within the same logit (ω, ω), that is, same matrix, is

$$I_{\omega j,\omega j'}(\mathbf{B}) = \frac{-\partial^2 \log L(\mathbf{B})}{\partial b_{\omega j} \partial b_{\omega j'}} = -\sum_{i=1}^{n} x_{ij'} x_{ij} \pi_{i\omega}(1 - \pi_{i\omega}), \tag{21.34}$$

whereas the second partial derivative for different variables (j, j') in different logits (ω, ω'), that is, different matrices, is

$$I_{\omega j,\omega' j'}(\mathbf{B}) = \frac{-\partial^2 \log L(\mathbf{B})}{\partial b_{\omega j} \partial b_{\omega' j'}} = \sum_{i=1}^{n} x_{ij'} x_{ij} \pi_{i\omega} \pi_{i\omega'}. \tag{21.35}$$

Similarly, maximum likelihood estimates of each $\beta_{\omega j}$ can be found by determining the matrix **B** iteratively with the matrix manipulation

$$\mathbf{B}_{i+1} = \mathbf{B}_i + \mathbf{I}(\mathbf{B})^{-1} \mathbf{S}(\mathbf{B}), \tag{21.36}$$

until convergence is reached when $||\mathbf{B}|| < \epsilon$. Values of ϵ are typically in the range $10^{-8} \leq \epsilon \leq 10^{-4}$.

Classification Procedures. Polytomous logistic regression accommodates all of the classes considered in a single run, and only $\Omega - 1$ logits are needed, where the general equation for the logit is

$$\begin{aligned} g_\omega(\mathbf{x}_i) &= \log \left(\frac{P(y_{i\omega} = \omega | \mathbf{x}_i)}{P(y_{i\omega} = 0 | \mathbf{x}_i)} \right) \\ &= b_{\omega 0} + b_{\omega 1} x_{i1} + b_{\omega 2} x_{i2} + \cdots + b_{\omega p} x_{ip}. \end{aligned} \tag{21.37}$$

Because y_{i0} drops out of the likelihood equation, only classes $\omega = 2, 3, \ldots, \Omega$ are needed for setting $y_{i\omega}$ for each microarray. For example, during analysis of a four-class problem, a training microarray from class 2 would require the coding $y_{i1} = 1, y_{i2} = 0$, and $y_{i3} = 0$ in order to represent classes 2-4. After the model is fit, the committee vote $(y = 0, 1, \ldots, \Omega - 1)$ for class 1 is

$$P(y = 0|\mathbf{x}) = \frac{1}{1 + \sum_{\omega=1}^{\Omega-1} e^{g_\omega(\mathbf{x})}}, \tag{21.38}$$

and for all other classes is

$$P(y = \omega|\mathbf{x}) = \frac{e^{g_\omega(\mathbf{x})}}{1 + \sum_{\omega=1}^{\Omega-1} e^{g_\omega(\mathbf{x})}}. \tag{21.39}$$

The final decision rule for class ω is

$$D(\mathbf{x} \rightsquigarrow \omega) = \underset{\omega \in \Omega - 1}{\arg\max}\{P(y = \omega|(\mathbf{x})\}. \tag{21.40}$$

21.4 CROSS-VALIDATION RESULTS

We used the nine datasets employed throughout this book for PLOG classification, based on input genes selected with a greedy plus-takeaway-1 method (see Table 12.10). Bootstrap accuracy ("CVB") was based on average 0.632 bootstrap accuracy (13.4) for 10 bootstraps. Table 21.1 lists cross-validation (CV) results, which indicate very little bias and variance of PLOG classification performance over the datasets employed. Evaluation runs for assessing

TABLE 21.1 Cross-Validation Results for Polytomous Logistic Regression (PLOG)

	Cross-validation method		
Dataset	CV5	CV10	CV-1
AMLALL2	1.00	1.00	1.00
Brain2	0.89	0.90	0.90
Breast2A	1.00	1.00	1.00
Breast2B	0.94	0.95	0.97
Colon2	0.91	0.90	0.92
Lung2	1.00	1.00	1.00
MLL_Leukemia3	1.00	1.00	1.00
Prostate2	0.97	0.97	0.98
SRBCT4	1.00	1.00	1.00

bias and variance during CV runs became problematic in terms of reaching convergence when running CVB and CV2, because smaller fractions ($\leq 60\%$) of the data were used for model training. However, there was no difficulty in reaching convergence for the nine datasets when running CV5, CV10, and CV $-$ 1.

Bootstrap bias and multiclass ROC curves were not generated for the PLOG classifier since the sample size of arrays used in training are based on fractions of 10-60 of the total input arrays for each dataset. For small sample sizes, there is a significant chance that convergence of PLOG will degenerate, whereby the likelihood will not be maximized.

21.5 DECISION BOUNDARIES

Polytomous logistic regression resulted in some of the smoothest decision boundaries generated for the classifiers covered thus far. As we pointed out above, however, when bootstrapping with varying sample sizes or running cross-validation with CV2 for which only half the data are used for training, there is potential for PLOG to run into convergence problems. Figure 21.1 illustrates 2D self-organizing maps of class decision boundaries for the PLOG based on the AMLALL2, Brain2, Breast2A, Breast2B, Colon2, Lung2, MLL_Leukemia3, Prostate2, and SRBCT4 datasets using mean-zero standardized input features. Figure 21.2 illustrates 2D self-organizing maps of class decision boundaries for the PLOG based on the AMLALL2, Brain2, Breast2A, Breast2B, Colon2, Lung2, MLL_Leukemia3, Prostate2, and SRBCT4 datasets using fuzzified input features.

21.6 SUMMARY

Polytomous logistic regression uses a maximum likelihood approach for solving regression parameters and can simultaneously determine the statistical significance of each model input feature. The Newton-Raphson method used is based on matrix operations on first- and second-order partial derivatives; therefore, the input features used should be normalized or mean-zero standardized in order to remove scale effects. It is also important to note that the information matrix $\mathbf{I}(\mathbf{B})$ and coefficient matrix $\mathbf{B}$ both have dimensions $(\Omega - 1)p \times (\Omega - 1)p$, so there must be enough objects in the database to minimize the chance of overfitting the model. As an example, if there are only $n = 30$ objects in a $\Omega = 5$ class dataset with $p = 5$ features, the information matrix $\mathbf{I}(\mathbf{B})$ and coefficient matrix $\mathbf{B}$ will each have dimensions 20×20, which will likely result in an overfitted or *overparametrized* model. Models that are overfitted assimilate too much information about any noise that is present in a dataset, causing bias in the generalization. This reduces the robustness of

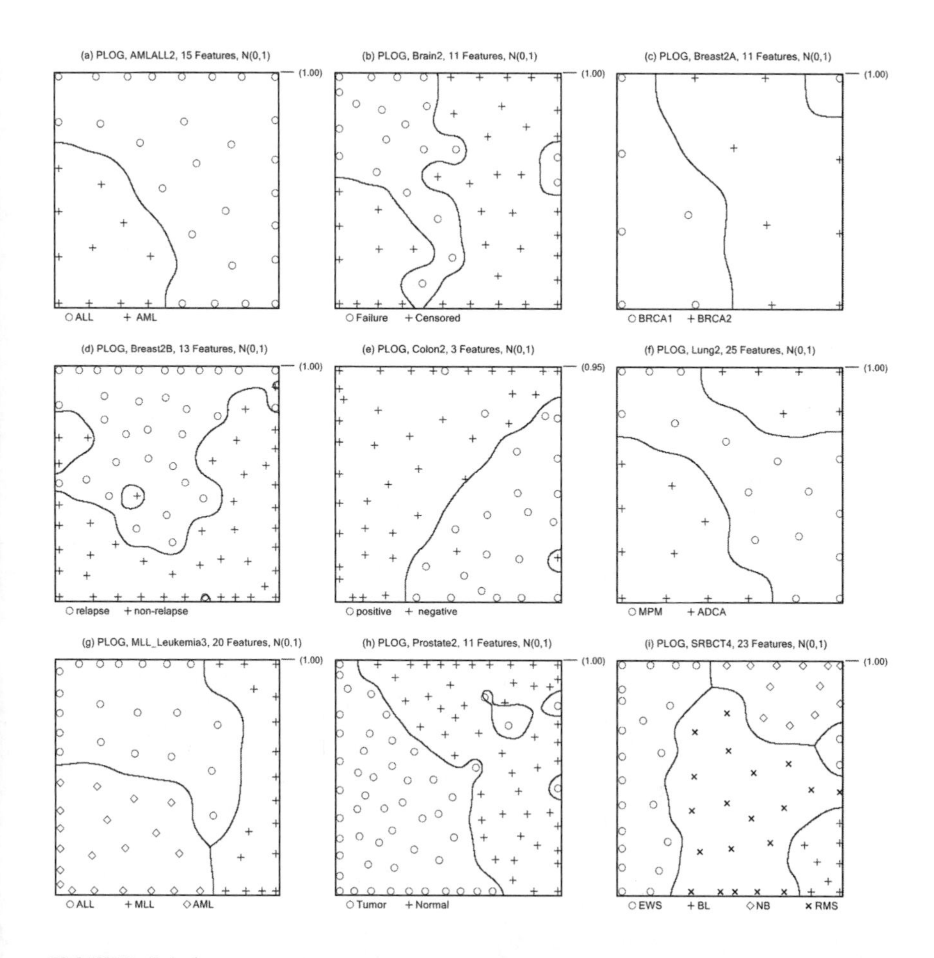

FIGURE 21.1 2D self-organizing maps of PLOG class decision boundaries and classification accuracy (upper right of each panel) for the AMLALL2, Brain2, Breast2A, Breast2B, Colon2, Lung2, MLL_Leukemia3, Prostate2, and SRBCT4 datasets based on mean-zero standardized input features: (a) AMLALL2; (b) Brain2; (c) Breast2A; (d) Breast2B; (e) Colon2; (f) Lung2; (g) MLL_Leukemia3; (h) Prostate2; (i) SRBCT4. The numbers of genes used for each dataset are listed in Table 12.10.

the findings. In the statistical literature, the rule of thumb for having enough *power* based on regression multiple R^2 is that one should have 10 times the number of cases (objects) as variables, that is, $n = 10p$. This does not wholly mean that a model will be invalid if $n < 10p$, but rather, that the power may not be sufficient for observing a significant F test for the multiple R^2 if the null hypothesis H_0: $R^2 = 0$ is false.

In summary, PLOG did not seem to be biased to the data to a large extent as the performance did not change dramatically over the three CV

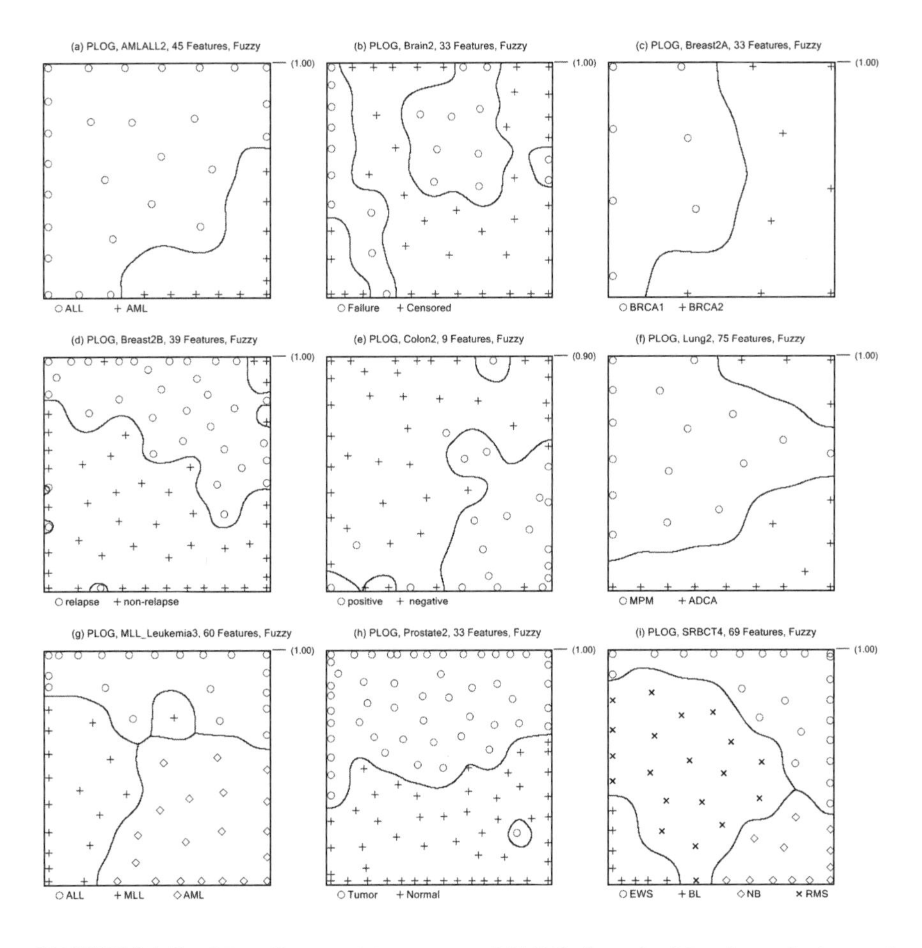

FIGURE 21.2 2D self-organizing maps of PLOG class decision boundaries and classification accuracy (upper right of each panel) for the AMLALL2, Brain2, Breast2A, Breast2B, Colon2, Lung2, MLL_Leukemia3, Prostate2, and SRBCT4 datasets based on fuzzified input features: (a) AMLALL2; (b) Brain2; (c) Breast2A; (d) Breast2B; (e) Colon2; (f) Lung2; (g) MLL_Leukemia3; (h) Prostate2; (i) SRBCT4. The numbers of genes used for each dataset are listed in Table 12.10.

methods that were employed. The decision boundaries were quite smooth, and PLOG assigned class membership of the tested microarrays quite well. The fact that PLOG is based on maximum likelihood methods indicates that it is more expensive computationally than, say, KNN, which is a lazy learner, but the performance of PLOG can exceed those many other classifiers.

REFERENCES

[1] D.W. Hosmer, S. Lemeshow. *Applied Logistic Regression*. Wiley, New York, 1989.

[2] E.T. Lee. A computer program for linear logistic regression. *Comput. Prog. Biomed.* **4**:80–92, 1974.

[3] A. Ben Hamida, I.S. Labidi, K. Mrad, E. Charafe-Jauffret, S. Ben Arab, B. Esterni, L. Xerri, P. Viens, F. Bertucci, D. Birnbaum, J. Jacquemier. Markers of subtypes in inflammatory breast cancer studied by immunohistochemistry: Prominent expression of P-cadherin. *BMC Cancer* **8**:28–, 2008.

[4] L. Dyrskjot, K. Zieger, F.X. Real, N. Malats, A. Carrato, C. Hurst, S. Kotwal, M. Knowles, P.U. Malmstrom, M. de la Torre, K. Wester, Y. Allory, D. Vordos, A. Caillault, F. Radvanyi, A.M. Hein, J.L. Jensen, K.M. Jensen, N. Marcussen, T.F. Orntoft. Gene expression signatures predict outcome in non-muscle-invasive bladder carcinoma: A multicenter validation study. *Clin. Cancer Res.* **13**(12):3545–3551, 2007.

[5] G. Weber, S. Vinterbo, L. Ohno-Machado. Multivariate selection of genetic markers in diagnostic classification. *Artif. Intell. Med.* **31**(2):155–167, 2004.

[6] M. Sanchez-Carbayo, N.D. Socci, J.J. Lozano, W. Li, E. Charytonowicz, T.J. Belbin, M.B. Prystowsky, A.R. Ortiz, G. Childs, C. Cordon-Cardo. Gene discovery in bladder cancer progression using cDNA microarrays. *Am. J. Pathol.* **163**(2):505–516, 2003.

[7] R. Orchekowski, D. Hamelinck, L. Li, E. Gliwa, M. vanBrocklin, J.A. Marrero, G.F. Vande Woude, Z. Feng, R. Brand, B.B. Haab. Antibody microarray profiling reveals individual and combined serum proteins associated with pancreatic cancer. *Cancer Res.* **65**(23):11193–11202, 2005.

[8] A.J. Stephenson, A. Smith, M.W. Kattan, J. Satagopan, V.E. Reuter, P.T. Scardino, W.L. Gerald. Integration of gene expression profiling and clinical variables to predict prostate carcinoma recurrence after radical prostatectomy. *Cancer* **104**(2):290–298, 2005.

[9] J. Zhu, T. Hastie. Classification of gene microarrays by penalized logistic regression. *Biostatistics* **5**(3):427–443, 2004.

[10] D.V. Nguyen, D.M. Rocke. Tumor classification by partial least squares using microarray gene expression data. *Bioinformatics* **18**(1):39–50, 2002.

CHAPTER 22

SUPPORT VECTOR MACHINES

22.1 INTRODUCTION

Support vector machines (SVMs) have been applied in many areas of investigation, including object classification with fMRI [1], B-cell epitopes [2], Alzheimer's diagnosis with PET images [3], arrhythmia detection with ECG signals [4], ECG analysis for hypoglycemia detection [5], recognition of white blood cells [6], cloud computing service architecture [7], fatigue monitoring [8], DNA sequence analysis [9], nanopore detection in chemoinformatics [10], enrofloxacin and ciprofloxacin in urine and plasma [11], and feed bagging in the animal feed industry [12]. Application of SVM to DNA microarray data has been quite popular; examples include use of gene pairs in classification [13], outcome prediction [14], ovarian cancer [15], outlier removal [16], intelligent parameter selection [17], prediction of BRCA1 and BRCA2 genotypes [18], alternate splicing in prostate cancer [19], regularization [20], neuroblastoma prognosis [21], discriminating toxicants [22], and recurrence of liver cancer [23].

The SVM algorithm was originally introduced by Boser et al. [24], and developed further over the next several years by Vapnik and colleagues [25,26]. SVM maps the input data into a higher-dimensional kernel space to find a hyperplane that optimally splits objects in different classes; therefore, SVM is recommended for classes that are not linearly separable. SVM usage with DNA microarray became popular in the early 2000s, when more and more classifiers were beginning to be applied in the field [27].

22.2 HARD-MARGIN SVM FOR LINEARLY SEPARABLE CLASSES

Let $(\mathbf{x}_1, y_1), \ldots, (\mathbf{x}_n, y_n)$ be a training dataset with $\mathbf{x}_i \in \mathbb{R}^p$ and class labels $y_i \in \{-1, +1\}$. For a two-class problem, class labels are assigned as follows

Classification Analysis of DNA Microarrays, First Edition. Leif E. Peterson.

$$y_i = \begin{cases} +1 & \mathbf{x}_i \in \omega = 1 \\ -1 & \mathbf{x}_i \in \omega = 2, \end{cases} \tag{22.1}$$

where $\omega = \{1, 2\}$ are the classes [24]. In addition, let $\mathbf{w}$ be a p-dimensional weight vector that is orthogonal to the hyperplane and b a bias term, both of which are determined during training. The decision rule for each training array is

$$D(\mathbf{x}_i) = \mathbf{w}^T \mathbf{x}_i + b. \tag{22.2}$$

The optimal hyperplane that splits the two classes of arrays is defined as $D(\mathbf{x}) = D(\mathbf{w}^T\mathbf{x} + b) = 0$ and the distance between each array $\mathbf{x}$ and the hyperplane is $|D(\mathbf{x})|/||\mathbf{w}||$. Figure 22.1 illustrates an *optimal hyperplane*, which splits two classes of arrays, centered between two *maximal margins* that are a distance M from the optimal hyperplane, and *support vectors* shown as the filled ▲ and ■ lying on the maximal margins. The goal of SVM is to minimize the Euclidean norm $||\mathbf{w}||$ to impose the inequality $y_i D(\mathbf{x}_i)/||\mathbf{w}|| > 1$ [24]. This is accomplished by using both the margin value M and the weight vector $\mathbf{w}$ and enforcing the constraint $M||\mathbf{w}|| = 1$, which when solving for the optimal margin, gives $M = 1/||\mathbf{w}||$. When the training data are linearly separable, the condition described above results in the following constraints

$$\begin{aligned} \mathbf{w}^T\mathbf{x}_i + b &\geq +1 \quad \text{for} \quad y_i = +1 \\ \mathbf{w}^T\mathbf{x}_i + b &\leq -1 \quad \text{for} \quad y_i = -1, \end{aligned} \tag{22.3}$$

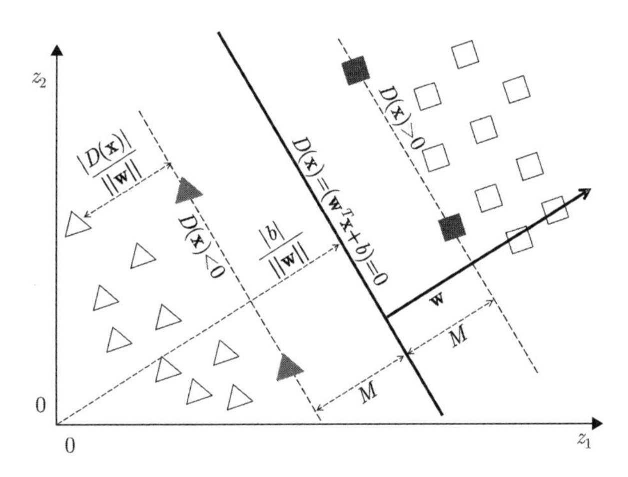

FIGURE 22.1 Geometric representation of central optimal hyperplane, maximal margins with support vectors (filled ▲ and ■), and two classes of objects; x values are transformed to z scores on the basis of mean-zero standardization; bold $\mathbf{x}$ are simply used to denote objects (microarrays).

which can be multiplied by y_i to give the inequality

$$y_i(\mathbf{w}^T \mathbf{x}_i + b) - 1 \geq 0. \tag{22.4}$$

We now introduce a Lagrange multiplier α_i for each inequality to obtain the unconstrained Lagrangian function

$$\begin{aligned} L(\mathbf{w}, b, \boldsymbol{\alpha}) &= \frac{1}{2}\mathbf{w}^T \mathbf{w} - \sum_{i=1}^{n} \alpha_i [y_i(\mathbf{w}^T \mathbf{x}_i + b) - 1] \\ \text{subject to} \quad & \alpha_i \geq 0 \\ y_i(\mathbf{w}^T \mathbf{x}_i + b) - 1 &= 0, \end{aligned} \tag{22.5}$$

for the which the Karush-Kuhn-Tucker (KKT) conditions are met [28, 29]. The KKT conditions $\alpha_i \geq 0$ and $y_i(\mathbf{w}^T \mathbf{x}_i + b) - 1 = 0$ ensure that the solutions $\mathbf{w}$, b, and $\boldsymbol{\alpha}$ exist and are optimal for the solution of $L(\mathbf{w}, b, \boldsymbol{\alpha})$. Values of $\alpha_i \neq 0$ are support vectors, which lie on the optimal margins. The objective function to optimize is found by first deriving the partial derivatives of $L(\mathbf{w}, b, \boldsymbol{\alpha})$ w.r.t. $\mathbf{w}$ and b. These partial derivatives are

$$\frac{\partial L(\mathbf{w}, b, \boldsymbol{\alpha})}{\partial \mathbf{w}} = \mathbf{w} - \sum_{i=1}^{n} \alpha_i y_i \mathbf{x}_i = \mathbf{0}, \tag{22.6}$$

$$\frac{\partial L(\mathbf{w}, b, \boldsymbol{\alpha})}{\partial b} = \sum_{i=1}^{n} \alpha_i y_i = 0, \tag{22.7}$$

and when solving each for $\mathbf{w}$ and b, b drops out and we are left with only

$$\mathbf{w} = \sum_{i=1}^{n} y_i \alpha_i \mathbf{x}_i. \tag{22.8}$$

We can now substitute $\mathbf{w}$ with $\sum_{i=1}^{n} y_i \alpha_i \mathbf{x}_i$ in the Lagrangian and then expand as follows:

$$\begin{aligned} L(\mathbf{w}, b, \boldsymbol{\alpha}) &= \frac{1}{2}\mathbf{w}^T \mathbf{w} - \sum_{i=1}^{n} \alpha_i [y_i(\mathbf{w}^T \mathbf{x}_i + b) - 1] \\ &= \frac{1}{2} \sum_{i,j=1}^{n} y_i y_j \alpha_i \alpha_j \mathbf{x}_i^T \mathbf{x}_j - \sum_{i,j=1}^{n} y_i y_j \alpha_i \alpha_j \mathbf{x}_i^T \mathbf{x}_j + \sum_{i}^{n} \alpha_i \\ &= \sum_{i}^{n} \alpha_i - \frac{1}{2} \sum_{i,j=1}^{n} y_i y_j \alpha_i \alpha_j \mathbf{x}_i^T \mathbf{x}_j. \end{aligned} \tag{22.9}$$

The final objective function is optimized in terms of α_i using either gradient ascent or least-squares matrix operations, and these are discussed in the next sections.

22.3 KERNEL MAPPING INTO NONLINEAR FEATURE SPACE

The abovementioned methods work optimally for linearly separable problems with decision boundaries, which are wholly linear. For nonlinear hyperplanes, however, this approach will suffer and deliver lower performance yields. To increase performance, a *kernel function* is incorporated into the Lagrangian as a means of mapping the input space into a nonlinear *feature space*, which potentially enhances linear separability. The kernel function $H(\mathbf{x}_i, \mathbf{x}_j)$ is essentially a microarray $\times$ microarray($n \times n$) dot product matrix, also known as a *gram matrix*, which provides the nonlinear mapping. The kernel function must meet Mercer's condition [30] for being positive semidefinite, such that there are no negative eigenvalues; this ensures that the problem will be concave quadratic. Some commonly used kernels are

$$\begin{aligned} H(\mathbf{x}_i, \mathbf{x}_j) &= \mathbf{x}_i^T \mathbf{x}_j && \text{(linear)} \\ &= (\mathbf{x}_i^T \mathbf{x}_j + 1)^d && \text{(polynomial)} \\ &= \exp\left(-\gamma ||\mathbf{x}_i - \mathbf{x}_j||\right) && \text{(RBF).} \end{aligned} \tag{22.10}$$

Gaussian RBF kernels are likely to yield the greatest class prediction accuracy, provided that a suitable choice of γ is used. To determine an optimum value of γ for use with RBF kernels, a grid search is carried out using incremental values of γ from $2^{-15}, 2^{-13}, \ldots, 2^3$ in order to evaluate accuracy for all training microarrays. Kernels must also be normalized; thus, polynomial kernels are normalized as

$$H(\mathbf{x}_i, \mathbf{x}_j) = \frac{(\mathbf{x}_i^T \mathbf{x}_j + 1)^d}{(p+1)^d}, \tag{22.11}$$

where p is the number of training features. Gaussian RBF kernels are normalized in the form

$$H(\mathbf{x}_i, \mathbf{x}_j) = \exp\left(-\frac{\gamma}{p} ||\mathbf{x}_i - \mathbf{x}_j||\right). \tag{22.12}$$

Figure 22.2 shows a nonlinear checkerboard pattern of objects in two classes that was solved straightforwardly using an RBF kernel function.

22.4 SOFT-MARGIN SVM FOR NONLINEARLY SEPARABLE CLASSES

Objects are rarely linearly separable and require complex decision boundaries for partitioning. Instead of minimizing $||\mathbf{w}||$ with a hard-margin SVM, we

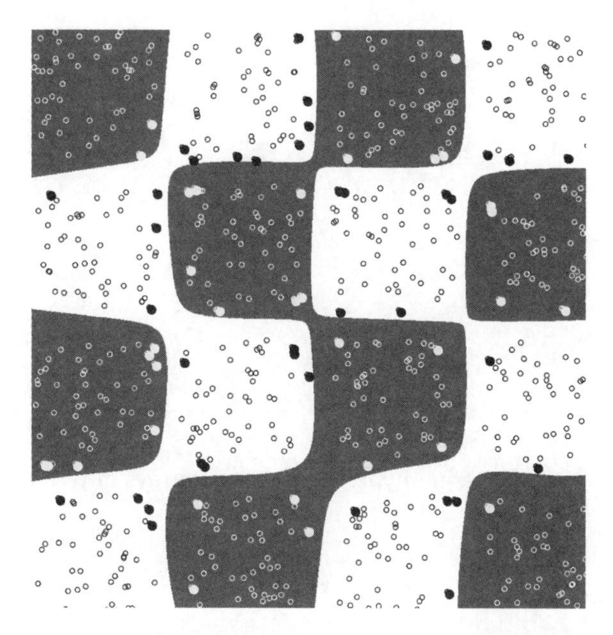

FIGURE 22.2 Nonlinearly separable objects for which an RBF kernel function is ideal for class assignment. There are 78 unbounded support vectors and zero bounded support vectors shown in the results. Parameter values were $C = 100$ and $\gamma = 3$ for the RBF kernel.

can employ a *soft-margin* SVM using preselected values for $||\mathbf{w}||$ and C and determine values of a *slack vector* $\boldsymbol{\xi}$ to "pick up the slack" during optimization. In this approach, the margin parameter C is varied for each fit, and during each run the norm $||\boldsymbol{\xi}||$ is minimized [25]. Figure 22.3 shows a soft-margin hyperplane with slack vector elements ξ_i. The *p-norm* soft-margin optimization problem can now be stated in the functional form

$$
\begin{aligned}
& \frac{1}{2}\mathbf{w}^T\mathbf{w} + \frac{C}{2}\sum_{i=1}^{n}\xi_i^p \\
\text{subject to} \quad & y_i(\mathbf{w}^T\mathbf{x}_i + b) \geq 1 - \xi_i, \\
& \xi_i \geq 0,
\end{aligned}
\tag{22.13}
$$

and on substitution of the Lagrange multipliers, this becomes

$$
L(\mathbf{w}, b, \boldsymbol{\alpha}, \boldsymbol{\xi}) = \frac{1}{2}\mathbf{w}^T\mathbf{w} + \frac{C}{2}\sum_{i=1}^{n}\xi_i^p - \sum_{i=1}^{n}\alpha_i[y_i(\mathbf{w}^T\mathbf{x}_i + b) - 1 + \xi_i]. \tag{22.14}
$$

This Lagrangian is often optimized by using either quadratic programming, gradient ascent, or least-squares approaches. In the next two sections, we introduce the gradient ascent and least-squares methods of optimization.

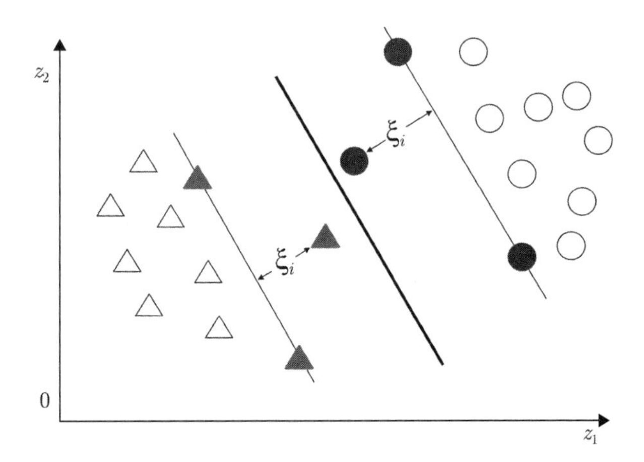

FIGURE 22.3 Soft-margin hyperplane split of arrays in two classes showing slack vector $\boldsymbol{\xi}$ elements.

22.5 GRADIENT ASCENT SOFT-MARGIN SVM

We now apply gradient ascent learning to solve the soft-margin Lagrangian objective function (22.14). First, set the partial derivatives of (22.14) with respect to the parameters equal to zero to obtain the following partial derivatives:

$$\frac{\partial L(\mathbf{w}, b, \boldsymbol{\alpha}, \boldsymbol{\xi})}{\partial \mathbf{w}} = \mathbf{w} - \sum_{i=1}^{n} \alpha_i y_i \mathbf{x}_i = \mathbf{0}, \tag{22.15}$$

$$\frac{\partial L(\mathbf{w}, b, \boldsymbol{\alpha}, \boldsymbol{\xi})}{\partial \xi_i} = C\boldsymbol{\xi} - \boldsymbol{\alpha} = \mathbf{0}, \tag{22.16}$$

$$\frac{\partial L(\mathbf{w}, b, \boldsymbol{\alpha}, \boldsymbol{\xi})}{\partial b} = \sum_{i=1}^{n} \alpha_i y_i = 0. \tag{22.17}$$

Resubstituting these relationships into the primal, we get the dual objective function

$$\begin{aligned} L(\boldsymbol{\alpha}) &= \sum_{i=1}^{n} \alpha_i - \frac{1}{2} \sum_{i,j=1}^{n} y_i y_j \alpha_i \alpha_j H(\mathbf{x}_i, \mathbf{x}_j) + \frac{1}{2C} \boldsymbol{\alpha}^T \boldsymbol{\alpha} - \frac{1}{C} \boldsymbol{\alpha}^T \boldsymbol{\alpha} \\ &= \sum_{i=1}^{n} \alpha_i - \frac{1}{2} \sum_{i,j=1}^{n} y_i y_j \alpha_i \alpha_j H(\mathbf{x}_i, \mathbf{x}_j) - \frac{1}{2C} \boldsymbol{\alpha}^T \boldsymbol{\alpha} \\ &= \sum_{i=1}^{n} \alpha_i - \frac{1}{2} \sum_{i,j=1}^{n} y_i y_j \alpha_i \alpha_j \left(H(\mathbf{x}_i, \mathbf{x}_j) + \frac{I(i=j)}{C} \right), \end{aligned} \tag{22.18}$$

where $I(i=j)/C$ is the indicator function equal to zero when $i \neq j$ and $1/C$ when $i=j$, which, when added to the diagonal element $H(\mathbf{x}_i, \mathbf{x}_i)$ of the gram matrix, ensures positive definiteness.

The partial derivative of (22.18) with respect to $\boldsymbol{\alpha}$ is

$$\frac{\partial L(\boldsymbol{\alpha})}{\partial \alpha_i} = 1 - y_i \sum_{j=1}^{n} y_j \alpha_j \left(H(\mathbf{x}_i, \mathbf{x}_j) + \frac{I(i=j)}{C} \right), \tag{22.19}$$

and α_1 is obtained iteratively in the form

$$\alpha_i^{g+1} = \alpha_i^g + \frac{1}{H(\mathbf{x}_i, \mathbf{x}_i)} \left(1 - y_i \sum_{j=1}^{n} y_j \alpha_j^g \left(H(\mathbf{x}_i, \mathbf{x}_j) + \frac{I(i=j)}{C} \right) \right), \tag{22.20}$$

where α_i^g is the value of α_i at the gth iteration. A grid search in the range of $10^{-2}, 10^{-1}, \ldots, 10^4$ for the margin parameter C is typically used in most implementations, and if k-fold cross-validation (CV) is employed, then the median value of C obtained from k grid searches for the training folds used provides a smooth value of C that works well when predicting class membership of arrays in the fold left out of training. The only difference between the $L^1(\xi)$ and $L^2(\xi^2)$ soft norm SVM is the constraint $0 \leq \alpha_i \leq C$ for L^1 norm and $\alpha_i \geq 0$ for L^2, which is monitored during the learning iterations performed in (22.20). The decision function for class l is

$$D_l(\mathbf{x}) = \sum_{m \neq l, m=1}^{\Omega} \operatorname{sign}(D_{lm}(\mathbf{x})), \tag{22.21}$$

where

$$D_{lm}(\mathbf{x}) = \sum_{i=1}^{n} \alpha_i y_i \mathbf{x}_i^T \mathbf{x} + b_{lm} \tag{22.22}$$

for a two-class problem involving classes l and m. The decision rule for test microarray $\mathbf{x}$ is

$$D_l(\mathbf{x} \leadsto \omega) = \underset{l=1,2,\ldots,\Omega}{\arg\max} \{D_l(\mathbf{x})\}. \tag{22.23}$$

22.5.1 Cross-Validation Results

The effect of CV on classification performance of SVMGA was assessed using the nine datasets and number of input genes listed in Table 12.10. Bootstrap accuracy ("CVB") was based on average 0.632 bootstrap accuracy (13.4) for 10 bootstraps. Table 22.1 lists the average accuracy based on 10 repartitions of

TABLE 22.1 Performance without Feature Transformation

Dataset	d^a	Cross-validation method CV2	CV5	CV10	CV − 1	CVB
AMLALL2	1	100	100	100	100	100
	2	100	100	100	100	100
	3	99.74	99.21	100	100	100
	4	100	100	100	100	100
	R	100	100	100	100	98.82
Brain2	1	92.17	89.50	86.67	88.33	94.86
	2	91.33	92.67	90.83	91.67	95.15
	3	91.50	87.83	89.83	90.00	95.14
	4	92.50	90.00	88.67	91.67	94.04
	R	91.33	91.83	89.67	93.33	95.04
Breast2A	1	100	100	100	100	100
	2	100	100	100	100	100
	3	100	100	100	100	100
	4	100	100	100	100	100
	R	100	100	100	100	100
Breast2B	1	90.90	93.08	95.38	97.44	95.82
	2	85.64	93.21	94.10	100	93.97
	3	90.77	93.59	93.59	98.72	93.94
	4	90.38	93.85	95.51	91.03	95.17
	R	91.41	92.31	94.62	97.44	93.16
Colon2	1	83.55	83.71	74.03	85.48	81.93
	2	81.45	76.77	83.39	77.42	82.62
	3	87.74	83.87	85.00	85.48	82.98
	4	87.26	86.13	77.90	66.13	83.80
	R	85.16	85.81	84.84	61.29	80.76
Lung2	1	100	100	100	100	100
	2	100	100	100	100	100
	3	100	100	100	100	100
	4	100	100	100	100	100
	R	100	100	100	100	100
MLL_Leukemia3	1	99.82	99.82	99.82	100	99.81
	2	99.65	98.42	99.65	100	96.91
	3	99.65	99.82	99.30	100	99.79
	4	99.82	99.65	99.82	100	99.83
	R	99.82	99.65	99.65	100	99.75
Prostate2	1	96.76	97.55	97.94	98.04	94.55
	2	96.27	96.47	97.84	98.04	97.61
	3	96.57	94.71	97.16	98.04	96.99
	4	95.39	95.49	96.96	98.04	97.83
	R	96.37	97.06	97.35	97.06	98.58
SRBCT4	1	99.84	100	100	100	99.95
	2	99.84	100	100	100	99.95
	3	100	99.84	100	100	99.97
	4	99.84	99.84	100	100	99.98
	R	99.84	100	100	100	99.95
Average		0.96	0.96	0.96	0.96	0.96

[a] The term d denotes the power of the kernel function and when equal to R, represents the RBF kernel.

the input arrays and subsequent CV, without any input feature transformations applied. Table 22.2 lists the average accuracy based on 10 repartitions of the input arrays and subsequent CV, using mean-zero standardized input feature values. Table 22.3 lists the average accuracy based on 10 repartitions of the input arrays and subsequent CV, using fuzzified input feature values. Figure 22.4 illustrates the average accuracy of SVMGA as a function of the CV method for each dataset without input feature transformations, and suggests somewhat strong bias of SVMGA due to the large change in accuracy of the Colon2, Brain2, and Breast2 datasets over the various CV methods. Figure 22.5 shows standard deviation of accuracy as a function of CV method for each dataset for 10 repartitions and indicates smaller variance for LOOCV because of the greater amount of training data used. Figure 22.6 reveals average accuracy for all datasets as a function of feature transformation and CV method, which shows that input feature fuzzification worsened mean accuracy, and mean-zero standardization did not markedly improve performance. Figure 22.7 reveals the SVMGA accuracy without transformations on features as a function of kernel power d in (22.10) and indicates that the Colon2 dataset is linearly separable since the accuracy was lower when $d > 1$. Figure 22.8 shows SVMGA accuracy with feature standardization as a function of kernel power d, which illustrates that accuracy for the Brain2 dataset was reduced when compared with no feature transformations. Figure 22.9 illustrates the SVMGA accuracy with feature fuzzification as a function of kernel power d, and indicates that feature fuzzification reduced the mean accuracy for the Breast2B dataset. Figure 22.10 shows average accuracy for all datasets as a function of kernel parameter d and feature transformation method. On average, there does not appear to be a large difference in the specific kernel used with the nine datasets; however, each specific dataset can result in drastically varying results over changing values of d.

22.5.2 Bootstrap Bias

Bootstrap bias was evaluated for SVMGA using the nine datasets listed in Table 12.10. Bootstrap bias runs were based on determining the 0.632 bootstrap accuracy (see Eq. 13.4) for each sampling fraction. Acc_0 in 13.4 was based on test accuracy of unsampled arrays, while Acc_b in 13.4 was based on test accuracy for sampled arrays. Figure 22.11 shows the SVMGA accuracy without transformations on features as a function of bootstrap sample size used during training, based on 10 resamplings from the input arrays. The x-axis values represent the number of arrays randomly sampled $B = 40$ times at fractions $f = 0.1, 0.2, 0.3, 0.4, 0.5$, or 0.6 of the available arrays. Not much bootstrap bias was presented by SVMGA, as there was little change in mean accuracy as more data were bootstrapped for training. Figure 22.12 illustrates the SVMGA accuracy with feature standardization

TABLE 22.2 Performance with Feature Standardization

		Cross-validation method				
Dataset	d^a	CV2	CV5	CV10	CV − 1	CVB
AMLALL2	1	100	100	100	100	100
	2	100	100	100	100	100
	3	100	100	100	100	100
	4	100	100	100	100	100
	R	100	100	100	100	100
Brain2	1	92.00	91.50	88.83	91.67	93.48
	2	91.33	92.33	87.50	90.00	95.08
	3	89.33	89.50	87.33	91.67	95.18
	4	91.17	88.50	89.17	93.33	95.17
	R	91.33	89.83	89.33	91.67	94.61
Breast2A	1	100	100	100	100	100
	2	100	100	100	100	100
	3	100	100	100	100	100
	4	100	100	100	100	100
	R	100	100	100	100	100
Breast2B	1	83.59	92.44	94.87	100	93.62
	2	90.64	93.59	94.74	98.72	91.71
	3	90.38	92.82	94.62	97.44	94.81
	4	89.62	93.85	95.38	97.44	96.18
	R	83.97	92.18	95.64	96.15	94.30
Colon2	1	85.16	84.19	82.42	66.13	85.82
	2	81.61	83.06	71.94	87.10	83.53
	3	83.55	82.26	76.94	82.26	85.14
	4	88.06	80.97	74.03	37.10	81.83
	R	86.29	81.61	82.58	72.58	83.17
Lung2	1	100	100	100	100	100
	2	100	100	100	100	100
	3	100	100	100	100	100
	4	100	100	100	100	100
	R	100	100	100	100	100
MLL_Leukemia3	1	98.77	99.82	100	100	99.85
	2	99.82	99.47	99.30	100	99.82
	3	99.65	99.65	99.47	100	99.90
	4	99.82	99.12	99.82	98.25	99.72
	R	99.65	99.65	97.72	100	99.84
Prostate2	1	96.76	96.57	97.75	97.06	97.07
	2	95.39	95.88	97.75	97.06	97.32
	3	96.27	96.37	97.35	97.06	98.33
	4	95.88	96.86	96.76	98.04	98.49
	R	96.57	96.57	97.55	97.06	98.45
SRBCT4	1	99.68	100	100	100	99.98
	2	99.68	100	100	100	99.97
	3	99.68	100	100	100	100
	4	99.84	100	100	100	99.98
	R	100	100	100	100	99.99
Average		0.95	0.96	0.95	0.95	0.97

[a]The term *d* denotes the power of the kernel function and when equal to *R*, represents the RBF kernel.

TABLE 22.3 Performance with Feature Fuzzification

Dataset	d^a	Cross-validation method				
		CV2	CV5	CV10	CV − 1	CVB
AMLALL2	1	100	100	100	100	100
	2	100	100	100	100	100
	3	100	100	100	100	100
	4	100	100	100	100	100
	R	100	100	100	100	100
Brain2	1	87.83	85.33	88.33	86.67	94.02
	2	84.50	85.33	86.67	90.00	86.55
	3	84.67	89.50	85.83	88.33	91.70
	4	81.33	85.17	83.00	91.67	93.59
	R	86.33	87.50	84.83	86.67	90.84
Breast2A	1	100	100	100	100	100
	2	100	100	100	100	100
	3	100	100	100	100	100
	4	100	100	100	100	100
	R	100	100	100	100	100
Breast2B	1	79.74	83.21	81.15	85.90	90.70
	2	82.56	81.54	84.36	82.05	91.94
	3	81.54	81.92	82.82	84.62	89.20
	4	82.18	80.77	81.92	84.62	90.14
	R	83.21	82.82	84.49	82.05	89.63
Colon2	1	79.52	77.26	79.68	64.52	85.18
	2	80.16	78.71	80.48	64.52	83.75
	3	82.90	82.42	77.26	40.32	84.79
	4	81.45	82.26	74.84	40.32	82.25
	R	84.84	81.29	76.29	67.74	84.73
Lung2	1	100	100	100	100	100
	2	100	100	100	100	100
	3	100	100	100	100	100
	4	100	100	100	100	100
	R	100	100	100	100	100
MLL_Leukemia3	1	97.72	97.19	95.79	98.25	99.29
	2	97.72	98.07	97.54	98.25	98.91
	3	95.96	97.37	97.89	98.25	95.69
	4	98.07	97.89	97.54	98.25	97.70
	R	96.32	95.44	96.67	96.49	99.11
Prostate2	1	92.06	93.92	91.96	90.20	93.53
	2	92.55	93.82	91.37	90.20	95.46
	3	92.84	94.02	92.16	85.29	96.81
	4	93.33	93.73	92.25	90.20	96.28
	R	93.14	92.65	92.35	88.24	95.98
SRBCT4	1	99.21	100	99.68	100	96.95
	2	98.73	99.84	99.84	100	99.79
	3	98.89	99.52	99.84	100	99.78
	4	99.52	99.68	99.52	100	99.55
	R	98.41	99.52	99.84	100	99.73
Average		0.93	0.93	0.93	0.91	0.95

[a]The term *d* denotes the power of the kernel function and when equal to *R*, represents the RBF kernel.

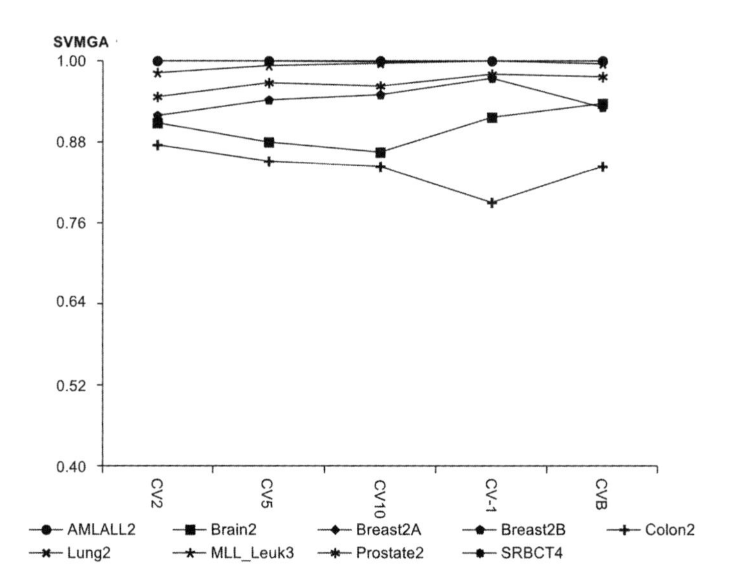

FIGURE 22.4 SVMGA accuracy as a function of cross-validation (CV) method for each dataset. Average based on CV accuracy for 10 repartitions of arrays. No feature transformations applied.

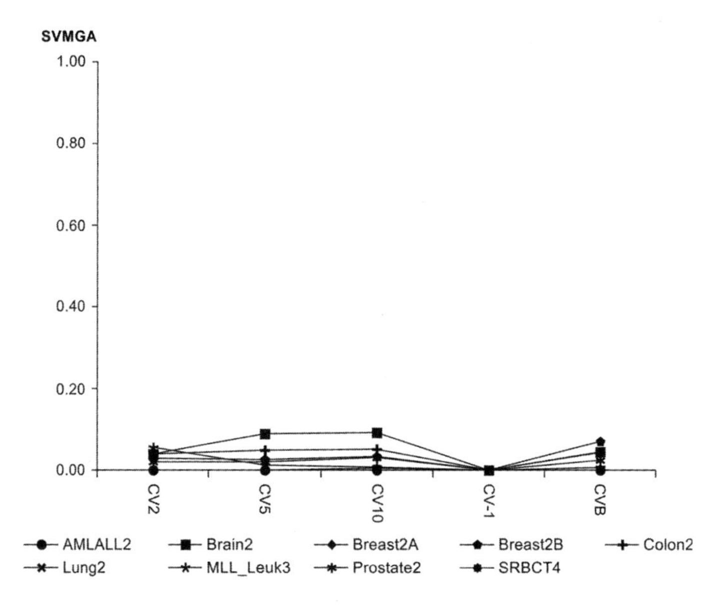

FIGURE 22.5 SVMGA standard deviation of accuracy as a function of CV method for each dataset. Standard deviation based on CV accuracy for 10 repartitions of arrays. No feature transformations applied.

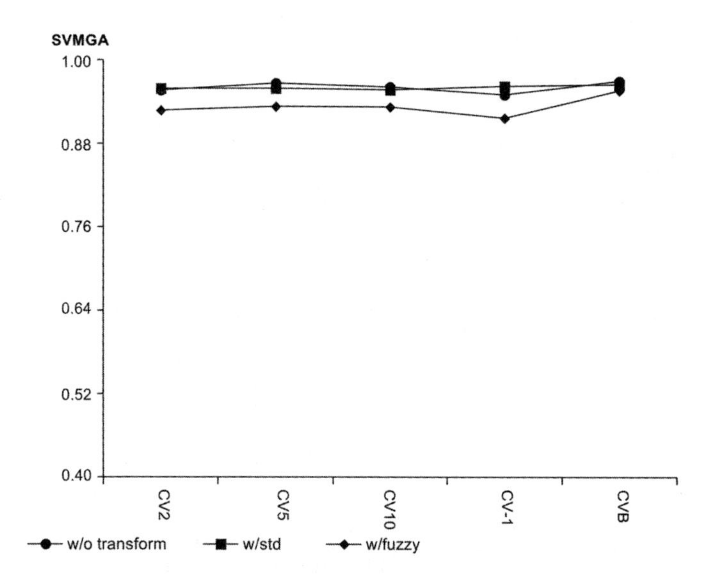

FIGURE 22.6 SVMGA average accuracy for all datasets as a function of feature transformation and CV method.

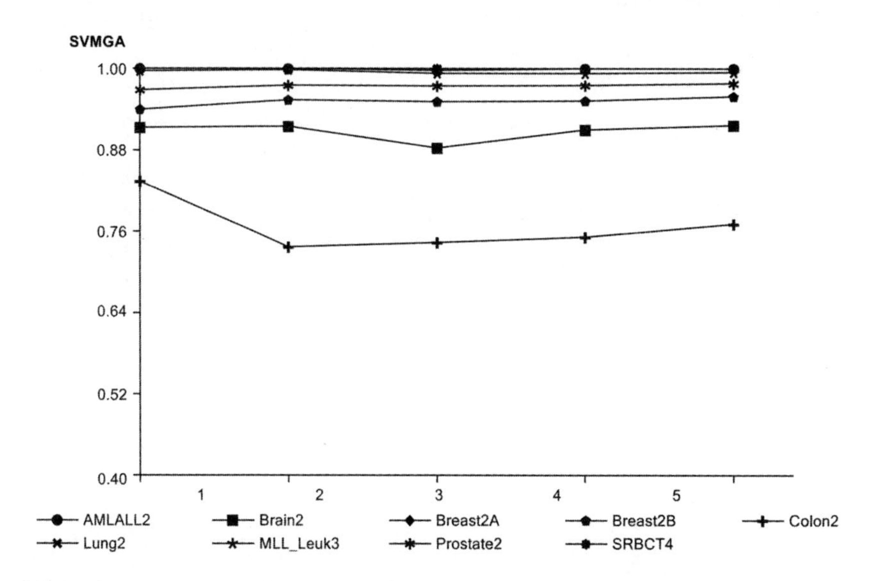

FIGURE 22.7 SVMGA accuracy without transformations on features as a function of kernel power d in (22.10). The Colon2 dataset is observed to be linearly separable since accuracy drops when $d > 1$. The x axis represents the kernel power d, and $d = 5$ represents an RBF kernel.

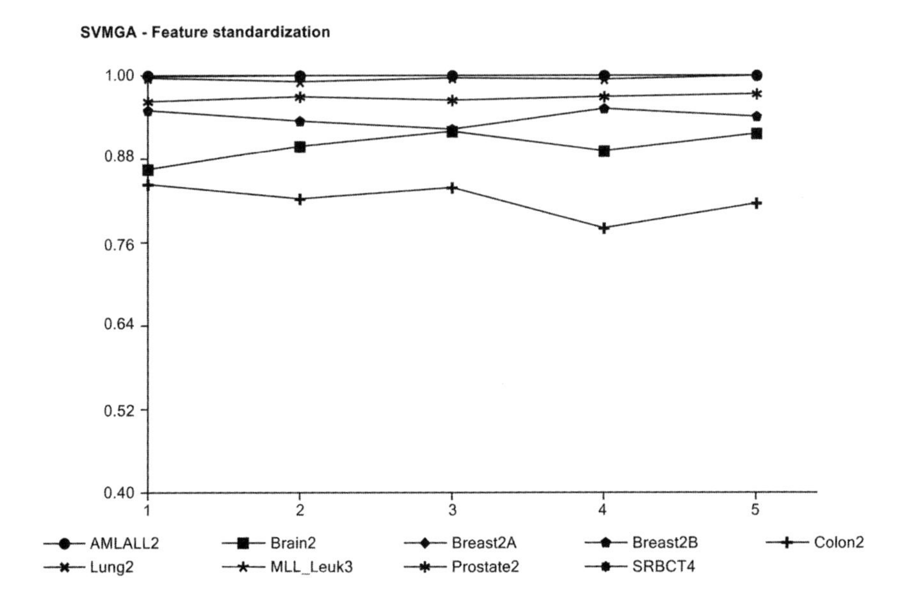

FIGURE 22.8 SVMGA accuracy with feature standardization as a function of kernel power d in (22.10). The x axis represents the kernel power d, and $d = 5$ represents an RBF kernel.

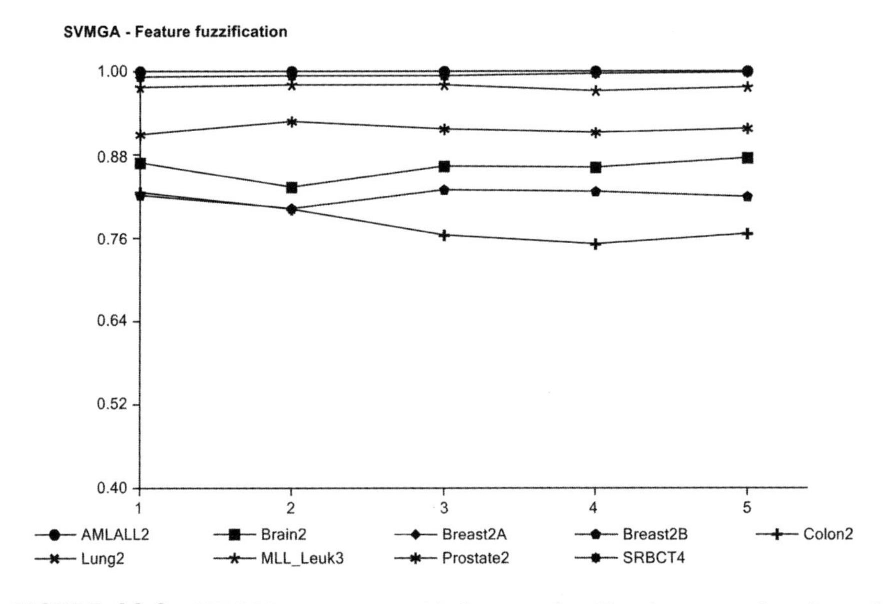

FIGURE 22.9 SVMGA accuracy with feature fuzzification as a function of kernel power d in (22.10). The x axis represents the kernel power d, and $d = 5$ represents an RBF kernel.

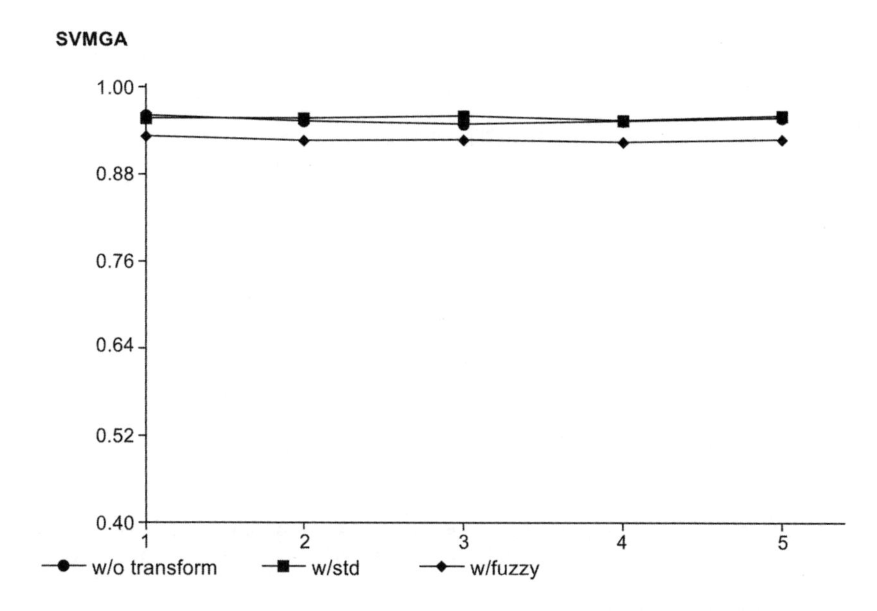

FIGURE 22.10 SVMGA average accuracy for all datasets as a function of kernel power d. The x axis represents the kernel power d, and $d = 5$ represents an RBF kernel.

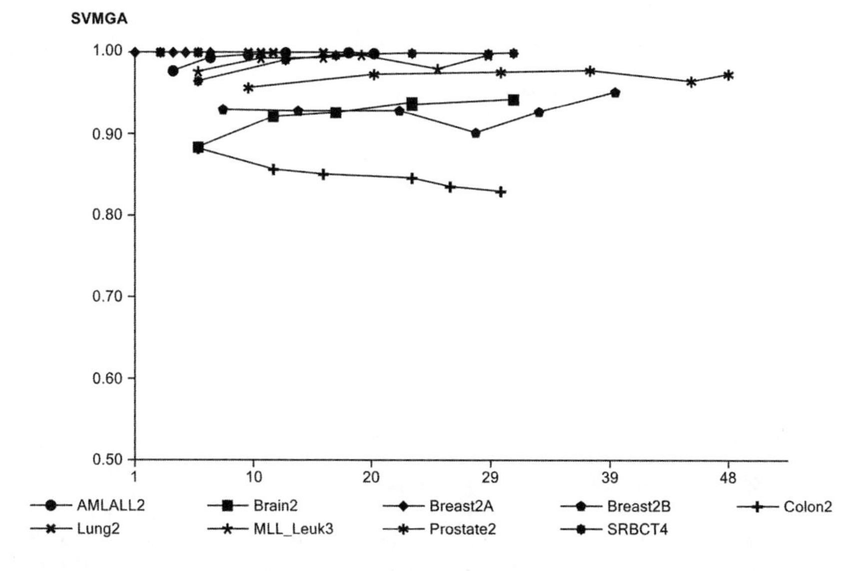

FIGURE 22.11 SVMGA accuracy without transformations on features as a function of bootstrap sample size used during training. The x axis values represent the number of arrays randomly sampled $B = 40$ times at fractions $f = 0.1$, 0.2, 0.3, 0.4, 0.5, or 0.6 of the available arrays.

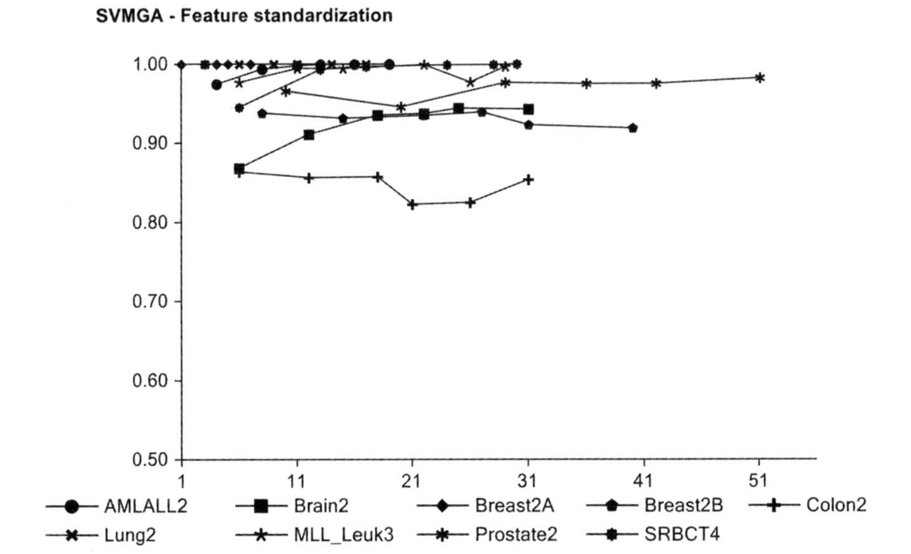

FIGURE 22.12 SVMGA accuracy with feature standardization as a function of bootstrap sample size used during training. The x axis values represent the number of arrays randomly sampled $B = 40$ times at fractions $f = 0.1$, 0.2, 0.3, 0.4, 0.5, or 0.6 of the available arrays.

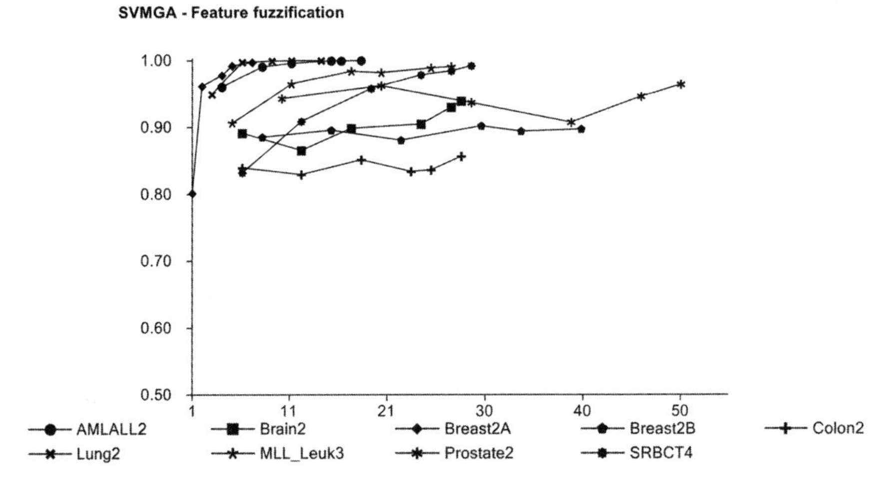

FIGURE 22.13 SVMGA accuracy with feature fuzzification as a function of bootstrap sample size used during training. The x axis values represent the number of arrays randomly sampled $B = 40$ times at fractions $f = 0.1$, 0.2, 0.3, 0.4, 0.5, or 0.6 of the available arrays.

as a function of bootstrap sample size used during training, and reflects minuscule changes in accuracy when compared with accuracy when no feature transformations were made. Figure 22.13 shows the SVMGA accuracy with feature fuzzification as a function of bootstrap sample size used during training and indicates that, for several datasets, accuracy at smaller sample sizes decreased when compared with mean-zero standardization and no use of transformations.

22.5.3 Multiclass ROC Curves

Multiclass ROC curves were generated using mean-zero standardized and fuzzified input features selected using greedy PTA (see Table 12.10) and reflect high performance levels of SVMGA for all datasets. Figure 22.14 illustrates ROC curves for the SVMGA classifier applied to the AMLALL2, Brain2, Breast2A, Breast2B, Colon2, Lung2, MLL_Leukemia3, Prostate2, and SRBCT4 datasets using mean-zero standardized input features. The legend values represent the number of arrays randomly sampled $B = 40$ times at fractions $f = 0.1, 0.2, 0.3, 0.4, 0.5$, or 0.6 of the available arrays. Figure 22.15 illustrates ROC curves for the SVMGA based on the same datasets using fuzzified input features. Fuzzification seemed to reduced performance levels slightly when 10-20% of the available training data were selected using bootstrapping.

22.5.4 Decision Boundaries

Figure 22.16 illustrates 2D self-organizing maps of class decision boundaries for the SVMGA based on the AMLALL2, Brain2, Breast2A, Breast2B, Colon2, Lung2, MLL_Leukemia3, Prostate2, and SRBCT4 datasets using mean-zero standardized input features. Figure 22.17 illustrates 2D self-organizing maps of class decision boundaries for the SVMGA based on the AMLALL2, Brain2, Breast2A, Breast2B, Colon2, Lung2, MLL_Leukemia3, Prostate2, and SRBCT4 datasets using fuzzified input features. SVMGA class decision boundaries based on input feature mean-zero standardization were smoother and less erratic than decision boundaries formed with SVMGA classification using fuzzified input features.

22.6 LEAST-SQUARES SOFT-MARGIN SVM

The least-squares support vector machine (SVM) is obtained by minimizing the function

$$\frac{1}{2}||\mathbf{w}||^2 + \frac{C}{2}\sum_{i=1}^{n}\xi_i^2. \tag{22.24}$$

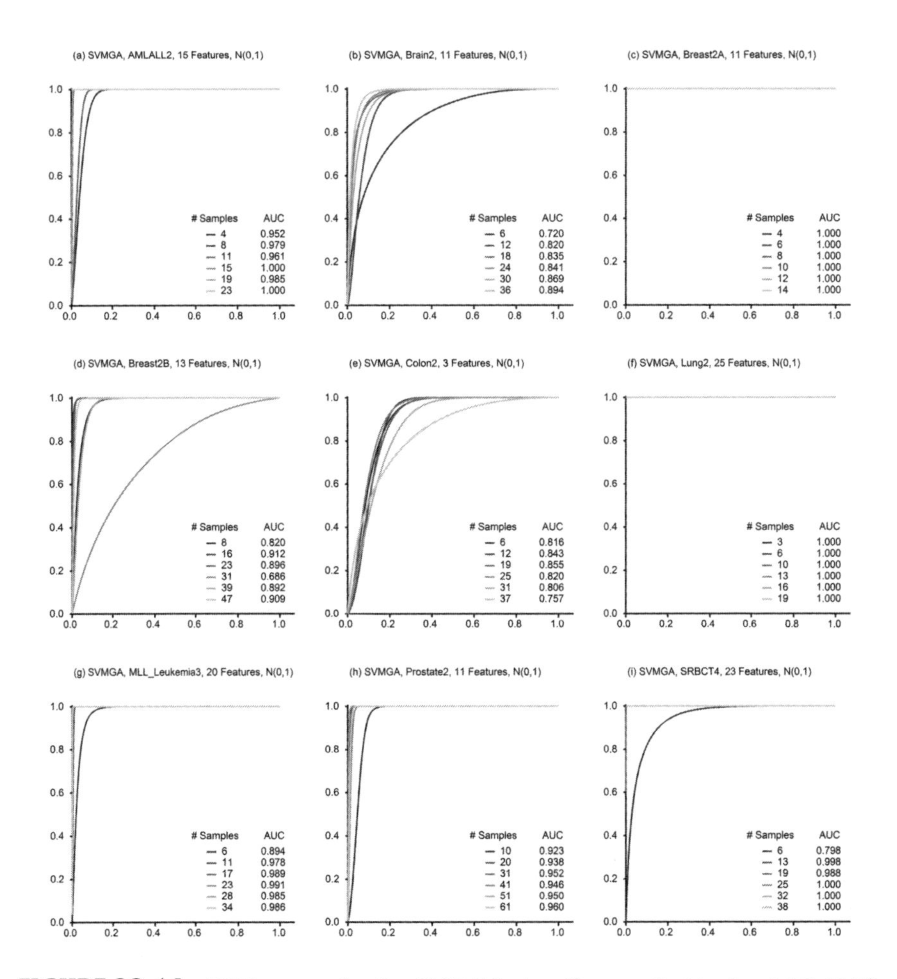

FIGURE 22.14 ROC curves for the SVMGA classifier applied to the AMLALL2, Brain2, Breast2A, Breast2B, Colon2, Lung2, MLL_Leukemia3, Prostate2, and SRBCT4 datasets based on mean-zero standardized input features: legend values represent the number of arrays randomly sampled $B = 40$ times at fractions $f = 0.1$, 0.2, 0.3, 0.4, 0.5, or 0.6 of the available arrays: (a) AMLALL2; (b) Brain2; (c) Breast2A; (d) Breast2B; (e) Colon2; (f) Lung2; (g) MLL_Leukemia3; (h) Prostate2; (i) SRBCT4. The numbers of genes used for each dataset are listed in Table 12.10.

With Lagrangian multipliers, the unconstrained objective function becomes

$$L(\mathbf{w}, b, \boldsymbol{\alpha}, \boldsymbol{\xi}) = \frac{1}{2}\mathbf{w}^T\mathbf{w} + \frac{C}{2}\sum_{i=1}^{n}\xi_i^2 - \sum_{i=1}^{n}\alpha_i[y_i(\mathbf{w}^T\mathbf{x}_i + b) - 1 + \xi_i]. \tag{22.25}$$

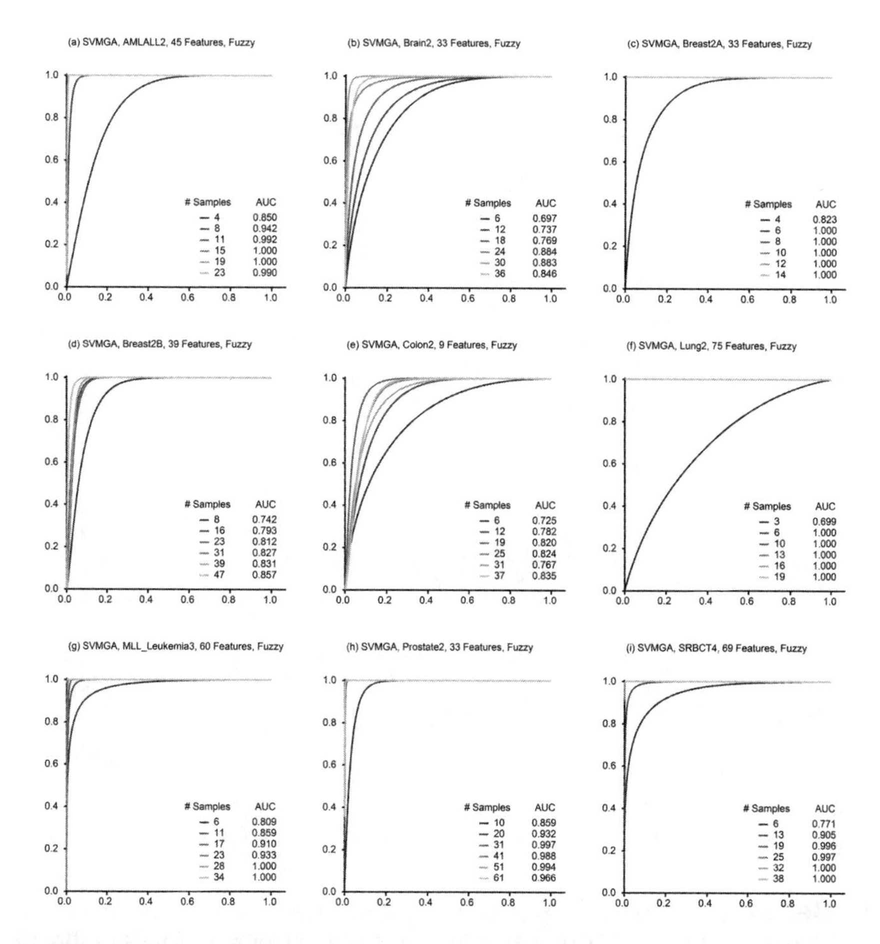

FIGURE 22.15 ROC curves for the SVMGA applied to the AMLALL2, Brain2, Breast2A, Breast2B, Colon2, Lung2, MLL_Leukemia3, Prostate2, and SRBCT4 datasets based on fuzzified input features; legend values represent the number of arrays randomly sampled $B = 40$ times at fractions $f = 0.1, 0.2, 0.3, 0.4, 0.5,$ or 0.6 of the available arrays: (a) AMLALL2; (b) Brain2; (c) Breast2A; (d) Breast2B; (e) Colon2; (f) Lung2; (g) MLL_Leukemia3; (h) Prostate2; (i) SRBCT4. The numbers of genes used for each dataset are listed in Table 12.10.

Setting the partial derivatives of (22.25) equal to zero, we obtain the following parameters for optimization:

$$\mathbf{w} = \sum_{i=1}^{n} \alpha_i y_i \mathbf{x}_i, \tag{22.26}$$

$$\sum_{i=1}^{n} \alpha_i y_i = 0, \tag{22.27}$$

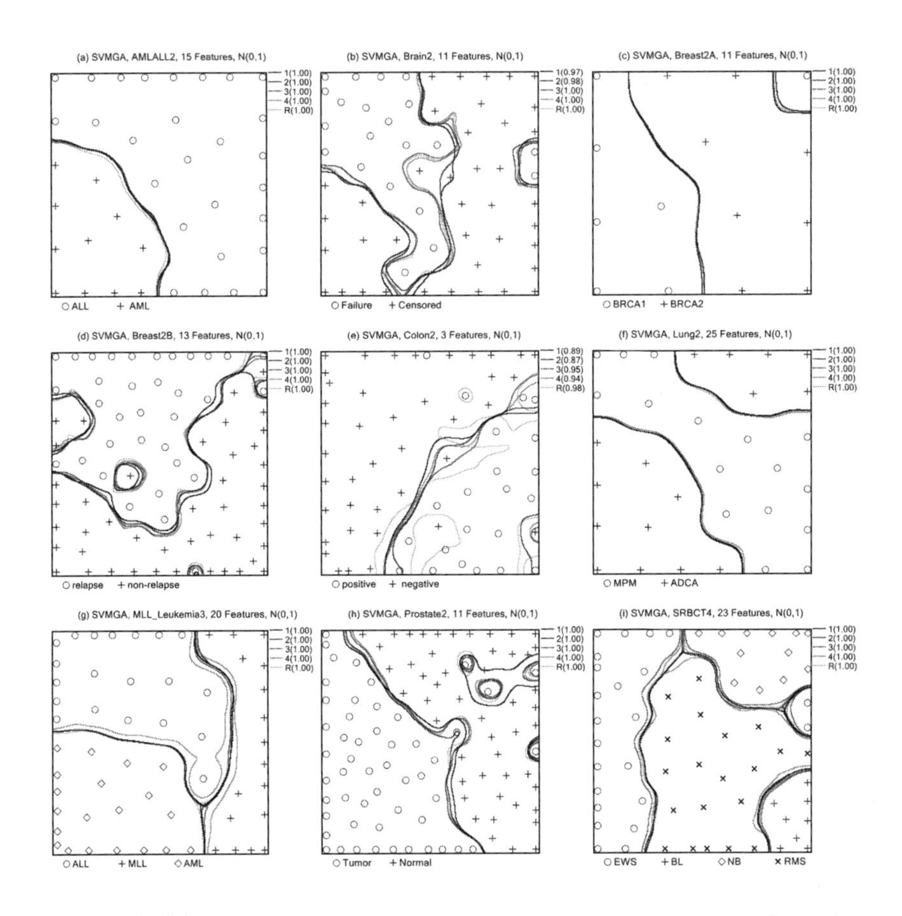

FIGURE 22.16 2D self-organizing maps of SVMGA class decision boundaries and classification accuracy (upper right of each panel) for the AMLALL2, Brain2, Breast2A, Breast2B, Colon2, Lung2, MLL_Leukemia3, Prostate2, and SRBCT4 datasets based on mean-zero standardized input features; the legend in each panel (upper right) contains accuracy as a function of kernel power d in (22.10); R represents radial basis function kernel: (a) AMLALL2; (b) Brain2; (c) Breast2A; (d) Breast2B; (e) Colon2; (f) Lung2; (g) MLL_Leukemia3; (h) Prostate2; (i) SRBCT4. The numbers of genes used for each dataset are listed in Table 12.10.

$$\alpha_i = C\xi_i, \tag{22.28}$$

$$y_i(\mathbf{w}^T\mathbf{x}_i + b) - 1 + \xi_i = 0. \tag{22.29}$$

By combining (22.26) and (22.29) and expressing them aggregately in matrix form [31], we obtain the following results:

$$\begin{pmatrix} \mathbf{0} & \mathbf{y}^T \\ \mathbf{y} & \mathbf{\Phi} \end{pmatrix} \begin{pmatrix} b \\ \boldsymbol{\alpha} \end{pmatrix} = \begin{pmatrix} \mathbf{0} \\ \mathbf{1} \end{pmatrix}, \tag{22.30}$$

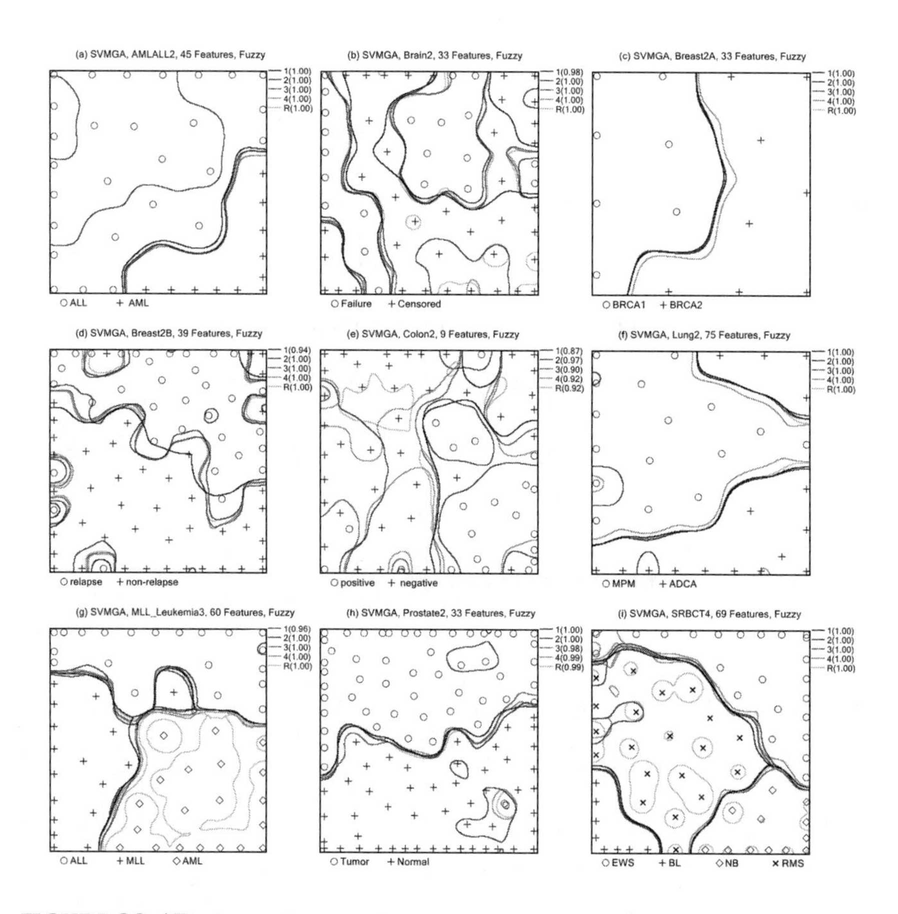

FIGURE 22.17 2D self-organizing maps of SVMGA class decision boundaries and classification accuracy (upper right of each panel) for the AMLALL2, Brain2, Breast2A, Breast2B, Colon2, Lung2, MLL_Leukemia3, Prostate2, and SRBCT4 datasets based on fuzzified input features; the legend in each panel (upper right) contains accuracy as a function of kernel power d in (22.10); R represents radial basis function kernel: (a) AMLALL2; (b) Brain2; (c) Breast2A; (d) Breast2B; (e) Colon2; (f) Lung2; (g) MLL_Leukemia3; (h) Prostate2; (i) SRBCT4. The numbers of genes used for each dataset are listed in Table 12.10.

Alternatively, this can be expressed as

$$\begin{aligned} \mathbf{y}^T\boldsymbol{\alpha} &= 0 \\ \mathbf{y}b + \boldsymbol{\Phi}\boldsymbol{\alpha} &= \mathbf{1}, \end{aligned} \tag{22.31}$$

where $\mathbf{y}$ is a microarray $\times$ 1 vector with $y_i = 1$ for the first class and $y_i = -1$ for the second class, $\boldsymbol{\alpha}$ is a microarray $\times$ 1 vector of support vectors (in least-squares SVM all α_i are support vectors), b is a constant, and $\boldsymbol{\Phi}$ is a microarray × microarray matrix, with elements

$$\phi_{ij} = y_i y_j H(\mathbf{x}_i, \mathbf{x}_j) + \frac{I(i=j)}{C}. \tag{22.32}$$

Solving for $\boldsymbol{\alpha}$ in the second equation of (22.31), we have

$$\boldsymbol{\alpha} = \boldsymbol{\Phi}^{-1}(\mathbf{1} - \mathbf{y}b), \tag{22.33}$$

and then, substituting in $\boldsymbol{\Phi}^{-1}(\mathbf{1} - \mathbf{y}b)$ for $\boldsymbol{\alpha}$ in the second equation of (22.31), we obtain

$$\mathbf{y}^T\boldsymbol{\Phi}^{-1}(\mathbf{1} - \mathbf{y}b) = \mathbf{0}. \tag{22.34}$$

Multiplying this equation out, we obtain

$$\mathbf{y}^T\boldsymbol{\Phi}^{-1}\mathbf{1} - \mathbf{y}^T\boldsymbol{\Phi}^{-1}\mathbf{y}b = \mathbf{0}, \tag{22.35}$$

and rearranging on opposite sides, we get

$$-\mathbf{y}^T\boldsymbol{\Phi}^{-1}\mathbf{y}b = -\mathbf{y}^T\boldsymbol{\Phi}^{-1}\mathbf{1}, \tag{22.36}$$

and finally, solving for b, we have

$$b = (\mathbf{y}^T\boldsymbol{\Phi}^{-1}\mathbf{y})^{-1}\mathbf{y}^T\boldsymbol{\Phi}^{-1}\mathbf{1}. \tag{22.37}$$

The decision functions used for SVMLS are the same as those used in Equations (22.21), (22.22), and (22.23).

22.6.1 Cross-Validation Results

Table 22.4 lists the average accuracy based on 10 repartitions of the input arrays and subsequent CV, without any input feature transformations applied. Table 22.5 lists the average accuracy based on 10 repartitions of the input arrays and subsequent CV, using mean-zero standardized input feature values. Table 22.6 lists the average accuracy based on 10 repartitions of the input arrays and subsequent CV, using fuzzified input feature values. Figure 22.18 illustrates accuracy as a function of CV method for each dataset based on 10 repartitions of arrays with no feature transformations applied. There was less bias and less variance of SVMLS when compared with SVMGA. Standard deviation of accuracy was also lower for SVMLS when compared with SVMGA (Figure 22.19). Figure 22.20 reveals average accuracy for all datasets as a function of feature transformation and CV method, and indicates that accuracy values with and without feature mean-zero standardization were identical and markedly greater than accuracy

TABLE 22.4 Performance without Feature Transformation

		Cross-validation method				
Dataset	*d*	CV2	CV5	CV10	CV − 1	CVB
AMLALL2	1	100	100	100	100	100
	2	99.74	100	100	100	100
	3	100	100	100	100	99.74
	4	98.95	100	100	100	100
	R	98.68	100	100	100	100
Brain2	1	89.00	89.67	91.00	91.67	91.81
	2	86.67	90.33	89.50	91.67	92.46
	3	89.67	89.67	89.33	91.67	92.03
	4	89.33	91.00	89.33	90.00	92.00
	R	89.33	90.67	91.00	91.67	91.53
Breast2A	1	100	100	100	100	100
	2	100	100	100	100	100
	3	100	100	100	100	100
	4	100	100	100	100	100
	R	100	100	100	100	100
Breast2B	1	93.72	93.33	94.23	94.87	96.88
	2	91.67	94.74	93.08	94.87	96.58
	3	92.82	92.69	95.26	94.87	97.11
	4	92.56	94.49	93.08	94.87	97.00
	R	92.05	94.49	94.10	94.87	96.56
Colon2	1	88.39	88.71	88.87	90.32	89.75
	2	88.87	89.35	89.52	90.32	89.68
	3	88.71	89.68	89.19	90.32	89.77
	4	89.19	89.35	89.03	90.32	89.46
	R	87.58	88.87	89.03	90.32	89.60
Lung2	1	100	100	100	100	100
	2	100	100	100	100	100
	3	100	100	100	100	100
	4	100	100	100	100	100
	R	100	100	100	100	100
MLL_Leukemia3	1	95.96	98.95	100	100	98.79
	2	95.44	99.47	100	100	99.44
	3	96.49	100	99.82	100	99.37
	4	95.79	98.95	100	100	98.38
	R	96.32	99.30	99.82	100	98.32
Prostate2	1	97.84	97.65	98.04	98.04	98.19
	2	97.65	97.25	98.04	98.04	98.02
	3	97.75	97.65	98.04	98.04	98.16
	4	97.55	97.75	98.04	98.04	98.49
	R	97.65	97.45	97.16	98.04	97.94
SRBCT4	1	97.62	94.76	96.51	96.83	98.83
	2	97.30	96.03	96.98	96.83	98.78
	3	97.46	96.03	96.98	96.83	98.96
	4	93.97	96.51	95.87	96.83	98.05
	R	97.78	94.76	96.19	96.83	98.20
Average		0.96	0.96	0.96	0.97	0.97

[a]The term *d* denotes the power of the kernel function and when equal to *R*, represents the RBF kernel.

TABLE 22.5 Performance with Feature Standardization

		Cross-validation method				
Dataset	d	CV2	CV5	CV10	CV − 1	CVB
AMLALL2	1	99.47	100	100	100	100
	2	99.47	100	100	100	100
	3	100	100	100	100	99.98
	4	99.21	100	100	100	100
	R	98.68	100	100	100	100
Brain2	1	88.00	89.33	89.83	90.00	91.64
	2	87.00	89.67	90.33	91.67	90.70
	3	87.17	89.83	90.00	91.67	92.12
	4	87.00	89.83	90.50	91.67	91.53
	R	88.17	88.83	89.83	90.00	91.39
Breast2A	1	100	100	100	100	100
	2	100	100	100	100	100
	3	100	100	100	100	100
	4	100	100	100	100	100
	R	100	100	100	100	100
Breast2B	1	93.59	93.08	92.44	94.87	96.31
	2	93.46	94.49	92.56	94.87	96.82
	3	91.41	93.72	93.72	94.87	97.26
	4	90.51	96.03	94.10	94.87	94.89
	R	92.95	95.26	93.97	94.87	96.24
Colon2	1	88.87	90.16	89.19	90.32	89.69
	2	88.87	89.03	88.55	90.32	89.55
	3	87.42	89.03	89.03	88.71	89.56
	4	88.39	88.55	88.55	90.32	89.78
	R	88.39	88.55	89.19	90.32	89.65
Lung2	1	100	100	100	100	100
	2	100	100	100	100	100
	3	100	100	100	100	100
	4	100	100	100	100	100
	R	100	100	100	100	100
MLL_Leukemia3	1	95.61	99.82	99.82	100	98.73
	2	97.89	99.47	99.65	100	98.10
	3	95.96	99.82	100	100	98.14
	4	94.21	99.65	99.82	100	99.15
	R	96.14	98.60	99.82	100	98.56
Prostate2	1	97.35	97.75	97.25	98.04	97.83
	2	97.35	97.65	98.14	98.04	98.12
	3	97.25	98.04	98.04	98.04	98.06
	4	97.16	98.14	97.55	98.04	98.25
	R	97.55	98.04	97.84	98.04	98.21
SRBCT4	1	98.41	95.08	96.83	96.83	99.08
	2	96.98	96.35	97.14	96.83	98.91
	3	96.67	96.51	96.67	96.83	98.95
	4	97.46	96.35	96.67	96.83	98.01
	R	97.78	97.46	97.46	96.83	98.66
Average		0.95	0.96	0.96	0.97	0.97

[a] The term d denotes the power of the kernel function and when equal to R, represents the RBF kernel.

TABLE 22.6 Performance with Feature Fuzzification

Dataset	d	Cross-validation method				
		CV2	CV5	CV10	CV − 1	CVB
AMLALL2	1	89.21	78.42	75.26	73.68	93.65
	2	89.21	76.58	75.00	73.68	93.42
	3	88.16	81.58	75.53	73.68	93.48
	4	88.16	80.26	72.63	73.68	93.38
	R	89.21	78.42	75.26	73.68	94.44
Brain2	1	57.33	52.83	53.83	38.33	65.12
	2	57.33	52.17	50.00	38.33	54.65
	3	62.50	51.67	49.17	38.33	61.06
	4	61.50	53.00	48.33	38.33	62.89
	R	59.00	52.17	46.83	38.33	60.68
Breast2A	1	100	100	100	100	100
	2	100	100	100	100	100
	3	100	100	100	100	100
	4	100	100	100	100	100
	R	100	100	100	100	100
Breast2B	1	78.08	83.97	79.23	78.21	85.54
	2	66.92	81.03	82.82	78.21	82.81
	3	70.51	80.26	83.21	78.21	81.35
	4	79.62	81.03	77.05	78.21	82.18
	R	83.46	78.97	77.69	78.21	86.14
Colon2	1	68.23	65.81	70.48	85.48	73.46
	2	65.65	65.16	69.84	85.48	68.37
	3	71.29	66.13	65.48	53.23	75.12
	4	76.94	67.74	68.23	85.48	64.88
	R	73.06	71.94	69.35	85.48	77.94
Lung2	1	100	100	100	100	100
	2	100	100	100	100	100
	3	100	100	100	100	100
	4	100	100	100	100	100
	R	100	100	100	100	100
MLL_Leukemia3	1	94.91	95.61	94.04	94.74	98.36
	2	93.86	95.61	95.09	94.74	97.31
	3	93.51	95.09	94.04	94.74	97.87
	4	93.86	95.79	95.09	94.74	97.96
	R	94.04	96.14	95.09	94.74	97.87
Prostate2	1	84.12	78.14	74.31	73.53	85.28
	2	82.94	77.45	73.14	73.53	73.47
	3	75.39	80.39	71.57	73.53	80.92
	4	85.29	71.08	72.35	73.53	83.70
	R	85.98	83.92	75.78	73.53	80.37
SRBCT4	1	92.22	93.17	94.13	93.65	96.87
	2	89.84	92.38	93.49	93.65	96.80
	3	89.84	92.86	92.22	93.65	96.80
	4	90.63	93.17	93.65	93.65	96.79
	R	90.63	92.70	93.17	93.65	96.98
Average		0.85	0.83	0.82	0.81	0.87

[a] The term d denotes the power of the kernel function and when equal to R, represents the RBF kernel.

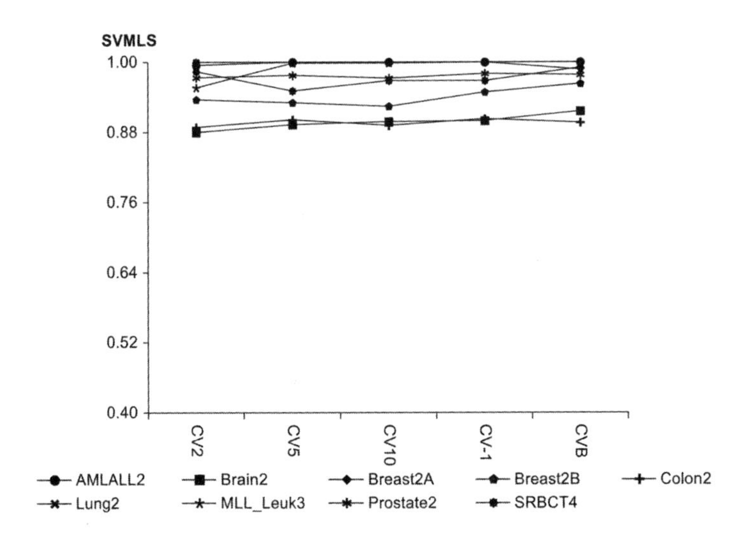

FIGURE 22.18 SVMLS accuracy as a function of CV method for each dataset. Average based on CV accuracy for 10 repartitions of arrays. No feature transformations applied.

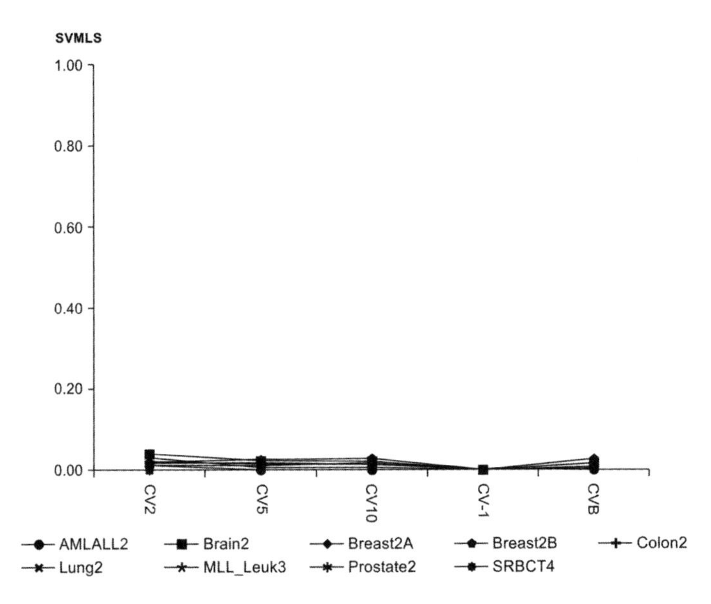

FIGURE 22.19 SVMLS standard deviation of accuracy as a function of CV method for each dataset. Standard deviation based on CV accuracy for 10 repartitions of arrays. No feature transformations applied.

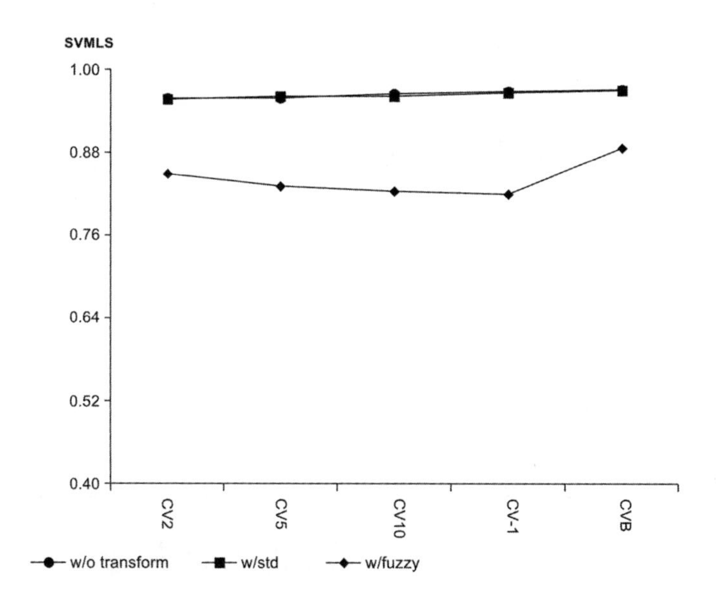

FIGURE 22.20 SVMLS average accuracy for all datasets as a function of feature transformation and CV method.

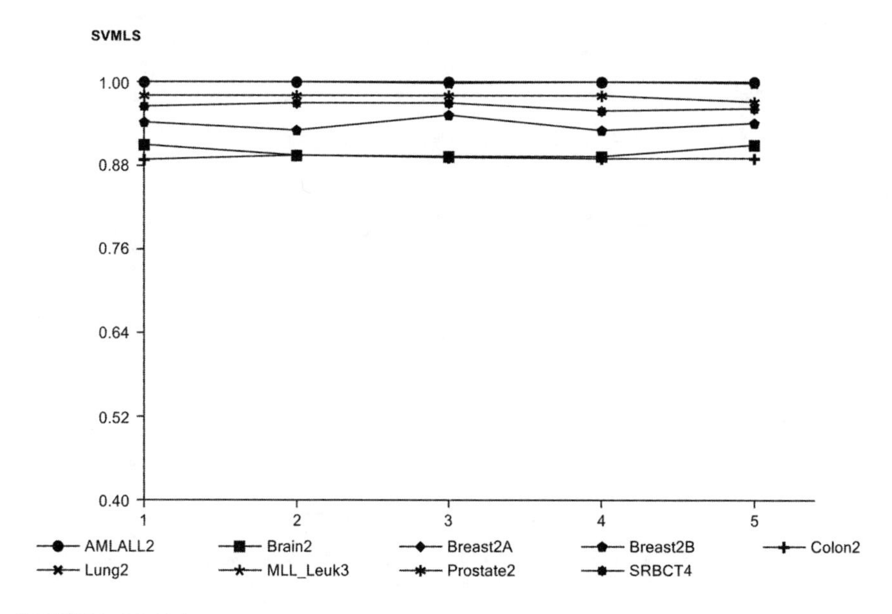

FIGURE 22.21 SVMLS accuracy without transformations on features as a function of d. The x axis represents the kernel power d, and $d = 5$ represents an RBF kernel.

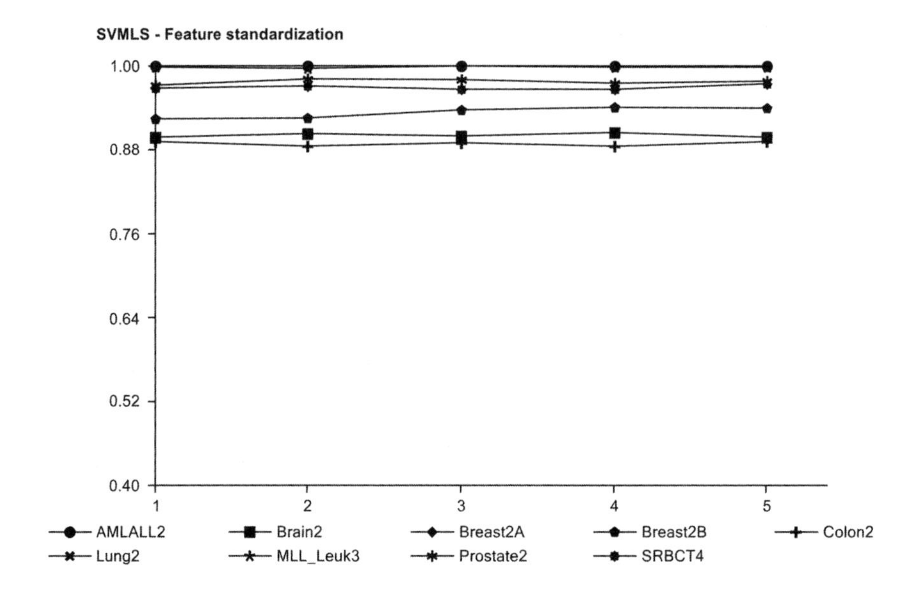

FIGURE 22.22 SVMLS accuracy with feature standardization as a function of d. The x axis represents the kernel power d, and $d = 5$ represents an RBF kernel.

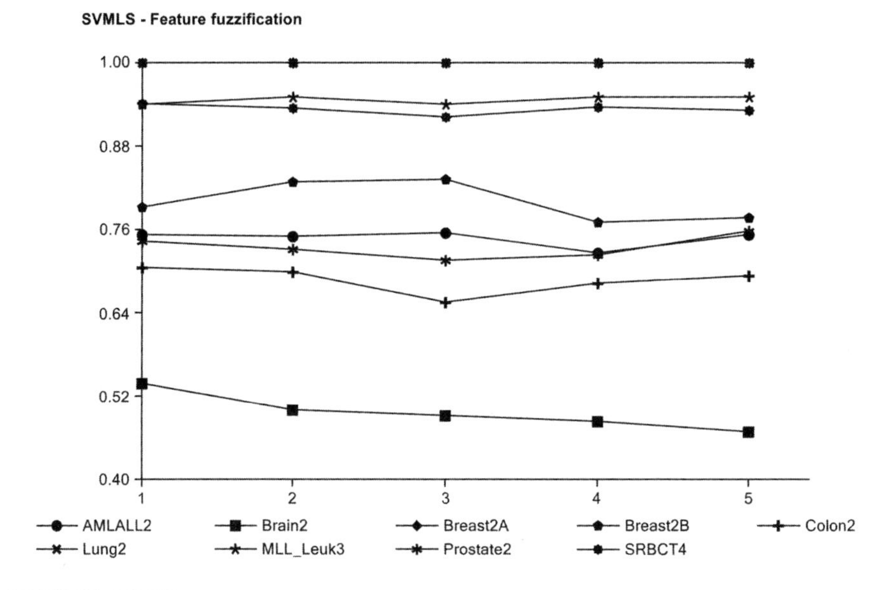

FIGURE 22.23 SVMLS accuracy with feature fuzzification as a function of d. The x axis represents the kernel power d, and $d = 5$ represents an RBF kernel.

when input features were fuzzified. Figure 22.21 reveals the SVMLS accuracy without transformations on features as a function of kernel power d. Figure 22.22 shows the SVMLS accuracy with feature standardization as a function of kernel power d. Figure 22.23 illustrates the SVMLS accuracy with feature fuzzification as a function of kernel power d. Figure 22.24 shows average accuracy for all datasets as a function of d. Results in Figures 22.21, 22.22, 22.23, and 22.24 reveal that input feature fuzzification substantially reduced performance, while mean-zero standardization of input features did not result in accuracy that underestimated accuracy when no feature transformations were applied.

22.6.2 Bootstrap Bias

Figure 22.5 shows the SVMLS accuracy without transformations on features as a function of bootstrap sample size used during training, based on 10 resamplings from the input arrays. The x axis values represent the number of arrays randomly sampled $B = 40$ times at fractions $f = 0.1, 0.2$, 0.3, 0.4, 0.5, or 0.6 of the available arrays. Bootstrap bias is greater for SVMLS when compared with SVMGA, as the accuracy at sampling fractions of 10% and 20% was lower when compared with SVMGA. Input feature mean-zero standardization did not reduce bias (Figure 22.26) when compared with accuracy when no feature transformation was performed. Fuzzification also appeared to reduce accuracy substantially (Figure 22.27).

22.6.3 Multiclass ROC Curves

The multivariate ROC curves shown in Figure 22.28 for SVMLS are quite similar to the SVMGA ROC curves. Input feature fuzzification decreased the AUC values (Figure 22.29) when compared with AUC values generated with mean-zero standardized input features.

22.6.4 Decision Boundaries

Figure 22.30 illustrates 2D self-organizing maps of class decision boundaries for the SVMLS based on the AMLALL2, Brain2, Breast2A, Breast2B, Colon2, Lung2, MLL_Leukemia3, Prostate2, and SRBCT4 datasets using mean-zero standardized input features. Figure 22.31 illustrates 2D self-organizing maps of class decision boundaries for the SVMLS based on the AMLALL2, Brain2, Breast2A, Breast2B, Colon2, Lung2, MLL_Leukemia3, Prostate2, and SRBCT4

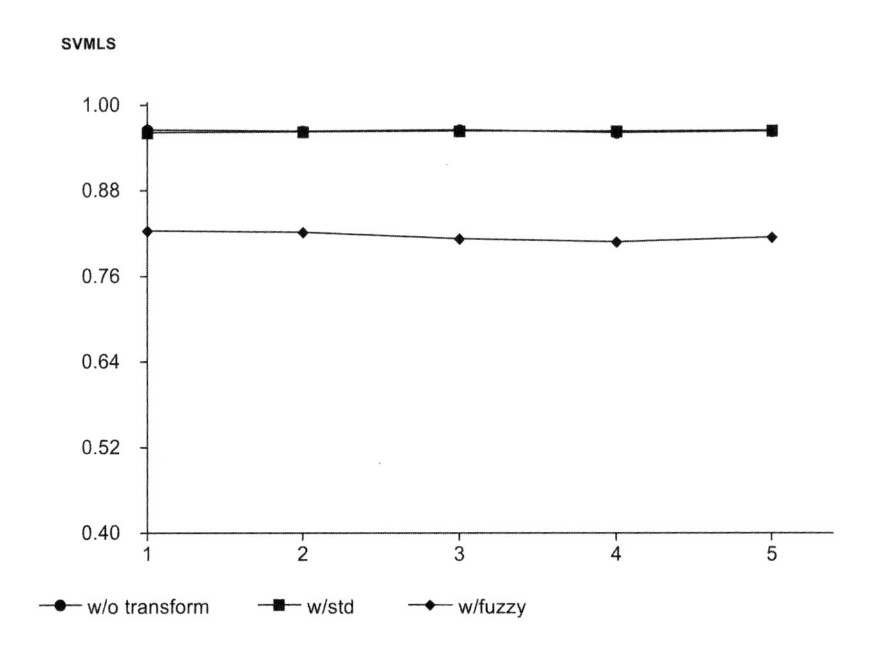

FIGURE 22.24 SVMLS average accuracy for all datasets as a function of d. The x axis represents the kernel power d, and $d = 5$ represents an RBF kernel.

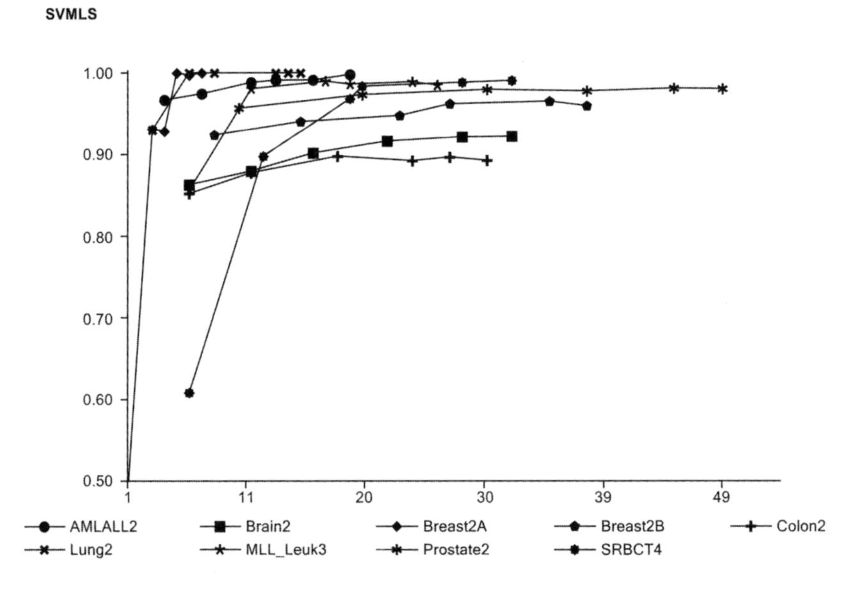

FIGURE 22.25 SVMLS accuracy without transformations on features as a function of bootstrap sample size used during training. The x axis values represent the number of arrays randomly sampled $B = 40$ times at fractions $f = 0.1$, 0.2, 0.3, 0.4, 0.5, or 0.6 of the available arrays.

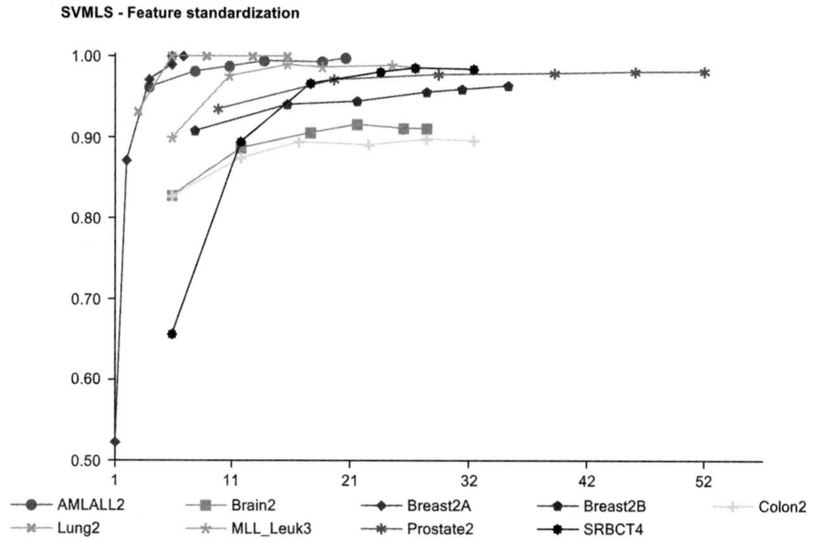

FIGURE 22.26 SVMLS accuracy with feature standardization as a function of bootstrap sample size used during training. The x axis values represent the number of arrays randomly sampled $B = 40$ times at fractions $f = 0.1$, 0.2, 0.3, 0.4, 0.5, or 0.6 of the available arrays. (*See insert for color representation of the figure.*)

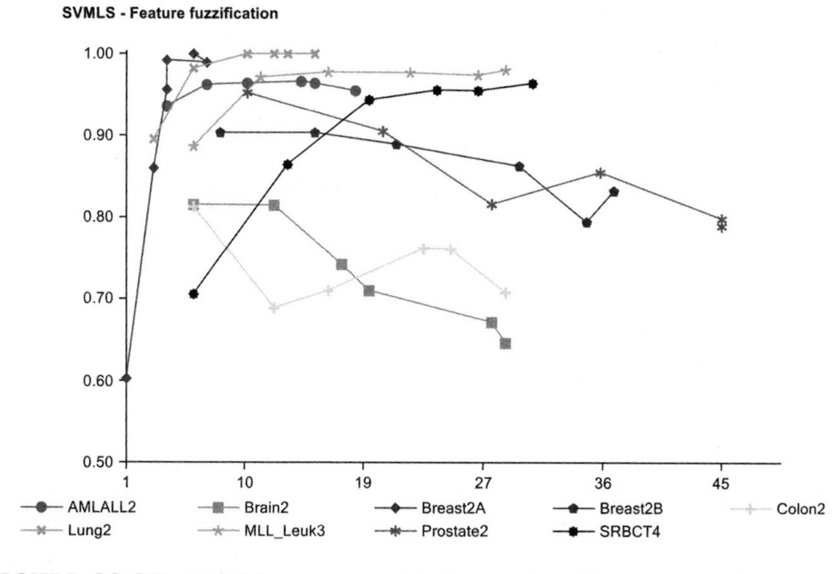

FIGURE 22.27 SVMLS accuracy with feature fuzzification as a function of bootstrap sample size used during training. The x axis values represent the number of arrays randomly sampled $B = 40$ times at fractions $f = 0.1$, 0.2, 0.3, 0.4, 0.5, or 0.6 of the available arrays.

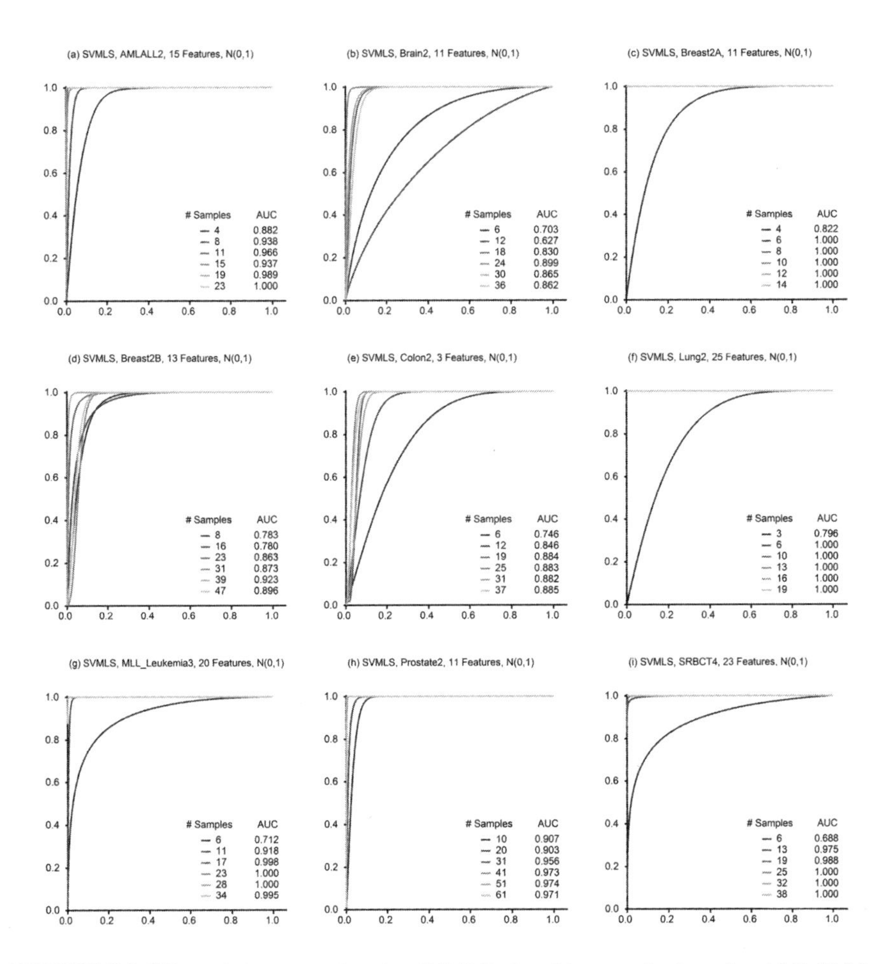

FIGURE 22.28 ROC curves for the SVMLS classifier applied to the AMLALL2, Brain2, Breast2A, Breast2B, Colon2, Lung2, MLL_Leukemia3, Prostate2, and SRBCT4 datasets based on mean-zero standardized input features; legend values represent the number of arrays randomly sampled $B = 40$ times at fractions $f = 0.1$, 0.2, 0.3, 0.4, 0.5, or 0.6 of the available arrays: (a) AMLALL2; (b) Brain2; (c) Breast2A; (d) Breast2B; (e) Colon2; (f) Lung2; (g) MLL_Leukemia3; (h) Prostate2; (i) SRBCT4. The numbers of genes used for each dataset are listed in Table 12.10.

datasets using fuzzified input features. The SVMLS decision boundaries formed for the nine datasets are quite similar to the decision boundaries generated during classification runs with SVMGA, shown in Figures 22.16 and 22.17.

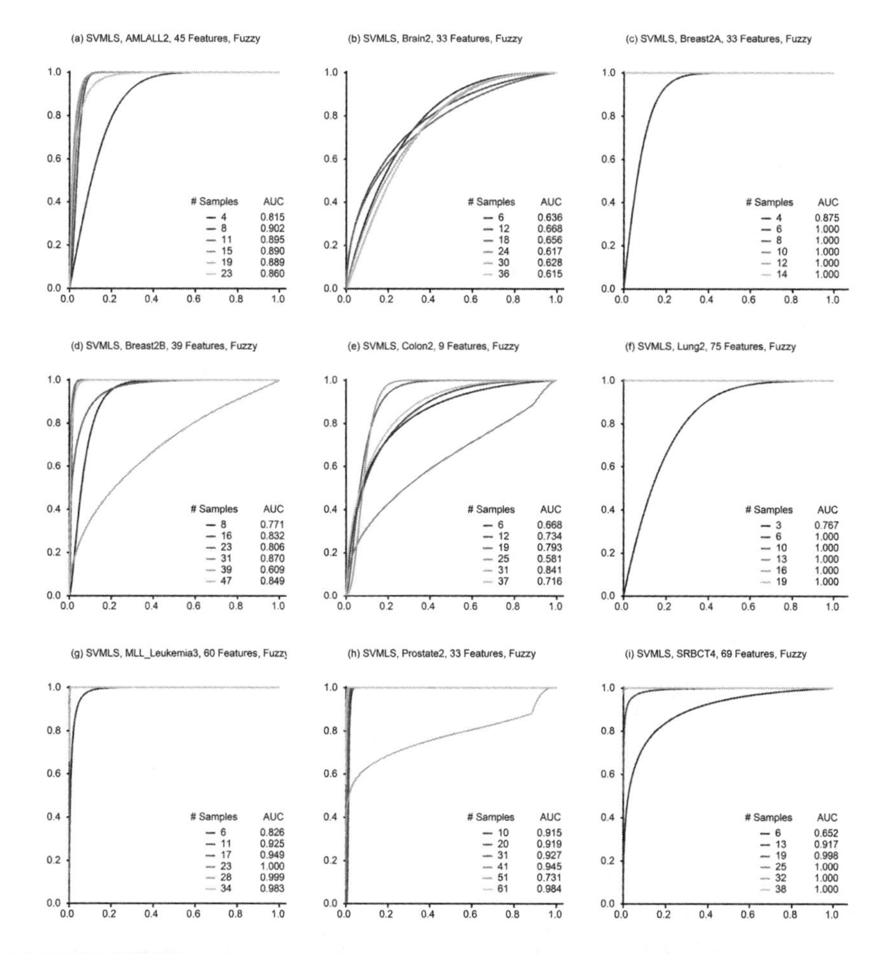

FIGURE 22.29 ROC curves for the SVMLS applied to the AMLALL2, Brain2, Breast2A, Breast2B, Colon2, Lung2, MLL_Leukemia3, Prostate2, and SRBCT4 datasets based on fuzzified input features; legend values represent the number of arrays randomly sampled $B = 40$ times at fractions $f = 0.1, 0.2, 0.3, 0.4, 0.5$, or 0.6 of the available arrays: (a) AMLALL2; (b) Brain2; (c) Breast2A; (d) Breast2B; (e) Colon2; (f) Lung2; (g) MLL_Leukemia3; (h) Prostate2; (i) SRBCT4. The numbers of genes used for each dataset are listed in Table 12.10.

22.7 SUMMARY

Support vector machines offer many advantages over other classifiers; for example, they maximize generalization ability, avoid local maxima, and are robust to outliers. However, they do not extend easily to multiclass problems,

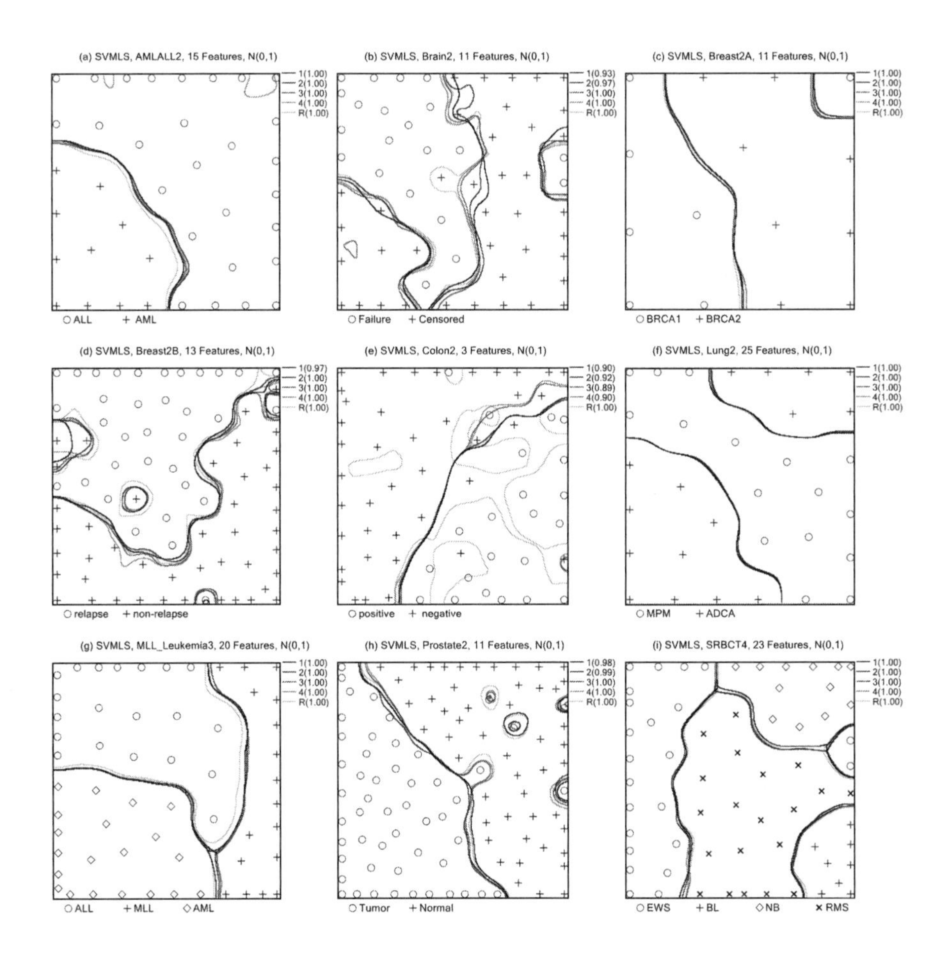

FIGURE 22.30 2D self-organizing maps of SVMLS class decision boundaries and classification accuracy (upper right of each panel) for the AMLALL2, Brain2, Breast2A, Breast2B, Colon2, Lung2, MLL_Leukemia3, Prostate2, and SRBCT4 datasets based on mean-zero standardized input features; the legend in each panel (upper right) contains accuracy as a function of kernel power d in (22.10); R represents radial basis function kernel: (a) AMLALL2; (b) Brain2; (c) Breast2A; (d) Breast2B; (e) Colon2; (f) Lung2; (g) MLL_Leukemia3; (h) Prostate2; (i) SRBCT4. The numbers of genes used for each dataset are listed in Table 12.10.

can require long training times when quadratic programming is used, and are sensitive to model parameters in the same way that artificial neural networks are sensitive to the number of hidden layers and the number of nodes at each hidden layer.

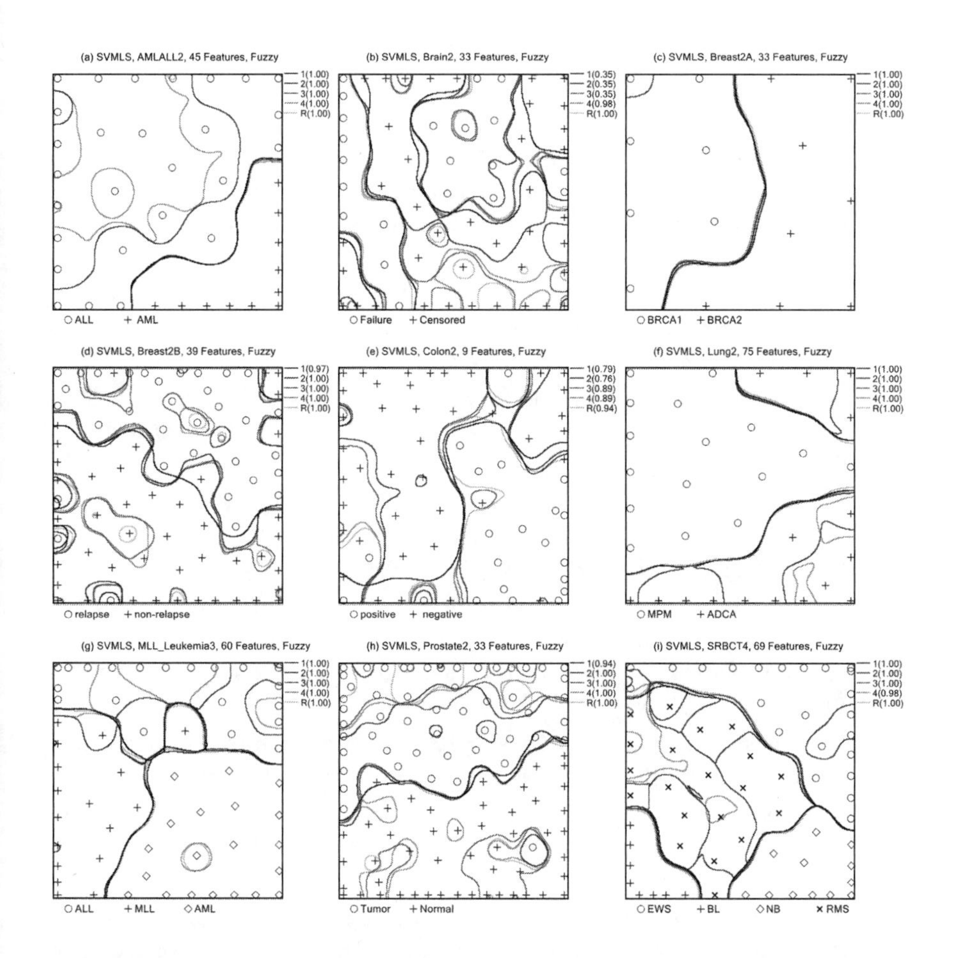

FIGURE 22.31 2D self-organizing maps of SVMLS class decision boundaries and classification accuracy (upper right of each panel) for the AMLALL2, Brain2, Breast2A, Breast2B, Colon2, Lung2, MLL_Leukemia3, Prostate2, and SRBCT4 datasets based on fuzzified input features; the legend in each panel (upper right) contains accuracy as a function of kernel power d in (22.10); R represents radial basis function kernel: (a) AMLALL2; (b) Brain2; (c) Breast2A; (d) Breast2B; (e) Colon2; (f) Lung2; (g) MLL_Leukemia3; (h) Prostate2; (i) SRBCT4. The numbers of genes used for each dataset are listed in Table 12.10.

REFERENCES

[1] S. Song, Z. Zhan, Z. Long, J. Zhang, L. Yao. Comparative study of SVM methods combined with voxel selection for object category classification on fMRI data. *PLoS One* **6**(2):e17191, 2011.

[2] L. J. Wee, D. Simarmata, Y. W. Kam, L. F. Ng, J. C. Tong. SVM-based prediction of linear B-cell epitopes using Bayes feature extraction. *BMC Genomics* **11**(Suppl 4):S21, 2010.

[3] D. Salas-Gonzalez, J. M. Gorriz, J. Ramirez, I. A. Illan, M. Lopez, F. Segovia, R. Chaves, P. Padilla, C. G. Puntonet. Feature selection using factor analysis for Alzheimer's diagnosis using 18F-FDG PET images. *Med. Phys.* **37**(11):6084-6095, 2010.

[4] C. Ye, M. T. Coimbra, B. K. Vijaya Kumar. Arrhythmia detection and classification using morphological and dynamic features of ECG signals. *Conf. Proc. IEEE Eng. Med. Biol. Soc.* **2010**:1918-1921, 2010.

[5] S. L. Nuryani, H. T. Nguyen. Electrocardiographic T-wave peak-to-end interval for hypoglycaemia detection. *Conf. Proc. IEEE Eng. Med. Biol. Soc.* **2010**:618-621, 2010.

[6] P. R. Tabrizi, S. H. Rezatofighi, M. J. Yazdanpanah. Using PCA and LVQ neural network for automatic recognition of five types of white blood cells. *Conf. Proc. IEEE Eng. Med. Biol. Soc.* **2010**:5593-5596, 2010.

[7] C. P. Shen, W. H. Chen, J. M. Chen, K. P. Hsu, J. W. Lin, M. J. Chiu, C. H. Chen, F. Lai. Bio-signal analysis system design with support vector machines based on cloud computing service architecture. *Conf. Proc. IEEE Eng. Med. Biol. Soc.* **2010**:1421-1424, 2010.

[8] D. Sommer, M. Golz. Evaluation of PERCLOS based current fatigue monitoring technologies. *Conf. Proc. IEEE Eng. Med. Biol. Soc.* **2010**:4456-4459, 2010.

[9] M. N. Akhtar, S. A. Bukhari, Z. Fazal, R. Qamar, I. A. Shahmuradov. POLYAR, a new computer program for prediction of poly(A) sites in human sequences. *BMC Genomics* **11**:646, 2010.

[10] A. M. Eren, I. Amin, A. Alba, E. Morales, A. Stoyanov, S. Winters-Hilt. Pattern recognition-informed feedback for nanopore detector cheminformatics. *Adv. Exp. Med. Biol.* **680**:99-108, 2010.

[11] A. A. Ensafi, M. Taei, T. Khayamian, F. Hasanpour. Simultaneous voltammetric determination of enrofloxacin and ciprofloxacin in urine and plasma using multiwall carbon nanotubes modified glassy carbon electrode by least-squares support vector machines. *Anal. Sci.* **26**(7):803-808, 2010.

[12] E. Fernandez-Ahumada, J. M. Roger, B. Palagos, J. E. Guerrero, D. Perez-Marin, A. Garrido-Varo. Multivariate near-infrared reflection spectroscopy strategies for ensuring correct labeling at feed bagging in the animal feed industry. *Appl. Spectros. c.* **64**(1):83-91, 2010.

[13] P. Chopra, J. Lee, J. Kang, S. Lee. Improving cancer classification accuracy using gene pairs. *PLoS One* **5**(12):e14305, 2010.

[14] M. Zervakis, M. E. Blazadonakis, G. Tsiliki, V. Danilatou, M. Tsiknakis, D. Kafetzopoulos. Outcome prediction based on microarray analysis: A critical perspective on methods. *BMC Bioinformatics* **10**:53, 2009.

[15] O. Gevaert, F. De Smet, T. Van Gorp, N. Pochet, K. Engelen, F. Amant, B. De Moor, D. Timmerman, I. Vergote. Expression profiling to predict the clinical behaviour of ovarian cancer fails independent evaluation. *BMC Cancer* **8**:18, 2008.

[16] R. Moffitt, J. Phan, S. Hemby, M. Wang. Effect of outlier removal on gene marker selection using support vector machines. *Conf. Proc. IEEE Eng. Med. Biol. Soc.* **1**:917–920, 2005.

[17] J. Phan, R. Moffitt, J. Dale, J. Petros, A. Young, M. Wang. Improvement of SVM algorithm for microarray analysis using intelligent parameter selection. *Conf. Proc. IEEE Eng. Med. Biol. Soc.* **5**:4838–4841, 2005.

[18] Z. Kote-Jarai, L. Matthews, A. Osorio, S. Shanley, I. Giddings, F. Moreews, I. Locke, D. G. Evans, R. D. Eccles, Carrier Clinic Collaborators, R. D. Williams, M. Girolami, C. Campbell, R. Eeles. Accurate prediction of BRCA1 and BRCA2 heterozygous genotype using expression profiling after induced DNA damage. *Clin. Cancer Res.* **12**(13):3896–3901, 2006.

[19] C. Zhang, H. R. Li, J. B. Fan, J. Wang-Rodriguez, T. Downs, X. D. Fu, M. Q. Zhang. Profiling alternatively spliced mRNA isoforms for prostate cancer classification. *BMC Bioinform.* **7**:202, 2006.

[20] N. Ancona, R. Maglietta, A. D'Addabbo, S. Liuni, G. Pesole. Regularized least squares cancer classifiers from DNA microarray data. *BMC Bioinform.* **6**(Suppl. 4):S2, 2005.

[21] A. Schramm, J. H. Schulte, L. Klein-Hitpass, W. Havers, H. Sieverts, B. Berwanger, H. Christiansen, P. Warnat, B. Brors, J. Eils, R. Eils, A. Eggert. Prediction of clinical outcome and biological characterization of neuroblastoma by expression profiling. *Oncogene* **24**(53):7902–7912, 2005.

[22] G. Steiner, L. Suter, F. Boess, R. Gasser, M. C. de Vera, S. Albertini, S. Ruepp. Discriminating different classes of toxicants by transcript profiling. *Environ. Health Perspect.* **112**(12):1236–1248, 2004.

[23] N. Iizuka, M. Oka, H. Yamada-Okabe, M. Nishida, Y. Maeda, N. Mori, T. Takao, T. Tamesa, A. Tangoku, H. Tabuchi, K. Hamada, H. Nakayama, H. Ishitsuka, T. Miyamoto, A. Hirabayashi, S. Uchimura, Y. Hamamoto. Oligonucleotide microarray for prediction of early intrahepatic recurrence of hepatocellular carcinoma after curative resection. *Lancet* **361**(9361):923–929, 2003.

[24] B. E. Boser, I. M. Guyon, V. N. Vapnik. A training algorithm for optimal margin classifiers. *Proc. 5th Annual Workshop on Computer Learning Theory (COLT'95)*, ACM Press, New York (NY), 1992, pp. 144–152.

[25] C. Cortes, V. N. Vapnik. Support-vector networks. *Machine Learn.* **20**(3):273–297, 1995.

[26] V. N. Vapnik. *The Nature of Statistical Learning Theory*. Springer-Verlag, Berlin, 1995.

[27] I. M. Guyon, J. Weston, S. Barnhill, V. N. Vapnik. Gene selection for cancer classification using support vector machines. *Machine Learn.* **46**:389–422, 2002.

[28] W. Karush. *Minima of Functions of Several Variables with Inequalities as Side Constraints*, master's thesis, Dept. Mathematics, Univ. Chicago, 1939.

[29] H. W. Kuhn, A. W. Tucker. Nonlinear programming, *Proc. 2nd Berkeley Symp.*, Univ. California Press, Berkeley, CA, 1951, pp. 481-492.

[30] J. Mercer. Functions of positive and negative type and their connection with the theory of integral equations. *Phil. Trans. Roy. Soc. A* **209**:415-446, 1909.

[31] T. Van Gestel, J. A. K. Suykens. Benchmarking least squares support vector machine classifiers. *Machine Learn.* **54**:5-32, 2004.

CHAPTER 23

ARTIFICIAL NEURAL NETWORKS

23.1 INTRODUCTION

Artificial neural networks (ANNs) offer several advantages over other classifiers for pattern searching among the increasingly complex genomic and biomolecular datasets currently being generated. Using the known outcome classes of arrays, ANNs perform supervised learning through coefficient updates that minimize error between the predicted and known outcome classes. ANNs are also capable of generalizing their learning results so that future test arrays not used during training can be classified.

Artificial neural networks are machine-based learning models that simulate information processing performed by the brain [1]. ANNs consist of neurons, or cells, interconnected by synaptic weights that filter and transmit information in a supervised fashion in order to acquire knowledge that can be stored in memory. After adapting to an environment in which an ANN is embedded, the stored knowledge can be generalized to future experiences to predict outcome based on input stimuli. In terms of modeling, the three most important advantages of ANNs in over other classifiers are that ANNs perform supervised learning, can be linear or nonlinear, and don't require fitting an objective function for which parameters are sought. Supervised (vs. unsupervised) learning, incorporates information on the known outcome classes into a classifier in order to guide the learning process. Supervised learning with ANNs involves the use of both the input and output vectors for each microarray, so that during training the ANN learns the relationship between the input vectors and output vectors. ANNs can also incorporate nonlinear transfer functions to reveal ultimately complex nonlinear mappings between input and output classes [2], since linear models cannot identify nonlinear

Classification Analysis of DNA Microarrays, First Edition. Leif E. Peterson.

relationships in data. ANNs are known to be universal approximators [3], as it has been shown mathematically that ANNs are able to approximate almost any continuous and most discontinuous functions, provided there are an adequate number of hidden nodes [4,5].

Another superior quality of ANNs is that they are data-driven and non-parametric in terms of a model, and therefore, do not require distributional assumptions. This eliminates the chance of model misspecification, resulting in poor goodness-of-fit (GOF) statistics. Multilayer ANNs incorporating a hidden layer between the input and output layers can model latent (unobservable) factors. Hidden layers also develop an internal representation of the relationship among variables, obviating the need to maintain hypotheses based on residual normality or lack of collinearity [6]. Overall, ANNs are capable of solving simultaneous systems of nonlinear and linear equations quite well [7]. As such, their performance is generally better than that using traditional techniques when the data exhibit nonlinear features [8].

23.2 ANN ARCHITECTURE

Figure 23.1 shows the basic structure of a three-layer gradient-based back-propagation ANN, with input, hidden, and output layers. Information flows between layers in a forward-reverse fashion, updating connection weights during the learning process.

23.3 BASICS OF ANN TRAINING

Before delving too deeply into the specifics of ANN training for gene expression analysis, we need to clarify several points. First, we assume that the user has already identified a smaller optimal set of genes that discriminate class, as it is very uncommon to use expression values for every gene on a large chip that passes quality control checks. However, an ANN can be used for recursive feature (gene) selection to identify the most important genes from a large set of genes. In the formulas that follow, we assume that an optimal set of p genes are selected before using an ANN for classification analysis.

Training a network proceeds as follows. Let $i = 1, 2, \ldots, p$ represent expression values for an optimal set of p genes on array $\mathbf{x}$, $j = 1, 2, \ldots, J$ represent the jth hidden node, and $k = 1, 2, \ldots, \Omega$ represent the kth output node or class. First, feature values are mean-zero standardized or normalized over the training arrays considered in each run to ensure the same scale. Next, the input p vector of feature values for training array $\mathbf{x}$ is clamped to the input nodes. All weights are then assigned random variates in the range $[-0.5, 0.5]$. The total input or *potential* received by neuron j in the hidden

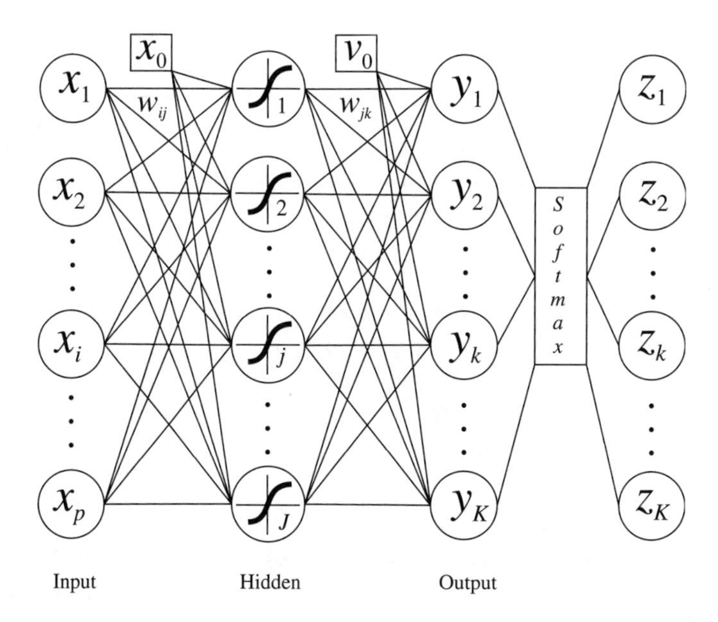

FIGURE 23.1 Basic architecture of an artificial neural network (ANN) showing the input layer to which the **x** vector is clamped, the hidden layer with activation functions, and output layer and formulas involved in connection weights and node inputs and outputs. By default, we set the number of nodes in the input layer equal to the number of input features (genes) or reduced dimensions and those in the output layer, equal to the number of classes to which the arrays are assigned.

layer is

$$u_j = \sum_i x_i w_{ij}^{ih}, \tag{23.1}$$

where x_i is the input value at the ith neuron in the input layer and w_{ij}^{ih} is the weight of the connection between the ith neuron in the input layer to the jth neuron in the hidden layer. The output of neuron j in the hidden layer is obtained by applying the following *activation* function to the potential in the form

$$v_j = \frac{1}{1 + e^{-u_j}}, \tag{23.2}$$

which is a logistic activation function. Table 23.1 lists some of the more common activation functions used. The RBF and Gaussian activation functions determine v_j as an exponentiated neighborhood function of the Euclidean distance between the entire **x** input vector and center $\mathbf{m}_j$ from K-means

TABLE 23.1 Common Activation Functions Used at ANN Hidden Layer Nodes

Activation	Formula[a]	Partial derivative[b] $(\partial v_j/\partial u_j)$
Linear	$v_j = u_j$	1
Logistic	$v_j = 1/[1+\exp(-u_j)]$	$\exp(-u_j)/(1+\exp(-u_j))^2$
tanh[c]	$v_j = 1.7159\left\{\frac{[\exp(\frac{2}{3}u_j)-\exp(-\frac{2}{3}u_j)]}{[\exp(\frac{2}{3})+\exp(-\frac{2}{3}u_j)]}\right\}$	$(1/\cosh(u_j))^2$
RBF	$v_j = \exp(-\lvert\lvert \mathbf{x}-\mathbf{m}_j \rvert\rvert)$	1^d
Gaussian	$v_j = \exp(-\lvert\lvert \mathbf{x}-\mathbf{m}_j \rvert\rvert/2\sigma^2)$	1^d
Hermite	$v_j = h_n(t), \quad t = u_j \quad (23.3)$	$dh_n(t)/dt$
Laguerre	$v_j = l_0^n \frac{t^2}{2}, \quad t = u_j \quad (23.6)$	$dl_0^n(t)/dt$

[a]Used in (23.2).
[b]Used in (23.22).
[c]$1.7159 \tanh\left(\frac{2}{3}x\right)$ suggested by LeCun et al. [9]
[d]Partial derivative $\partial v_j/\partial u_j$ set to unity, since u_j is not an element of v_j.

cluster analysis, where the number of centers is equal to the number of hidden nodes J. Hermite activation functions belong to the general class of orthonormal functions, and use the product of a Hermite polynomial $H_n(t)$ and a Gaussian function t, which is set equal to u_j

$$h_n(t) = \frac{H_n(t)}{\sqrt{2^n n! \sqrt{\pi}}} \exp\left(\frac{-t^2}{2}\right), \tag{23.3}$$

where the first Hermite polynomial $H_0(t) = 1$ and the remaining polynomials $(n \geq 1)$ are based on the recurrence relationship

$$H_n(t) = 2tH_n(t) - 2nH_{n-1}(t). \tag{23.4}$$

Figure 23.2 shows Hermite activation functions for $n = 0, 1, \ldots, 5$. First-order derivatives for the activation functions are

$$\begin{aligned} \frac{dh_n(t)}{dt} &= \sqrt{2n}h_{n-1}(t) - th_n(t), \qquad n \geq 1 \\ \frac{dh_0(t)}{dt} &= -th_0(t), \qquad n = 0 \end{aligned} \tag{23.5}$$

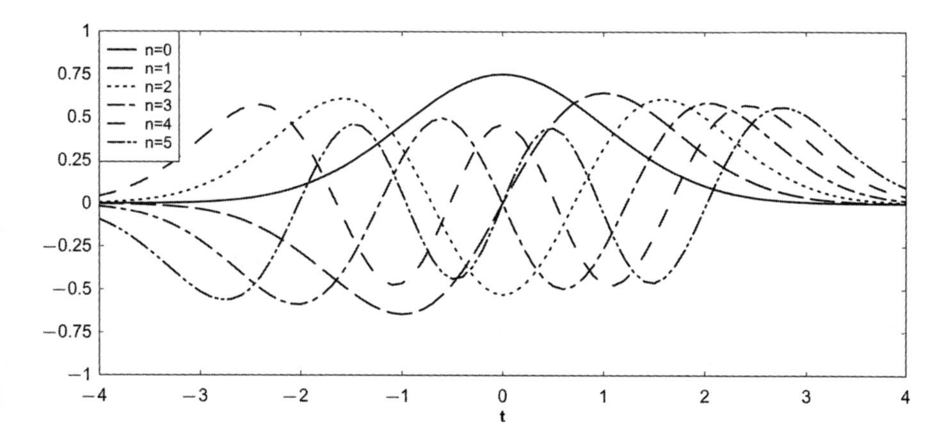

FIGURE 23.2 Hermite polynomials of order $0, 1, \ldots, 5$.

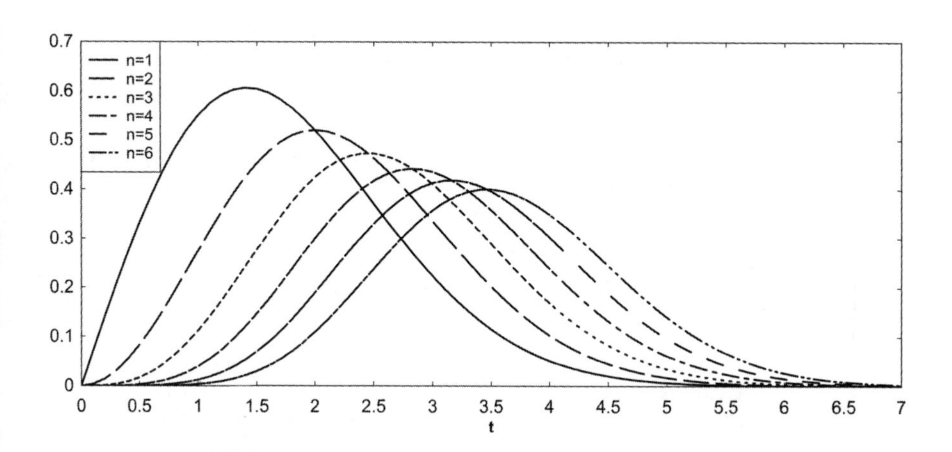

FIGURE 23.3 Laguerre polynomials of order $1, 2, \ldots, 6$.

Laguerre polynomials, which are strictly positive, are shown for $n = 1, 2, \ldots, 6$ in Figure 23.3. Laguerre activation functions at the hidden level are

$$l_0^n \frac{t^2}{2} = \frac{t^{n/2} \exp(-t/2)}{\sqrt{n!}}. \tag{23.6}$$

The first-order derivative of $l_0^n(t^2/2)$ with respect to t is

$$\frac{dl_0^n(t)}{dt} = \frac{(2n - t^2)\sqrt{\left(\frac{1}{2}\right)^n (t^2)^n} \exp\left(-t^2/4\right)}{2t\sqrt{n!}}. \tag{23.7}$$

Using the output of neuron j in the hidden layer, we obtain the total input to the kth neuron in the output layer as

$$y_k = \sum_j v_j w_{jk}^{ho}, \tag{23.8}$$

where w_{jk}^{ho} is the connection weight between neuron j in the hidden layer and neuron k in the output layer. Note that $k = 1, 2, \ldots, \Omega$, where Ω is the number of classes. The output of neuron k in the output layer is obtained using the *softmax function* in the form

$$z_k = \frac{\exp(y_k)}{\sum_l \exp(y_l)}, \tag{23.9}$$

which normalizes all the z_k so that they sum to unity. The softmax function is commonly used for classification when the arrays are partitioned into Ω classes.

Finally, the *mean-square error* (MSE) is determined as the total sum of squares based on the difference between the output vector and the "target" vector representing the true output, given as

$$\text{MSE} = E = \frac{1}{2}\sum_{k=1}^{\Omega}(z_k - c_k)^2. \tag{23.10}$$

where c_k is the kth element in the binary vector with known class for this training microarray. For example, assuming a four-class problem, the four possible target vectors $\mathbf{c}$ for training arrays known to be in classes 1, 2, 3, or 4 are $\mathbf{c} = (1, 0, 0, 0)$, $\mathbf{c} = (0, 1, 0, 0)$, $\mathbf{c} = (0, 0, 1, 0)$, and $\mathbf{c} = (0, 0, 0, 1)$, respectively. Thus, a given microarray known to be in class 2 could have class prediction of $\mathbf{z} = (0.17, 0.79, 0.04, 0)$ and when compared with its truth table $\mathbf{c} = (0, 1, 0, 0)$, would result in

$$\begin{aligned}\text{MSE} &= \tfrac{1}{2}[(0 - 0.17)^2 + (1 - 0.79)^2 + (0 - 0.04)^2 + (0 - 0)^2] \\ &= \tfrac{1}{2}[0.0289 + 0.0441 + 0.0016 + 0] \\ &= \frac{0.0746}{2} \\ &= 0.0373.\end{aligned} \tag{23.11}$$

If the predicted class membership for this same microarray in class 2 instead leaned toward membership in, say, class 4, as shown as $\mathbf{z} = (0, 0.04, 0.17, 0.79)$, then the MSE would be

$$\begin{aligned}
\text{MSE} &= \tfrac{1}{2}(0-0)^2 + (1-0.04)^2 + [(0-0.17)^2 + (0-0.79)^2] \\
&= \tfrac{1}{2}[0 + 0.9216 + 0.0289 + 0.6241] \\
&= \frac{1.5746}{2} \\
&= 0.7873. \qquad (23.12)
\end{aligned}$$

So it is clear that the greater the degree to which predicted class membership is off, the greater the MSE. An alternate form of error for a network is the *cross-entropy*, which for a 2-class problem is given in the form

$$\text{CE} = -\sum_{k=1}^{\Omega}[c_k \log(z_k) + (1-c_k)\log(1-z_k)], \qquad (23.13)$$

where c_k is assumed to be a random binary variable for which $P(c_k = 1) = z_k$. For multiclass problems, the CE is

$$\text{CE} = -\sum_{k=1}^{\Omega} c_k \log(z_k). \qquad (23.14)$$

Greater values of CE represent disagreement between c_k and z_k, and therefore CE is minimized via gradient descent. The minimum value of CE occurs when all $c_k = z_k$.

The next section, on backpropagation learning, describes how an ANN learns to correct for misclassification via an iterative gradient descent approach for adjusting connection weights.

23.3.1 Backpropagation Learning

Now that the total error is known for one microarray, we use the delta (δ) rule for gradient descent-based back-propagation learning to update the connection weights in the output and hidden layers. According to the chain rule, the partial derivative of the error with respect to the weight at the output level is

$$\frac{\partial E}{\partial w_{jk}^{ho}} = \frac{\partial E}{\partial z_k}\frac{dz_k}{dy_k}\frac{\partial y_k}{\partial w_{jk}^{ho}}. \qquad (23.15)$$

By parts, we have

$$\frac{\partial E}{\partial z_k} = \begin{cases} z_k - c_k & \text{MSE} \\ [(1-c_k)/(1-z_k)] - (c_k/z_k) & \text{CE} \end{cases}$$

$$\frac{dz_k}{dy_k} = \frac{\exp(y_k)\left(\sum_l \exp(y_l) - \exp(y_k)\right)}{\left(\sum_l \exp(y_l)\right)^2}, \tag{23.16}$$

$$\frac{\partial y_k}{\partial w_{jk}^{ho}} = v_j,$$

and then substitute to obtain

$$\begin{aligned} \frac{\partial E}{\partial w_{jk}^{ho}} &= \frac{\partial E}{\partial z_k}\frac{dz_k}{dy_k}\frac{\partial y_k}{\partial w_{jk}^{ho}} \\ &= (z_k - c_k)\left(\frac{\exp(y_k)\left(\sum_l \exp(y_l) - \exp(y_k)\right)}{\left(\sum_l \exp(y_l)\right)^2}\right) v_j. \end{aligned} \tag{23.17}$$

The chain rule is used to determine $\partial E/\partial w_{jk}^{ho}$ for weights between all hidden nodes and all output nodes. At each cycle, the weight update between the hidden and output nodes is

$$\Delta w_{jk}^{ho}(t) = -\varepsilon\gamma \frac{\partial E}{\partial w_{jk}^{ho}} + \alpha \Delta w_{jk}^{ho}(t-1), \tag{23.18}$$

where ε is the *learning rate*, γ is the weight decay equal to 1/(#total sweeps), and α is the *momentum*. The learning rate at each sweep is determined as

$$\varepsilon = \varepsilon_0 \left(1 - \frac{\text{sweep} - 1}{\text{\#sweeps}}\right), \tag{23.19}$$

where $\varepsilon_0 = 0.01$. Weight updates at cycle t are then applied to weights for the previous cycle at $t-1$ in the form

$$w_{jk}^{ho}(t) = w_{jk}^{ho}(t-1) + \Delta w_{jk}^{ho}(t). \tag{23.20}$$

With the sum-of-squares error (SSE) in the form $z_k - c_k$, whenever the predicted output exceeds the target (e.g., 1-0), the prediction is too large, requiring reduction in weights, hence; the negative sign precedes $\partial E/\partial w_{jk}^{ho}$ in the computation of $\Delta w_{jk}^{ho}(t)$. Analogously, when the prediction is too small, the error $z_k - c_k$ takes the form 0-1 and the double negative will increase the weights. With regard to weights between the input and hidden layer, the chain rule is used again to determine the partial derivative of error with respect to the weight at the hidden layer:

$$\frac{\partial E}{\partial w_{ij}^{ih}} = \frac{\partial E}{\partial v_j}\frac{dv_j}{du_j}\frac{\partial u_j}{\partial w_{ij}^{ih}}. \tag{23.21}$$

At the hidden layer, the output of neuron j is transmitted to each output node via $v_j w_{jk}^{ho}$, so a summation is required for the partial derivative $\partial E/\partial v_j$, shown as

$$\begin{aligned}\frac{\partial E}{\partial v_j} &= \sum_k \frac{\partial E}{\partial z_k}\frac{dz_k}{dy_k}\frac{\partial y_k}{\partial v_j}\\ &= \sum_k \frac{\partial E}{\partial z_k}\frac{dz_k}{dy_k} w_{jk}^{ho}\\ &= \sum_k \left((z_k - c_k)\left(\frac{\exp(y_k)(\sum_l \exp(y_l) - \exp(y_k))}{(\sum_l \exp(y_l))^2}\right) w_{jk}^{ho}\right),\end{aligned} \tag{23.22}$$

whereas the other derivatives are

$$\frac{dv_j}{du_j} = \frac{e^{-u_j}}{(1+e^{-u_j})^2}, \tag{23.23}$$

and

$$\frac{\partial u_j}{\partial w_{ij}^{ih}} = x_i, \tag{23.24}$$

which are substituted into the expression for the chain rule as

$$\begin{aligned}\frac{\partial E}{\partial w_{ij}^{ih}} &= \frac{\partial E}{\partial v_j}\frac{dv_j}{du_j}\frac{\partial u_j}{\partial w_{ij}^{ih}}\\ &= \sum_k \left((z_k - c_k)\left(\frac{\exp(y_k)(\sum_l \exp(y_l) - \exp(y_k))}{(\sum_l \exp(y_l))^2}\right) w_{jk}^{ho}\right)\frac{e^{-u_j}}{(1+e^{-u_j})^2}x_i.\end{aligned} \tag{23.25}$$

Again, weight updates are similar for the hidden layer, and are given as

$$\Delta w_{ij}^{ih}(t) = -\varepsilon\gamma\frac{\partial E}{\partial w_{ij}^{ih}} + \alpha\Delta w_{ij}^{ih}(t-1), \tag{23.26}$$

and

$$w_{ij}^{ih}(t) = w_{ij}^{ih}(t-1) + \Delta w_{ij}^{ih}(t). \tag{23.27}$$

During training, an ANN will learn both the features of interest and the noise present in the input data. This phenomenon is called *overfitting*, and, although it results in better predictions for the within-sample training data, it degrades generalization of results when applied to out-of-sample test data. Several methods have been proposed for improving network generalization such as weight elimination [10,11], network pruning [12,13], Bayesian regularization [4], early stopping [15,16], and cross-validation [17,18].

23.3.2 Resilient Backpropagation (RPROP) Learning

Backpropagation developed as a natural outgrowth of Hebbian learning, Widrow-Hoff learning, and delta rule learning [19]. Several improvements to backpropagation have been introduced. Incorporation of a momentum term in the weight update of ANNs was introduced as a way to smooth oscillations in the step direction [20]. However, because backpropagation is a *first-order* method, it uses information only on the first partial derivatives, which are independent of the curvature of the gradient. To overcome this limitation of first-order techniques, *quickprop* learning was introduced, which assumes a locally quadratic error surface and attempts to jump to the minimum error in one step [21]. Resilient backpropagation (RPROP) learning was another first-order method that operates in batch mode to minimize the effects of large partial derivatives on the step length [22]. RPROP assumes a changing error surface and uses the sign of the derivatives instead of the magnitude to determine whether a local minimum has been overstepped. During training, an *update value* $\Delta_{ij}(t)$ specific to each connection weight is adapted. This approach results in accelerated convergence in shallow regions of the gradient, causing RPROP to be one of the fastest first-order methods [23]. Algorithm 23.1 lists the computational steps required for implementing RPROP. Network training error was compared for backpropagation and RPROP using a variety of input- and output-side activation functions; results are shown in Section 23.12.

23.3.3 Cycles and Epochs

The calculations described above apply only to training with the first array, from the time the input vector data are clamped to the input nodes up to the last step used in backpropagation learning to update all weights. A *cycle* constitutes a sequence of forward and backward passes between (23.1) and (23.27) for one array. A complete cycle of training using all of the arrays is

Algorithm 23.1: RPROP

$\Delta_{min} \leftarrow 1.0E-6$, $\Delta_{max} \leftarrow 50$, $\Delta_0 \leftarrow 0.1$
$\eta^- \leftarrow 0.5$, $\eta^+ \leftarrow 1.2$
$\Delta_{ij}(t) = \Delta_0$
$(\partial E/\partial w_{ij})(t-1) = 0$
foreach *Weight* **do**
  if $(\partial E/\partial w_{ij})(t-1) \times (\partial E/\partial w_{ij})(t) > 0$ **then**
    $\Delta_{ij}(t) = \min\{\Delta_{ij}(t-1) \times \eta^+, \Delta_{max}\}$
    $w_{ij}(t) = -\text{sign}\left((\partial E/\partial w_{ij})(t)\right) \times \Delta_{ij}(t)$
    $w_{ij}(t+1) = w_{ij}(t) + \Delta_{ij}(t)$
    $(\partial E/\partial w_{ij})(t-1) = (\partial E/\partial w_{ij})(t)$
  else if $(\partial E/\partial w_{ij})(t-1) \times (\partial E/\partial w_{ij})(t) < 0$ **then**
    $\Delta_{ij}(t) = \max\{\Delta_{ij}(t-1) \times \eta^-, \Delta_{min}\}$
    $(\partial E/\partial w_{ij})(t-1) = 0$
  else if $(\partial E/\partial w_{ij})(t-1) \times (\partial E/\partial w_{ij})(t) = 0$ **then**
    $w_{ij}(t) = -\text{sign}(\partial/\partial w_{ij}(t)) \times \Delta_{ij}(t)$
    $w_{ij}(t+1) = w_{ij}(t) + \Delta_{ij}(t)$
    $(\partial E/\partial w_{ij})(t-1) = (\partial E/\partial w_{ij})(t)$

endfch
Note: w_{ij} is a generic term for weight-RPROP is applied to all weights.

termed an *epoch* or *sweep*. After the initial assignment of random weights and several sweeps through the training arrays, classification error will begin to decrease. As this occurs, the ANN will be learning the relationship between the input and output vectors. For most data when using an optimal set of preselected features, a marked reduction in error usually occurs within 25 sweeps, so by default one can likely use 100 sweeps and reach a considerable decrease in MSE. However, when recurrent feature selection is sought by using the ANN to identify optimal features, or the problem being solved is very complex, thousands of sweeps may be necessary to ensure that overfitting does not occur beyond, say, 5000 sweeps.

23.4 ANN TRAINING METHODS

23.4.1 Method 1: Gene Dimensional Reduction and Recursive Feature Elimination for Large Gene Lists

K-Means Cluster Analysis. K-means cluster analysis can be used to collapse expression of genes with similar profiles (over the arrays) into K

groups [24]. Let $\mathbf{X}$ be an $n \times p$ expression matrix for n arrays (in rows) and a large number of p genes (in columns), given in the form

$$\underset{n\times p}{\mathbf{X}} = \begin{pmatrix} x_{11} & x_{12} & \cdots & x_{1p} \\ x_{21} & x_{22} & \cdots & x_{2p} \\ \vdots & \vdots & \ddots & \vdots \\ x_{n1} & x_{n2} & \cdots & x_{np} \end{pmatrix}. \tag{23.28}$$

After K-means cluster analysis, we obtain the reduced matrix $\mathbf{M}$ with n-length column vectors $\mathbf{m}_k$, shown as

$$\underset{n\times K}{\mathbf{M}} = \begin{pmatrix} m_{11} & m_{12} & \cdots & m_{1K} \\ m_{21} & m_{22} & \cdots & m_{2K} \\ \vdots & \vdots & \ddots & \vdots \\ m_{n1} & m_{n2} & \cdots & m_{nK} \end{pmatrix}. \tag{23.29}$$

Each row of $\mathbf{M}$ represents an array (object), and the ANN is trained by clamping each $1 \times K$ row of $\mathbf{M}$ while using the class label of each row as the basis for incrementing elements in the confusion matrix. Following ANN training, we determine the K-means score, which maps the genes back to the center k as in the form

$$z_{gk} = \frac{\|\mathbf{x}_g - \mathbf{m}_k\| - \mu_k}{\sigma_k} \qquad k = 1, 2, \ldots, K, \tag{23.30}$$

where $\mathbf{x}_g$ is the standardized expression vector for gene g, $\mathbf{m}_k$ is the mean vector for center k, $\|\mathbf{x}_g - \mathbf{m}_k\|$ is the Euclidean distance between expression for gene g and center k, and μ_k and σ_k are the average and standard deviation of distances $\|\mathbf{x}_g - \mathbf{m}_k\|$ between all genes and center k. This procedure is repeated for each cluster center to yield a $p \times K$ $\mathbf{Z}$ matrix of K-means scores. The $\mathbf{Z}$ matrix has the form

$$\underset{p\times K}{\mathbf{Z}} = \begin{pmatrix} z_{11} & z_{12} & \cdots & z_{1K} \\ z_{21} & z_{22} & \cdots & z_{2K} \\ \vdots & \vdots & \ddots & \vdots \\ z_{p1} & z_{p2} & \cdots & z_{pK} \end{pmatrix}. \tag{23.31}$$

Since the scores are standard normal distributed, the bulk of scores will be centered around zero, and genes with the smallest or greatest distance from the cluster center will yield greater scores. Each row of $\mathbf{Z}$ representing a gene is presented to the trained ANN using the last known weights, and class membership is predicted for each.

Principal Components Analysis (PCA). During PCA, the top 10 eigenvalues are extracted from the $p \times p$ gene-by-gene correlation matrix $\mathbf{R}$. The array-by-component ($n \times p*$) $\mathbf{F}$ matrix of *PC scores* is determined with the matrix of standardized expression values (standardized with mean and standard deviation over the genes) as follows:

$$\underset{n \times p*}{\mathbf{F}} = \underset{n \times p}{\mathbf{Z}} \underset{p \times p*}{\mathbf{W}} . \tag{23.32}$$

In order to map genes back to the arrays via the principal components, the matrix of *PC score coefficients* is obtained using the matrix operation

$$\underset{p \times p*}{\mathbf{W}} = \underset{p \times p*}{\mathbf{L}} \underset{p* \times p*}{(\mathbf{L}'\mathbf{L})^{-1}} . \tag{23.33}$$

where $\mathbf{L}$ is the loading matrix reflecting the correlation between each gene expression profile and the extracted PCs. The top 10 PC's are always extracted from the gene-by-gene correlation matrix and used for training. Orthogonal rotations are not performed.

Training the ANN with K-Means Centers or Principal Components. When ANN training is based on K-means centers, each row (array) of the $n \times K$ $\mathbf{M}$ matrix of cluster centers is clamped to the input nodes. Whereas when ANN training is based on PCA, each row (array) of the $n \times p*$ $\mathbf{F}$ matrix of PC scores is clamped to the input nodes. After each sweep (epoch) through all of the arrays, ANN training results in input-side connection weights w_{ij}^{ih} and output-side weights w_{jk}^{ho}.

Recursive Feature Elimination. Once the input-side and output-side connection weights are obtained after a full sweep through all the arrays, we next sweep through the gene-specific rows of $\mathbf{Z}$ or $\mathbf{W}$ and obtain gene-specific target outputs $\hat{t}_{\omega}^{g}$. It warrants noting that the ANN is not retrained here, but rather gene-specific values of $\hat{t}_{\omega}^{g}$ are determined by applying the last known weights to gene-specific row vectors of $\mathbf{Z}$ or $\mathbf{W}$, which map the genes back to the original $\mathbf{M}$ and $\mathbf{F}$ matrices used for training. The information regarding outcome classes is embedded in the learned values of the connections weights when the ANN was trained with data matrices $\mathbf{M}$ and $\mathbf{F}$.

Gene Selection with Maximum Sensitivity. The average gene-class-specific sensitivity [26] of each gene is then determined as

$$S_{\omega}^{g} = \frac{1}{n} \sum_{i=1}^{n} \frac{\partial \hat{t}_{\omega}^{g}}{\partial x_i}, \tag{23.34}$$

where g is the gene, ω is the class, n is the number of input nodes based on $n = K$, and $\mathbf{x} = \mathbf{z}_g$ if the ANN is trained with K-means centers based on $\mathbf{Z}$, or $n = p*$ and $\mathbf{x} = \mathbf{w}_g$ if the ANN is trained with PC score coefficients based on $\mathbf{W}$. The partial derivative $\partial \hat{t}_\omega^g / \partial x_i$ is determined via the chain rule, by first differentiating $\hat{t}_\omega^g$ with respect to hidden layer outputs, v_j, and then input row values x_i, given by

$$\begin{aligned}
\frac{\partial \hat{t}_\omega^g}{\partial x_i} &= \sum_j \frac{\partial \hat{t}_\omega^g}{\partial v_j} \frac{\partial v_j}{\partial x_i} \\
&= \sum_j \frac{d\hat{t}_\omega^g}{dy_c} \frac{\partial y_\omega}{\partial v_j} \frac{dv_j}{du_j} \frac{\partial u_j}{\partial x_i} \\
&= \sum_j \left(\frac{\exp(y_\omega)\left(\sum_l \exp(y_l) - \exp(y_\omega)\right)}{\left(\sum_l \exp(y_l)\right)^2} w_{j\omega}^{ho} \frac{e^{-u_j}}{(1+e^{-u_j})^2} w_{ij}^{ih} \right).
\end{aligned} \tag{23.35}$$

Class-specific sensitivities for each gene are summed over all models and then sorted in descending order. Genes at the top of the sort are selected as the best predictors based on gene-class-specific sensitivity. The list of genes is divided equally into genes with the greatest sensitivity for discriminating each class. For example, for a list of eight genes and two classes, the four genes with the greatest sensitivity for discriminating class 1 are used along with the four genes with the greatest sensitivity for discriminating class 2.

Gene Selection with Minimum Error. In addition to RFE based on sensitivity, we also calculate the gene-class-specific mean-square error (MSE) during the last sweep, using the recomputed values of $\hat{t}_\omega^g$ described above [25]. Analogously, we derive lists of genes for which each class is represented equally by genes having the lowest gene-class-specific MSE. Recall that the predicted class target $\hat{t}_\omega^g$ for each gene is class-specific; however, genes do not have a true class membership (i.e., t_ω^g), so we calculate error as $E_\omega^g = \sum_j^\Omega 0.5(\hat{t}_\omega^g - I(j))^2$, where $I(j)$ is one if $j = \omega$ and zero if $j \neq \omega$.

Generating Lists of Selected Genes. A modular approach is employed for generating the list of genes identified during RFE. Lists are divided uniformly into genes that best discriminate each outcome class, depending on whether the selection criterion is minimum gene-class-specific MSE or maximum gene-class-specific sensitivity. The total number of genes in a list is based on powers of 2 multiplied by the number of classes, such that the list is uniformly loaded with genes that best discriminate each class. As an example, in a two-class study, gene selection lists contain a total of 2, 4, 8, 16, 32, or 64 genes; half of these genes have the total maximum sensitivity (or least error) for the first class, and half have the maximum sensitivity (or least error) for the second class.

ANN Learning Curves from Selected Genes. After recursive feature identification, the ANN models are trained with the actual standardized values of expression for each list of 2, 4, 8, 16, 32, or 64 genes identified. Expression for each gene is standardized over the arrays (using array-specific mean and s.d.). ANN clamping involves use of the rows $\mathbf{x}_i$ of the standardized expression matrix $\mathbf{X}$ (Table 23.2). The number of hidden nodes can be set to, for example, 40% of the number of input nodes. For example, for 64 genes (features) and two outcome classes, a 64-26-2 network is employed. During runs with actual gene expression profiles, we assess accuracy and the proportion of between-gene correlation coefficients that are significant ($P \leq 0.01$).

In clinical settings where large streams of daily gene expression data are available, it is advantageous to use large sets of genes for classification because of the *concept drift* problem [26]. As physician referral patterns change through time, the race, ethnicity, environmental/occupational exposures, and unknown gene-environment determinants among patients screened

TABLE 23.2 Data Used in ANN Training, Recursive Feature Elimination (RFE), and Clamping Criteria

		Reduction method	
Data	ANN usage	K means	PCA
Matrices	Train[a]	$\underset{n\times K}{\mathbf{M}}$	$\underset{n\times p*}{\mathbf{F}}$
	RFE[b]	$\underset{p\times K}{\mathbf{Z}}$	$\underset{p\times p*}{\mathbf{W}}$
	Prediction[c]	$\underset{n\times 2,4,8,...}{\mathbf{X}}$	$\underset{n\times 2,4,8...}{\mathbf{X}}$
ANN clamping vectors	Train	$\mathbf{m}_i(1 \times K)$	$\mathbf{f}_i(1 \times p*)$
	RFE	$\mathbf{z}_j(1 \times K)$	$\mathbf{w}_j(1 \times p*)$
	Prediction	$\mathbf{x}_i(1 \times 2, 4, 8\ldots)$	$\mathbf{x}_i(1 \times 2, 4, 8, \ldots)$

[a]Matrices **M** and **F** used for training ANN.
[b]Apply weights after last training epoch against matrices **Z** and **W** looping over all genes once to determine E^g_ω and S^g_ω.
[c]During RFE, all p genes are used for training, whereas during prediction sets of the p top genes are used ($p = 2, 4, 8, 16, 32, 64$). After RFE, network is trained using small sets of the top genes ($p = 2, 4, 8, 16, 32, 64$).
Notation: M—matrix of k-means centers; F—matrix of PC scores; Z—matrix of K-means scores; W—matrix of PC score coefficients; X—matrix of standardized expression; n—number of arrays ($i = 1, 2, \ldots, n$); K—number of K-means cluster centers ($k = 1, 2, \ldots, K$); $p*$—number of PCs ($j = 1, 2, \ldots, p*$); p—number of genes ($j = 1, 2, \ldots, p$).

will also begin to vary. Genes that are informative for classifying cancer subtypes during one year may not be the same as genes that are informative another year. Therefore, gene expression-based intelligent systems used for large-scale patient classification must be retrained daily or weekly in order to minimize bias from concept drift.

23.4.2 Method 2: Gene Filtering and Selection

Dimensional reduction and recursive feature elimination for ANN training described above is useful when the number of features greatly outweighs the number of arrays. However, there are other ways to train an ANN when there are a large number of features. One of the most common methods, called *filtering*, identifies genes that best discriminate the various classes, followed by *selection* to identify genes that are jointly statistically significant for discrimination. (Chapter 12 describes several gene filtering and selection schemes.) The greedy plus takeaway-one (greedy PTA) and "best ranked" selection methods are discussed in the approach in which features are ranked according to a score or statistic. For two-class problems, many groups use *t*-tests, Mann-Whitney tests, and for multiclass problems, the F test or Kruskal-Wallis tests can be used. Moreover, information gain and Gini diversity index also work well for both two-class and multiclass problems. The choice of how many features to use for the best ranked approach depends on the goal of the ANN application. Here, learning curves can be generated for ANN training based on features added singly or in groups such as the modular approach.

At each step of training with a given number of features, a grid search can be employed to find the lowest MSE as a function of varying the both the learning rate ϵ and momentum α over the range $2^{-9}, 2^{-8}, \ldots, 2^{-1}$. The grid search should include an evaluation of error for a variable number of hidden nodes in the single hidden layer, which ranges from the number of output nodes (number of classes) up to the number of training features (i.e., the length of input vector for each microarray) incremented by 2. A grid search does not have to be performed, but it will accelerate training by reducing *MSE* faster during earlier sweeps.

23.5 ALGORITHM

Algorithm 23.2 lists the computational steps required for implementing ANN classification analysis via an online learning approach, which assumes that genes have already been selected prior to analysis (i.e., the ANN analysis proper is not used for gene selection through recursive feature elimination).

Algorithm 23.2: ANN (online learning)

Data: $n \times p$ data matrix $\mathbf{X}$ with n arrays and p genes
Result: Classification Accuracy
$m \leftarrow 10$, $\text{Acc}_{\text{total}} = 0$, $\varepsilon_0 \leftarrow 0.01$, $\gamma \leftarrow 1/\#\text{sweeps}$, $\alpha \leftarrow 0.7$
foreach *Repartition* $= 1 \leftarrow 10$ **do**
    Set $\mathbf{C} \leftarrow \mathbf{0}$, $n_{\text{test}} \leftarrow 0$
    Randomly partition input arrays into $\mathcal{D}_1, \mathcal{D}_2, \ldots, \mathcal{D}_m$ folds
    for *each cross-validation fold* $\mathcal{D}_m$ **do**
        Initialize $\{w_{ij}^{ih}\}$ and $\{w_{jc}^{ho}\}$ in the range [−0.5,0.5]
        Select all arrays in $\mathcal{D}_m$ for testing
        Use all remaining arrays not in $\mathcal{D}_m$ for training
        $\Delta w_{ij}^{ih} \leftarrow 0, \forall i, j$, $\Delta w_{jc}^{ho} \leftarrow 0, \forall j, c$
        foreach *sweep* $= 1 \leftarrow \#\text{sweeps}$ **do**
            $\varepsilon = \varepsilon_0 \left(1 - \frac{\text{sweep}-1}{\#\text{sweeps}}\right)$
            for *each training array* $\mathbf{x}_i \notin \mathcal{D}_m$ **do**
                $u_j = \sum_{i=1}^{I} x_i w_{ij}^{ih}, \forall j$
                $v_j = 1/(1 + e^{-u_j}), \forall j$
                $y_k = \sum_{j=1}^{J} v_j w_{jk}^{ho}, \forall k$
                $\hat{z}_k = \exp(y_k)/\{\sum_{l=1} \exp(y_l)\}, \forall k$
                $E = \frac{1}{2} \sum_{k=1}^{\Omega} (\hat{z}_k - t_k)^2$
                $\Delta w_{jk}^{ho} = -\varepsilon\gamma(\partial E/\partial w_{jk}^{ho}) + \alpha \Delta w_{jk}^{ho}, \forall j, k$
                $w_{jk}^{ho} = w_{jk}^{ho} + \Delta w_{jk}^{ho}, \forall j, k$
                $\Delta w_{ij}^{ih} = -\varepsilon\gamma(\partial E/\partial w_{ij}^{ih}) + \alpha \Delta w_{ij}^{ih}, \forall i, j$
                $w_{ij}^{ih} = w_{ij}^{ih} + \Delta w_{ij}^{ih}, \forall i, j$
            endfor
        endfch
        for *each test array* $\mathbf{x}_i \in \mathcal{D}_m$ **do**
            $n_{\text{test}} += 1$
            for *each class* $k \in \Omega$ **do**
                Determine $\hat{z}_k$
            endfor
            Predict class $\hat{\omega} = \arg\max_{k \in \Omega}\{\hat{z}_k\}$
            Increment confusion matrix $\mathbf{C}$ by 1 in element $c_{\omega,\hat{\omega}}$
        endfor
    endfor
    $\text{Acc}_{\text{total}} += (\sum_i c_{ii})/n_{\text{test}}$
endfch
$\text{Acc} = \text{Acc}_{\text{total}}/10$

23.6 BATCH VERSUS ONLINE TRAINING

Backpropagation training can be accomplished in several ways, depending on how the partial derivatives and weight updates are calculated. Batch training maintains running sums of the partial derivatives through all cycles within a sweep and then updates connection weights after the sweep. Online training, on the other hand, updates weights during each cycle as each array is presented to the network. Batch training is faster than online; however, there can be differences in the training error based on the type of functions used on both the input and output sides. Another main difference between batch and online learning is that gradient descent step lengths are larger for batch training, due to the accumulation of derivative values during each sweep; hence, the learning rate and momentum may need to be reduced during batch processing. Online training, however, uses smaller array-specific derivatives during each cycle, causing much smaller step sizes during gradient descent; greater learning rates and momentum have a lesser impact. Users need to determine via trial and error which combination of network functions and learning approach results in the least error.

23.7 ANN TESTING

After training is complete and MSE has been minimized to a satisfactory level, the connection weights are saved from the last sweep. During testing, the feature values of each test microarray **x** are clamped to the input nodes and propagated through the connections using the last known weights to derive the target vector **z**. The decision rule $D(\mathbf{x} \rightsquigarrow \omega)$ is to assign **x** into the class for which z_k is the greatest.

23.8 CROSS-VALIDATION RESULTS

Cross-validation (CV) was run using the nine datasets listed in Table 12.10, for which genes were filtered and selected using greedy PTA. Bootstrap accuracy ("CVB") was based on average 0.632 bootstrap accuracy (13.4) for 10 bootstraps. Batch processing was used with 50 sweeps, and the learning rate was set to $\varepsilon = 0.01$, while momentum was set to $\alpha = 0.7$. Weight decay was not employed. The effect of the number of hidden nodes (HNs) used on performance was evaluated using values of 2, 4, 6, 8, and 10. Table 23.3 lists the average accuracy for various fold sizes of CV with no any input feature transformations applied. As the number of hidden nodes increased, accuracy did not improve for the majority of datasets. This was attributable to the fact that the preselected features (see Table 12.10) are highly discriminating for the classes, requiring a small number of hidden nodes to achieve high performance. In fact, if 100% accuracy is obtained using two hidden nodes in the hidden layer, then more hidden nodes would truly result in overfitting.

TABLE 23.3 ANN Cross-Validation (CV) Performance without Input Feature Transformation

		Cross-validation method				
Dataset	HN[a]	CV2	CV5	CV10	CV − 1	CVB
AMLALL2	2	100	100	100	100	100
	4	100	100	100	100	100
	6	100	100	100	100	100
	8	100	100	100	100	100
	10	100	100	100	100	100
Brain2	2	88.33	95.00	95.00	95.00	96.67
	4	90.00	93.33	96.67	96.67	97.75
	6	91.67	93.33	95.00	96.67	97.64
	8	91.67	93.33	93.33	96.67	96.14
	10	95.00	95.00	96.67	96.67	96.41
Breast2A	2	100	100	100	100	100
	4	100	100	100	100	100
	6	100	100	100	100	100
	8	100	100	100	100	100
	10	100	100	100	100	100
Breast2B	2	92.31	94.87	93.59	94.87	97.43
	4	87.18	93.59	93.59	94.87	97.66
	6	92.31	97.44	96.15	93.59	96.23
	8	91.03	94.87	96.15	96.15	97.43
	10	91.03	93.59	96.15	94.87	97.37
Colon2	2	77.42	88.71	88.71	90.32	88.66
	4	85.48	90.32	88.71	88.71	89.63
	6	88.71	87.10	88.71	90.32	89.97
	8	88.71	88.71	88.71	88.71	90.39
	10	87.10	90.32	91.94	90.32	90.18
Lung2	2	100	100	100	100	100
	4	100	100	100	100	100
	6	100	100	100	100	100
	8	100	100	100	100	100
	10	100	100	100	100	100
MLL_Leukemia3	2	98.25	98.25	100	100	99.12
	4	96.49	98.25	98.25	100	99.69
	6	100	100	100	98.25	99.61
	8	100	98.25	100	100	100
	10	100	100	100	100	99.49
Prostate2	2	97.06	98.04	98.04	98.04	98.67
	4	97.06	98.04	98.04	98.04	98.90
	6	98.04	98.04	97.06	98.04	98.77
	8	98.04	98.04	98.04	98.04	99.22
	10	97.06	97.06	98.04	98.04	98.58
SRBCT4	2	93.65	100	90.48	92.06	91.53
	4	100	100	100	100	99.30
	6	100	100	100	100	100
	8	100	100	100	100	100
	10	96.83	100	100	100	99.82
Average		0.96	0.97	0.97	0.97	0.98

[a]HN represents the number of hidden nodes employed in the three-layer feed forward ANN.

Mean-zero standardization did not improve accuracy (Table 23.4) and input feature fuzzification reduced accuracy (Table 23.5).

Figure 23.4 illustrates accuracy as a function of CV method for each dataset with no feature transformations applied. ANN was quite biased for the Breast2A and SRBCT4 datasets, as accuracy ramped up quickly after CV2 when more data were used for training. Figure 23.5 reveals average accuracy for all datasets as a function of feature transformation and CV method. Average accuracy increased with increasing data used in training, that is, while ramping through CV2, CV5, and CV10, since these use $\frac{1}{2}$, $\frac{4}{5}$, and $\frac{9}{10}$ of the available data for training, respectively.

Figure 23.6 reveals the ANN accuracy without transformations on features as a function of the number of hidden nodes. Overall, there was not much change in performance as the number of hidden nodes increased. Input feature mean-zero standardization did not seem to improve performance by any appreciable amount (Figure 23.7), and input feature fuzzification worsened performance (Figure 23.8). Figure 23.9 shows that average performance for all datasets as a function of input feature transformation did not seem to change with increasing number of hidden nodes.

23.9 BOOTSTRAP BIAS

Figure 23.10 shows the ANN accuracy based on runs without transformations on features as a function of bootstrap sample size used during training, based on 10 resamplings from the input arrays. Bootstrap bias runs were based on determining the 0.632 bootstrap accuracy (see Eq. 13.4) for each sampling fraction. Acc_0 in 13.4 was based on test accuracy of unsampled arrays, while Acc_b in 13.4 was based on test accuracy for sampled arrays. The x-axis values represent the number of arrays randomly sampled $B = 40$ times at fractions $f = 0.1$, 0.2, 0.3, 0.4, 0.5, or 0.6 of the available arrays. Figure 23.11 illustrates the ANN accuracy with feature standardization as a function of bootstrap sample size used during training, and Figure 23.12 shows the ANN accuracy with feature fuzzification. ANN was slightly biased when 10% of the training data was bootstrapped before learning.

23.10 MULTICLASS ROC CURVES

Figure 12.13 illustrates ROC curves for the ANN classifier applied to the AMLALL2, Brain2, Breast2A, Breast2B, Colon2, Lung2, MLL_Leukemia3, Prostate2, and SRBCT4 datasets using mean-zero standardized input features. The legend values represent the number of arrays randomly sampled $B = 40$ times at fractions $f = 0.1, 0.2, 0.3, 0.4, 0.5$, or 0.6 of the available arrays. Figure 23.14 illustrates ROC curves for the ANN based on the AMLALL2, Brain2, Breast2A, Breast2B, Colon2, Lung2, MLL_Leukemia3, Prostate2, and SRBCT4 datasets using fuzzified input features.

TABLE 23.4 ANN CV Performance with Input Feature Mean-Zero Standardization

		Cross-validation method				
Dataset	HN[a]	CV2	CV5	CV10	CV − 1	CVB
AMLALL2	2	100	100	100	100	98.86
	4	100	100	100	100	100
	6	100	100	100	100	100
	8	100	100	100	100	100
	10	100	100	97.37	100	100
Brain2	2	93.33	93.33	98.33	96.67	96.97
	4	88.33	95.00	98.33	96.67	98.19
	6	96.67	95.00	95.00	95.00	97.37
	8	91.67	95.00	90.00	96.67	97.20
	10	96.67	98.33	95.00	95.00	97.10
Breast2A	2	80.00	100	100	100	100
	4	100	100	100	100	100
	6	100	100	100	100	100
	8	100	100	100	100	100
	10	100	100	100	100	100
Breast2B	2	88.46	92.31	93.59	92.31	96.62
	4	89.74	93.59	92.31	94.87	98.03
	6	93.59	96.15	97.44	93.59	97.61
	8	89.74	94.87	96.15	93.59	97.03
	10	92.31	97.44	96.15	96.15	98.28
Colon2	2	88.71	85.48	87.10	87.10	86.56
	4	90.32	88.71	88.71	88.71	88.59
	6	90.32	90.32	88.71	88.71	90.67
	8	88.71	88.71	90.32	88.71	89.86
	10	90.32	87.10	88.71	88.71	90.59
Lung2	2	100	100	100	100	100
	4	100	100	100	100	100
	6	100	100	100	100	100
	8	100	100	100	100	100
	10	100	100	100	100	100
MLL_Leukemia3	2	92.98	100	100	100	98.98
	4	96.49	98.25	100	100	99.85
	6	98.25	100	100	100	99.84
	8	98.25	100	100	100	99.82
	10	100	100	100	100	99.43
Prostate2	2	98.04	98.04	98.04	98.04	98.81
	4	97.06	98.04	98.04	98.04	98.90
	6	97.06	98.04	98.04	98.04	99.33
	8	97.06	98.04	98.04	98.04	97.98
	10	97.06	98.04	98.04	98.04	99.16
SRBCT4	2	77.78	90.48	90.48	92.06	91.49
	4	100	100	100	100	97.33
	6	100	100	100	100	99.85
	8	100	100	100	100	100
	10	100	100	100	100	99.86
Average		0.96	0.97	0.97	0.97	0.98

[a]HN represents the number of hidden nodes employed in the three-layer feed forward ANN.

TABLE 23.5 ANN CV Performance with Input Feature Fuzzification

		Cross-validation method				
Dataset	HN[a]	CV2	CV5	CV10	CV − 1	CVB
AMLALL2	2	100	100	100	100	99.54
	4	100	100	100	100	100
	6	89.47	100	100	100	100
	8	100	100	100	100	100
	10	100	100	100	100	100
Brain2	2	86.67	88.33	81.67	90.00	89.38
	4	78.33	90.00	90.00	88.33	92.03
	6	90.00	86.67	88.33	90.00	95.03
	8	78.33	90.00	90.00	91.67	94.41
	10	85.00	86.67	90.00	88.33	94.26
Breast2A	2	60.00	80.00	100	100	92.32
	4	93.33	100	100	100	100
	6	100	100	100	100	99.47
	8	100	100	100	100	100
	10	100	100	100	100	100
Breast2B	2	79.49	84.62	82.05	83.33	91.65
	4	84.62	82.05	82.05	84.62	92.87
	6	82.05	80.77	80.77	85.90	93.05
	8	84.62	80.77	84.62	85.90	92.95
	10	84.62	82.05	80.77	83.33	92.79
Colon2	2	79.03	77.42	87.10	85.48	86.77
	4	72.58	83.87	87.10	85.48	87.25
	6	87.10	85.48	85.48	87.10	88.19
	8	88.71	88.71	88.71	87.10	88.21
	10	83.87	83.87	83.87	87.10	89.02
Lung2	2	100	100	100	100	100
	4	100	100	100	100	100
	6	100	100	100	100	100
	8	100	100	100	100	100
	10	100	100	100	100	100
MLL_Leukemia3	2	98.25	100	98.25	98.25	95.63
	4	92.98	98.25	100	96.49	99.00
	6	98.25	98.25	98.25	98.25	99.43
	8	98.25	96.49	100	98.25	98.95
	10	94.74	98.25	100	98.25	99.17
Prostate2	2	89.22	95.10	94.12	95.10	96.09
	4	93.14	94.12	95.10	95.10	96.87
	6	91.18	95.10	95.10	95.10	97.21
	8	94.12	95.10	94.12	95.10	97.38
	10	85.29	91.18	94.12	95.10	97.03
SRBCT4	2	71.43	77.78	92.06	84.13	82.20
	4	90.48	96.83	100	100	96.96
	6	100	92.06	100	100	100
	8	96.83	100	100	100	99.25
	10	98.41	98.41	100	100	99.68
Average		0.91	0.93	0.94	0.95	0.96

[a]HN represents the number of hidden nodes employed in the three-layer feed forward ANN.

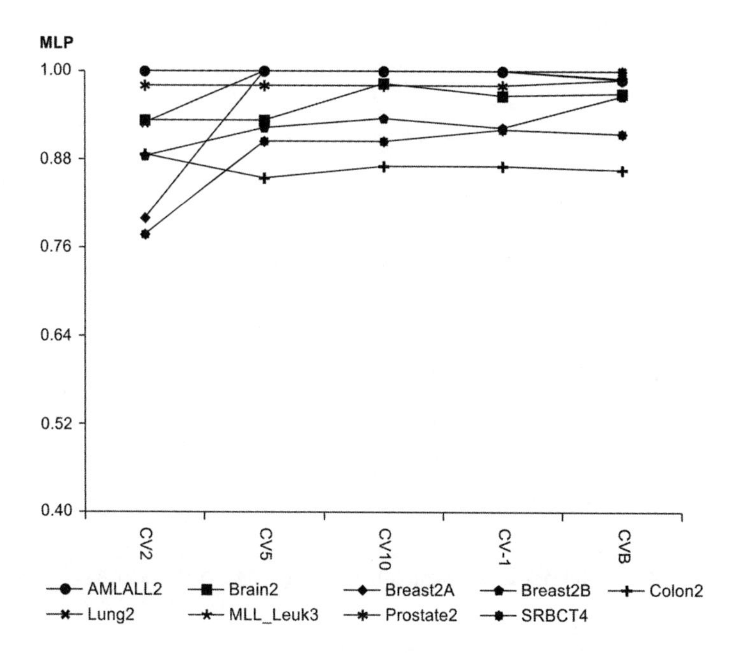

FIGURE 23.4 ANN accuracy as a function of CV method for each dataset without feature transformations applied.

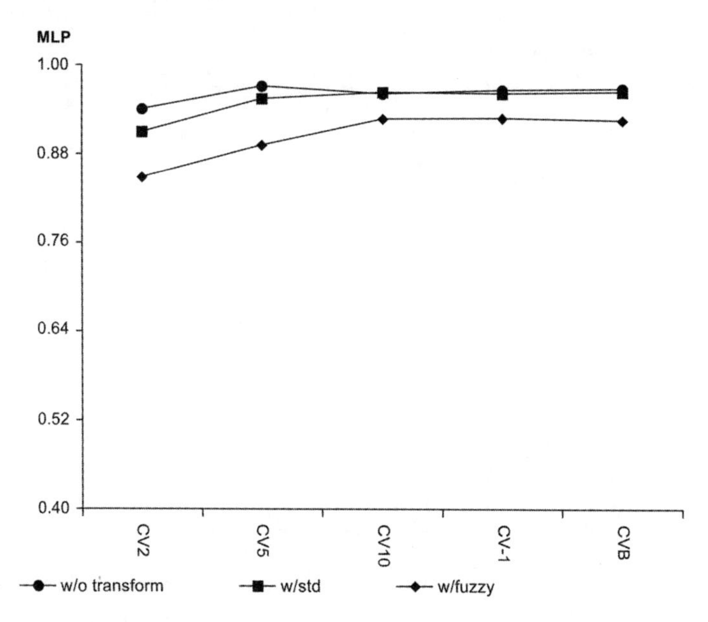

FIGURE 23.5 ANN average accuracy for all datasets as a function of feature transformation and CV method.

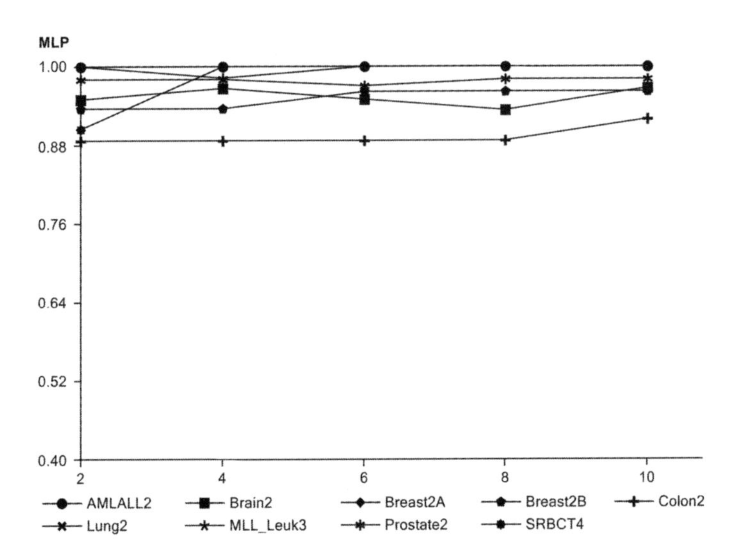

FIGURE 23.6 ANN accuracy without feature transformation as a function of the number of hidden nodes.

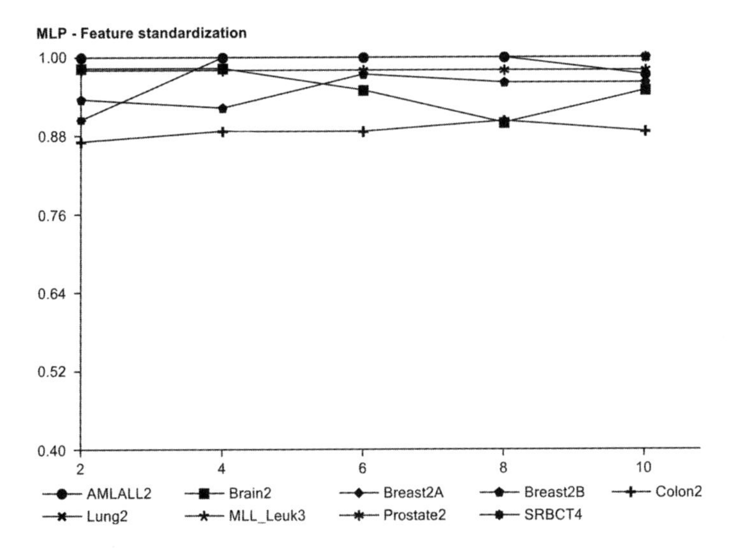

FIGURE 23.7 ANN accuracy with feature standardization as a function of the number of hidden nodes.

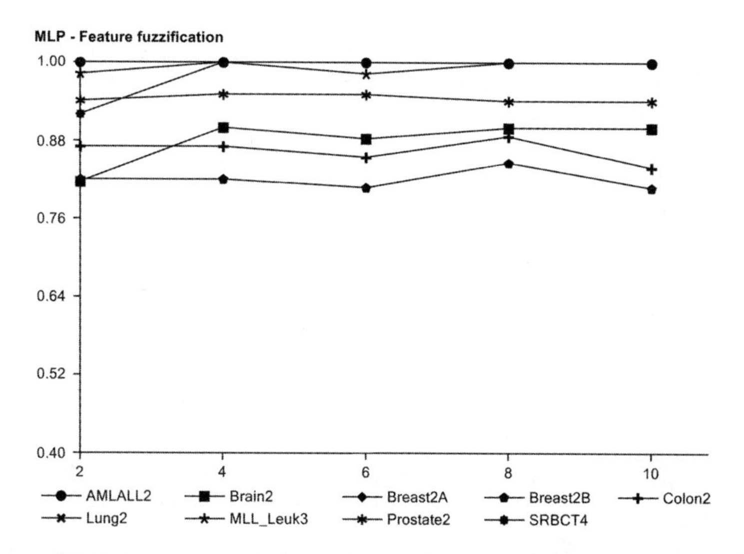

FIGURE 23.8 ANN accuracy with feature fuzzification as a function of the number of hidden nodes.

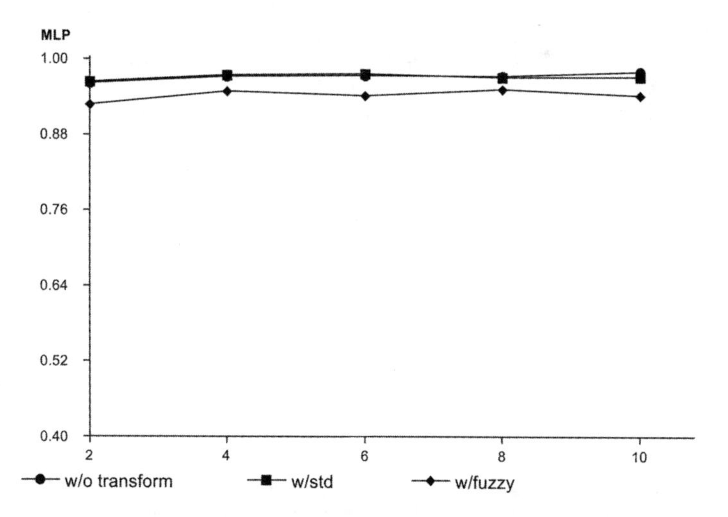

FIGURE 23.9 ANN average accuracy for all datasets as a function of the number of hidden nodes. Ten 10-fold cross-validation used.

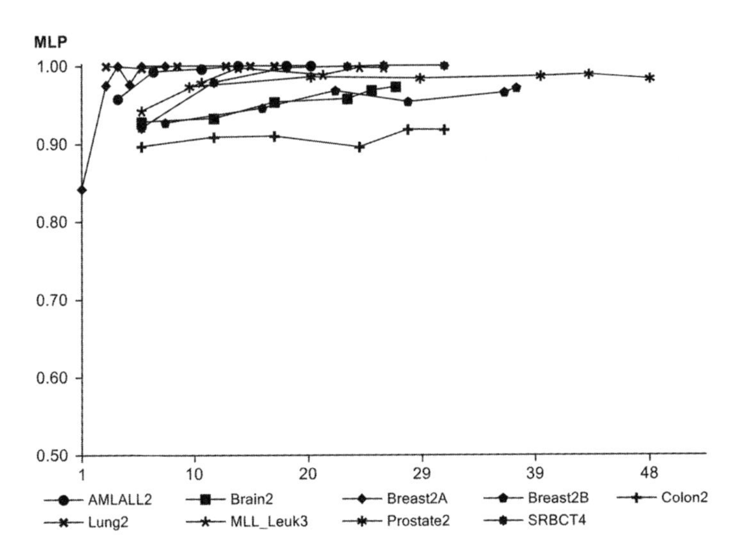

FIGURE 23.10 ANN accuracy without transformations on features as a function of bootstrap sample size used during training. The x-axis values represent the number of arrays randomly sampled $B = 40$ times at fractions $f = 0.1$, 0.2, 0.3, 0.4, 0.5, or 0.6 of the available arrays.

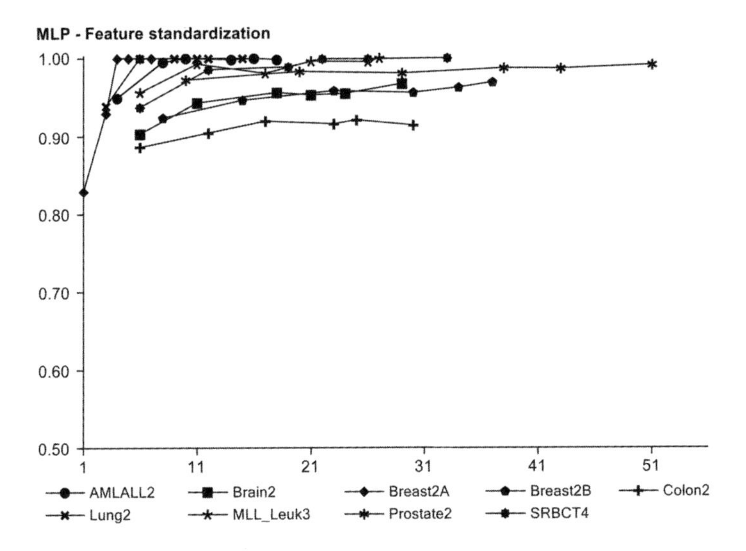

FIGURE 23.11 ANN accuracy with feature standardization as a function of bootstrap sample size used during training. The x-axis values represent the number of arrays randomly sampled $B = 40$ times at fractions $f = 0.1$, 0.2, 0.3, 0.4, 0.5, or 0.6 of the available arrays.

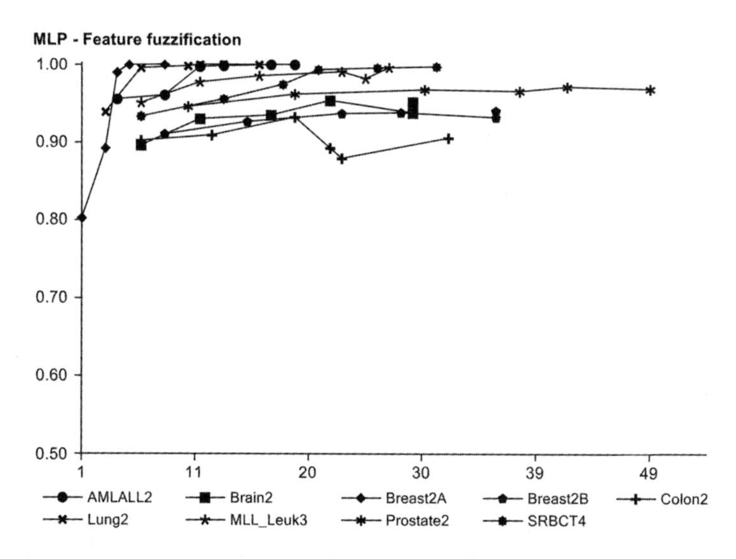

FIGURE 23.12 ANN accuracy with feature fuzzification as a function of bootstrap sample size used during training. The x-axis values represent the number of arrays randomly sampled $B = 40$ times at fractions $f = 0.1, 0.2, 0.3, 0.4, 0.5$, or 0.6 of the available arrays.

23.11 DECISION BOUNDARIES

Figure 23.15 illustrates 2D self-organizing maps of class decision boundaries for the ANN based on the AMLALL2, Brain2, Breast2A, Breast2B, Colon2, Lung2, MLL_Leukemia3, Prostate2, and SRBCT4 datasets using mean-zero standardized input features. Figure 23.16 illustrates 2D self-organizing maps of class decision boundaries for the ANN based on the AMLALL2, Brain2, Breast2A, Breast2B, Colon2, Lung2, MLL_Leukemia3, Prostate2, and SRBCT4 datasets using fuzzified input features. The ANN classifier exploited the feature fuzzification and provided very smooth decision boundaries. A drawback with input feature fuzzification is that it creates 3 times as many input features, causing an increase in runtimes. However, when compared with the majority of other classifiers presented in this book, the ANN seemed to be able to yield smoother decision boundaries from feature value fuzzification prior to input.

23.12 RPROP VERSUS BACKPROPAGATION

The four-class SRBCT dataset was used for ANN training to investigate the effect of hidden node activation function and backpropagation method on network training error. In total, 200 sweeps were used for $n = 63$ arrays

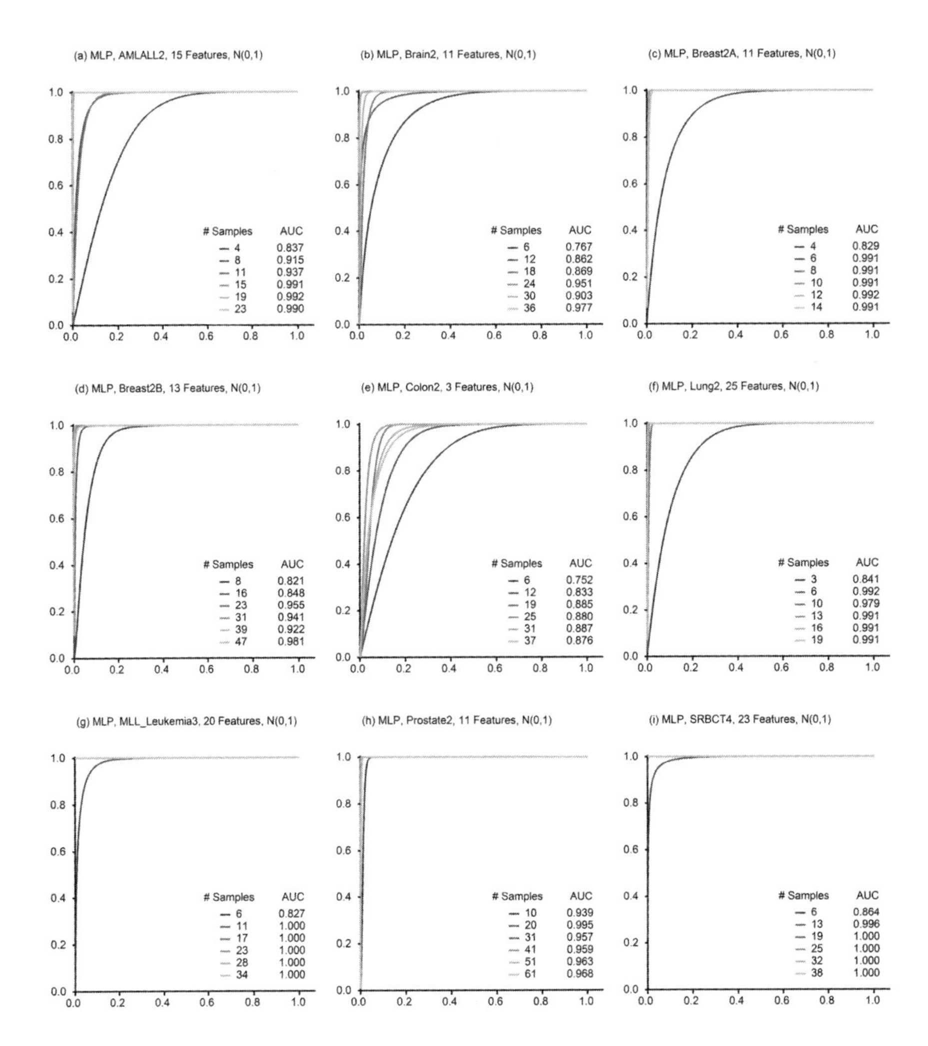

FIGURE 23.13 ROC curves for the ANN classifier applied to the AMLALL2, Brain2, Breast2A, Breast2B, Colon2, Lung2, MLL_Leukemia3, Prostate2, and SRBCT4 datasets based on mean-zero standardized input features: legend values represent the number of arrays randomly sampled $B = 40$ times at fractions $f = 0.1$, 0.2, 0.3, 0.4, 0.5, or 0.6 of the available arrays: (a) AMLALL2; (b) Brain2; (c) Breast2A; (d) Breast2B; (e) Colon2; (f) Lung2; (g) MLL_Leukemia3; (h) Prostate2; (i) SRBCT4. The numbers of genes used for each dataset are listed in Table 12.10.

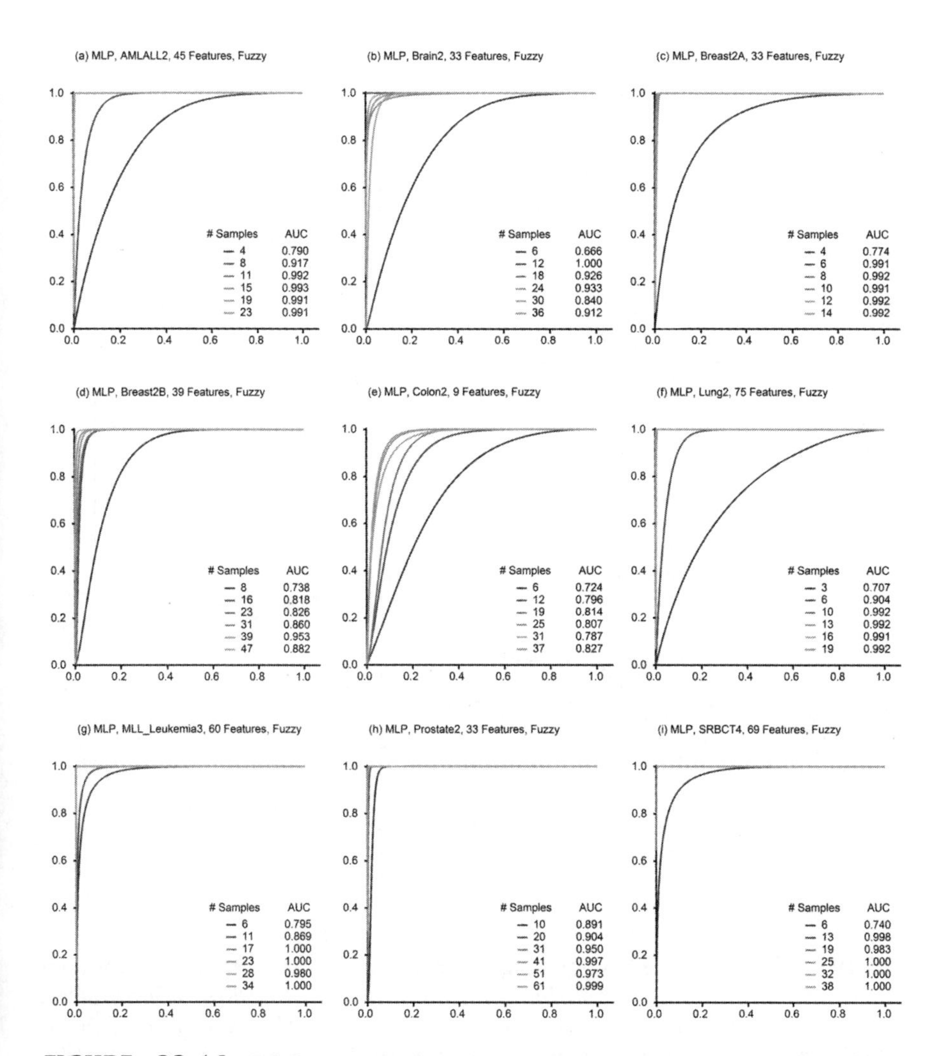

FIGURE 23.14 ROC curves for the ANN applied to the AMLALL2, Brain2, Breast2A, Breast2B, Colon2, Lung2, MLL_Leukemia3, Prostate2, and SRBCT4 datasets based on fuzzified input features; legend values represent the number of arrays randomly sampled $B = 40$ times at fractions $f = 0.1, 0.2, 0.3, 0.4, 0.5$, or 0.6 of the available arrays: (a) AMLALL2; (b) Brain2; (c) Breast2A; (d) Breast2B; (e) Colon2; (f) Lung2; (g) MLL_Leukemia3; (h) Prostate2; (i) SRBCT4. The numbers of genes used for each dataset are listed in Table 12.10.

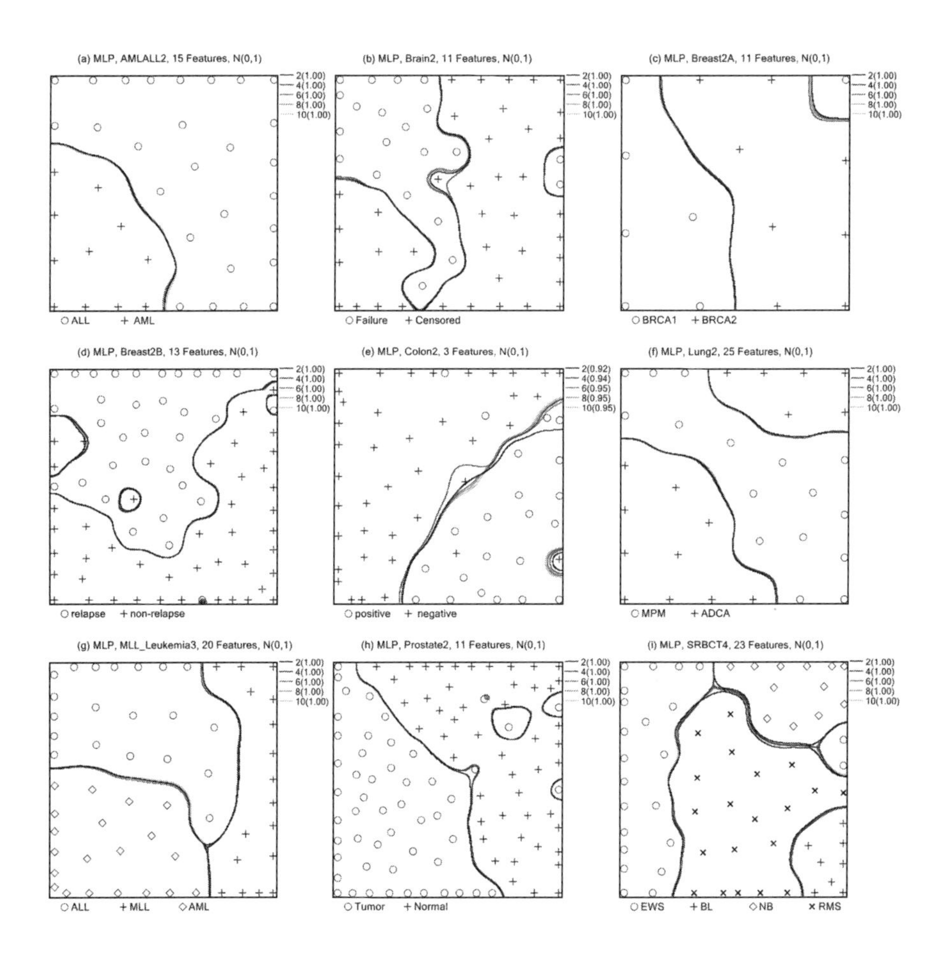

FIGURE 23.15 2D self-organizing maps of ANN class decision boundaries and classification accuracy as a function of the number of hidden nodes (upper right of each panel) for the AMLALL2, Brain2, Breast2A, Breast2B, Colon2, Lung2, MLL_Leukemia3, Prostate2, and SRBCT4 datasets based on mean-zero standardized input features: (a) AMLALL2; (b) Brain2; (c) Breast2A; (d) Breast2B; (e) Colon2; (f) Lung2; (g) MLL_Leukemia3; (h) Prostate2; (i) SRBCT4. The numbers of genes used for each dataset are listed in Table 12.10.

and $p = 23$ input features. The softmax function was always used on the output-side, while the choices for hidden node activation function were logistic, tanh, and Hermite. Batch and online learning were evaluated, as well as backpropagation versus RPROP. Figures 23.17 and 23.18 show the variation in MSE and CE for the various training combinations as a function of sweep. Similarly, Tables 23.6 and 23.7 list in ascending order the network training error values of MSE and CE at the 50th iteration on Figures 23.17 and 23.18. For MSE in Figure 23.17, the use of the

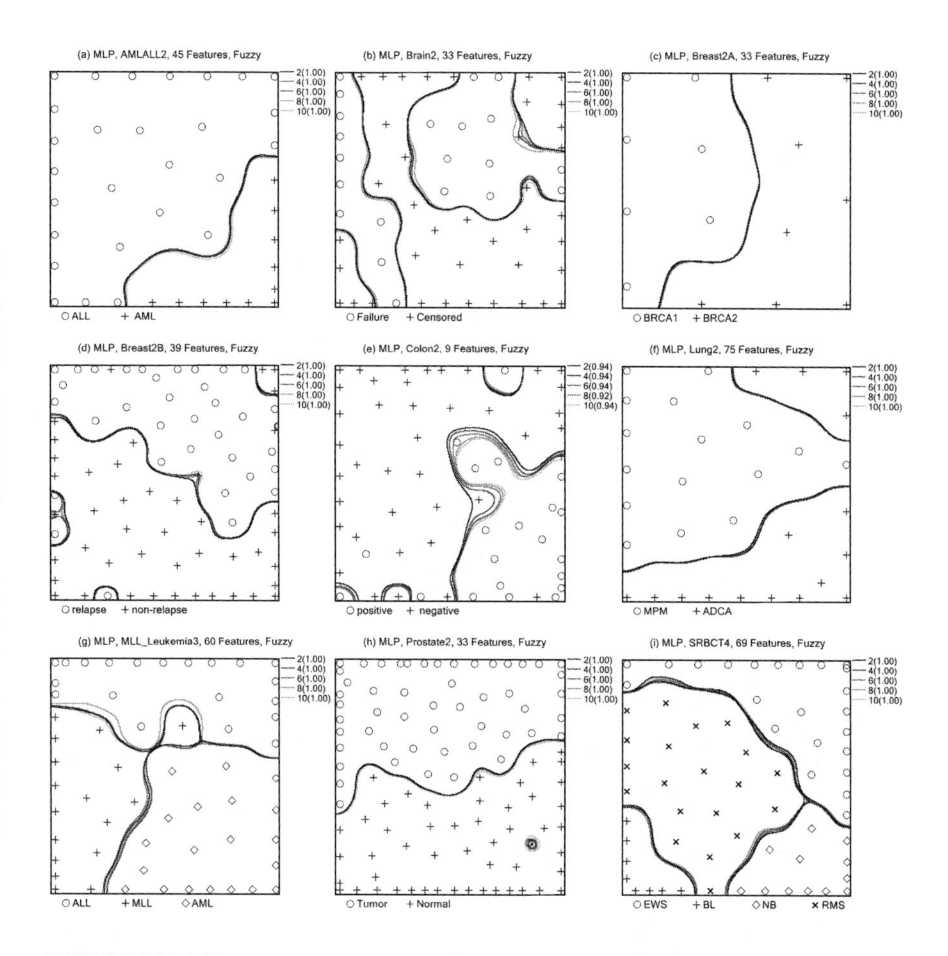

FIGURE 23.16 2D Self-organizing map of ANN class decision boundaries and classification accuracy as a function of the number of hidden nodes (upper right of each panel) for the AMLALL2, Brain2, Breast2A, Breast2B, Colon2, Lung2, MLL_Leukemia3, Prostate2, and SRBCT4 datasets based on fuzzified input features: (a) AMLALL2; (b) Brain2; (c) Breast2A; (d) Breast2B; (e) Colon2; (f) Lung2; (g) MLL_Leukemia3; (h) Prostate2; (i) SRBCT4. The numbers of genes used for each dataset are listed in Table 12.10.

tanh activation function at the hidden nodes (softmax at output nodes) resulted in the lowest MSE—and convergence was much faster than convergence based on other configurations, whereas for CE in Figure 23.18, the logistic activation function at the hidden nodes resulted in the least CE.

Differences in learning rates due to the use of either the logistic or tanh activation function at hidden nodes is due largely to differences in magnitude of the derivatives of these functions. Figure 23.19 illustrates that

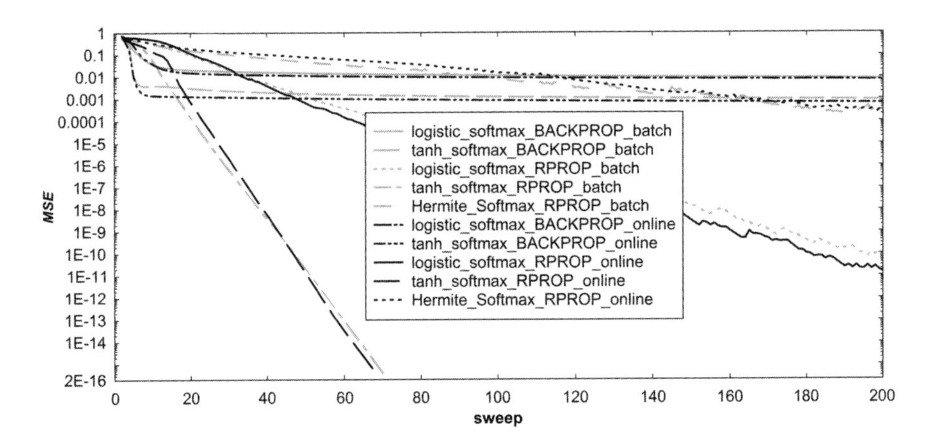

FIGURE 23.17 Network mean-square error MSE and effects of hidden node activation function and network training method for the four-class SRBCT dataset with $n = 63$ arrays and $p = 23$ genes. In total, 23 hidden nodes were used and the softmax function was used on the output side.

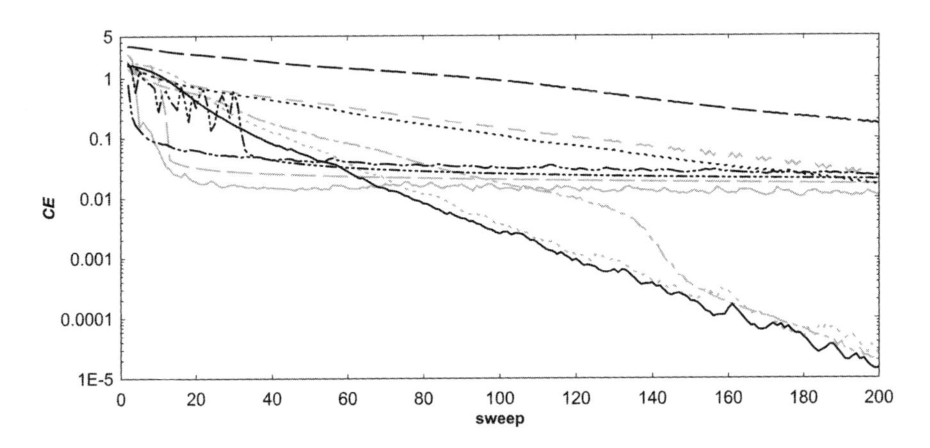

FIGURE 23.18 Network cross-entropy, (CE) and effects of hidden node activation function and network training method for the four-class SRBCT dataset with $n = 63$ arrays and $p = 23$ genes. In total, 23 hidden nodes were used and the softmax function was employed on the output side.

the magnitude of $y = \tanh(x)$ and its derivative dy/dx in the x range of $[-3, 3]$ is much greater than the derivative values for the logistic function. While $y = \text{logistic}(x)$ spans a range $[0, 1]$, $y = \tanh(x)$ spans $[-1.65, 1.65]$. The maximum value of dy/dx for the logistic function is only 0.25 at $x = 0$, while the maximum dy/dx of tanh at $x = 0$ is unity. The logistic function

TABLE 23.6 Network Mean Square Error (MSE) at 50th Training Iteration[a] for the Four-Class SRBCT Dataset with n = 63 Arrays and p = 23 Genes

Method	MSE
tanh_softmax_RPROP_online	1.33×10^{-11}
tanh_softmax_RPROP_batch	2.05×10^{-11}
logistic_softmax_RPROP_online	0.00065
tanh_softmax_BACKPROP_online	0.001057
logistic_softmax_RPROP_batch	0.001261
tanh_softmax_BACKPROP_batch	0.001725
logistic_softmax_BACKPROP_online	0.011053
logistic_softmax_BACKPROP_batch	0.014631
Hermite_Softmax_RPROP_batch	0.045988
Hermite_Softmax_RPROP_online	0.077206

[a]See Figure 23.17.

TABLE 23.7 Network Cross-Entropy Error (CE) at 50th Training Iteration[a] for the Four-Class SRBCT Dataset with n = 63 Arrays and p = 23 Genes[b]

Method	CE
logistic_softmax_BACKPROP_batch	0.014336
tanh_softmax_BACKPROP_batch	0.024029
tanh_softmax_BACKPROP_online	0.035997
logistic_softmax_BACKPROP_online	0.041378
logistic_softmax_RPROP_online	0.049626
logistic_softmax_RPROP_batch	0.071932
tanh_softmax_RPROP_batch	0.118491
Hermite_Softmax_RPROP_online	0.349621
Hermite_Softmax_RPROP_batch	0.402449
tanh_softmax_RPROP_online	1.597736

[a]See Figure 23.18.
[b]In total, 23 hidden nodes were used.

clearly provides a lesser degree of updating, due to the overall smaller magnitude of its function value and derivative. When activation function values are near zero when in a smooth region of the error, the greater values of dy/dx for tanh will result in greater step lengths —which should reduce the convergence rate. The issue of stagnation, or getting stuck in a region where the update needs to be large but is constrained to lower values of dy/dx in the tails of the dy/dx distribution, can be overcome by using

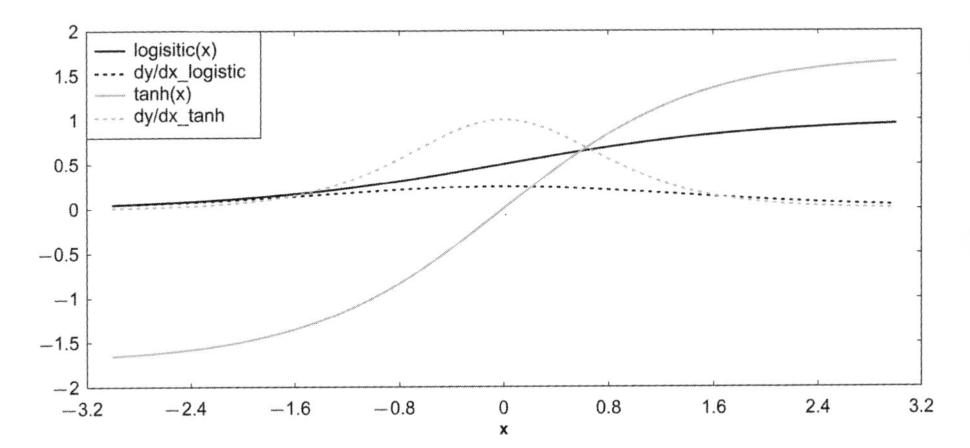

FIGURE 23.19 Function and derivative (dy/dx) values of $y = \text{logistic}(x)$ and $y = \tanh(x)$.

cross-entropy error. However, for the SRBCT dataset that was used, we did not observe improved convergence when compared with mean-square error.

A more thorough understanding of how RPROP works when compared with backpropagation can be gleaned from batch learing results shown in Figures 23.20, 23.21, and 23.22. Backpropagation alone with logistic and

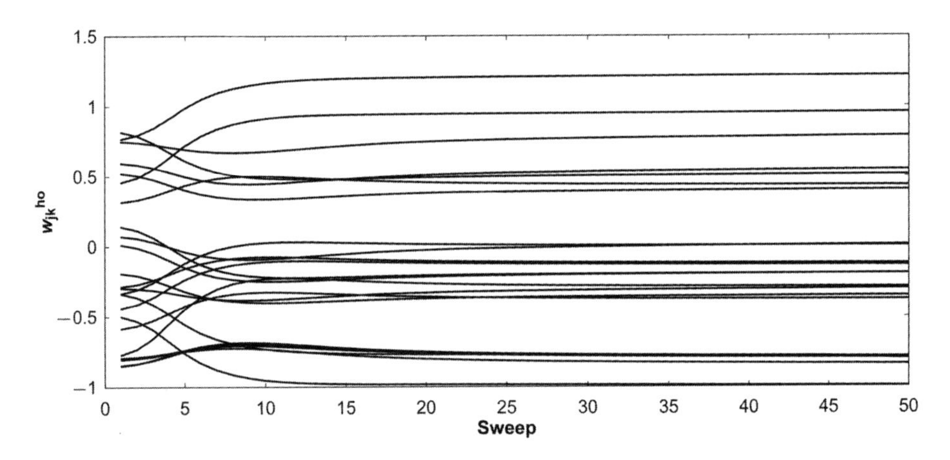

FIGURE 23.20 Logistic function results: ANN connection weights between hidden nodes and output nodes w_{jk}^{ho} for each cycle when using backpropagation, logistic activation function at hidden nodes, and softmax function at output nodes. Two-class Breast2A dataset was used.

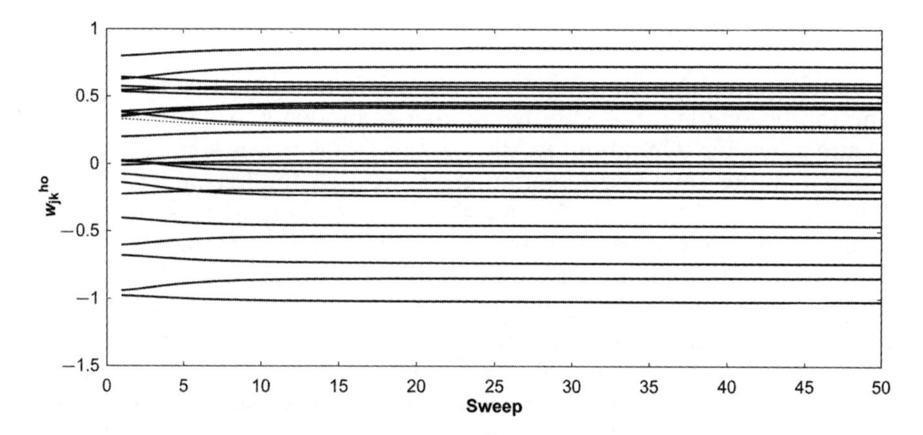

FIGURE 23.21 tanh function results: ANN connection weights between hidden nodes and output nodes w_{jk}^{ho} for each cycle when using backpropagation, tanh activation function at hidden nodes, and softmax function at output nodes. Two-class Breast2A dataset was used.

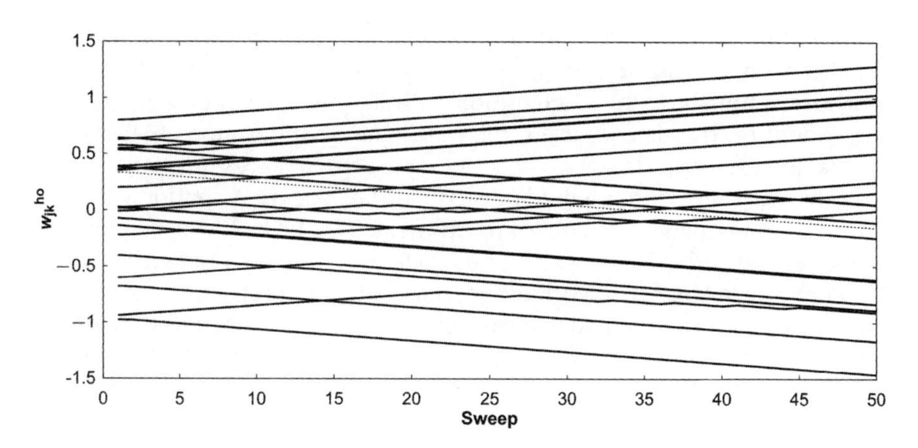

FIGURE 23.22 RPROP results: ANN connection weights between hidden nodes and output nodes w_{jk}^{ho} for each cycle when using RPROP, tanh activation function at hidden nodes, and softmax function at output nodes. Two-class Breast2A dataset was used.

tanh activation function works up to a point when no additional updates are made. On the other hand, RPROP assesses indirectly the curvature of the error by detecting a sign change of the derivative for each sweep and continues to update beyond 50 iterations. As a result, RPROP does not cease correction of error when the indirect measure of curvature is still present.

23.13 SUMMARY

Artifical neural networks offer superior classification performance if overfitting can be avoided. Keeping the number of hidden nodes smaller should ensure a lower potential for overfitting, but this is not guaranteed as overfitting is quite complex. In addition to pruning hidden nodes, pruning connection weights, and using weight decay schemes, Bayesian learning methods can be implemented to help evaluate and minimize overfitting. Network regularization via Bayesian learning and *automatic relevance determination* (ARD) can help the practitioner understand the uncertainty distribution of the predictions, the effects of network structure, and characteristics of the data effects on network generalization. ARD first uses regularization [27] during training to find exact [28] or approximate [29] solutions of the most probable weights and hyperparameters, and then determines the distribution of predictions over the weight space [30]. Markov chain Monte Carlo (MCMC) methods have also been implemented, which can result in more reliable estimates at the expense of longer network training times [31].

Successful employment of ANNs often depends on trial and error regarding the learning rate, momentum, weight decay, number of hidden nodes, and input- and output-side functions employed. It was also reported in this chapter that when RPROP is used with the tanh and softmax activation functions on the input and output sides, respectively, the MSE dropped very precipitously within 50 iterations. While this observation is obviously dependent on the dataset used, the particular configurations that resulted in the lowest error levels may serve as useful choices for other analyses. The weight update methods described [i.e., backpropagation (RPROP)] are first-order methods, which are known to result in lower convergence rates. Second-order learning methods such as conjugate gradient (CG) and Levenberg-Marquardt (LM) methods use both the first- and second-order partial derivatives of network error with respect to each connection weight, and therefore achieve must shorter convergence times [19,32-34]. Second-order methods do have greater computational costs when compared with first-order methods; however, their performance yield is typically better. Certainly, large networks can be trained faster using first-order methods.

REFERENCES

[1] R. Hecht-Nielsen. Theory of the back propagation neural network. *Proc. Int. Joint Conf. Neural Networks (IJCNN'89)*, IEEE Press, Piscataway, NJ, 1989, pp. 593-608.

[2] K. Funahashi. On the approximate realization of continuous mappings by neural networks. *Neural Networks* **2**:183-192, 1989.

[3] K. Hornik, M. Stinchcombe, H. White. Multilayer feedforward networks are universal approximators. *Neural Networks* **2**:359-366, 1989.

[4] M. Leshno, V. Ya Lin, A. Pinkus, S. Schoken. Multilayer feedforward networks with a nonpolynomial activation function can approximate any function. *Neural Networks* **6**:861-867, 1993.

[5] G. Cybenko. Approximations by superpositions of sigmoid function. *Math. Control Signals Sys.* **2**:303-314, 1989.

[6] G.P. Zhang. *Neural Networks in Business Forecasting*. IRM Press, London, 2004.

[7] C.M. Kaun, H. White. Artificial neural networks: An econometric perspective. *Econometr. Rev.* **13**(1):1-91; 1994.

[8] W.P. Gorr, D. Nagin, J. Szczpula. Comparative study of artificial neural network and statistical models for predicting student gradepoint averages. *Int. J. Forecast.* **10**(1):17-34, 1994.

[9] Y. LeCun, L. Bottou, G. Orr, K. Muller. Efficient BackProp. In G. Orr, K. Muller (eds.), *Neural Networks: Tricks of the Trade, Lecture Notes in Computer Science, 1524*, Springer, Berlin, 1998, pp 9-50.

[10] A.S. Weigend, D.E. Rumelhart, B.A. Huberman. Generalization by weight-elimination applied to currency exchange rate prediction. *Proc. IEEE Int. Joint Conf. Neural Networks (IJCNN'91)* 1991, vol. 3, pp. 2374-2379.

[11] M. Cottrell, B. Girard, M. Mangeas, C. Muller. Neural modeling for time series: A statistical stepwise method for weight elimination. *IEEE Trans. Neural Networks* **6**:1355-1364, 1995.

[12] B. Hassibi, D.G. Stork, G.J. Wolff. Optimal brain surgeon and general network pruning. *Proc. IEEE Int. Joint Conf. Neural Networks (IJCNN'93)*, 1993, vol. 1, pp. 293-299.

[13] R. Reed. Pruning algorithms—a survey. *IEEE Trans. Neural Networks* **4**:740-747, 1993.

[14] D.J.C MacKay. Bayesian interpolation. *Neural Comput.* **4**:415-447, 1992.

[15] H. Demuth, M. Beale. *Neural Network Toolbox: For Use with MATLAB*. The Math Works, Natick, MA, 1998.

[16] S. Amari, N. Murata, K.R. Muller, M. Finke, H.H. Yang. *IEEE Trans. Neural Networks* **8**:985-996, 1997.

[17] G. Peterson, D.C. St. Clair, S.R. Aylward, W.E. Bond. Using Taguchi's method of experimental design to control errors in layered perceptrons. *IEEE Trans. Neural Networks* **6**:949-961, 1995.

[18] F. Leisch, L.C. Jain, K. Hornik. Cross-validation with active pattern selection for neural-network classifiers. *IEEE Trans. Neural Networks* **9**:35-41, 1998.

[19] B.D. Wilamowski. Neural network architectures and learning. *Proc. IEEE Int. Conf. Industrial Technology*, IEEE Press, Piscataway, NJ, 2003, pp. TU1-TU11.

[20] D.E. Rumelhart, G.E. Hinton, R.J. Williams. Learning representations by back-propagating errors. *Nature* **323**:533-536, 1986.

[21] S.E. Fahlman. Faster-learning variations on back-propagation: An empirical study. *Proc. Connectionist Models Summer School*, Morgan-Kaufmann, Los Altos, CA, 1988.

[22] M. Riedmiller, H. Braun. A direct adaptive method for faster backpropagation learning: The RPROP algorithm. *Proc. IEEE Int. Conf. Neural Networks*, IEEE Press, Piscatawy, NJ, 1993, pp. 586-591.

[23] S.L. Phung, A. Bouzerdoum. A pyramidal neural network for visual pattern recognition. *IEEE Trans. Neural Networks* **18**(2):329-343, 2007.

[24] L.E. Peterson, M. Ozen, H. Erdem, A. Amini, L. Gomez, C.C. Nelson, M. Ittmann. Artificial neural network analysis of DNA microarray-based prostate cancer recurrence. *Proc. 2005 IEEE Symp. Computational Intelligence in Bioinformatics and Computational Biology (CIBCB05)*, IEEE Press, Piscataway, NJ, 2005, pp. 1-8.

[25] L.E. Peterson, M.A. Coleman. Comparison of gene identification based on artificial neural network pre-processing with k-means cluster and principal component analysis. *Lecture Notes Artif. Intell.*, **3849**:267-276. 2006.

[26] G. Widmer, M. Kubat. Learning in the presence of concept drift and hidden contexts. *Machine Learn.* **23**(1):69-101, 1996.

[27] F. Girosi, M. Jones, T. Poggio. Regularization theory and neural network architectures. *Neural Comput.* **7**:216-269, 1993.

[28] W.L. Buntine, A.S. Weigend. Bayesian back-propagation. *Complex Sys.* **5**:603-643, 1991.

[29] D.J.C. MacKay. A practical Bayesian framework for backpropagation networks. *Neural Computation* **4**:448-472, 1992.

[30] D.H. Wolpert. On the use of evidence in neural networks. In C.L. Giles, S.J. Hanson, J.D. Cowan, eds., *Advances in Neural Information Processing Systems*, Morgan Kaufmann, San Mateo, CA, 1993, vol. 5, pp. 539-546.

[31] R.M. Neal. Bayesian learning via stochastic dynamics. In C.L. Giles, S.J. Hanson, J.D. Cowan, eds., *Advances in Neural Information Processing Systems*, Morgan Kaufmann, San Mateo, CA, 1993, vol. 5, pp. 475-482.

[32] B.D. Wilamowski, N. Cotton, J. Hewlett. Neural network trainer with second order learning algorithms. *Proc. 11th Int. Conf. Intelligent Engineering Systems*, IEEE Press, Piscataway, NJ, 2007, pp. 127-132.

[33] M.T. Hagan, M.B. Menhaj. Training feedforward networks with the Marquardt algorithm. *IEEE Trans. Neural Networks* **5**(6):989-993, 1994.

[34] C. Charalambous. A conjugate gradient algorithm for the efficient training of artificial neural networks. *IEE Proc. Part G.* **139**(3):301-310, 1992.

CHAPTER 24

KERNEL REGRESSION

24.1 INTRODUCTION

Kernel regression (KREG) employs kernel tricks in a least-squares fashion to determine coefficients, which reliably predict class membership when multiplied against kernels for test objects (arrays) [1]. The majority of applications of kernel regression focus on a function approximation problem of the form $y = f(x)$ and use the Nadarya-Watson kernel weighted average of y [2,3], deconvoluting kernels [4], or some other form of kernel density estimation for deriving weights used in regression [5]. In this chapter, we shall be concerned with kernels of the form $H(\mathbf{x}_i, \mathbf{x}_j)$ for two arrays $\mathbf{x}_i$ and $\mathbf{x}_j$, based on several kernel functions. The combination of linear regression and kernels creates a nonlinear alternative to straightforward linear regression classification (using LREG), which can be especially useful if the data are not entirely linearly separable.

We will assume that the genes of interest for classification have been selected through filtering prior to use of KREG, and that we are not using KREG for gene selection via wrapping. First, the polynomial kernel $H(\mathbf{x}_i, \mathbf{x}_j) = (\mathbf{x}_i^T \mathbf{x}_j + 1)^d$ is determined for each pair of arrays i and j, where d is the power of the polynomial. When $d = 1$, the kernel is purely linear. Power values of d typically range from 1 to 4 or 5, since values of $d > 5$ are seldom employed. Radial basis function (RBF) or Gaussian kernels can also be used, where $H(\mathbf{x}_i, \mathbf{c}_m) = \exp(-\|\mathbf{x}_i - \mathbf{c}_m\|)$ for RBF kernels and $H(\mathbf{x}_i, \mathbf{c}_m) = \exp(-\|\mathbf{x}_i - \mathbf{c}_m\|/2\sigma_m^2)$ for Gaussian kernels, where σ_m is the width of center $\mathbf{c}_m$. Candidate values of σ_m can be determined via Gutta's method [6], given as

$$\sigma_m = \max\{\|\mathbf{c}_m - \mathbf{x}_a\|, \|\mathbf{c}_m - \mathbf{x}_b\|\}, \tag{24.1}$$

Classification Analysis of DNA Microarrays, First Edition. Leif E. Peterson.

where $\mathbf{x}_a$ is the farthest member of cluster $\mathbf{c}_m$ based on

$$a = \underset{i \in \mathbf{c}_m}{\arg\max}\{\|\mathbf{c}_m - \mathbf{x}_i\|\} \tag{24.2}$$

and $\mathbf{x}_b$ is the closest microarray from all other clusters based on

$$b = \underset{j \in \mathbf{c}_k, k \neq m}{\arg\min}\{\|\mathbf{c}_k - \mathbf{x}_j\|\}. \tag{24.3}$$

An alternative approach for determining σ_m is to use $\sigma = (d_1 + d_2)/2$, where $d_1 = \|\mathbf{c}_m - \mathbf{c}_1\|$ is the distance to the closest cluster and $d_2 = \|\mathbf{c}_m - \mathbf{c}_2\|$ is the distance to the second closest cluster. Coefficients for kernel regression are determined using the least-squares model

$$\boldsymbol{\alpha}_{kl} = (\mathbf{H}^T\mathbf{H})^{-1}\mathbf{H}^T\mathbf{y}, \tag{24.4}$$

where $\mathbf{H}$ is an $n \times n$ (array-by-array) matrix with elements $h_{ij} = H(\mathbf{x}_i, \mathbf{x}_j)$ or a $n \times M$ matrix with elements $h_{im} = H(\mathbf{x}_i, \mathbf{c}_m)$ when M centers are used, and $\mathbf{y}$ is a $n \times 1$ vector with y_i set to 1 for training arrays in class ω_k and y_i set to -1 for arrays in class ω_l. We have learned that using RBF kernels typically results in better classification accuracy when compared with LREG and LDA when classes are not linearly separable–hence, we focus on matrix $\mathbf{H}$ having dimensions $n \times M$. Because $\mathbf{H}^T$ is an $M \times n$ matrix, the inverse $(\mathbf{H}^T\mathbf{H})^{-1}$ of the dispersion matrix $\mathbf{H}^T\mathbf{H}$ is an $M \times M$ matrix. When cross-multiplied by $\mathbf{H}^T\mathbf{y}$, which is $M \times 1$, the resulting vector $\boldsymbol{\alpha}_{kl}$ is an $M \times 1$ vector. The predicted y for a test microarray is based on the sum product of the kernel $K(\mathbf{x}, \mathbf{c}_m)$ for the test microarray and each center by the respective α_{klm} for the mth center, shown as

$$\hat{y}_{kl} = \sum_{m}^{M} H(\mathbf{x}, \mathbf{c}_m)\alpha_{klm}. \tag{24.5}$$

As pointed out above, all possible two-class problems (pairs of classes) are used to determine class membership for each test array $\mathbf{x}$. The decision function for class k is

$$D_k(\mathbf{x}) = \underset{l \neq k, l=1}{\arg\min}\ \{D_{kl}(\mathbf{x})\}, \tag{24.6}$$

where

$$D_{kl}(\mathbf{x}) = \hat{y}_{kl} \tag{24.7}$$

for classes k and l. The decision rule for test microarray $\mathbf{x}$ is

$$D(\mathbf{x} \rightsquigarrow \omega) = \underset{l=1,2,\ldots,\Omega}{\arg\max}\ \{D_k(\mathbf{x})\}. \tag{24.8}$$

```
Algorithm 24.1: KREG

Data: n × p data matrix X with n microarrays and p genes
Result: Predicted class value of test array x
u ← 10, Acc_total = 0
foreach Repartition = 1 ← 10 do
    Set C ← 0, n_test ← 0
    Randomly partition input arrays into D_1, D_2, ... ,D_10 folds
    for u = 1 to 10 do
        for k = 1 to Ω − 1 do
            for l = k + 1 to Ω do
                for each training array x_i ∉ D_u do
                    if x_i ∈ ω_k then
                        y = −1
                    endif
                    if x_i ∈ ω_l then
                        y = 1
                    endif
                endfor
                Determine α_kl
            endfor
        endfor
        for each test array x ∈ D_u do
            n_test+ = 1
            Calculate ŷ and D_k(x)
            Predict class ω̂ = D(x ⇝ ω)
            Increment confusion matrix C by 1 in element c_{ω,ω̂}
        endfor
    endfor
    Acc_total+ = Σ_i c_ii/n_test
endfch
Acc = Acc_total/10
```

24.2 ALGORITHM

Algorithm 24.1 lists the computational steps required for implementing KREG classification analysis.

24.3 CROSS-VALIDATION RESULTS

Cross-validation (CV) analysis was performed with the nine datasets for which filtered genes were preselected. Bootstrap accuracy ("CVB") was based on average 0.632 bootstrap accuracy (13.4) for 10 bootstraps. Runs were made for linear and polynomial kernels using $H(\mathbf{x}_i, \mathbf{x}_j) = (\mathbf{x}_i^T \mathbf{x}_j + 1)^d$ and an RBF kernel for which $H(\mathbf{x}_i, \mathbf{c}_m) = \exp(-\|\mathbf{x}_i - \mathbf{c}_m\|)$, with K-means cluster centers using $K = \Omega$. Table 24.1 lists the average accuracy based on 10 repartitions of the input arrays and subsequent CV, with no input feature transformations applied. Note that for the majority of runs, the RBF kernel resulted in much greater performance levels when compared with the linear ($d = 1$) and polynomial kernels ($d > 1$). Table 24.2 lists the average accuracy based on 10 repartitions of the input arrays and subsequent CV, using mean-zero standardized input feature values. Table 24.3 lists the average accuracy based on 10 repartitions of the input arrays and subsequent CV, using fuzzified input feature values. Overall, feature transformations involving mean-zero standardization and fuzzification did not yield substantial performance increases.

The remaining figures in this section on CV summarize average classification accuracy of KREG over all datasets. Recall that for $d > 0$, the kernel is defined as $H(\mathbf{x}_i, \mathbf{x}_j) = (\mathbf{x}_i^T \mathbf{x}_j + 1)^d$, whereas when $d = 0$, the RBF kernel $H(\mathbf{x}_i, \mathbf{c}_m) = \exp(-\|\mathbf{x}_i - \mathbf{c}_m\|)$, where K-means cluster analysis is used to identify $K = \Omega$ centers. Figure 24.1 illustrates accuracy as a function of CV method for each dataset when RBF kernels are used, without input feature transformations and averages based on CV accuracy for 10 repartitions of arrays. Figure 24.2 shows standard deviation of accuracy as a function of CV method for each dataset when RBF kernels are used, with no feature transformations, and with standard deviation based on CV accuracy for 10 repartitions of arrays. Figure 24.3 reveals average accuracy for all datasets as a function of feature transformation and CV method when RBF kernels are used. Figure 24.4 reveals the KREG accuracy without transformations on features as a function of d. Figure 24.5 shows the KREG accuracy with feature standardization as a function of d. Figure 24.6 illustrates the KREG accuracy with feature fuzzification as a function of d. Figure 24.7 shows average accuracy for all datasets as a function of d. In conclusion, only the RBF kernel provided a satisfactory level of classification accuracy, and there was no input feature transformation that appreciably increased performance.

24.4 BOOTSTRAP BIAS

Bootstrap bias was determined using arrays randomly sampled $B = 40$ times at fractions $f = 0.1, 0.2, 0.3, 0.4, 0.5$, or 0.6 of the available arrays. Bootstrap bias runs were based on determining the 0.632 bootstrap accuracy (see Eq. 13.4) for each sampling fraction. Acc_0 in 13.4 was based on test accuracy of unsampled arrays, while Acc_b in 13.4 was based on test accuracy for sampled arrays. RBF kernels were used exclusively, and classification accuracy was calculated for each bth iteration and averaged over the B iterations at each sample size. Figure 24.8 shows the KREG accuracy (RBF

TABLE 24.1 KREG Classification Performance without Feature Transformation[a]

Dataset	d	Cross-validation method				
		CV2	CV5	CV10	CV − 1	CVB
AMLALL2	0	85.09	91.23	92.98	92.11	95.20
	1	73.68	67.54	61.40	65.79	80.94
	2	70.18	71.05	67.54	65.79	69.13
	3	99.12	73.68	53.51	60.53	97.26
	4	70.18	68.42	71.05	71.05	69.21
Brain2	0	76.11	76.67	82.78	85.00	86.24
	1	48.33	53.89	49.44	78.33	49.25
	2	50.00	51.11	58.33	65.00	50.03
	3	58.89	47.78	57.22	68.33	48.20
	4	52.78	46.11	35.00	43.33	50.41
Breast2A	0	100	100	100	100	100
	1	15.56	77.78	46.67	0.00	47.73
	2	48.89	46.67	40.00	73.33	54.37
	3	53.33	35.56	6.67	46.67	37.95
	4	62.22	62.22	46.67	46.67	49.54
Breast2B	0	86.75	85.47	89.32	94.87	90.82
	1	40.60	55.56	40.17	64.10	41.61
	2	55.56	57.26	53.85	52.56	52.80
	3	44.44	50.00	57.69	75.64	41.31
	4	56.41	55.98	44.44	61.54	54.59
Colon2	0	89.78	88.71	89.78	90.32	89.20
	1	56.99	59.14	65.05	70.97	67.90
	2	61.83	66.13	64.52	62.90	65.21
	3	52.69	50.00	55.91	48.39	50.97
	4	54.30	59.14	48.92	46.77	47.79
Lung2	0	100	100	100	100	100
	1	58.33	51.04	42.71	34.38	56.07
	2	53.13	42.71	43.75	46.88	50.73
	3	52.08	20.83	45.83	3.13	40.52
	4	41.67	57.29	45.83	62.50	50.15
MLL_Leukemia3	0	99.42	98.25	99.42	100	99.75
	1	23.39	20.47	38.01	56.14	37.30
	2	22.81	36.84	21.64	21.05	28.71
	3	28.65	38.01	24.56	28.07	30.01
	4	26.90	25.73	29.82	29.82	26.71
Prostate2	0	97.39	97.71	98.04	98.04	97.58
	1	51.31	37.58	53.27	32.35	45.21
	2	48.69	50.33	44.77	50.00	48.42
	3	69.28	68.95	54.25	85.29	57.40
	4	60.46	62.09	43.46	57.84	51.59
SRBCT4	0	100	100	100	100	100
	1	37.57	23.28	26.98	20.63	25.46
	2	35.45	38.62	34.92	26.98	36.06
	3	32.80	28.57	24.34	12.70	37.47
	4	32.28	34.92	29.10	25.40	32.00
Average		0.59	0.58	0.55	0.58	0.59

[a]For $d > 0$, the kernel is defined as $H(\mathbf{x}_i, \mathbf{x}_j) = (\mathbf{x}_i^T \mathbf{x}_j + 1)^d$, whereas when $d = 0$, the RBF kernel $H(\mathbf{x}_i, \mathbf{c}_m) = \exp(-\|\mathbf{x}_i - \mathbf{c}_m\|)$ is used with K-means cluster centers $K = \Omega$.

TABLE 24.2 KREG Classification Performance with Feature Standardization[a]

Dataset	d	Cross-validation method				
		CV2	CV5	CV10	CV − 1	CVB
AMLALL2	0	94.74	92.11	92.11	92.11	95.15
	1	72.81	60.53	64.04	23.68	73.25
	2	70.18	70.18	71.05	71.05	67.21
	3	98.25	70.18	64.91	78.95	96.64
	4	69.30	70.18	70.18	68.42	68.76
Brain2	0	78.33	82.78	83.33	90.00	85.82
	1	37.22	50.00	46.11	51.67	53.69
	2	59.44	47.22	47.22	46.67	54.22
	3	50.00	49.44	56.11	38.33	54.86
	4	56.67	53.89	51.11	73.33	54.38
Breast2A	0	100	100	100	100	100
	1	60.00	44.44	51.11	60.00	59.37
	2	46.67	44.44	51.11	40.00	54.57
	3	4.44	55.56	53.33	86.67	29.61
	4	44.44	46.67	44.44	66.67	56.19
Breast2B	0	86.75	85.04	87.61	93.59	90.35
	1	32.05	29.49	55.56	44.87	51.16
	2	51.71	42.74	48.29	66.67	54.54
	3	43.59	54.70	51.28	46.15	43.38
	4	49.15	54.70	55.98	55.13	55.07
Colon2	0	88.71	88.71	90.32	90.32	89.68
	1	56.99	64.52	61.29	67.74	64.34
	2	68.82	66.67	66.13	64.52	63.42
	3	46.77	49.46	54.30	75.81	49.89
	4	39.25	43.55	42.47	24.19	51.61
Lung2	0	100	100	100	100	100
	1	41.67	51.04	55.21	90.63	45.04
	2	45.83	51.04	57.29	37.50	54.65
	3	43.75	59.38	36.46	21.88	56.81
	4	57.29	53.13	42.71	34.38	50.33
MLL_Leukemia3	0	98.83	100	100	100	99.70
	1	62.57	26.32	47.37	22.81	31.84
	2	22.81	35.09	24.56	14.04	22.25
	3	22.81	23.98	26.32	26.32	24.82
	4	25.15	29.82	33.92	36.84	27.35
Prostate2	0	96.73	98.04	98.04	98.04	97.48
	1	35.62	53.27	70.26	64.71	46.60
	2	55.56	41.18	46.73	44.12	48.33
	3	60.78	59.15	39.22	66.67	57.05
	4	43.46	35.29	49.35	70.59	53.52
SRBCT4	0	100	100	100	100	100
	1	24.87	26.98	26.46	17.46	26.00
	2	34.92	30.16	24.87	44.44	32.18
	3	46.03	31.75	31.22	12.70	35.82
	4	33.86	31.22	43.92	34.92	34.07
Average		0.57	0.57	0.58	0.59	0.59

[a]For $d > 0$, the kernel is defined as $H(\mathbf{x}_i, \mathbf{x}_j) = (\mathbf{x}_i^T \mathbf{x}_j + 1)^d$, whereas when $d = 0$, the RBF kernel $H(\mathbf{x}_i, \mathbf{c}_m) = \exp(-\|\mathbf{x}_i - \mathbf{c}_m\|)$ is used with K-means cluster centers $K = \Omega$.

TABLE 24.3 KREG Classification Performance with Feature Fuzzification[a]

Dataset	d	Cross-validation method CV2	CV5	CV10	CV − 1	CVB
AMLALL2	0	100	100	100	100	100
	1	50.88	48.25	66.67	63.16	51.40
	2	68.42	85.09	57.89	84.21	71.21
	3	69.30	67.54	75.44	71.05	71.24
	4	70.18	68.42	72.81	71.05	71.45
Brain2	0	82.78	82.22	81.67	80.00	82.69
	1	60.00	52.22	45.00	66.67	51.70
	2	66.11	61.67	58.33	45.00	59.22
	3	57.22	49.44	56.11	51.67	55.76
	4	58.89	62.78	65.56	51.67	61.92
Breast2A	0	100	100	100	100	99.30
	1	60.00	86.67	57.78	80.00	41.69
	2	46.67	64.44	28.89	40.00	44.50
	3	55.56	75.56	48.89	73.33	67.40
	4	60.00	51.11	55.56	0.00	57.32
Breast2B	0	82.91	82.91	81.62	85.90	86.34
	1	67.09	52.56	44.44	33.33	47.27
	2	52.99	50.00	46.15	46.15	51.95
	3	51.28	41.88	67.09	69.23	44.93
	4	54.70	45.73	49.15	35.90	54.81
Colon2	0	80.11	84.41	83.33	85.48	87.06
	1	57.53	63.98	63.98	59.68	57.35
	2	45.70	53.23	48.92	32.26	52.89
	3	43.55	64.52	40.32	56.45	44.15
	4	60.22	47.85	62.90	59.68	48.84
Lung2	0	100	100	100	100	100
	1	50.00	22.92	48.96	84.38	63.66
	2	32.29	72.92	72.92	6.25	50.80
	3	31.25	42.71	61.46	21.88	42.12
	4	59.38	46.88	27.08	53.13	47.17
MLL_Leukemia3	0	98.25	98.25	98.25	98.25	99.30
	1	21.64	19.88	25.15	8.77	36.61
	2	38.60	39.77	25.73	63.16	44.39
	3	34.50	42.69	38.01	26.32	34.51
	4	36.26	32.75	25.15	12.28	32.36
Prostate2	0	93.14	93.46	93.14	94.12	94.29
	1	29.08	61.44	41.18	42.16	55.69
	2	42.48	55.23	48.04	48.04	58.54
	3	52.94	71.24	42.81	45.10	56.02
	4	40.20	34.64	37.58	62.75	49.33
SRBCT4	0	99.47	100	100	100	99.28
	1	37.57	36.51	27.51	15.87	31.19
	2	35.98	33.86	25.40	23.81	34.21
	3	30.16	29.10	32.80	30.16	35.35
	4	27.51	35.98	27.51	26.98	34.46
Average		0.58	0.60	0.57	0.56	0.59

[a]For $d > 0$, the kernel is defined as $H(\mathbf{x}_i, \mathbf{x}_j) = (\mathbf{x}_i^T \mathbf{x}_j + 1)^d$, whereas when $d = 0$, the RBF kernel $H(\mathbf{x}_i, \mathbf{c}_m) = \exp(-\|\mathbf{x}_i - \mathbf{c}_m\|)$ is used with K-means cluster centers $K = \Omega$.

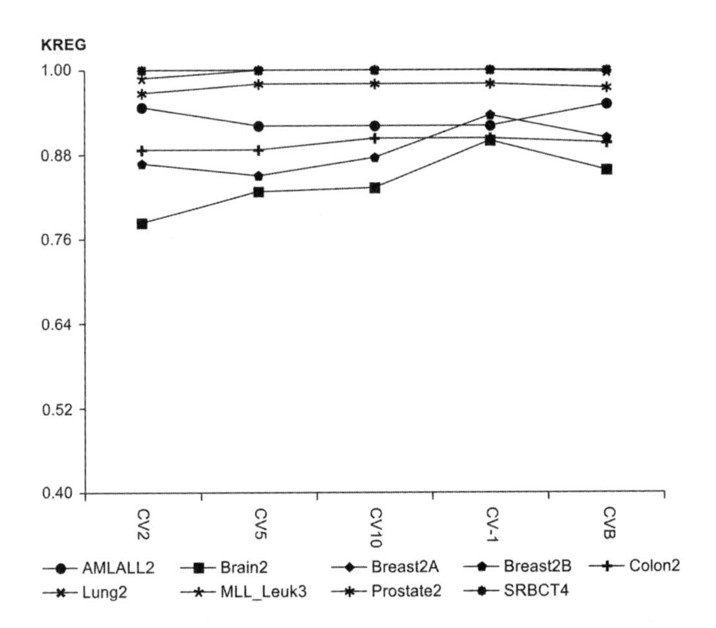

FIGURE 24.1 KREG accuracy as a function of cross-validation (CV) method for each dataset when RBF kernels were used. Average based on CV accuracy for 10 repartitions of arrays. No feature transformations applied.

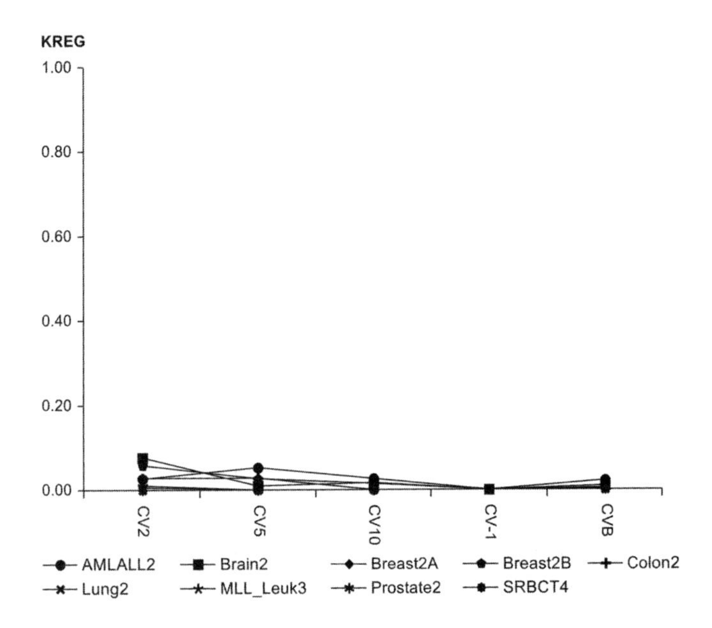

FIGURE 24.2 KREG standard deviation of accuracy as a function of CV method for each dataset when RBF kernels were used. Standard deviation based on CV accuracy for 10 repartitions of arrays. No feature transformations applied.

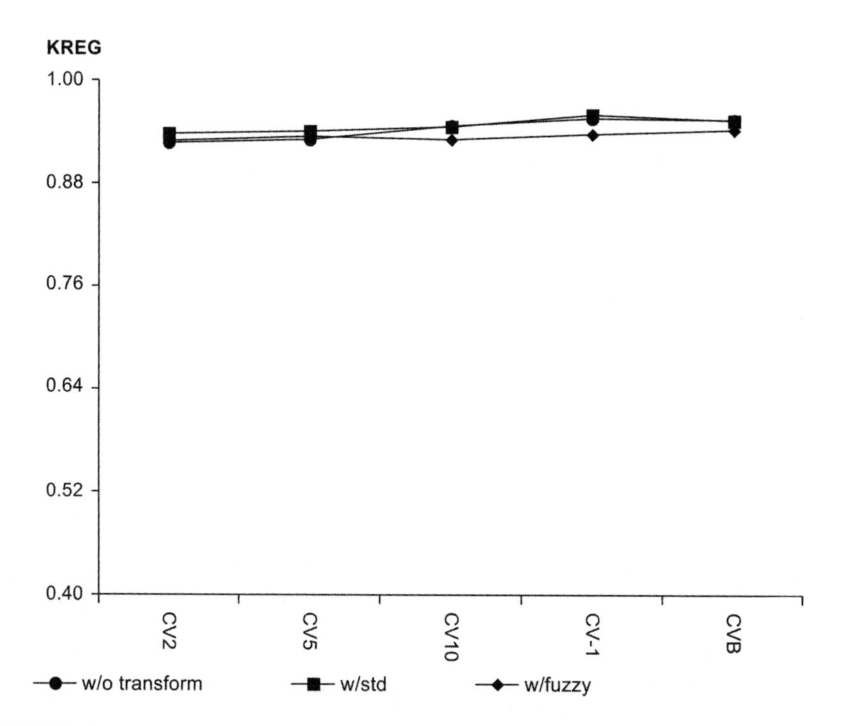

FIGURE 24.3 KREG average accuracy for all datasets as a function of feature transformation and CV method when RBF kernels were used.

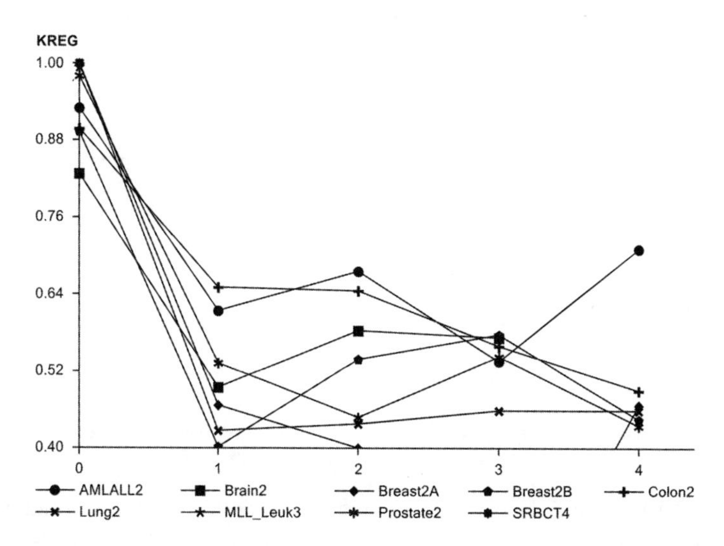

FIGURE 24.4 KREG accuracy without transformations on features as a function of d. Integer values on the x axis represent values of d; when $d > 0$, the kernel is defined as $H(\mathbf{x}_i, \mathbf{x}_j) = (\mathbf{x}_i^T \mathbf{x}_j + 1)^d$, whereas when $d = 0$, the RBF kernel $H(\mathbf{x}_i, \mathbf{c}_m) = \exp(-\|\mathbf{x}_i - \mathbf{c}_m\|)$ is used with K-means cluster centers $K = \Omega$.

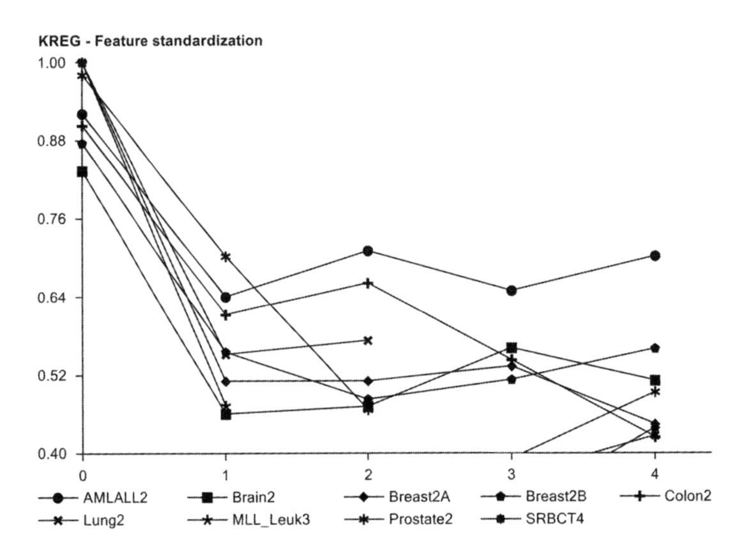

FIGURE 24.5 KREG accuracy with feature standardization as a function of d. Integer values on the x axis represent values of d; when $d > 0$, the kernel is defined as $H(\mathbf{x}_i, \mathbf{x}_j) = (\mathbf{x}_i^T \mathbf{x}_j + 1)^d$, whereas when $d = 0$, the RBF kernel $H(\mathbf{x}_i, \mathbf{c}_m) = \exp(-\|\mathbf{x}_i - \mathbf{c}_m\|)$ is used with K-means cluster centers $K = \Omega$.

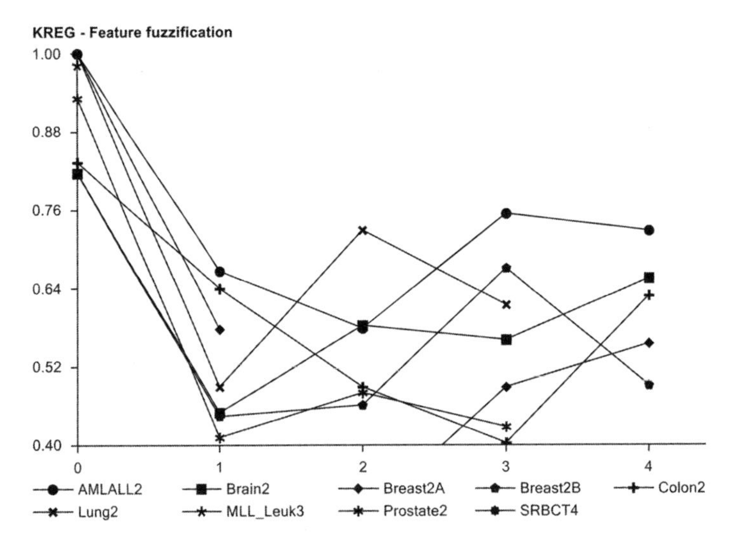

FIGURE 24.6 KREG accuracy with feature fuzzification as a function of d. Integer values on the x axis represent values of d; when $d > 0$, the kernel is defined as $H(\mathbf{x}_i, \mathbf{x}_j) = (\mathbf{x}_i^T \mathbf{x}_j + 1)^d$, whereas when $d = 0$, the RBF kernel $H(\mathbf{x}_i, \mathbf{c}_m) = \exp(-\|\mathbf{x}_i - \mathbf{c}_m\|)$ is used with K-means cluster centers $K = \Omega$.

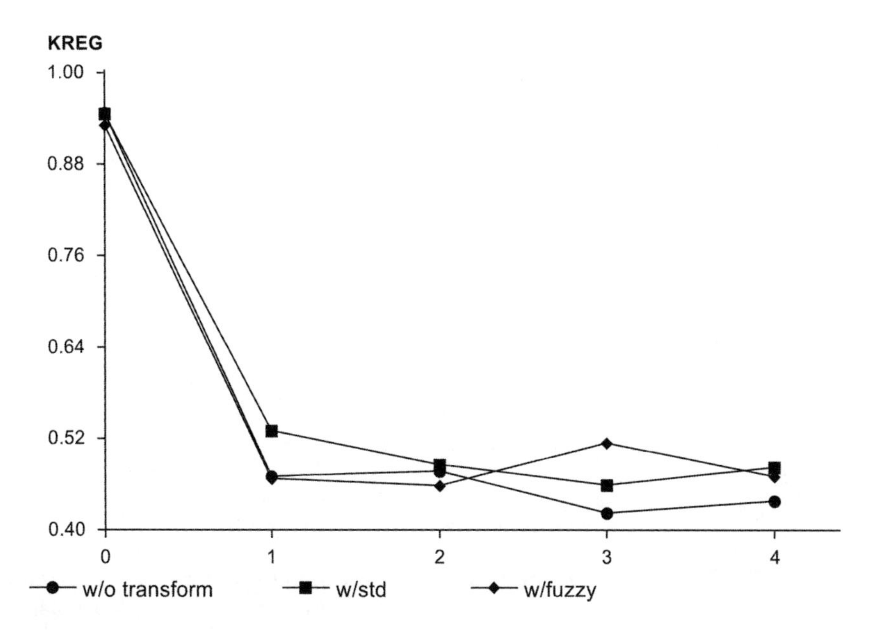

FIGURE 24.7 KREG average accuracy for all datasets as a function of d. Integer values on the x axis represent values of d; when $d > 0$, the kernel is defined as $H(\mathbf{x}_i, \mathbf{x}_j) = (\mathbf{x}_i^T \mathbf{x}_j + 1)^d$, whereas when $d = 0$, the RBF kernel $H(\mathbf{x}_i, \mathbf{c}_m) = \exp(-\|\mathbf{x}_i - \mathbf{c}_m\|)$ is used with K-means cluster centers $K = \Omega$.

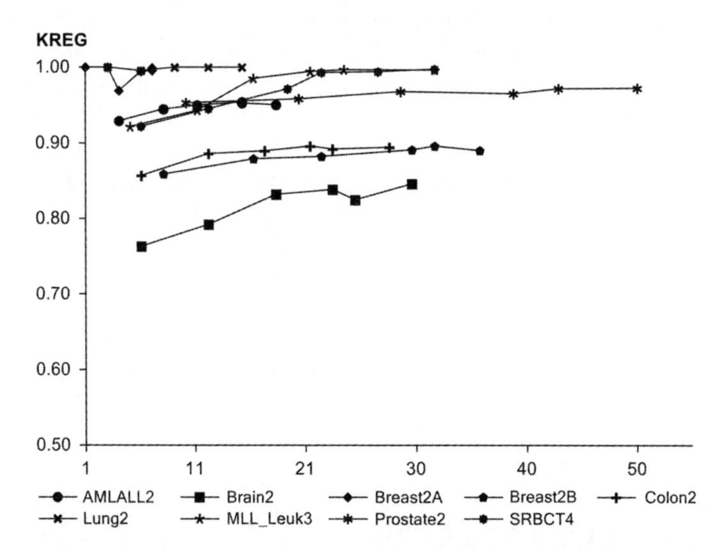

FIGURE 24.8 KREG accuracy (RBF kernel used) without transformations on features as a function of bootstrap sample size used during training. The x-axis values represent the number of arrays randomly sampled $B = 40$ times at fractions $f = 0.1, 0.2, 0.3, 0.4, 0.5$, or 0.6 of the available arrays.

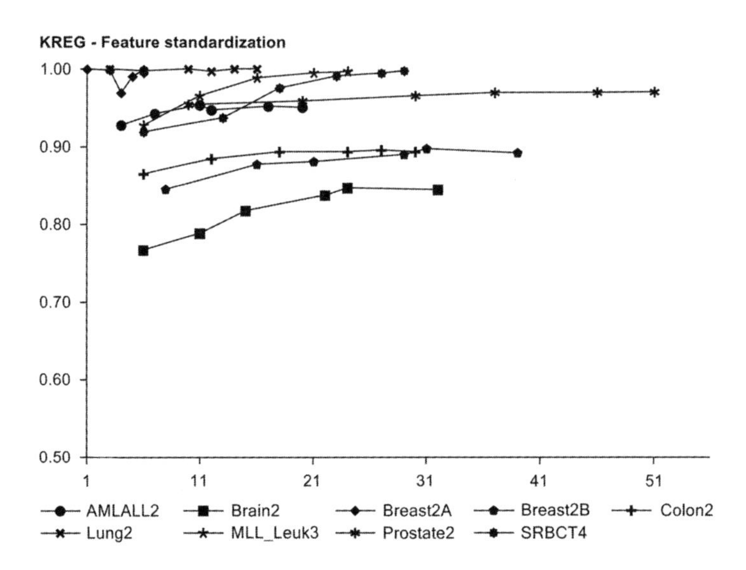

FIGURE 24.9 KREG accuracy (RBF kernel used) with feature standardization as a function of bootstrap sample size used during training. The x-axis values represent the number of arrays randomly sampled $B = 40$ times at fractions $f = 0.1, 0.2, 0.3, 0.4, 0.5$, or 0.6 of the available arrays.

kernel used) without transformations on features as a function of bootstrap sample size used during training, based on 10 resamplings from the input arrays. The x-axis values represent the number of arrays randomly sampled $B = 40$ times at fractions $f = 0.1, 0.2, 0.3, 0.4, 0.5$, or 0.6 of the available arrays. Figure 24.9 illustrates the KREG accuracy (RBF kernel used) with feature standardization as a function of bootstrap sample size used during training, based on 10 resamplings from the input arrays. Figure 24.10 shows the KREG accuracy (RBF kernel used) with feature fuzzification as a function of bootstrap sample size used during training, based on 10 resamplings from the input arrays. Performance results for bootstrap bias suggest that, with and without mean-zero standardization of input features, KREG was relatively unbiased by the amount of data used during training. However, input feature fuzzification seemed to introduce a greater degree of bias, such that more training data were required in order to achieve greater performance.

24.5 MULTICLASS ROC CURVES

Multiclass receiver-operator characteristic (ROC) curves were generated for the nine datasets used in previous sections. During ROC curve generation, arrays were randomly sampled using $f = 0.1, 0.2, 0.3, 0.4, 0.5$, or 0.6 of the available arrays $B = 40$ times. The numbers of genes used for each dataset are listed in Table 12.10. Figure 24.11 illustrates ROC curves for the KREG

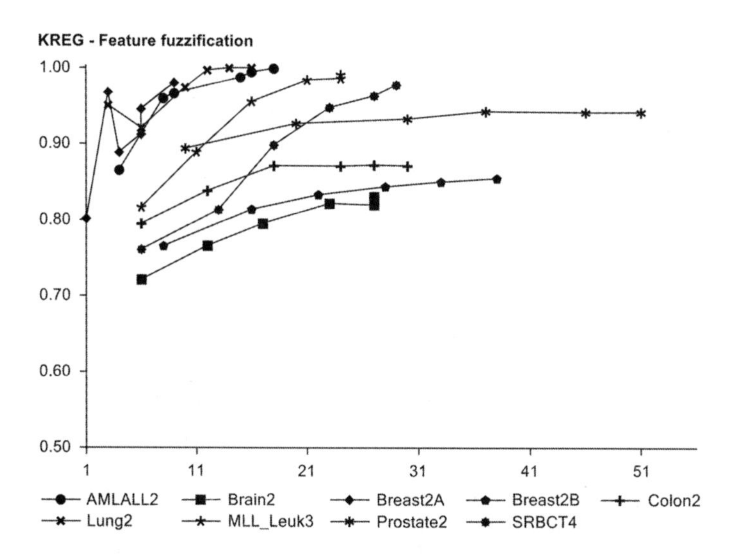

FIGURE 24.10 KREG accuracy (RBF kernel used) with feature fuzzification as a function of bootstrap sample size used during training. The x-axis values represent the number of arrays randomly sampled $B = 40$ times at fractions $f = 0.1, 0.2, 0.3, 0.4, 0.5$, or 0.6 of the available arrays.

classifier based on RBF kernels applied to the AMLALL2, Brain2, Breast2A, Breast2B, Colon2, Lung2, MLL_Leukemia3, Prostate2, and SRBCT4 datasets using mean-zero standardized input features. The legend values represent the number of arrays randomly sampled $B = 40$ times at fractions $f = 0.1, 0.2, 0.3, 0.4, 0.5$, or 0.6 of the available arrays. Figure 24.12 illustrates ROC curves for KREG based on RBF kernels for the same data using fuzzified input features.

24.6 DECISION BOUNDARIES

Class prediction results showing decision boundaries via self-organizing maps for KREG are presented in this section. The numbers of genes used for each dataset are listed in Table 12.10. Integers in the upper right of each panel represent values d: when $d > 0$, the kernel is defined as $H(\mathbf{x}_i, \mathbf{x}_j) = (\mathbf{x}_i^T \mathbf{x}_j + 1)^d$, whereas when $d = 0$ the RBF kernel $H(\mathbf{x}_i, \mathbf{c}_m) = \exp(-\|\mathbf{x}_i - \mathbf{c}_m\|)$ is used with K-means cluster centers, where $K = \Omega$. Figure 24.13 illustrates 2D self-organizing maps of class decision boundaries for the KREG based on the AMLALL2, Brain2, Breast2A, Breast2B, Colon2, Lung2, MLL_Leukemia3, Prostate2, and SRBCT4 datasets using mean-zero standardized input features. Figure 24.14 illustrates 2D self-organizing maps of class decision boundaries for the same data using fuzzified input features. The RBF kernel resulted in the best performance, while the linear and polynomial kernels yielded lower performance levels.

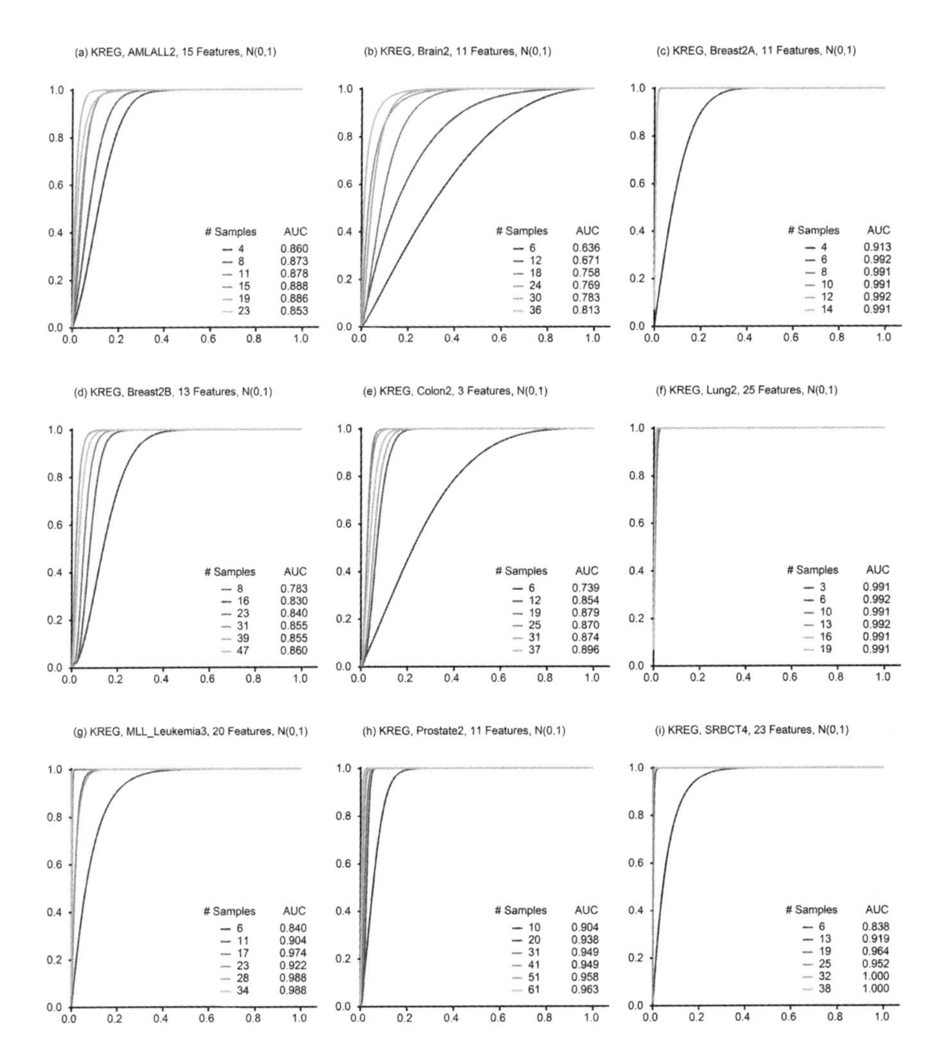

FIGURE 24.11 ROC curves for the KREG classifier based on RBF kernels applied to the AMLALL2, Brain2, Breast2A, Breast2B, Colon2, Lung2, MLL_Leukemia3, Prostate2, and SRBCT4 datasets based on mean-zero standardized input features; legend values represent the number of arrays randomly sampled $B = 40$ times at fractions $f = 0.1, 0.2, 0.3, 0.4, 0.5$, or 0.6 of the available arrays: (a) AMLALL2; (b) Brain2; (c) Breast2A; (d) Breast2B; (e) Colon2; (f) Lung2; (g) MLL_Leukemia3; (h) Prostate2; (i) SRBCT4. The numbers of genes used for each dataset are listed in Table 12.10.

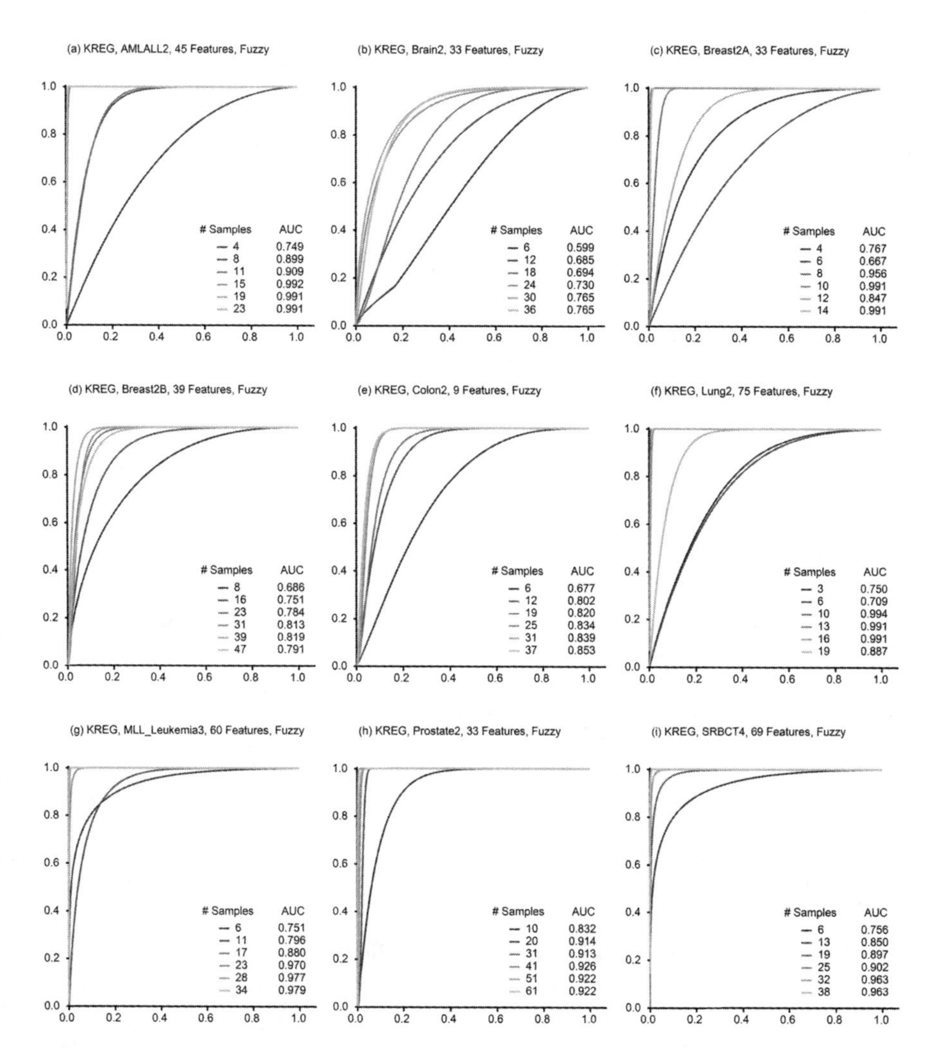

FIGURE 24.12 ROC curves for the KREG based on RBF kernels applied to the AMLALL2, Brain2, Breast2A, Breast2B, Colon2, Lung2, MLL_Leukemia3, Prostate2, and SRBCT4 datasets based on fuzzified input features; legend values represent the number of arrays randomly sampled $B = 40$ times at fractions $f = 0.1, 0.2, 0.3, 0.4, 0.5$, or 0.6 of the available arrays: (a) AMLALL2; (b) Brain2; (c) Breast2A; (d) Breast2B; (e) Colon2; (f) Lung2; (g) MLL_Leukemia3; (h) Prostate2; (i) SRBCT4. The numbers of genes used for each dataset are listed in Table 12.10.

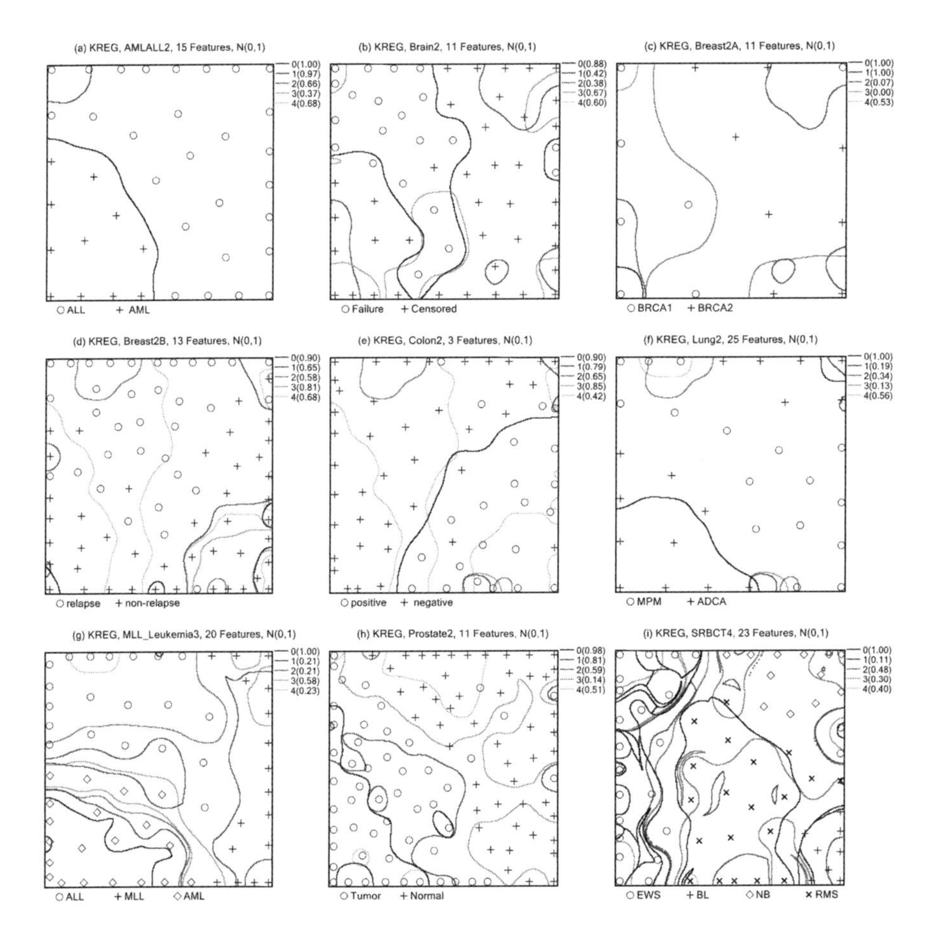

FIGURE 24.13 2D self-organizing maps of KREG class decision boundaries and classification accuracy (upper right of each panel) for the AMLALL2, Brain2, Breast2A, Breast2B, Colon2, Lung2, MLL_Leukemia3, Prostate2, and SRBCT4 datasets based on mean-zero standardized input features: (a) AMLALL2, (b) Brain2, (c) Breast2A, (d) Breast2B, (e) Colon2, (f) Lung2, (g) MLL_Leukemia3, (h) Prostate2, and (i) SRBCT4. The number of genes used for each dataset are listed in Table 12.10. Integers in upper right of each panel represent values d; when $d > 0$, the kernel is defined as $H(\mathbf{x}_i, \mathbf{x}_j) = (\mathbf{x}_i^T \mathbf{x}_j + 1)^d$, whereas when $d = 0$, the RBF kernel $H(\mathbf{x}_i, \mathbf{c}_m) = \exp(-\|\mathbf{x}_i - \mathbf{c}_m\|)$ is used with K-means cluster centers $K = \Omega$.

24.7 SUMMARY

Results of the runs reported in the previous sections indicate that KREG was more biased for the Brain2 and Breast2B datasets, since performance changed considerably over the various CV methods used. Average performance of

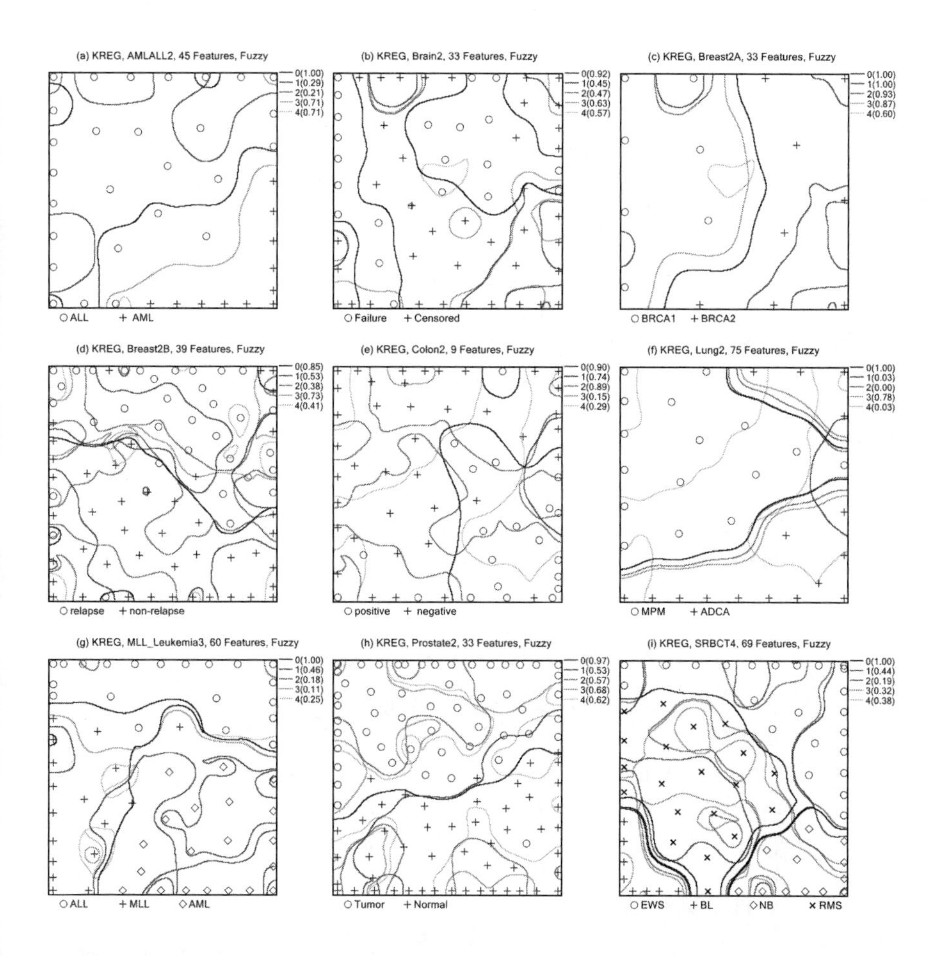

FIGURE 24.14 2D self-organizing maps of KREG class decision boundaries and classification accuracy (upper right of each panel) for the AMLALL2, Brain2, Breast2A, Breast2B, Colon2, Lung2, MLL_Leukemia3, Prostate2, and SRBCT4 datasets based on fuzzified input features: (a) AMLALL2; (b) Brain2; (c) Breast2A; (d) Breast2B; (e) Colon2; (f) Lung2; (g) MLL_Leukemia3; (h) Prostate2; (i) SRBCT4. The numbers of genes used for each dataset are listed in Table 12.10. Integers in upper right of each panel represent values d; when $d > 0$, the kernel is defined as $H(\mathbf{x}_i, \mathbf{x}_j) = (\mathbf{x}_i^T \mathbf{x}_j + 1)^d$, whereas when $d = 0$, the RBF kernel $H(\mathbf{x}_i, \mathbf{c}_m) = \exp(-\|\mathbf{x}_i - \mathbf{c}_m\|)$ is used with K-means cluster centers $K = \Omega$.

KREG for all datasets across CV methods reflected that, generally speaking, KREG was not biased. The variance across CV method for all datasets was also quite low. Despite the low bias and variance of KREG, we observed that the RBF kernel resulted in much greater performance levels when compared with linear ($d = 1$) and polynomial kernels ($d > 1$). The bootstrap bias of KREG was quite low, and fuzzification of input feature data degraded performance.

Recall from Chapter 11, on nonlinear manifold learning, that kernel methods form an essentially nonlinear mapping. If the sample space is wholly linearly separable, then the LREG and LDA classifiers should be applied initially. If LREG and/or LDA does (do) not perform well, then apply KREG with a linear kernel ($d = 1$), and if this proves unsuccessful, then employ polynomial kernels ($d > 1$) and an RBF kernel. A last resort, in terms of kernel-based methods, would be to employ support vector machines, which are discussed in Chapter 22.

REFERENCES

[1] U. Holst, O. Hössjer, C. Björklund, P. Ragnarson, H. Edner. Locally weighted least squares kernel regression and statistical evaluation of lidar measurements. *Environmetrics* **7**:401-416, 1998.

[2] E.A. Nadaraya. On estimating regression. *Theory Probab. Appl.* **10**:186-190, 1964.

[3] G.S. Watson. Smooth regression analysis. *Sankhyā: Indian J. Stat.* **26**(4): 359-372, 1964.

[4] L.A. Stefansky, R.J. Carroll. Deconvoluting kernel density estimators. *Statistics* **21**:169-184, 1990.

[5] H. Takeda, S. Farsiu, P. Milanfar. Kernel regression for image processing and reconstruction. *IEEE Trans. Image Proc.* **16**(2):349-366, 2007.

[6] S. Gutta, J.R.J. Huang, P. Jonathon, H. Weschler. Mixture of experts for classification of gender, ethnic origin, and pose of human faces. *IEEE Trans. Neural Networks* **11**(4):948-960, 2000.

CHAPTER 25

NEURAL ADAPTIVE LEARNING WITH METAHEURISTICS

The application of machine learning techniques in biomedical data analysis is rapidly increasing as the constraints of traditional techniques become more important. The majority of challenging problems in numerical analyses are high-dimensional and involve NP-hard combinatorial optimization such as timetabling, vehicle routing, quadratic assignment, and maximum satisfiability problems. Algorithms developed for solving combinatorial optimization problems are termed "exact" or "approximate." Exact algorithms can find an optimal solution within a given runtime and are usually limited to solving small problems. For larger problems requiring much longer runtimes, approximate methods are typically used for deriving suboptimal solutions via short runtimes. A *heuristic* is a type of labor-intensive approximation algorithm developed for a particular cost function and search space, which can successfully solve a problem based on sound principles. More specifically, heuristics involve permutations of solution vectors tuned via machine learning in order to find "good" solutions. The field of *metaheuristics* was introduced to address development of problem-independent high-level search strategies for optimization problems [1]. Examples of early metaheuristics include genetic algorithms [2,3], evolutionary programming [4,5], evolution strategies [6-9], simulated annealing [10], tabu search [11], and iterated local search [12]. More recently, particle swarm optimization [13] and ant colony optimization [14,15] were introduced as new swarm intelligence variants of metaheuristics.

This chapter focuses on the employment of metaheuristics such as genetic algorithms (GA), covariance matrix self-adaptation evolution strategies (CMSA-ES), particle swarm optimization (PSO), and ant colony optimization (ACO) for adaptive learning of neural network connection weights as an alternative to back-propagation learning.

Classification Analysis of DNA Microarrays, First Edition. Leif E. Peterson.

25.1 MULTILAYER PERCEPTRONS

Multilayer perceptrons (MLPs), or artificial neural networks, are biologically motivated parallel connectionist models designed to mimic synaptic information processing performed by neurons in the brain. MLPs are universal approximators [16] that can exploit nonlinearities and memorize patterns using unobservable constructs called *hidden layers*. The connection weights between nodes of a feedforward MLP are trained using gradient-descent-based backpropagation. Weight training can also be accomplished with biologically inspired evolutionary algorithms, which have been the topic of considerable interest in artificial intelligence research over the last several decades [17-19].

Example 25.1 Datasets listed in Table 1.3 whose features were generated with greedy PTA (see Table 12.10) were used for all MLP runs. Data preprocessing prior to use of the MLP involved mean-zero standardization of input features. The number of hidden nodes used varied within $2, 4, \ldots, 10$. The $\tanh$ activation was used for hidden nodes, while the output-side function was the softmax with cross-entropy error accumulation. MLP connection weights were initialized using random variates uniformly sampled from the range $[-2,2]$, and 500 epochs were used for training. The classification run results are identical to results shown in Figure 23.15. Classification accuracy as a function of the number of hidden nodes was near perfect for all except the Colon2 dataset. This was expected, as an MLP should provide high classification performance levels when the best genes are filtered prior to analysis.

25.2 GENETIC ALGORITHMS

Evolutionary algorithms are employed primarily for solving direct search and optimization problems. Originally, there were three variants of evolutionary algorithms: genetic algorithms (GA) introduced by Holland in Ann Arbor [2,3], evolutionary programming (EP) introduced by Fogel in San Diego [4,5], and evolution strategies (ESS) introduced by students Rechenberg and Schwefel in Berlin [6-9]. This section focuses only on GAs and ES and therefore, does not evaluate the application of EP.

Goldberg's building-block hypothesis and Holland's schema theorem are fundamental principles for the way GAs operate [3,17]. Under the building-block hypothesis, GAs reassemble solutions to a problem by successive juxtapositioning of low-order high-performance building blocks to attain a near optimal solution. The schema theorem is based on templates of n-bit binary $\{0,1\}$ strings such as (0101010), and (111010010101011000), which undergo generational selection proportional to fitness. Note that the schema theorem is, however, independent of genetic crossover, mutation, and parallelism of wildcards such as [*,1,0], which are essentially considered disturbances [20]. An important characteristic of GAs is that fitness proportional selection does not ensure convergence to a local or global maximum because of the nonzero chance of losing good solutions. Convergence is guaranteed mostly by use of *elitism*, or the selection and unaltered transformation (via crossover, mutation) of the most elite fitness-valued chromosomes for each successive generation [18]. The most common workflow for GAs involving the sequential steps for solving a problem with N unknowns is as follows:

1. Simulate N binary coded $\{0, 1\}$ genes on λ parent chromosomes.
2. Use binary-to-decimal translation to recode binary strings of N genes into real-value parameters.
3. Determine fitness for each parent chromosome using the real-value N parameter values.
4. Select several elite members to go forward in the next generation (elitism).
5. Select the best half of parent chromosomes with the best fitness to go forward.
6. Randomly select pairs of parent chromosomes for mating (crossover) of their binary strings to form child chromosomes.
7. Randomly apply mutations to the binary child chromosomes.
8. Go to step 2.

Genetic algorithms have found widespread use in biomedicine since the 1980s, and there is a voluminous literature on the application of GAs in biomedicine. Applications include dyad pattern finding in DNA sequences [21], chemical structure-activity relationship analysis [22], biomarkers for risk stratification [23], colonic polyp detection in CT colonography [24], structural properties of nanoalloys [25], automated identification of dementia using images [26], automated protein crystal recognition [27], biochemical networks [28], and evolving K nearest neighbor classifiers [29].

Chromosome Structure. In the field of genetics, a *chromosome* is defined as a biochemical structure that contains genetic information in the form of genes. The notation we use for a chromosome and its genes is $\mathcal{C} = (g_1, g_2, \ldots, g_N)$. Mammalian chromosomes commonly contain thousands of genes, and this varies according to species. In a GA, each chromosome serves as a trial, or permutation, in which the genes carried in the chromosome represent the N parameters being solved. Therefore, each chromosome carries N genes representing the N parameters. In other words, a chromosome essentially represents a solution vector. The number of chromosomes used in a GA is denoted as λ.

Gene Structure and Length. A single gene represents a single parameter value. Genes are represented by binary strings of zeroes and ones from the set $\{0,1\}$. Examples of binary strings for genes on chromosome l are $\mathcal{S}_{gl} = (01001)$, and $\mathcal{S}_{gl} = (10101010101010)$, where the full set of genes on chromome l is $\{\mathcal{S}_{1l}, \mathcal{S}_{2l}, \ldots, \mathcal{S}_{gl}, \ldots, \mathcal{S}_{Nl}\}$. The length of a gene's binary string dictates the level of precision of the parameter that it represents. A common level of precision is 10^{-6}, based on a *gene length* of $L = 20$. Using decimal notation for a parameter in the form x_{10} and binary notation x_2, and assuming a parameter range of $[x_{10,\min}, x_{10,\max}]$, the following binary to decimal encoding is used

$$x_{10} = x_{10,\min} + \frac{x_{10,\max} - w_{10,\min}}{2^L - 1} x_2 \tag{25.1}$$

where x_{10} is the calculated result in decimal code, x_2 is the integer value of the binary string (e.g., 10101101), and L is the gene length. Binary bits $\{0,1\}$ are initialized by use of a uniform random variate $\mathcal{U}(0,1)$ which is rounded down to 0 if $\mathcal{U}(0,1) \leq 0.5$ and rounded up to 1 if $\mathcal{U}(0,1) > 0.5$. As a result, each chromosome $\mathcal{C}$ has length $N \times L$, representing the N L-length strings. Using the simulated binary strings, binary to decimal encoding is then performed during an iterative learning procedure in order to determine decimal values, which are substituted into an objective function to obtain fitness.

Fitness. Historically, the majority of applications that have employed GA have focused on function minimization or maximization. The simplest form of fitness is the value of a mathematical function evaluated with various parameter values. Consider, for example, a function maximization problem where $f(x) = x^3$ is to be maximized over the parameter range $0 \leq x \leq 10$. We already know that $x = 10$ is the solution, since the greatest value is obtained when $f(10) = 1000$. In this example, the function value of 1000 is the fitness, and the greatest fitness value is obtained when $x = 10$.

Fitness f is determined for each chromosome by substituting gene-specific parameter values into the function being evaluated. Fitness portrays the quality of a solution vector. A descending sort of chromosome-specific fitness values will yield an array of permutations, the best of which are at the top of the stack. Fitness can be derived in a number of ways. Mean-square error (MSE) of a predicted function value is commonly used and determined as

$$\text{MSE}_l = \frac{1}{n} \sum_{i}^{n} (y_i - \hat{y}_i)^2, \tag{25.2}$$

where y_i is the observed y value for the ith array and $\hat{y}_i$ is the predicted y value based on the objective function or method being used. The inverse of MSE_l (i.e., $1/\text{MSE}_l$) should be used since a lower MSE represents a better fit of the objective function. We will see later in the chapter how a GA can be embedded into a neural network architecture to solve function approximation and multiclass classification problems. Fitness can also be represented by the value of log-likelihood, χ^2, R^2, or other goodness-of-fit (GOF) criteria for a statistical model. There is no limit to the various forms of fitness, and any technique using iterative learning for problem solving can equate intermediate iteration-specific results to a fitness value.

Mutation. Mutation is used to induce random fluctuations in the 0-1 bits of a chromosome in order to minimize genetic drift. Mutation is applied at a rate of P_m. Each bit is evaluated for mutation, and undergoes mutation if a randomly drawn uniform variate $\mathcal{U}(0,1)$ is less than P_m. It is customary to use low values of mutation in the range 0.001-0.005.

Selection. Genetic selection is used to determine which of the fittest chromosomes are considered for the next generation. The two most common types of selection are *roulette selection* and *tournament selection*.

In roulette selection, fitness values for chromosomes are sorted in descending order $[f_{(1)}, f_{(2)}, \ldots, f_{(l)}, \ldots, f_{(\lambda)}]$ and normalized into weights for which $\sum w_i = 1$. The cumulative sum of weights for each ranked chromosome are added in the form $s_l = \sum_i^l w_i$. A random uniform variate $\mathcal{U}(0,1)$ is then drawn, and the unselected chromosome whose s_l is closest to $\mathcal{U}(0,1)$ is selected as $\mathcal{C}_1$. Next, another uniform variate is drawn, and a second unselected chromosome $\mathcal{C}_2$ is selected. The selected pair of chromosomes then undergo crossover, if specified, before proceeding forward to the next generation.

In tournament selection, two pairs of chromosomes $(\mathcal{C}_1, \mathcal{C}_2)$, $(\mathcal{C}_3, \mathcal{C}_4)$ are randomly selected from the pool of unselected chromosomes, followed by a comparison of fitness values. For the first pair $(\mathcal{C}_1, \mathcal{C}_2)$, if $f(\mathcal{C}_1) > f(\mathcal{C}_2)$, then chromosome $\mathcal{C}_1$ is selected; otherwise $\mathcal{C}_2$ is selected. For the second pair $(\mathcal{C}_3, \mathcal{C}_4)$, if $f(\mathcal{C}_3) > f(\mathcal{C}_4)$, then chromosome $\mathcal{C}_3$ is selected; otherwise $\mathcal{C}_4$ is selected. The final pair of winning chromosomes undergo crossover (if specified) before being used in the next generation.

Elitism. Elitism is a process in which one or more of the most fit chromosomes are always used in the next generation. Elitism has been known to improve the fit of a GA; however, the dilemma is that once the fittest chromosome(s) is used, the solution path always follows the direction of the fittest chromosome. Let f_e be the fitness of the most elite chromosome $\mathcal{C}_e$, and let $\bar{f}$ be the average fitness. If $f_e \gg \bar{f}$, then there may be little chance that fitness of other chromosomes exceed f_e, causing the solution vector for chromosome $\mathcal{C}_e$ to become *nested* in the problem. The generic term for this phenomenon is *nesting*, which occurs when a set of model parameters becomes stuck during an iterative learning procedure, which may not be beneficial to the model being trained.

Crossover. Genetic crossover occurs during meiosis in which genetic material is exchanged between the paternal (father) and maternal (mother) gametes. Consider the selection of a pair of parent chromosomes and assume that they are paternal and maternal copies of a gamete, for which the labels are *parent A* and *parent B*. If single-point crossover is used, and, for example, the last 4 bits of the pair of chromosomes are selected for crossover, then the ends of the chromosomes are swapped (traded) resulting in two child chromosomes containing the exchanged genetic material, with the last 4 bits of the chromosomes selected for crossover shown in bold:

Parent A	1	0	0	0	1	**1**	**0**	**1**	**1**
Parent B	0	1	0	1	0	**0**	**1**	**0**	**1**

The child chromosomes containing the exchanged genetic material are denoted as follows:

Child A	1	0	0	0	1	**0**	**1**	**0**	**1**
Child B	0	1	0	1	0	**1**	**0**	**1**	**1**

Single-point crossover is commonly performed at a randomly selected bit on a pair of *selected* chromosome if a randomly drawn uniform variate $\mathcal{U}(0,1)$ is below the crossover probability P_c. Common values for P_c are 0.5-0.9. It is important to note that crossover is initiated at a certain bit location on the parental chromosomes, and is independent of gene location.

As stated above, GA is usually employed for optimization of a known function for which fitness can be determined. In classification analysis, a classifier is needed for determining class prediction error or accuracy, which can serve as a source of information for fitness. GAs can be embedded in most classifiers, and have proven quite useful for parameter learning in MLPs. This approach is called *adaptive learning with metaheuristics*. Figure 25.1 shows an MLP architecture with nodes linked to a metaheuristic that performs connection weight learning. In the following example, we apply a GA as the heuristic for MLP parameter learning. Algorithm 25.1 lists the pseudocode for the GA-MLP algorithm.

Example 25.2 This example uses datasets listed in Table 1.3 whose features were generated with greedy PTA (see Table 12.10). Preprocessing data before use in the MLP involved mean-zero standardization of input features. For GA weight training, MLP connection weights were uniformly sampled from the range $[-2, 2]$. We used 500 generations for training the MLP. The number of

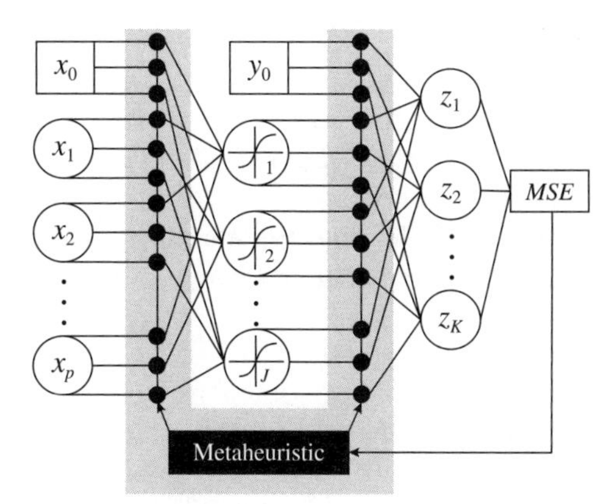

FIGURE 25.1 A metaheuristic adaptive learning method embedded into a MLP for connection weight learning.

chromosomes employed was 40, and the number of N genes per chromosome was equal to the number of p MLP input features (plus 1 for bias), times the number of hidden nodes in the hidden layer (plus 1 for bias), times the number of connection weights between the activation functions and the $K = \Omega$ outputs. Overall, the total number of weights in the MLP was $N = (p+1)J + (J+1)K$. The process of (1) binary-to-decimal conversion for obtaining MLP weights, (2) sweeping arrays through the MLP with the current connection weights, (3) determination of fitness for each chromosome, (4) tournament selection, (5) crossover, and (6) point mutation was repeated for the specified number of generations. Figure 25.2 illustrates 2D self-organizing maps of class decision boundaries for the GA-MLP based on the AMLALL2, Brain2, Breast2A, Breast2B, Colon2, Lung2, MLL_Leukemia3, Prostate2, and SRBCT4 datasets using mean-zero standardized input features. The list of integers (with corresponding classification accuracies in parentheses) in the upper right of each panel represent the number of chromosomes employed during training. Results suggest that GA-MLP performance was not that good, and in fact, classification accuracy for the multiclass datasets (i.e., three-class MLL_Leukemia3 and four-class SRBCT4) was in the 70-90% range. The two-class datasets AMLALL2, Breast2A, and Lung2 resulted in perfect classification accuracy values of 100%.

25.3 COVARIANCE MATRIX SELF-ADAPTATION-EVOLUTION STRATEGIES

The second type of evolutionary algorithm addressed in this chapter is covariance matrix adaptation-evolution strategies (CMA-ES). CMA-ES has

Algorithm 25.1: GA-MLP

```
Data: n training arrays, p features, J hidden nodes, K output nodes, λ
      chromosomes, connection weights w_ij^ih and w_jk^ho
Result: Trained GA-MLP network
Specify N = (p + 1)J + (J + 1)K genes (parameters)
Generate 𝓈_gl ∈ {0, 1}  ∀g, l 16-bit binary strings
Set crossover probability P_c = 0.6
Set mutation probability P_m = 0.005
Scaling constant c_m = 2
for gen ← 1 to #generations do
    Decode binary string 𝓈_gl into x_gl  ∀g, l
    for l ← 1 to λ do
        MSEtrain ← 0, MSEtest ← 0
        w_ij^ih = x(g, l)   g ≤ (p + 1)J
        w_jk^ho = x(g, l)   (p + 1)J + 1 ≤ g ≤ (p + 1)J + (J + 1)K
        for i ← 1 to n do
            Determine ẑ using x_i
            MSEtrain ← MSEtrain + Σ_k ½(z_k − ẑ_k)²
        endfor
        Determine MSEtest
        MSEtrain = MSEtrain/ntrain
        MSEtest = MSEtest/ntest
        E_l = MSEtrain + MSEtest
        fitness_l = 1/E_l
    endfor
    Perform tournament selection
    Perform crossover for each pair of parent chromosomes
    Mutate within each child chromosome
    Select pair of elite chromosomes
    Inject new genes
endfor
Perform binary-to-decimal conversion of most elite chromosome's binary
string to obtain solution vector
```

been found useful for difficult nonlinear optimization problems involving rugged multimodal fitness landscapes with discontinuities and numerous local maxima, since it does not use or approximate gradients [30]. At present, however, there is a sparse biomedical literature on the application of CMA-ES. The limited number of reports focused on chemical crystal spectral interferometry [31], detection of bacterial gene starts [32], multiobjective optimization [33], optimization of conductance parameters of cortical neural

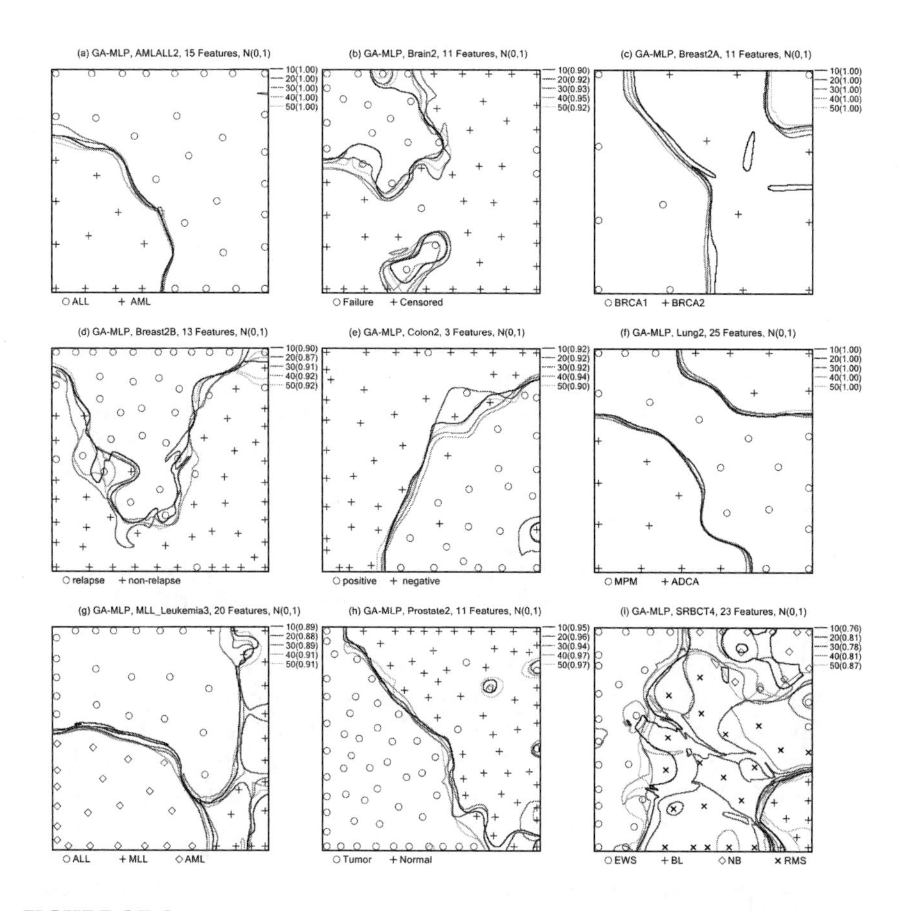

FIGURE 25.2 2D self-organizing maps of GA-MLP class decision boundaries and classification accuracy (upper right of each panel) for the AMLALL2, Brain2, Breast2A, Breast2B, Colon2, Lung2, MLL_Leukemia3, Prostate2, and SRBCT4 datasets based on mean-zero standardized input features; integers in the upper right of each panel represent the number of chromosomes employed during training: (a) AMLALL2; (b) Brain2; (c) Breast2A; (d) Breast2B; (e) Colon2; (f) Lung2; (g) MLL_Leukemia3; (h) Prostate2; (i) SRBCT4. The numbers of genes used for each dataset are listed in Table 12.10.

models [34], self-organizing nets for optimization [35], and publication on refinements to CMA-ES [36-38].

The earliest form of ES was called *mutative strategy parameter control* (MSPC), in which search steps are taken by stochastic variation in the form of a mutation involving the addition of a realization of a normally distributed vector [7]. Let λ be the number of chromosomes (individuals), μ be the fittest offspring selected for the next generation, $\mathbf{x}_l \in \mathbb{R}^N$ be a real-valued solution vector represented by the lth chromosome, and let $\sigma_l \in \mathbb{R}^+$ be the step size taken for the lth chromosome ($l = 1, 2, \ldots, \lambda$). The mutation workflow for $l = 1, \ldots, \lambda$ chromosomes for $(1, \lambda)$-ES, the simplest form of CMA-ES, is as follows

$$\sigma_l^{(g+1)} = \sigma^{(g)} \exp(\xi_l) \tag{25.3}$$

$$\mathbf{x}_l^{(g+1)} = \mathbf{x}_l^{(g)} + \sigma_l^{(g+1)} \mathbf{z}_l, \tag{25.4}$$

where $\xi_l \in \mathbb{R}$ is a random vector of variates distributed $\mathcal{N}(0,1/\sqrt{2p})$ and $\mathbf{z}_l$ is a random vector of variates distributed $\mathcal{N}(0,1)$. Following the mutation update, fitness is determined for each chromosome using the current values of $\mathbf{x}_l$, and the μ best chromosomes are selected for the next generation. A noted drawback of the $(1, \lambda)$-ES method for adapting individual step sizes is that the number of parameters N has to scale linearly with the population size λ [37].

The most prominent development of ES algorithms since the introduction of the original $(1, \lambda)$-ES has focused on *adaptation*, an ES principle stating that the strategy parameters adapt to the objective function. The approach taken by $(1, \lambda)$-ES was for global and individual step sizes. Individual step sizes do not ensure adaptation, and thus the use of arbitrary normal mutation distributions was sought. In 1981, Schwefel developed $(\mu/\rho_I, \lambda)$, an expanded version that employs a rotation angle adaptation [39] and mean-zero arbitrary normal mutations. The $(\mu/\rho_I, \lambda)$ algorithm included using $(N^2 - N)/2$ rotations, which undergo selection, recombination, and mutation. While MSPC was still preserved, $(\mu/\rho_I, \lambda)$ could not satisfy the ES adaptation principle, which guarantees that progress on a spherical objective function ($\sum_i x_i^2$) must be similar to that for any convex quadratic objective function.

The covariance matrix adaptation variant of ES (CMA-ES) was introduced by Hansen and Ostermeier [40] as a completely derandomized self-adaptation approach. CMSA-ES implements MSPC and uses the covariance matrix of the mutation distribution to increase the probability of repeated successful mutation steps. A brief history of the computational developments of ES is provided as follows. The basic model for ES includes a series of learning iterations, expressed by the relationship

$$\mathbf{x}^{(g+1)} = \mathbf{x}^{(g)} + \text{recombination} + \text{mutation}. \tag{25.5}$$

An early variant of CMA-ES [37] called (μ_W, λ)-CMA-ES updates parameters using the relationship

$$\mathbf{x}^{(g+1)} = \langle \mathbf{x} \rangle_w^{(g)} + \sigma^{(g)} \mathbf{E}^{(g)} \mathbf{D}^{(g)} \mathbf{z}^{(g+1)}, \tag{25.6}$$

where $\langle \mathbf{x} \rangle_w^{(g)}$ are the parameter means based on the μ most fit chromosomes, $\mathbf{E}^{(g)}$ is the matrix of eigenvectors of $\mathbf{C}$, $\mathbf{D}^{(g)}$ is the diagonal square-root matrix of eigenvalues of $\mathbf{C}$, $\mathbf{C}$ is defined as the covariance matrix of the normally distributed matrix $\mathbf{E}^{(g)}\mathbf{D}^{(g)}\mathcal{N}(\mathbf{0},\mathbf{I})$, and $\mathbf{z}^{(g+1)}$ is a vector of random standard normal variates distributed $\mathcal{N}(0,1)$ [40]. The vector $\mathbf{E}^{(g)}\mathbf{D}^{(g)}\mathbf{z}^{(g+1)}$ forms the *mutation* distribution used in learning, and is multiplied by the global step size $\sigma^{(g)}$. Initially, $\mathbf{C} = \mathbf{I}$, with later updates [40] based on

$$\mathbf{C}^{(g+1)} = \mathbf{C}^{(g)} + \mathbf{E}^{(g)}\mathbf{D}^{(g)}\mathbf{z}^{(g)}\left[\mathbf{E}^{(g)}\mathbf{D}^{(g)}\mathbf{z}^{(g)}\right]^T. \quad (25.7)$$

The multivariate normal distribution of the mutation vector $\mathbf{E}^{(g)}\mathbf{D}^{(g)}\mathbf{z}^{(g+1)}$ is distributed $\mathcal{N}(\mathbf{0},\mathbf{C}^{(g)})$ for $\mathbf{z}^{(g+1)} \sim \mathcal{N}(\mathbf{0},\mathbf{I})$, and the only random stochastic feature of this process is the use of $\mathbf{z}$ [40].

The matrix $\mathbf{ED}$ is the square root matrix of $\mathbf{C}$ since

$$\mathbf{D} = \mathrm{diag}(\sqrt{\lambda_1}, \sqrt{\lambda_2}, \ldots, \sqrt{\lambda_p}), \quad (25.8)$$

and $\mathbf{C} = \mathbf{ED}(\mathbf{ED})^T$. Simulation of a multivariate normal random vector $\mathbf{y}$ is accomplished by performing a Cholesky factorization $\mathbf{C} = \mathbf{LL}^T$ when covariance is provided or $\mathbf{DRD} = \mathbf{LL}^T$ when the correlation matrix $\mathbf{R}$ is input, where $\mathbf{D} = \mathrm{diag}(\sigma_1, \sigma_2, \ldots, \sigma_p)$. Realization of standard normal variates is then based on the operation $\mathbf{Y} = \mathbf{LZ} + \boldsymbol{\mu}$. However, it is irrelevant whether $\mathbf{y}$ is based on $\mathbf{ED}$ from eigenanalysis or $\mathbf{L}$ from Cholesky factorization. The computational complexity of the eigenanalysis in (μ_W, λ)-CMA-ES described above was quite high, so naturally further development resulted in a Cholesky update [41] called $(1, \lambda)$-CMA-ES, which used $\mathbf{L}$ from the Cholesky factorization of $\mathbf{C}^{(g)}$. Its parameter vector update was in the form

$$\mathbf{x}^{(g+1)} = \langle \mathbf{x} \rangle_w^{(g)} + \sigma^{(g)}\mathbf{L}^{(g)}\mathbf{z}^{(g+1)}, \quad (25.9)$$

with the associated covariance matrix update

$$\mathbf{C}^{(g+1)} = \mathbf{C}^{(g)} + \mathbf{L}^{(g)}\mathbf{z}^{(g)}\left[\mathbf{L}^{(g)}\mathbf{z}^{(g)}\right]^T. \quad (25.10)$$

Several criteria must be maintained for the mutation distribution and step size: (1) the condition number of $\mathbf{C}$ must remain below 10^{14} [40]; (2) $\mathbf{C}$ must remain positive definite–this is accomplished through use of the positive definite $\mathbf{Lz}[\mathbf{Lz}]^T$ for the covariance matrix update; and (3) because the update in (25.10) shifts the mutation distribution toward the line distribution $\mathcal{N}(\mathbf{0}, \mathbf{Lz}[\mathbf{Lz}]^T)$, the line distribution has to be shortened during each step [40]. Shortening is accomplished by using the relationships

$$\begin{aligned} \mathbf{s}^{(g+1)} &= (1-c)\cdot \mathbf{s}^{(g)} + c_u \cdot \mathbf{L}^{(g)}\mathbf{z} \\ \mathbf{C}^{(g+1)} &= (1-c_{\text{cov}})\cdot \mathbf{C}^{(g)} + c_{\text{cov}} \cdot \mathbf{s}^{(g)}(\mathbf{s}^{(g)})^T, \end{aligned} \tag{25.11}$$

where $\mathbf{s}$ is a line distribution with variance $||\mathbf{s}||^2$, $c \in (0,1]$ determines the accumulation time for $\mathbf{s}$, $c_u = \sqrt{c\cdot(2-c)}$ normalizes the variance of $\mathbf{s}$ by solving the equation $1^2 = (1-c)^2 + c_u^2$, and $c_{\text{cov}} \in (0,1]$ determines the rate of change of $\mathbf{C}$. The last adjustment for the iterative model is the global step size based on the relationship

$$\sigma = \sigma \exp\{\tau(\|\mathbf{s}\| - \hat{\chi})\} \tag{25.12}$$

in order to reduce the difference between the step size taken and $\hat{\chi}$, the expected length of a p-dimensional random vector distributed $\mathcal{N}(\mathbf{0}, \mathbf{I})$. The parameter τ is a learning parameter that dampens the step size variation.

Example 25.3 We used the more efficient Cholesky factorization-based version of CMA-ES known as CMSA (covariance matrix self-adaptation)-ES [42] to classify the DNA microarrays in the same nine datasets used in previous examples. Note that CMSA-ES has only two external parameters, τ and $\tau_c = 1/c_{\text{cov}}$. During each training generation, fitness was determined as $1/\text{MSE}_l$, and the top μ chromosomes were used for calculating revised values of solution parameters $\mathbf{y}$. Algorithm 25.2 lists the pseudocode for the CMSA-ES-MLP, the neural network version. Figure 25.3 illustrates 2D self-organizing maps of class decision boundaries for the CMSA-ES-MLP based on the AMLALL2, Brain2, Breast2A, Breast2B, Colon2, Lung2, MLL_Leukemia3, Prostate2, and SRBCT4 datasets using mean-zero standardized input features. The list of integers (with corresponding classification accuracies in parentheses) in the upper right of each panel represents the number of generations used during training. Classification accuracy from CMSA-ES-MLP as a function of the number of generations increased to 100% when more than 200 generations were employed—and this occurred for all datasets with the exception of the Colon2 dataset. CMSA-ES was not designed to be a fast learner, because the constraints on step direction and step size during gradient descent prevent the algorithm from overstepping the solution. However, in order to attain a reasonably good solution, at least 1000 generations should be used. This tends to be a pitfall for CMSA-ES when applied to simple problems, and fast results are expected. Rather, CMSA-ES works best on complex multimodal fitness landscapes in which many local maxima are present, providing several thousand training generations are used.

Algorithm 25.2: CMSA-ES-MLP

Data: n arrays, p features, J hidden nodes, K output nodes, λ chromosomes, connection weights w_{ij}^{ih} and w_{jk}^{ho}
Result: Trained CMSA-ES-MLP network
Specify $G = (p+1)J + (J+1)K$ genes (parameters)
$\lambda = 4 + \lfloor(3\log(N))\rfloor$
if $N < 8$ **then**
  $\lambda = 8$
endif
$\mu = \lfloor\lambda/4\rfloor$; set learning rate, $\tau = 1/\sqrt{2N}$
Set time constant, $\tau_c = 1 + (N(N+1))/(2\mu)$
$\mathbf{C} = \mathbf{I}$
$\mathbf{L} = \mathbf{I}$
$\bar{\sigma} = 0.5$
$\sigma_l = \mathcal{N}(0,1) \quad \forall l$
$y_m = \mathcal{N}(0,1) \quad \forall m$
for *gen* $\leftarrow$ 1 **to** *#generations* **do**
  for $l \leftarrow 1$ **to** λ **do**
    $\text{MSE}_{\text{train}} = 0$
    $\sigma_l = \bar{\sigma}\exp\{\tau\mathcal{N}(0,1)\}$
    $\mathbf{s}_l = \mathbf{Lz}$
    $\mathbf{v}_l = \sigma_l \mathbf{s}_l$
    $\mathbf{y}_l \leftarrow \mathbf{y} + \mathbf{v}_l$
    $w_{ij} = y(g,l) \quad g \leq (p+1)J$
    $w_{jk} = y(g,l) \quad (p+1)J + 1 \leq g \leq (p+1)J + (J+1)K$
    for $i \leftarrow 1$ **to** n **do**
      Determine $\hat{z}$ using $\mathbf{x}_i$
      $\text{MSE}_{\text{train}} \leftarrow \text{MSE}_{\text{train}} + \sum_k \frac{1}{2}(z_k - \hat{z}_k)^2$
    endfor
    $\text{MSE}_{\text{train}} = \text{MSE}_{\text{train}}/n_{\text{train}}$
    $\text{fitness}_l = 1/\text{MSE}_{\text{train}}$
  endfor
  Select top μ fitness_l
  Determine $\bar{\mathbf{v}}$
  Determine $\bar{\mathbf{s}}$
  Determine $\bar{\sigma}$
  $\mathbf{y} \leftarrow \mathbf{y} + \bar{\mathbf{z}}$
  $\mathbf{C} \leftarrow [1 - (1/\tau_c)] \cdot \mathbf{C} + (1/\tau_c) \cdot \bar{\mathbf{s}}\bar{\mathbf{s}}^T$
  $\mathbf{C} = \mathbf{LL}^T$ (**L** from Cholesky factorization)
endfor
Perform testing by use of $\mathbf{y}_l$ for solution vector

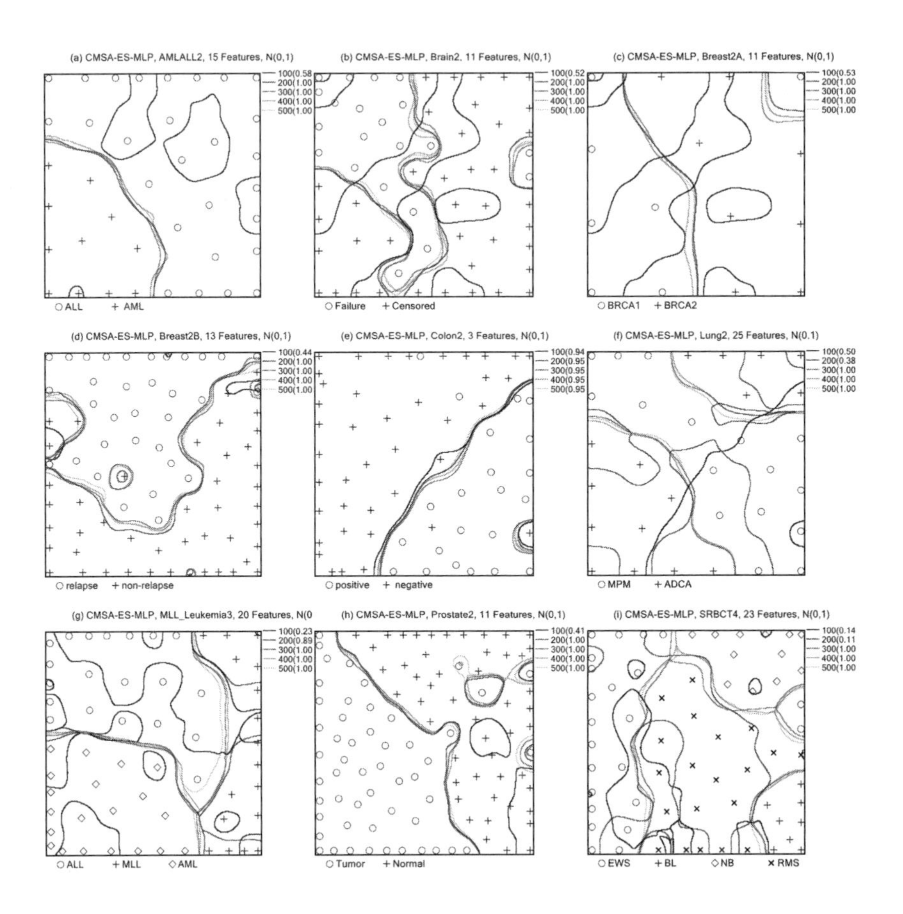

FIGURE 25.3 2D self-organizing maps of CMSA-ES-MLP class decision boundaries and classification accuracy (upper right of each panel) for the AMLALL2, Brain2, Breast2A, Breast2B, Colon2, Lung2, MLL_Leukemia3, Prostate2, and SRBCT4 datasets based on mean-zero standardized input features; integers in the upper right of each panel represent the number of generations employed during training: (a) AMLALL2; (b) Brain2; (c) Breast2A; (d) Breast2B; (e) Colon2; (f) Lung2; (g) MLL_Leukemia3; (h) Prostate2; (i) SRBCT4. The numbers of genes used for each dataset are listed in Table 12.10.

25.4 PARTICLE SWARM OPTIMIZATION

Particle swarm optimization (PSO) was introduced by Kennedy and Eberhart in 1995 [13] as a new optimization tool for stochastic searching in multimodal space. PSO is modeled after the behavior of migrating flocks of birds or feeding behaviors of schools of fish, in which "particles" fly through the multidimensional space, exchanging information along the way in order to influence each other's movements to find a global maximum. Each particle has a cognitive memory about the position where the best fitness occurred, as well as a social memory on where the best fitness occurred among all members of the swarm. Particle travel is hinged to the velocity based on the last position, the cognitive and social memories, and randomness.

To date, PSO has been applied to numerous applications, including classification and regression trees [44], Huber fractal image compression [45], ventricular contractions [46], tongue motion in speech disorders [47], white blood cell image segmentation [48], flow shop scheduling [49], and minimum spanning trees [50]. DNA microarray studies have focused on biclustering [51], constricted PSO [52], and semisupervised ellipsoid ARTMAP PSO [53].

For the lth particle, let the position matrix $\mathbf{R}$ and velocity matrix $\mathbf{V}$ have dimension $\Omega \times p$, where Ω is the number of classes and p is the number of features. Let the p-dimensional position and velocity vectors for particle l be $(\mathbf{r}_l^1, \ldots, \mathbf{r}_l^\Omega)$ and $(\mathbf{v}_l^1, \ldots, \mathbf{v}_l^\Omega)$. The class-specific particle position is $\mathbf{r}_l^\omega = \{r_{1,l}^\omega, \ldots, r_{p,l}^\omega\}$, and the class-specific particle velocity is $\mathbf{v}_l^\omega = \{v_{1,l}^\omega, \ldots, v_{p,l}^\omega\}$. At each iteration, the fitness [54] of each particle is based on

$$f(l) = \frac{1}{n} \sum_{i=1}^{n} |\mathbf{x}_i - \mathbf{r}_l^{\omega(\mathbf{x}_i)}|, \tag{25.13}$$

where n is the number of training arrays and $\omega(\mathbf{x}_i)$ is the class of array $\mathbf{x}_i$. The velocity update is

$$\begin{aligned} v_l(t+1) = {} & w v_l(t) + c_1 \mathcal{U}(0,1) \otimes (\mathbf{b}_l(t) - \mathbf{r}_l(t)) \\ & + c_2 \mathcal{U}(0,1) \otimes (\mathbf{b}_g(t) - \mathbf{r}_l(t)), \end{aligned} \tag{25.14}$$

where w is the *inertia factor*, c_1 is the cognitive parameter and c_2 is the social parameter, $\mathbf{b}_l(t)$ is the best historical position for particle l, and $\mathbf{b}_g(t)$ is the global best position. The inertia at iteration t is $w(t) = w_{\max} - (w_{\max} - w_{\min}) * (t/T_{\max})$. The particle position update is $\mathbf{r}_l(t+1) = \mathbf{r}_l(t) + \mathbf{v}_l(t+1)$. The decision rule for class j of a test array $\mathbf{x}$ is

$$D(\mathbf{x} \rightsquigarrow \omega) = \underset{j=1,2,\ldots,\Omega}{\arg\min}\{|\mathbf{x} - \mathbf{b}_g^j|\}. \tag{25.15}$$

Parameter values for PSO were set to #numparticles $= 50, T_{\max} = 300$, with the following constraints: $v_{\min} = -0.05$, $v_{\max} = 0.05$, $c_1 = 2$, $c_2 = 2$, $w_{\min} = 0.4$, and $w_{\max} = 0.9$. Algorithm 25.3 lists the pseudocode for the PSO-MLP algorithm.

Example 25.4 Classification runs were performed for the nine datasets employed in previous examples in this chapter. Figure 25.4 illustrates 2D

Algorithm 25.3: PSO-MLP

Data: n arrays, p features, J hidden nodes, K output nodes, λ chromosomes, connection weights w_{ij}^{ih} and w_{jk}^{ho}

Result: Trained PSO-MLP network

$\lambda \leftarrow \#chrom$

Dimension chromerr(λ), chromfitness(λ), $\mathbf{p}(N,\lambda)$

$\mathbf{v}(N,\lambda)$, $\mathbf{b}(N,\lambda)$, $\mathbf{b}_{gf}(N)$, $\mathbf{b}_f(\lambda)$

$\mathbf{p}(j,i) \leftarrow \mathcal{U}(0,1)$

$\mathbf{v}(j,i) \leftarrow 0$

$v_{\min} \leftarrow -0.05$, $v_{\max} \leftarrow 0.05$, $c_1 \leftarrow 2$, $c_2 \leftarrow 2$, $w_{\min} \leftarrow 0.4$, $w_{\max} \leftarrow 0.9$

for $gen \leftarrow 1$ **to** $\#generations$ **do**

- $w = w_{\max} - (w_{\max} - w_{\min})(gen/\#generations)$
- **for** $l \leftarrow 1$ **to** λ **do**
  - $\text{MSE}_{\text{train}} \leftarrow 0$
  - $w_{ij}^{ih} = p(m,l) \quad m \le (p+1)J$
  - $w_{jk}^{ho} = p(m,l) \quad (p+1)J + 1 \le m \le (p+1)J + (J+1)K$
  - **for** $i \leftarrow 1$ **to** n **do**
    - Determine $\hat{z}$ using $\mathbf{x}_i$
    - $\text{MSE}_{\text{train}} \leftarrow \text{MSE}_{\text{train}} + \sum_k \frac{1}{2}(z_k - \hat{z}_k)^2$
    - $E_{\text{train}} += \text{MSE}_{\text{train}}$
  - **endfor**
  - Determine MSE_{test}
  - $\bar{E}_{\text{train}} = E_{\text{train}}/n_{\text{train}}$
  - $\bar{E}_{\text{test}} = E_{\text{test}}/n_{\text{test}}$
  - $E_l = \bar{E}_{\text{train}} + \bar{E}_{\text{test}}$
  - **if** $\bar{E}_{\text{train}} < b_f(l)$ **then**
    - $b_f(l) = \bar{E}_{\text{train}}$
    - $b(j,l) = p(j,l) \quad \forall j$
  - **endif**
  - **if** $\bar{E}_{\text{train}} < E_{\min}$ **then**
    - $E_{\min} = \bar{E}_{\text{train}}$
    - $b_{gf}(j) = p(j,l) \quad \forall j$
  - **endif**
  - **for** $i \leftarrow 1$ **to** n **do**
    - $v(j,l) = wv(j,l) + c_1\mathcal{U}(0,1)(b(j,l) - p(j,l)) + c_2 *$ $\mathcal{U}(0,1)(b_{gf}(j) - p(j,l))$
  - **endfor**
- **endfor**
- $v(j,l) = \max\{v(j,l), v_{\min}\} \quad \forall j,l$
- $v(j,l) = \min\{v(j,l), v_{\max}\} \quad \forall j,l$
- $p(j,l) = p(j,l) + v(j,l) \quad \forall j,l$

endfor

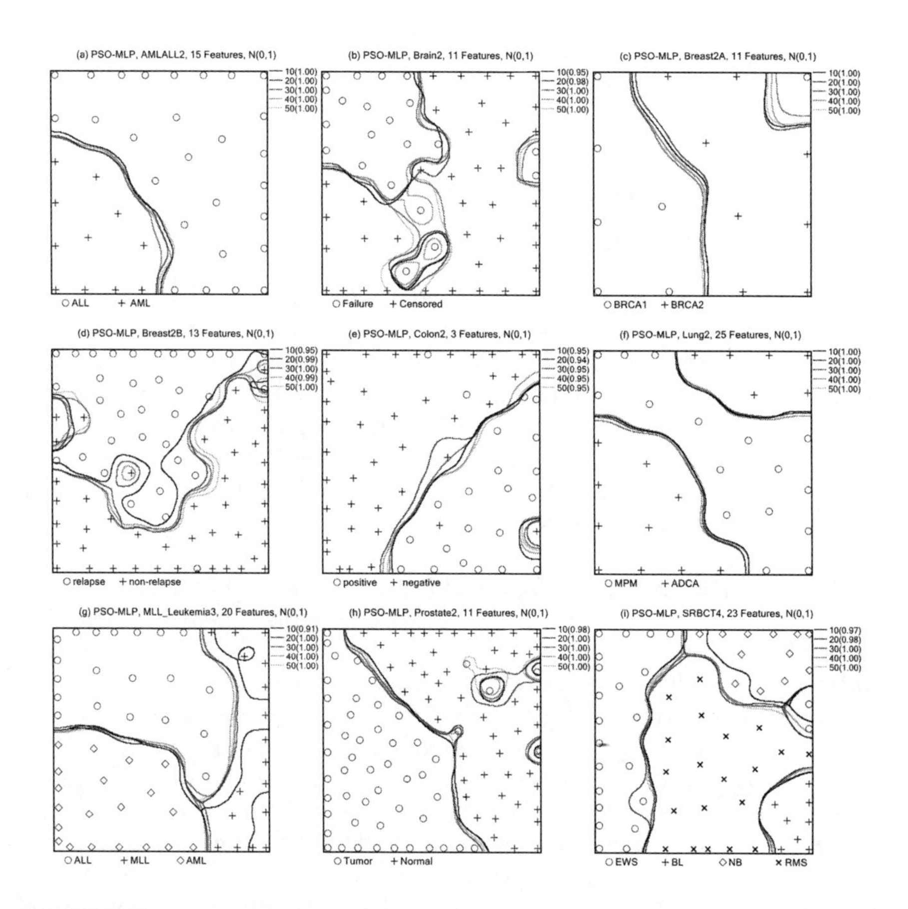

FIGURE 25.4 2D self-organizing maps of PSO-MLP class decision boundaries and classification accuracy (upper right of each panel) for the AMLALL2, Brain2, Breast2A, Breast2B, Colon2, Lung2, MLL_Leukemia3, Prostate2, and SRBCT4 datasets based on mean-zero standardized input features; integers in the upper right of each panel represent the number of chromosomes employed during training: (a) AMLALL2; (b) Brain2; (c) Breast2A; (d) Breast2B; (e) Colon2; (f) Lung2; (g) MLL_Leukemia3; (h) Prostate2; (i) SRBCT4. The numbers of genes used for each dataset are listed in Table 12.10.

self-organizing maps of class decision boundaries for the PSO-MLP based on the AMLALL2, Brain2, Breast2A, Breast2B, Colon2, Lung2, MLL_Leukemia3, Prostate2, and SRBCT4 datasets using mean-zero standardized input features. The list of integers (with corresponding classification accuracies in parentheses) in the upper right corner of each panel represents the number of chromosomes employed during training. The PSO-MLP classifier achieved 100% accuracy for all datasets except Colon2 when 50 chromosomes were used. Below 50 chromosomes, the pool of potential solutions was too small for seeding a reliable solution.

25.5 ANT COLONY OPTIMIZATION

25.5.1 Classification

Ant colony optimization (ACO) is a swarm intelligence method inspired by the foraging behavior of ants [14]. The principle behind ACO is based on pheromone deposition and decay by ants during their return from a food source. Any time food is found, on the return trip each ant carrying food deposits pheromone, which decays over time. A path with a high level of pheromone signifies one of two possibilities: the path is long and a lot of ants traversed it because a large food source was found, or the path is short and a moderate number of ants recently traversed it.

Pathway Structure. ACO for classification [55] begins with defining the path structure for p features in the form of a table, given in Table 25.1. Table 25.1 column headings $A_j (j = 1, 2, \ldots, p)$ represent p features (attributes), and rows represent the categorical values for each feature. Table 25.1 elements a_{jk} represent the kth value of the jth feature and serve as a pathway *node* to which ants can travel. If feature values are not nominal, categorical, or ordinal, then continuously scaled features are discretized into a fixed number of categories, such as quartiles. Henceforth, a_{jk} is the kth quartile ($k = 1, 2, 3, 4$) of the jth feature. What is important to realize in this relationship is that ACO for classification assumes discretized values of each feature that are at least nominal. For continuously scaled features, ordinal scaling is used in which quartiles are implemented. Looking at Table 25.1, we see that all ant travel is rightward, and each ant must visit one node of each attribute. (Later, we will discuss node pruning, which removes nodes visited that do not increase classification "quality" singly.) In Table 25.1 the boxed elements reflect an example pathway trip made by an ant to several nodes in the order $a_{12} \rightarrow a_{21} \rightarrow a_{j4} \rightarrow a_{p1}$.

TABLE 25.1 Pathway Structure for Ant Travel Based on Features and Their Quartile Values

A_1	A_2	$\cdots$	A_j	$\cdots$	A_p
a_{11}	$\boxed{a_{21}}$	$\cdots$	a_{j1}	$\cdots$	$\boxed{a_{p1}}$
$\boxed{a_{12}}$	a_{22}	$\cdots$	a_{j2}	$\cdots$	a_{p2}
a_{13}	a_{23}	$\cdots$	a_{j3}	$\cdots$	a_{p3}
a_{14}	a_{24}	$\cdots$	$\boxed{a_{j4}}$	$\cdots$	a_{p4}

Pheromone Initialization. Before any ants can travel, an equal amount of pheromone is assigned to all nodes in the pathway, given by

$$\tau_{jk} = \frac{1}{\sum_{j=1}^{p} b_j}, \tag{25.16}$$

where p is the total number of features and b_j is the number of feature values for the jth feature. When quartiles are used for continuously scaled features, $b_j = 4$.

Rule Construction. The nodes traversed by each ant form a classification rule R_t at iteration t. Feature quartiles visited by an ant form the basis of the rule. The majority class of arrays having the same quartiles as the rule constructed by the ant determines the following rule: IF $\langle a_{1k}$ AND a_{2k} AND $\ldots a_{pk}\rangle$ THEN $\langle$ CLASS $\rangle$. Before continuing, it is noteworthy to define several definitions used in rule construction. The phrase *arrays covered by the rule* constructed by an ant varies and refers to arrays whose feature quartile values are the same as those selected by the ant. The *class predicted by the rule* remains fixed and is defined as the majority class label among arrays covered by the rule.

During the beginning of rule construction, the ant needs to know which node (quantile) to travel to for the next feature in the pathway [56]. Assume that feature j is the next feature to step to during rule construction. The level of feature j chosen is based on the greatest value of

$$P_{jk}(t) = \frac{\tau_{jk}\eta_{jk}}{\sum_{j=1}^{p} x_j \sum_{k=1}^{b_j} (\eta_{jk}\tau_{jk}(t))}, \tag{25.17}$$

among all b_j values of feature j, where η_{jk} is an information theoretic heuristic reflecting how relevant a node is for classification, $x_j = 1$ if the feature has been used in the rule and $x_j = 0$ otherwise, and b_j is the number of values for feature j. Let n_{jk} be the number of arrays having feature value a_{jk} for the jth feature and n_{jk}^{ω} be the number of similar arrays having class label $\omega\,(\omega = 1, 2, \ldots, \Omega)$. The density-based heuristic is

$$\eta_{jk} = \frac{\log_2(\Omega) - I(\Omega))}{\sum_{j=1}^{p} \sum_{k=1}^{b_j} \log_2(\Omega) - I(\Omega)}, \tag{25.18}$$

where the information gain is

$$I(\Omega) = -\sum_{\omega=1}^{\Omega} \left(\frac{n_{jk}^{\omega}}{n_{jk}}\right) \log_2 \left(\frac{n_{jk}^{\omega}}{n_{jk}}\right). \tag{25.19}$$

Rule Pruning and Pheromone Update. The quality of a rule is defined by the relationship

$$Q = \left(\frac{\text{TP}}{\text{TP}+\text{FN}}\right)\left(\frac{\text{TN}}{\text{FP}+\text{TN}}\right), \tag{25.20}$$

where TP is the number of arrays covered by the rule whose class is the same as the class predicted by the rule; FP is the number of false positive arrays, which are arrays covered by the rule whose class differs from the class predicted by the rule; FN represents the number of false negative arrays not covered by the rule but with the same class as the class predicted by the rule; and TN is the number of true negative arrays not covered by the rule whose class differs from the class predicted by the rule. The range of rule quality is $0 \leq Q \leq 1$.

After an ant traverses all nodes (features) and constructs a rule, the rule is then pruned by dropping each node singly while recalculating Q, until Q no longer increases. For example, if there are 10 features in an original rule, then 10 values of Q are determined each time that one of the 10 nodes is dropped singly. Nodes for which Q values increase when dropped are dropped from the rule. This process continues until there is at least one node remaining in the rule.

After rule pruning, the pheromone level is adjusted for only the nodes a_{jk} that were retained in the rule, using the form

$$\tau_{jk}(t) = \tau_{jk}(t-1) + \tau_{jk}(t-1)Q \qquad \forall j,k \in R_t. \tag{25.21}$$

When an evaporation rate is used, the update becomes

$$\tau_{jk}(t) = (1-\rho)\tau_{jk}(t-1) + \left(1 - \frac{1}{1+Q}\right)\tau_{jk}(t-1) \quad \forall j,k \in R_t, \tag{25.22}$$

where ρ is the pheromone evaporation rate. Large values of ρ increase evaporation, while low values ease the evaporation. Liu et al. [57] used an evaporation rate of $\rho = 0.1$ when comparing ACO methods.

25.5.2 Continuous-Function Approximation

The ACO classification methods described above can be used only for classification and are not well suited for MLP connection weight learning. In 2008, Socha and Dorigo [15] introduced a continuous-function approximation variant of ACO, which happens to be ideal for MLP connection weight learning. This approach employs kernel density estimation to improve solutions that yield the best fitness among a population of solution vectors.

Let $\mathbf{S} \in \mathbb{R}^{N \times \lambda}$ be a storage array for $\lambda (l = 1, 2, \ldots, \lambda)$ solution vectors having N parameters $(j = 1, 2, \ldots, N)$. An MLP incorporating solution vectors from $\mathbf{S}$ as the input-side and output-side connection weights is used to determine fitness of a predicted outcome. When classification is performed with the MLP network, the number of outputs is K, which represents the number of classes, namely, Ω. The total number of parameters N in a solution vector $\mathbf{s}_l$ is equal to the number of input features $I = p$ plus one (for bias) times J plus (J+1)(K), where J is the number of hidden nodes in the hidden layer. Thus, $N = (I + 1)J + (J + 1)K$. Let $\text{MSE}_l = \sum_i (1/2) \sum_k (z_k - \hat{z}_k)^2$ be the error for the lth solution vector when applied to network connection weights and all training arrays, and let $1/\text{MSE}_l$ represent the fitness. Solution vectors are ranked $R_{(1)}, R_{(2)}, \ldots, R_{(l)}, \ldots, R_{(\lambda)}$, and the weight at generation t for each solution vector in $\mathbf{S}$ is defined as

$$w_l(t) = \frac{1}{qp\sqrt{(2\pi)}} \exp\left\{ \frac{(R_{(l)} - 1)^2}{2q^2p^2} \right\}, \tag{25.23}$$

where q is a fixed parameter for the algorithm. The weight w defines a Gaussian variate with mean 1 and $\sigma = qp$. Small values of q provide greater weight to the best fitting solution vectors, while larger values distribute the weight more uniformly. We used a value of $q = 0.1$ for all runs, which seemed to provide the best fits by solution vectors. The probability of choosing solution vector l is

$$P_l(t) = \frac{w_l}{\sum_l^{\lambda} w_l}. \tag{25.24}$$

For each successive learning generation, KDE is used to simulate a normally distributed quantile for each parameter, with the mean equal to the current parameter value, that is, $\mu_l = s_{jl}$, and the standard deviation, determined as

$$\sigma_l(t) = \xi \sum_e^{\lambda} \frac{d(\mathbf{s}_e, \mathbf{s}_l)}{N - 1}, \tag{25.25}$$

where $\mathbf{s}_e$ are each of the λ solution vectors. The term ξ is similar to the pheromone evaporation rate and controls the overall learning rate of the ACO system. We have used $\xi = 0.3$, which seems to yield the best results after performing a grid search with increments of $\Delta\xi = 0.05$ in the range $0 < \xi \leq 1$. For each generation, the PDF for a new parameter value s_{jl} is approximated by using KDE [58] with $M = 100$ equally spaced bins over the range $(s_{jl} - 4, s_{jl} + 4)$, based on the bin frequencies

$$f(m) = \frac{1}{N_s h} \sum_{i=1}^{N_s} K\left(\frac{s_i - s_m}{h} \right), \tag{25.26}$$

where $f(m)$ is the smoothed bin count for the mth bin, $N_s = 1000$ simulated variates for constructing the PDF, s_i is the simulated quantile, s_m is the lower wall of the mth bin, and $h = 1.06\sigma_l N_s^{-0.2}$ is the optimal bandwidth for a Gaussian [59], and σ_l is as determined in (25.25). K is the Epanechnikov kernel function [60], defined as

$$K(u) = \begin{cases} \frac{3}{4}(1-u^2) & |u| \leq 1 \\ 0 & \text{otherwise,} \end{cases} \tag{25.27}$$

where $u = (s_i - s_m)/h$ and s_m is the lower bound of the mth bin. Once the smooth PDF is derived from KDE, the rejection-acceptance method is used for obtaining a single variate. Under the rejection method, bins in the simulated Gaussian distribution are randomly selected with the formula

$$m = (M_{\max} - 1)\mathcal{U}(0,1)_1 + 1, \tag{25.28}$$

where $\mathcal{U}(0,1)_1$ is a pseudorandom uniform distributed variate. For each m, a second pseudorandom variate is obtained, and if the following criterion is met,

$$\mathcal{U}(0,1)_2 < \text{pdf}(m)/\max\{\text{PDF}(i)\}, \qquad i = 1, 2, \ldots, M_{\max} \tag{25.29}$$

then the histogram bin wall s_m is used as the Gaussian variate s_{jl} in the current generation. Features of ACO-MLP are listed in Algorithm 25.4.

Example 25.5 Prior to training, **S** was initialized with standard normal variates distributed as $\mathcal{N}(0,1)$. During each generation, the fitness of each solution vector $1/\text{MSE}_l$ was determined by applying each solution vector to the input-side and output-side network connection weights, and feeding all arrays through the network. Fitness is then sorted in descending order. Next, selection weights w_l, probabilities P_l, and standard deviations σ_l for KDE were determined for each solution vector. For each generation, two new solution vectors were simulated. The first new solution vector was simulated by using KDE N times, based on the single value of σ_l and N values of $\mu = s_{jl}$ for the lth existing solution vector for which fitness rank was $R_{(l)} = 1$. This was repeated for the second new solution vector using values from the existing solution vector for which rank was $R_{(l)} = 2$. If the fitness values of either (both) of the new solution vectors was greater than the worst fitness values, then the solution vectors with the worst fitness were replaced with these new solution vectors. The process of simulating two new solution vectors per generation represents two ants, which travel through the solution space. Replacement of the solution vectors whose fitness is the worst is similar to pheromone update of a potential pathway through the solution space. Figure 25.5 illustrates 2D self-organizing maps of

Algorithm 25.4: ACO-MLP

Data: n arrays, p features, J hidden nodes, K output nodes, λ chromosomes, connection weights w_{ij}^{ih} and w_{jk}^{ho}
Result: Trained ACO-MLP network
$\lambda \leftarrow \#chrom$
Dimension $\mathbf{p}(N, \lambda)$
$\mathbf{S}(N, \lambda)$, $\mathbf{w}(N)$, $\mathbf{P}(N)$,$\sigma(N)$
Initialize $\mathbf{S} \leftarrow \mathcal{N}(0, 1)$
for $gen \leftarrow 1$ **to** $\#generations$ **do**
  for $l \leftarrow 1$ **to** λ **do**
    $\text{MSE}_{\text{train}} \leftarrow 0$
    $w_{ij}^{ih} = S(m, l) \quad m \leq (p+1)J$
    $w_{jk}^{ho} = S(m, l)xs \quad (p+1)J + 1 \leq m \leq (p+1)J + (J+1)K$
    for $i \leftarrow 1$ **to** n **do**
      Determine $\hat{z}$ using $\mathbf{x}_i$
      $\text{MSE}_{\text{train}} \leftarrow \text{MSE}_{\text{train}} + \sum_k \frac{1}{2}(z_k - \hat{z}_k)^2$
      $E_{\text{train}} += \text{MSE}_{\text{train}}$
    endfor
    Determine MSE_{test}
    $\bar{E}_{\text{train}} = E_{\text{train}}/n_{\text{train}}$
    $\bar{E}_{\text{test}} = E_{\text{test}}/n_{\text{test}}$
    $E_l = \bar{E}_{\text{train}} + \bar{E}_{\text{test}}$
  endfor
  $R_{(l)} = 1 \Rightarrow a = l$
  $R_{(l)} = 2 \Rightarrow b = l$
  $R_{(l)} = \lambda - 1 \Rightarrow c = l$
  $R_{(l)} = \lambda - 2 \Rightarrow d = l$
  $E_a < E_c \Rightarrow \mathbf{s}_c = \mathbf{s}_a; E_a < E_d \Rightarrow \mathbf{s}_d = \mathbf{s}_a$
  $E_b < E_c \Rightarrow \mathbf{s}_c = \mathbf{s}_b; E_b < E_d \Rightarrow \mathbf{s}_d = \mathbf{s}_b$
  for $j \leftarrow 1$ **to** N **do**
    $s_{j,a} = \mathcal{N}(s_{j,a}, \sigma_a)$
    $s_{j,b} = \mathcal{N}(s_{j,b}, \sigma_b)$
  endfor
endfor
Perform testing using solution vector $\mathbf{s}_a$;

class decision boundaries for the ACO-MLP based on the AMLALL2, Brain2, Breast2A, Breast2B, Colon2, Lung2, MLL_Leukemia3, Prostate2, and SRBCT4 datasets using mean-zero standardized input features. The list of integers (with corresponding classification accuracies in parentheses) in the upper right corner of each panel represents the number of generations employed

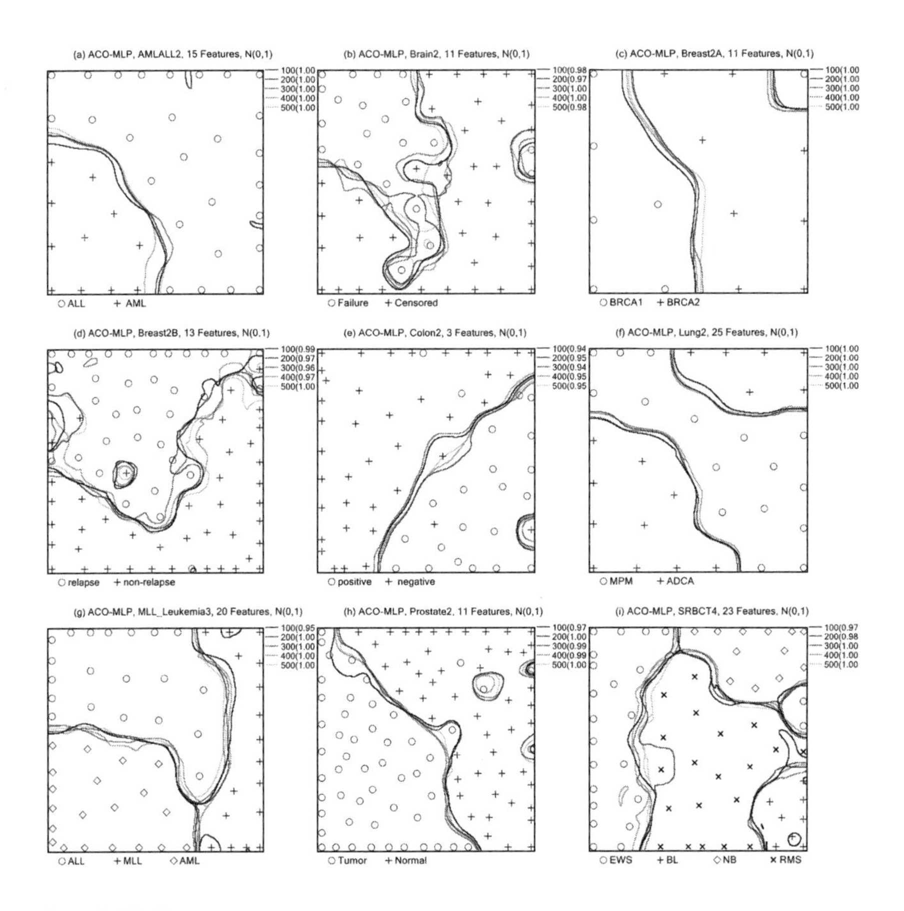

FIGURE 25.5 2D self-organizing maps of ACO-MLP class decision boundaries and classification accuracy (upper right of each panel) for the AMLALL2, Brain2, Breast2A, Breast2B, Colon2, Lung2, MLL_Leukemia3, Prostate2, and SRBCT4 datasets based on mean-zero standardized input features; integers in the upper right of each panel represent the number of generation employed during training: (a) AMLALL2; (b) Brain2; (c) Breast2A; (d) Breast2B; (e) Colon2; (f) Lung2; (g) MLL_Leukemia3; (h) Prostate2; (i) SRBCT4. The numbers of genes used for each dataset are listed in Table 12.10.

during training. Classification accuracy results obtained from the ACO-MLP classifier were consistently (100%) for all of the datasets with the exception of the Brain2 and Colon2 datasets. On several occasions, performance results improved when the number of generations approached 500. Recall, however, that the continuous version of ACO that was used for connection weight learning was designed to employ only two chromosomes, and thus, the greater the number of generations used, the better the expected results.

25.6 SUMMARY

Evolutionary algorithm optimizers are global optimization methods and scale well to higher-dimensional problems. They are robust with respect to noisy evaluation functions, and the handling of evaluation functions, which do not yield a sensible result in given period of time. Although ES has existed as long as GA, there has been considerably more systematic investigation of GA in biomedicine. GA essentially involves random generation of binary strings, conversion from binary to decimal to obtain real-coded parameters, followed by selection, recombination, and mutation of the most fit chromosomes. On the other hand, ES requires calculation of the covariance matrix either by diagonalization with eigenanalysis or factorization with Cholesky factorization, simulation of normally distributed mutation vectors, followed by recombination of real-valued parameters from the most fit chromosomes, adaptation of the covariance matrix, and long-term cumulation. ES also requires an in-depth understanding of matrix algebra and associated geometric perception of the quasi-Newton methods implemented for adapting the covariance matrix; hence, more computational tools from linear algebra are required. GA does not involve matrix multiplication, transpose, eigenanalysis, factorization, and simulation of normally distributed vectors, and can therefore be implemented more rapidly than ES. At most, the simulation in GA merely involves rounding a uniform random variate in the range $[0,1]$ into the set $\{0,1\}$. PSO also provided strongly positive performance results and fundamentally learned at a greater rate when compared with the 2-chromosome ACO metaheuristic that was employed.

REFERENCES

[1] C. Blum, A. Roli. Metaheuristics in combinatorial optimization: Overview and conceptual comparison. *ACM (Association for Computing Machinery) Comput. Surv.* **35**(3):268–308, 2003.

[2] J.H. Holland. Outline for a logical theory of adaptive systems. *J. ACM* **9**:297–314, 1962.

[3] J.H. Holland. *Adaptation in Natural and Artificial Systems.* Univ. Michigan Press, Ann Arbor, 1975.

[4] L.J. Fogel. Autonomous automata. *Industr. Res.* **4**:14–19, 1962.

[5] L.J. Fogel, A.J. Owens, M.J. Walsh. *Artificial Intelligence through Simulated Evolution*, Wiley, New York, 1966.

[6] I. Rechenberg. *Cybernetic Solution of Path of an Experimental Problem.* Royal Aircraft Establishment, Farnborough, Library Translation 1122, 1965.

[7] I. Rechenberg. *Evolution Strategy: Optimization of Technical Systems According to Principles of Biological Evolution.* Frommann-Holzboog Verlag, Stuttgart, 1973.

[8] H-P. Schwefel. *Cybernetic Evolution as Experimental Research Strategy in Fluid Mechanics*, Master's Thesis, Technical Univ. Berlin, 1965.

[9] H-P. Schwefel. *Evolutionary Strategy and Numerical Optimization.* Dissertation, Techincal Univ. Berlin, 1975.

[10] S. Kirkpatrick, C.D. Gelatt, M.P. Vecchi. Optimization by simulated annealing. *Science* **220**:671-680, 1983.

[11] F. Glover. Future paths for integer programming and links to artificial intelligence. *Comput. Oper. Res.* **13**(5):533-549, 1986.

[12] H. Ramalhinho-Lourenco, O. Martin, T. Stützle. Iterated local search. In F. Glover, G. K. Herausgeber, eds., *Handbook of Metaheuristics*, Kluwer Academic Publishers, Norwell, MA, 2002, pp. 321-353.

[13] J. Kennedy, R.C. Eberhart. Particle swarm optimization. *Proc. IEEE Int. Conf. Neural Networks*, IEEE Press, Piscataway, NJ, 1995, pp. 1942-1948.

[14] M. Dorigo. *Optimization, Learning and Natural Algorithms* (in Italian). Dissertation, Dept. Electronics, Politecnico di Milano, 1992.

[15] K. Socha, M. Dorigo, Ant colony optimization for continuous domains. *Eur. J. Oper. Res.* **185**:1155-1173, 2008.

[16] K. Hornik, M. Stinchcombe, H. White. Multilayer feedforward networks are universal approximators. *Neural Networks* **2**(5):359-366, 1989.

[17] D.A. Goldberg. *Genetic Algorithms in Search, Optimization, and Machine Learning.* Addison-Wesley, 1989.

[18] K.A. De Jong. Are genetic algorithms function optimizers? In R. Manner B. Manderick, eds., *Parallel Problem Solving from Nature*, vol. 2, North-Holland, Amsterdam, 1992, pp. 3-13.

[19] H.-G. Beyer, H.-P. Schwefel. Evolution strategies: A comprehensive introduction. *J. Nat. Comput.* **1**(1):3-52, 2002.

[20] J.H. Holland. *Adaptation in Natural and Artificial Systems.* MIT Press, Cambridge, MA, 1992.

[21] F. Zare-Mirakabad, H. Ahrabian, M. Sadeghi, S. Hashemifar, A. Nowzari-Dalini, B. Goliaei. Genetic algorithm for dyad pattern finding in DNA sequences. *Genes Genet. Syst.* **84**(1):81-93, 2009.

[22] A.M. Abu Hammad, M.O. Taha. Pharmacophore modeling, quantitative structure-activity relationship analysis, and shape-complemented in silico screening allow access to novel influenza neuraminidase inhibitors. *J. Chem. Inform. Model.* **49**(4):978-996, 2009.

[23] X. Zhou, H. Wang, J. Wang, Y. Wang, G. Hoehn, J. Azok, M.L. Brennan, S.L. Hazen, K. Li, S.F. Chang, S.T. Wong. Identification of biomarkers for risk stratification of cardiovascular events using genetic algorithm with recursive local floating search. *Proteomics* **9**(8):2286-2294, 2009.

[24] N. Kilic, O.N. Ucan, O. Osman. Colonic polyp detection in CT colonography with fuzzy rule based 3D template matching. *J. Med. Syst.* **33**(1):9-18, 2009.

[25] F. Chen, R.L. Johnston. Energetic, electronic, and thermal effects on structural properties of Ag-Au nanoalloys. *ACS Nano.* **2**(1):165–175, 2008.

[26] Y. Xia, L. Wen, S. Eberl, M. Fulham, D. Feng. Genetic algorithm-based PCA eigenvector selection and weighting for automated identification of dementia using FDG-PET imaging. *Conf. Proc. IEEE Eng. Med. Biol. Soc.* **2008**:4812–4815, 2008.

[27] M.J. Po, A.F. Laine. Leveraging genetic algorithm and neural network in automated protein crystal recognition. *Conf. Proc. IEEE Eng. Med. Biol. Soc.* **2008**:1926–1929, 2008.

[28] M.R. Maurya, S.J. Bornheimer, V. Venkatasubramanian, S. Subramaniam. Mixed-integer nonlinear optimisation approach to coarse-graining biochemical networks. *IET Syst. Biol.* **3**(1):24–39, 2009.

[29] R. Gil-Pita, X. Yao. Evolving edited k-nearest neighbor classifiers. *Int. J. Neural Syst.* **18**(6):459–67, 2008.

[30] A. Ostermeier, A. Gawelczyk, N. Hansen. A derandomized approach to self-adaptation of evolution strategies. *Evol. Comput.* **4**(2):369–380, 1994.

[31] J.W. Wilson, P. Schlup, M. Lunacek, D. Whitley, R.A. Bartels. Calibration of liquid crystal ultrafast pulse shaper with common-path spectral interferometry and application to coherent control with a covariance matrix adaptation evolutionary strategy. *Rev. Sci. Instrum.* **79**(3):033103, 2008.

[32] B. Mersch, T. Glasmachers, P. Meinicke, C. Igel. Evolutionary optimization of sequence kernels for detection of bacterial gene starts. *Int. J. Neural Syst.* **17**(5):369–381, 2007.

[33] C. Igel, N. Hansen, S. Roth. Covariance matrix adaptation for multi-objective optimization. *Evol. Comput.* **15**(1):1–28, 2007.

[34] K. Bush, J. Knight, C. Anderson. Optimizing conductance parameters of cortical neural models via electrotonic partitions. *Neural Networks* **18**(5–6):488–496, 2005.

[35] M. Milano, P. Koumoutsakos, J. Schmidhuber. Self-organizing nets for optimization. *IEEE Trans. Neural Networks*, **15**(3):758–765, 2004.

[36] K. Deb, A. Anand, D. Joshi. A computationally efficient evolutionary algorithm for real-parameter optimization. *Evol. Comput.*, **10**(4):371–395, 2002.

[37] N. Hansen, A. Ostermeier. Completely derandomized self-adaptation in evolution strategies. *Evol. Comput.*, **9**(2):159–195, 2001.

[38] N. Hansen, S.D. Muller, P. Koumoutsakos. Reducing the time complexity of the derandomized evolution strategy with covariance matrix adaptation (CMA-ES). *Evol. Comput,.* **11**(1):1–18, 2003.

[39] H-P. Schwefel. *Numerical Optimization of Computer Models.* John Wiley, Chichester (UK), 1981.

[40] N. Hansen, A. Ostermeier. Adapting arbitrary normal mutation distributions in evolutionary strategies. The covariance matrix adaptation. *Proc. IEEE Int. Conf. Evolutionary Computation*, IEEE Press, Piscataway, NJ, 1996, pp. 312-317.

[41] C. Igel, T. Suttorp, N. Hansen. A computational efficient covariance matrix update and a (1+1)-CMA for evolution strategies. *Proc. Genetic Evolutionary Computation Conf.* (GECCO-2006), ACM Press, New York, 2006.

[42] H-G. Beyer, B. Sendhoff. Covariance matrix adaptation revisted: The CMSA evolution strategy. *Lecture Notes Comput. Sci.* **5199**:123-132, 2008.

[43] J. Kennedy, R.C. Eberhart, Particle swarm optimization. *Proc. IEEE Int. Conf. Neural Networks.* IEEE Press, Piscataway, NJ, 1995, pp. 1942-1948.

[44] Y.P. Zhou, L.J. Tang, J. Jiao, D.D. Song, J.H. Jiang, R.Q. Yu. Modified particle swarm optimization algorithm for adaptively configuring globally optimal classification and regression trees. *J. Chem. Inform. Model.* **49**(5):1144-1453, 2009.

[45] J.H. Jeng, C.C. Tseng, J.G. Hsieh. Study on Huber fractal image compression. *IEEE Trans. Image Process.* **18**(5):995-1003, 2009.

[46] T. Ince, S. Kiranyaz, M. Gabbouj. Automated patient-specific classification of premature ventricular contractions. *Conf. Proc. IEEE Eng. Med. Biol. Soc.* **2008**:5474-5477, 2008.

[47] J. Wang, X. Huo, M. Ghovanloo. A quadratic particle swarm optimization method for magnetic tracking of tongue motion in speech disorders. *Conf. Proc. IEEE Eng. Med. Biol. Soc.* **2008**:4222-4225, 2008.

[48] F. Yi, Z. Chongxun, P. Chen, L. Li. White blood cell image segmentation using on-line trained neural network. *Conf. Proc. IEEE Eng. Med. Biol. Soc.* **6**:6476-6479, 2005.

[49] B. Liu, L. Wang, Y.H. Jin. An effective PSO-based memetic algorithm for flow shop scheduling. *IEEE Trans. Syst. Man Cybernet. B Cybernet.* **37**(1):18-27, 2007.

[50] Q. Shen, J.H. Jiang, C.X. Jiao, S.Y. Huan, G.L. Shen, R.Q. Yu. Optimized partition of minimum spanning tree for piecewise modeling by particle swarm algorithm. QSAR studies of antagonism of angiotensin II antagonists. *J. Chem. Inform. Comput. Sci.* **44**(6):2027-2031, 2004.

[51] J. Liu, Z. Li, X. Hu, Y. Chen. Biclustering of microarray data with MOSPO based on crowding distance. *BMC Bioinform.* **10**(Suppl. 4):S9, 2009.

[52] L.E. Peterson, M.A. Coleman. Machine learning-based receiver operating characteristic (ROC) curves for crisp and fuzzy classification of DNA microarrays in cancer research. *Int. J. Approx. Reason.* **47**(1):17-36, 2008.

[53] R. Xu, G.C. Anagnostopoulos, D.C. Wunsch, II. Multiclass cancer classification using semisupervised ellipsoid ARTMAP and particle swarm optimization with gene expression data. *IEEE/ACM Trans. Comput. Biol. Bioinform.* **4**(1):65-77, 2007.

[54] I. De Falco, A. Della Cioppa, E. Tarantino. Evaluation of particle swarm optimization effectiveness in classification *Lecture Notes Comput. Sci.* **3849**:164-171, 2006.

[55] D. Martens, M. De Backer, R. Haesen, J. Vanthienen, M. Snoeck, B. Baesens. Classification with ant colony optimization. *IEEE Trans. Evol. Comput.* **11**(5):651-665; 2007.

[56] R.S. Parpinelli, H.S. Lopes, A.A. Freitas. Data mining with an ant colony optimization algorithm. *IEEE Trans. Evol. Comput.* **6**(4):321-332, 2002.

[57] B. Liu, H.A. Abbass, R. McKay. Classification rule discovery with ant colony optimization. *IEEE Comput. Intell. Bull.* **3**(1):31-35; 2004.

[58] D. Fadda, E. Slezak, A. Bijaoui. *Astron. Astrophys. Suppl. Ser.* **127**:335-352, 1998.

[59] B.W. Silverman. *Density Estimation for Statistics and Data Analysis*, Chapman & Hall, New York, 1986.

[60] V.A. Epanechnikov. *Theor. Probab. Appl.* **14**:163, 1969.

CHAPTER 26

SUPERVISED NEURAL GAS

26.1 INTRODUCTION

Supervised neural gas (SNG) is an extension of UNG, which employs microarray class labels for class prediction. More recent implementations of SNG include histologic grading of astrocytomas [1], remote sensing [2], handwritten digit sampling [3], proteomics mass spectroscopy [4,5], and clustering [6]. Heinke has evaluated and benchmarked SNG against other classifiers [7]. Class prediction with SNG requires two modifications of the UNG algorithm. The first modification requires that K prototypes be generated for each class. One of the most common ways to create K prototypes per class is to use, for example, K-means cluster analysis on arrays within each class in order to yield K centers. It is also assumed that K is the same for all classes. The second fundamental change requires that adaptive learning occurs when a prototype has the same class label as the training array being considered. This stipulation is carried out by introducing the class of the prototype $\mathbf{v}_m$ into the neighborhood function $h_\lambda(\omega_m, \mathbf{v}_m, \mathbf{x}_i)$. Recall that the neighborhood function of UNG for class discovery used only $h_\lambda(\mathbf{v}_m, \mathbf{x}_i)$, which did not depend on the class of the prototypes or training arrays.

Let $\mathbf{x}_i$ be the ith array $(i = 1, 2, \ldots, n)$ and $\mathbf{v}_m$ $(m = 1, 2, \ldots, M)$ be a prototype with the same number of features as input arrays. In addition, let Ω be the number of classes with individual class labels $(\omega = 1, 2, \ldots, \Omega)$, ω_i be the class label for array $\mathbf{x}_i$, and ω_m be the class label for prototype $\mathbf{v}_m$ obtained with K-means cluster analysis of arrays in class ω. Here, the M prototypes are based on the use of K-means cluster analysis of training arrays within each class, yielding K centers per class, times the number of classes: $M = K\Omega$.

During learning, each array is selected and the prototypes $\mathbf{v}$ are rank-ordered according to their distance from the selected array. Adaptive learning occurs only if the class label of the rank-ordered prototype is the same as the

Classification Analysis of DNA Microarrays, First Edition. Leif E. Peterson.

class label of the array, based on the relationship

$$\mathbf{v}_m = \mathbf{v}_m + \alpha(t)h_\lambda(\omega_m, \mathbf{v}_m, \mathbf{x}_i)(\mathbf{x}_i - \mathbf{v}_m) \qquad \omega_m = \omega_i, \tag{26.1}$$

where $\alpha(t)$ is the learning rate at the tth iteration $(t = 1, 2, \ldots, T)$ and $(\mathbf{v}_m - \mathbf{x}_i)$ is the vector difference between prototype $\mathbf{v}_m$ and array $\mathbf{x}_i$. The SNG *neighborhood function* $h_\lambda(\omega_m, \mathbf{v}_m, \mathbf{x}_i)$ is functionally expressed as

$$h_\lambda(\omega_m, \mathbf{v}_m, \mathbf{x}_i) = \exp\left(\frac{-R_k}{\lambda(t)}\right), \tag{26.2}$$

where R_k $(R_k = 0, 1, \ldots, K-1)$ is the rank index of prototype $\mathbf{v}_m$ with class label ω_m based on distance from array $\mathbf{x}_i$ and $\lambda(t)$ is the width. For example, $R_k = 0$ is the closest prototype whose class label ω_m equals the class label ω_i of array $\mathbf{x}_i$, $R_k = 1$ is the second closest prototype whose class label ω_m equals the class label ω_i of array $\mathbf{x}_i$, and so on. Suitable choices for the learning rate and width are

$$\alpha(t) = \alpha_0 \left(\frac{\alpha_f}{\alpha_0}\right)^{t/T}, \tag{26.3}$$

and

$$\lambda(t) = \lambda_0 \left(\frac{\lambda_f}{\lambda_0}\right)^{t/T}. \tag{26.4}$$

This is then carried out over T total iterations. Martinetz et al. [8] suggested using values of $\alpha_0 = 0.5$, $\alpha_f = 0.005$, $\lambda_0 = 10$, $\lambda_f = 0.01$, $T = 100{,}000$.

26.2 ALGORITHM

Algorithm 26.1 lists the procedural calculations of the SNG algorithm.

26.3 CROSS-VALIDATION RESULTS

The effects of different cross-validation (CV) runs were evaluated for SNG using the nine datasets and specified number of arrays, classes, and number of filtered genes listed in Table 12.10. Bootstrap accuracy ("CVB") was based on average 0.632 bootstrap accuracy (13.4) for 10 bootstraps. Table 26.1 lists the average accuracy as a function of K prototypes used per class with no input feature transformations applied. Table 26.2 lists the average accuracy using mean-zero standardized input feature values. Table 26.3 lists the average accuracy using fuzzified input feature values. Overall, the mean accuracy values with and without input feature mean-zero standardization were similar, and fuzzification reduced performance only slightly. The number of prototypes K employed per class also did not seem to strongly impact SNG classification performance. In addition, Figure 26.1 illustrates accuracy as a function of CV method for each dataset when no feature transformations were applied. Figure 26.2 shows standard deviation of accuracy as a function

Algorithm 26.1: Supervised Neural Gas

Data: p-dimensional input arrays $\mathbf{x}_i$ ($i = 1, 2, \ldots, n$), M prototypes $\mathbf{V} = \{\mathbf{v}_1, \mathbf{v}_2, \ldots, \mathbf{v}_M\}$
Result: Trained prototypes
Specify the number of prototypes $M = K\Omega$
Initialize $\alpha_0 = 0.5$, $\alpha_f = 0.005$, $\lambda_0 = 10$, $\lambda_f = 0.01$, $T = 100{,}000$
Initialize M reference vectors $\mathbf{V} = \{\mathbf{v}_1, \mathbf{v}_2, \ldots, \mathbf{v}_M\}$
for $t \leftarrow 1$ **to** T **do**
  $\alpha(t) = \alpha_0(\alpha_f/\alpha_0)^{t/T}$
  $\lambda(t) = \lambda_0(\lambda_f/\lambda_0)^{t/T}$
  for $i \leftarrow 1$ **to** n **do**
    Determine prototype ranks $R_1 < R_2 < \cdots < R_K$ based on ascending $d(\mathbf{v}_m, \mathbf{x}_i)$ for prototypes whose class label ω_m is equal to the array class label ω_i (*Note*: $R_1 = 0, R_2 = 1, \ldots, R_K = K - 1$)
    for $m \leftarrow 1$ **to** M **do**
      $h_\lambda(\omega_m, \mathbf{v}_m, \mathbf{x}_i) = \exp(-R_k/\lambda(t)) \quad \omega_m = \omega_i$
      $\mathbf{v}_m = \mathbf{v}_m + \alpha(t)h_\lambda(\omega_i, \mathbf{v}_m, \mathbf{x}_i)(\mathbf{x}_i - \mathbf{v}_m)$
    endfor
  endfor
endfor
Assign class of each test array to the class of its closest trained prototype

of CV method for each dataset with no feature transformations applied. CV results suggest that there is little bias in SNG because the performance did not change appreciably with varying levels of data used for training and testing. However, there is considerable variance because of the wide variation in results across the nine datasets used. Figure 26.3 reveals average accuracy for all datasets as a function of feature transformation and CV method, and reveals that there was a negligible effect of feature transformation method on performance. Figure 26.4 reveals the SNG accuracy without transformations on features as a function of K, the number of prototypes used per class. Figure 26.5 shows the SNG accuracy with feature standardization as a function of K, the number of prototypes used per class. Figure 26.6 illustrates the SNG accuracy with feature fuzzification as a function of K, the number of prototypes used per class. Figure 26.7 shows average accuracy for all datasets as a function of K, the number of prototypes used per class. There were negligible effects observed for the number of prototypes K employed per class. The genes used in the datasets considered were also preselected and offered the best level of class separability, so we did not expect significant changes in performance with increasing values of K. In other words, use of a smaller number of prototypes per class was sufficient for achieving high performance.

TABLE 26.1 SNG Performance without Feature Transformation

		Cross-validation method				
Dataset	K^a	CV2	CV5	CV10	CV – 1	CVB
AMLALL2	2	100	100	100	100	100
	3	99.74	100	100	100	99.97
	4	100	100	100	100	100
	5	100	100	100	100	99.98
	6	100	100	100	100	100
Brain2	2	81.00	81.33	80.50	85.00	87.93
	3	77.67	82.00	82.33	81.67	88.22
	4	82.33	82.17	81.00	85.00	88.36
	5	81.83	81.00	83.33	78.33	88.46
	6	80.83	82.17	81.33	81.67	88.32
Breast2A	2	100	100	100	100	100
	3	100	100	100	100	100
	4	100	100	100	100	100
	5	100	100	100	100	99.95
	6	100	100	100	100	100
Breast2B	2	86.92	86.67	86.92	89.74	89.63
	3	85.77	86.41	87.44	88.46	89.68
	4	86.79	87.05	87.31	85.90	89.26
	5	86.79	87.31	87.56	88.46	88.95
	6	86.67	85.13	87.05	87.18	89.62
Colon2	2	87.74	87.26	87.90	90.32	88.16
	3	86.61	87.58	87.10	88.71	88.20
	4	86.61	88.23	87.58	88.71	88.24
	5	87.10	87.10	87.74	88.71	88.30
	6	87.74	87.74	86.94	87.10	88.11
Lung2	2	100	100	100	100	100
	3	100	100	100	100	100
	4	100	100	100	100	100
	5	100	100	100	100	100
	6	100	100	100	100	100
MLL_Leukemia3	2	99.12	98.95	98.77	100	99.56
	3	99.12	98.95	98.60	100	99.67
	4	98.95	98.77	98.77	98.25	99.48
	5	99.30	98.60	98.95	98.25	99.73
	6	98.95	98.42	98.95	100	99.51
Prostate2	2	94.31	94.80	94.90	95.10	94.92
	3	94.22	94.71	94.71	95.10	94.74
	4	94.51	94.31	95.10	95.10	94.91
	5	94.71	94.71	95.00	94.12	94.75
	6	93.53	94.71	94.41	95.10	94.82
SRBCT4	2	99.84	100	100	100	99.95
	3	99.21	100	100	100	99.97
	4	99.37	100	100	100	99.94
	5	99.37	100	100	100	99.97
	6	100	100	100	100	99.97
Average		0.94	0.94	0.94	0.95	0.96

[a]The term K represents the number of prototypes used per class (here and in Tables 26.2 and 26.3 also).

TABLE 26.2 SNG Performance with Feature Standardization

		Cross-validation method				
Dataset	K	CV2	CV5	CV10	CV − 1	CVB
AMLALL2	2	100	100	100	100	100
	3	100	100	100	100	100
	4	100	100	100	100	100
	5	99.74	100	100	100	100
	6	100	100	100	100	100
Brain2	2	82.00	82.00	83.00	78.33	89.20
	3	81.00	82.17	81.50	81.67	88.66
	4	80.67	81.50	81.83	76.67	88.44
	5	79.67	81.67	82.50	78.33	87.94
	6	78.67	84.33	82.50	83.33	88.38
Breast2A	2	100	100	100	100	100
	3	100	100	100	100	100
	4	100	100	100	100	100
	5	100	100	100	100	100
	6	100	100	100	100	100
Breast2B	2	86.79	86.03	86.79	88.46	89.44
	3	84.23	85.13	87.44	89.74	89.53
	4	84.23	85.13	87.44	89.74	89.53
	5	84.23	85.13	87.44	89.74	89.53
	6	84.23	85.13	87.44	89.74	89.53
Colon2	2	86.94	87.10	87.26	88.71	88.15
	3	85.97	86.94	88.39	88.71	88.14
	4	87.74	86.94	87.58	88.71	88.18
	5	86.94	86.94	87.58	88.71	88.11
	6	87.42	87.58	87.90	88.71	87.95
Lung2	2	100	100	100	100	100
	3	100	100	100	100	100
	4	100	100	100	100	100
	5	100	100	100	100	100
	6	100	100	100	100	100
MLL_Leukemia3	2	98.42	98.95	98.77	98.25	99.56
	3	98.42	98.95	98.95	98.25	99.44
	4	98.60	98.77	98.77	98.25	99.57
	5	99.30	98.07	98.77	98.25	99.70
	6	99.30	98.42	98.95	98.25	99.51
Prostate2	2	93.92	94.41	95.10	96.08	94.93
	3	94.22	94.61	94.90	96.08	94.96
	4	94.71	94.71	94.71	95.10	94.95
	5	94.51	94.71	94.51	95.10	94.80
	6	93.53	94.41	94.90	95.10	94.64
SRBCT4	2	100	100	100	100	99.98
	3	99.84	100	100	100	99.95
	4	99.84	100	100	100	99.98
	5	100	100	100	100	99.99
	6	99.84	100	100	100	99.97
Average		0.94	0.94	0.95	0.95	0.96

TABLE 26.3 SNG Performance with Feature Fuzzification

		Cross-validation method				
Dataset	K	CV2	CV5	CV10	CV − 1	CVB
AMLALL2	2	100	100	100	100	100
	3	100	100	100	100	100
	4	100	100	100	100	100
	5	100	100	100	100	100
	6	100	100	100	100	100
Brain2	2	80.00	78.17	75.67	78.33	86.03
	3	78.33	78.50	77.50	78.33	85.72
	4	77.17	77.00	78.67	76.67	85.91
	5	80.83	77.17	77.00	76.67	85.78
	6	79.33	79.17	78.17	78.33	86.19
Breast2A	2	100	100	100	100	100
	3	100	100	100	100	100
	4	100	100	100	100	100
	5	100	100	100	100	100
	6	100	100	100	100	100
Breast2B	2	82.18	83.59	84.62	84.62	88.02
	3	80.90	83.21	84.74	82.05	87.78
	4	81.03	83.33	84.36	78.21	87.54
	5	83.21	84.36	83.33	88.46	87.57
	6	81.67	83.72	84.10	87.18	87.63
Colon2	2	85.32	87.42	86.45	87.10	87.52
	3	85.00	87.58	86.61	85.48	87.61
	4	86.13	85.48	86.77	88.71	87.61
	5	83.55	86.94	86.77	87.10	87.69
	6	85.65	86.29	86.77	87.10	87.52
Lung2	2	100	100	100	100	100
	3	100	100	100	100	100
	4	100	100	100	100	100
	5	100	100	100	100	100
	6	100	100	100	100	100
MLL_Leukemia3	2	98.42	98.25	98.25	98.25	98.66
	3	98.07	98.25	98.25	98.25	98.49
	4	98.42	98.25	98.07	98.25	98.75
	5	98.60	98.25	98.25	98.25	98.60
	6	98.07	98.25	98.25	98.25	98.80
Prostate2	2	93.63	94.41	94.51	95.10	94.01
	3	93.63	94.12	94.41	94.12	94.13
	4	93.33	94.22	94.31	95.10	94.13
	5	93.92	94.02	94.02	94.12	94.23
	6	94.22	93.92	94.31	95.10	94.33
SRBCT4	2	98.73	100	100	100	99.97
	3	98.73	100	100	100	99.95
	4	99.84	99.84	100	100	99.92
	5	99.68	99.84	100	100	100
	6	99.84	99.84	100	100	99.94
Average		0.93	0.93	0.93	0.94	0.95

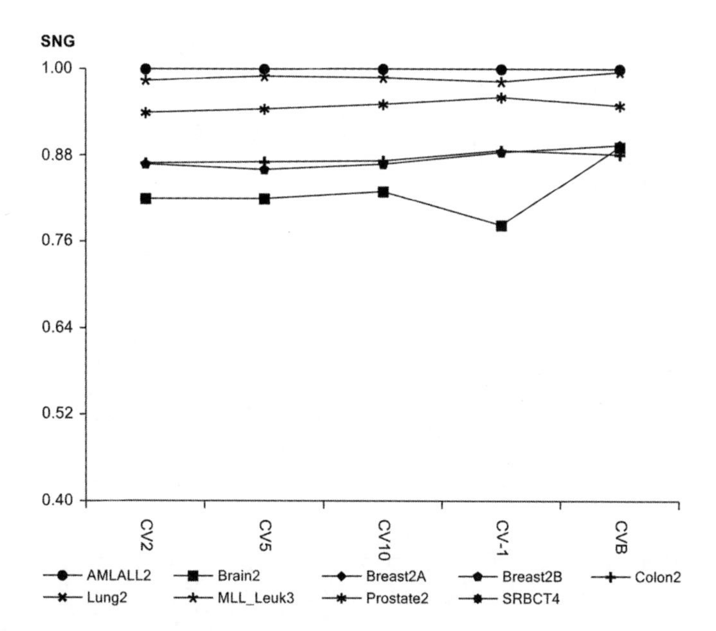

FIGURE 26.1 SNG accuracy as a function of cross-validation (CV) method for each dataset. Average based on CV accuracy for 10 repartitions of arrays. No feature transformations applied.

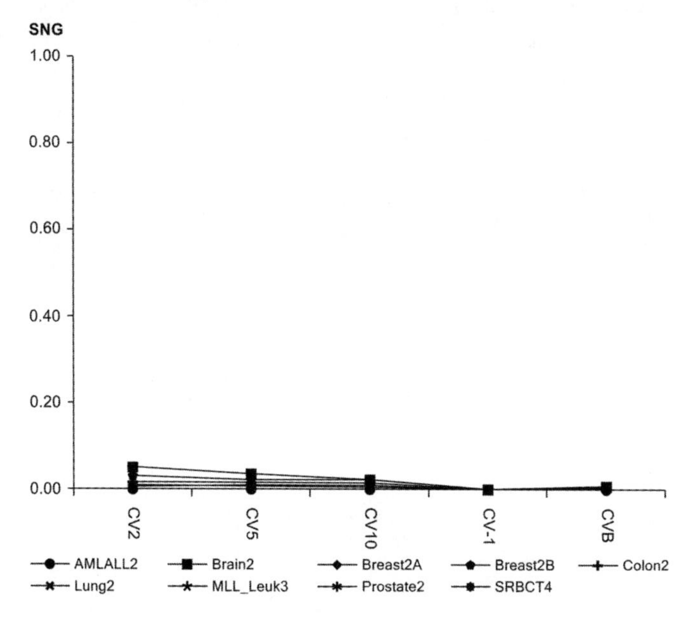

FIGURE 26.2 SNG standard deviation of accuracy as a function of CV method for each dataset. Standard deviation based on CV accuracy for 10 repartitions of arrays. No feature transformations applied.

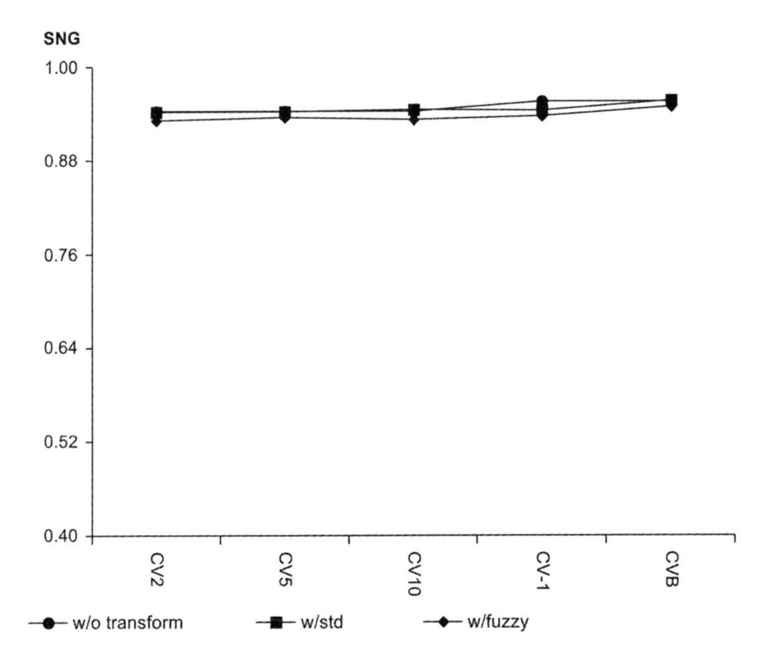

FIGURE 26.3 SNG average accuracy for all datasets as a function of feature transformation and CV method.

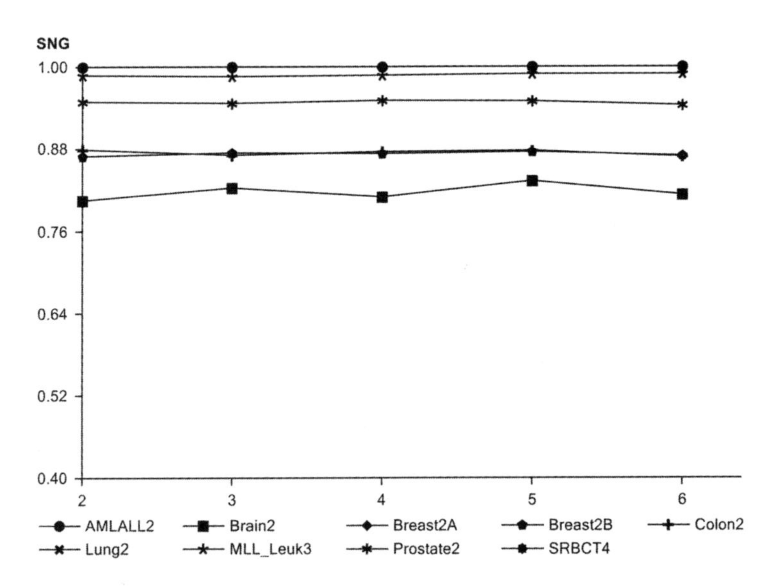

FIGURE 26.4 SNG accuracy without transformations on features as a function of K, the number of prototypes per class.

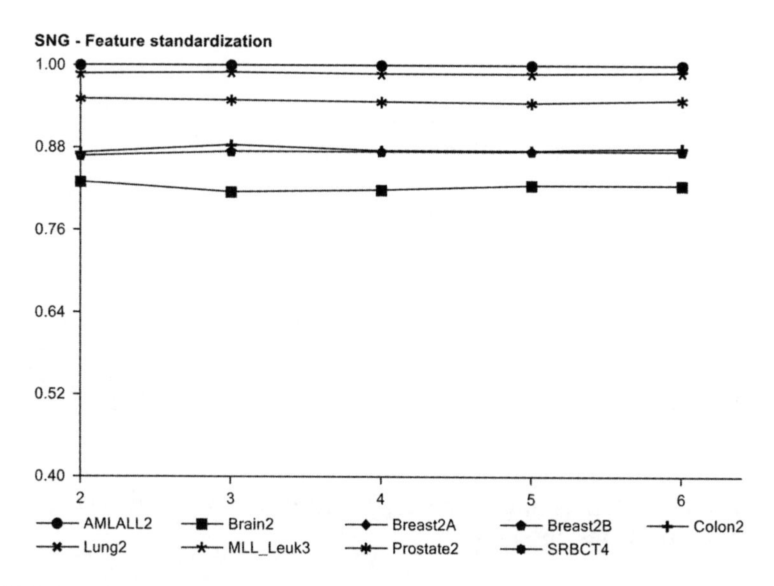

FIGURE 26.5 SNG accuracy with feature standardization as a function of *K*, the number of prototypes per class.

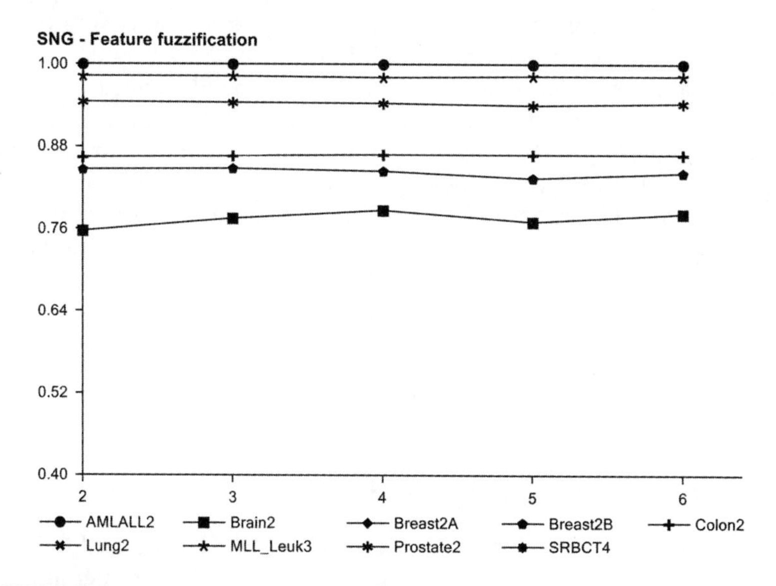

FIGURE 26.6 SNG accuracy with feature fuzzification as a function of *K*, the number of prototypes per class.

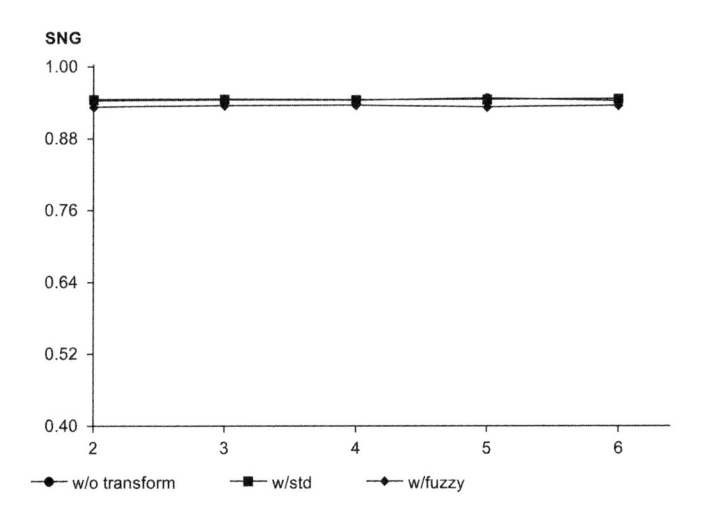

FIGURE 26.7 SNG average accuracy for all datasets as a function of K, the number of prototypes per class.

26.4 BOOTSTRAP BIAS

The bootstrap bias of SNG was determined using the same nine datasets. Bootstrap bias runs were based on determining the 0.632 bootstrap accuracy (see Eq. 13.4) for each sampling fraction. Acc_0 in 13.4 was based on test accuracy of unsampled arrays, while Acc_b in 13.4 was based on test accuracy for sampled arrays. Figure 26.8 shows the SNG accuracy without transformations on features as a function of bootstrap sample size used during training. The x-axis values represent the number of arrays randomly sampled $B = 40$ times at fractions $f = 0.1$, 0.2, 0.3, 0.4, 0.5, or 0.6 of the available arrays. Figure 26.9 illustrates the SNG accuracy with feature standardization as a function of bootstrap sample size used during training, and Figure 26.10 shows the SNG accuracy with input feature fuzzification. The bootstrap results observed indicate that feature fuzzification increased performance for lower sampling fractions of $f = 0.1$, 0.2, and 0.3. Nevertheless, there is still a noticeable degree of variance, due to the range of performance observed across the datasets.

26.5 MULTICLASS ROC CURVES

The area under the curve (AUC) was assessed for SNG using multiclass ROC curves. Figure 26.11 illustrates ROC curves for the SNG classifier applied to the AMLALL2, Brain2, Breast2A, Breast2B, Colon2, Lung2, MLL_Leukemia3, Prostate2, and SRBCT4 datasets using mean-zero standardized input features. The legend values represent the number of arrays randomly sampled

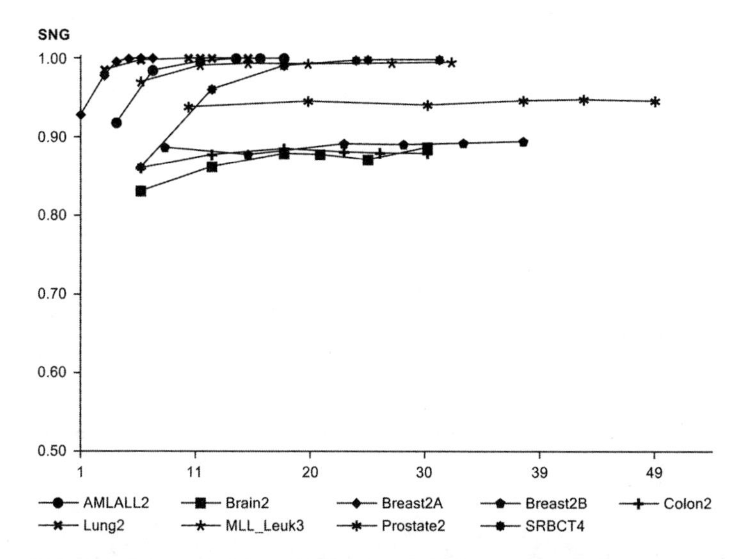

FIGURE 26.8 SNG accuracy without transformations on features as a function of bootstrap sample size used during training. The x-axis values represent the number of arrays randomly sampled $B = 40$ times at fractions $f = 0.1$, 0.2, 0.3, 0.4, 0.5, or 0.6 of the available arrays.

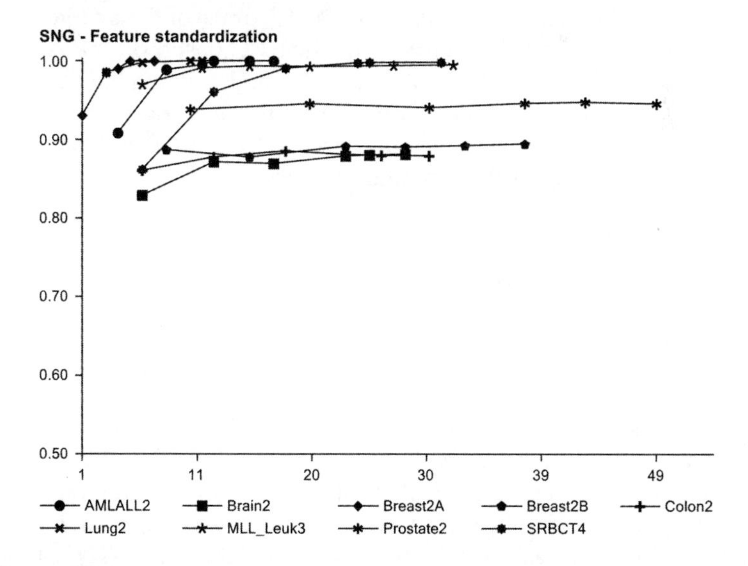

FIGURE 26.9 SNG accuracy with feature standardization as a function of bootstrap sample size used during training. The x-axis values represent the number of arrays randomly sampled $B = 40$ times at fractions $f = 0.1$, 0.2, 0.3, 0.4, 0.5, or 0.6 of the available arrays.

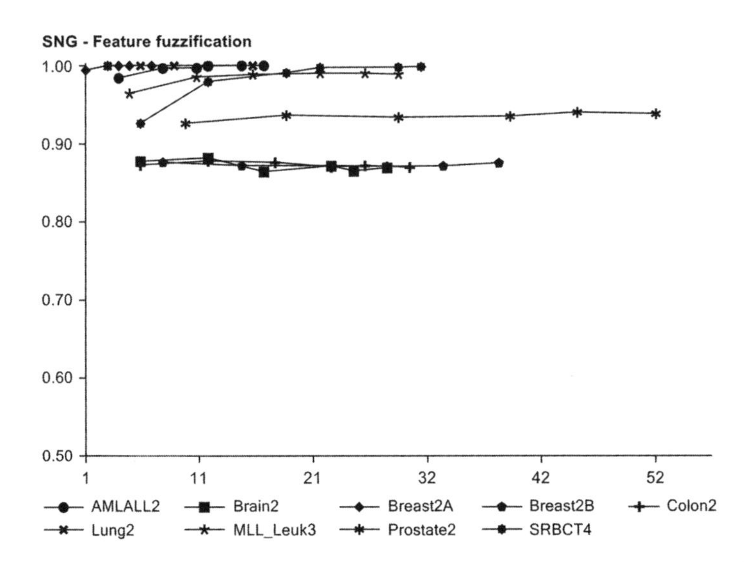

FIGURE 26.10 SNG accuracy with feature fuzzification as a function of bootstrap sample size used during training. The *x*-axis values represent the number of arrays randomly sampled $B = 40$ times at fractions $f = 0.1, 0.2, 0.3, 0.4, 0.5$, or 0.6 of the available arrays.

$B = 40$ times at fractions $f = 0.1, 0.2, 0.3, 0.4, 0.5$, or 0.6 of the available arrays. Figure 26.12 illustrates a ROC curves for the SNG based on the AMLALL2, Brain2, Breast2A, Breast2B, Colon2, Lung2, MLL_Leukemia3, Prostate2, and SRBCT4 datasets using fuzzified input features. SNG had lower values of AUC in the 80% range for the Brain2, Breast2B, and Colon2 datasets, and AUC in the 90% range for the remaining datasets.

26.6 CLASS DECISION BOUNDARIES

Class decision boundaries were identified by first running SNG on the arrays to extract the decision rules. The decision rules were then applied to each array as well as each reference vector of a 300 × 300 2D SOM in order to predict class membership. The class decision boundaries were then displayed on the SOM along with the arrays using class-specific symbols for the true class labels. Figure 26.13 illustrates 2D self-organizing maps of class decision boundaries for the SNG based on the AMLALL2, Brain2, Breast2A, Breast2B, Colon2, Lung2, MLL_Leukemia3, Prostate2, and SRBCT4 datasets using mean-zero standardized input features. Figure 26.14 illustrates decision boundaries for the same datasets when input features were fuzzified. A major conclusion concerning the data used is that use of additional prototypes did not increase SNG classification performance. Input feature fuzzification resulted in smoother decision boundaries, but without an accompanying increase in performance.

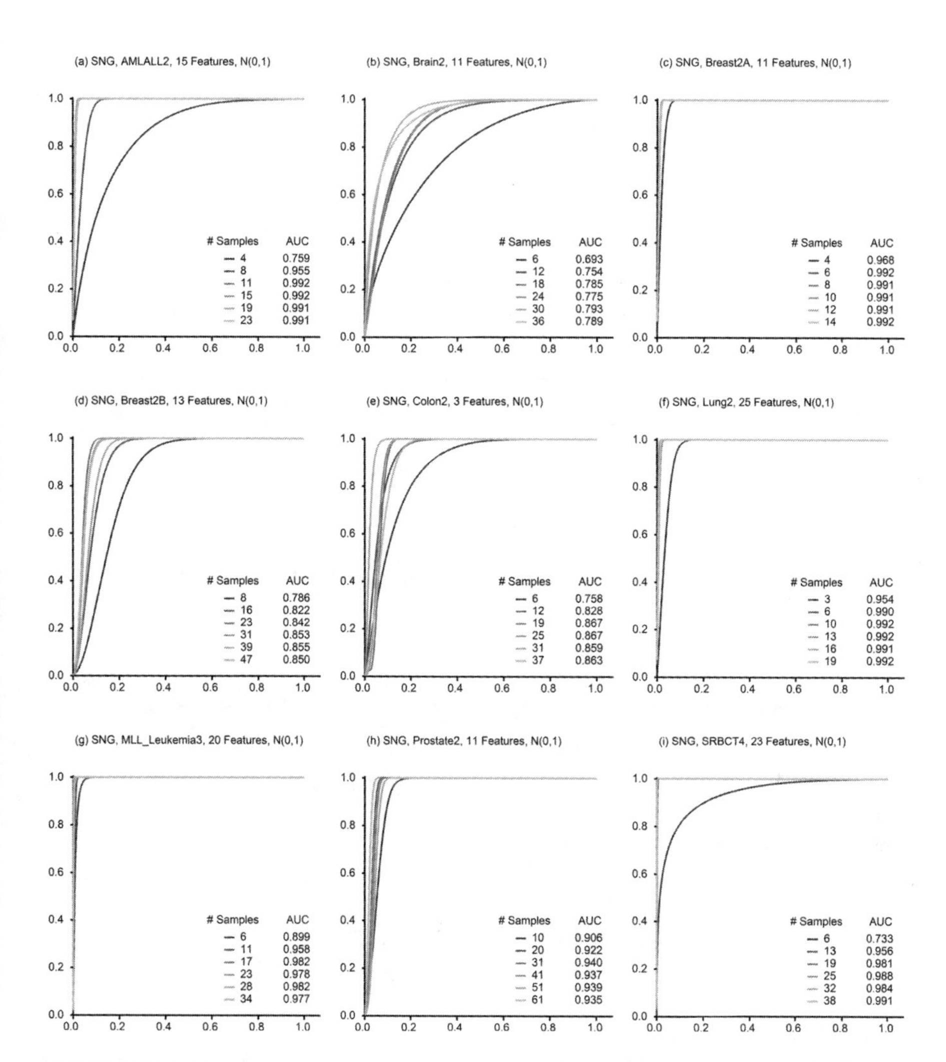

FIGURE 26.11 ROC curves for the SNG classifier applied to the AMLALL2, Brain2, Breast2A, Breast2B, Colon2, Lung2, MLL_Leukemia3, Prostate2, and SRBCT4 datasets based on mean-zero standardized input features; legend values represent the number of arrays randomly sampled $B = 40$ times at fractions $f = 0.1$, 0.2, 0.3, 0.4, 0.5, or 0.6 of the available arrays: (a) AMLALL2; (b) Brain2; (c) Breast2A; (d) Breast2B; (e) Colon2; (f) Lung2; (g) MLL_Leukemia3; (h) Prostate2; (i) SRBCT4. The numbers of genes used for each dataset are listed in Table 12.10.

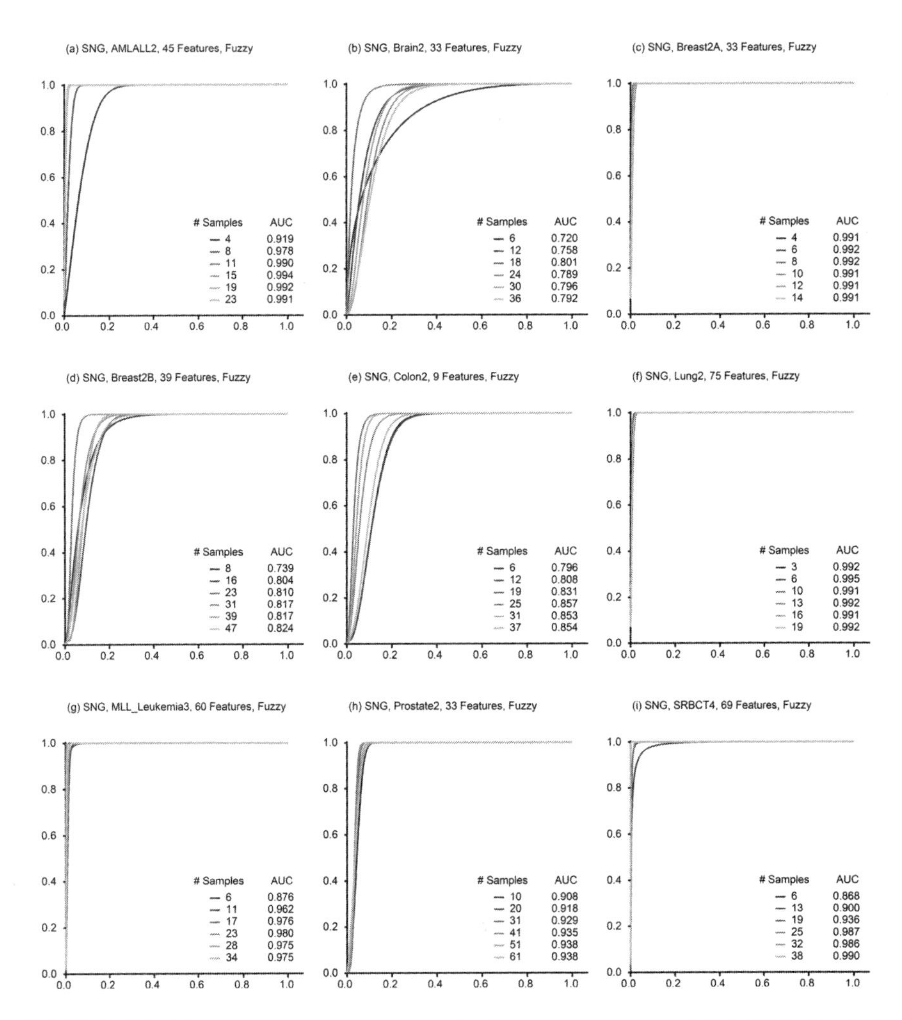

FIGURE 26.12 ROC curves for the SNG applied to the AMLALL2, Brain2, Breast2A, Breast2B, Colon2, Lung2, MLL_Leukemia3, Prostate2, and SRBCT4 datasets based on fuzzified input features; legend values represent the number of arrays randomly sampled $B = 40$ times at fractions $f = 0.1, 0.2, 0.3, 0.4, 0.5$, or 0.6 of the available arrays: (a) AMLALL2; (b) Brain2; (c) Breast2A; (d) Breast2B; (e) Colon2; (f) Lung2; (g) MLL_Leukemia3; (h) Prostate2; (i) SRBCT4. The numbers of genes used for each dataset are listed in Table 12.10.

26.7 SUMMARY

We observed the SNG classifier to have little bias and a large degree of variance, as the classification accuracy varied widely across the different

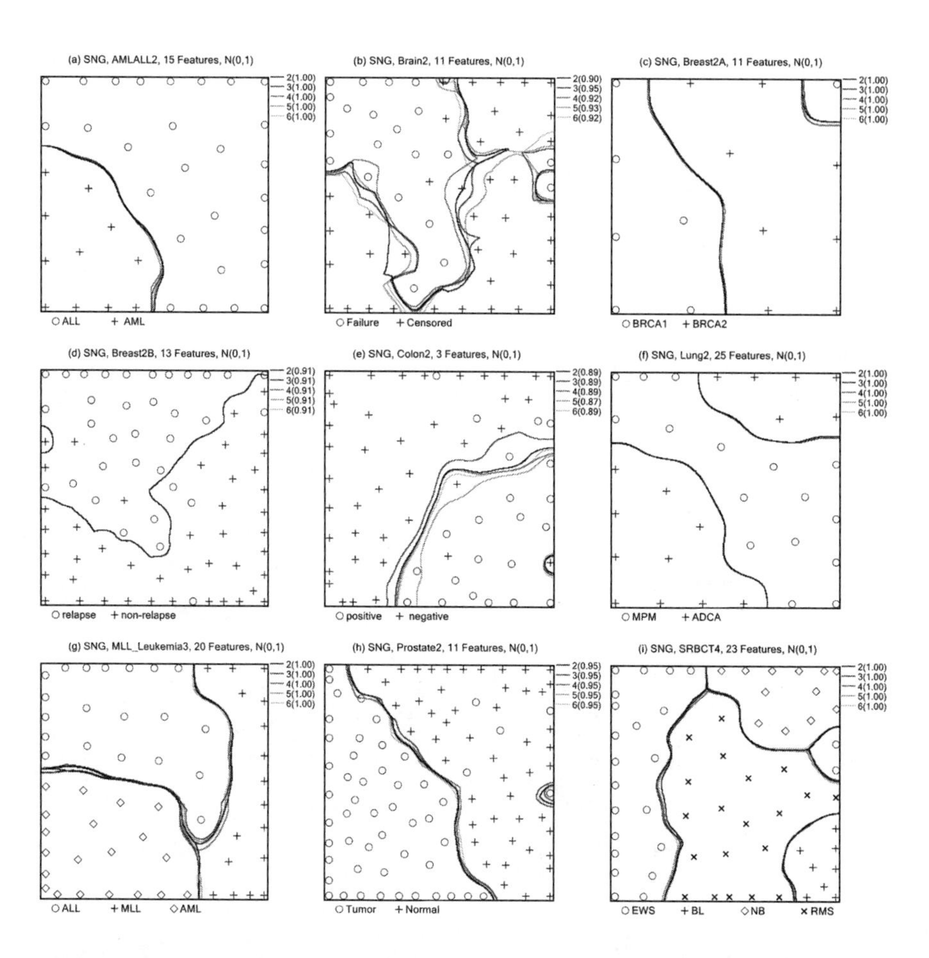

FIGURE 26.13 2D self-organizing maps of SNG class decision boundaries and classification accuracy (upper right of each panel) for the AMLALL2, Brain2, Breast2A, Breast2B, Colon2, Lung2, MLL_Leukemia3, Prostate2, and SRBCT4 datasets based on mean-zero standardized input features: (a) AMLALL2; (b) Brain2; (c) Breast2A; (d) Breast2B; (e) Colon2; (f) Lung2; (g) MLL_Leukemia3; (h) Prostate2; (i) SRBCT4. The numbers of genes used for each dataset are listed in Table 12.10. Integer values in upper right legend represent K number of prototypes used per class.

datasets used. The prefiltered input features that were used also provided maximum class separation between arrays, so we did not expect the addition of prototypes to enhance classification performance. Readers are encouraged to explore other approaches for SNG, such as fuzzy labeled neural gas [9] and divergence-based vector quantization [10].

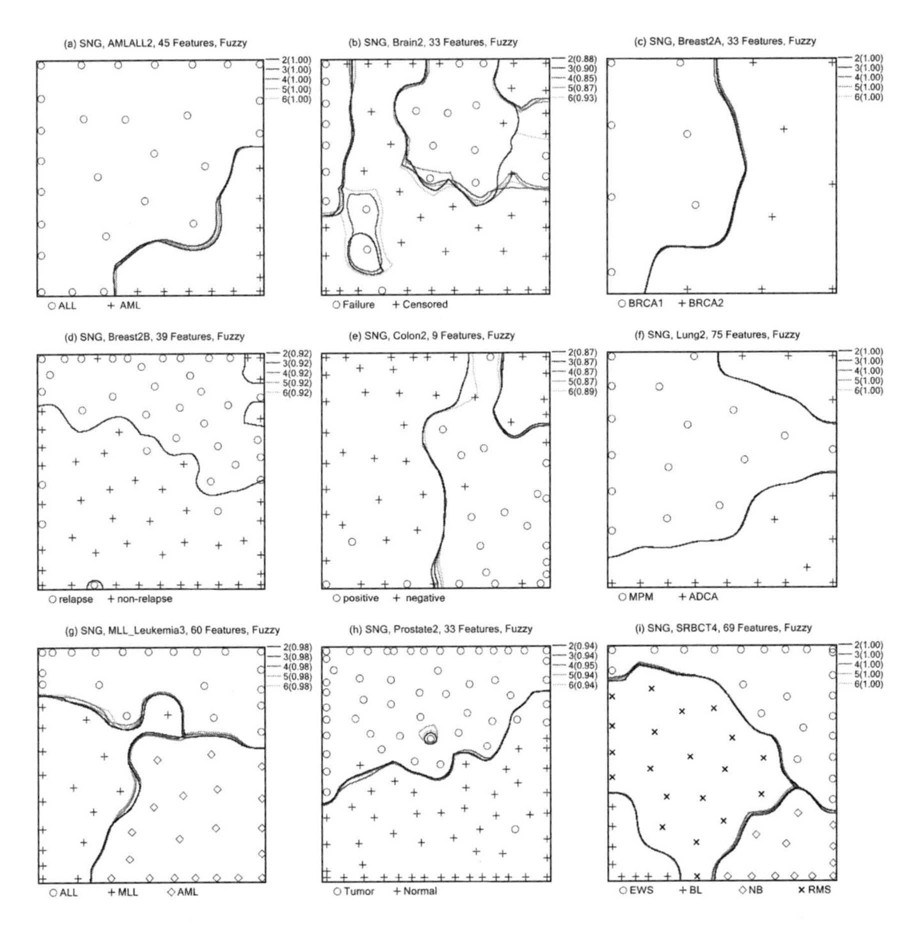

FIGURE 26.14 2D self-organizing maps of SNG class decision boundaries and classification accuracy (upper right of each panel) for the AMLALL2, Brain2, Breast2A, Breast2B, Colon2, Lung2, MLL_Leukemia3, Prostate2, and SRBCT4 datasets based on fuzzified input features: (a) AMLALL2; (b) Brain2; (c) Breast2A; (d) Breast2B; (e) Colon2; (f) Lung2; (g) MLL_Leukemia3; (h) Prostate2; (i) SRBCT4. The numbers of genes used for each dataset are listed in Table 12.10. Integer values in upper right legend represent *K* number of prototypes used per class.

REFERENCES

[1] N. Rohrl, J.R. Iglesias-Rozas, G. Weidl. A modern reproducible method for the histologic grading of astrocytomas with statistical classification tools. *Anal. Quant. Cytol. Histol.* **30**(1):33–38, 2008.

[2] G.S. Ruppert, M. Schardt, G. Balzuweit, M. Hussain. A hybrid classifier for remote sensing applications. *Int. J. Neural Syst.* **8**(1):63–68, 1997.

[3] A.S. Atukorale, P.N. Sugantham. An efficient neural gas network for classification. *Proc. Int. Conf. Control Automation Robotics and Vision (ICARCV98)*, 1998, pp. 1152–1156.

[4] T. Villmann, F-M. Schleif, M. Kostrzewa, A. Walch, B. Hammer. Classification of mass-spectrometric data in clinical proteomics using learning vector quantization methods. *Brief. Bioinform.* **9**(2):129–143, 2008.

[5] F.-M. Schleif, T. Villmann, B. Hammer. Supervised neural gas for functional data and its application to the analysis of clinical proteom spectra. *Proc. 9th Int. Conf. Artificial Neural Networks (IWANN2007)*, San Sebastian, Spain, June 20–22, 2007, Springer, Berlin, 2007, pp. 1036–1044.

[6] A. Jirayusakul, S. Auwatanamongkol. A supervised growing neural gas algorithm for cluster analysis. *Int. J. Hybrid Intell. Syst.* **4**(2):129–141, 2007.

[7] D. Heinke, F.H. Hamker. Comparing neural networks: A benchmark on growing neural gas, growing cell structures, and fuzzy ARTMAP. *IEEE Trans. Neural Networks* **9**(6):1279–1291, 1998.

[8] T. Martinetz, S. Berkovich, K. Schulten. Neural gas network for vector quantization and its application to time-series prediction. *IEEE Trans. on Neural Networks* **4**(4):558–568; 1993.

[9] T. Villmann, B. Hammer, F.-M. Schleif, T. Geweniger, W. Herrmann. Fuzzy classification by fuzzy labeled neural gas. *Neural Networks* **19**(6-7):772–779, 2006.

[10] T. Villmann, S. Haase. Divergence-based vector quantization. *Neural Comput.* **23**(5):1343–1392, 2011.

CHAPTER 27

MIXTURE OF EXPERTS

27.1 INTRODUCTION

Mixture of experts (MOE) networks perform classification by a "divide and conquer" approach to solve large difficult problems by breaking them up into smaller easier-to-solve problems [1-3]. MOE networks have been applied to biomolecular sequence similarity [4]; autoregressive time series [5]; diagnosis of leukemia [6]; breast cancer survival [7] and diagnosis [8]; EEG signal classification [9]; bloodflow signals [10]; classification of gender, ethnicity, and pose of human faces [11]; and clinical factors and gene markers [12].

In principle, MOE networks are structured, in the same way Kohonen's self-organizing maps are organized, where nodes establish a primacy over other nodes in certain areas of the input sample space. Depending on the region of the input sample space, the network output gives greater weight to results if the microarrays are closer to the specific "expert" or modular component that focuses on a specific area of the sample space. The MOE architecture consists of a *gating* network and an *expert* network (see Figure 7.1). An MOE classifier first partitions the input sample space with, for example, K-means cluster analysis and assigns an expert classifier to each region. Each expert network is responsible for class prediction using the assigned microarrays and the posterior class probabilities weighted by gating probabilities.

The basic structure of an MOE network is decomposed into the conditional probability for each microarray

$$p(\mathbf{x}|g,f) = \sum_{j}^{G} g_j(\mathbf{v},\mathbf{x}) f_i(\mathbf{w},\mathbf{x}), \tag{27.1}$$

where $g_j(\mathbf{v},\mathbf{x})$ is the gating probability for expert j and $f_i(\mathbf{w},\mathbf{x})$ is the probability of microarray $\mathbf{x}_i$ for expert j. Each gating probability is determined by use of

Classification Analysis of DNA Microarrays, First Edition. Leif E. Peterson.

the softmax function

$$g_j = \frac{g_j(\mathbf{x}^T\mathbf{v})}{\sum_k^G g_k(\mathbf{x}^T\mathbf{v})}. \tag{27.2}$$

For a linear regression *expert* model, $f_i(\mathbf{w}, \mathbf{x})$ is equal to

$$f_i(\mathbf{x}_i^T\mathbf{w}_j) = \frac{1}{(2\pi)^{\Omega/2}} \exp\left(-\frac{1}{2}||\hat{\mathbf{y}} - \mathbf{y}||\right), \tag{27.3}$$

where Ω is the number of output classes and $\hat{\mathbf{y}} = \mathbf{x}_i^T\mathbf{w}_j$.

The expectation-maximization (EM) algorithm [13,14] has been useful for maximizing the *complete* likelihood function

$$L(\mathbf{y}, \mathbf{z}, \mathbf{v}, \mathbf{w}) = \prod_{i=1}^{n}\prod_{j=1}^{G}[g_j(\mathbf{x}^T\mathbf{v})f_i(\mathbf{x}_i^T\mathbf{w}_j)]^{z_{ij}}, \tag{27.4}$$

where $z_{ij} = 1$ if microarray $\mathbf{x}_i$ is generated by expert j and $z_{ij} = 0$ otherwise. A less trivial and more easily interpretable form of $L(\mathbf{y}, \mathbf{z}, \mathbf{v}, \mathbf{w})$ is the loglikelihood in the form

$$\log(L(\mathbf{y}, \mathbf{z}, \mathbf{v}, \mathbf{w})) = \sum_{i=1}^{n}\sum_{j=1}^{G} z_{ij}[\log g_j(\mathbf{x}^T\mathbf{v}) + \log f_i(\mathbf{x}_i^T\mathbf{w}_j)]. \tag{27.5}$$

Since the z_{ij} are unobservable, the expectation of the complete loglikelihood is

$$\log(L((\mathbf{y}, \mathbf{z}, \mathbf{v}, \mathbf{w})) = \sum_{i=1}^{n}\sum_{j=1}^{G} \mathbf{E}(z_{ij}|\mathbf{y}_i)[\log g_j(\mathbf{x}^T; \mathbf{v}) + \log f_i(\mathbf{x}_i^T\mathbf{w}_j)], \tag{27.6}$$

where the expected values of the missing z_{ij} are determined with Bayes' rule as

$$\begin{aligned}
\mathbf{E}(z_{ij}|\mathbf{y}_i) &= \Pr(z_{ij} = 1|\mathbf{y}_i) \\
p_{ij} &= \frac{\Pr(\mathbf{y}_i|z_{ij} = 1)\Pr(z_{ij} = 1)}{\sum_k^G \Pr(\mathbf{y}_i|z_{ik} = 1)\Pr(z_{ik} = 1)} \\
&= \frac{g_j(\mathbf{x}^T\mathbf{v})f_i(\mathbf{x}_i^T\mathbf{w}_j)}{\sum_k g_k(\mathbf{x}^T\mathbf{v})f_i(\mathbf{x}_i^T\mathbf{w}_j)} \\
&= p_{ij}.
\end{aligned} \tag{27.7}$$

Substituting the expected values into the expected loglikelihood equation yields

$$\log(L(\mathbf{y},\mathbf{z},\mathbf{v},\mathbf{w})) = \sum_{i=1}^{n}\sum_{j=1}^{G} p_{ij}\log g_j(\mathbf{x}^T\mathbf{v}) + \sum_{i=1}^{N}\sum_{j=1}^{G} p_{ij}\log f_i(\mathbf{x}_i^T\mathbf{w}_j). \tag{27.8}$$

Maximization of the loglikelihood requires the partial derivatives of the $f_i(\mathbf{x}_i^T\mathbf{w}_j)$ with respect to the weight vector $\mathbf{w}$. Recall, each microarray has membership in one of Ω classes, so the y outcomes are formed by use of a matrix $\mathbf{Y}$ with dimensions $n \times \Omega$. For the expert networks, we therefore have

$$\begin{aligned}
\frac{\partial \log(L)}{\partial \mathbf{w}_j} &= \sum_{i=1}^{n}\sum_{j=1}^{G} p_{ij}\frac{1}{f_i(\mathbf{x}_i^T\mathbf{w}_j)}\frac{\partial f_i(\mathbf{x}_i^T\mathbf{w}_j)}{\mathbf{w}_j} \\
&= \sum_{i=1}^{n}\sum_{j=1}^{G} p_{ij}\frac{1}{\exp\left(-||\hat{\mathbf{y}}_i - \mathbf{y}_i||\right)}\frac{\partial \exp\left(-||\hat{\mathbf{y}}_i - \mathbf{y}_i||\right)}{\mathbf{w}_j} \\
&= \sum_{i=1}^{n}\sum_{j=1}^{G} p_{ij}\frac{1}{\exp\left(-||\hat{\mathbf{y}}_i - \mathbf{y}_i||\right)} \\
&\quad \exp\Big(-(x_{i1}w_{i1} - y_{i1})^2 + (x_{i2}w_{i2} - y_{i2})^2 + \cdots \\
&\quad + (x_{i\Omega}w_{i\Omega} - y_{i\Omega})^2\Big)(\hat{\mathbf{y}}_i - \mathbf{y}_i)\mathbf{x}_i \\
&= p_{ij}(\hat{\mathbf{y}}_i - \mathbf{y}_i)\mathbf{x}_i,
\end{aligned} \tag{27.9}$$

and for the gating networks, we have

$$\begin{aligned}
\frac{\partial \log(L)}{\partial \mathbf{v}} &= \sum_{i=1}^{n}\sum_{j=1}^{G} \frac{p_{ij}}{g_j}(\delta_{jk}g_j - g_j g_k)\mathbf{x}_i \\
&= (g_j - p_{ij})\mathbf{x}_i.
\end{aligned} \tag{27.10}$$

The procedures necessary for performing the EM algorithm can be solved using an iteratively reweighted least-squares (IRLS) approach. First, let's review the normal equations used in ordinary least squares $\hat{\mathbf{Y}} = \mathbf{X}\boldsymbol{\beta}^T$. Recall that the residuals for each observation are determined as the difference between the fitted (predicted) value and the observed value in the form $e_i = \hat{y}_i - y_i$. If the residual is zero, the model fits the particular observation with zero error. Using matrix algebra, we typically solve for the weights (regression coefficients) in a single pass:

$$\begin{aligned} \mathbf{X}^T(\hat{\mathbf{Y}} - \mathbf{Y}) &= 0 \\ \mathbf{X}^T(\mathbf{X}\boldsymbol{\beta}^T - \mathbf{Y}) &= 0 \\ \mathbf{X}^T\mathbf{X}\boldsymbol{\beta}^T - \mathbf{X}^T\mathbf{Y} &= 0 \\ \mathbf{X}^T\mathbf{X}\boldsymbol{\beta}^T &= \mathbf{X}^T\mathbf{Y} \\ \boldsymbol{\beta}^T &= (\mathbf{X}^T\mathbf{X})^{-1}\mathbf{X}^T\mathbf{Y}. \end{aligned} \tag{27.11}$$

We now develop the matrix-based solution for solving the EM problem for an MOE network, based on the least-squares algorithm introduced by Jordan and Jacobs [1] and Moerland [15]. Let n be the number of training microarrays for a particular problem, p be the number of features (variables), and Ω be the number of classes. For matrix definitions, let $\mathbf{X}$ be an $n \times p$ matrix of input data for all microarrays, $\mathbf{Y}$ be an $n \times \Omega$ matrix of true class memberships of training arrays, $\mathbf{Y}_j$ and $\hat{\mathbf{Y}}_j$ be $n \times \Omega$ matrices of predicted class memberships, $\mathbf{W}_j$ be an $\Omega \times p$ weight (coefficient) matrix containing weights for the jth expert network, $\mathbf{V}_j$ be a $p \times n$ weight matrix for the jth gating network, $\mathbf{G}_j$ be an $n \times n$ gate probability matrix, and $\boldsymbol{\Pi}_j$ be an $n \times n$ diagonal matrix with p_{ij} for expert j in the diagonal elements. By assuming $\hat{\mathbf{Y}} = \mathbf{X}\mathbf{W}_j^T$ and using the partial derivative $p_{ij}(\hat{\mathbf{y}}_i - \mathbf{y}_i)\mathbf{x}_i$ from (27.9), the maximum likelihood estimates of coefficients for our Gaussian MOE are now obtained for expert j as

$$\begin{aligned} \mathbf{X}^T\boldsymbol{\Pi}_j(\hat{\mathbf{Y}} - \mathbf{Y}) &= 0 \\ \mathbf{X}^T\boldsymbol{\Pi}_j(\mathbf{X}\mathbf{W}_j^T - \mathbf{Y}) &= 0 \\ \mathbf{X}^T\boldsymbol{\Pi}_j\mathbf{X}\mathbf{W}_j^T - \mathbf{X}^T\boldsymbol{\Pi}_j\mathbf{Y} &= 0 \\ \mathbf{X}^T\boldsymbol{\Pi}_j\mathbf{X}\mathbf{W}_j^T &= \mathbf{X}^T\boldsymbol{\Pi}_j\mathbf{Y} \\ \Delta\mathbf{W}_j^T &= (\mathbf{X}^T\boldsymbol{\Pi}_j\mathbf{X})^{-1}\mathbf{X}^T\boldsymbol{\Pi}_j\mathbf{Y}. \end{aligned} \tag{27.12}$$

For the gating networks, we assume $\hat{\boldsymbol{\Pi}} = \mathbf{X}\mathbf{V}^T$ and use $(g_j - p_{ij})\mathbf{x}_i$ from the partial derivative in (27.10), and for gating network j we get

$$\begin{aligned} \mathbf{X}^T(\hat{\boldsymbol{\Pi}}_j - \boldsymbol{\Pi}_j) &= 0 \\ \mathbf{X}^T(\mathbf{X}\mathbf{V}^T - \boldsymbol{\Pi}_j) &= 0 \\ \mathbf{X}^T\mathbf{X}\mathbf{V}^T - \mathbf{X}^T\boldsymbol{\Pi}_j &= 0 \\ \mathbf{X}^T\mathbf{X}\mathbf{V}^T &= \mathbf{X}^T\boldsymbol{\Pi}_j \\ \Delta\mathbf{V}_j^T &= (\mathbf{X}^T\mathbf{X})^{-1}\mathbf{X}^T\boldsymbol{\Pi}_j. \end{aligned} \tag{27.13}$$

The relationship $\Delta\mathbf{W}_j^T = (\mathbf{X}^T\mathbf{\Pi}_j\mathbf{X})^{-1}\mathbf{X}^T\mathbf{\Pi}_j\mathbf{Y}$ above is used during the first iteration. When the iteration is greater than one, we use $\Delta\mathbf{W}_j^T = (\mathbf{X}^T\mathbf{\Pi}_j\mathbf{X})^{-1}\mathbf{X}^T\mathbf{\Pi}_j\mathbf{E}$, where $\mathbf{E} = \mathbf{Y} - \hat{\mathbf{Y}}_j$. In addition, we also set $\mathbf{\Pi}_j = \mathbf{G}_j$ after the first iteration.

27.2 ALGORITHM

The iterative steps used in MOE training are listed in Algorithm 27.1. The result of iterative training is a weight matrix $\mathbf{W}_j^T$ for each expert, which is saved in memory for class prediction of test arrays. Once training is complete, the set of test objects (arrays) is assembled into a data matrix $\mathbf{X}$. Next, the distances between each test array and the K-means centers derived prior to iterative training are compared, and test arrays are assigned to the closest center. Class prediction for test arrays is carried out using a one-step approach without any iterations, which is outlined below. Class-specific weights for test arrays based on feature values are first determined, using the matrix operation

$$\hat{\mathbf{Y}} = \mathbf{X}\mathbf{W}_j^T. \tag{27.14}$$

Next, diagonal values Π_{ii} of the $\mathbf{\Pi}_j$ matrix are set to one if the cluster assignment is equal to the particular jth expert being evaluated. Using the test arrays in matrix format $\mathbf{X}$, we next calculate the weight matrix for the jth gating network $\mathbf{V}_j^T$, based on K-means cluster membership, represented by $\mathbf{\Pi}_j$ as follows:

$$\Delta\mathbf{V}_j^T = (\mathbf{X}^T\mathbf{X})^{-1}\mathbf{X}^T\mathbf{\Pi}_j. \tag{27.15}$$

We now cross-multiply with the test array data matrix to obtain the gating probability matrix in the form

$$\mathbf{G}_j = \mathbf{X}\Delta\mathbf{V}_j^T. \tag{27.16}$$

Finally, the class-specific weights based on features are multiplied by gating probabilities to yield the final class memberships for the jth expert:

$$\mathbf{Y}_j = \mathbf{G}_j^T\hat{\mathbf{Y}}_j \tag{27.17}$$

As we loop through the experts and perform the preceding matrix operations, a running total is maintained on the predicted class membership probabilities using the relationship $\mathbf{YTOT} = \mathbf{YTOT} + \mathbf{Y}_j$. The decision rule for test microarray $\mathbf{x}$ is determined using the row-specific vector $\mathbf{ytot}$ of the matrix $\mathbf{YTOT}$, in the form

$$D(\mathbf{x} \leadsto \omega) = \underset{l=1,2,\ldots,\Omega}{\arg\max}\{ytot_l\}. \tag{27.18}$$

Algorithm 27.1: Mixture of Experts (MOE) Network, Training Algorithm

Data: p-dimensional input arrays $\mathbf{x}_i$ ($i = 1, 2, \ldots, n$), K experts
Result: $\mathbf{W}_j^T$ for each expert
Perform K-means cluster, set $c_i \rightarrow c \quad c = 1, 2, \ldots, K$
for $j \leftarrow 1$ **to** K **do**
  Set $Y_{i\omega_i} = 1$
  for $t \leftarrow 1$ **to** T **do**
    if $t = 1$ **then**
      $\Pi_{ii} = 0$
      if $c_i = j$ **then**
        $\Pi_{ii} = 1$
      endif
      $\mathbf{Y}_j = \mathbf{Y}$
    endif
    if $t > 1$ **then**
      if $\|\Delta\mathbf{W}_j^T\| < \epsilon$ *and* $\|\Delta\mathbf{V}_j^T\| < \epsilon$ **then**
        goto next j
      endif
      $\mathbf{W}_j^T += \Delta\mathbf{W}_j^T$
      $\mathbf{V}_j^T += \Delta\mathbf{V}_j^T$
      $\mathbf{Y}_j = \mathbf{Y} - \hat{\mathbf{Y}}_j$
      $\Pi_{ik} = 0 \quad i \neq k = 1, 2, \ldots, n$
      $\Pi_{ii} = G_{ii} \quad \forall i$
    endif
    $\Delta\mathbf{W}_j^T = (\mathbf{X}^T\Pi_j\mathbf{X})^{-1}\mathbf{X}^T\Pi_j\mathbf{Y}_j$
    $\hat{\mathbf{Y}}_j = \mathbf{X}\Delta\mathbf{W}_j^T$
    $\Delta\mathbf{V}_j^T = (\mathbf{X}^T\mathbf{X})^{-1}\mathbf{X}^T\Pi_j$
    $\mathbf{G}_j = \mathbf{X}\Delta\mathbf{V}_j^T$
    $\mathbf{Y}_j = \mathbf{G}_j^T\hat{\mathbf{Y}}_j$
  endfor
endfor

27.3 CROSS-VALIDATION RESULTS

A MOE classification analysis was performed on the nine datasets used in all other chapters on class prediction. Bootstrap accuracy ("CVB") was based on average 0.632 bootstrap accuracy (13.4) for 10 bootstraps. Table 27.1 lists the classification accuracy for the nine datasets based on ten 10-fold cross-validation (CV). Results suggest that MOE resulted in quite high levels of performance, and that accuracy was greater when the number of experts employed was equal to the number of classes. This is essentially an effect due wholly to the use of optimized feature sets, which maximize the between-class

TABLE 27.1 Classification Performance (Accuracy) of MOE Classifier[a]

			# Experts			
Dataset	#Samples	#Features	2	3	4	5
AMLALL2	38	15	**1.00**	0.97	0.98	0.98
Brain2	60	11	**0.86**	0.78	0.76	0.69
Breast2A	15	11	**1.00**	1.00	1.00	1.00
Breast2B	78	13	**0.95**	0.88	0.84	0.81
Colon2	62	3	**0.88**	0.87	0.86	0.85
Lung2	32	25	**1.00**	1.00	1.00	1.00
MLL_Leukemia3	57	20	0.97	**1.00**	1.00	1.00
Prostate2	102	11	**0.96**	0.94	0.93	0.91
SRBCT4	63	23	0.98	0.98	**1.00**	1.00

[a]Based on ten 10-fold cross-validation (CV) and mean-zero standardization of input feature values prior to classification. The number of experts (# Experts) used along with the number of arrays and filtered features selected from each dataset are listed. Entries in bold reflect the accuracy when the number of experts equaled the number of classes.

Mahalanobis distance of arrays used. Initiating the assignment of training arrays to K-means clusters based on optimized features will nominally result in the greatest performance when the number of K-means clusters equals the number of classes, as the assignment of training arrays to the closest cluster (expert) will almost assuredly result in the assignment of arrays with the same true class label to the same cluster. MOE is nevertheless a supervised class prediction classifier, so the true number of classes is always known beforehand. We would have expected different results if optimal gene sets were not employed.

27.4 DECISION BOUNDARIES

The MOE classifier was also employed for obtaining class decision boundaries. Figure 27.1 illustrates 2D self-organizing maps of class decision boundaries for the MOE based on the AMLALL2, Brain2, Breast2A, Breast2B, Colon2, Lung2, MLL_Leukemia3, Prostate2, and SRBCT4 datasets using mean-zero standardized input features. Figure 27.2 illustrates decision boundaries for the same datasets when input feature values are fuzzified.

27.5 SUMMARY

The MOE classifier consists of experts and gating functions. The goal of each expert is to find an optimal gating function, while the goal of the gating functions are to train each expert for maximal performance based on the data chosen by the gating function. It is through this decomposition of learning

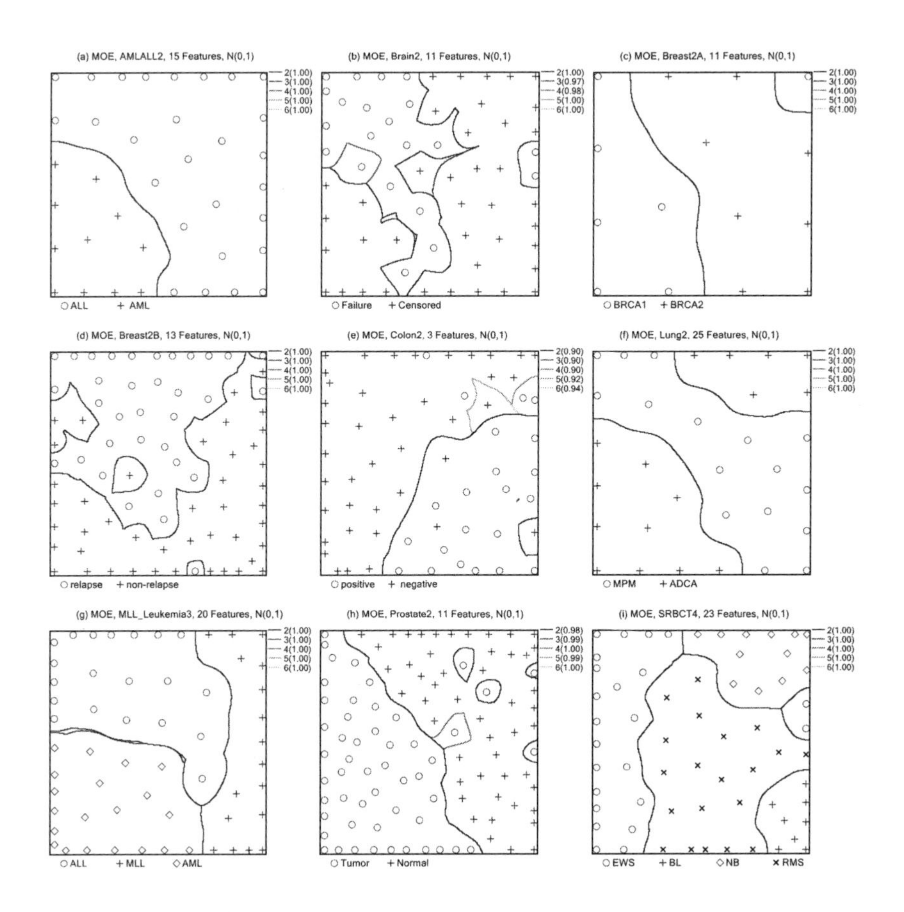

FIGURE 27.1 2D self-organizing maps of MOE class decision boundaries and classification accuracy (upper right of each panel) for the AMLALL2, Brain2, Breast2A, Breast2B, Colon2, Lung2, MLL_Leukemia3, Prostate2, and SRBCT4 datasets based on mean-zero standardized input features: (a) AMLALL2; (b) Brain2; (c) Breast2A; (d) Breast2B; (e) Colon2; (f) Lung2; (g) MLL_Leukemia3; (h) Prostate2; (i) SRBCT4. The numbers of genes used for each dataset are listed in Table 12.10.

for which the EM algorithm is exploited. The main advantage of the MOE classifier is that the experts are localized to a given portion of the data and then combined dynamically to provide an optimal output that generalizes quite well [16]. A disadvantage of the MOE classifier presented in this chapter is that the computational complexity can be high becuase of the degree of matrix multiplication used. Another potential disadvantage of our MOE algorithm is possible pathology of the $\mathbf{X}^T\mathbf{X}$ matrix in terms of singularity and non-positive definiteness due to multicollinearity. Multicollinearity can result in zero eigenvalues of $\mathbf{X}^T\mathbf{X}$, rendering inversion intractable.

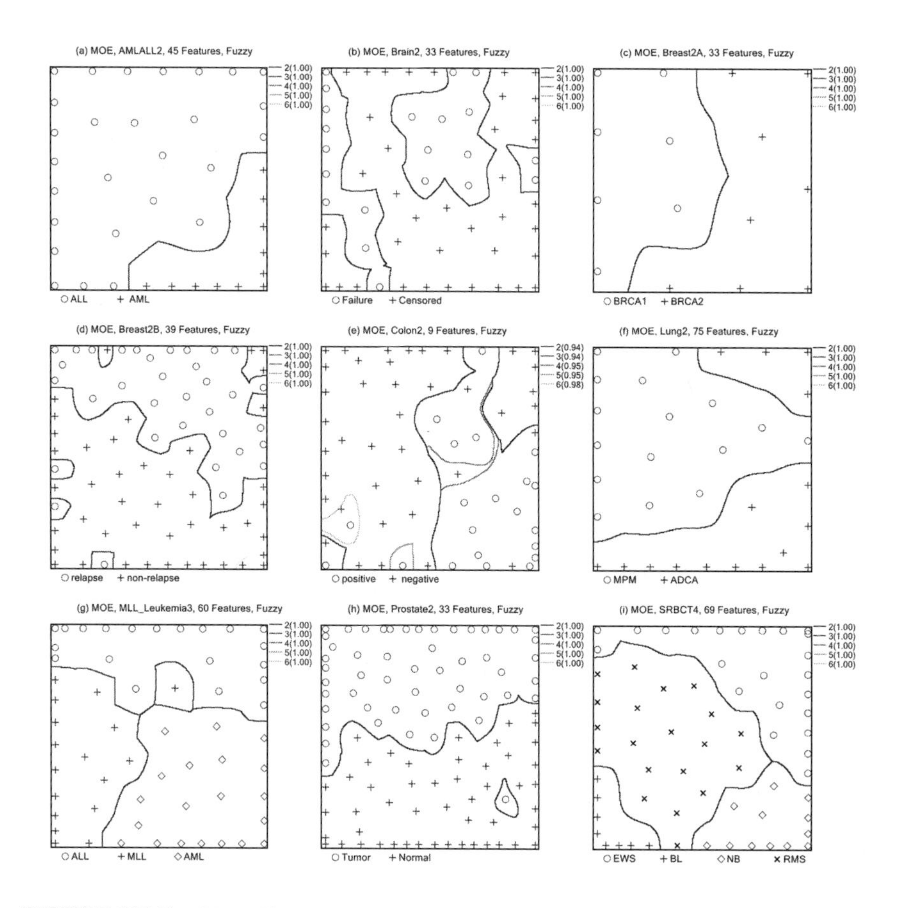

FIGURE 27.2 2D self-organizing maps of MOE class decision boundaries and classification accuracy (upper right of each panel) for the AMLALL2, Brain2, Breast2A, Breast2B, Colon2, Lung2, MLL_Leukemia3, Prostate2, and SRBCT4 datasets based on fuzzified input features: (a) AMLALL2; (b) Brain2; (c) Breast2A; (d) Breast2B; (e) Colon2; (f) Lung2; (g) MLL_Leukemia3; (h) Prostate2; (i) SRBCT4. The numbers of genes used for each dataset are listed in Table 12.10.

REFERENCES

[1] M.I. Jordan, R.R. Jacobs. *Hierarchical Mixtures of Experts and the EM Algorithm.* Artificial Intelligence Laboratory Memo 1440, MIT Press, Cambridge, MA, 1993.

[2] S.R. Waterhouse, A.J. Robinson. Classification using hierarchical mixtures of experts. *Proc. 1994 IEEE Workshop on Neural Networks for Signal Processing*, IEEE Press, Piscataway, NJ, 1994, pp. 177-186.

[3] K. Chen, L. Xu, H. Chi. Improved learning algorithms for mixture of experts in multiclass classification. *Neural Networks* **12**:1229-1252, 1999.

[4] C. Caragea, J. Sinapov, D. Dobbs, V. Honavar. Mixture of experts models to exploit global sequence similarity on biomolecular sequence labeling. *BMC Bioinform.* **10**(S4):S4, 2009.

[5] A.X. Carvalho, M.A. Tanner. Mixtures-of-experts of autoregressive time series: Asymptotic normality and model specification. *IEEE Trans. Neural Networks* **16**:39–56, 2005.

[6] J.M. Corchado, J.F. De Paz, S. Rodriguez, J. Bajo. Model of experts for decision support in the diagnosis of leukemia patients. *Artif. Intell. Med.* **46**:179–200, 2009.

[7] S. Raman, T.J. Fuchs, P.J. Wild, E. Dahl, J.M. Buhmann, V. Roth. Infinite mixture-of-experts model for sparse survival regression with application to breast cancer. *BMC Bioinform.* **11**(S8):S8, 2010.

[8] E.D. Ubeyli. A mixture of experts network structure for breast cancer diagnosis. *J. Med. Syst.* **29**:569–579, 2005.

[9] I. Gule, U.E. Derya, G.N. Fatma. A mixture of experts network structure for EEG signals classification. *Proc. IEEE Eng. Med. Biol. Soc.* **3**:2707–2710, 2005.

[10] I. Guler, E.D. Ubeyl. A mixture of experts network structure for modelling Doppler ultrasound blood flow signals. *Comput. Biol. Med.* **35**:565–582, 2005.

[11] S. Gutta, J.J. Huang, P. Jonathon, H. Wechsler. Mixture of experts for classification of gender, ethnic origin, and pose of human faces. *IEEE Trans. Neural Networks* **11**:948–960, 2000.

[12] K.A. Le Cao, E. Meugnier, G.J. McLachlan. Integrative mixture of experts to combine clinical factors and gene markers. *Bioinformatics* **26**:1192–1198, 2010.

[13] A.P. Dempster, N.M. Laird, D.B. Rubin. Maximum likelihood from incomplete data via the EM algorithm. *J. Roy. Stat. Soc.* **39**:1–38, 1977.

[14] S-K. Ng, G.J. McLachlan. Using the EM algorithm to train neural networks: Misconceptions and a new algorithm for multiclass classification. *IEEE Trans. Neural Networks* **15**(3):738–749, 2004.

[15] P. Moerland. *Mixture Models for Unsupervised and Supervised Learning*. Thesis 2189, Dept. Informatics, École Polytechnique Fédérale de Lausanne (EPFL), Lausanne, 2000.

[16] R. Avnimelech, N. Intrator. Boosted mixture of experts: An ensemble learning scheme. *Neural Comput.* **11**(2):483–497, 1999.

CHAPTER 28

COVARIANCE MATRIX FILTERING

28.1 INTRODUCTION

This chapter focuses on covariance matrix filtering (CMF) techniques to identify the region of noise within the eigenvalues of a covariance or correlation matrix. The fundamental problem posed by sampling is that as the number of rows n of $\mathbf{X}$ tend to infinity ($n \rightarrow \infty$), the covariance between the p columns becomes more reliable and approaches the true value. However, if both n and p approach infinity ($n, p \rightarrow \infty$), correlation becomes inconsistent and not all the eigenvalues λ are true. Another problem posed by dealing with covariance matrices is that whenever $p = n$, there is by definition one zero eigenvalue, and when $p > n$ there will be $p - n$ zero eigenvalues. To tackle this problem, we will employ random matrix theory (RMT) techniques to generate various types of random matrices for which we assume that there is zero correlation between columns (rows). This will allow us to compare eigenvalues from a random matrix with eigenvalues from a sample covariance matrix to identify a noise threshold among eigenvalues. By knowing which eigenvalues lie in the noise subspace, we will be able to directly observe which eigenvalues are informative through the process of elimination. Another technique we will use involves moment-based *shrinkage* procedures to transform the covariance matrix so that it can be used again in filtered form.

28.2 COVARIANCE AND CORRELATION MATRICES

By definition, the population covariance matrix is

$$\boldsymbol{\Sigma} \stackrel{\partial ef}{=} \mathbb{E}[(\mathbf{x} - \boldsymbol{\mu})^T (\mathbf{x} - \boldsymbol{\mu})], \tag{28.1}$$

Classification Analysis of DNA Microarrays, First Edition. Leif E. Peterson.

where $\mathbf{x}$ is a p-tuple $(x_{i1}, x_{i2}, \ldots, x_{ip})$ for the ith object and $\boldsymbol{\mu}$ is a p-tuple with means for the p features. Because it is impossible to measure $\boldsymbol{\Sigma}$, we are limited to working with the sample covariance matrix, given as

$$\mathbf{S} = \frac{1}{n}\sum_{i=1}^{n}(\mathbf{x}_i - \bar{\mathbf{x}})^T(\mathbf{x}_i - \bar{\mathbf{x}}). \tag{28.2}$$

Let $\mathbf{D} = \text{diag}(\mathbf{S})$, the sample correlation matrix is

$$\mathbf{R} = \mathbf{D}^{-1}\mathbf{s.d.}^{-1}. \tag{28.3}$$

During computation, elements r_{jk} of $\mathbf{R}$ can be determined individually using the relationship

$$\begin{aligned} r_{jk} &= \frac{s_{jk}}{\sqrt{s_{jj}s_{kk}}} \\ &= \frac{s_{jk}}{s_j s_k}. \end{aligned} \tag{28.4}$$

If the means of columns of $\mathbf{X}$ are known, covariance elements s_{jk} of $\mathbf{S}$ can be directly calculated using the functional form

$$s_{jk} = \frac{1}{n}\sum_{i=1}^{n}(x_{ij} - \bar{x}_j)(x_{ik} - \bar{x}_k), \tag{28.5}$$

whereas, if the standard deviations and correlations are already known, then covariance elements can be calculated using

$$s_{jk} = r_{jk}s_j s_k. \tag{28.6}$$

It is assumed that the sample covariance matrix $\mathbf{S}$ is an accurate and reliable estimate of the population covariance matrix $\boldsymbol{\Sigma}$. When the sample size n is infinitely large, there are fewer problems with this assumption, but as $n \ll p$, the eigenvalues of $\mathbf{S}$ are biased where small eigenvalues are too small and the large eigenvalues are too large. Moreover, several of the eigenvalues become zero, so $\mathbf{S}$ becomes singular by losing full rank and therefore, is not positive definite and cannot be inverted.

28.3 RANDOM MATRICES

Random matrices are commonly used in probability theory, mathematical statistics, theoretical physics, and chaos and complexity theory to study eigenvalue limit densities for various types of matrices derived from probability distributions [1-5]. With growing interest in fusion and integration of large datasets, there is an ever-increasing chance that the assumption $n \gg p$ is becoming less true, especially since the dimension p occasionally dwarfs the sample size n. In this chapter, we draw information from the field of RMT to explore the consistency and noise subspace of eigenvalues of random matrices and covariance (correlation) matrices as a function of the ratio $\gamma = p/n$ and probability distribution used for generating the data. Random matrices are generated via an ensemble of multivariate Gaussian independent and identically distributed (i.i.d.) normal variates. As such, their rows and columns are orthogonal, since the correlation between them is assumed to be zero. We shall be interested in two kinds of Gaussian ensembles: the Gaussian orthogonal ensemble (GOE) and the Wishart ensemble $[W_p(n, \mathbf{\Sigma})]$.

Gaussian Orthogonal Ensemble (GOE). A GOE is a $p \times p$ symmetric matrix $\mathbf{A}_p = (\mathbf{X} + \mathbf{X}^T)/\sqrt{2p}$, where $\mathbf{X}$ is a square $p \times p$ matrix with elements $x_{ij} \sim \mathcal{N}(0, 1)$ [6]. Diagonal elements of $\mathbf{A}_p$ are always $a_{ii} \sim \mathcal{N}(0, \sigma^2/p)$, with off-diagonal elements $a_{ij} \sim \mathcal{N}(0, 2\sigma^2/p)$. The obvious choice for obtaining eigenvalues of eigenvectors of $\mathbf{A}_p$ is to use matrix factorization for the symmetric eigenvalue problem. The main point about GOEs is that they are not a covariance or correlation matrix, but rather a simple random symmetric matrix.

Wishart Ensemble $W_p(n, \Sigma)$. A Wishart ensemble $W_p(n, \mathbf{\Sigma}) = \mathbf{X}\mathbf{X}^T$ is also a square symmetric $p \times p$ matrix, and has n degrees of freedom based on the number of rows of $\mathbf{X}$. The elements of $\mathbf{X}$ are i.i.d. standard normal variates, so the crossproduct of $\mathbf{X}$ and its transpose $\mathbf{X}^T$ are assumed to be a random covariance or correlation matrix, for which the null hypothesis is $H_0 : \mathbf{\Sigma} = \mathbf{I}$ or $H_0 : \mathbf{R} = \mathbf{I}$. The key point about Wishart ensembles is that, when $\mathbf{X}$ has normally distributed elements that are i.i.d., the resulting Wishart matrix is assumed to form the identity matrix $\mathbf{I}$, for which all the off-diagonals are assumed to be zero. Additionally, since $|\mathbf{I}| = 1$, all the eigenvalues of $W_p(n, \mathbf{I})$ are supposed to be unity. We also know that by definition, the determinant of a square symmetric Wishart matrix is $|W_p(n, \mathbf{I})| = \prod_j^p \lambda_j$, and since the eigenvalues $\lambda_1, \lambda_2, \ldots, \lambda_p$ of $\mathbf{I}$ are all unity, the determinant is also $|W_p(n, \mathbf{I})| = 1$ when $W_p(n, \mathbf{I}) = \mathbf{I}$. However, as we shall see later, the eigenvalues of random matrices are not all equal to one and instead have a certain level of spread. For GOEs, the empirical eigenvalue distribution (e.e.d.) always forms a semicircle distribution, which is defined by the Wigner semi-circle law. Whereas, for Wishart matrices, the e.e.d. follows the Marčenko-Pastur law and exhibits a variety of forms depending on the p/n ratio γ.

Wigner Semicircle Law for GOE Matrices. RMT was popularized by Wigner's research in the field of nuclear physics [3], for which the limit density in $n \rightarrow \infty$ for $n \times n$ random matrices was observed to describe resonance fluctuations of compound nuclei. For any $n \times n$ GOE matrix $\mathbf{A}_p$, the limit density of eigenvalues is on the order of

$$f(\lambda) = \begin{cases} \sqrt{4\sigma^2 - \lambda^2}, & |\lambda| \leq 2\sigma \\ 0, & |\lambda| > 2\sigma \end{cases}$$

with results shown in Figure 28.1, in which the semicircle density can be observed. The intent of including the Wigner semicircle law for GOE matrices is mainly to show the reader the eigenvalue density for a random square symmetric GOE matrix.

Marčenko-Pastur Law for Wishart Matrices. While the Wigner semicircle law addresses real symmetric random matrices, we now delve deeper into the limiting density of eigenvalues for random Wishart matrices. The Marčenko-Pastur (MP) law states that for i.i.d. columns in $\mathbf{X}$ and $(n, p \rightarrow \infty, \gamma = p/n)$, the minimum and maximum eigenvalues of $W_p(n, \mathbf{\Sigma})$ almost surely converge to $\lambda^- = \sigma^2(1 - \sqrt{\gamma})^2$ and $\lambda^+ = \sigma^2(1 + \sqrt{\gamma})^2$, respectively. The e.e.d. for $W_p(n, \mathbf{\Sigma})$ based on $\mathbf{X}$ with i.i.d. elements is given by

$$f(\lambda, \gamma) = \max\left(0, 1 - \frac{1}{\gamma}\right)\delta(\lambda) + \frac{\sqrt{(\lambda^+ - \lambda)(\lambda - \lambda^-)}}{2\pi\gamma\lambda\sigma^2} I(\lambda^- \leq \lambda \leq \lambda^+), \tag{28.7}$$

where the $\max()\delta(\lambda)$ term represents the density component at $\lambda = 0$ for the $p - n = p(1 - 1/\gamma)$ zero eigenvalues when $p > n$ (i.e., $\gamma > 1$) and the $I(\#)$ represents the density when λ is between λ^- and λ^+. Figure 28.2 shows Marčenko-Pastur limit densities of the eigenvalues of $\mathbf{W}_p(n, \mathbf{I})$ with $n = 1000$ and $p = 1000$, that is, $\gamma = p/n = 1$. Overlaid on the histogram are MP eigenvalue density lines representing $f(\lambda, \gamma)$ for other Wishart matrices for which $n = 1000$ and $\gamma = 1, 0.5, 0.25, 0.1$, and 0.05. Note that when $\gamma = 0.05 = \frac{50}{1000}$, the number of variables is $\frac{1}{20}$ the number of rows and the MP eigenvalue distribution begins to converge to unity, when $n \rightarrow \infty$. The MP law applies strictly to random Wishart matrices. Later in this chapter, we will see how the MP law can be applied to a sample covariance (correlation) matrix to filter out the eigenvalues falling within the noise region.

Tracy-Widom Law for Wishart Matrices. A threshold for the noise subspace of eigenvalues can be determined by use of the Tracy-Widom (TW) law [8], which provides the mean and standard deviation of λ_1, the largest eigenvalue of an i.i.d. random Wishart matrix. The first-order variates of the

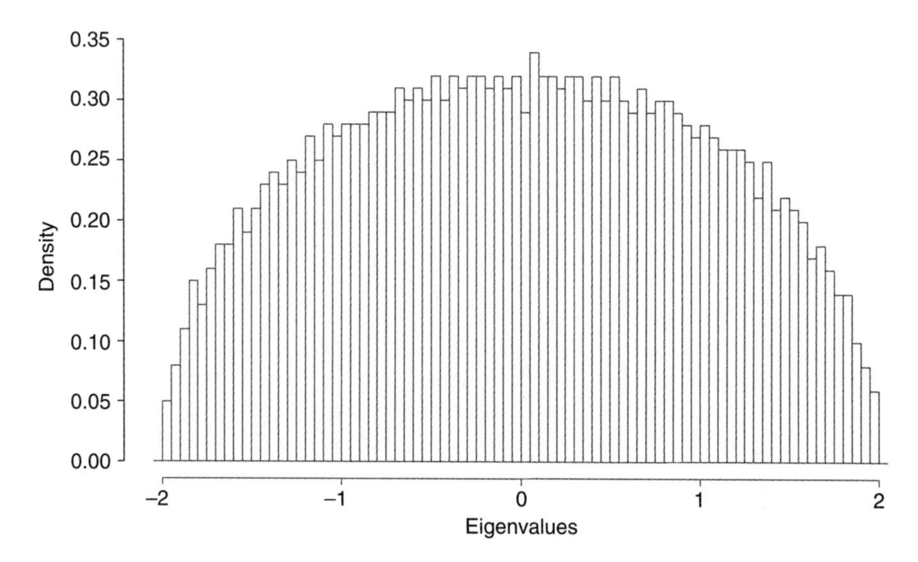

FIGURE 28.1 Limiting eigenvalue density for an $n \times n$ GOE matrix $\mathbf{A}_p$ showing the characteristic Wigner semicircle shape.

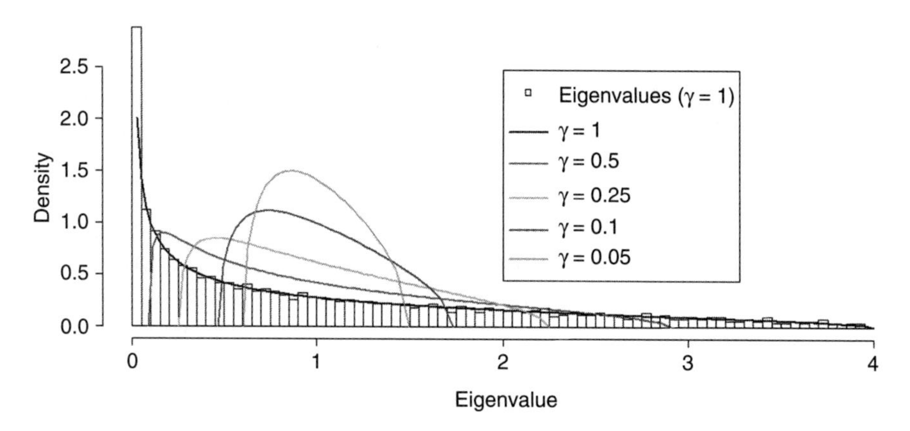

FIGURE 28.2 Marčenko-Pastur limiting eigenvalue densities for Wishart ensemble matrices $W_p(n, \mathbf{I})$ for $n = 1000$, $p = 1000(\gamma = 1)$, $p = 500(\gamma = 0.5)$, $p = 250(\gamma = 0.25)$, $p = 100(\gamma = 0.1)$, and $p = 50(\gamma = 0.05)$. (*See insert for color representation of the figure.*)

TW law are defined as

$$F_1 \sim \frac{\lambda_1 - \mu_{np}}{\sigma_{np}}, \tag{28.8}$$

where

$$\mu_{np} = \left(\sqrt{n - \tfrac{1}{2}} + \sqrt{p - \tfrac{1}{2}}\right)^2 \tag{28.9}$$

and

$$\sigma_{np} = \sqrt{\mu_{np}} \left(\frac{1}{\sqrt{n - \frac{1}{2}}} + \frac{1}{\sqrt{p - \frac{1}{2}}} \right)^{1/3}. \tag{28.10}$$

The TW law provides the central estimate of the maximum eigenvalue of a random covariance matrix. When using sample covariance matrices, Baik [9] and Johnstone [13] suggest that as soon as the largest eigenvalue λ_1 in the bulk of eigenvalues exceeds $1 + \sqrt{\gamma}$, F_1 jumps to a Gaussian distribution with mean increasing with λ_1.

Covariance Matrix Filtering with RMT Results. There are several key points for employing RMT results for filtering a covariance matrix; these are introduced using the following conjectures:

Conjecture 1: The MP and TW laws apply mainly to Wishart matrices $W_p(n, \mathbf{\Sigma})$ where the covariance matrix $\mathbf{\Sigma}$ is assumed to be equal to the identity matrix **I**. Information surrounding MP and TW densities of eigenvalues stem from Wishart matrices that are either the identity matrix or Wishart matrices based on a data matrix **X** whose x values are random i.i.d variates. The latter does not ensure an identity matrix, but does help enforce zero correlation between columns.

Conjecture 2: The eigenvalues of a covariance or correlation matrix derived from x values of **X** *that are random i.i.d. and distributed $\mathcal{N}(0, 1)$ will follow the MP and TW density laws for the particular value of γ.* The MP and TW eigenvalue densities can be simulated by constructing an $n \times p$ data matrix **X** whose values are standard normal random variates.

Conjecture 3: An eigenvalue distribution that purely follows the MP and TW laws can be obtained from any sample covariance **S** *or correlation matrix* **R** *by permuting (shuffling) the columns of the* **X** *matrix from which* **S** *or* **R** *were derived.* Permuting or shuffling the columns of **X** helps to enforce zero correlation between columns, causing the off-diagonals of **S** or **R** to approach zero. The closer the resulting **S** or **R** matrices are to **I**, the more the MP and TW laws will be satisfied.

Conjecture 4: Covariance and correlation matrices are Wishart matrices, so the MP and TW laws apply to their eigenvalue distributions. The MP and TW densities of eigenvalues from covariance and correlation matrices can be used for CMF. The improved matrices can then be substituted back into the data analytic problem(s) being considered.

Conjecture 5: The MP-based maximum eigenvalue λ^+ will usually fall within the TW distribution of the greatest eigenvalue λ_1 as long as λ_1 in the bulk of eigenvalues is less than the theoretical value of $1+\sqrt{\gamma}$. When λ_1 exceeds the theoretical value of $1+\sqrt{\gamma}$, the TW distribution F_1 becomes a Gaussian distribution with mean increasing with λ_1 [9,13].

Conjecture 6: The mean value μ_{np} of λ_1 based on the TW law is typically a reliable estimate of λ_1, but can break down depending on the empirical value of λ_1. The TW law for obtaining the mean value of λ_1 tends to be a reliable estimate, but can be biased after the phase transition occurs when the empirical observed λ_1 exceeds the theoretical value of $1+\sqrt{\gamma}$.

Conjecture 7: As $n \to \infty$, the greatest eigenvalue λ_1 of **S** *or* **R** *for a data matrix* **X** *whose columns are permuted will approach the theoretical value of $\lambda^+ = \sigma^2(1+\sqrt{\gamma})^2$ and fall within the TW distribution of λ_1.* Permuting columns of **X** is a reliable approach for finding the greatest eigenvalue of λ_1 of the sample **S** or **R** when it is assumed that **S** or **R** are purely random.

Conjecture 8: The theoretical value of λ^+ from the MP law or the maximum observed value of λ_1 from **S** *or* **R** *after permuting columns in the data matrix* **X** *is one of the most commonly used estimators of the cutpoint below which eigenvalues are considered noninformative or noise.* Eigenvalues above λ^+ or λ_1 based on a permuted **X** are considered to be informative—so only their eigenvectors (principal components, factors) are typically used in later analyses. However, this is not the most reliable approach for finding the maximum eigenvalue.

Conjecture 9: The most reliable estimator of the noise threshold in eigenvalues is based on λ^+ obtained from fitting MP distributions while varying parameter values for γ, σ, λ^-, and λ^+. When models of the MP density are fitted to the empirical eigenvalues and parameters are allowed to vary, the resulting value of λ^+ from the best fit will tend to be greater than the theoretical value of $\lambda^+ = \sigma^2(1+\sqrt{\gamma})^2$ based on only p and n, as well as greater than the maximum λ_1 after permuting the data matrix. When permuting columns of **X** and then extracting eigenvalues of **S** or **R**, the maximum obtained eigenvalue will be conservatively too low, and this also assumes that the data are random and not empirical. The maximum eigenvalue obtained after permuting the data columns in **X** are typically similar to the theoretical value of λ^+, which assumes that the data are random. However, the input data are not random realizations, so a better representation of λ^+ can be obtained from fitting MP distributions.

Conjecture 10: Covariance or correlation matrix eigenvalues above the MP-based λ^+ are considered informative, while eigenvalues below λ^+ are believed to fall in the noise region. We will show how to filter a covariance (correlation) matrix for noise removal by regressing the input data matrix on principal components whose eigenvalues fall below λ^+. The subsequent filtered covariance matrix can be used directly in the intended analysis being considered. Use of a filtered covariance matrix will result in less bias, since the noise is removed from the data.

Conjecture 11: Informative eigenvalues above the fitted value of λ^+ should coincide with eigenvectors whose elements are normally distributed. Eigenvector elements are, by definition, supposed to be normally distributed, but this is not always the case. A rich cluster structure of objects (features) can result in eigenvector elements that are not normally distributed, so histograms of eigenvector elements for all informative eigenvalues should be evaluated for departures of normality. Inferential tests of hypotheses can also be performed to rule out eigenvectors whose elements are not normally distributed.

These conjectures describe properties of RMT results based on the TW and MP laws, which can be applied to CMF to identify which eigenvalues are considered to be noise and noninformative.

28.4 COMPONENT SUBTRACTION

***Filtering out the First Principal Component from* X.** Occasionally the value of the greatest eigenvalue of the covariance matrix is much larger than the remaining eigenvalues. This phenomenon is observed quite frequently, where, for example, $\lambda_1 = 200$, $\lambda_2 = 80$, and $\lambda_3 = 50$, and so on. When the greatest eigenvalue far exceeds the other eigenvalues, it suggests that many of the dimensions or variables (genes) are highly correlated with one another and that the principal component associated with the eigenvalue represents a large degree of the variance explanation for all variables. Unfortunately, this may not always be desirable, especially if the component or driving factor is too overwhelming, such as a genomic regulatory control pathway that masks more subtle regulatory control patterns that otherwise would not be discernible. A way to effectively unmask or "subtract" this component of highly correlated variables is to regress each variable's data values on the eigenvector or principal component and replace the original data values with the regression residuals. Any variable whose values (i.e., gene whose expression values) are predicted well with the model will have small or near-zero residual values, which, when used to replace the original data entries, will force the given variable (gene) to be less informative. Subtraction of the first principal component from the entire dataset is accomplished with the following multivariate regression model

$$\underset{n\times p}{\mathbf{X}} = \underset{n\times 2}{\mathbf{F}} \underset{2\times p}{\boldsymbol{\beta}}, \tag{28.11}$$

where $\mathbf{X}$ is the data matrix, $\mathbf{F}$ is a matrix with the first column containing all ones and the second column containing the n principal component scores for the first component, and $\boldsymbol{\beta}$ is the $2 \times p$ regression coefficient matrix. Once the model is fitted, we predict the data matrix with $\hat{\mathbf{X}} = \mathbf{F}\boldsymbol{\beta}$. The new data matrix consists of the residuals and is formed by the relationship

$$\mathbf{X}^{\text{res}} = \hat{\mathbf{X}} - \mathbf{X}, \tag{28.12}$$

which can be used to obtain a new covariance matrix $\mathbf{S}_{\text{res}}$ or correlation matrix $\mathbf{R}_{\text{res}}$. Results can be compared with results based on $\mathbf{X}$ to inspect the participation [10] of each n row in each eigenvector, using the inverse participation ratio

$$\text{IPR}_j = \sum_{k=1} |e_{jk}|^4, \tag{28.13}$$

where e_{jk} is the eigenvector element of the jth eigenvector. Large values of IPR indicate participation of only several dimensions, while small values reflect that rows participate equally.

The methods described above assume that a variable-by-variable ($p \times p$) covariance matrix $\mathbf{S}$ is being investigated. When running eigenanalysis on an object-by-object ($n \times n$) $\mathbf{S}$ matrix, we can determine the average class-specific strength of participation in eigenvectors. First, we set an indicator to denote which class each object is a member of, using

$$\Delta_{l\omega} = \begin{cases} 1 & \text{if} \quad \mathbf{x}_l \in \omega \\ 0 & \text{otherwise,} \end{cases} \tag{28.14}$$

and then loop over all objects to obtain the mean strength of participation in the ith eigenvector by objects in class ω:

$$\text{IPR}_{i\omega} = \frac{1}{n_\omega} \sum_{l=1}^{n} \Delta_{l\omega} |e_{il}|^2. \tag{28.15}$$

We then construct plots of $\text{IPR}_{i\omega}$ showing the class-specific participation of objects in each eigenvector.

Filtering out m Noise Principal Components from X after Fitting a Marčenko-Pastur Density. When filtering a covariance matrix for the MP distribution, we first fit the MP density to the empirical eigenvalue distribution to obtain parameter values for γ, σ, λ^-, and λ^+. We employed particle swarm optimization for fitting the MP parameters and observed this approach to be more stable than, for example, Newton-Raphson gradient descent. The influence of variables that load on noise eigenvectors whose eigenvalues are less than the fitted λ^+ is filtered out of the data matrix **X** by use of a multivariate linear regression of the form

$$\underset{n\times p}{\mathbf{X}} = \underset{n\times(m+1)}{\mathbf{F}}\underset{(m+1)\times p}{\boldsymbol{\beta}}, \tag{28.16}$$

where **X** is the data matrix, **F** is the principal component score matrix for n objects and $m+1$ columns (a column of ones precedes the **F** matrix), $\boldsymbol{\beta}$ is the $(m+1)\times p$ regression coefficient matrix, and $m = \#\{j : \lambda_j \leq \lambda^+\}$, that is, the number of covariance matrix eigenvalues that are less than the MP-fitted value of λ^+. After running this regression, procedure, we simply determine the predicted data matrix by $\hat{\mathbf{X}} = \mathbf{F}\boldsymbol{\beta}$. The new data matrix $\mathbf{X}^{\mathrm{MP}}$ without the effects of the m eigenvectors in the noise region is determined from the relationship

$$\mathbf{X}^{\mathrm{MP}} = \hat{\mathbf{X}} - \mathbf{X}, \tag{28.17}$$

which is then used to obtain the filtered covariance matrix $\mathbf{S}_{\mathrm{MP}}$ and correlation matrix $\mathbf{R}_{\mathrm{MP}}$. The newly determined $\mathbf{X}^{\mathrm{MP}}$ matrix is filtered and does not contain the noise component associated with eigenvectors whose eigenvalues are less than λ^+, i.e., noise eigenvalues. There are typically many covariance matrix eigenvalues below λ^+, so the resulting removal of m components from **X** will result in a covariance matrix with $p-m$ zero eigenvalues. All the dimensions will now be highly correlated with the remaining informative principal eigenvectors having the greatest eigenvalues.

28.5 COVARIANCE MATRIX SHRINKAGE

One of the earliest reports on covariance bias from high-dimensional problems was published by Stein [11], who stated that maximum likelihood estimation can be improved with tremendous success. Efron and Morris [12] reported that inference includes both maximum likelihood estimation and maximum likelihood summarization–and it is under the former where problems occur. Covariance shrinkage methods attempt to stabilize **S** by forcing it to (1) be positive definite with nonzero eigenvalues, (2) become wellconditioned so that the ratio of the largest to the smallest eigenvalue is not too large, (3) have reduced variance regarding its bias-variance decomposition. Most shrinkage methods either shrink the eigenvalues or shrink an unstructured estimator of $\boldsymbol{\Sigma}$ toward a structured one. We discuss three shrinkage methods introduced by Daniels and Kass [14], Ledoit and Wolf [15], and Schäfer and Strimmer [16].

Daniels-Kass (DK) Shrinkage. The DK shrinkage method [14] is an eigenvalue shrinkage method that assumes the eigenvalues are log-normally distributed, and forms a maximum likelihood estimator based on the lognormal priors. The simple closed form version of the eigenvalue estimator is

$$\hat{\lambda}_i = \exp\left(\frac{2/n}{2/n+\hat{\tau}^2}\log(\hat{\lambda}_i) + \frac{\hat{\tau}^2}{2/n+\hat{\tau}^2}\log(\hat{\lambda}_i)\right), \tag{28.18}$$

where $\log(\hat{\lambda}_i)$ is the natural logarithm of each eigenvalue, and $\hat{\tau}^2$ is determined using

$$\hat{\tau}^2 = \frac{\sum_{i=1}^{p}\left(\log(\hat{\lambda}_i) - \log(\lambda)\right)^2}{p+4} - \frac{2}{n}, \tag{28.19}$$

where $\log(\lambda)$ is the mean of the logarithm of eigenvalues.

Ledoit-Wolf (LW) Shrinkage. Ledoit-Wolf shrinkage [15] is based on minimum squared error of an optimal combination of the unconstrained high-dimensional covariance matrix and a constant correlation prior matrix **F**. First, determine both the sample covariance **S** and sample correlation **R** matrices. Next, estimate the prior matrix for constant correlation using mean correlation $\bar{r}$ and determine elements of **F** in the form

$$f_{jk}, = \begin{cases} s_{jj} & \text{if } j = k \\ \bar{r}\sqrt{s_{jj}s_{kk}} & \text{if } j \neq k. \end{cases} \tag{28.20}$$

Using the definition of asymptotic variance of s_{jk} as

$$\text{AsyVar}[s_{jk}] = \frac{1}{n}\lim_{n\to\infty} \text{E}[\{s_{jk} - \lim_{n\to\infty}\text{E}[s_{jk}]\}^2], \tag{28.21}$$

we now scale the covariance matrix elements by n and find the individual asymptotic variances of covariance elements for diagonal elements using

$$\hat{\pi}_{jj} = \frac{1}{n}\sum_{i=1}^{n}[\Delta_{ij}^2 - s_{jj}]^2, \tag{28.22}$$

where $\Delta_{ij} = (x_{ij} - \bar{x}_j)$ and $\Delta_{ik} = (x_{ik} - \bar{x}_k)$, and then for off-diagonal elements using the form

$$\hat{\pi}_{jk} = \frac{1}{n}\sum_{i=1}^{n}[\Delta_{ij}\Delta_{ik} - s_{jk}]^2. \tag{28.23}$$

We then find the sum using

$$\hat{\pi} = \sum_{j=1}^{p}\sum_{k=1}^{p}\hat{\pi}_{jk}. \tag{28.24}$$

Next, we calculate the asymptotic covariances for $[s_{jj}, s_{jk}]$ and $[s_{kk}, s_{jk}]$ using the functional forms

$$\sigma_{jj,jk} = \frac{1}{n}[\Delta_{ij}^2 - s_{jj}][\Delta_{ij}\Delta_{ik} - s_{jk}] \tag{28.25}$$

and

$$\sigma_{kk,jk} = \frac{1}{n}[\Delta_{ik}^2 - s_{kk}][\Delta_{ij}\Delta_{ik} - s_{jk}]. \tag{28.26}$$

The sum total asymptotic covariance between the covariance matrix elements and shrinkage targets then becomes

$$\hat{\rho} = \sum_{j=1}^{p} \pi_{jj} + \sum_{j=1}^{p}\sum_{k=1}^{p} \frac{\bar{r}}{2}\left[\sqrt{\frac{s_{kk}}{s_{jj}}}\sigma_{jj,jk} + \sqrt{\frac{s_{jj}}{s_{kk}}}\sigma_{kk,jk}\right] I(j \neq k). \tag{28.27}$$

Finally, the shrinkage misspecification parameter is found with the equation

$$\hat{\psi} = \sum_{j=1}^{p}\sum_{k=1}^{p}(f_{jk} - s_{jk})^2. \tag{28.28}$$

The final shrinkage form of covariance is now

$$\mathbf{S}_{LW} = \delta^*\mathbf{F} + (1 - \delta^*)\mathbf{S}, \tag{28.29}$$

where the shrinkage intensity is

$$\delta^* = \max\left\{0, \min\left\{\frac{\kappa}{n}, 1\right\}\right\} \tag{28.30}$$

and

$$\kappa = \frac{\hat{\pi} - \hat{\rho}}{\hat{\psi}}. \tag{28.31}$$

Schäfer-Strimmer (SS) Shrinkage. Covariance matrix shrinkage based on the Schäfer-Strimmer method [16] is slightly similar to LW shrinkage, and employs the correlation matrix entries using the mean-zero standardized data such that Δ_{ij} and Δ_{ik} described previously are based on Z scores instead of the input x values. The asymptotic variance of r_{jk} is determined as

$$\mathrm{Var}(r_{jk}) = \frac{n}{(n-1)^3}\sum_{i=1}^{n}(\Delta_{ij}\Delta_{ik} - \mu_{\Delta\Delta})^2, \tag{28.32}$$

where

$$\mu_{\Delta\Delta} = \frac{1}{n}\Delta_{ij}\Delta_{ik}. \tag{28.33}$$

The shrinkage intensity is determined as

$$\lambda^* = \frac{\sum_{j \neq k} \text{Var}(r_{jk})}{\sum_{j \neq k} r_{jk}^2}, \tag{28.34}$$

and finally, elements of the shrinkage correlation matrix $\mathbf{R}_{SS}$ become

$$r^* = \begin{cases} 1, & \text{if } j = k \\ r_{jk} \min(1, \max(0, 1 - \lambda^*)) & \text{if } j \neq k. \end{cases} \tag{28.35}$$

In the next section, we apply shrinkage and filtering to datasets to evaluate the induced changes and characterize the results.

28.6 COVARIANCE MATRIX FILTERING

Covariance matrix filtering (CMF) is a multistep procedure that first involves extraction of eigenvalues from the covariance (correlation) matrix being considered and then filtering out eigenvalues assumed to be in the noise region based on RMT or shrinkage. Once CMF is applied and a new covariance matrix is obtained, the intent is to use the results in any analyses requiring a covariance (correlation) matrix.

We begin our CMF analysis by comparing several CMF results for a 100×100 random Wishart matrix $\mathbf{W}_{100}(400, \mathbf{1})$ with 400 degrees of freedom. We then evaluate CMF for a 59×59 array-by-array covariance matrix for the NCI-60 cell line cDNA expression dataset [17]. We dropped one microarray from the analyses since its diagnostic class is listed as unknown, leaving $n = 59$ arrays. All analyses were based on the first 500 genes in the NCI-60 expression dataset, which originally consisted of $p = 1375$ genes. We also applied CMF to the 280×280 (array-by-array) covariance matrix for 280 arrays in the GCM expression dataset for 27 classes of tumors and normal tissue types [18].

Six CMF runs were performed to obtain eigenvalues from the following matrices:

1. $\mathbf{S}_{\text{raw}}$—covariance matrix for raw input expression values
2. $\mathbf{S}_{\text{res}}$—filtered covariance matrix based on data matrix $\mathbf{X}^{\text{res}}$
3. $\mathbf{S}_{\text{MP}}$—filtered covariance matrix based on data matrix $\mathbf{X}^{\text{MP}}$
4. $\mathbf{S}_{\text{DK}}$—shrunken covariance matrix derived after DK eigenvalue shrinkage
5. $\mathbf{S}_{\text{LW}}$—shrunken covariance matrix derived after LW shrinkage
6. $\mathbf{S}_{\text{SS}}$—shrunken covariance matrix derived after SS shrinkage

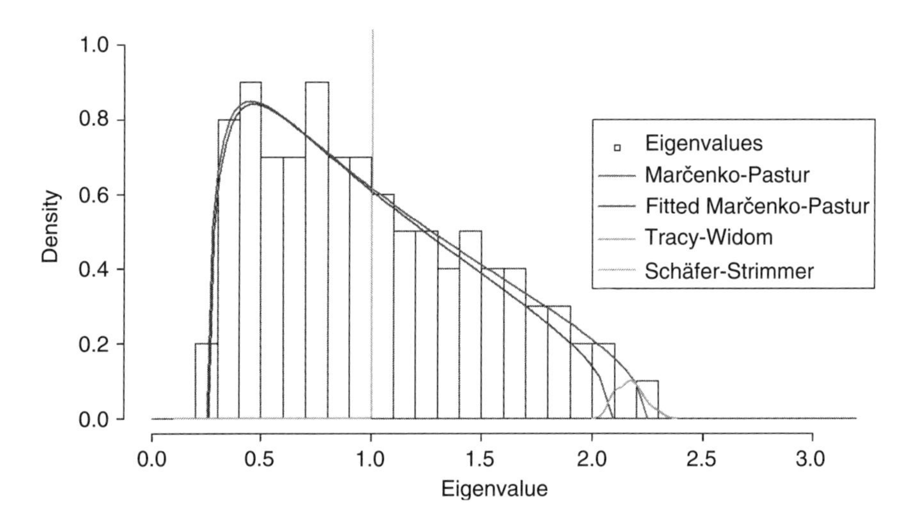

FIGURE 28.3 Histogram of eigenvalues for a $\mathbf{W}_{100}(400, \mathbf{1})$ Wishart matrix ($100 \times 100\,\mathbf{R}$ matrix) showing the theoretical Marčenko-Pastur density, the fitted Marčenko-Pastur density, the Tracy-Widom density of the largest eigenvalue, and the eigenvalues after shrinking **R** with the Shäfer-Strimmer shrinkage method.

During CMF, we were interested mainly in calculating the eigenvalues for the numbered sets of genes as they appeared in the datasets, and were not interested in identifying genes that were predictive of class and how their eigenvalues differed, even though optimally discriminating gene sets would likely yield different eigenvalues. Thus, the CMF analysis was not discriminative, but rather focused on eigenvalues obtained after CMF.

Results for several of the CMF methods are shown in Figure 28.3 for a random $\mathbf{W}_{100}(400, \mathbf{1})$ Wishart matrix, that is, a 100×100 $\mathbf{R}$ matrix, based on a random 400×100 $\mathbf{X}$ matrix with random elements distributed $\mathcal{N}(0,1)$. Note that the TW density of the largest eigenvalue encapsulates both the greatest empirical and greatest MP-fitted eigenvalue. The fitted MP density also happens to overlap the empirical density, but this phenomenon changes with empirical nonrandom data.

Fitted MP eigenvalue densities overlaid on histograms of the empirical eigenvalue densities for the 59×59 array-by-array covariance matrix $\mathbf{S}_{\text{raw}}$ for the NCI-60 arrays and 280×280 covariance matrix $\mathbf{S}_{\text{raw}}$ for the 280 array in the GCM dataset are shown in Figures 28.4 and 28.5. The maximum eigenvalue of $\lambda^+ = 1.296$ was obtained for the NCI-60 array covariance matrix, while $\lambda^+ = 0.4107$ was obtained for the GCM array covariance matrix. Eigenvalues that exceed the fitted λ^+ based on the MP density are considered informative, whereas eigenvalues below λ^+ are considered noise.

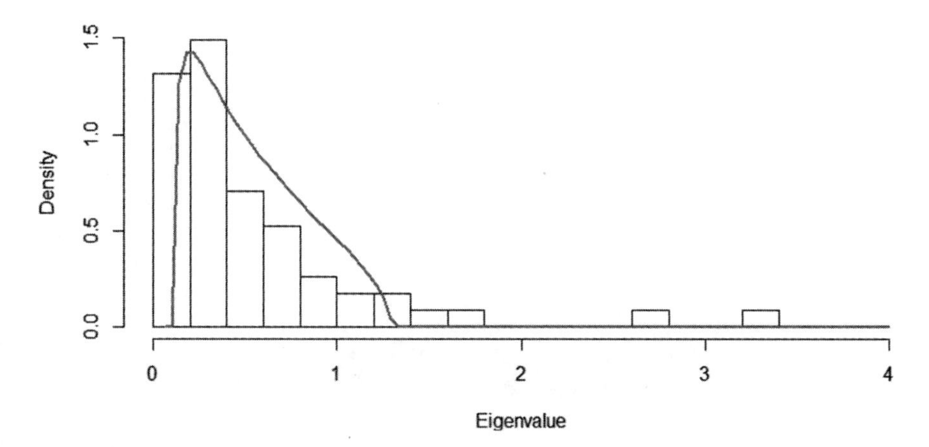

FIGURE 28.4 Histogram of eigenvalues from a 59 × 59 array-by-array covariance matrix $\mathbf{S}_{\text{raw}}$ based on the mean-zero standardized NCI-60 expression dataset using the first 500 genes in the 1375 gene set. Fitted line is the MP density fitted to the empirical eigenvalue density, based on fitted values of $\gamma = 0.2079$, $\sigma = 0.998$, $\lambda^- = 0.1045$, and $\lambda^+ = 1.296$. Eigenvalues above $\lambda^+ = 1.296$ are considered informative, while those below are considered to be in the noise region. (*See insert for color representation of the figure.*)

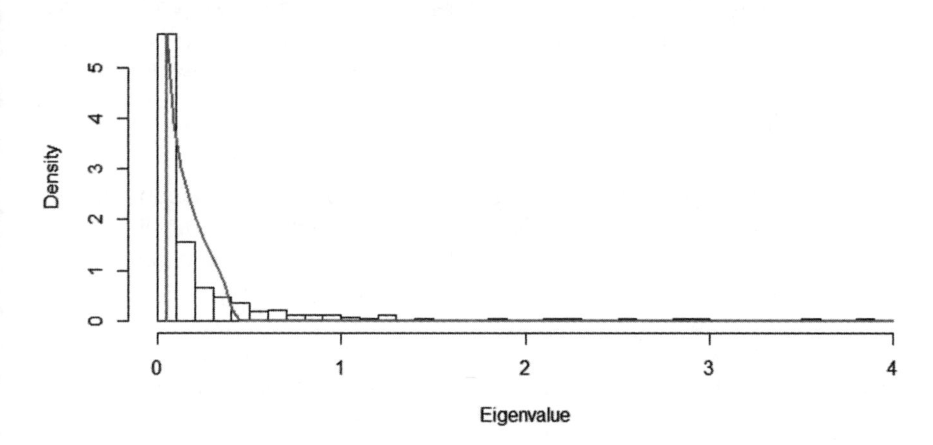

FIGURE 28.5 Histogram of eigenvalues from a 280 × 280 array-by-array covariance matrix $\mathbf{S}_{\text{raw}}$ based on the GCM expression dataset using the 999 informative genes with more than 80% present calls (log-transformed, mean-zero standardized). Fitted line is the MP density fitted to the empirical eigenvalue density, based on fitted values of $\gamma = 0.2392$, $\sigma = 1.2387$, $\lambda^- = 0$, and $\lambda^+ = 0.4107$. Eigenvalues above $\lambda^+ = 0.4107$ are considered informative, while those below are considered to be in the noise region. (*See insert for color representation of the figure.*)

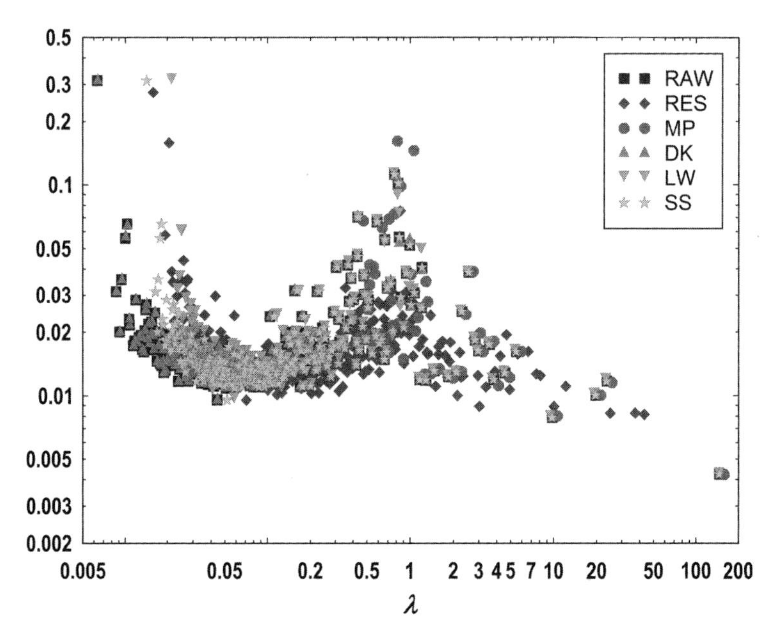

FIGURE 28.6 Inverse participation ratios (28.13) for eigenvectors as a function of eigenvalue for the 280 × 280 (array-by-array) covariance matrix ($p = 280$) for the GCM expression dataset with all informative $n = 999$ genes ($\gamma = 0.28 = 280/999$) having more than 80% present calls. Covariance matrices evaluated include $\mathbf{S}_{\text{raw}}$, $\mathbf{S}_{\text{res}}$, $\mathbf{S}_{\text{MP}}$, $\mathbf{S}_{\text{DK}}$, $\mathbf{S}_{\text{LW}}$, and $\mathbf{S}_{\text{SS}}$. (*See insert for color representation of the figure.*)

Plots of IPR (28.13) for each eigenvector as a function of eigenvalue are shown in Figure 28.6 for the 280 × 280 array-by-array covariance matrices $\mathbf{S}_{\text{raw}}$, $\mathbf{S}_{\text{res}}$, $\mathbf{S}_{\text{MP}}$, $\mathbf{S}_{\text{DK}}$, $\mathbf{S}_{\text{LW}}$, and $\mathbf{S}_{\text{SS}}$, based on the GCM expression dataset with all informative $n = 999$ genes ($\gamma = 0.28 = 280/999$) having more than 80% present calls. Figure 28.7 shows IPRs for the 280 × 280 covariance matrices $\mathbf{S}_{\text{raw}}$, $\mathbf{S}_{\text{res}}$, $\mathbf{S}_{\text{MP}}$, $\mathbf{S}_{\text{DK}}$, $\mathbf{S}_{\text{LW}}$, $\mathbf{S}_{\text{SS}}$ with the exception that $n = 70$ genes were used in the input dataset. Looking at these figures, it is instructive to note that IPR values for eigenvalues below the MP eigenvalues are not considered informative, so the values of IPR and participation of arrays within them are not credible. When $\gamma = 0.28$ as in Figure 28.6, at $\lambda = 1$, where noise levels pick up, the IPRs are observed to increase because there are fewer participating arrays (dimensions), which explains the loss of variation. When $\gamma = 0.28$, the sample size n outweighs the number of dimensions of the covariance matrix, and covariance shrinkage methods are not very helpful in improving stability of eigenvalues. However, when $\gamma = 4$ as in Figure 28.7, the IPR values increase rapidly near the smallest values of λ, and are more dramatic for the shrunken covariance matrices.

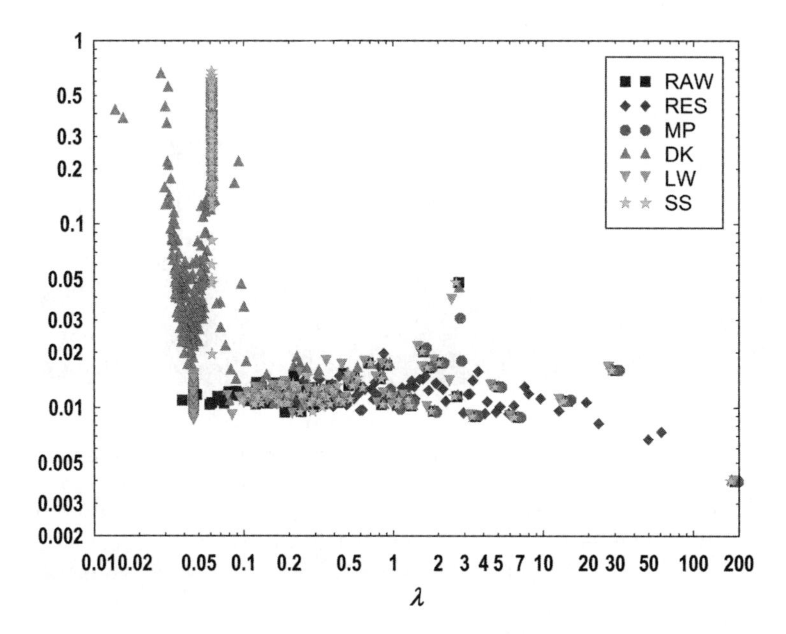

FIGURE 28.7 Inverse participation ratios for eigenvectors as a function of eigenvalue for the 280 × 280 (array-by-array) covariance matrix ($p = 280$) for the GCM expression dataset with $n = 70$ genes ($\gamma = 4 = 280/70$) having more than 80% present calls. Covariance matrices evaluated include $\mathbf{S}_{\text{raw}}$, $\mathbf{S}_{\text{res}}$, $\mathbf{S}_{\text{MP}}$, $\mathbf{S}_{\text{DK}}$, $\mathbf{S}_{\text{LW}}$, and $\mathbf{S}_{\text{SS}}$.

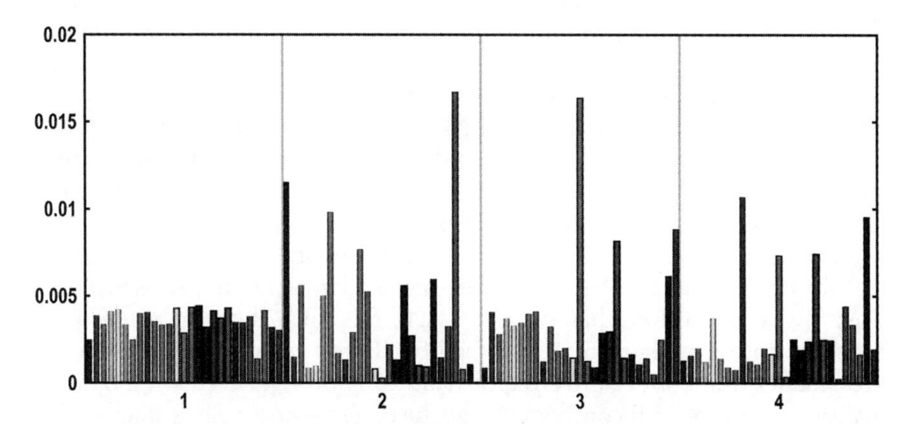

FIGURE 28.8 Class-specific inverse participation ratios for eigenvectors as a function of eigenvalue for the 280 × 280 (array-by-array) covariance matrix ($p = 280$) for the GCM expression dataset with $n = 70$ genes having more than 80% present calls ($\gamma = 4 = 280/70$). The x-axis values of 1,2,3,4 represent the first, second, third, and fourth eigenvectors, respectively. Covariance matrix evaluated includes $\mathbf{S}_{\text{raw}}$.

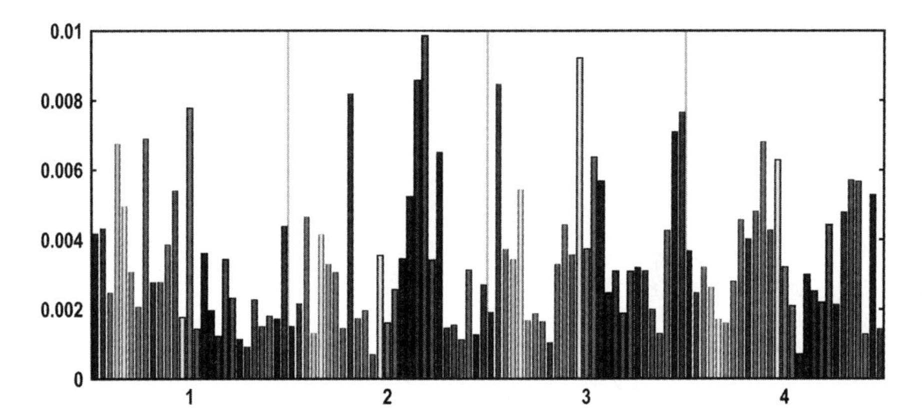

FIGURE 28.9 Class-specific inverse participation ratios for eigenvectors as a function of eigenvalue for the 280×280 (array-by-array) covariance matrix ($p = 280$) for the GCM expression dataset with $n = 70$ genes having more than 80% present calls ($\gamma = 4 = 280/70$). The x-axis values of 1,2,3,4 represent the first, second, third, and fourth eigenvectors, respectively. Covariance matrix evaluated includes $\mathbf{S}_{\text{res}}$.

Class-specific IPRs are illustrated for the GCM expression set array-by-array (280×280) covariance matrix $\mathbf{S}_{\text{raw}}$ based on 70 genes (Figure 28.8) using the raw $\mathbf{X}$ data matrix and the same 70 genes using the residual matrix $\mathbf{X}^{\text{res}}$ (Figure 28.9) with the first principal component "subtracted" via regression analysis. As expected, the raw data result in similar values of the majority of the 27 class-specific IPRs for the first eigenvector (greatest eigenvalue), whereas the class-specific IPRs of the first eigenvalue based on the $\mathbf{X}^{\text{res}}$ data matrix are much more heterogeneous. Subtraction of factor effects for the first principal component removes widespread correlation among many of the arrays, which may be due to global regulatory transcriptional control, or any batch (systematic) effects that are present.

Figures 28.10, 28.11, 28.12, and 28.13 show the empirical eigenvalue distribution for the 280×280 (array-by-array) covariance matrices $\mathbf{S}_{\text{raw}}$, $\mathbf{S}_{\text{res}}$, $\mathbf{S}_{\text{MP}}$, $\mathbf{S}_{\text{DK}}$, $\mathbf{S}_{\text{LW}}$, and $\mathbf{S}_{\text{SS}}$ based on 999, 280, 140, and 70 genes, respectively. First, we note that eigenvalues for the MP-filtered covariance matrix results in a smaller number of nonzero eigenvalues. This is due to the removal of a majority of orthogonal components whose eigenvalues are less than λ^{+}. The effects of covariance matrix shrinkage methods are not really observable until $\gamma = 1$ and higher, when the covariance matrix off-diagonals are shrunken to keep the matrix positive definite. At the greatest value of $\gamma = 4$, the DK shrinkage estimator was observed to make a smooth transition among eigenvalues, whereas the LW and SS shrinkage techniques portray jumpy transitions among eigenvalues.

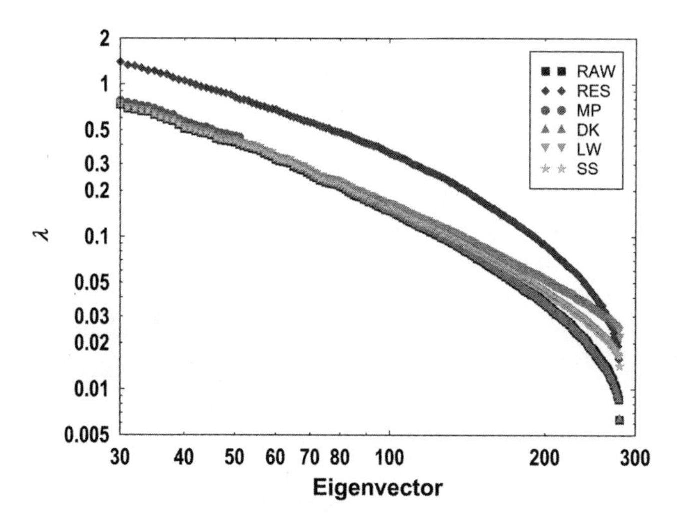

FIGURE 28.10 Empirical eigenvalue distribution for 280 × 280 (array-by-array) covariance matrix ($p = 280$) using the GCM expression dataset with $n = 999$ genes ($\gamma = 0.28 = 280/999$) having more than 80% present calls. Covariance matrices evaluated include $\mathbf{S}_{\text{raw}}$, $\mathbf{S}_{\text{res}}$, $\mathbf{S}_{\text{MP}}$, $\mathbf{S}_{\text{DK}}$, $\mathbf{S}_{\text{LW}}$, and $\mathbf{S}_{\text{SS}}$.

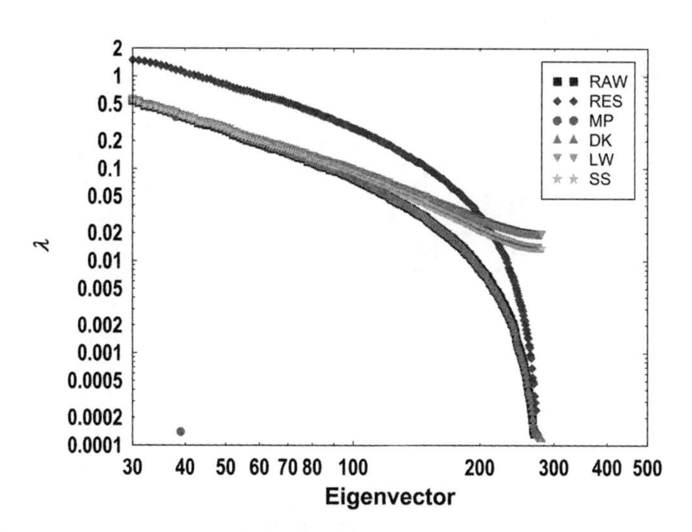

FIGURE 28.11 Empirical eigenvalue distribution for 280 × 280 (array-by-array) covariance matrix ($p = 280$) using the GCM expression dataset with $n = 280$ genes ($\gamma = 1 = 280/280$) having more than 80% present calls. Covariance matrices evaluated include $\mathbf{S}_{\text{raw}}$, $\mathbf{S}_{\text{res}}$, $\mathbf{S}_{\text{MP}}$, $\mathbf{S}_{\text{DK}}$, $\mathbf{S}_{\text{LW}}$, and $\mathbf{S}_{\text{SS}}$. (*See insert for color representation of the figure.*)

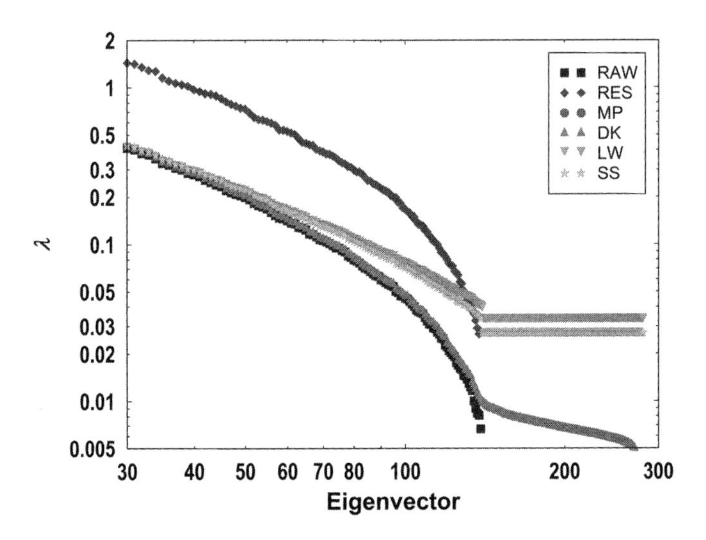

FIGURE 28.12 Empirical eigenvalue distribution for 280 × 280 (array-by-array) covariance matrix ($p = 280$) using the GCM expression dataset with $n = 140$ genes ($\gamma = 2 = 280/140$) having more than 80% present calls. Covariance matrices evaluated include $\mathbf{S}_{\text{raw}}$, $\mathbf{S}_{\text{res}}$, $\mathbf{S}_{\text{MP}}$, $\mathbf{S}_{\text{DK}}$, $\mathbf{S}_{\text{LW}}$, and $\mathbf{S}_{\text{SS}}$.

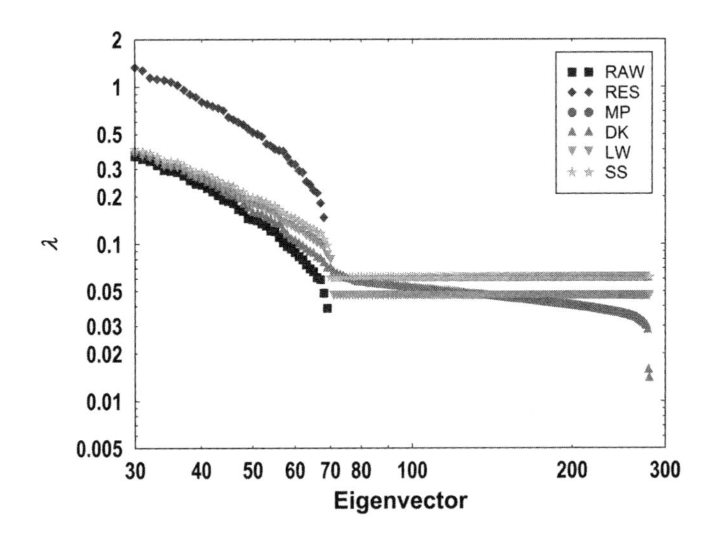

FIGURE 28.13 Empirical eigenvalue distribution for 280 × 280 (array-by-array) covariance matrix ($p = 280$) using the GCM expression dataset with $n = 70$ genes ($\gamma = 4 = 280/70$) having more than 80% present calls. Covariance matrices evaluated include $\mathbf{S}_{\text{raw}}$, $\mathbf{S}_{\text{res}}$, $\mathbf{S}_{\text{MP}}$, $\mathbf{S}_{\text{DK}}$, $\mathbf{S}_{\text{LW}}$, and $\mathbf{S}_{\text{SS}}$.

28.7 SUMMARY

In this chapter, we explored RMT methods for filtering noise from a covariance matrix. There are essentially two major problems inherent in the eigenanalysis of covariance matrices. The first is related to a widespread global influence on many of the variables, which is characterized by the greatest eigenvalue, λ_1. When λ_1 far exceeds the remaining eigenvalues, many variables are highly correlated with one another, suggesting redundancy of information. When the effect of the greatest eigenvector (eigenvalue) was subtracted from the data matrix by replacing the data with residuals obtained from regressing each variable on the first principal component, we could observe greater heterogeneity among class-specific participation ratios for the first eigenvector. The eigenvector participation ratios were instrumental in determining which eigenvalues are dominated by fewer dimensions, ultimately reflecting whether the within-eigenvector contribution to the eigenvalue is homogeneous or heterogeneous.

The second problem surrounds the stability and noise region of eigenvalues as the ratio $\gamma = p/n$ increases and decreases or whether both $p, n \to \infty$. For noise removal, we showed how to filter the noise component out of a covariance matrix by subtracting all eigenvectors from the data whose eigenvalues were less than the fitted value of the maximum eigenvalue (λ^+) of the Marčenko-Pastur density. A key point to remember when filtering out effects of many of the smaller eigenvectors from the data matrix is that after rerunning eigenanalysis on the filtered data, the smaller eigenvalues will become zero, since the variables were enforced not to load on these eigenvectors. This is not viewed as a problem, however, because the intent of filtering out the noise eigenvectors is to work with only those eigenvalues that are considered informative, namely, the nonzero eigenvalues of the filtered covariance matrix. Our results from filtering covariance matrices by fitting MP densities and then subtracting effects of noise eigenvectors from the data show that the commonly used rule of thumb to use eigenvalues greater than unity may lead to biased results. The covariance shrinkage methods that were employed were observed to exert their greatest effect when γ was close to and greater than unity, when $p > n$. Covariance shrinkage techniques attempt to ensure positive definiteness by enforcing none of the eigenvalues to become zero, and also improve the condition index of the covariance matrix such that the ratio of the greatest to the smallest eigenvalue is not too large. No significant differences were observed in the effects of the three shrinkage methods employed, except that the DK method seemed to result in a smooth transition of eigenvalues in regions of criticality. A major finding for the shrinkage methods was that, although the smallest eigenvalues lie in the noise region, shrinkage nevertheless transforms them to acquire greater values, which ultimately minimizes the condition number.

Filtered covariance matrices are beneficial in PCA, regression, risk estimation, and prediction and may lead to reduced bias and improved performance. In other fields such as financial engineering, quantitative finance, and portfolio optimization, eigenvalues are viewed as risk since they represent the variance explained by the respective eigenvectors, so every effort is commonly made to remove noise to ensure greater profitability. Regardless of whether the application is scientific or financial, covariance matrix filtering techniques will likely play a central role in data mining for the next several decades.

REFERENCES

[1] J. Wishart. Generalized product moment distributions in samples. *Biometrika* **1**(2):32-52, 1928.

[2] A.T. James. Normal multivariate analysis and the orthogonal group. *Ann. Math. Stat.* **1**:40-75, 1954.

[3] E.P. Wigner. Characteristic vectors of bordered matrices with infinite dimensions. *Ann. Math.* **62**(3):548-564, 1955.

[4] L.A. Pastur. On the spectrum of random matrices. *Teor. Mat. Fiz.*, **10**:102-111, 1973.

[5] A. Khorunzhy. Sparse random matrices: Spectral edge and statistics of rooted trees. *Adv. Appl. Probab.* **33**:1-18, 2001.

[6] A. Edelman, N.R. Rao. Random matrix theory. *Acta Numerica* **14**:233-297, 2005.

[7] V.A. Marčenko, L.A. Pastur. Distribution of eigenvalues for some sets of random matrices. *Mat. Sb.* (NS) **72**(114):4, 507-536, 1967.

[8] C.A. Tracy, H. Widom. Correlation functions, cluster functions, and spacing distributions for random matrices. *J. Stat. Phys.* **92**:809-835, 1998.

[9] J. Baik, G. BenArous, S. Péché. Phase transition of the largest eigenvalue for nonnull complex sample covariance matrices. *Ann. Probab.* **33**(5):1643-1697, 2005.

[10] C. Biely, S. Thurner. Random matrix ensembles of time-lagged correlation matrices: Derivation of eigenvalue spectra and analysis of financial time-series. *Quant. Finance* **8**:705-722, 2008.

[11] C. Stein. Inadmissibility of the usual estimator for the mean of a multivariate normal distribution. *Proc. 3rd Berkeley Symp. Mathematical and Statistical Probability,* Univ. California Press, Berkeley, 1956, vol. 1, 197 pp. 197-206.

[12] B. Efron, C. Morris. Stein's paradox in statistics. *Sci. Am.* **237**:119-127, 1977.

[13] I.M. Johnstone. High dimensional statistical inference and random matrices. *Proc. 2006 Int. Congress of Mathematicians,* Madrid, European Mathematical Society, Zürich, 2007.

[14] M.J. Daniels, R.E. Kass. Shrinkage estimators for covariance matrices. *Biometrics* **57**(4):1173-1184, 2001.

[15] O. Ledoit, M. Wolf. Honey, I shrunk the sample covariance matrix. *J. Portfolio Manage.* **30**(4):110-119, 2004.

[16] J. Schäfer, K. Strimmer. A shrinkage approach to large-scale covariance matrix estimation and implications for functional genomics. *Stat. Appl. Genet. Mol. Biol.* **4**(1):1-30, 2005.

[17] D.T. Ross, U. Scherf, M.B. Eisen, C.M. Perou, C. Rees, P. Spellman, V. Iyer, S.S. Jeffrey, M. Van de Rijn, M. Waltham, A. Pergamenschikov, J.C. Lee, D. Lashkari, D. Shalon, T.G. Myers, J.N. Weinstein, D. Botstein, P.O. Brown. Systematic variation in gene expression patterns in human cancer cell lines. *Nat. Genet.* **24**(3):227-235, 2000.

[18] S. Ramaswamy, P. Tamayo, R. Rifkin, S. Mukherjee, C.-H. Yeang, M. Angelo, C. Ladd, M. Reich, E. Latulippe, J.P. Mesirov, T. Poggio, W. Gerald, M. Loda, E.S. Lander, T.R. Golub. Multiclass cancer diagnosis using tumor gene expression signatures. *Proc. Natl. Acad. Sci. USA* **98**(26):15149-15154, 2001.

APPENDIXES

APPENDIX A

PROBABILITY PRIMER

A.1 CHOICES

Definition A.1 When two choices must be made, and there are m choices for the first and n choices for the second, the total number of choices (i.e., options) is equal to

$$N = mn. \tag{A.1}$$

Example A.1 How many ways can a new computer be selected if there are four hard-disk options ($m = 4$) and three RAM memory options ($n = 3$)?

Solution: The number of ways is

$$N = mn = 4 \times 3 = 12. \tag{A.2}$$

■

Definition A.2 When a series of k choices must be made, and the number of choices at each step are $r_1, r_2, \ldots, r_k$, then the total number of ways of making the $r_1, r_2, \ldots, r_k$ choices from n distinct objects $a_1, a_2, \ldots, a_n$ is equal to

$$N = r_1 \times r_2 \times \cdots \times r_k. \tag{A.3}$$

Example A.2 A new sequencer is available with four capillary designs, three computer options, and two printer options. How many ways can a customer order the sequencer?

Solution:

$$N = r_1 \times r_2 \times r_3 = 4 \times 3 \times 2 = 24. \tag{A.4}$$

■

Classification Analysis of DNA Microarrays, First Edition. Leif E. Peterson.

Definition A.3 Whenever we choose r objects out of n distinct objects and the selected objects are returned to their originating population, the number of ordered samples is

$$N = n^r = \text{\# distinct objects}^{\text{\# trials}}. \tag{A.5}$$

Example A.3 (Sampling with Replacement) There are four bases: A, G, C, and T (distinct objects, $n = 4$). How many 1-megabase sequences are possible ($r = 1,000,000$)?

Solution:

$$N = n^r = 4^{1,000,000}. \tag{A.6}$$

■

Example A.4 (Sampling with Replacement) Using the four bases ($n = 4$), how many three-letter codons are possible ($r = 3$)?

Solution:

$$N = n^r = 4^3 = 64. \tag{A.7}$$

Note that there are only 20 amino acids, since each amino acid is represented by more than one codon. ■

Example A.5 (Sampling with Replacement) A fair coin is flipped 8 times. Given that there are $n = 2$ distinct objects [heads (h) and tails (t)], how many distinct outcomes are possible?

Solution:

$$N = n^r = 2^8 = 2 \times 2 \times 2 \times 2 \times 2 \times 2 \times 2 \times 2 = 256. \tag{A.8}$$

■

A.2 PERMUTATIONS

Permutations arise whenever $r \leq n$ objects are selected from a set of n distinct objects in a particular order, and describe the number of ways in which r objects can be arranged (ordered). For example, the letters `AE` and `EA` are permutations of two letters from the 26-letter alphabet. Consider the first word `AE`; once `A` was selected, it was not available during the second selection. The same rule applies for the word `EA`; once `E` was selected, it was not available for selection during the second selection. Thus, after the first object is chosen, there are $n - 1$ objects available for selection. After the second selection, there are $n-2, n-3, \ldots,$ until there is one object remaining. This is another form of sampling without replacement.

Definition A.4 The total number of permutations of all n distinct objects $a_1, a_2, \ldots, a_n$ when chosen objects are not replaced back into the population is

$$N = n(n-1)(n-2) \times \cdots \times 3 \times 2 \times 1 = n! \tag{A.9}$$

Example A.6 (Sampling without Replacement) A biologist needs to conduct five new experiments and has five technicians who are each capable of performing them. How many ways can the five technicians be assigned (ordered) to the five experiments ($r = n$)?

Solution:

$$N = 5! = 5 \times 4 \times 3 \times 2 \times 1 = 120. \tag{A.10}$$

■

Definition A.5 The total number of permutations of assigning n distinct objects when r_1 are alike, r_2 are alike, up to r_k alike ($i = 1, 2, \ldots, k$) is

$$N = \frac{n!}{r_1! r_2! \times \cdots \times r_k!}. \tag{A.11}$$

Example A.7 A company has 10 desktop computers with different (i.e., distinct) capabilities. One is a Dell, two are HPs, three are Alienware, and four are Macs. How many ways can the 10 computers be ordered?

Solution:

$$N = \frac{10!}{1!2!3!4!} = \frac{3,628,800}{288} = 12{,}600. \tag{A.12}$$

■

Definition A.6 The number of permutations of r objects selected from a population of n distinct objects $a_1, a_2, \ldots, a_n$ when chosen objects are not replaced back into the population is

$${}_nP_r = n(n-1)(n-2) \times \cdots \times (n-r+1) = \frac{n!}{(n-r)!}. \tag{A.13}$$

Example A.8 (Sampling without Replacement) How many three-letter ordered permutations ($r = 3$) are there from the 26-letter alphabet ($n = 26$)?

Solution:

$$_nP_r = \frac{26!}{(26-3)!} = \frac{\exp\{\log(26) + \log(25) + \cdots + \log(1)\}}{\exp\{\log(23) + \log(22) + \cdots + \log(1)\}} = 15{,}600. \quad \text{(A.14)}$$

This is also equal to $26 \times 25 \times 24 = 15,600$. Note that results of this sampling scheme would produce all of the 15,600 permutations

```
ABC  ACB  BAC  BCA  CAB  CBA
ABD  ADB  BAD  BDA  DAB  DBA
ABE  AEB  BAE  BEA  EAB  EBA
.
.
.
XYZ  XZY  ZXY  ZYX ZXY  ZYX
```

Finally, one notes above that when multiplying large integers, computer integer overflow can be avoided by first taking the logarithm of the integers, summing them up, and then exponentiating the sum, since $\log(A \times B) = \log(A) + \log(B)$. Thus, $\exp(\sum_i \log(X_i)) = \prod_i X_i$. ■

A.3 COMBINATIONS

Each row in the text output in the example above contains $r! = 3! = 3 \times 2 \times 1 = 6$ three-letter permutations and forms a *combination*. Combinations are unordered arrangements of r objects selected from n distinct objects. As an example, the word ABC lists three letters in no particular order, while ABC ACB BAC BCA CAB CBA lists the $r! = 3! = 6$ ways (permutations) in which the three letters can be ordered. ■

Definition A.7 The number of combinations of r objects selected from a population of n distinct objects $a_1, a_2, \ldots, a_n$ is equal to

$$\begin{aligned} _nC_r = \binom{n}{r} &= \frac{n(n-1)(n-2) \times \cdots \times (n-r+1)}{r!} \\ &= \frac{n!}{r!(n-r)!}. \end{aligned} \quad \text{(A.15)}$$

where $\binom{n}{r}$ is the *binomial coefficient*.

Example A.9 How many three-letter combinations are there ($r = 3$) from the 26-letter alphabet ($n = 26$)?

Solution:

$$ {}_nC_r = \binom{26}{3} = \frac{26!}{3!(26-3)!} = \frac{\exp\{\log(26) + \log(25) + \cdots + \log(1)\}}{6\exp\{\log(23) + \log(22) + \cdots + \log(1)\}} = 2600. \tag{A.16} $$

Note that, by using results of Example A.8, each combination represents $r!$ permutations as shown below

```
ABC ACB BAC BCA CAB CBA
ABD ADB BAD BDA DAB DBA
ABE AEB BAE BEA EAB EBA
```

(combination 1, $r! = 3! = 6$ permutations)
(combination 2, $r! = 3! = 6$ permutations)
(combination 3, $r! = 3! = 6$ permutations)

.
.
.

`XYZ XZY ZXY ZYX ZXY ZYX` (combination 2600, $r! = 3! = 6$ permutations)

We already know that there were $3! = 6$ permutations in each combination, so we can check the results by multiplying the number of permutations per combination by the number of combinations to obtain the total number of permutations. This gives $6 \times 2600 = 15,600$. These results illustrate why there are $r!$ more ordered permutations of size r than there are combinations of size r. Hence, to determine the number of combinations, we divide by $r!$ ■

Example A.10 Assume a single-locus two-allele model at the "A" locus. What is the total number of possible genotypes given the two alleles A and a?

Solution: There are n homozygotes (AA, aa), and the number of heterozygotes is

$$ N = \binom{2}{2} = \frac{2!}{2!(2-2)!} = \frac{2}{2} = 1. \tag{A.17} $$

Adding together the two homozygote genotypes AA and aa with the one heterozygote Aa gives us three possible genotypes. ■

Definition A.8 The total number of genotypes for a locus with m alleles is

$$ N = \binom{m+1}{2} = \frac{m!}{r!(m-r)!} = \frac{m(m+1)}{2}. \tag{A.18} $$

Example A.11 A locus has four alleles: 1,2,3,4. How many possible genotypes are there for this locus?

Solution:

$$N = \binom{4+1}{2} = \frac{5!}{2!(5-2)!} = \frac{120}{12} = 10 = \frac{4(4+1)}{2}, \qquad \text{(A.19)}$$

and they are the sampled heterozygotes 12, 13, 14, 23, 24, and 34, and the homozygotes (11, 22, 33, and 44), which are essentially added because they can't be drawn from the population when there is no replacement (i.e., when permutations are the goal). The number of homozygotes (11, 22, 33, and 44) is added to the resulting number for heterozygotes by use of the +1 in the top of the binomial coefficient. If we had to determine how many permutations there are for selecting two alleles at a time (without replacement) from the four available alleles, the solution would be

$$N = \frac{m!}{(m-r)!} = \frac{4!}{(4-2)!} = \frac{24}{2} = 12, \qquad \text{(A.20)}$$

and the resulting 12 permutations would be

12 21
13 31
14 41
23 32
24 42
34 43

Note that homozygotes are not sampled since there is no replacement when dealing with permutations. For genotype counting, distinction between maternal and paternal copies of alleles is rarely necessary unless there is interest in identity by descent, paternal or maternal influences, and other factors. Therefore, we are more interested in the number of combinations that could be sampled. ■

A.4 PROBABILITY

Definition A.9 The sample space Ω contains all of the ordered samples (with replacement) of r objects from n distinct objects.

Example A.12 What is the sample space of genotypes for a locus with four alleles: 1,2,3,4?

Solution: There are $n = 4$ distinct objects and $r = 2$ selected, so there are $4^2 = 16$ ordered samples in the sample space:

$$\Omega = \{11, 12, 13, 14, 21, 22, 23, 24, 31, 32, 33, 34, 41, 42, 43, 44\}. \qquad \text{(A.21)}$$

Note that when sampling with replacement, each sequence has a $\frac{1}{16}$ probability of being selected. ■

A.4.1 Addition Rule

Events that are *mutually exclusive* cannot occur at the same time. For example, individuals with genotype AA cannot have genotype Aa or aa.

Definition A.10 For two mutually exclusive events A and B, the probability that one or the other will occur is

$$P(A \text{ or } B) = P(A \cup B) = P(A) + P(B). \tag{A.22}$$

Example A.13 A group of patients were genotyped at the A locus. The genotype frequency for homozygotes AA, denoted as P_{AA}, was found to be 0.2, while the genotype frequency for Aa individuals was $P_{Aa} = 0.3$. What is the probability that at least one patient will have genotype AA or Aa?

Solution: The probability for two mutually exclusive genotypes is

$$P(AA \text{ or } Aa) = P_{AA} + P_{Aa} = 0.2 + 0.3 = 0.5. \tag{A.23}$$

■

Many loci have more than two alleles. Let m represent the number of alleles at a locus ($m > 2$).

Definition A.11 The probability of at least one among all mutually exclusive genotypes being observed at a locus is

$$\begin{aligned} P(A_1A_1 \text{ or } A_1A_2 \text{ or } \cdots \text{ or } A_mA_m) &= P(A_1A_1) + P(A_1A_2) + \cdots + P(A_mA_m) \\ &= \sum_{i=1}^{m}\sum_{j=1}^{m} P(A_iA_j) = P(\Omega) = 1. \end{aligned} \tag{A.24}$$

Example A.14 Consider again the genotype frequencies $P_{AA} = 0.2$ and $P_{Aa} = 0.3$. What is the probability of observing at least one of the mutually exclusive genotypes?

Solution: Since genotypes at the same locus are mutually exclusive, $P_{aa} = 1 - (P_{AA} + P_{Aa})$, which is 0.5. The probability for at least one of the mutually exclusive genotype occurring is

$$P(AA \text{ or } Aa \text{ or } aa) = P_{AA} + P_{Aa} + P_{aa} = 0.2 + 0.3 + 0.5 = 1. \tag{A.25}$$

■

Definition A.12 The general addition rule for the probability of two events A and B is

$$P(A \cup B) = P(A \text{ or } B) = P(A) + P(B) - P(A \cap B), \tag{A.26}$$

where $P(A \cap B)$ represents $P(A$ and $B)$. This relationship holds whether events A and B are mutually exclusive. If A and B are mutually exclusive events, then $P(A \cap B) = 0$.

Example A.15 The probability of "finding" a new gene by performing experiment A is 0.25; experiment B, 0.1; and both A and B, 0.06. You perform both. What is the probability of finding the gene?

Solution: The general addition rule can be used as follows:

$$P(A \cup B) = P(A \text{ or } B) = P(A) + P(B) = 0.25 + 0.1 - 0.06 = 0.29. \tag{A.27}$$

■

A.4.2 Multiplication Rule and Conditional Probabilities

Definition A.13 The probability that events A and B occur is the probability that A occurs times the probability that B occurs given that A occurs or has already occurred.

$$P(A \text{ and } B) = P(A \cap B) = P(A)P(B|A). \tag{A.28}$$

Analogously, the probability that A and B occur is the probability that B occurs times the probability that A occurs given B occurs or has already occurred, given as

$$P(A \cap B) = P(B)P(A|B). \tag{A.29}$$

Thus, we obtain the equivalence

$$P(A)P(B|A) = P(B)P(A|B). \tag{A.30}$$

The *conditional probability* of A occurring given that B occurs is then found by solving for $P(A|B)$:

$$P(A|B) = \frac{P(A)P(B|A)}{P(B)}. \tag{A.31}$$

Here we note that the numerator involving events A and B cannot be mutually exclusive, where $P(A \cap B) = 0$.

Example A.16 Penetrance of a genotype represents the probability of being affected given a particular genotype, and is represented by $P(D|g)$. If 36 subjects with genotype AA are affected, and there are 360 individuals in a study with genotype AA, what is the penetrance of the AA genotype?

Solution: We obtain penetrance by calculating the proportion of affecteds with genotype AA by the total number of individuals with genotype AA:

$$P(D|AA) = \frac{36}{360} = 0.10. \tag{A.32}$$

If there were 1000 total subjects, then the probability of being affected and having genotype AA would be $P(D \text{ and } AA) = \frac{36}{1000} = 0.036$, whereas the genotype frequency for AA is $P_{AA} = \frac{360}{1000} = 0.36$. In terms of probabilities, our results would be

$$P(D|AA) = \frac{P(D \text{ and } AA)}{P_{AA}} = \frac{0.036}{0.36} = 0.10. \tag{A.33}$$

Penetrance $P(D|G_i)$ can also be described using the form

$$P(A|B) = \frac{P(A)P(B|A)}{P(B)}, \tag{A.34}$$

rewritten as

$$P(D|G_i) = \frac{P(D)P(G_i|D)}{P(G_i)}. \tag{A.35}$$

One may think that penetrance can now be determined solely by finding the values of the constituent parts $P(D)$, $P(G_i|D)$, and $P(G_i)$. This must be approached with caution, however, since $P(D)$ itself is a conditional probability that depends on all of the genotype frequencies and their respective penetrances. Further, it is difficult to get a handle on what the "reverse probability" $P(G_i|D)$ is. In later sections, we show how $P(D)$ is determined using the *elimination rule*, and how $P(G_i|D)$ is determined using Bayes' rule. ■

A.4.3 Multiplication Rule for Independent Events

Definition A.14 The probability of two independent events is

$$P(A \cap B) = P(A \text{ and } B) = P(A)P(B). \tag{A.36}$$

Example A.17 Assume two loci A and B are unlinked. If the frequency of genotype A_1A_4 is 0.25 and frequency of B_2B_5 is 0.10, what is the probability that someone has both independent genotypes.

Solution:

$$P(A_1A_4 \text{ and } B_2B_5) = P(A_1A_4)P(B_2B_5) = (0.25)(0.10) = 0.025. \quad \text{(A.37)}$$

■

A.4.4 Elimination Rule (Disease Prevalence)

Definition A.15 For k mutually exclusive pathways $G_1, G_2, \ldots, G_k$, the probability of event E occurring

$$\begin{aligned} P(E) &= P(G_1)P(E|G_1) + P(G_2)P(E|G_2) + \cdots + P(G_k)P(E|G_k) \\ &= \sum_{i=1}^{k} P(G_i)P(E|G_i). \end{aligned} \quad \text{(A.38)}$$

It is easy to understand how the elimination rule can be used to rule out probabilities with the process of elimination.

Example A.18 Assume five genotypes $G_1, G_2, \ldots, G_5$ that confer risk for developing disease D. We know that the penetrances are $P(D|G_1), P(D|G_2), \ldots, P(D|G_5)$. What is the *prevalence* of disease given the five penetrances?

Solution:

$$P(D) = P(G_1)P(D|G_1) + P(G_2)P(D|G_2) + \cdots + P(G_5)P(D|G_5). \quad \text{(A.39)}$$

■

Figure A.1 illustrates the concept of disease prevalence based on genotypic probabilities and genotype-specific penetrances.

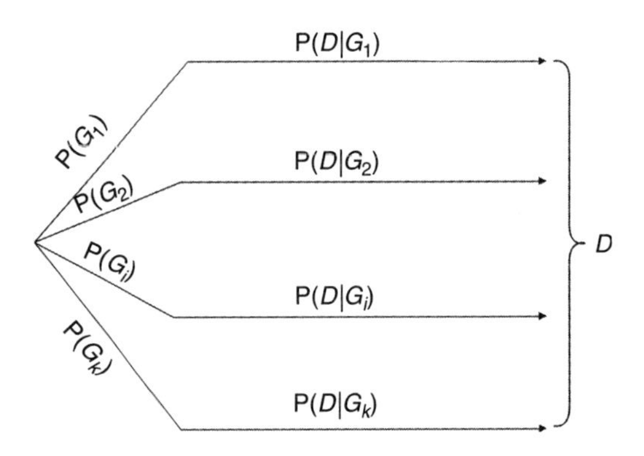

FIGURE A.1 Prevalence of disease based on genotypic probabilities and genotype penetrances.

A.4.5 Bayes' Rule (Pathway Probabilities)

We already mentioned conditional probabilities described by the relationship

$$P(A \cap B) = P(B)P(A|B) = P(A)P(B|A). \tag{A.40}$$

Knowing that penetrance is $P(D|G_i)$, we can easily show that

$$P(D)P(G_i|D) = P(G_i)P(D|G_i). \tag{A.41}$$

By dividing both sides by $P(D)$, we get

$$P(G_i|D) = \frac{P(G_i)P(D|G_i)}{P(D)}. \tag{A.42}$$

Earlier, it was shown by the elimination rule that for k penetrances, the prevalence of disease D was

$$P(D) = \sum_{i=1}^{k} P(G_i)P(D|G_i). \tag{A.43}$$

Bayes' rule simply places prevalence in the denominator of the relationship

$$\begin{aligned} P(G_i|D) &= \frac{P(G_i)P(D|G_i)}{P(D)} \\ &= \frac{P(G_i)P(D|G_i)}{\sum_{i=1}^{k} P(G_i)P(D|G_i)}. \end{aligned} \tag{A.44}$$

Thus, Bayes' rule tells us the probability that disease occurred as a result of a specific genotype's frequency multiplied by its penetrance ($P(G_i)P(D|G_i)$), divided by the prevalence. Another way of looking at disease D is by assuming each product $P(G_i)P(D|G_i)$ as a pathway to disease. If there are k pathways to disease, then Bayes' rules gives the contribution of pathway $P(G_i)P(D|G_i)$ out of total prevalence [i.e., $\sum_{i=1}^{k} P(G_i)P(D|G_i)$].

APPENDIX B

MATRIX ALGEBRA

B.1 VECTORS

Distance between Two Vectors. Let P be a point in n dimensional space based on n dimensions, where, for example, n is the number of patients and x_i is the value of the random variable x for the ith patient ($i = 1, 2, \ldots, n$). The column vector $\mathbf{x}$ represents the values of n patients (objects, arrays) for a variable

$$\mathbf{x} = \begin{bmatrix} x_1 \\ x_2 \\ \vdots \\ x_n \end{bmatrix} \quad \text{or} \quad \mathbf{x}^T = \begin{bmatrix} x_1 & x_2 & \cdots & x_n \end{bmatrix}, \tag{B.1}$$

where the ^{T} denotes transpose of a row vector. Consider a similar n-dimensional vector for another variable, denoted $\mathbf{y}$, given as

$$\mathbf{y} = \begin{bmatrix} y_1 \\ y_2 \\ \vdots \\ y_n \end{bmatrix} \quad \text{or} \quad \mathbf{y}^T = \begin{bmatrix} y_1 & y_2 & \cdots & y_n \end{bmatrix}. \tag{B.2}$$

The *distance* between two points P and Q represented by vectors $\mathbf{x}$ and $\mathbf{y}$ is

$$d(P, Q) = \sqrt{(x_1 - y_1)^2 + (x_2 - y_2)^2 + \cdots + (x_n - y_n)^2}. \tag{B.3}$$

Euclidean Norm of a Vector. Now consider the point 0, represented by an n-dimensional vector whose elements are all equal to zero. The distance $d(P, 0)$ between point P, based on vector $\mathbf{x}$ and the origin 0, is called the

Classification Analysis of DNA Microarrays, First Edition. Leif E. Peterson.

Euclidean norm of **x**, given as

$$
\begin{aligned}
||\mathbf{x}|| &= \sqrt{(x_1 - 0)^2 + (x_2 - 0)^2 + \cdots + (x_n - 0)^2} \\
&= \sqrt{(x_1 - 0)^2 + (x_2 - 0)^2 + \cdots + (x_n - 0)^2} \\
&= \sqrt{x_1^2 + x_2^2 + \cdots + x_n^2} \\
&= \sqrt{x_1 x_1 + x_2 x_2 + \cdots + x_n x_n} \\
&= \sqrt{\sum_{i=1}^{n} x_i^2} \\
&= \sqrt{\mathbf{x}^T \mathbf{x}}.
\end{aligned}
\tag{B.4}
$$

In this solution, the term $\mathbf{x}^T\mathbf{x}$ is known as the *inner product* between two vectors, in this case $\mathbf{x}^T$ and **x**. The vector operations needed for calculating the inner product are discussed below.

***p*-Norm of a Vector.** The *p-norm* of a vector is defined as

$$
\|\mathbf{x}\| = \left(\sum_{i=1}^{n} (x_i)^p \right)^{1/p}, \tag{B.5}
$$

and forms the basis of the L^1-norm and L^2-norm (Euclidean norm).

Unit Vectors. The features (attributes) of an object commonly have varying range and scale, and if not transformed to be scale invariant, may become difficult when used in matrix operations. Unit vectors are *normalized* by dividing a vector's components by its length. Assume that object **x** has three feature values (x_1, x_2, x_3). The length of **x** is

$$
\|\mathbf{x}\| = \sqrt{x_1^2 + x_2^2 + x_3^2}. \tag{B.6}
$$

The unit form of **x** is

$$
\mathbf{x}_u = \left(\frac{x_1}{\|\mathbf{x}\|}, \frac{x_2}{\|\mathbf{x}\|}, \frac{x_3}{\|\mathbf{x}\|} \right), \tag{B.7}
$$

where each element (i.e., $x_j/\|\mathbf{x}\|$) is a *direction cosine*. The length of all unit vectors is unity such that $\|\mathbf{x}_u\| = 1$. Direction cosines offer an additional metric for feature transformation, and transforming features into direction cosines prior to analysis can result in remarkable performance of a model.

Concept Vectors. The center vector $\mathbf{v}$ defines the centroid vector of mean feature values for objects assigned to a particular cluster. Center vectors are a fundamental component of crisp K-means cluster analysis, kernel regression, and radial basis function networks. Assume that there are K centers of the form $\mathbf{v}_1, \mathbf{v}_2, \ldots, \mathbf{v}_K$. The concept vector for each center $\mathbf{v}_k$ is

$$\mathbf{c}_k = \frac{\mathbf{v}_k}{\|\mathbf{v}_k\|}. \tag{B.8}$$

Within a single cluster, concept vectors have the greatest correlation with all of the objects. Concept clusters are widely employed in text mining for document clustering.

Inner Product. The inner product between two vectors $\mathbf{x}$ and $\mathbf{y}$ is given by

$$\begin{aligned}\underbrace{\underset{(1\times n)}{\mathbf{x}^T}\underset{(n\times 1)}{\mathbf{y}}}_{1\times 1} &= \begin{bmatrix} x_1 & x_2 & \cdots & x_n \end{bmatrix} \begin{bmatrix} y_1 \\ y_2 \\ \vdots \\ y_n \end{bmatrix} \\ &= x_1 y_1 + x_2 y_2 + \cdots + x_n y_n. \end{aligned} \tag{B.9}$$

The inner product is determined by the following procedure:

Step 1:

$$\begin{bmatrix} \boxed{x_1} & x_2 & \cdots & x_n \end{bmatrix} \begin{bmatrix} \boxed{y_1} \\ y_2 \\ \vdots \\ y_n \end{bmatrix} \rightarrow \text{gives } x_1 y_1. \tag{B.10}$$

Step 2:

$$\begin{bmatrix} x_1 & \boxed{x_2} & \cdots & x_n \end{bmatrix} \begin{bmatrix} y_1 \\ \boxed{y_2} \\ \vdots \\ y_n \end{bmatrix} \rightarrow \text{gives } x_2 y_2. \tag{B.11}$$

Step n:

$$\begin{bmatrix} x_1 & x_2 & \cdots & \boxed{x_n} \end{bmatrix} \begin{bmatrix} y_1 \\ y_2 \\ \vdots \\ \boxed{y_n} \end{bmatrix} \rightarrow \text{gives } x_n y_n. \tag{B.12}$$

The inner product always results in a scalar value.

Outer Product. The outer product places the transpose on the rightmost vector, causing the outer dimension of the product to be greater than one:

$$\underbrace{\underset{(n\times 1)}{\mathbf{x}}\underset{(1\times n)}{\mathbf{y}^T}}_{n\times n} = \begin{bmatrix} x_1 \\ x_2 \\ \vdots \\ x_n \end{bmatrix} \begin{bmatrix} y_1 & y_2 & \cdots & y_n \end{bmatrix}$$

$$= \begin{bmatrix} x_1y_1 & x_1y_2 & \cdots & x_1y_n \\ x_2y_1 & x_2y_2 & \cdots & x_2y_n \\ \vdots & \vdots & \ddots & \vdots \\ x_ny_1 & x_ny_2 & \cdots & x_ny_n \end{bmatrix}. \tag{B.13}$$

This first row in the $n \times n$ matrix is obtained in the following manner:

Step 1:

$$\begin{bmatrix} \boxed{x_1} \\ x_2 \\ \vdots \\ x_n \end{bmatrix} \begin{bmatrix} \boxed{y_1} & y_2 & \cdots & y_n \end{bmatrix} \rightarrow \text{gives } x_1y_1, \tag{B.14}$$

Step 2:

$$\begin{bmatrix} \boxed{x_1} \\ x_2 \\ \vdots \\ x_n \end{bmatrix} \begin{bmatrix} y_1 & \boxed{y_2} & \cdots & y_n \end{bmatrix} \rightarrow \text{gives } x_1y_2, \tag{B.15}$$

Step n:

$$\begin{bmatrix} \boxed{x_1} \\ x_2 \\ \vdots \\ x_n \end{bmatrix} \begin{bmatrix} y_1 & y_2 & \cdots & \boxed{y_n} \end{bmatrix} \rightarrow \text{gives } x_1y_n. \tag{B.16}$$

These steps are repeated n times to yield the $n \times n$ outer product matrix. The outer product operation always results in a matrix if the dimensions of both vectors is greater than one.

B.2 MATRICES

Design Matrix. In linear regression, the independent (predictor) variables are arranged into what is called the *design matrix*. The design matrix has

n rows equal to the number of records in the dataset used, and p columns representing the total number of variables. An example of a design matrix is

$$\underset{(n\times p)}{\mathbf{X}} = \begin{bmatrix} x_{11} & x_{12} & x_{13} & \cdots & x_{1j} & \cdots & x_{1p} \\ x_{21} & x_{22} & x_{23} & \cdots & x_{2j} & \cdots & x_{2p} \\ \vdots & & & \vdots & & \vdots & \\ x_{i1} & x_{i2} & x_{i3} & \cdots & x_{ij} & \cdots & x_{ip} \\ \vdots & & & \vdots & & \vdots & \\ x_{n1} & x_{n2} & x_{n3} & \cdots & x_{nj} & \cdots & x_{np} \end{bmatrix}. \tag{B.17}$$

As we will see later, the design matrix can be manipulated in a number of ways to solve a system of linear equations, determine the covariance and correlation between variables, and extract important information from the variables used. The following sections describe several matrix operations that are commonly used in matrix operations for statistical analysis.

Transpose. Before a matrix can be used in multiplication, it must be transposed, such that rows become columns, and columns become rows. In addition, the dimensions of the matrix are swapped during the transpose. For example, if a matrix has dimensions $n \times p$, then after a transpose operation, the new dimensions are $p \times n$. The transpose of the $n \times p$ design matrix $\mathbf{X}$ is

$$\underset{(p\times n)}{\mathbf{X}^T} = \begin{bmatrix} x_{11} & x_{12} & x_{13} & \cdots & x_{1i} & \cdots & x_{1n} \\ x_{21} & x_{22} & x_{23} & \cdots & x_{2i} & \cdots & x_{2n} \\ \vdots & & & \vdots & & \vdots & \\ x_{j1} & x_{j2} & x_{j3} & \cdots & x_{ji} & \cdots & x_{jn} \\ \vdots & & & \vdots & & \vdots & \\ x_{p1} & x_{p2} & x_{p3} & \cdots & x_{pj} & \cdots & x_{pn} \end{bmatrix}. \tag{B.18}$$

An instructive example of the transpose is shown below. In the example matrix below the array elements across row 1 are 123, 33, and 3.5. When considering the transposed version, the first column now has the entries 123, 33, and 3.5

$$\underset{(5\times 4)}{\mathbf{X}} = \begin{bmatrix} 1 & 123 & 33 & 3.5 \\ 1 & 99 & 30 & 2.7 \\ 1 & 102 & 45 & 1.4 \\ 0 & 212 & 39 & 4.6 \\ 0 & 187 & 57 & 2.8 \end{bmatrix} \tag{B.19}$$

$$\underset{(4\times5)}{\mathbf{X}^T} = \begin{bmatrix} 1 & 1 & 1 & 0 & 0 \\ 123 & 99 & 102 & 212 & 187 \\ 33 & 30 & 45 & 39 & 57 \\ 3.5 & 2.7 & 1.4 & 4.6 & 2.8 \end{bmatrix}. \tag{B.20}$$

Addition. Matrix addition simply adds together the matrix elements on an element-by-element basis:

$$\underset{(2\times3)}{\mathbf{X}} = \begin{bmatrix} 4 & 1 & 7 \\ 3 & 2 & 6 \end{bmatrix} \tag{B.21}$$

$$\underset{(2\times3)}{\mathbf{Y}} = \begin{bmatrix} 2 & 3 & 6 \\ 5 & 4 & 1 \end{bmatrix} \tag{B.22}$$

$$\underset{(2\times3)}{\mathbf{X}} + \underset{(2\times3)}{\mathbf{Y}} = \begin{bmatrix} 4+2 & 1+3 & 7+6 \\ 3+5 & 2+4 & 6+1 \end{bmatrix} = \begin{bmatrix} 6 & 4 & 13 \\ 8 & 6 & 7 \end{bmatrix}. \tag{B.23}$$

Vector-Matrix Multiplication. Vector-matrix multiplication treats the vector as if it were a matrix, with the exception that one of the vector's dimensions is one:

$$\underset{(p\times n)}{\mathbf{X}^T}\underset{(n\times 1)}{\mathbf{y}} = \underbrace{\begin{bmatrix} x_{11} & x_{12} & \cdots & x_{1n} \\ x_{21} & x_{22} & \cdots & x_{2n} \\ \vdots & \vdots & \ddots & \vdots \\ x_{p1} & x_{p2} & \cdots & x_{pn} \end{bmatrix} \begin{bmatrix} y_1 \\ y_2 \\ y_3 \\ \vdots \\ y_i \\ \vdots \\ y_n \end{bmatrix}}_{(n\times1)(p\times1)} \tag{B.24}$$

$$\underset{(2\times3)}{\mathbf{X}^T}\underset{(3\times 1)}{\mathbf{y}} = \begin{bmatrix} 2 & 1 & 4 \\ 5 & 3 & 7 \end{bmatrix} \begin{bmatrix} 3 \\ 2 \\ 5 \end{bmatrix}. \tag{B.25}$$

Step 1:

$$\underset{(2\times3)}{\mathbf{X}^T}\underset{(3\times 1)}{\mathbf{y}} = \begin{bmatrix} \boxed{2} & \boxed{1} & \boxed{4} \\ 5 & 3 & 7 \end{bmatrix} \begin{bmatrix} \boxed{3} \\ \boxed{2} \\ \boxed{5} \end{bmatrix} \rightarrow 2(3) + 1(2) + 4(5)$$

$$= 6 + 2 + 20 = 28 \rightarrow \begin{bmatrix} 28 \\ \ \end{bmatrix}, \tag{B.26}$$

Step 2:

$$\underset{(2\times3)}{\mathbf{X}^T}\underset{(3\times1)}{\mathbf{y}} = \begin{bmatrix} 2 & 1 & 4 \\ \boxed{5} & \boxed{3} & \boxed{7} \end{bmatrix} \begin{bmatrix} \boxed{3} \\ \boxed{2} \\ \boxed{5} \end{bmatrix} \rightarrow 5(3) + 3(2) + 7(5)$$

$$= 15 + 6 + 35 = 56 \rightarrow \begin{bmatrix} 28 \\ 56 \end{bmatrix}. \tag{B.27}$$

Matrix-Matrix Multiplication. Analogously, matrix-matrix multiplication is identical to vector-matrix multiplication, as shown in the example below:

$$\underset{(p\times n)(n\times p)}{\mathbf{X}^T\ \mathbf{X}} = \underbrace{\begin{bmatrix} x_{11} & x_{12} & \cdots & x_{1n} \\ x_{21} & x_{22} & \cdots & x_{2n} \\ \vdots & \vdots & \ddots & \vdots \\ x_{p1} & x_{p2} & \cdots & x_{pn} \end{bmatrix} \begin{bmatrix} x_{11} & x_{12} & \cdots & x_{1p} \\ x_{21} & x_{22} & \cdots & x_{2p} \\ \vdots & \vdots & \ddots & \vdots \\ x_{n1} & x_{n2} & \cdots & x_{np} \end{bmatrix}}_{(p\times n)(n\times p)}. \tag{B.28}$$

Consider the 4×2 matrix with four rows and two columns:

$$\underset{(4\times2)}{\mathbf{X}} = \begin{bmatrix} 2 & 6 \\ 1 & 4 \\ 3 & 2 \\ 4 & 7 \end{bmatrix}. \tag{B.29}$$

Let's cross-multiply the matrix **X** with itself. First, we have to transpose **X** in order to get $\mathbf{X}^T$. Then $\mathbf{X}^T$ is placed in front of **X** in the following fashion:

$$\underset{(2\times4)(4\times2)}{\mathbf{X}^T\ \mathbf{X}} = \begin{bmatrix} 2 & 1 & 3 & 4 \\ 6 & 4 & 2 & 7 \end{bmatrix} \begin{bmatrix} 2 & 6 \\ 1 & 4 \\ 3 & 2 \\ 4 & 7 \end{bmatrix}. \tag{B.30}$$

Step 1:

$$\underset{(2\times4)(4\times2)}{\mathbf{X}^T\ \mathbf{X}} = \begin{bmatrix} \boxed{2} & \boxed{1} & \boxed{3} & \boxed{4} \\ 6 & 4 & 2 & 7 \end{bmatrix} \begin{bmatrix} \boxed{2} & 6 \\ \boxed{1} & 4 \\ \boxed{3} & 2 \\ \boxed{4} & 7 \end{bmatrix}$$

$$= 2(2) + 1(1) + 3(3) + 4(4)$$

$$= 4 + 1 + 9 + 16 = 30$$

$$= \begin{bmatrix} 30 & \\ & \end{bmatrix}. \tag{B.31}$$

Step 2:

$$\underset{(2\times4)(4\times2)}{\mathbf{X}^T\ \mathbf{X}} = \begin{bmatrix} \boxed{2} & \boxed{1} & \boxed{3} & \boxed{4} \\ 6 & 4 & 2 & 7 \end{bmatrix} \begin{bmatrix} 2 & \boxed{6} \\ 1 & \boxed{4} \\ 3 & \boxed{2} \\ 4 & \boxed{7} \end{bmatrix}$$

$$= 2(6) + 1(4) + 3(2) + 4(7)$$

$$= 12 + 4 + 6 + 28 = 50$$

$$= \begin{bmatrix} 30 & 50 \\ & \end{bmatrix}. \tag{B.32}$$

Step 3:

$$\underset{(2\times4)(4\times2)}{\mathbf{X}^T\ \mathbf{X}} = \begin{bmatrix} 2 & 1 & 3 & 4 \\ \boxed{6} & \boxed{4} & \boxed{2} & \boxed{7} \end{bmatrix} \begin{bmatrix} \boxed{2} & 6 \\ \boxed{1} & 4 \\ \boxed{3} & 2 \\ \boxed{4} & 7 \end{bmatrix}$$

$$= 6(2) + 4(1) + 2(3) + 7(4)$$

$$= 12 + 4 + 6 + 28 = 50$$

$$= \begin{bmatrix} 30 & 50 \\ 50 & \end{bmatrix}. \tag{B.33}$$

Step 4:

$$\underset{(2\times4)(4\times2)}{\mathbf{X}^T\ \mathbf{X}} = \begin{bmatrix} 2 & 1 & 3 & 4 \\ \boxed{6} & \boxed{4} & \boxed{2} & \boxed{7} \end{bmatrix} \begin{bmatrix} 2 & \boxed{6} \\ 1 & \boxed{4} \\ 3 & \boxed{2} \\ 4 & \boxed{7} \end{bmatrix}$$

$$= 6(6) + 4(4) + 2(2) + 7(7)$$

$$= 36 + 16 + 4 + 49 = 105$$

$$= \begin{bmatrix} 30 & 50 \\ 50 & 105 \end{bmatrix}. \tag{B.34}$$

B.3 SAMPLE MEAN, COVARIANCE, AND CORRELATION

We begin this section by setting up a mean vector $\bar{\mathbf{x}}$, and show how this vector is used for subtraction with the data matrix $\mathbf{X}$ to first obtain the covariance matrix, followed by the correlation matix. The mean vector is determined by cross-multiplying each $\mathbf{x}$ vector's values with a vector of ones (i.e., $\mathbf{1}$) and dividing by the sample size n in the following manner:

$$\bar{\mathbf{x}} = \begin{bmatrix} \bar{x}_1 \\ \bar{x}_2 \\ \vdots \\ \bar{x}_p \end{bmatrix} = \begin{bmatrix} \frac{\mathbf{x}_1^T\mathbf{1}}{n} \\ \frac{\mathbf{x}_2^T\mathbf{1}}{n} \\ \vdots \\ \frac{\mathbf{x}_p^T\mathbf{1}}{n} \end{bmatrix} = \frac{1}{n} \begin{bmatrix} x_{11} & x_{12} & \cdots & x_{1n} \\ x_{21} & x_{22} & \cdots & x_{2n} \\ \vdots & \vdots & \ddots & \vdots \\ x_{p1} & x_{p2} & \cdots & x_{pn} \end{bmatrix} \begin{bmatrix} 1 \\ 1 \\ 1 \\ 1 \end{bmatrix} = \frac{1}{n}\mathbf{X}^T\mathbf{1}. \tag{B.35}$$

Next, we replicate the mean vector into n rows, given as

$$\underset{(n\times 1)(1\times p)}{\mathbf{1}\ \bar{\mathbf{x}}^T} = \underbrace{\begin{bmatrix} \bar{x}_1 & \bar{x}_2 & \cdots & \bar{x}_p \\ \bar{x}_1 & \bar{x}_2 & \cdots & \bar{x}_p \\ \vdots & \vdots & \ddots & \vdots \\ \bar{x}_1 & \bar{x}_2 & \cdots & \bar{x}_p \end{bmatrix}}_{(n\times p)}, \tag{B.36}$$

and then subtract the replicate mean vector matrix from the data matrix using the operation

$$\underset{(n\times p)}{\mathbf{X}} - \underset{(n\times 1)(1\times p)}{\mathbf{1}\ \bar{\mathbf{x}}^T} = \begin{bmatrix} x_{11}-\bar{x}_1 & x_{12}-\bar{x}_2 & \cdots & x_{1p}-\bar{x}_p \\ x_{21}-\bar{x}_1 & x_{22}-\bar{x}_2 & \cdots & x_{2p}-\bar{x}_p \\ \vdots & \vdots & \ddots & \vdots \\ x_{n1}-\bar{x}_1 & x_{n2}-\bar{x}_2 & \cdots & x_{np}-\bar{x}_p \end{bmatrix}. \tag{B.37}$$

The *covariance matrix* $\mathbf{S}$ is obtained by cross-multiplying the preceding matrix with its transpose and dividing by the sample size, using the following operations:

$$\underset{p\times p}{\mathbf{S}} = \frac{1}{n}\left(\mathbf{X} - \mathbf{1}\bar{\mathbf{x}}^T\right)^T\left(\mathbf{X} - \mathbf{1}\bar{\mathbf{x}}^T\right)$$

$$= \frac{1}{n}\begin{bmatrix} x_{11}-\bar{x}_1 & x_{21}-\bar{x}_1 & \cdots & x_{n1}-\bar{x}_1 \\ x_{12}-\bar{x}_2 & x_{22}-\bar{x}_2 & \cdots & x_{n2}-\bar{x}_2 \\ \vdots & \vdots & \ddots & \vdots \\ x_{1p}-\bar{x}_p & x_{2p}-\bar{x}_p & \cdots & x_{np}-\bar{x}_p \end{bmatrix}$$

$$\times \begin{bmatrix} x_{11}-\bar{x}_1 & x_{12}-\bar{x}_2 & \cdots & x_{1p}-\bar{x}_p \\ x_{21}-\bar{x}_1 & x_{22}-\bar{x}_2 & \cdots & x_{2p}-\bar{x}_p \\ \vdots & \vdots & \ddots & \vdots \\ x_{n1}-\bar{x}_1 & x_{n2}-\bar{x}_2 & \cdots & x_{np}-\bar{x}_p \end{bmatrix}$$

$$= \begin{bmatrix} s_{11} & s_{12} & \cdots & s_{1p} \\ s_{21} & s_{22} & \cdots & s_{2p} \\ \vdots & \vdots & \ddots & \vdots \\ s_{p1} & s_{p2} & \cdots & s_{pp} \end{bmatrix}. \tag{B.38}$$

As noted, $\mathbf{S}$ is a $p \times p$ square symmetric matrix, with equal upper and lower triangulars. The *diagonal* of $\mathbf{S}$ consists of the variance of each variable, while the *off-diagonals* are the covariances.

B.4 DIAGONAL MATRICES

The *sample standard deviation matrix* is a diagonal matrix with elements only on the diagonal shown as

$$\underset{p\times p}{\mathbf{D}}^{1/2} = \begin{bmatrix} \sqrt{s_{11}} & 0 & \cdots & 0 \\ 0 & \sqrt{s_{22}} & \cdots & 0 \\ \vdots & \vdots & \ddots & \vdots \\ 0 & 0 & \cdots & \sqrt{s_{pp}} \end{bmatrix} = \begin{bmatrix} s_1 & 0 & \cdots & 0 \\ 0 & s_2 & \cdots & 0 \\ \vdots & \vdots & \ddots & \vdots \\ 0 & 0 & \cdots & s_p \end{bmatrix}. \tag{B.39}$$

When the inverse of the diagonal standard deviation matrix is of interest, we simply invert the diagonal elements and obtain

$$\underset{p\times p}{\mathbf{D}}^{-1/2} = \begin{bmatrix} \frac{1}{\sqrt{s_{11}}} & 0 & \cdots & 0 \\ 0 & \frac{1}{\sqrt{s_{22}}} & \cdots & 0 \\ \vdots & \vdots & \ddots & \vdots \\ 0 & 0 & \cdots & \frac{1}{\sqrt{s_{pp}}} \end{bmatrix} = \begin{bmatrix} \frac{1}{s_1} & 0 & \cdots & 0 \\ 0 & \frac{1}{s_2} & \cdots & 0 \\ \vdots & \vdots & \ddots & \vdots \\ 0 & 0 & \cdots & \frac{1}{s_p} \end{bmatrix}. \tag{B.40}$$

We can then sandwich the covariance matrix between inverse standard deviation matrices to obtain the correlation matrix in the following form:

$$\mathbf{R} = \mathbf{D}^{-1/2}\mathbf{s.d.}^{-1/2}$$

$$= \begin{bmatrix} \frac{1}{\sqrt{s_{11}}} & 0 & \cdots & 0 \\ 0 & \frac{1}{\sqrt{s_{22}}} & \cdots & 0 \\ \vdots & \vdots & \ddots & \vdots \\ 0 & 0 & \cdots & \frac{1}{\sqrt{s_{pp}}} \end{bmatrix} \begin{bmatrix} s_{11} & s_{12} & \cdots & s_{1p} \\ s_{21} & s_{22} & \cdots & s_{2p} \\ \vdots & \vdots & \ddots & \vdots \\ s_{p1} & s_{p2} & \cdots & s_{pp} \end{bmatrix} \begin{bmatrix} \frac{1}{\sqrt{s_{11}}} & 0 & \cdots & 0 \\ 0 & \frac{1}{\sqrt{s_{22}}} & \cdots & 0 \\ \vdots & \vdots & \ddots & \vdots \\ 0 & 0 & \cdots & \frac{1}{\sqrt{s_{pp}}} \end{bmatrix}$$

$$= \begin{bmatrix} \frac{s_{11}}{\sqrt{s_{11}}\sqrt{s_{11}}} & \frac{s_{12}}{\sqrt{s_{11}}\sqrt{s_{22}}} & \cdots & \frac{s_{1p}}{\sqrt{s_{11}}\sqrt{s_{pp}}} \\ \frac{s_{21}}{\sqrt{s_{11}}\sqrt{s_{22}}} & \frac{s_{22}}{\sqrt{s_{22}}\sqrt{s_{22}}} & \cdots & \frac{s_{2p}}{\sqrt{s_{22}}\sqrt{s_{pp}}} \\ \vdots & \vdots & \ddots & \vdots \\ \frac{s_{p1}}{\sqrt{s_{pp}}\sqrt{s_{11}}} & \frac{s_{p2}}{\sqrt{s_{pp}}\sqrt{s_{22}}} & \cdots & \frac{s_{pp}}{\sqrt{s_{pp}}\sqrt{s_{pp}}} \end{bmatrix}$$

$$= \begin{bmatrix} 1 & r_{12} & \cdots & r_{1p} \\ r_{21} & 1 & \cdots & r_{2p} \\ \vdots & \vdots & \ddots & \vdots \\ r_{p1} & r_{p2} & \cdots & 1 \end{bmatrix}. \tag{B.41}$$

Thus, the *correlation matrix* **R** is also a square symmetric matrix with equal upper and lower triangulars. The diagonal elements of **R** consists of unity, since the correlation between a variable and itself is always one. The off-diagonals of **R** are the bivariate correlation coefficients.

B.5 IDENTITY MATRICES

Identity matrices are square symmetric matrices with ones on the diagonal, and zeroes on the off-diagonals. The structure of an identity matrix is shown below:

$$\mathbf{I} = \begin{bmatrix} 1 & 0 & \cdots & 0 \\ 0 & 1 & \cdots & 0 \\ \vdots & \vdots & \ddots & \vdots \\ 0 & 0 & \cdots & 1 \end{bmatrix}. \tag{B.42}$$

Any matrix multiplied by the identity matrix is equal to itself:

$$\underset{(n\times p)(n\times p)}{\mathbf{I}\ \mathbf{X}} = \underset{(n\times p)(n\times p)}{\mathbf{X}\ \mathbf{I}} = \underset{(n\times p)}{\mathbf{X}}. \tag{B.43}$$

As an example, when the 2×3 matrix below is multiplied by the identity matrix, the result is the same matrix:

$$\begin{bmatrix} 1 & 0 \\ 0 & 1 \end{bmatrix} \begin{bmatrix} 3 & 1 & -2 \\ 6 & 7 & 4 \end{bmatrix} = \begin{bmatrix} 3 & 1 & -2 \\ 6 & 7 & 4 \end{bmatrix}. \tag{B.44}$$

Identity matrices are quite useful in theorems, especially when working with orthogonal matrices, because by definition, the product of an orthogonal matrix and its transpose is equal to the identity matrix. If $\mathbf{X}$ is orthogonal, then $\mathbf{X}\mathbf{X}^T = \mathbf{I}$.

B.6 TRACE OF A MATRIX

The *trace* of square-symmetric matrix is the sum of the diagonals.

$$\text{tr}(\mathbf{X}) = x_{11} + x_{22} + \cdots + x_{nn} = \sum_{i=1}^{n} x_{ii}, \tag{B.45}$$

$$\text{tr}\begin{bmatrix} 2 & 9 & 7 \\ 5 & -1 & 3 \\ 3 & 4 & 7 \end{bmatrix} = 2 - 1 + 7 = 8, \tag{B.46}$$

$$\text{tr}(\mathbf{X}^T) = \text{tr}(\mathbf{X}), \tag{B.47}$$

$$\text{tr}\begin{bmatrix} 1 & 5 & 6 \\ 4 & 6 & 7 \\ -2 & 4 & 9 \end{bmatrix} = \text{tr}\begin{bmatrix} 1 & 4 & -2 \\ 5 & 6 & 4 \\ 6 & 7 & 9 \end{bmatrix} = 1 + 6 + 9 = 16. \tag{B.48}$$

B.7 EIGENANALYSIS

Let $\mathbf{R} \in \mathbb{R}^{p \times p}$ be a real square symmetric matrix. Eigenanalysis of $\mathbf{R}$ is accomplished by use of the characteristic polynomial function of a matrix. Eigenvalues of $\mathbf{R}$ are the roots of the characteristic polynomial function $p(\lambda) = \det(\mathbf{R} - \lambda\mathbf{I})$. The eigenvector $\mathbf{e} \neq 0$ for each eigenvalue λ is determined by solving the linear system $(\mathbf{R} - \lambda\mathbf{I})\mathbf{e} = 0$. The set of eigenvectors $\mathbf{e}$ satisfying $\mathbf{Re} = \lambda\mathbf{e}$ is called the eigenspace of $\mathbf{R}$ corresponding to λ.

B.8 SYMMETRIC EIGENVALUE PROBLEM

Eigenanalysis is commonly performed on square, real symmetric, positive definite matrices. In statistics, the sample correlation matrix $\mathbf{R}$ and sample covariance matrix $\mathbf{S}$ are real square symmetric matrices, that is, $\mathbf{R}, \mathbf{S} \in \mathbb{R}^{p \times p}$. Positive definiteness of $\mathbf{R}$ and $\mathbf{S}$ is guaranteed only if none of their eigenvalues are zero. The *principal form* of $\mathbf{R}$ is defined as

$$\mathbf{R} = \mathbf{E}\boldsymbol{\Lambda}\mathbf{E}^{-1}, \tag{B.49}$$

where $\mathbf{E}$ is an orthogonal *eigenvector* matrix and $\boldsymbol{\Lambda} = \text{diag}\{\lambda_1, \lambda_2, \ldots, \lambda_p\}$, the diagonal matrix of *eigenvalues*. Likewise, the principal form of $\mathbf{S}$ is

$$\mathbf{S} = \mathbf{E}\boldsymbol{\Lambda}\mathbf{E}^{-1}. \tag{B.50}$$

Because $\mathbf{E}$ is orthogonal, $\mathbf{E}\mathbf{E}^T = \mathbf{E}\mathbf{E}^T = \mathbf{I}$ and $\mathbf{E}^T = \mathbf{E}^{-1}$. If every eigenvector of $\mathbf{E}$ has length one, then $\mathbf{E}$ is also said to be orthonormal. Orthonormal eigenvectors can be obtained by normalizing each eigenvector, that is, dividing by the length in the form $\mathbf{e}/||\mathbf{e}||$. There is tremendous utility in the principal form of a matrix, since it can be exploited to solve many problems in matrix algebra.

Theorem (Power Matrix)
If $\mathbf{R}$ is a real square symmetric positive definite matrix, then the power matrix is equal to $\mathbf{R}^n = \mathbf{E}\mathbf{D}^n\mathbf{E}^{-1}$.

Proof: Letting $\mathbf{R} = \mathbf{R}^{1/2}\mathbf{R}^{1/2} = (\mathbf{R}^{1/2})^2$, we may show that

$$\begin{aligned}
\mathbf{R} &= \mathbf{R}^{1/2}\mathbf{R}^{1/2} \\
&= (\mathbf{E}\mathbf{D}^{1/2}\mathbf{E}^{-1})(\mathbf{E}\mathbf{D}^{1/2}\mathbf{E}^{-1}) \\
&= (\mathbf{E}\mathbf{D}^{1/2})(\mathbf{E}^{-1}\mathbf{E})(\mathbf{D}^{1/2}\mathbf{E}^{-1}) \\
&= (\mathbf{E}\mathbf{D}^{1/2})(\mathbf{I})(\mathbf{D}^{1/2}\mathbf{E}^{-1}) \\
&= (\mathbf{E}\mathbf{D}^{1/2})(\mathbf{D}^{1/2}\mathbf{E}^{-1}) \\
&= \mathbf{E}(\mathbf{D}^{1/2}\mathbf{D}^{1/2})\mathbf{E}^{-1} \\
&= \mathbf{E}(\mathbf{D})\mathbf{E}^{-1}.
\end{aligned} \tag{B.51}$$

■

B.9 GENERALIZED EIGENVALUE PROBLEM

In Section 12.6, on feature selection methods using the greedy PTA approach, we used eigenanalysis to calculate Wilks' lambda for the nonsymmetric matrix $\mathbf{W}^{-1}\mathbf{B}$. The generalized eigenvalue problem can be used for eigenanalysis of a nonsymmetric matrix, and is typically solved with the QR algorithm. The QR algorithm starts by rearranging all elements in upper Hessenberg form, followed by iterative production of a series of Hessenberg matrices that converge to a triangular matrix. During each iteration a bulge is created and *chased* down using a Givens transformation. We favor an alternative approach that allows use of the symmetric eigenvalue problem and *power matrix theorem* to extract eigenvalues from $\mathbf{W}^{-1}\mathbf{B}$. This is based on the fact that the eigenvalues of the non-symmetric matrix $\mathbf{W}^{-1}\mathbf{B}$ are equal to the eigenvalues of the symmetric matrix $(\mathbf{W}^{1/2})^{-1}\mathbf{B}(\mathbf{W}^{1/2})^{-1}$, in accordance with the following theorem.

Theorem (Equivalent Eigenvalues)
The eigenvalues of the square nonsymmetric matrix $\mathbf{W}^{-1}\mathbf{B}$ are equal to the eigenvalues of the square symmetric matrix $(\mathbf{W}^{1/2})^{-1}\mathbf{B}(\mathbf{W}^{1/2})^{-1}$.

Proof: Letting $(\mathbf{W}^{-1}\mathbf{B} - \mathbf{\Lambda I})\mathbf{E} = \mathbf{0}$, we may show that

$$\begin{aligned}
(\mathbf{W}^{-1}\mathbf{B} - \mathbf{\Lambda I})\mathbf{E} &= \mathbf{0} \\
(\mathbf{B} - \mathbf{\Lambda W})\mathbf{E} &= \mathbf{0} \quad [\text{multiply by } \mathbf{W}] \\
(\mathbf{B} - \mathbf{\Lambda W}^{1/2}\mathbf{W}^{1/2})\mathbf{E} &= \mathbf{0} \quad [\text{substitute } \mathbf{W} \text{ with } \mathbf{W}^{1/2}\mathbf{W}^{1/2}] \\
(\mathbf{W}^{1/2})^{-1}(\mathbf{B} - \mathbf{\Lambda W}^{1/2}\mathbf{W}^{1/2})\mathbf{E} &= (\mathbf{W}^{1/2})^{-1}\mathbf{0} \\
&= \mathbf{0} \quad [\text{multiply by } (\mathbf{W}^{1/2})^{-1}] \\
[(\mathbf{W}^{1/2})^{-1}\mathbf{B} - \mathbf{\Lambda W}^{1/2}](\mathbf{W}^{1/2})^{-1}\mathbf{W}^{1/2}\mathbf{E} &= \mathbf{0} \quad [\text{insert } (\mathbf{W}^{1/2})^{-1}\mathbf{W}^{1/2} = \mathbf{I}] \\
[(\mathbf{W}^{1/2})^{-1}\mathbf{B}(\mathbf{W}^{1/2})^{-1} - \mathbf{\Lambda I}]\mathbf{W}^{1/2}\mathbf{E} &= \mathbf{0}. \qquad (B.52)
\end{aligned}$$

■

This proof shows that the eigenvalues of $\mathbf{W}^{-1}\mathbf{B}$ are the same as the eigenvalues of $(\mathbf{W}^{1/2})^{-1}\mathbf{B}(\mathbf{W}^{1/2})^{-1}$. Because $(\mathbf{W}^{1/2})^{-1}\mathbf{B}(\mathbf{W}^{1/2})^{-1}$ is square symmetric, extracting the eigenvalues becomes a symmetric eigenvalue problem, which can be solved using, for example, the Jacobi method or singular value decomposition. The last theorem, which we used to obtain the eigenvalue matrix $\mathbf{\Lambda}$ from the nonsymmetric matrix $\mathbf{W}^{-1}\mathbf{B}$ satisfies our need in order to calculate Wilks' lambda in Section 12.6. One should also recognize, however, that the eigenvectors of $\mathbf{W}^{-1}\mathbf{B}$ are not equal to the eigenvectors of $(\mathbf{W}^{1/2})^{-1}\mathbf{B}(\mathbf{W}^{1/2})^{-1}$.

B.10 MATRIX PROPERTIES

Finally, we list several properties of matrices that were utilized in this book:

1. A square symmetric matrix $\mathbf{A}$ can be constructed by adding any nonzero square matrix to its transpose: $\mathbf{X} + \mathbf{X}^T$.
2. A $p \times p$ square symmetric matrix $\mathbf{A}$ can be constructed from an $n \times p$ matrix $\mathbf{X}$ by multiplying the transpose of $\mathbf{X}$ by $\mathbf{X}$: $\underset{p\times p}{\mathbf{A}} = \underset{p\times n}{\mathbf{X}^T}\underset{n\times p}{\mathbf{X}}$.
3. An $n \times n$ square symmetric matrix $\mathbf{A}$ can be constructed from an $n \times p$ matrix $\mathbf{X}$ by multiplying $\mathbf{X}$ by its transpose: $\underset{n\times n}{\mathbf{A}} = \underset{n\times p}{\mathbf{X}}\underset{p\times n}{\mathbf{X}^T}$.
4. An $n \times p$ matrix whose columns are correlated according to the correlation matrix $\mathbf{R}$ can be simulated using the matrix operation

$\mathbf{Y} = \mathbf{X}\mathbf{R}^{1/2}$, where $\mathbf{R}^{1/2}$ is obtained using Cholesky factorization. Note that $\mathbf{R}$ must be positive definite. $\mathbf{R}^{1/2}$ is also equal to $\mathbf{E}\mathbf{\Lambda}^{1/2}\mathbf{E}^T$, where $\mathbf{\Lambda}^{1/2} = \text{diag}\{\sqrt{\lambda_1}, \sqrt{\lambda_2}, \ldots, \sqrt{\lambda_p}\}$ is the diagonal square root matrix of eigenvalues of $\mathbf{R}$, and $\mathbf{E}$ is the eigenvector matrix.

5. All of the p eigenvalues $(\lambda_1, \lambda_2, \ldots, \lambda_p)$ of a $p \times p$ identity matrix $\mathbf{I}$ are equal to one.
6. The determinant of a matrix is equal to the product of all of its eigenvalues: $|\mathbf{A}| = \prod_{j=1}^{p} \lambda_j$.
7. The determinant of the identity matrix is one ($|\mathbf{I}|$=1), since all of its eigenvalues are equal to one.
8. Assume an $n \times p$ data matrix $\mathbf{X}$ and its $p \times p$ correlation matrix $\mathbf{R}$ representing the correlation between all possible pairs of columns. Whenever there are more columns than rows in $\mathbf{X}$ $(p > n)$, there will be $p - n$ zero eigenvalues (n nonzero eigenvalues) of $\mathbf{R}$.
9. Assume an $n \times p$ data matrix $\mathbf{X}$ and its $p \times p$ correlation matrix $\mathbf{R}$. If there are q columns that are identical, then there will be $q - 1$ zero eigenvalues of $\mathbf{R}$.
10. A real square symmetric matrix $\mathbf{A}$ can be inverted by finding its principal form, and sandwiching the inverse eigenvalue matrix between the eigenvector matrix: $\mathbf{A}^{-1} = \mathbf{E}\mathbf{\Lambda}^{-1}\mathbf{E}^T$, where $\mathbf{\Lambda}^{-1} = \text{diag}\{1/\lambda_1, 1/\lambda_2, \ldots, 1/\lambda_p\}$ is the diagonal inverse matrix of eigenvalues of $\mathbf{R}$ and $\mathbf{E}$ is the eigenvector matrix.

APPENDIX C

MATHEMATICAL FUNCTIONS

The following mathematical review sections are fundamental in applied statistics. Of particular importance is the development of maximum likelihood models for multiplicative (and additive) regression models.

C.1 INEQUALITIES

$$(-\infty, a) = \{x < a\} \tag{C.1}$$

$$(a, \infty) = \{x > a\} \tag{C.2}$$

$$(a, b) = \{a < x < b\} \tag{C.3}$$

$$[a, b] = \{a \leq x \leq b\} \tag{C.4}$$

$$[a, b) = \{a \leq x < b\} \tag{C.5}$$

$$(a, b] = \{a < x \leq b\} \tag{C.6}$$

C.2 LAWS OF EXPONENTS

$$x^n = x \times x \times \cdots \times x \times \cdots \qquad (n \text{ times}) \tag{C.7}$$

$$x^m x^n = x^{m+n} \tag{C.8}$$

$$(x^m)^n = x^{mn} \tag{C.9}$$

$$(xy)^n = x^n y^n \tag{C.10}$$

$$\left(\frac{x}{y}\right)^n = \frac{x^n}{y^n} \qquad y \neq 0 \tag{C.11}$$

Classification Analysis of DNA Microarrays, First Edition. Leif E. Peterson.

$$\frac{x^m}{x^n} = x^{m-n} \qquad m > n, x \neq 0 \tag{C.12}$$

$$\frac{x^m}{x^n} = \frac{1}{x^{n-m}} \qquad n > m, x \neq 0 \tag{C.13}$$

$$\frac{x^m}{x^n} = 1 \qquad m = n, \ x \neq 0 \tag{C.14}$$

$$x^0 = 1 \quad \text{since} \quad \frac{x^0}{x^0} = 1 \qquad m = n, x \neq 0 \tag{C.15}$$

$$x^{-n} = \frac{1}{x^n} \qquad n > 0, x \neq 0 \tag{C.16}$$

$$x^{1/n} = \sqrt[n]{x} \tag{C.17}$$

$$x^{m/n} = \sqrt[n]{x^m} \tag{C.18}$$

C.3 LAWS OF RADICALS

$$\left(\sqrt[n]{x}\right)^n = x \tag{C.19}$$

$$\sqrt[n]{x}\,\sqrt[n]{y} = \sqrt[n]{xy} \tag{C.20}$$

$$\frac{\sqrt[n]{x}}{\sqrt[n]{y}} = \sqrt[n]{\frac{x}{y}} \qquad x \neq y \tag{C.21}$$

$$\left(\sqrt[n]{x}\right)^m = \sqrt[n]{x^m} \tag{C.22}$$

$$\sqrt[m]{\sqrt[n]{x}} = \sqrt[mn]{x} \tag{C.23}$$

C.4 ABSOLUTE VALUE

$$|ab| = |a||b| \tag{C.24}$$

$$\left|\frac{a}{b}\right| = \frac{|a|}{|b|} \tag{C.25}$$

C.5 LOGARITHMS

$$a^{\log_a u} = u \tag{C.26}$$

$$\log_a 1 = 0 \tag{C.27}$$

$$\log_a a = 1 \tag{C.28}$$

$$\log_a(uw) = \log_a(u) + \log_a(w) \tag{C.29}$$

$$\log_a\left(\frac{u}{w}\right) = \log_a(u) - \log_a(w) \tag{C.30}$$

$$\log_a(u^c) = c\log_a(u) \tag{C.31}$$

$$\log_b(u) = \frac{\log_a(u)}{\log_a(b)} \tag{C.32}$$

Consider the following example:

$$\log_{10}(x) = \frac{\log_e(x)}{\log_e(10)} \tag{C.33}$$

$$\log_b(a) = \frac{1}{\log_a(b)} \tag{C.34}$$

C.6 PRODUCT AND SUMMATION OPERATORS

We begin this section by looking at product and summation operators that are commonly used in likelihood equations. Recall that most likelihood functions equate to the product of many probabilities in the form

$$f(x) = \prod_{i=1}^{n} p_i. \tag{C.35}$$

We already know that $\log(AB) = \log(A) + \log(B)$, so, for example, we can rewrite Equation (C.35) as $\log(p_1p_2) = \log(p_1) + \log(p_2)$. The product operator then becomes the summation symbol over the same range of x values

$$\log f(x) = \sum_{i=1}^{n} \log(1 + p_i), \tag{C.36}$$

where one is added to p_i in case $p_i = 0$, since the logarithm of zero is undefined. The value of $f(x)$ can now be determined as

$$f(x) = \exp\left(\sum_{i=1}^{n} \log(1 + p_i)\right). \tag{C.37}$$

C.7 PARTIAL DERIVATIVES

In order to review partial derivatives, it is essential to have a strong foothold in solving derivatives of functions. Let's assume the function

$$f(x) = 3x^2 + 4y^5. \tag{C.38}$$

The first derivative is

$$f(x)' = 6x + 20y^4, \tag{C.39}$$

and the second derivative is

$$f(x)'' = 6 + 80y^3. \tag{C.40}$$

Now, the partial derivative of the function with respect to y is

$$\frac{\partial(f(x))}{\partial y} = 20y^4. \tag{C.41}$$

As one can note, partial derivatives focus on taking the derivative of whatever term is in the denominator. As another example, assume that the function

$$f(x) = \beta_0 + \beta_1 x_1 + \beta_2 x_2, \tag{C.42}$$

the partial derivative of the function with respect to β_1 is

$$\frac{\partial(f(x))}{\partial \beta_1} = x_1. \tag{C.43}$$

C.8 LIKELIHOOD FUNCTIONS

Consider the exponential function that commonly appears in likelihood functions for many of the multiplicative regression models. The derivative of any exponentiated function is simply the derivative of the function times the original exponentiated function. In notation, this is written in functional form as

$$\frac{\partial e^u}{\partial u} = e^u du. \tag{C.44}$$

Let us now work with a likelihood "kernel" function replacing the function u with

$$p(x_i) = e^{\beta_0 + \beta_1 x_{i1} + \beta_2 x_{i2} + \cdots + \beta_p x_{ip}} \qquad i = 1, 2, \ldots, n; j = 1, 2, \ldots, p \tag{C.45}$$

where x_{ij} are the covariate values for p covariates ($j = 1, 2, \ldots, p$) for the ith subject ($i = 1, 2, \ldots, n$). We already know that the complete likelihood function for all subjects in the dataset is

$$f(x) = \prod_{i=1}^{n} p_i. \tag{C.46}$$

A general property of likelihood functions is that they are not linear. A commonly used method for "linearizing" likelihood functions is to take the natural logarithm of the kernel, which makes it more amenable to solution.

Taking the log of our function now results in the form

$$\begin{aligned} \log(f(x)) &= \sum_{i=1}^{n} \log(p_i) \\ &= \sum_{i=1}^{n} \log\left(e^{\beta_0+\beta_1 x_{i1}+\beta_2 x_{i2}+\cdots+\beta_p x_{ip}}\right) \\ &= \sum_{i=1}^{n} \beta_0 + \beta_1 x_{i1} + \beta_2 x_{i2} + \cdots + \beta_p x_{ip}. \end{aligned} \tag{C.47}$$

Next, recall that our goal is to maximize the likelihood, so we will need to find values of the β parameters at the maxima of the loglikelihood function. First, to find the maximum of the likelihood due to the contribution of β_1, we must find the first partial derivative of the function with respect to (w.r.t.) β_1; this is known as the *score*, which is expressed as

$$\frac{\partial(f(x))}{\partial\beta_1} = x_{i1} e^{\beta_0+\beta_1 x_{i1}+\beta_2 x_{i2}+\cdots+\beta_p x_{ip}}, \tag{C.48}$$

and the second derivative of $f(x)$ w.r.t. β_2, called the *information*, is

$$\frac{\partial(f(x))}{\partial\beta_1 \partial\beta_2} = x_{i1} x_{i2} e^{\beta_0+\beta_1 x_{i1}+\beta_2 x_{i2}+\cdots+\beta_p x_{ip}}. \tag{C.49}$$

Scores, or first partial derivatives of the loglikelihood function w.r.t. coefficients, are typically arranged in a p-dimension vector $U(\beta_0, \beta_1, \ldots, \beta_p)$, whereas the second derivatives form a $p \times p$ information matrix $\mathbf{I}(\beta_j, \beta_k)$.

Finally, another common element seen in likelihood functions involves the partial derivatives of the log function:

$$f(x) = \log(y). \tag{C.50}$$

The partial derivative of log function is

$$\frac{\partial \log(y)}{\partial y} = \frac{1}{y}. \tag{C.51}$$

Putting everything discussed, together we close with solving the likelihood function for Cox proportional hazards regression. To begin, assume the ordered unique failure times $t_{(1)} < t_{(2)} < \cdots < t_{(k)}$. We recall that subjects in the risk set at time $t_{(i)}$ [i.e., $R(t_{(i)})$] are those subjects with survival times that are at least $t_{(i)}$. In other words, the group of subjects at risk at time $t_{(i)}$ are subjects that survived up to time $t_{(i)}$, and subjects who failed or were censored after $t_{(i)}$ [since they too were also at risk before failure time $t_{(i)}$]. The probability contribution of each failure at time $t_{(i)}$ conditional on those subjects at risk $R(t_{(i)})$ is

$$p(\text{fail}|R(t_{(i)}, \beta) = \frac{\exp\left(\sum_{j=1}^{p} \beta_j x_{ij}\right)}{\sum_{\ell \in R(t_{(i)})} \exp\left(\sum_{j=1}^{p} \beta_j x_{\ell j}\right)}. \tag{C.52}$$

For k total failures, the conditional likelihood becomes

$$L(\beta) = \prod_{i=1}^{k} \frac{\exp\left(\sum_{j=1}^{p} \beta_j x_{ij}\right)}{\sum_{\ell \in R(t_{(i)})} \exp\left(\sum_{j=1}^{p} \beta_j x_{\ell j}\right)}. \tag{C.53}$$

We first take the log of the likelihood above. Taking the log of the numerator gives us

$$\log\left[\exp\left(\sum_{j=1}^{p} \beta_j x_{ij}\right)\right] = \sum_{j=1}^{p} \beta_j x_{ij}. \tag{C.54}$$

The denominator term in the likelihood cannot be reduced by taking the log and remains as

$$\log\left[\sum_{\ell \in R(t_{(i)})} \exp\left(\sum_{j=1}^{p} \beta_j x_{\ell j}\right)\right]. \tag{C.55}$$

Looking at the likelihood kernel $p(\text{fail}|R(t_{(i)}, \beta)$, we also notice that it is a ratio, so we have to use the property for logarithms $\log(A/B) = \log(A) - \log(B)$. Combining the logarithms of the numerator and denominator of the kernel yields the loglikelihood (LL) function

$$\text{LL}(\beta) = \sum_{i=1}^{k} \left\{ \sum_{j=1}^{p} \beta_j x_{ij} - \log\left[\sum_{\ell \in R(t_{(i)})} \exp\left(\sum_{j=1}^{p} \beta_j x_{\ell j}\right)\right]\right\}. \tag{C.56}$$

Next, let's only work with partial derivatives for each kth risk set, so for now we can drop the summation symbol $\sum_{i=1}^{k}$. Therefore, we are dealing with the loglikelihood for each risk set in the form

$$\text{LL}(\beta, R(t_{(i)})) = \sum_{j=1}^{p} \beta_j x_{ij} - \log\left[\sum_{\ell \in R(t_{(i)})} \exp\left(\sum_{j=1}^{p} \beta_j x_{\ell j}\right)\right]. \tag{C.57}$$

Now we take the first partial derivative of the left term $\sum_{j=1}^{p} \beta_j x_{ij}$ in the loglikelihood function w.r.t. each β term to get the score vector. We already know that for the function

$$f(x) = \beta_0 + \beta_1 x_1 + \beta_2 x_2, \tag{C.58}$$

the partial derivative of the function w.r.t. β_1 is

$$\frac{\partial(f(x))}{\partial\beta_1} = x_1. \tag{C.59}$$

Therefore the first partial derivative for each β in the left term of the loglikelihood is

$$\frac{\partial(\sum_{j=1}^{p} \beta_j x_{ij})}{\beta_j} = x_{ij}. \tag{C.60}$$

At this point, we know that the first part of the score function is

$$s(\beta_j) = x_{ij} \ldots. \tag{C.61}$$

We are left with the right term of the loglikelihood in the form

$$\log\left[\sum_{\ell \in R(t_{(i)})} \exp\left(\sum_{j=1}^{p} \beta_j x_{\ell j}\right)\right]. \tag{C.62}$$

Notice that this is a log function. We know that the derivative of the logarithm of some function u, $\log(u)$ is the inverse of the function, du/u. This suggests that part of the right term of the score contains the inverse

$$\frac{\partial u/\partial\beta_j}{\log\left[\sum_{\ell \in R(t_{(i)})} \exp\left(\sum_{j=1}^{p} \beta_j x_{\ell j}\right)\right]}. \tag{C.63}$$

Since we already handled the logarithm, we can now attack the function itself. Our target is β_j, and we note that the subscript j indicates that the coefficients are not a function of each subject ℓ in the risk set $R(t_{(i)})$ considered. Thus, β values don't change across risk sets, so their addition across risk sets (and anywhere they appear in the summation) has to be recognized and left intact. Given this, we can now work with the terms for a single individual ℓ in the risk set who contributes the factor

$$\exp\left(\sum_{j=1}^{p} \beta_j x_{\ell j}\right). \tag{C.64}$$

Taking the first partial derivative of this function w.r.t. β_j gives

$$\frac{\partial\left(\exp\left(\sum_{j=1}^{p} \beta_j x_{\ell j}\right)\right)}{\partial\beta_j} = x_{\ell j} \exp\left(\sum_{j=1}^{p} \beta_j x_{\ell j}\right). \tag{C.65}$$

For the entire group of subjects in the risk set, we can now include the first partial derivative into the summation and rewrite this as

$$\sum_{\ell \in R(t_{(i)})} x_{\ell j} \exp\left(\sum_{j=1}^{p} \beta_j x_{\ell j}\right). \tag{C.66}$$

Going back to our original log function on the right side of the loglikelihood, we now have

$$\frac{\sum_{\ell \in R(t_{(i)})} x_{\ell j} \exp\left(\sum_{j=1}^{p} \beta_j x_{\ell j}\right)}{\sum_{\ell \in R(t_{(i)})} \exp\left(\sum_{j=1}^{p} \beta_j x_{\ell j}\right)}. \tag{C.67}$$

For the risk set considered, we now have the score

$$s(\beta_1) = x_{ij} - \frac{\sum_{\ell \in R(t_{(i)})} x_{\ell 1} \exp\left(\sum_{j=1}^{p} \beta_1 x_{\ell 1}\right)}{\sum_{\ell \in R(t_{(i)})} \exp\left(\sum_{j=1}^{p} \beta_1 x_{\ell 1}\right)}. \tag{C.68}$$

Now that the first partial derivative (score) of the loglikelihood function w.r.t. β_1 is solved, we can calculate the second partial derivative of the loglikelihood function w.r.t. β_2. First, we drop the term x_{ij} on the left of the first partial derivative since it is independent of β_2. Looking at the remaining terms, we note that the first partial derivative is a quotient in the form

$$\frac{\sum_{\ell \in R(t_{(i)})} x_{\ell 1} \exp\left(\sum_{j=1}^{p} \beta_1 x_{\ell 1}\right)}{\sum_{\ell \in R(t_{(i)})} \exp\left(\sum_{j=1}^{p} \beta_1 x_{\ell 1}\right)}, \tag{C.69}$$

so we use the quotient rule for differentiation. The quotient rule is

$$\frac{f(x)}{g(x)} = \frac{g(x)\frac{\partial f(x)}{\partial x} - f(x)\frac{\partial g(x)}{\partial x}}{g(x)^2}, \tag{C.70}$$

so we have

$$\begin{aligned}\frac{f(x)}{g(x)} = &\sum_{\ell \in R(t_{(i)})} \exp\left(\sum_{j=1}^{p} \beta_1 x_{\ell 1}\right) \frac{\partial \sum_{\ell \in R(t_{(i)})} x_{\ell 1} \exp\left(\sum_{j=1}^{p} \beta_1 x_{\ell 1}\right)}{\partial \beta_2} \\ &- \sum_{\ell \in R(t_{(i)})} x_{\ell 1} \exp\left(\sum_{j=1}^{p} \beta_1 x_{\ell 1}\right) \frac{\partial \sum_{\ell \in R(t_{(i)})} \exp\left(\sum_{j=1}^{p} \beta_1 x_{\ell 1}\right)}{\partial \beta_2} \\ &\div \sum_{\ell \in R(t_{(i)})} \exp\left(\sum_{j=1}^{p} \beta_1 x_{\ell 1}\right)^2,\end{aligned} \tag{C.71}$$

and after rearranging this gives

$$\frac{f(x)}{g(x)} = \sum_{\ell \in R(t_{(i)})} \exp\left(\sum_{j=1}^{p} \beta_1 x_{\ell 1}\right) \sum_{\ell \in R(t_{(i)})} x_{\ell 1} x_{\ell 2} \exp\left(\sum_{j=1}^{p} \beta_1 x_{\ell 1}\right)$$
$$- \sum_{\ell \in R(t_{(i)})} x_{\ell 1} \exp\left(\sum_{j=1}^{p} \beta_1 x_{\ell 1}\right) \sum_{\ell \in R(t_{(i)})} x_{\ell 2} \exp\left(\sum_{j=1}^{p} \beta_1 x_{\ell 1}\right) \quad \text{(C.72)}$$
$$\div \sum_{\ell \in R(t_{(i)})} \exp\left(\sum_{j=1}^{p} \beta_1 x_{\ell 1}\right)^2,$$

and finally

$$\frac{f(x)}{g(x)} = \frac{\sum_{\ell \in R(t_{(i)})} \exp\left(\sum_{j=1}^{p} \beta_1 x_{\ell 1}\right) \sum_{\ell \in R(t_{(i)})} x_{\ell 1} x_{\ell 2} \exp\left(\sum_{j=1}^{p} \beta_1 x_{\ell 1}\right)}{\sum_{\ell \in R(t_{(i)})} \exp\left(\sum_{j=1}^{p} \beta_1 x_{\ell 1}\right)^2}$$
$$- \frac{\sum_{\ell \in R(t_{(i)})} x_{\ell 1} \exp\left(\sum_{j=1}^{p} \beta_1 x_{\ell 1}\right) \sum_{\ell \in R(t_{(i)})} x_{\ell 2} \exp\left(\sum_{j=1}^{p} \beta_1 x_{\ell 1}\right)}{\sum_{\ell \in R(t_{(i)})} \exp\left(\sum_{j=1}^{p} \beta_1 x_{\ell 1}\right)^2}. \quad \text{(C.73)}$$

Canceling out terms, we get

$$\mathbf{I}(1,2) = \frac{\sum_{\ell \in R(t_{(i)})} x_{\ell 1} x_{\ell 2} \exp\left(\sum_{j=1}^{p} \beta_1 x_{\ell 1}\right)}{\sum_{\ell \in R(t_{(i)})} \exp\left(\sum_{j=1}^{p} \beta_1 x_{\ell 1}\right)}$$
$$- \frac{\sum_{\ell \in R(t_{(i)})} x_{\ell 1} \exp\left(\sum_{j=1}^{p} \beta_1 x_{\ell 1}\right) \sum_{\ell \in R(t_{(i)})} x_{\ell 2} \exp\left(\sum_{j=1}^{p} \beta_1 x_{\ell 1}\right)}{\sum_{\ell \in R(t_{(i)})} \exp\left(\sum_{j=1}^{p} \beta_1 x_{\ell 1}\right)^2}, \quad \text{(C.74)}$$

which forms element $(1,2)$ in the information matrix (second partial derivative *Hessian* matrix). Each element in the information matrix is calculated using the equation above and by substituting in appropriate values for subjects in all risk sets. A running total of elements in the score vector and information matrix is kept as the covariates (e.g., $x_{\ell 1}$) for each subject are looped through within each risk set and over each β_j term.

APPENDIX D

STATISTICAL PRIMITIVES

D.1 RULES OF THUMB

Always Induce Scale Invariance among Feature Values. The range and scale of feature values input into a classifier can be a significant problem if not dealt with appropriately. *Scale invariance* simply means that there is no difference in scale among the features used for an analysis. Whenever there is more than an order of magnitude (factor of 10) difference in the scale of feature values, then there is potential for problems during learning or generalization of results. For example, if the unit of one feature is μg/mL and the unit of another in ng/mL, and no unit conversion is made before modeling, the results will be biased. In this case, feature values in units of μg/mL should be multiplied by 1000 if the final working unit is ng/mL, or the ng/mL values should be divided by 1000 to convert into μg/mL. Either way, all features used should have the same scale (unit of measurement) before attempting to train an algorithm.

Always Ensure a Gaussian Shape of Feature Distributions. Histograms should be generated for each feature so that the Gaussian shape and skewness can be visually assessed. Numerous transformations can be performed on feature values, such as $1/x$, $\sqrt{x}$, $\log(x)$, depending on the original shape of the distribution. For lognormally distributed feature values, either the natural logarithm $\log_e(x)$ or base-10 logarithm $\log_{10}(x)$ can be used to log-transform feature values. When log transforming, be sure to add a one before taking the log, that is, $\log_e(1 + x)$, since the log of zero is undefined.

Always Ensure a Similar Range Among Feature Values. Once skewness is removed, the mean and standard deviation of each distribution representing an array can still be different, and therefore, mean-zero standardization should be performed to ensure that the range is within

Classification Analysis of DNA Microarrays, First Edition. Leif E. Peterson.

expectation of the standard normal distribution. For a large number of genes, the $n - 1$ correction to degrees of freedom in the denominator of the standard deviation will have a negligible effect, so it can often be avoided. Normalizing each feature to ensure that the range of its values is [0,1] is another approach that can ensure a similar range among all features used. The choice of whether to mean-zero-standardize or normalize feature values depends on the performance of the filter used for gene selection or classifier employed. Below is example code that can be used for mean-zero standardizing each feature, where `NumObjects` represents the number of objects, or arrays; `NumFeatures` is the number of features, or genes; `x(NumObjects,NumFeatures)` is the data array; and `mean` and `sigma` represent the mean and standard deviation:

```
For j = 1 To NumFeatures
  sum = 0
  For i = 1 To NumObjects
    sum += x(i, j)
  Next i
  mean = sum / NumObjects
  sum = 0
  For i = 1 To NumObjects
    diff = x(i, j) - mean
    sum += diff * diff
  Next i
  sigma = Math.Sqrt(sum / (NumObjects - 1))
  For i = 1 To NumObjects
    If sigma > 0 Then x(i, j) = (x(i, j) - mean) / sigma
  Next i
Next j
```

For normalizing each feature into a range of [0,1], the following example code can be used:

```
For j = 1 To NumFeatures
  max = -1.0E+30
  min = 1.0E+30
  For i = 1 To NumObjects
    If x(i, j) < min Then min = x(i, j)
    If x(i, j) > max Then max = x(i, j)
  Next i
  range = max - min
  For i = 1 To NumObjects
    x(i, j) = (x(i, j) - min) / range
  Next i
Next j
```

Don't Mean-Zero-Standardize or Normalize Feature Values within Each Class. Mean-zero standardization or normalization for a feature should be done for all objects in all the classes. In other words, don't use class-specific means and standard deviations when mean-zero-standardizing a feature. For normalization, don't use the minimum feature value from each class, but rather, use the single minimum feature value from all objects. An informative feature will have values that are consistently different within each class. If mean-zero standardization or normalization are performed on feature values within each class, then any mean differences will likely be removed, resulting in feature values that are much less distinctive (noninformative) across the classes.

Finally, another transformation that can be used to ensure equal scale of feature values is the direction cosine transform, which divides each vector element by the norm (length) of the vector:

```
For i = 1 To NumObjects
  sum = 0
  For j = 1 To NumFeatures
    sum += x(i, j) * x(i, j)
  Next j
  norm = Math.Sqrt(sum)
  For j = 1 To NumFeatures
    dircosines(j) = x(i, j) / norm
  Next j
Next i
```

Shuffle Array Order Before Cluster Initialization. During initialization of CKM cluster analysis, each object needs to be randomly assigned to one of the K clusters. When assigning cluster labels, apply them repeatedly to the objects in the order $1, 2, \ldots, K$ until you reach the end of the list of arrays, and then shuffle the cluster labels among the objects, so that they are in random order. Simply applying labels randomly in the range $1, 2, \ldots, K$ is inefficient, and will often result in an imbalance in the number of objects per class–in other words, the class priors will be markedly different. Assigning ordered cluster labels and shuffling objects will ensure an even distribution and equal opportunity for all objects to be distributed among the clusters. Below is example code that lists how to first assign cluster labels to the `vector clusterassigned(NumObjects)` sequentially to the objects, and then how to shuffle the cluster labels among the objects:

```
'Fill clusterassigned() with consecutive values of k
'  (i.e., if K=3, then 1,2,3,1,2,3,... until filled
j = 0
For i = 1 To NumObjects
  For k = 1 To NumClusters
    j += 1
    If j > NumObjects Then GoTo continue1
    clusterassigned(j) = k
```

```
  Next k
Next i

continue1:
'Permute cluster labels among objects
Nperm = NumObjects
Do While Nperm >= 1
  Mperm = Math.Round((Nperm - 1) * Rnd()) + 1
  buffer = clusterassigned(Nperm)
  clusterassigned(Nperm) = clusterassigned(Mperm)
  clusterassigned(Mperm) = buffer
  Nperm = Nperm - 1
Loop
```

Use Normalized Feature Values for Cluster Validity. When determining the appropriate number of clusters for a dataset via *cluster validity*, it is often better to use normalized feature values in the range [0,1] instead of mean-zero standardization.

Repartition Objects After Shuffling during Cross-Validation. Assessing the stability of a classifier by using cross validation requires that multiple repartitions are performed when randomly assigning objects to folds prior to cross validation (CV). It is recommended that ten 10-fold CVs be performed to ensure a reliable measure of stability. Naturally, the preferred approach for assigning objects into the respective folds is to first shuffle them and then assign them to folds in the order $1, 2, \ldots, 10$. Each new CV run will repeat these steps using a new shuffling of objects prior to assigning objects to the folds again. This is the process of *repartitioning* via shuffling.

D.2 PRIMITIVES

Below are listed some of the most commonly used statistical primitives for classification analysis.

Correlation and Covariance Matrices. Determination of the variable-by-variable (feature-by-feature) covariance matrix **S** and correlation matrix **R** requires calculation of the standard deviation s and mean $\bar{x}$ of expression for each variable (feature). The output arrays `cov(,)` and `cor(,)` have dimensions $p \times p$ when p genes are being correlated and dimensions $n \times n$ when arrays are being correlated:

```
For j = 1 To NumFeatures
  sum = 0
  For i = 1 To NumObjects
```

```
    sum += x(i, j)
  Next i
  avg(j) = sum / NumObjects
Next j
For j = 1 To NumFeatures
  sum = 0
  For i = 1 To NumObjects
    sum += (x(i, j) - avg(j)) * (x(i, j) - avg(j))
  Next i
  sd(j) = Math.Sqrt(sum / (NumObjects - 1))
Next j
For j = 1 To NumFeatures -1
  For k = j + 1 to NumFeatures
    sum = 0
  For i = 1 To NumObjects
    sum += (x(i, j) - avg(j)) * (x(i, k) - avg(k))
  Next
  cov(j, k) = sum / NumObjects
  cov(k, j) = cov(j, k)
  Next
Next
For j = 1 To NumFeatures
  cov(j, j) = sd(j) * sd(j)
Next j
For j = 1 To NumFeatures - 1
  For k = j + 1 To NumFeatures
  sum = 0
  For i = 1 To NumObjects
    sum += (x(i, j) - avg(j)) / sd(j) * (x(i, k) - avg(k)) / sd(k)
  Next i
  cor(j, k) = sum / (NumObjects - 1)
  cor(k, j) = cor(j, k)
  Next k
Next j
For j = 1 To NumFeatures
  cor(j, j) = 1.0
Next j
```

Matrix Transpose. The matrix transpose operation is employed in the majority of techniques that involve matrix operations. The input matrix `a(,)` has dimensions $m \times n$ while the output matrix `b(,)` has dimensions $n \times m$:

```
'Input: matrix a(m,n)
'Output: matrix b(n,m)
For i = 1 To m
  For j = 1 To n
    b(j, i) = 0
    b(j, i) = a(i, j)
  Next j
Next i
```

Matrix-Vector Multiplication. This routine is used extensively in regression when we cross-multiply $(\mathbf{X}^T\mathbf{X})^{-1}$ by $\mathbf{X}^T\mathbf{y}$ in kernel regression, $(\mathbf{X}^T\mathbf{X})^{-1}$ by $\mathbf{X}^T\mathbf{Y}$ in the LREG classifier and covariance matrix filtering, and $\mathbf{I}(\boldsymbol{\beta})^{-1}\mathbf{s}(\boldsymbol{\beta})$ in the PLOG classifier. The input matrix `a(,)` has dimensions $l \times m$, while the input vector `b()` is an $m \times 1$ vector, and the corresponding output is a vector `c()` with dimensions $l \times 1$, where l is the number of rows of `a(,)`.

```
'Input: matrix a(l,m)
'        vector b(m)
'Output: vector c(l)
For i = 1 To l
  d = 0
  For j = 1 To m
    d += a1(i, j) * b1(j)
  Next j
  c(i) = d
Next i
```

Matrix-Matrix Multiplication. Matrix-matrix multiplication is used extensively in classification analysis. The input matrices `a(,)` and `b(,)`, have dimensions $l \times m$ and $m \times n$, respectively. The output matrix `c(,)` has dimensions $l \times n$:

```
'Input: matrix a(l,m), with l rows, m columns
'       matrix b(m,n), with m rows, n columns
'Output: vector c(l,n), with l rows, n columns
For i = 1 To l
  For j = 1 To n
    d = 0
    For k = 1 To m
      d += a1(i, k) * b1(k, j)
    Next k
    c(i, j) = d
  Next j
Next i
```

Indexed Sort. The indexed sort is one of the better methods to use when there is a need to keep the original order intact for recordkeeping. A noteworthy example is when the original index of an eigenvector is needed based on the sort order of all the eigenvalues. Using the notation below, `arr()` is the unsorted input vector of length `size`, `ndx()` is the original index for each input value (typically 1,2,...,`size`), and `descending` is a Boolean variable to specify whether an ascending or descending sort order is desired. In the specific code example below, the array `arr()` will contain the values sorted in the order requested, while `ndx()` will contain the original index for each value prior to sorting. For example, if the original unsorted values of `arr()` are (7.2, 3.4, 23.0) and original `ndx()` values are (1, 2, 3), then for a descending sort the output values for `arr()` will be (23.0, 7.2, 3.4) and for `ndx()` will be (3,1,2):

```
'Inputs: size  -- length of arr(), ndx() vectors
         'arr() -- unsorted input vector
         'ndx() -- index value (1,2,...,size)
         'descending -- Boolean
If Not descending Then
  System.Array.Sort(arr, ndx, 1, size)
  Exit Sub
End If
If descending = True Then
  System.Array.Sort(arr, ndx, 1, size)
  System.Array.Reverse(arr, 1, size)
  System.Array.Reverse(ndx, 1, size)
End If
```

Kernel Density Estimation. The following code can be used for kernel density estimation (KDE), which was described in Chapter 13. The variant of KDE used here is to determine a smoothed PDF of a distribution of x values. For inputs, `numbins` is the number of equally spaced nonoverlapping bins of the histogram (100 is a reliable default value when thousands of x values are used), `sampsize` is the number of x values used, `xinput(sampsize)` is the input vector of x values, `xmin` is the minimum value of x, `xmax` is the maximum value of x, and `sd` is the standard deviation of x. There are two output vectors: `xvalue(numbins+3)`, which contains the x values at the histogram bin walls; and `f_x(numbins+3)`, containing the PDF values for the respective histogram bins. The results can be plotted in an x-y scatterplot using the contents of `xvalue()`for the x axis and `f_x()` for the y axis.

There are five options of KDE available in the code: uniform, triangle, Epanechnikov, quartic, triweight, cosinus, and Gaussian; Epanechnikov is used as a default. If the distribution of x tends to be Gaussian, then the Gaussian option will provide a smoother PDF when compared with Epanechnikov. However, Epanechnikov is likely a better choice if the PDF is multimodal:

```
precisionfactor = (xmax - xmin) / numbins
bandwidth = 2 * 1.06 * sd * sampsize ^ (-0.2) '1.06 is nominal
   bandwidth
fac1 = 1 / (sampsize * bandwidth)
fac2 = 1 / Math.Sqrt(2 * Math.PI)
fac3 = 1 / bandwidth
For i = 1 To numbins + 3
  x = xmin + ((i - 1) * precisionfactor)
```

```
  If i <= numbins Then xvalue(i) = x
  sum = 0
  For j = 1 To sampsize
    y1 = (xinput(j) - x) * fac3
    'Uniform
    'sum += 0.5
    'Triangle
    'sum += (1 - Math.Abs(y1))
    'Epanechnikov
     sum += 0.75 * (1 - y1 * y1)
    'Quartic
    'sum += 15.0# / 16.0# * (1 - y1 * y1) ^ 3
    'Triweight
    'sum += 35.0# / 32.0# * (1 - y1 * y1) ^ 3
    'Cosinus
    'sum += 3.1415 / 4.0# * Cos(3.1415 / 2 * y1)
    'Gaussian
    'sum += fac2 * Math.Exp(-0.5 * y1 * y1)
  Next j
  f_x(i) = sum * fac1
Next i
'Normalize f_x() to total sample size
sum = 0
For j = 1 To numbins
  sum += f_x(j)
Next j
fac = 1 / sum
sum = 0
For j = 1 To numbins
  f_x(j) *= fac
sum += f_x(j)
Next j
```

Rejection-Acceptance Method. The rejection-acceptance method can be used to simulate a large number of quantiles (x values) when the function values $f(x)$ in `numbin` equally spaced nonoverlapping intervals are known for a specified range of x. The function $f(x)$ can be a PDF for a distribution, for which only the histogram bin values are known. The purpose of using the acceptance-rejection method is to generate via simulation the individual x values, or quantiles, that will result in a histogram that is similar to the input histogram. The rejection-acceptance method works as follows. First, obtain the empirical function values $f(x)$ at M equally spaced nonoverlapping intervals or "bins" for the specified range and find the maximum value $f_{\max}$. Next, to generate each simulated quantile value, draw two random uniform variates $\mathcal{U}_1(0,1)$ and $\mathcal{U}_2(0,1)$, and set the quantile value in the function value in the mth interval ($m = \mathcal{U}_1(0,1)(M-1)+1$) to ("accept") if

$\mathcal{U}_2(0,1) < f_x(m)/f_{\max}$, otherwise "reject" and draw another pair of uniform variates and continue until `sampsize` quantile values have been generated.

For inputs, `sampsize` is the the number of quantile values desired, `numbins`$= M$ is the number of equally spaced nonoverlapping intervals, and `f_xin(numbins)` is the input vector of $f(x)$ values, or PDF values. For outputs, `quantile(sampsize)` is the output vector containing `sampsize` simulated quantile values for the input PDF, and `hist(numbins)` is the output vector of histogram bin counts for the simulated quantiles:

```
totsamp = 0
fmax = -1E+30
For j = 1 To numbins
 If f_xin(j) > fmax Then fmax = f_xin(j)
Next j
Do
  m = (numbins - 1) * Rnd + 1
  ran2 = Rnd
  If ran2 < f_xin(m) / fmax Then
    totsamp += 1
    If totsamp > sampsize Then Exit Do
    quantile(totsamp) = f_xin(m)
    hist(m) += 1
   End If
Loop
```

Nonlinear Regression. In nonlinear regression, the objective function to be minimized is

$$\mathrm{SSE} = \sum_{i=1}^{n} (y_i - \hat{y}_i)^2, \tag{D.1}$$

where y_i is the observed value for the ith patient and $\hat{y}_i$ is the predicted value. The consistent equations for solving the objective function is

$$\boldsymbol{\delta} = (\mathbf{D}^T\mathbf{D})^{-1}\mathbf{D}^T\mathbf{e}, \tag{D.2}$$

where $\boldsymbol{\delta}$ is the update vector, $\mathbf{D}$ is the Jacobian matrix of first partial derivatives of $f(x)$ w.r.t. each β parameter, and $\mathbf{e}$ is the vector of residuals (i.e., $y_i - \hat{y}_i$). At the (i+1)th iteration, the values of the parameters are

$$\beta_j^{(i+1)} = \beta_j^{(i)} + \delta_j^{(i)}. \tag{D.3}$$

Convergence is reached at the point when $||\boldsymbol{\delta}|| \leq 10^{-4}$. We focus on a published example from the text by Neter et al. [1], for which nonlinear regression was performed to iteratively solve for the coefficients or parameters of the nonlinear function $f(x) = \beta_1 \exp(\beta_2 x_i)$. The data represent n=15 hospitalized patients for which x_i is the length of stay in days and $y_i = f(x_i)$ is a prognosis index. Starting values used are $\beta_1 = 50$ and $\beta_2 = 0$. After several iterations when convergence is reached, the parameter solutions are $\beta_1 = 58.607$ and $\beta_2 = -0.04$, with a MSE of 3.8. Setting up the regression involved creation of the 15×1 **x** vector of length-of-stay values, and **e** containing the residuals, based on the starting parameter values. `nrow` is set to 15 and `ncol` is set to 2 since we are solving two parameters: β_1 and β_2.

There are two options for reaching convergence in the `nonlin()` routine listed below. The first involves the typical *method of scoring* that directly uses the first partial derivatives in the Jacobian matrix, and the second involves *finite differencing*, which approximates derivatives by exploiting machine precision [2]. Finite differencing assumes the derivative is approximated by

$$f(x)' = \frac{f(x+h) - f(x)}{h}, \tag{D.4}$$

where $h \approx \sqrt{\epsilon_{\text{mach}}}$, where $\epsilon_{\text{mach}} \approx 10^{-16}$, and therefore $h \approx 10^{-8}$. For p variables, finite differencing is about p times as expensive as computing $f(x)$. There are also two options for obtaining the update vector, `g()`=$\boldsymbol{\delta}$. The first option employs the Newton–Raphson approach, which inverts the matrix $\mathbf{D}^T\mathbf{D}$; and then multiplies its inverse $(\mathbf{D}^T\mathbf{D})^{-1}$ by $\mathbf{D}^T\mathbf{e}$; the second uses the *conjugate gradient* method, which does not perform matrix inversion or factorization:

```
Sub nonlin(nrow,ncol,x,y,coef)
'(C) 2001 Leif E. Peterson
'Subroutine to perform non-linear regression
'Algorithm is capable of solving f(x) via the second-order
'method of scoring, which requires first and second partial
'derivatives of the f(x) w.r.t. parameters, or the first-order
'conjugate gradient method.
'Example partial derivatives are for the non-linear function:
'   f(x) = coef(1) * Math.Exp(coef(2) * x(i)), from Chap. 14
'Nonlinear Regression, in Neter, J., Wasserman, W.,
   Kutner, M.H.
'Applied Linear Statistical Models, 2nd Edition.
'Homewood (IL), Richard Irwin Publishers, 1985.
'Dim g(ncol), d(ncol), r(ncol), p(ncol), s(ncol, ncol),
   sinv(ncol, ncol)
'Dim u(ncol), e(nrow), gnorm As Double
'Starting values for coef(1) and coef(2) are below:
coef(1) = 50
coef(2) = 0
For iter = 1 To 100
  gnorm = 0
```

```
If iter > 1 Then
  For j = 1 To ncol
    coef(j) += g(j)
    gnorm += Math.Abs(g(j))
  Next j
  If gnorm < 0.00001 Then
    GoTo converged
  End If
End If
For l = 1 To ncol
  For m = 1 To ncol
    s(l, m) = 0
  Next m
  U(l) = 0
  D(l) = 0
Next l
SSE = 0
For i = 1 To nrow
  yhat = coef(1) * Math.Exp(coef(2) * x(i))
  e(i) = y(i) - yhat
  SSE += e(i) * e(i)
  '***Use derivatives directly in the method of scoring****
  'remove comments below and add comments to finite
  'differencing section to use partial derivatives
  ' D(1) = Math.Exp(coef(2) * x(i))
  ' D(2) = coef(1) * x(i) * Math.Exp(coef(2) * x(i))
  '*****************************************************
  '***Use finite differencing to approximate derivatives***
  'comment section below and remove comments above to
  'use partial derivatives
  For j = 1 To ncol
    temp = coef(j)
    h = 0.00000001 * Math.Abs(temp)
    If h = 0 Then h = 0.00000001
    coef(j) = temp + h
    h = coef(j) - temp
    ynew = coef(1) * Math.Exp(coef(2) * x(i))
    coef(j) = temp
    d(j) = (ynew - yhat) / h
  Next j
  '*****************************************************
  For l = 1 To ncol
    For m = 1 To ncol
      s(l, m) += d(l) * d(m)
    Next m
  Next l
  For l = 1 To ncol
    u(l) += d(l) * e(i)
```

```
    Next l
  Next i
  '*********************************************************
  '***Use Newton-Raphson to obtain update vector g(), by
  'first inverting s(,) to obtain sinv(,), and
     cross-multiplying
  'sinv(,) with u() to obtain g().
  Call invertmatrix(ncol, s, sinv)
  Call matmul1(sinv, u, g, ncol, ncol)
  '*********************************************************
  '***Use conjugate gradient to obtain g(), by avoiding matrix
  'inversion.
  rtrold = 0
  For j = 1 To ncol
    r(j) = u(j)
    rtrold += r(j) * r(j)
    p(j) = 0
  Next j
  Call conjug(ncol, s, r, p, rtrold, g)
  '*********************************************************
Next iter
converged:
Call invertmatrix(ncol, s, sinv) 'get sinv for
   variance-covariance matrix MSE = SSE / (nrow - ncol)
'Note: every element of sinv must be multiplied by MSE to
   obtain the variance-covariance matrix
```

Conjugate Gradient Method. The conjugate gradient method does not work explicitly with the coefficient matrix, but rather splits up a multi-dimensional minimization problem into multiple univariate minimization problems [3]. Since matrix inversion and factorization are not used, conjugate gradient is one of the faster "low-storage" methods for solving a system of equations. Conjugate gradient methods are also advantageous when the second partial derivatives are difficult to determine in closed form or are computationally expensive to solve. The particular code provided in subroutine `conjug()` below is called in the *non-linear regression routine* described earlier in this appendix:

```
Sub conjug(ncol, s, r, p, rtrold, g)
'Design & code: (C) 2001, Leif E. Peterson
'Reference - Conjugate gradient method for solving linear
   systems of equations
'based on equations in:
'Nash, S.G., Ariela, S.  Low-storage Methods for
'Unconstrained Problems, Chap 12. In: Linear and Nonlinear
'Programming. New York, McGraw-Hill, 1996.
```

```
Dim a, B, rtr, pap, gnorm, ap(ncol) As Double
For j = 1 To ncol
  g(j) = 0
Next j
For iter = 1 To 100
  For j = 1 To ncol
    ap(j) = 0
  Next j
  gnorm = 0
  For j = 1 To ncol
    gnorm += r(j) * r(j)
  Next j
  gnorm = Math.Sqrt(gnorm)
  If gnorm < 0.0001 Then
    GoTo conv
  End If
  If citer > 1 Then
    rtr = 0
    For j = 1 To ncol
      rtr += r(j) * r(j)
    Next j
    B = rtr / rtrold
    rtrold = rtr
  End If
  For j = 1 To ncol
    p(j) = r(j) + B * p(j)
  Next j
  For j = 1 to ncol
    sum =0
    For k = 1 to ncol
      sum + = s(j,k) * p(k)
    Next k
    ap(j) = sum
  Next j
  pap = 0
  For j = 1 To ncol
   pap += p(j) * ap(j)
  Next j
  a = rtrold / pap
  For j = 1 To ncol
    g(j) += a * p(j)
  Next j
  For j = 1 To ncol
   r(j) -= a * ap(j)
  Next j
Next iter
conv:
```

REFERENCES

[1] J. Neter, W. Wasserman, M.H. Kutner. Nonlinear regression. In *Applied Linear Statistical Models*, 2nd ed., Richard Irwin Publishers, Homewood, IL, 1985, chap. 14.

[2] A.R. Curtis, M.J.D. Powell, J.K. Reid. On the estimation of sparse Jacobian matrices. *J. Inst. Math. Appl.* **13**:117–119, 1974.

[3] S.G. Nash, S. Ariela. Low-storage methods for unconstrained problems. In *Linear and Nonlinear Programming*, McGraw-Hill, New York, 1996, chap. 12.

APPENDIX E

PROBABILITY DISTRIBUTIONS

E.1 BASICS OF HYPOTHESIS TESTING

Probability values are the primary objective of statistical hypothesis testing, and their use enables us to determine the degree of significance for a test statistic. This chapter covers the numerical methods used for obtaining P values for several tests of hypotheses. Some commonly used tests are listed below along with the distribution for the test statistic and function used for determining the P value.

Before moving on, let's define some terminology used in hypothesis testing, so that we understand where P values fit in. Examples are provided for several areas of statistics in order to expand the horizons of the reader.

Null Hypothesis. The null hypothesis is the formal statement of the research issue, establishing the basis for conducting the experiment. The goal is to discredit the null hypothesis in order to confirm whether a statistically significant effect is present in the sampled data. The notation used for the null hypothesis of, for example, equal sample means is $H_0 : \bar{x}_1 = \bar{x}_2$. The following are some commonly used null hypotheses:

Null hypothesis	Application	Remarks
$P(Z_i \leq Z_\alpha)$	Extreme value, outlier	$P(x \leq x_0)$
$\bar{x}_1 = \bar{x}_2$	t test	Means are equal
$P(\chi^2 \leq \chi^2_\nu)$	χ^2 test	χ^2 less than critical value
$\beta_j = 0$	Linear regression	Slope is equal to zero

Alternative Hypothesis. The alternative hypothesis is true when the null hypothesis is false, and embraces the belief that a statistically significant signal or effect exists in the sampled data. The reason for pursuing a research project is due to the alternative hypothesis, since it forms the belief.

Classification Analysis of DNA Microarrays, First Edition. Leif E. Peterson.

Some examples of alternative hypotheses for the above mentioned null hypotheses are listed here:

Alternative hypothesis	Application	Remarks
$P(Z_i > Z_\alpha)$	Extreme value, outlier	x_i exceeds x_0
$\bar{x}_1 \neq \bar{x}_2$	t test	Means are not equal
$P(\chi^2 > \chi_\nu^2)$	χ^2 test	χ^2 exceeds critical value
$\beta_j \neq 0$	Linear regression	Slope is not zero

Test Statistic. The test statistic is what is calculated during hypothesis testing. Examples of test statistics are as follows:

Notation	Statistic	Distribution
Z	Z score	Standard normal distribution
$t_{\nu,1-\alpha/2}$	t statistic	Student's t distribution
F_{ν_1,ν_2}	F statistic	F (variance ratio) distribution
χ_ν^2	χ^2 statistic	χ^2 distribution
$(\beta_j/s.e.(\beta_j))^2$	Wald statistic (regression)	χ^2 distribution

Level of Significance, $\alpha = 0.05$. The level of significance, termed α, is the percentage of total probability in the tail of a probability distribution used for selecting the critical value, against which the test statistic is compared. For example, at the $\alpha = 0.05$ level of significance, the standard normal variate is $Z_{\alpha=0.05} = 1.645$, as shown in Figure 12.3. At the 2.5% level, $Z_{\alpha=0.05/2} = 1.96$. The level of significance also reflects the proportion of repeated experiments in which the null hypothesis will be falsely rejected, that is, the false positive rate. By specifying a level of significance of $\alpha = 0.05$, we are accepting a 5% false positive rate of a significant result (based on the test statistic used), which translates to 1 out of 20 experiments being significant by chance alone.

Critical Value. Critical values are what a test statistic is compared with. Typically, if the test statistic exceeds the tabled critical value from the same distribution, then the test is significant. For example, if a Z score Z_i for the ith observation from a normally distributed set of sample data is equal to 1.7, then the original observation x_i [before the transformation $Z_i = (x_i - \mu)/\sigma$ is made] lies 1.7 standard deviations from the mean. We know from the standard normal distribution that, for a one-tailed test, the probability of Z exceeding 1.645 is 0.05, or 5%, and thus the probability of exceeding 1.7 is less than 5%. In this case, $Z_\alpha = 1.645$ is the critical value for a tail probability of 5%. That $Z_i = 1.7$ exceeded the critical value of 1.65 means that the test was significant at the 0.05 level.

Rejection Region. When the test statistic determined from the data exceeds the tabled critical value, the test is said to lie in the rejection region above the tabled critical value. Figure 12.3 shows the 5% and 10% rejection regions for the standard normal distribution.

P Values (Tail Probabilities). The P value is equal to total probability in the tail of the probability distribution above (or below) the value of the test statistic. As you can see, the total probability in the tails shown in Figure 12.3 can be determined for any value of Z. If the test statistic Z happens to equal 1.645, then, for a one-tailed test, the tail probability or P value is 0.05. On the other hand, if the test statistic Z happens to be 1.96, then the P value is 0.025, assuming a one-tailed test. However, in most cases, P values will not take on exact probabilities of 5% or 2.5%, but rather other values. Let us assume that the sampling distribution is distributed as a standard normal distribution, and that we wish to use a significance level of $\alpha = 0.05$ for a one-tailed hypothesis test. In this case, the critical value for the standard normal distribution for the 5% level is $Z_{\alpha=0.05} = 1.645$. If, after calculating the test statistic Z, we find that it is equal to 1.96, then we know that the P value is equal to 0.025, since 0.025 is the total probability in the tail above a value of 1.96. Moreover, we also know that 0.025 is less than 0.05, so we can state that $P < 0.05$. In other words, since the test statistic $Z = 1.96$ exceeded the tabled critical value of $Z_{\alpha=0.05} = 1.645$ (based on a level of significance of $\alpha = 0.05$), we can directly conclude that $P < 0.05$.

Reporting P Values. The reporting of P values in the literature is usually done the same way by most researchers. What matters in the end, however, is that the reader can understand which tests are significant and which are not. Here are a few examples for reporting P values and what they mean:

Statement	Translation
$P < 0.05$	Test statistic exceeded critical value
$P = 0.023$	Tail probability equal to 0.023
$P < 0.001$	Test was very significant
"S"	Test was significant
"NS"	Test was not significant
"Reject"	Test was significant

It is customary to never report a test that is not significant as $P > 0.05$. If you want to report a P value that was not significant, then use, for example, $P = 0.13$, assuming that $\alpha = 0.05$.

E.2 PROBABILITY FUNCTIONS: SOURCE OF *P* VALUES

The main intent of this appendix is to discuss how P values are generated through the use of numerical methods. In the remaining sections, computation of P values for various distributions are provided. We will see that P values computation for the normal distribution is directly based on the standard normal distribution. However, for the χ^2, Student's t, and F distributions, we will link our computation to variate relationships with the incomplete gamma and incomplete beta functions for determining probability. These are listed in the following table:

Test	Distribution of test statistic	P value function
Large-sample t test ($n_1, n_2 > 30$)	Standard normal	Cumulative distribution function
t test ($n_1, n_2 < 30$)	Student's t	Incomplete β function
F test	F(Variance ratio)	Incomplete β function
Wald	χ^2	Incomplete γ function
Mann-Whitney	Standard normal	Cumulative distribution function
Kruskall-Wallis	χ^2	Incomplete γ function
χ^2 test	χ^2	Incomplete γ function

E.3 NORMAL DISTRIBUTION

One of the most common distributions in statistics is the normal distribution. Whenever we think of an average, we are assuming that the data on which the average is based are normally distributed. The definition of the normal distribution rests primarily on the *central limit theorem,* which is defined as follows:

Theorem (Central Limit Theorem)

If a variable is distributed with mean μ and standard deviation σ, and if random samples of size n are drawn, then the mean of the samples will be approximately normally distributed with mean μ and standard deviation $\sigma/\sqrt{n}$ for sufficiently large n.

The central limit theorem tells us that multiple means and standard deviations from many sampling distributions converge to a single normal distribution that has its own mean and standard deviation. The primary importance of the central limit theorem is that the means from many samples with size n>30 is normally distributed, irrespective of the original distribution from which the samples were drawn.

Notation. In order to understand the characteristics of a distribution, it is essential to know the notation used for describing its parameters, statistics based on the distribution, and range. For the normal distribution, these are as follows:

Notation	Remarks
x	Normal variate
Z	Standard normal variate, Z score (statistic)
μ	Population mean (parameter)
σ	Population standard deviation (parameter)
$\bar{x}$	Sample mean (statistic)
s	Sample standard deviation (statistic)
$\phi(x), p(x)$	Probability density function
$\Phi, P(x)$	Cumulative distribution function
$1 - \Phi, Q(x), \alpha$	P value ("tail probability")
$(-\infty < x < \infty)$	Range

Probability Functions. In order to obtain P values, we have to work with the probability functions for the normal distribution. Probability functions allow us to determine how much total probability lies within a certain area of the distribution. For most probability distributions such as the normal distribution, essentially three functions are commonly employed:

Function	Notation	Description
PDF	$\phi(x), p(x)$	Probability density function, probability of occurrence of the normal variate x
	$Z(x)$	Used by Abramowitz and Stegun [1] to denote the PDF
CDF	$\Phi, P(x)$	Cumulative distribution function, total probability under the curve from $-\infty$ to x
1 – CDF	$1 - \Phi, Q(x), \alpha$	P value, total probability under the curve from x to ∞, also known as the (*tail probability*)

The *probability density function* of the normal distribution is

$$\phi(x) = \frac{1}{\sigma\sqrt{2\pi}} \exp\left[\frac{-(x-\mu)^2}{2\sigma^2}\right], \tag{E.1}$$

where x is a normal variate, μ is the mean, and σ is the standard deviation. For a standard normal distribution, each x value is standardized with the relationship $Z_i = (x_i - \mu)/\sigma$. After standardization, the Z-score distribution has mean zero ($\mu = 0$) and variance one ($\sigma^2 = 1$), and thus the standard deviation is one ($\sigma = 1$). By substitution, we have

$$\phi(Z) = \frac{1}{\sqrt{2\pi}} \exp[-0.5Z^2], \tag{E.2}$$

where Z is a standard normal variate. For the sake of efficiency, let's use x to represent Z, assuming, of course, that we are dealing with a standard normal distribution. We are interested in determining the *cumulative distribution function* $\Phi(x)$ of the standard normal distribution. The cumulative distribution function (CDF) is equal to the total probability under the curve from $-\infty$ up to x and has the form

$$\Phi(x) = \int_{-\infty}^{x} \phi(x)dx. \tag{E.3}$$

Because $\Phi(x)$ is the cumulative distribution defining the total probability under the normal curve from $-\infty$ up to x, its complement is $1 - \Phi(x)$ (or the *tail probability* representing the total probability under the curve from x to ∞) and is functionally composed as

$$1 - \Phi(x) = \int_{x}^{\infty} \phi(x)dx. \tag{E.4}$$

The use of Φ is specific to the normal distribution, so a more commonly used notation for representing values of the cumulative distribution for any distribution is $P(x)$. Accordingly, the commonly used notation for the tail probability is $Q(x)$. Collectively, these probabilities can be functionally composed as follows:

$$\begin{aligned} P(x) &= 1 - Q(x) = \Phi(x) = \int_{-\infty}^{x} \phi(x)dx \qquad \text{(CDF)} \\ Q(x) &= 1 - P(x) = 1 - \Phi(x) = \int_{x}^{\infty} \phi(x)dx \quad (P \text{ value, or tail probability}). \end{aligned} \tag{E.5}$$

Abramowitz and Stegun [1] provided *continued fraction* series expansions for calculating the tail probability $Q(x)$ for a given x value of a standard normal distribution. For $x < 0.6$, we use their [1] Equation (26.2.15), given as

$$Q(x) = \phi(x)\left\{\frac{x}{1-}\frac{x^2}{3+}\frac{2x^2}{5-}\frac{3x^2}{7+}\frac{4x^2}{9-}\cdots\right\}, \tag{E.6}$$

where the continued fraction is

$$\left\{\frac{x}{1-}\frac{x^2}{3+}\frac{2x^2}{5-}\frac{3x^2}{7+}\frac{4x^2}{9-}\cdots\right\} = \cfrac{x}{1-\cfrac{x^2}{3+\cfrac{2x^2}{5-\cfrac{3x^2}{7+\cfrac{4x^2}{9+\cdots}}}}}. \tag{E.7}$$

However, if $x \geq 0.6$, then their Equation (26.2.14) is used as

$$Q(x) = 0.5 - \phi(x)\left\{\frac{1}{x+}\frac{1}{x+}\frac{2}{x+}\frac{3}{x+}\frac{4}{\cdots}\right\}, \tag{E.8}$$

where the continued fraction is

$$\left\{\frac{1}{x+}\frac{1}{x+}\frac{2}{x+}\frac{3}{x+}\frac{4}{\cdots}\right\} = 1 + \cfrac{1}{x\cfrac{1}{x+\cfrac{2}{x+\cfrac{3}{x+\cfrac{4}{x+\cdots}}}}}. \tag{E.9}$$

The function normpvalue listed below returns the P value $Q(x)$, where x is a standard normal variate. The function normcf returns $Q(x)$ based on continued fraction series described in (our) Equations (E.6) and (E.8):

```
Function stdnormpvalue(ByVal x) As Double
    Dim phi As Double
    x = Math.Abs(x)
    phi = Math.Exp(-0.5 * x ^ 2) / Math.Sqrt(2 *
       3.1415926535897931)
    If x > 0.6 Then stdnormpvalue = phi * normcf(x)
    If x <= 0.6 Then stdnormpvalue = 0.5 - phi * normcf(x)
End Function

Function normcf(ByVal x) As Double
    Dim af, bf, quot, func, funcold, eps As Double
    Dim k, j, quitafterbestk, bestk As Integer
    quot = 0
    quitafterbestk = 0
    eps = 0.0000001
```

```
        For k = 1 To 1000
            quot = 0
            bestk = k
redoatbestk:
            For j = bestk To 1 Step -1
                If x <= 0.6 Then 'Abramowitz & Stegun Eq.
                   26.2.15, pp. 932
                    af = j * x ^ 2
                    bf = (j + 1) * 2 - 1
                    If j Mod 2 = 0 Then quot = (af / (bf -
                       quot))
                    If j Mod 2 <> 0 Then quot = (af / (bf +
                       quot))
                End If
                If x > 0.6 Then 'Abramowitz & Stegun Eq.
                   26.2.14, pp. 932
                    af = j
                    bf = x
                    quot = (af / (bf + quot))
                End If
            Next j
            If x > 0.6 Then func = 1 / (x + quot)
            If x <= 0.6 AndAlso quot <> 1 Then func = x /
               (1 - quot)
            If quot = 0 Then func = 0
            If quitafterbestk = 1 Then Exit For
            If k > 1 AndAlso Math.Abs(func - funcold) /
               Math.Abs(func) < eps Then
                bestk = k
                quitafterbestk = 1
                GoTo redoatbestk
            End If
            funcold = func
        Next k
        normcf = func
    End Function
```

E.4 GAMMA FUNCTION

Unlike normal distribution applications, applications requiring P values for the χ^2 distribution are rarely based on the χ^2 distribution. There may be any number of reasons why tail probabilities are not based on the cumulative density function (CDF) from the same distribution. The primary reason is that there is usually no closed-form analytic solution of the CDF. In such cases, one has to solve the CDF numerically, or employ some other distribution for which the numerical solution to the CDF is more straightforward and therefore, less expensive computationally. In consideration of the above, we introduce two functions, the *incomplete gamma γ function* and the *gamma γ function*, which are used for computing tail probabilities for the χ^2 distribution. The tail probability of the χ^2 distribution based on the gamma function is

$$Q(\chi^2|\nu) = 1 - \frac{\Gamma(a,x)}{\Gamma(a)} = 1 - \frac{\Gamma\left(\frac{\nu}{2}, \frac{\chi^2}{2}\right)}{\Gamma\left(\frac{\nu}{2}\right)} \qquad \left(a = \frac{\nu}{2}, x = \frac{\chi^2}{2}\right), \tag{E.10}$$

where χ^2 is the value of the χ^2 test statistic, ν is the degrees of freedom (d.f.), $\Gamma(\nu/2, \chi^2/2)$ is the incomplete γ function evaluated with parameters representing $\nu/2$ and $\chi^2/2$, $\Gamma(\nu/2)$ is the γ function evaluated at $\nu/2$, and a is a calling parameter used for the subroutine `gammaln`. The γ function can be best described as the factorial of the argument plus one, given as

$$\Gamma(x+1) = x\Gamma(x) = 1 \times 2 \times 3 \times \cdots \times (x-1) \times x = x! \tag{E.11}$$

The definition of the incomplete gamma function, however, is less straightforward than $\Gamma(a)$ and is given in Equation (6.5.2) of Abramowitz and Stegun [1] as

$$\Gamma(a,x) = \int_0^x e^{-t} t^{a-1} dt. \tag{E.12}$$

Gautschi [2] provided several forms for $\Gamma(a,x)$ based on the relationship between a and x. When $a < x$, the continued fraction is recommended

$$\Gamma(a,x) = \cfrac{x^a e^{-x} \frac{1}{(x+1-a)}}{1 + \cfrac{\alpha_1}{1 + \cfrac{\alpha_2}{1 + \cfrac{\alpha_3}{1 + \cfrac{\alpha_4}{1 + \cdots}}}}}, \tag{E.13}$$

where

$$\alpha_j = \frac{j(a-j)}{(x+2j-1-a)(x+2j+1+a)}. \tag{E.14}$$

However, when $a \geq x$, the series expansion is recommended

$$1 - \frac{\Gamma(a,x)}{\Gamma(a)} = x^a e^{-x} \sum_0^\infty \frac{x^j}{\Gamma(a+j+1)}. \tag{E.15}$$

The P value for a χ^2 distribution is returned from the function `chisqpvalue`. The `gammaln` returns the natural log of $\Gamma(a)$, which must be exponentiated as $\exp[\Gamma(a)]$ before use. Finally, the function `gammaprob` is called within function `chisqpvalue` and evaluates the continued fraction [Equation (E.13)] and series expansion [Equation (E.15)]:

```
    Function chisqpvalue(ByVal chisq, ByVal df) As Double
        Dim a, x As Double
        Dim ifault As Integer
        a = df / 2
        x = chisq / 2
        chisqpvalue = probgamma(a, x, ifault)
    End Function

    Function probgamma(ByVal a, ByVal x, ByVal ifault) As
       Double
        Dim af, bf, quot, funcold As Double
        Dim j, k, bestk, quitafterbestk As Integer
        Dim logx, sum As Double
        Dim eps, func As Double
        eps = 0.0000001
        If a < x Then 'continued fraction
            quot = 0
            quitafterbestk = 0
            For k = 1 To 100
                quot = 0
                bestk = k
redoatbestk:
                For j = bestk To 1 Step -1
                    af = (j * (a - j)) / ((x + 2 *
                       j - 1 - a) * _
                      (x + 2 * j + 1 + a))
                    bf = 1
                    quot = (af / (bf + quot))
                Next j
                func = ((1 / (x + 1 - a)) * (x ^ a) *
                   Math.Exp(-x)) / _
                  (1 + quot)
                If quitafterbestk = 1 Then Exit For
                If k > 1 AndAlso Math.Abs(func - funcold) / _
                      Math.Abs(func) < eps Then
                    bestk = k
                    quitafterbestk = 1
                    GoTo redoatbestk
                End If
                funcold = func
            Next k
            probgamma = func / Math.Exp(gammaln(a))
```

```
    Else '
        sum = 0
        For j = 0 To 20
            sum += (x ^ j) / Math.Exp(gammaln(a + j + 1))
        Next
        probgamma = x ^ a * Math.Exp(-x) * sum
        If probgamma > 1 Then probgamma = 1
        probgamma = 1 - probgamma
    End If
End Function

Function gammaln(ByVal a) As Double
    Dim ac(6), quot As Double
    Dim j As Integer
    ac(0) = 1 / 12
    ac(1) = 1 / 30
    ac(2) = 53 / 210
    ac(3) = 195 / 371
    ac(4) = 22999 / 22737
    ac(5) = 29944523 / 19733142
    ac(6) = 109535241009 / 48264275462 / a

    quot = ac(6)
    For j = 5 To 0 Step -1
        quot = ac(j) / (a + quot)
    Next j
    gammaln = quot + 0.5 * Math.Log(2 *
       3.1415926535897931) + _
      (a - 0.5) * Math.Log(a) - a
End Function
```

E.5 BETA FUNCTION

The beta (β) function is commonly used for computing tail probabilities for the F distribution and Student's t distribution. The tail probability for the F distribution based on the incomplete beta function is

$$Q(F|\nu_1, \nu_2) = I_x(a, b)$$
$$= I_x\left(\frac{\nu_2}{2}, \frac{\nu_1}{2}\right) \qquad \left(a = \frac{\nu_2}{2}, b = \frac{\nu_1}{2}\right), \tag{E.16}$$

where

$$x = \frac{\nu_2}{\nu_2 + F\nu_1}. \tag{E.17}$$

Abramowitz and Stegun [1] provide in Equations (26.5.8) and (26.5.9) continued fraction solutions for the incomplete β function. For $x < (a - 1)/$

$(a+b-2)$, Equation (26.5.8) is used as

$$I_x(a,b) = \frac{x^a(1-x)^b}{aB(a,b)} \left\{ \frac{1}{1+} \frac{d_1}{1+} \frac{d_2}{1+} \cdots \right\}, \tag{E.18}$$

where

$$B(a,b) = \frac{\Gamma(a)\Gamma(b)}{\Gamma(a+b)}, \tag{E.19}$$

$$d_{2j+1} = \frac{(a+j)(a+b+j)}{(a+2j)(a+2j+1)} x, \tag{E.20}$$

and

$$d_{2j} = \frac{j(b-j)}{(a+2j-1)(a+2j)} x. \tag{E.21}$$

For $x < 1$, we use their Equation (26.5.9) in the form

$$I_x(a,b) = \frac{x^a(1-x)^{b-1}}{aB(a,b)} \left\{ \frac{1}{1+} \frac{e_1}{1+} \frac{e_2}{1+} \cdots \right\}, \tag{E.22}$$

where

$$e_{2j} = -\frac{(a+j-1)(b-j)}{(a+2j-2)(a+2j-1)} \frac{x}{1-x}, \tag{E.23}$$

and

$$e_{2j+1} = \frac{j(a+b-1+j)}{(a+2j-1)(a+2j)} \frac{x}{1-x}. \tag{E.24}$$

Function `Fpvalue` returns the tail probability for an F statistic with degrees of freedom ν_1 and ν_2. Function `beta` returns the value of the β function $B(a,b)$. `betaprob` is called from within `Fpvalue` and returns the solution to the incomplete β function $I_x(a,b)$:

```
Function Fpvalue(ByVal F, ByVal nDF, ByVal dDF) As Double
    Dim x, a, b As Double
    Dim ifault As Integer
    a = dDF / 2
    b = nDF / 2
    x = dDF / (dDF + F * nDF)
    Fpvalue = probbeta(x, a, b, ifault)
End Function

Function beta(ByVal a, ByVal b, ByVal ifault)
    'Eq. 6.2.2 in Abramowitz & Stegun (pp. 258)
    beta = (Math.Exp(gammaln(a)) * Math.
       Exp(gammaln(b))) / _
      Math.Exp(gammaln(a + b))
End Function
```

```
    Function probbeta(ByVal x, ByVal a, ByVal b, ByVal ifault)
       As Double
        Dim af, bf, quot, funcold As Double
        Dim j, k, bestk, quitafterbestk As Integer
        Dim sum As Double
        Dim eps, func As Double
        eps = 0.0000001
        If x < (a - 1) / (a + b - 2) Then ''Abramowitz &
           Stegun 26.5.8, pp. 944
            quot = 0
            quitafterbestk = 0
            For k = 1 To 1000
                quot = 0
                bestk = k
redoatbestk:
                For j = bestk To 1 Step -1
                    If j Mod 2 = 0 Then af = -((a + j) *
                      (a + b + j)) / ((a + 2 * j) *
                         (a + 2 * j + 1)) * x
                    If j Mod 2 <> 0 Then af = (j * (b - j)) / _
                      ((a + 2 * j - 1) * (a + 2 * j)) * x
                    bf = 1
                    quot = (af / (bf + quot))
                Next j
                func = 1 / (1 + quot)
                If quitafterbestk = 1 Then Exit For
                If k > 1 AndAlso Math.Abs(func - funcold) / _
                     Math.Abs(func) < eps Then
                    bestk = k
                    quitafterbestk = 1
                    GoTo redoatbestk
                End If
                funcold = func
            Next k
            probbeta = (x ^ a * (1 - x) ^ b) / (a * beta(a, b,
               ifault)) * func
        End If
        If x < 1 Then 'Abramowitz & Stegun 26.5.9, pp. 944
            quot = 0
            quitafterbestk = 0
            For k = 1 To 1000
                quot = 0
                bestk = k
redoatbestk1:
                For j = bestk To 1 Step -1
                    If j Mod 2 = 0 Then af = (j * (a + b -
                       1 + j)) / _
                      ((a + 2 * j - 1) * (a + 2 * j)) *
```

```
                    x / (1 - x)
                If j Mod 2 <> 0 Then af = -((a + j - 1) * _
                  (b - j)) / ((a + 2 * j - 2) * (a + 2 *
                      j - 1)) *_
                  x / (1 - x)
                bf = 1
                quot = (af / (bf + quot))
            Next j
            func = 1 / (1 + quot)
            If quitafterbestk = 1 Then Exit For
            If k > 1 AndAlso Math.Abs(func - funcold) / _
                  Math.Abs(func) < eps Then
                bestk = k
                quitafterbestk = 1
                GoTo redoatbestk1
            End If
            funcold = func
        Next k
        probbeta = (x ^ a * (1 - x) ^ (b - 1)) / (a * _
          beta(a, b, ifault)) * func
    End If
End Function
```

E.6 PSEUDO-RANDOM-NUMBER GENERATION

Let $x \in \mathbb{R}^x$ be a random variate that takes on the quantiles, $x_1, x_2, \ldots, x_n$ such that $p(x_i) = P(X = x_i)$ and $\sum p(x_i) = 1$. Recall the cumulative distribution function $F(x)$, which maps X into the uniform distribution $\mathcal{U}(0,1)$ and inverse cumulative distribution $F(x)^{-1}$, which performs inverse mapping of $\mathcal{U}(0,1)$ into x. Because most computer algorithms can generate pseudorandom numbers from $\mathcal{U}(0,1)$, we can directly obtain values for x by substituting $\mathcal{U}(0,1)$ into $F(x)^{-1}$ and then solve for x. By repeatedly generating numerous quantiles, we can simulate the desired statistical distribution with a frequency distribution consisting of the binned quantiles. We describe some basic methods for generating quantiles for various statistical distributions in the following paragraphs.

E.6.1 Standard Uniform Distribution

Standard uniform variates [$\mathcal{U}(0,1)$] can be obtained directly from a pseudo random number generator using a fixed seed or one that varies with the clock cycle of the CPU. The result is a standard uniform variate in the range [0,l]. Figure E.1 illustrates a frequency histogram of 50,000 uniformly distributed pseudorandom variates. Uniform variates whose range is other than 0 and 1

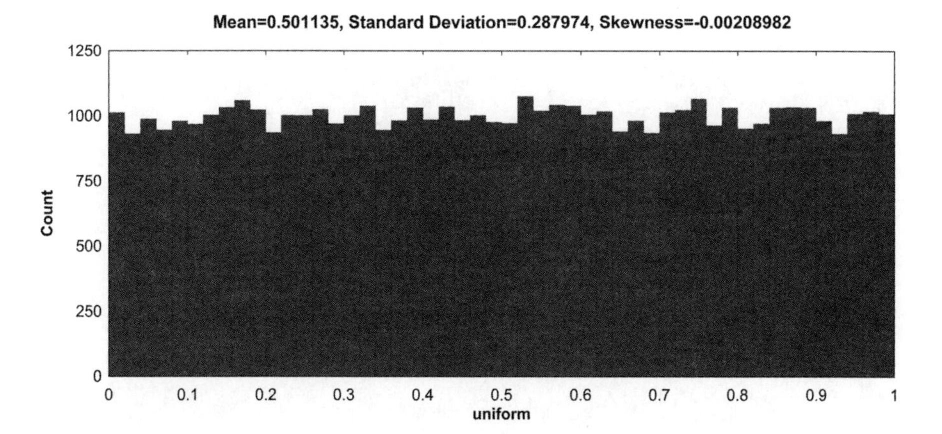

FIGURE E.1 Frequency distribution of 50,000 pseudorandom uniform variates distributed $\mathcal{U}(0, 1)$.

can be obtained with the ''floor'' function as

$$\mathcal{U}(a,b) = F^{-1}(x) = a + [\mathcal{U}(0,1)(b-a)]. \tag{E.25}$$

Finally, pseudorandom numbers for the log-uniform distribution are obtained with the functional form

$$\mathcal{U}(a,b) = \ln(a) + [\mathcal{U}(0,1)(\ln(b) - \ln(a))]. \tag{E.26}$$

E.6.2 Normal Distribution

Pseudo-random-number generation for a normal quantile involves generation of a pseudorandom variate that is distributed $\mathcal{N}(0,1)$. Figure E.2 illustrates a frequency histogram of 50,000 standard normal variates distributed $\mathcal{N}(0,1)$. To obtain a normal quantile that is not $\mathcal{N}(0,1)$ but rather $\mathcal{N}(\mu,\sigma^2)$, the $\mathcal{N}(0,1)$ variate is first multiplied by the standard deviation σ and then shifted by the mean μ. A tried-and-true method for generating normally distributed pseudorandom quantiles is as follows:

$$x_{\mathcal{N}(\mu,\sigma^2)} = F^{-1}(x) = \left[\left(\sum_{i=1}^{12} \mathcal{U}_i(0,1)\right) - 6\right]\sigma + \mu. \tag{E.27}$$

Another method, called the by Box-Muller (1958) method [3], transforms two uniform random variates into two normal variates in the form

$$x_{1,\mathcal{N}(\mu,\sigma^2)} = F^{-1}(x) = \sqrt{-2\ln \mathcal{U}_1(0,1)}\cos(2\pi\,\mathcal{U}_2(0,1))\sigma + \mu$$

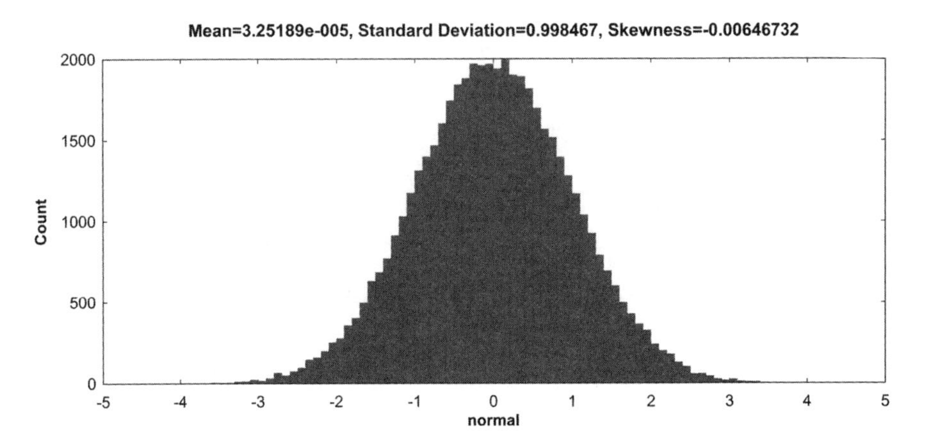

FIGURE E.2 Frequency distribution of 50,000 pseudorandom standard normal variates distributed $\mathcal{N}(0, 1)$.

$$x_{2,\mathcal{N}(\mu,\sigma^2)} = F^{-1}(x) = \sqrt{-2\ln \mathcal{U}_1(0,1)}\sin(2\pi\,\mathcal{U}_2(0,1))\sigma + \mu, \qquad \text{(E.28)}$$

which is one of the most common methods used. Figure E.3 illustrates a frequency histogram of 50,000 normal variates distributed $\mathcal{N}(15, 2.25)$.

E.6.3 Lognormal Distribution

Pseudorandom numbers from the lognormal distribution based on the natural logarithm are obtained from the relationship

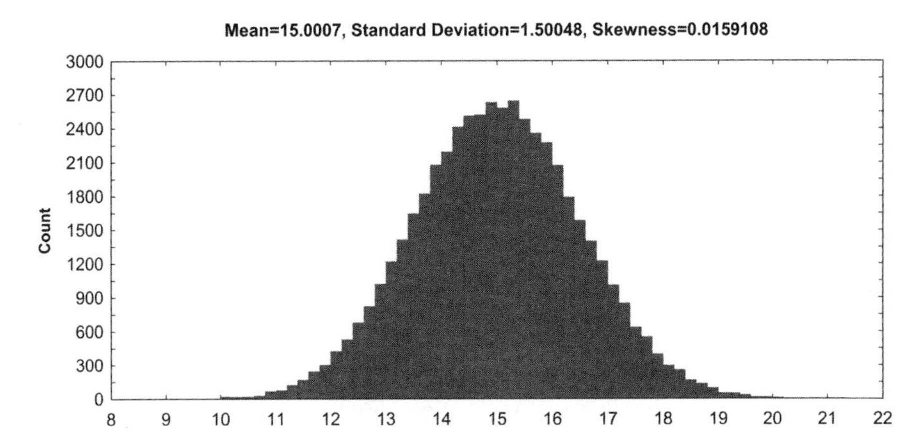

FIGURE E.3 Frequency distribution of 50,000 pseudorandom normal variates distributed $\mathcal{N}$. Note, the standard deviation of $1.5 = \sqrt{2.25}$.

$$x_{\mathcal{LN}(\mu,\sigma^2)} = F^{-1}(x) = \exp\left[\left(\left(\sum_{i=1}^{12} \mathcal{U}_i(0,1)\right) - 6\right)\sigma + \mu\right], \tag{E.29}$$

where μ is the mean of the lognormal distribution with variance σ^2, whereas pseudorandom numbers from the common logarithm-based lognormal distribution are calculated in the form

$$x_{\mathcal{LN}(\mu,\sigma^2)} = F^{-1}(x) = 10\left[\left(\left(\sum_{i=1}^{12} \mathcal{U}_i(0,1)\right) - 6\right)\sigma + \mu\right], \tag{E.30}$$

where μ is the mean of the lognormal distribution with variance σ^2 on the arithmetic scale.

For $\mathcal{LN}$ on the natural log scale, one can exponentiate the Box-Muller (1958) standard normal variates

$$\begin{aligned} x_{1,\mathcal{LN}(\mu,\sigma^2)} &= F^{-1}(x) = \exp[\sqrt{-2\ln \mathcal{U}_1(0,1)}\cos(2\pi\,\mathcal{U}_2(0,1))\sigma + \mu] \\ x_{2,\mathcal{LN}(\mu,\sigma^2)} &= F^{-1}(x) = \exp[\sqrt{-2\ln \mathcal{U}_1(0,1)}\sin(2\pi\,\mathcal{U}_2(0,1))\sigma + \mu], \end{aligned} \tag{E.31}$$

and for the common logarithm, one can use

$$\begin{aligned} x_{1,\mathcal{LN}(\mu,\sigma^2)} &= F^{-1}(x) = 10^{[\sqrt{-2\ln \mathcal{U}_1(0,1)}\cos(2\pi\,\mathcal{U}_2(0,1))\sigma + \mu]} \\ x_{2,\mathcal{LN}(\mu,\sigma^2)} &= F^{-1}(x) = 10^{[\sqrt{-2\ln \mathcal{U}_1(0,1)}\sin(2\pi\,\mathcal{U}_2(0,1))\sigma + \mu]}, \end{aligned} \tag{E.32}$$

where σ and μ are on the log scale. Figure E.4 illustrates a frequency histogram of 50,000 lognormal variates distributed $\mathcal{N}(15, 1.5)$, that is, GM = 15 and GSD = 1.5.

E.6.4 Binomial Distribution

One of the least expensive (in terms of computing time and memory constraints) ways to simulate a binomial deviate $\mathcal{B}(n,p)$ is to use the *geometric distribution method* introduced by Evans et al. [4]. This involves adding x geometric random variates as in

$$P = \sum_{j=1}^{x} \frac{\log(\mathcal{U}(0,1))}{\log(1-p) - 1} \tag{E.33}$$

until P exceeds $n - x$, which will yield $\mathcal{B}(n,p) = j = F^{-1}(x)$. Figure E.5 illustrates a frequency histogram of 50,000 binomial variates distributed $\mathcal{B}(10, 0.4)$.

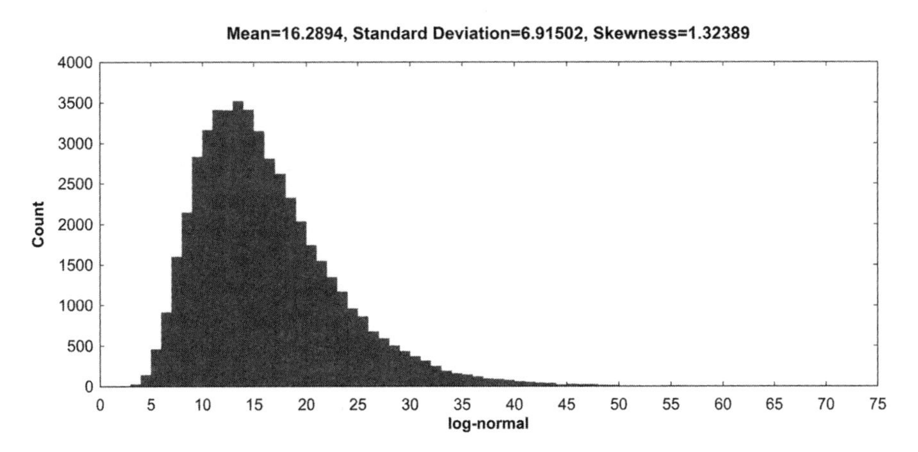

FIGURE E.4 Frequency distribution of 50,000 pseudorandom lognormal variates distributed $\mathcal{N}(15, 1.5)$, that is, GM = 15 and GSD = 1.5.

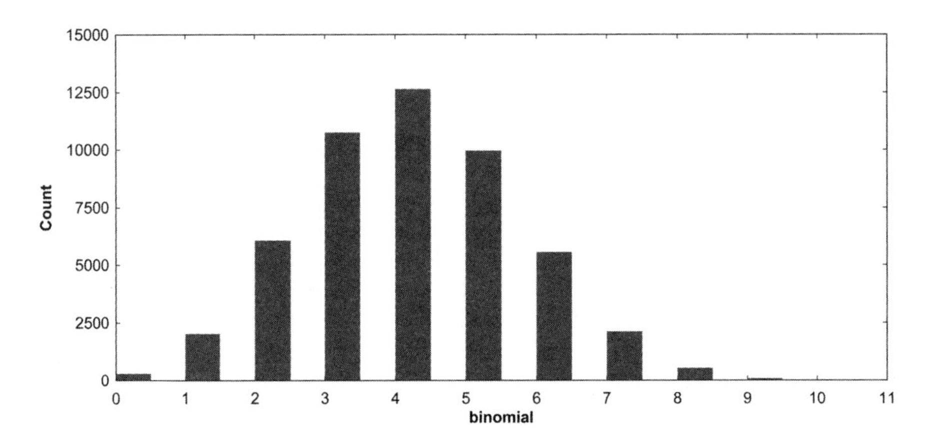

FIGURE E.5 Frequency distribution of 50,000 pseudorandom binomial variates distributed $\mathcal{B}(10, 0.4)$.

E.6.5 Poisson Distribution

Poisson variates $\mathcal{P}(\mu)$ are also generated using the method described by Evans et al. [3]. To start, the cumulative distribution function $F(x)$, $x = 1, 2, \ldots, N$ is calculated for arbitrary values of N. Next, successive values of uniform ($\mathcal{U}_i(0,1)$) variates are generated until the criterion $F(x) \leq \mathcal{U}_i(0, 1) < F(x + 1)$ is true, at which time x is the Poisson random deviate. Figure E.6 illustrates a frequency histogram of 50,000 Poisson variates distributed $\mathcal{P}(10)$.

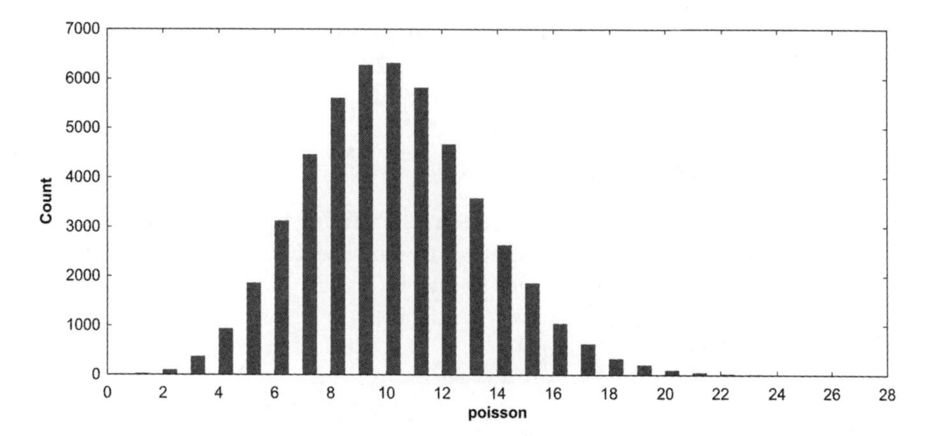

FIGURE E.6 Frequency distribution of 50,000 pseudorandom Poisson variates distributed $\mathcal{P}(10)$.

E.6.6 Triangle Distribution

Using a, b, and mode c random quantiles for the triangular distribution, we can find $\mathcal{TRI}(a, c, b)$ by using the equation

$$x_{\mathcal{TRI}(a,c,b)} = F^{-1}(x) = \begin{cases} a + \sqrt{\mathcal{U}(0,1)(b-a)(c-a)} & \text{if } a \leq y \leq c \\ b - \sqrt{(1-\mathcal{U}(0,1))(b-a)(b-c)} & \text{if } c \leq y \leq b \end{cases} \tag{E.34}$$

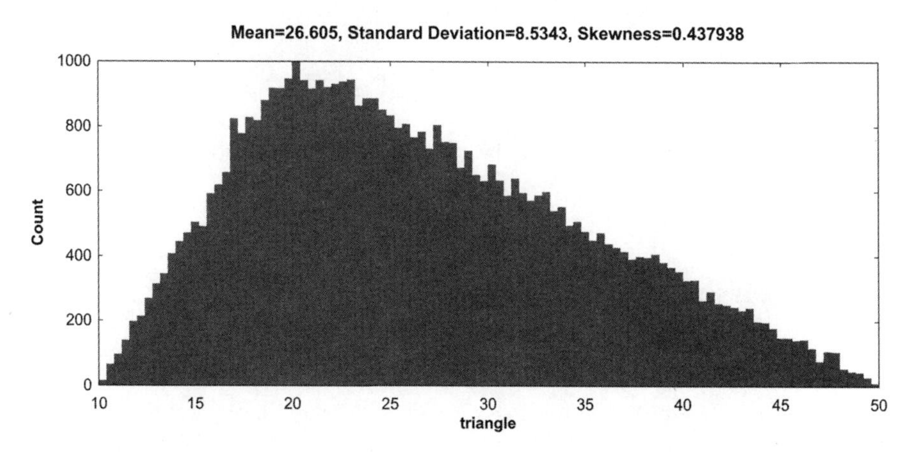

FIGURE E.7 Frequency distribution of 50,000 pseudorandom triangle variates distributed $\mathcal{TRI}(10, 20, 50)$.

where $y = a + \mathcal{U}(0,1)(b - a)$. Figure E.7 illustrates a frequency histogram of 50,000 triangle variates distributed $\mathcal{TRI}(10, 20, 50)$.

E.6.7 Log-Triangle Distribution

Log-triangle pseudorandom numbers are obtained by feeding the inverse cumulative distribution function for the triangle distribution with lower bound $\ln(a)$, mode $\ln(c)$, and upper bound $\ln(b)$.

REFERENCES

[1] M. Abramowitz, I.A. Stegun. *Handbook of Mathematical Functions with Formulas, Graphs, and Mathematical Tables*. Dover Press, New York, 1964.

[2] W. Gautschi, A computational procedure for incomplete gamma functions. *ACM Trans. Math. Software* **5**:466–481, 1979.

[3] G.E.P. Box and M.E. Muller. A note on the generation of random normal deviates. *The Annals of Mathematical Statistics* **29**(2):610–611, 1958.

[4] M. Evans, N. Hastings, B. Peacock. *Statistical Distributions*. Wiley, New York, 1993.

APPENDIX F

SYMBOLS AND NOTATION

Greek Symbols

α	Alpha
β	Beta
γ	Gamma
δ	Delta
ε	Varepsilon
ζ	Zeta
η	Eta
θ, Θ	Theta
ϑ	Vartheta
ι	Iota
κ	Kappa
λ, Λ	Lambda
μ	Mu
ν	Nu
ξ, Ξ	Xi
o	Omicron
π, Π	Pi
ϖ	Varpi
ρ	Rho
ϱ	Varrho
σ, Σ	Sigma
ς	Varsigma
τ	Tau
υ, Υ	Upsilon
ϕ, Φ	Phi
φ	Varphi
χ, X	Chi
ψ, Ψ	Psi
ω, Ω	Omega

Classification Analysis of DNA Microarrays, First Edition. Leif E. Peterson.

Machine Learning Notation

c_k	True membership probability for class k
$\mathcal{C}$	Chromosome
d	Power of kernel
$d(\mathbf{x}_i, \mathbf{x}_j)$	Distance between objects $\mathbf{x}_i$ and $\mathbf{x}_j$
$D(\mathbf{x} \rightsquigarrow \omega)$	Class ω decision rule for object $\mathbf{x}$
f_e	Fitness of most elite chromosomes
$\bar{f}$	Average fitness
$K(\mathbf{x}_i, \mathbf{x}_j)$	Kernel for objects $\mathbf{x}_i$ and $\mathbf{x}_j$
λ	Number of chromosomes
$\mathbf{L}$	Square-root matrix from Cholesky factorization
n	Sample size of objects (microarrays)
N	Number of unknowns in function approximation problem
ω_k	Class label for object k
Ω	Total number of classes for a sample of objects
p	Number of features (variables or genes)
$\mathcal{S}_{gl}$	Binary string for gene g on chromosome l
$\mathbf{v}_m$	Prototype m ($m = 1, 2, \ldots, M$)
$\mathbf{v}_k$	Center (mean vector) for cluster k
$\mathbf{x}_i$	Object (microarray) i ($i = 1, 2, \ldots, n$)
x_{ij}	jth feature value for ith object (microarray).
z_k	Predicted membership probability for class k

Mathematical Symbols

$\leq$	Less than or equal to
$\geq$	Greater than or equal to
$\neq$	Not equal to
$\pm$	Plus or minus error
$\simeq, \approx$	Is approximately equal to
$\rightarrow$	Approaches
∞	Infinity
$\sum_i x_i$	Sum, $x_1 + x_2 + \cdots + x_p$
$\prod_i x_i$	Product, $x_1 \times x_2 \times \cdots \times x_p$
$\int_a^b dx$	Definite integral over a to b
$\ln, \log_e$	Natural logarithm
$\log_{10}$	Common logarithm, base 10
$n!$	Factorial, $n(n-1)(n-2)\cdots 1$
$\min\{a, b\}$	Minimum of a and b
$\max\{a, b\}$	Maximum of a and b
$\mathbb{R}^p$	Real numbers with p dimensions

Statistical Symbols

$\stackrel{\partial ef}{=}$	By definition
X	Random variable
x	Quantile, F^{-1}
$\mathbb{E}[X]$	Expectation of X
μ_X	Mean of X
σ_X	Standard deviation of X
σ_X^2	Variance of X
σ_{XY}	Covariance between X and Y
$\rho_{X,Y}$	Correlation between X and Y
$\mathrm{m}(x)$	Median of quantiles
β_j	Regression coefficient
$\mathrm{s.e.}(\beta_j)$	Standard error of regression coefficient
$\binom{n}{x}$	Binomial coefficient, "n choose x"
$L(\theta)$	Likelihood function of parameter θ
$\ell(\theta)$	Loglikelihood function of parameter θ, using ln
P	P value, tail probability
$P(A)$	Probability of event A
$\cup$	Union
$\cap$	Intersection
$\#\{j : P_j \leq 0.05\}$	Number of times $P_j \leq 0.05$
e.e.d.	Empirical eigenvalue distribution
i.i.d.	Independent and identically distributed
$f(\lambda)$	Wigner eigenvalue density
$f(\lambda, \gamma)$	Marčenko-Pastur eigenvalue density
F_1	Tracy-Widom eigenvalue density
d.f.	degrees of freedom
s.d.	standard deviation

Matrix Notation

$\mathbf{x}$	Vector
$\mathbf{X}$	Matrix
$\mathbf{C}$,$\mathbf{S}$	Sample covariance matrix
$\mathbf{\Sigma}$	Population covariance matrix
$\mathbf{R}$	Correlation matrix
$\mathbf{A}_p$	Gaussian orthogonal ensemble (GOE) matrix
$\mathbf{W}_p(n, \mathbf{\Sigma})$	Wishart ensemble matrix

Probability Distributions

$\mathcal{U}(0,1)$	Standard uniform distribution
$\mathcal{N}(0,1)$	Standard normal distribution

$\mathcal{N}(\mu, \sigma^2)$	Normal distribution
$\mathcal{B}(n, p)$	Binomial distribution
$\mathcal{P}(\mu)$	Poisson distribution
$\mathcal{L}\mathcal{N}(\mu, \sigma^2)$	Lognormal distribution
$\mathcal{T}\mathcal{R}\mathcal{I}(a, b, c)$	Asymmetric triangle distribution

INDEX

AAO, all-at-once, 250
Accuracy, 278
Adjacency matrix, 68
Agglomerative cluster analysis, 4, 91
AID algorithm, 311
ALLAML2 dataset, 260
Analysis of variance (ANOVA), 230
Ant colony optimization, 560
APP, all-possible-pairs, 247
Area under the curve (AUC), 293
Artificial neural network, 487
Associative learning, 57
Asymptotic covariance, 611
Asymptotic variance, 611
Attributes, 277
Automatic relevance determination (ARD), 522
Average, 214
Average quantization error, 73
Average silhouette index, 31

Bartlett's sphericity test, 261
Batch training, 504
Bayes' rule, 637
Behrens–Fisher problem, 222
Best matching unit, 57, 68
`beta`, 689
Beta function, 689
Bias, 284
Binary logistic regression (BLOG), 433, 434
Binomial distribution, 695
Bonferroni adjustment to *P* value, 237
Bootstrap bias, 291
Bootstrap sampling, 290, 332
Brain2 dataset, 260
Breast2A dataset, 354
Breast2B dataset, 340
Business intelligence (BI), 119, 120

C4.5 algorithm, 311
C5.0 algorithm, 311
Canberra distance, 23
Canonical discriminant function, 408
Cardinality, 17, 26, 28
CART algorithm, 311
Cauchy–Schwarz inequality, 124
Cross-entropy error, 493
Centers, 7, 16
Central limit theorem, 682
CHAID algorithm, 311
Chebyshev distance, 22
Chemoinformatics, 145
`chisqpvalue`, 686
Cholesky factorization, 553, 567
Chromosome, 545
City block (Manhattan) distance, 22
Class labels, definition, 6
Class prior probability, 244
Classifier
  ant colony optimization (ACO), 560
  artificial neural network (ANN), 487
  covariance matrix self-adaptation evolution strategies (CMSA-ES), 549
  decision tree classifier (DTC), 311
  Fisher's discriminant analysis (FDA), 406
  genetic algorithms (GA), 544
  K nearest neighbor (KNN), 361
  kernel regression (KREG), 525
  learning vector quantization (LVQ), 415
  linear discriminant analysis (LDA), 396

Classification Analysis of DNA Microarrays, First Edition. Leif E. Peterson.

Classifier (*Continued*)
linear regression (LREG), 297
mixture of experts (MOE), 591
naïve Bayes classifier (NBC), 379
particle swarm optimization (PSO), 556
polytomous logistic regression (PLOG), 433, 439
quadratic discriminant analysis (QDA), 403
random forests (RF), 331
supervised neural gas (SNG), 573
support vector machines (SVM), 449
Cluster
center, 20
crisp *K* means, 15
fuzzy *K* means, 47
initialization, 37
random partitions, 40
prototype splitting, 41
random *K*, 37
outliers, 44
V-fold cross-validation (VFCV), 35
validity, 24
Cluster initialization, 37
Cluster validity, 3
Coefficient of variation (CV), 217
Colon2 dataset, 260
Combinations, 630
Comorbidity, 2
Component map, 71
Concept drift, 2, 501
Concept vectors, 119, 124, 641
Conditional probability, 634
Conflation, 121
Confusion matrix, 278, 332
Conjugate gradient method, 674, 676
Conscience, self-organizing map, 61
Continuous responses, 210
Correlation, 338
simulating correlated variables, 553
Correlation matrix, 91, 163, 395
Correlation-based PCA (CPCA), 190
Covariance matrix, 395, 410, 552
Covariance matrix filtering (CMF), 601
Covariance matrix self-adaptation, 549
Covariance matrix shrinkage, 610
Credit risk, 120
Cross-entropy error (CE), 493
Cross-validation (CV), 279
Customer churn, 120
CV, 217
Cycles, 496
Daniels–Kass shrinkage, 610
Datasets used, 9
Davies–Bouldin index, 25
Decision tree classification (DTC), 311
child node, 312
feature cutpoints, 312
features used, 314
Gini index, 312
information gain, 312
node purity, 312
parent node, 312
stopping criteria, 314
terminal nodes, 314
Degrees of freedom (d.f.), 219, 221–223, 232–234, 260, 261, 408, 438, 439, 603, 613
Delta-rule learning, 496
Deviance residuals, 438
d.f., *see* Degrees of Freedom
Diffusion maps (DM), 192
Direction cosines, 640
Discrete responses, 209
Discretization, 242
Discriminant analysis
Fisher's, 406
linear, 396
quadratic, 403
Dissimilarity coefficient, 20
Distance
average diameter, 26
average linkage, 28
average-to-centroids linkage, 28
Canberra, 23
centroid diameter, 26
centroid linkage, 28
Chebyshev, 22
city block (Manhattan), 22
complete diameter, 26
complete linkage, 27
Euclidean, 18, 22, 91
Hausdorff, 28
Mahalanobis, 396, 409
Manhattan, 22
Pearson product moment correlation, 21
Pitman correlation, 20
silhouette, 30
single linkage, 27
Tanimoto, 23
Distribution
binomial, 695
chi-square (χ^2), 236
F ratio, 235
lognormal, 694

normal, 213, 693
Poisson, 696
standard normal, 218, 225
triangular, 697
uniform, 692
Divisive cluster analysis, 4
Document length normalization, 123
DTC, *see* Decision tree classification
Dunn's index, 25
Duo-mining, 119

Eigenanalysis, 162, 257, 342
Eigenvalue, 163, 553, 650
Eigenvector, 163, 553, 650
Elitism, genetic, 547
Empirical eigenvalue distribution (e.e.d.), 603
Ensemble classifier fusion, 280
ensemble diversity, 281
majority vote, 281
random oracles, 281
hyperplane splits, 281
kappa-error plot, 281
miniclassifier, 281
principal direction linear, 282
spherical linear, 283
weighted majority vote, 281
Entropy, 242, 312
Epanechnikov kernel, 563
Epochs, 496
Equivalent eigenvalues theorem, 651
Error sum of squares (SSE), 234
Euclidean distance, 18, 22, 91
Evolutionary algorithms, 9
Exchangeability, 37
Expert network, 591

F distribution, 689
F test, 230
False discovery rates
SAM, 240
Storey *q* values, 239
Westfall–Young method, 242
False discovery rate (FDR), 239
False positive rate, 286
Feature selection, 207
Feature space, 452
Features, definition, 6
Filtering, 208
APP, all-possible-pairs, 247
AAO, all-at-once, 250
OAA, one-against-all, 249
Filtering (features), 6

Finite differencing, 674
Fitness function, 557, 564
Fitness, genetic, 546
fMRI, *see* Functional MRI
`Fpvalue`, 689
Function words, 124
Functional MRI (fMRI), 81
Fuzzy *K*-means cluster analysis, 47

Gamma function, 686
`gammaln`, 686
Gating network, 591
Gaussian mixture model, 107
Gaussian orthogonal ensemble (GOE), 603
Gene, 546
Generalized eigenvalue problem, 651
Genetic algorithm (GA)
chromosome, 545
crossover, 548
elitism, 547
fitness, 546
gene, 546
mutation , 547
selection, 547
Geometric mean, 217
Gini index, 246, 312
Gold standard, 283
Goodness of fit (GOF), 438
Gram matrix, 452

Harmonic mean, 217
Hebbian learning, 9, 496
Hierarchical cluster analysis (HCA)
agglomeration sequence, 98
distance matrices, 98
drawing dendograms, 104
heatmap color control, 96
user specifications, 97
Hierarchical data format (HDF), 274
Holdout testing, 280
Homogeneity of variance test, 220
Homoscedasticity, 220
HSV color, 69
Hyperparameters, 522
Hypothesis tests
F test, 230
homogeneity of variance, 220
Kruskal–Wallis, 235
Levene, 220
Mann–Whitney *U*, 223
t test, 220
assuming equal variances, 221

Hypothesis tests (*Continued*)
assuming unequal variances, 222
independent samples, 221
paired samples, 222
variances assumed unequal, 221, 222

ID3 algorithm, 311
Importance, 331
Incomplete beta function, 689
Incomplete gamma function, 685, 686
Independent component analysis (ICA), 81
Information gain, 245, 312
Instance-based model, 361
Instances, definition, 6
Integers, 209
Interval scale, 210
Intractable solution, 598
Inverse document frequency, 123
Irrational numbers, 209

J48 algorithm, 312
Jacobi method, 163, 299

k-fold cross-validation, 279
K-means cluster analysis, 7, 15
K-means scores, 498
K-nearest neighbor, 361
Kappa error plot, 281
Kernel density estimation (KDE), 291, 563
Kernel function, 452
Kernel PCA, 191
Kernel regression (KREG), 525
Knowledge discovery, 2
Kruskal–Wallis test, 235
Kurtosis, 217

Lagrange multipliers, 450
Laplacian eigenmaps (LEM), 192
Lazy learner, 377
Learning vector quantization, 415
Leave-one-out cross-validation (LOOCV), 279
Ledoit–Wolf shrinkage, 611
Lexical analysis, 120
Linear discriminant analysis (LDA), 364, 393
Linear regression (LREG), 297, 364
Linearly separable, 297
Local linear embedding (LLE), 193
Locality preserving projections (LPP), 194
Logistic regression, 433
Lognormal distribution, 694
Lung2 dataset, 260

Machine learning methods, 9
Mahalanobis distance, 396, 409
Manhattan distance, 22
Mann–Whitney *U* test, 223
`mannwhitney`, 223
Marčenko–Pastur law, 604
Markov chain Monte Carlo (MCMC), 522
Matrix
correlation, 91, 163, 395
covariance, 395, 410, 552
data, multivariate **X**, 213
determinant, 650
doubly-centered, 193, 343
double-centered eigenvalue, 163, 553, 650
double-centered eigenvector, 163, 553, 650
Gram, 191, 193
Hessian, 674
Jacobian, 674
orthogonal, 650
orthonormal, 650
positive definite, 299, 598
random, 602
singularity, 299, 598
Wishart, 604
Matrix setup of data, 213
Maximum likelihood method, 446
MCMC, *see* Markov chain Monte Carlo
Mean, 214
Mean absolute deviation, 348
Mean-square error (MSE), 492
Mean-zero standardization, 59
Median, 216
Membership functions, 51, 52
Mercer's condition, 452
Metaheuristics, 543
Method of scoring, 674
Misclassification, 3
Mixture models, 591
Mixture of expert network, 591
MLL_Leukemia3, 260
Model-based cluster analysis, 107
MSE, *see* Mean-square error
Multicollinearity, 259, 299, 598
Multinomial logistic regression, 439
Multiple testing problem, 237
Multiple Mutation, genetic, 547

N grams, 139
NP-hard problems, 543
Naïve Bayes, 379
Neighborhood function, 489
Nesting, 548
Nesting problem, 203, 254

Neural adaptive learning, 543
Neural gas, 81, 573
Newton–Raphson procedure, 434, 437, 674
No-free-lunch theorem, 8
Nominal scale, 210
Nonlinear manifold learning, 7, 189
  correlation-based PCA (CPCA), 190
  diffusion map (DM), 192
  kernel PCA, 191
  Laplacian eigenmaps (LEM), 192
  local linear embedding (LLE), 193
  locality preserving projections (LPP), 194
  Sammon mapping, 195
Nonlinear regression, 673
Norm, L^1,L^2, 640
Normal distribution, 213, 693
Normal variates, 213
Normalization, 59
Normalized vector, 640
Normalizing, 59
normcf, 685
Numbers
  integers, 209
  irrational, 209
  rational, 209

OAA, one-against-all, 249
Objects, 6, 277
  definition, 6
Occam's razor, 8, 208
Online training, 504
Optimal gene set, 254, 258
Oracles, random, 281
Ordinal scale, 210
Ordinary least squares, 593
Out-of-bag (OOB) sample, 332
Out-of-place distance, 153
Outliers, 3, 348
Overfitting, 207, 311

P values
  P Bonferroni adjustment, 237
  chi-square distribution, 686
  F distribution, 689
  Sidak adjustment, 237
  standard normal distribution, 685
  Student's *t* distribution, 689
Paired *t* test, 222
Parsimony, 207, 208
Particle swarm optimization, 556
Partitional clustering, 3
Patient satisfaction, 120
Pearson product moment correlation, 21
Pearson residuals, 438
Penetrance, 635, 636
Performance
  accuracy, 278
  area under the curve (AUC), 293
  bias, 284
  bootstrap bias, 291
  bootstrap sampling, 290
  confusion matrix, 278
  cross-validation, 279
  false positive rate, 286
  holdout testing, 280
  leave-one-out cross-validation, 279
  plotting ROC curves, 292
  ROC curves, 286
  sensitivity, 283
  specificity, 283
  statistical power, 295
  testing, 277
  training, 277
  true positive rate, 286
  validation, 277
  variance, 285
  0.632 Bootstrap, 279
Permutation tests
  between-gene, 237
  genomewide test, 207
  t test, 237
Pitman correlation, 20
Plotting ROC curves, 292
Poisson distribution, 696
Polytomous logistic regression (PLOG), 439
Power, 295
Power matrix theorem, 651
Predictive value negative (PV^-), 284
Predictive value positive (PV^+), 284
Prevalence, 636
Principal axis theorem, 163
Principal component analysis (PCA), 7, 161
  communalities, 174
  correlation matrix, 163, 171
  eigenvalues, 163, 172
  eigenvectors, 163, 172
  loadings, 163, 168, 174
  principal component score coefficients, 168
  principal component scores, 169, 176
  principal components, 164
  Q-mode, 170
  R-mode, 170
  solution, 163
  varimax rotation, 166, 174

Prior, 244
Probability, 632
  addition rule, 633
  Bayes' rule, 637
  class prior, 244
  conditional, 634
  elimination rule, 636
  multiplication rule, 634
    for independent events, 635
`probbeta`, 689
`probgamma`, 686
Profitability, 120
Programs
  `conjug`, 676
  `matmul1`, 670
  `matmul`, 670
  `nonlin`, 674
  `beta`, 689
  `chisqpvalue`, 686
  `Fpvalue`, 689
  `gammaln`, 686
  `mannwhitney`, 223
  `normcf`, 685
  `probbeta`, 689
  `probgamma`, 686
  `stdnormpvalue`, 685
Prostate2 dataset, 260
Prototype learning, 57, 415
Prototypes, 7, 16, 415
Proximity, 342
Pruning, 311
Punishment–reward approach, 9

Q-mode PCA, 170
Quantile, 217
Quickprop adaptive learning, 496

R-mode PCA, 170
Random forests (RF), 331
  bootstrapping, 332
  confusion matrices, 332
  correlation, 338
  eigenanalysis, 342
  importance, 331
  out-of-bag (OOB) sample, 332
  outliers, 348
  proximity, 342
  scores, 342
  strength, 338
  supervised clustering, 345
  unsupervised clustering, 345
Random matrix, 602
Random matrix theory (RMT), 604
Random oracles, 281
Randomization tests
  between-gene, 237
  genome-wide test, 237
  t test, 237
Range, 216
Ratio scale, 211
Rational numbers, 209
Receiver–operator characteristic (ROC) curves, 286
Regular expressions, 127
Regularization, 522
Rejection–acceptance method, 564, 673
Repartitioning, 275, 279, 668
Resemblance coefficient, 20
Resilient backpropagation (RPROP), 496
Responses, 209
  continuous, 210
  discrete, 209
Return on investment (ROI), 2, 120
RGB color, 69
ROC curves, *see* Receiver-operator characteristic ROC curves
ROI, *see* Return on investment

Sammon mapping, 195
Sample space, 6
Sampling
  bootstrapping, 331
  with replacement, 628
  without replacement, 629
Scale
  interval, 210
  nominal, 210
  ordinal, 210
  ratio, 211
Scale invariance, 665
Schäfer–Strimmer shrinkage, 612
Scores, 342
s.d., *see* Standard deviation
Selection, genetic, 547
Self-organizing map (SOM), 9, 57, 591
  average quantization error, 73
  cluster visualization, 67
    adjacency matrix, 68
    best matching unit (BMU), 57, 61, 68
    cluster connectivity, 69
    crisp *k*-means cluster, 67
    HSV color normalization, 69
  component map, 71
  goodness, 73
  neighborhood function, 57

nonlinearity, 75
preprocessing data, 75
U matrix, 71
Sensitivity, 283
Shuffling, 668
Sidak adjustment to *p* value, 237
Silhouette index, 30
Similarity coefficient, 20
Simulating correlated variables, 553
Simulation
binomial probability distribution, 695
lognormal probability distribution, 694
normal probability distribution, 693
Poisson probability distribution, 696
triangular probability distribution, 697
uniform probability distribution, 692
Singular value decomposition, 299, 652
Skewness, 217
Slack vector, 453
Small-sample problem, 7
SMILES strings, 145
Softmax function, 492, 592
Specificity, 283
SSE, error sum of squares, 234
SST, treatment sum of squares, 234
SSTO, total sum of squares, 234
Standard deviation (s.d.), 21, 31, 32, 59, 63, 69, 92–95, 162, 171, 191, 215, 259, 285, 286, 291, 301–303, 310, 318, 323, 339, 363, 364, 368, 380, 383, 394, 398–400, 417, 422, 457, 460, 470, 474, 498, 491, 563–564 574, 579, 602, 604
Standard normal distribution, 218, 685
Standardization, 59
Statistical analysis of microarrays (SAM), 240
Statistical power, 295
`stdnormpvalue`, 685
Stemming, 121
Stepwise regression, 7
Stopwords, 121
Storey *q* values, 239
Strength, 338
Sum of ranks, 224, 235
Sum of squares (ANOVA), 232
Supervised classification
ant colony optimization, 560
artificial neural network, 487
covariance matrix self-adaptation, 549
decision trees, 311
ensemble classifier fusion, 280
genetic algorithms, 544
k nearest neighbor, 361
kernel regression, 525
learning vector quantization, 415
linear discriminant analysis, 393
linear regression, 297
mixture of expert network, 591
naïve Bayes classifier, 379
neural gas, 573
particle swarm optimization, 556
polytomous logistic regression, 433
random forests, 331
support vector machines
gradient ascent, 454
least squares, 465
Supervised clustering, 345
Supervised neural gas (SNG), 573
Support vector machines (SVM), 449
feature space, 452
gradient ascent approach, 453
gram matrix, 452
hard-margin hyperplane, 449, 452
Karush–Kuhn–Tucker (KKT) condition, 451
kernel functions, 451
Lagrange multipliers, 450
least-squares approach, 465
Mercer's condition, 451
slack vectors, 452
Support vectors, 451
Swarm intelligence, 556, 560
Sweep, 496
Symmetric eigenvalue problem, 603

t distribution, 689
t test, 220
assuming equal variances, 221
assuming unequal variances, 222
paired, 222
Tanimoto distance, 23
Term frequency, 123
Testing, 277
Text mining, 119
Cauchy–Schwartz inequality, 124
concept vectors, 124
document length normalization, 123
duomining, 119
function words, 124
duo-mining inverse document frequency, 123
lexical analysis, 120
N grams, 139
out-of-place distance, 153
stemming, 121

Text mining (*Continued*)
  stopwords, 121
  term frequency, 123
Theorem
  central limit theorem, 682
  equivalent eigenvalues, 651
  no free lunch, 8
  power matrix, 651
  principal axis, 163
  ugly duckling, 8
Total sum of squares (SSTO), 234
Tracy–Widom law, 604
Training, 277
Treatment sum of squares (SST), 234
Triangular distribution, 697
True positive rate, 286
Truth table, definition, 6
Tuples, 311

U-matrix, 67
Ugly duckling theorem, 8
Unified distance matrix, 71
Uniform distribution, 692
Unit vector, 640
Unsupervised classification
  crisp *k* means, 15
  fuzzy *k* means, 47
  Gaussian mixture model, 107
  hierarchical cluster analysis, 91
  neural gas, 81
  principal component analysis, 161
  random forests, 331
  self-organizing map, 57
Unsupervised clustering, 345
Unsupervised neural gas (UNG)
  nonlinearity, 85
  preprocessing data, 85

V-fold cross-validation (VFCV), 35
Validation, 277
Variables, 211
  qualitative, 211
  quantitative, 211
Variance, 215, 285
Varimax orthogonal rotation, 166
Vector
  L^1 norm, 22, 640
  L^2 norm, 22, 640
  concept, 641
  direction cosines, 640
  normalization, 640
  unit, 640

Westfall–Young, 242
Widrow–Hoff learning, 496
Winner-take-all approach, 9
Wishart ensemble, 603
Wishart matrix, 604
Wrapping, 6, 208

0.632 bootstrap, 279